T0356649

Monoidal Category Theory

Monoidal Category Theory

Unifying Concepts in Mathematics, Physics, and Computing

Noson S. Yanofsky

The MIT Press
Cambridge, Massachusetts
London, England

© 2024 Massachusetts Institute of Technology

All rights reserved. No part of this book may be reproduced in any form by any electronic or mechanical means (including photocopying, recording, or information storage and retrieval) without permission in writing from the publisher.

The MIT Press would like to thank the anonymous peer reviewers who provided comments on drafts of this book. The generous work of academic experts is essential for establishing the authority and quality of our publications. We acknowledge with gratitude the contributions of these otherwise uncredited readers.

This book was set in LaTeX by Westchester Publishing Services. Printed and bound in the United States of America.

Library of Congress Cataloging-in-Publication Data

Names: Yanofsky, Noson S., 1967– author.
Title: Monoidal category theory : a unifying concept in mathematics, physics, and computing / Noson S. Yanofsky.
Description: Cambridge, Massachusetts : The MIT Press, [2024] | Includes bibliographical references and index.
Identifiers: LCCN 2023054563 (print) | LCCN 2023054564 (ebook) | ISBN 9780262049399 (hardcover) | ISBN 9780262380799 (epub) | ISBN 9780262380782 (pdf)
Subjects: LCSH: Monoids. | Categories (Mathematics)
Classification: LCC QA169 .Y36 2024 (print) | LCC QA169 (ebook) | DDC 512/.62—dc23/eng/20240324
LC record available at https://lccn.loc.gov/2023054563
LC ebook record available at https://lccn.loc.gov/2023054564

10 9 8 7 6 5 4 3 2 1

Dedicated to the memory of my father
Rabbi Moshe Yanofsky
(1942–2022)
and to my mother
Sharon Yanofsky

C'est l'harmonie des diverses parties, leur symétrie, l'eur heureux balancement; c'est en un mot tout ce qui y met de l'ordre, tout ce qui leur donne de l'unité, ce qui nous permet par conséquent d'y voir clair et d'en comprendre l'ensemble en même temps que les détails.

It is the harmony of the diverse parts, their symmetry, their happy balance; in a word it is all that introduces order, all that gives unity, that permits us to see clearly and to comprehend at once both the ensemble and the details.

Henri Poincaré, *Science et Méthode* [211], page 25, and, *Science and Method* [212], page 30

Contents

Preface

If we expand our experience into wilder and wilder regions of experience–every once in a while, we have these integrations when everything's pulled together into a unification, in which it turns out to be simpler than it looked before.
Richard P. Feynman, *The Pleasure of Finding Things Out: The Best Short Works of Richard P. Feynman* [73], page 15

Over the past few decades, category theory has been used in many areas of mathematics, physics, and computers. The applications of category theory have arisen in (to name just a few) quantum field theory, database theory, abstract algebra, formal language theory, quantum algebra, theoretical biology, knot theory, universal algebra, string theory, quantum computing and self-referential paradoxes. This book will introduce the category theory that is necessary to understand large parts of these different areas.

Category theory studies categories, which are collections of structures and ways of changing those structures. Categories have been used to describe many phenomena in mathematics and science. Our central focus will be monoidal categories, which are enhanced categories that allow one to describe even more phenomena. The theory of monoidal categories has emerged as a theory of structures and processes.

Category theory is a simple, extremely clear, and concise language in which various fields of science can be discussed. It is also a unifying language. Different fields are expressed in this single language so one can see common themes and properties. Category theory also brings together different fields by actually establishing connections between them. It is particularly suited to show the big picture. Once this language is understood, one is capable of easily learning an immense amount of science, mathematics, and computing.

This book is an introduction to category theory. It begins with the basic definitions of category theory and takes the reader all the way up to cutting-edge research topics. Rather than going "down the rabbit hole" with a lot of very technical, pure category theory, our central focus will be examples and applications. In fact, an alternative title of this text could be *Category Theory by Example*. A major goal is to show the ubiquity of category theory and finally put an end to the silly canard that category theory is "general abstract nonsense." Another important goal is to show how a multitude of fields are related through category theory.

Within each chapter, whenever there is a definition or theorem, it is immediately followed by examples and exercises that clarify the important categorical idea and makes it come alive. The fun is in the examples, and our examples come from many diverse disciplines.

This text contains twelve self-contained mini-courses on various fields. They are short introductions to major fields such as quantum computing, self-referential paradoxes, and quantum algebra. These sections do not introduce new categorical ideas. Rather, they use the category theory already presented to describe an entire field. The point we are making with these mini-courses is that, with the language of category theory in your toolbox, you can master totally new and diverse fields with ease.

This book is different from other books on category theory. In contrast to other books, we do not assume that the reader is already a mathematician, a physicist, or a computer scientist. Rather, this book is for anyone who wants to learn the wonders of category theory. We assume that the reader is broad-minded, is interested in many areas, and wants to see how diverse areas are related to each other. The reader will not only learn all about category theory, they will also learn an immense amount of science, mathematics, and computers.

Another major difference between this book and other introductory category theory books is the way that the book is organized. While in most books, the concepts of category, functor, and natural transformation are introduced in the first few pages, here we slowly present each idea separately and in its correct time. With such a presentation, the novice will not be overwhelmed.

Organization

Chapter 1 is an introduction that places category theory in its historical and philosophical context. It ends with a discussion of constructions on sets, which will be useful for the rest of the text. Chapters 2, 3, and 4 give a simple introduction to category theory. Chapter 2 contains the basic definitions and properties of categories. Chapter 3 deals with special structures within a category. The real magic begins in Chapter 4, where we see how various categories relate to each other.

Chapters 5, 6, and 7 discuss monoidal categories. Chapter 5 describes monoidal categories, which are categories that have extra structure. Chapter 6 deals with the relationships between monoidal categories. The core of the book is Chapter 7, where several variations of monoidal categories are presented with many of their properties and applications.

The final three chapters contain some advanced topics. Many categorical ways of describing structures are explored in Chapter 8. Chapter 9 offers a sampling of research areas in advanced category theory. We conclude with Chapter 10, which is a collection of more mini-courses from several areas.

At the end of each chapter, there is a self-contained mini-course on a single topic. Every chapter and mini-course end with several pointers to where you can learn more about the particular topic.

Appendix A contains Venn diagrams that describe the relationships of various structures used throughout the text.

Appendix B is an index of the categories that appear in the text.

Appendix C has some suggestions for further study of category theory and its applications.

Appendix D has answers to selected exercises.

Ancillaries

This text does not stand alone. I maintain a web page for the text at
www.sci.brooklyn.cuny.edu/~noson/MCTtext.html
There will be links to important resources and on-line lectures. Instructors will find classroom slides for lectures.

Acknowledgments

There are many people who have been instrumental in the production of this book.

Brooklyn College has been a large part of my life since I entered as a freshman in September 1985. I am grateful for the intellectual environment that it provides. My colleagues and friends are always there for a chat, to help develop an idea, and for editing a draft. In particular, I would like to thank David Arnow, Eva Cogan, James Cox, Scott Dexter, Lawrence Goetz, Jackie Jones, Keith Harrow, Yedidyah Langsam, Michael Mandel, Simon Parsons, Ira Rudowsky, Charles Schnabolk, Bridget Sheridan, Alexander Sverdlov, Joseph Thurm, Gerald Weiss, and Paula Whitlock.

A large part of this book was written during a sabbatical. I am thankful to President Michelle J. Anderson, Dean Louise Hainline, Dean Peter Tolias, and Chairman Yedidyah Langsam for making that sabbatical possible.

This book owes much to Professor Rohit Parikh who, besides teaching me many ideas, showed me that "you can teach anything to anyone ... You just have to teach them the prerequisites first." How true! I am also in debt to Jamie Lennox for repeatedly reminding me that he is my target audience. I hope he is satisfied. Ralph Wojtowicz suggested the mini-course about sets and functions, which is critical for the novice.

My education in category theory owes much to many people. At the end of my first year in graduate school at The Graduate Center of the City University of New York, Alex Heller posted a sign stating that he would be teaching a class called "Categorical Logic" the following semester. He wrote that the prerequisites were the first four chapters of Saunders Mac Lane's *Categories for the Working Mathematician* [180]. Until then, I had never heard of category theory. Over that summer, Mirco Mannucci and I struggled through those chapters. Gradually, the ideas started taking shape. Eventually, Heller became my thesis adviser. Between Heller's clarity and Mirco's infectious enthusiasm, I was hooked on category theory. I am forever in their debt.

After receiving my degree, I had the good fortune of doing a postdoc at McGill University. The main professors who organized the category theory group were Michael Barr, Marta Bunge, Jim Lambek, and Michael Makkai. I learned much from them. Many people were part of the category theory group and regularly joined the weekly seminar. I gained much from Rick Blute, Thomas Fox, Jonathan Funk, André Lebel,

Jean-Pierre Marquis, Prakash Panangaden, Robert Paré, Phill Scott, Robert Seely, and others. Many became lifelong friends.

On returning to New York, I continued to meet Alex Heller at the Graduate Center. In addition to being a brilliant polymath who could easily discuss myriad topics, he was an extremely kind, gentle man who always had a warm and encouraging word. His vision of mathematics was astonishing and unique. While chatting with him, one was able to get a fleeting glimpse of the clarity, interconnectedness, and broadness of that vision. From my graduation in 1996 until his passing in 2008, we met and chatted once or twice a week. In essence, he gave me twelve years of a postdoctoral research fellowship. I am forever indebted to him for all his teaching and his sincere friendship.

My category theory education continues up to today. Since 2009, I have been organizing the New York City Category Theory Seminar out of The Graduate Center of the City University of New York. We have a great group of regular participants, including Gershom Bazerman and Raymond Puzio. World-class category theorists from around the globe come and give talks at the seminar. They explain their research and answer our questions. I have gained much from the participants and speakers.

I am grateful to the many people who helped with editing the book: Rio Alvarado, Edward Arroyo, Michael Barr, Dominic Barraclough, Tai-Danae Bradley, Miriam Briskman, Leo Caves, Ellen Cooper, Thierry Coquand, Eugene Dorokhin, Hindy Drillick, Vasili I. Galchin, Michael Goldenberg, Jonathan Hanon, Shai Hershfeld, Mark Hillery, Rick Jardine, Samantha Jarvis, Yigal Kamel, Deric Kwok, Moshe Lach, Steve Lack, Klaas Landsman, Jamie Lennox, Armando Matos, William Mayer, Ieke Moerdijk, Micah Miller, Elina Nourmand, Jason Parker, Arthur Parzygnat, Brian Porter, Saeed Salehi, Lorenzo Sauras, Nathan Schor, Dan Shiebler, Michael Shulman, Evan Siegel, David I. Spivak, Ross Street, Joshua Sussan, Karl Svozil, Walter Tholen, Do Anh Tuan, Hadassah Yanofsky, Sharon Yanofsky, Shayna Leah Yanofsky, Joel Zablow, and Marek Zawadowski. Thank you all!

Miriam Briskman did a magnificent job carefully editing the entire book. She also used her exceptional LaTeX skills to make most of the more complicated Ti*k*Z diagrams in the text. Thank you, Miriam! Adina Scheinfeld edited every chapter of this book. She made many great suggestions. Thank you, Adina!

I would also like to thank Elizabeth Swayze, Emma Donovan, Madhulika Jain, Malerie Lovejoy, Janet E Rossi, Matthew Valades, and everyone else at the MIT Press for helping me get this book into shape.

I am thankful to my children Hadassah, Rivka, Boruch, and Miriam for giving me much joy and the strength to complete this text. Essentially, this book was written because my wife, Shayna Leah, took care of everything else in my life. It is her love and encouragement that made this book possible. I am forever indebted to her.

This book is dedicated to the memory of my father, Rabbi Moshe Yanofsky, and to my mother, Sharon Yanofsky. My father was a professor of mathematics at The City University of New York for many years and concurrently a prominent figure in Jewish education. At the age of twenty-five, he became the principal of a Jewish girls, high school with over 1,000 students. Over the next fifty years, he continued educating and eventually established his own high school. His wisdom and warmth had a profound influence on everyone who met him. He was a wonderful, caring father who was always there for us. My father led by example and showed us what it means to live a meaningful life and to have a strong moral code. At the time of his death, he was predeceased by a son and was survived by my mother, four children, thirty-eight grandchildren, tens of great-grandchildren, and thousands of loving students. His warm smile is painfully missed.

My mother is a woman of many talents. She was a teacher, a principal, and a bank vice president, and is currently a school administrator. Her main hobby is being best friends with every one of her grandchildren and great-grandchildren. She is excellent at everything she does, and she does it all with class and a certain panache. My mother raised us with the understanding that anything is possible when you put enough perseverance into it. Everything that I am and all that I do are because of all the effort and love she put into raising me.

1

Introduction

So the problem is not so much to see what nobody has yet seen, as to think what nobody has yet thought concerning that which everybody sees.
Arthur Schopenhauer

We gently introduce the world of category theory here. In Section 1.1, a little of the historical and philosophical context is provided. It shows how unification has always been a motivating factor of category theory. Section 1.2 explains the motivation for monoidal categories. We then lay out the structure of this book in Section 1.3 and also present some standard notation. Section 1.4 is a mini-course that uses constructs on sets and functions to teach many categorical ideas.

1.1 Categories

Category theory began with the intention of relating and unifying two different areas of study. The aim was to characterize and classify certain types of geometric objects by assigning to each of them certain types of algebraic objects as depicted in Figure 1.1. (In detail, the geometric objects are structures called "topological spaces," "manifolds," "bundles," etc. The algebraic objects are called "groups," "rings," "abelian groups," etc. The assignments have exotic names like "homology," "cohomology," "homotopy," and "K-theory," etc.) Researchers realized that if they were going to relate geometric objects with algebraic objects, they needed a language that is specialized to neither geometric content nor algebraic content. Only with such a general language can one discuss both fields.

Category theory was invented by Samuel Eilenberg and Saunders Mac Lane [67]. They described various collections of mathematical objects. Each collection was called a **category**. There was a collection of geometric objects and a collection of algebraic objects. Eilenberg and Mac Lane were interested in many different categories, and in order to relate one category to another, they formulated the notion of a **functor** which—like a function—assigns to each entity in one category an entity in another category. They went further and formulated the notion of a **natural transformation**, which is a

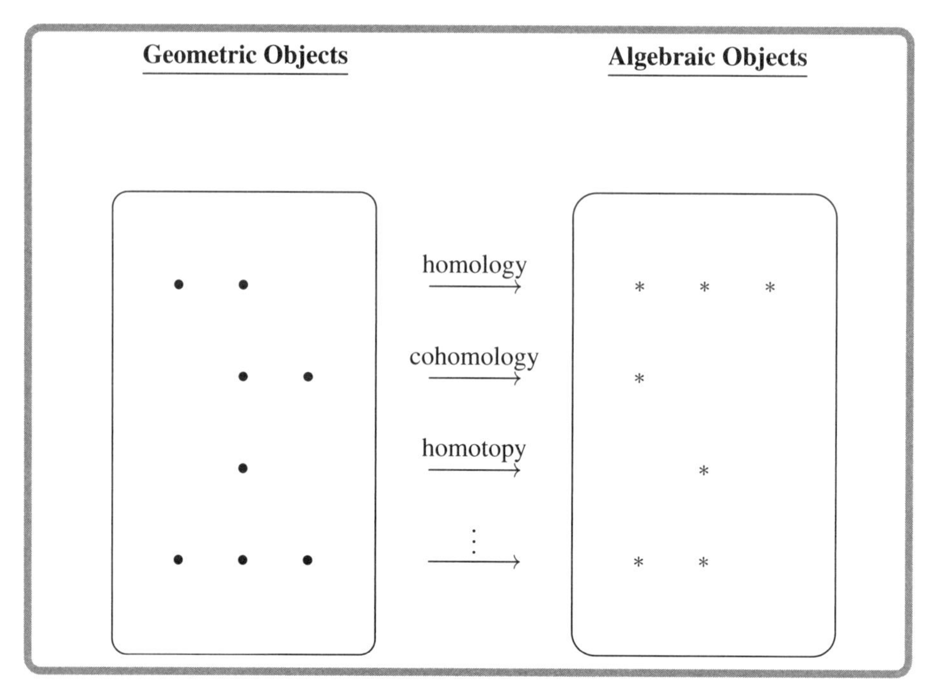

Figure 1.1. Relating geometric objects to algebraic objects.

way of relating one functor to another. (In a sense, a natural transformation *transfers* the results of one functor to the results of another functor.) These structures can be visualized as follows:

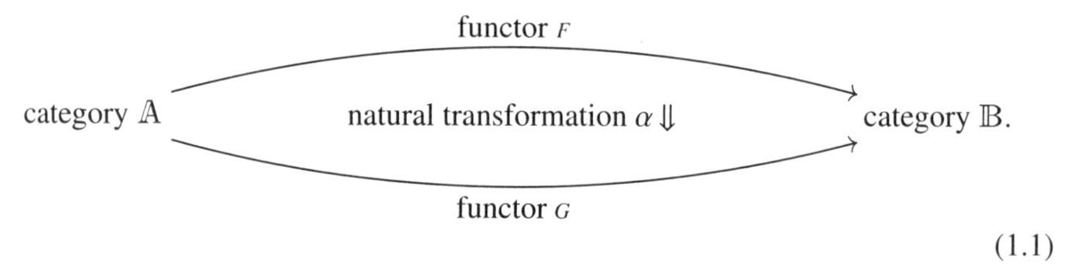

(1.1)

There is category $\mathbb{A}$ and category $\mathbb{B}$. These categories are related by functor F and functor G. And, finally, these functors are related by the natural transformation α.

What is a category? It refers to a collection of structures called **objects** of a particular type, and a collection of transformations or processes between the objects. We call these transformations or processes **morphisms** or **maps**. If a and b are objects and f is a morphism from a to b, we write it as $f\colon a \longrightarrow b$ or $a \xrightarrow{f} b$. We can visualize part of a category as Figure 1.2. The morphisms are to be thought of as ways of transforming objects. As time went on, the morphisms between objects took center stage. Category theory became not only the study of structures but also the study of transformations or processes between structures. One of the main properties of processes is that they

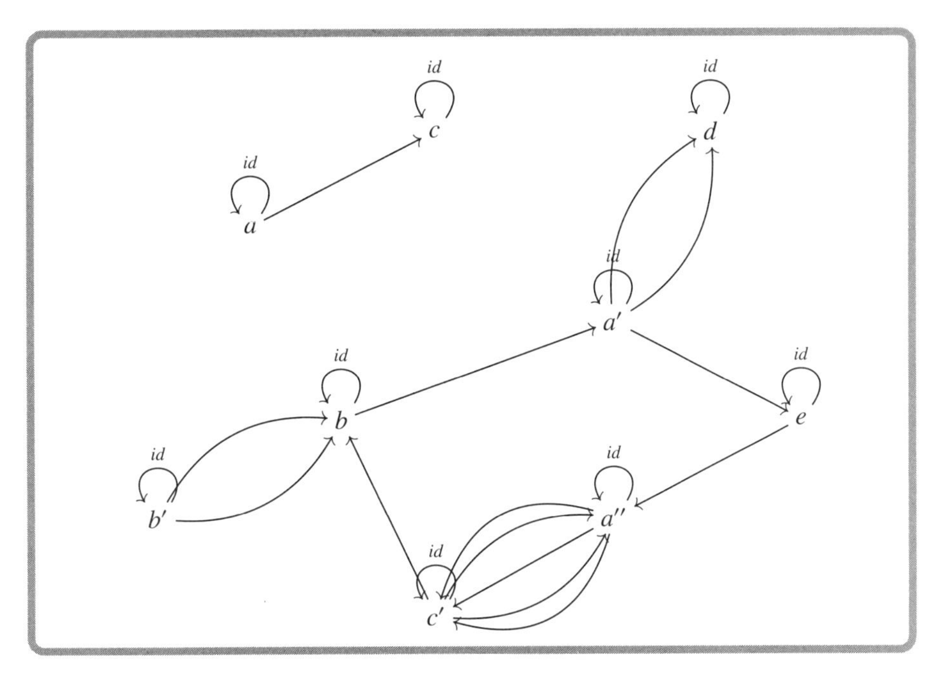

Figure 1.2. An example of part of a category. Every object has an identity morphism (*id*).

can be combined. That is, one process followed by another process can be combined into a single process. In a category, if there is a morphism from object a to object b called f and a morphism from object b to object c called g, then there is an associated morphism from object a to object c written as $g \circ f$, which is called "g composed f," or "g following f," or "g after f." This can be drawn as follows:

$$a \xrightarrow{f} b \xrightarrow{g} c. \quad (\text{with } g \circ f : a \to c \text{ drawn as an arc above}) \tag{1.2}$$

This composition is the most fundamental part of a category.

Categories are related to more familiar structures, namely directed graphs. A **directed graph** is a structure that has objects (also called "vertices," "nodes," or "points") and morphisms (also called "arrows" or "directed edges") between them. One can view a category as a souped-up directed graph. Categories, like directed graphs, also have objects and morphisms, but within categories, one morphism after another can be composed. A directed graph is used to deal with various phenomena of interconnectivity. A category, with its extra structure, deals with more sophisticated notions of interconnectivity (such as reachability). A category can also be seen as a generalization of a group. A **group** is a set where one can combine elements to form other elements. In a category, one can combine morphisms that follow one another. Graphs and groups are ubiquitous in modern science and mathematics. Categories—as generalizations of both structures—are even more pervasive.

Since categories are disassociated from any specific field or area, category theory received the reputation of being a language without content. Because of this, the field was derided by some as "general abstract nonsense." However, it is precisely this independence from any field that gives category theory its power. By not being formulated for one particular field, it is capable of dealing with *any* field. At first, category theory was extremely successful in dealing with various fields of mathematics. As time went on, researchers realized that many branches of science that deal with structures or processes can be discussed in the language of category theory. Computer science is the study of computational processes, and hence it has included a deep interest in category theory. More recently, category theory has been shown to be very adept at discussing structures and processes in physics. Researchers have also demonstrated that category theory is great at discussing the structures and processes of chemistry, biology, artificial intelligence, and linguistics.

Many diverse fields are shown to be related because they are discussed in the single language of category theory. Researchers have found similar theorems and patterns in areas that were thought to be unrelated. Moreover, in the past few decades, category theory has further unified different fields by revealing amazing relationships among them. There are functors from a category in one field to a category in a totally different field that preserve properties and structures. Such property-preserving functors show that the two fields are similar. For example, quantum algebra is a field that uses categorical language to show how certain algebraic structures are related to geometric structures like knot theory. Another prominent example is topological quantum field theory, which is a branch of math and physics that uses functors to unite relativity and quantum theory. Quantum computing is a field that sits at the intersection of computer science, physics, and mathematics and can easily be understood using various categorical structures.

1.2 Monoidal Categories

In the early 1960s, Jean Bénabou and Saunders Mac Lane described categories that have more structure, called **monoidal categories** or **tensor categories**. In these categories, one can "multiply" or "combine" objects. Symbolically, within a monoidal category, object a and object b can be combined to form object $a \otimes b$ (read as "a tensor b"). As always in category theory, one is interested not only in combining objects but also in combining morphisms. With morphisms $f\colon a \longrightarrow a'$ and $g\colon b \longrightarrow b'$, there are objects $a \otimes b$, $a' \otimes b'$, and there also is a morphism $f \otimes g$, which we write as

$$\begin{array}{c} a \xrightarrow{\quad f \quad} a' \\ \otimes \\ b \xrightarrow[\quad g \quad]{} b' \end{array} \qquad \text{or} \qquad a \otimes b \xrightarrow{\quad f\otimes g \quad} a' \otimes b'. \tag{1.3}$$

Notice that there are two ways of combining morphisms in a monoidal category: $f \circ g$ and $f \otimes g$. In physics, the combination $f \circ g$ corresponds to performing one process after another, while the combination $f \otimes g$ corresponds to performing two independent processes. In computers, the combination $f \circ g$ corresponds to sequential processes, while $f \otimes g$ corresponds to parallel processes. In mathematics, the interplay of the two combinations of morphisms is very important.

Classical algebra is the branch of mathematics that deals with sets and operations on those sets. For sets of numbers and the addition operation, we have the rule that $x + y = y + x$, while in general, for subtraction, $x - y \neq y - x$. In the theory of monoidal categories, there are rules that govern the relationship between $a \otimes b$ and $b \otimes a$. What about the relationship between $(a \otimes b) \otimes c$ and $a \otimes (b \otimes c)$? Within monoidal categories, there are many possible relationships when dealing with these combined objects. For each rule relating these operations, there will be a corresponding type of monoidal category. In Chapter 7, we will see many types of monoidal categories. This variability allows many phenomena to be modeled by monoidal categories. The area that deals with the various types of rules among operations is called **coherence theory** (i.e., how the various operations *cohere* with each other) or **higher-dimensional algebra**. This area of study has become pervasive, and it is believed that higher-dimensional algebra will arise even more frequently in the science and mathematics of the coming decades.

1.3 The Examples and the Mini-Courses

This text is centered on the examples. Our goal is to show the pervasiveness of categories, and in particular monoidal categories. We also want to emphasize how categories can reveal the interconnectedness of various fields. We do so by introducing many examples from many different areas. Immediately following a definition or a theorem of category theory, there are lots of examples that illustrate the idea. There are also some examples that are left to the reader as exercises. It is important to realize that although this book is chock-full of examples, we have barely scratched the surface. The literature of category theory has many more examples. For this book, we chose the examples that arise most frequently or are the easiest to understand. The reader will be directed to places in the literature where other examples are described. We are showing the beauty of category theory but revealing only the tip of the iceberg.

Most of the examples can be loosely split into three broad groupings: mathematics, physics, and computers. The problem is that the boundaries between these areas are hazy. For example, is quantum computing part of computer science, physics, or abstract mathematics? Is knot theory part of mathematics or physics? There are no firm boundaries.

Since most readers are familiar with sets and functions between sets, we usually try to first show an idea or definition in terms of sets. In later chapters, it will become apparent that sets and functions between sets are not the right contexts to examine certain phenomena. This is where category theory really gets interesting.

The examples are spread throughout the book. To illustrate, in Chapter 2, a category will be introduced. In Chapter 3, some properties of this category will be described. This category will be related to other categories in Chapter 4. In Chapter 5, we will show that the category has a monoidal structure, and we will see how that monoidal structure relates to the monoidal structure of other categories in Chapter 6. This same category and variations of this category will be shown to have even more structure in Chapter 7. We will also see how this category arises in various mini-courses. By the time the reader finishes the book, the category will be an old friend.

Not all categories are introduced early on. In order not to overwhelm the reader in the beginning, we will introduce many categories in later chapters as well. Our aim is readability and understanding.

These examples will take the reader rather far. In mathematics, the reader will meet lots of algebra and topology. In physics, we will see the basics of quantum theory. In computers, we will see how categories are good for describing certain models of computation and some advanced logic.

Due to space limitations, and by concentrating on examples, we are going to omit some results in pure category theory. We only describe the category theory required to understand the examples. In Appendix C, we point out various places where one can learn more about (pure) category theory.

Category theory is a language that can deal with many different areas of science. The really fun part of category theory is that once we have this language in our toolbox, we can easily pick up whole new branches of science. We show this flexibility with little mini-courses. At the end of every chapter is a little self-contained section that describes a whole field in the context of the category theory already learned. Mini-courses in later chapters depend on the knowledge of earlier mini-courses. In Chapter 10, we offer several other mini-courses.

One of the intended mini-courses for this book was on the topic of theoretical computer science. Within that mini-course, there were sections on models of computation, computability theory, complexity theory, and several other topics. Some of the deepest ideas and theorems of modern computers and mathematics, such as Turing machines, unsolvable problems, the P = NP question, Kurt Gödel's incompleteness theorem, intractable problems, and Turing's Halting problem, among others,were met and explained using simple categorical language. While writing about these topics, I was surprised at the ease in which complicated ideas of theoretical computer science can be expressed and proved using category theory. That mini-course took on a life of its own, growing out of this book to become its own book, *Theoretical Computer Science for the Working Category Theorist* [274]. Consider that book a companion volume of this text for readers interested in those fascinating topics.

This book owes a tremendous debt to previous works, including the following:

- I cut my teeth learning category theory from Saunders Mac Lane's *Categories for the Working Mathematician* [180]. This is *the* classic text by one of the founders of category theory. It influenced my thinking and this book in the most profound way. As the title implies, Mac Lane assumed that the reader knows large parts of mathematics before opening his book. My goal with this book is to show the beauty of category theory as Mac Lane did, but to a larger audience.
- John Baez and Michael Stay wrote a wonderful paper called "Physics, Topology, Logic and Computation: A Rosetta Stone" [24], which highlights connections

between many different fields. I would like to think of this book as an explanation and expansion of that paper.
- I learned much from Christian Kassel's textbook *Quantum Groups* [134]. His clarity and exactness is an inspiration.
- This book attempts to be as readable as Michael Barr and Charles Wells's textbook *Category Theory for Computing Science* [33]. Their work goes through large segments of category theory, with many examples along the way. We try to do the same, but with examples from physics and mathematics as well.

It must be noted that this is not a history book. We are not going to say who thought of some particular construction or example first. Some of the examples in this book came from other books and papers. Some examples are just known in the folklore of category theory. And we made up other examples. The history is too complicated for us to disentangle and is of absolutely no pedagogical use to the novice. We name some places to learn about the history of category theory in Appendix C.

There are many potential topics that could have gone into this book. Painful choices had to be made. In the end, topics were chosen based on a desire to provide as diverse a set of examples as possible to satisfy a broad readership, with a natural bias to those areas which I feel more confident to address. I would like to believe that the topics chosen will be important as we march into the unfathomable future.

Finally, I would like to apologize to all my friends in the category theory community if I neglected their favorite example or did not discuss an area in which they did great work. It was not my intention to omit anyone's work.

To improve readability, for the most part, we keep to the following notation:

- Categories are in blackboard bold font: $\mathbb{A}, \mathbb{B}, \mathbb{C}, \mathbb{D}, \mathbb{Circuit}, \mathbb{Set}, \ldots$
- Objects in general categories are the first few lowercase Latin letters: a, b, c, d, a', $b', a'' \ldots$
- Morphisms in general categories are later lowercase Latin letters: f, g, h, i, j, k, f', $g'', \ldots$
- Functors are capital Latin letters: $F, G, H, I, J, \ldots$
- Natural transformations are lowercase Greek letters: $\alpha, \beta, \gamma, \delta, \eta, \kappa, \ldots$
- Higher-dimensional morphisms are capital Greek letters: $\Gamma, \Delta, \Theta, \Phi, \Psi, \ldots$
- Sets of numbers are denoted as follows: $\mathbf{N}, \mathbf{Z}, \mathbf{Q}, \mathbf{R}, \mathbf{C}$.
- 2-categories are in blackboard bold font with a line above: $\overline{\mathbb{A}}, \overline{\mathbb{B}}, \overline{\mathbb{C}}, \overline{\mathbb{D}}$, $\overline{\mathbb{Cat}} \ldots$
- 3-categories are in blackboard bold font with a two lines above: $\overline{\overline{\mathbb{A}}}, \overline{\overline{\mathbb{B}}}, \overline{\overline{\mathbb{C}}}, \overline{\overline{\mathbb{D}}}$, $\overline{\overline{2\mathbb{Cat}}} \ldots$

There are several types of arrows in this book:

- Morphism, map or functor: $\longrightarrow$
- The input and output of a function or a functor: $\longmapsto$ or $\leadsto$
- Inclusion or injection: $\hookrightarrow$
- Surjection or full functor: $\twoheadrightarrow$
- Natural transformation: $\Longrightarrow$

1.4 Mini-Course: Sets and Categorical Thinking

Category theory is not just a language that is capable of describing an immense amount of science and mathematics. Rather, it is a *new and innovative way of thinking.* One of its central ideas is that we define properties of objects by the way that they interact with other objects.

Important Categorical Idea 1.4.1. Morphisms Are Central. Properties and structures in a category can be described by the morphisms of the category. That is, the objects do not stand alone. One must see how the objects relate to each other with morphisms. The objects have to be seen in the context of the morphisms.

In particular, many properties of object b can be understood by looking at the collections of morphisms $a \longrightarrow b$ for various simple objects a. Similarly, many properties of b are described by looking at collections of morphisms $b \longrightarrow c$ for various simple objects c. ○

To get a feel for this, we take an in-depth look at the familiar world of sets and functions between sets. We show that many of the usual ideas and constructions about sets can be described with functions between sets. This mini-course will also be a gentle reminder of many concepts that are needed in the rest of the text. Throughout this book, we will point back to equations, diagrams, and ideas found in this section.

Sets and Operations

Definition 1.4.2. A **set** is a collection of elements. If S is a set and x is an element of S, we write $x \in S$. If x is not an element of S, we write $x \notin S$.

Example 1.4.3. We will deal with both infinite sets and finite sets. Some of the most important infinite sets of numbers are

- The natural numbers, $\mathbf{N} = \{0, 1, 2, 3, \ldots\}$
- The integers, $\mathbf{Z} = \{\ldots, -3, -2, -1, 0, 1, 2, 3, \ldots\}$
- The rational numbers, $\mathbf{Q} = \{\frac{m}{n} : m$ and n in $\mathbf{Z}$ and $n \neq 0\}$
- The real numbers, $\mathbf{R}$ (i.e., all numbers on the real number line)
- The complex numbers, $\mathbf{C} = \{a + bi : a$ and b in $\mathbf{R}\}$. □

We begin by discussing several operations on sets. Let S and T be sets. If s is in S and t is in T, we write an **ordered pair** of the elements as (s, t). The set of all ordered pairs is called the **Cartesian product of sets** S and T:

$$S \times T = \{(s, t) : s \in S, t \in T\}. \tag{1.4}$$

Example 1.4.4. If *Pants* = {black, blue1, blue2, gray} is the set of pants that you own, and *Shirts* = {white, blue, orange} is the set of shirts that you own, then the set of

Outfits is

$$Pants \times Shirts = \left\{ \begin{array}{c} \text{(black, white), (black, blue), (black, orange),} \\ \text{(blue1, white), (blue1, blue), (blue1, orange),} \\ \text{(blue2, white), (blue2, blue), (blue2, orange),} \\ \text{(gray, white), (gray, blue), (gray, orange)} \end{array} \right\} \tag{1.5}$$

(True scholarly category theorists do not care if their clothes don't match!) □

Technical Point 1.4.5. The most important aspect of an ordered pair is its order. In contrast, sets are just collections, and as such they have no preferred order. The set $\{s, t\}$ is considered to be the same set as $\{t, s\}$. In contrast, the pair (s, t) is not considered the same as (t, s). Hence, we cannot simply use two curly brackets to describe ordered pairs. There are other ways of describing an ordered pair of elements from S and T. For example, we could write them as $\langle s, t\rangle$ or $\{s, t, \{s\}\}$ (where we collect the elements but indicate the first element by putting it in a set by itself), or even $\{s, t, \{t\}\}$ (where we indicate the second element by putting it in a set by itself). There is nothing special about the notation (s, t). (We will see this again in the beginning of Section 3.1) ♡

The most interesting property about a finite set is the number of elements in such a set. For every finite set S, we write $|S|$ to denote the number of elements in S. If there are m elements in S and n elements in T, then there are mn elements in $S \times T$. In symbols, we write this as

$$|S \times T| = |S| \cdot |T|. \tag{1.6}$$

Exercise 1.4.6. How many elements are in *Pants*? How many elements are in *Shirts*? How many elements in *Outfits*? Show that the above formula works. ■

We can generalize the notion of ordered pairs to **ordered triples**, **ordered 4-tuples**, **ordered 5-tuples**, etc. If there are n sets, $S_1, S_2, \ldots, S_n$, then an **ordered n-tuple** is written as $(s_1, s_2, \ldots, s_n)$, where s_i is in S_i. The set of all n-tuples is $S_1 \times S_2 \times \cdots \times S_n$. The number of n-tuples follows a generalization of Equation (1.6):

$$|S_1 \times S_2 \times \cdots \times S_n| = |S_1| \cdot |S_2| \cdots |S_n|. \tag{1.7}$$

Exercise 1.4.7. In addition to pants and shirts, an outfit might consist of a hat, socks, and shoes. How many outfits are there if there are m hats, n pairs of socks, p pairs of shoes, q pants, and r shirts? ■

Another operation performed on sets is the union. Let S and T be sets. The **union** of S and T is the set $S \bigcup T$, which contains those elements that are in S or in T:

$$S \bigcup T = \{x\colon\ x \in S \text{ or } x \in T\}. \tag{1.8}$$

It is important to notice that if there is some element that is in both S and T, then it will occur only once in $S \bigcup T$. This is because when dealing with a set, repetition does not matter. The set $\{a, b, c, b\}$ is considered the same as $\{a, b, c\}$.

A related operation is the disjoint union. Given sets S and T, one forms the **disjoint union** $S \amalg T$, which contains the elements from S and T but considers elements that are in both sets as different elements. One way in which this is done is by tagging every element with extra information that says which set it comes from. This way, an element that is in both S and T would be considered two different elements. For example, if $S = \{a, b, c, x, y\}$ and $T = \{q, w, b, x, e, r\}$, then

$$S \amalg T = \{(a,0),(b,0),(c,0),(x,0),(y,0),(q,1),(w,1),(b,1),(x,1),(e,1),(r,1)\}, \quad (1.9)$$

where the elements of S are tagged with a 0 and the elements of T are tagged with a 1. In general for sets S and T, we have

$$S \amalg T = (S \times \{0\}) \bigcup (T \times \{1\}). \quad (1.10)$$

The formula for the number of elements in the disjoint union is $|S \amalg T| = |S| + |T|$.

Exercise 1.4.8. When does the union of two sets have the same number of elements as the disjoint union of those same sets? ■

Functions

The central idea of this mini-course is the notion of functions between sets and how they determine properties of sets.

Important Categorical Idea 1.4.9. Not Entities, But Morphisms Between Entities. In category theory, whenever we have a notion (e.g., a set), the next immediate task is to consider how these notions relate to each other (e.g., functions are ways for sets to relate to each other.) As we have stated, category theory is not about "things," but rather about how "things" relate to other "things." Relations between objects are usually described by morphisms or functions between the objects.

Following this rule, once we describe the morphisms between the objects, we must immediately ask what is between the morphisms. Usually, the answer will be other morphisms. The computer scientist might protest that this recursive procedure will lead into an infinite loop. It will! We will see this in Section 9.4 when we describe infinite levels of morphisms in higher-dimensional category theory. Such structures are a reflection of this important idea that is at the center of category theory. ○

Definition 1.4.10. Let S and T be sets. A **function** f from S to T, written $f: S \longrightarrow T$, is an assignment of an element of T to every element of S. The value of f on the element s is written as $f(s)$ (stated as "f of s"). If $f(s) = t$, we write $s \mapsto t$ ("s maps to t").

It is important to understand the difference between the symbol $\longrightarrow$ and the symbol $\mapsto$. The symbol $\longrightarrow$ goes between two sets. It describes a function from one set to another. In contrast, the symbol $\mapsto$ goes from an element in the first set to an element in the second set. It describes how the function is defined.

Example 1.4.11. For every set S, there is an **identity function** $id_S : S \longrightarrow S$, which takes every element to itself. In symbols, it is defined as $id_S(s) = s$ or $s \mapsto s$. □

Following Important Categorical Idea 1.4.1, let us look at how morphisms to a set determine the properties of that set. Functions can be used as a way of describing or choosing elements of a set. Consider a one-element set, $\{*\}$. (There are many one-element sets, such as $\{a\}$, $\{b\}$, and $\{\text{Bill}\}$) For set S, function $f\colon \{*\} \longrightarrow S$ picks out one element of S. The single element $*$ goes to the selected element s in S. In symbols, $f(*) = s$ or $* \mapsto s$.

Example 1.4.12. Let S be the set $\{\text{Jack, Jill, Joan, June, Joe, John}\}$. The element Joe in S can be described as a function $f\colon \{*\} \longrightarrow S$ where $f(*) = \text{Joe}$. We might want to distinguish this function by calling it $f_{\text{Joe}}\colon \{*\} \longrightarrow S$. There will be other functions like $f_{\text{Jill}}\colon \{*\} \longrightarrow S$, where $f_{\text{Jill}}(*) = \text{Jill}$. For this set of six elements, there are six different functions from $\{*\}$ to S. □

If we are interested in choosing two elements of S, we can look at functions from a two-element set to S. So $f\colon \{0,1\} \longrightarrow S$ will choose two elements of S. The first element is $f(0)$ and the second element is $f(1)$. If $f(0) \neq f(1)$, then f will choose two *different* elements of S. Every such function chooses two elements of S. Functions from $\{a,b,c\}$ to S will choose three elements of S. This can go on; if we wanted to choose n elements of a set, we would look at functions of the form

$$\{1, 2, \ldots, n\} \longrightarrow S. \tag{1.11}$$

Definition 1.4.13. A set T is a **subset** of S if every element of T is an element of S. We write this as $T \subseteq S$. If T is a subset of S but not equal to S, we call T a **proper subset** and write $T \subsetneq S$ or $T \subset S$. This is the case when there is at least one element in S that is *not* in T. If T is a subset of S, there is an **inclusion function** that takes every element of T to its corresponding element of S, which is written as $inc\colon T \hookrightarrow S$. There is a special set that has no elements, called the **empty set** and denoted as $\emptyset$. Since it is true that whatever is in $\emptyset$ (nothing) is in any other set, the empty set is a subset of every set. Furthermore, for every set S, there is a unique function from the empty set to S. We sometimes denote this function as $!\colon \emptyset \longrightarrow S$.

Subsets of a set are of fundamental importance. We shall be interested in the collection of all subsets of a particular set.

Definition 1.4.14. For set S, the set of all subsets of S is called the **powerset** or **power set** of S and is denoted $\mathcal{P}(S)$. In other words,

$$\mathcal{P}(S) = \{T : T \text{ is a subset of } S\}. \tag{1.12}$$

Example 1.4.15. For set $\{a\}$, the powerset is $\mathcal{P}(\{a\}) = \{\emptyset, \{a\}\}$. The powerset of a two-element set $\{a,b\}$ is $\mathcal{P}(\{a,b\}) = \{\emptyset, \{a\}, \{b\}, \{a,b\}\}$. The powerset of a three-element set $\{a,b,c\}$ is $\mathcal{P}(\{a,b,c\}) = \{\emptyset, \{a\}, \{b\}, \{c\}, \{a,b\}, \{a,c\}, \{b,c\}, \{a,b,c\}\}$. Whenever we add an element to a set, we double the number of elements in the powerset. We have following rule: if S has n elements, then $\mathcal{P}(S)$ has 2^n elements. In symbols, $|S| = n$ implies $|\mathcal{P}(S)| = 2^n$. We can also write this as $|\mathcal{P}(S)| = 2^{|S|}$. □

Functions can be used to describe subsets.

Definition 1.4.16. For any set S and subset $T \subseteq S$, there is an associated **characteristic function** $\chi_T : S \longrightarrow \{0,1\}$. This function assigns either 1 or 0 to every element s of S. If s is in T, the characteristic function assigns 1 to s, and if s is not in T, it assigns 0 to s; that is,

$$\chi_T(s) = \begin{cases} 1 & : s \in T \\ 0 & : s \notin T. \end{cases} \tag{1.13}$$

(The Greek letter χ is pronounced "chi" and is supposed to remind you of the first syllable of "characteristic.") Function χ_T tells which elements of S are in T and which elements of S are not in T. Characteristic functions establish a correspondence between subsets of S and functions from S to $\{0,1\}$.

Example 1.4.17. Let S be the set {Jack, Jill, Joan, June, Joe, John}. Consider the subset $T =$ {Jack, Joe, John} of S that contains all the boys in S. This subset can be described by the function $\chi_T : S \longrightarrow \{0,1\}$, which can be visualized as

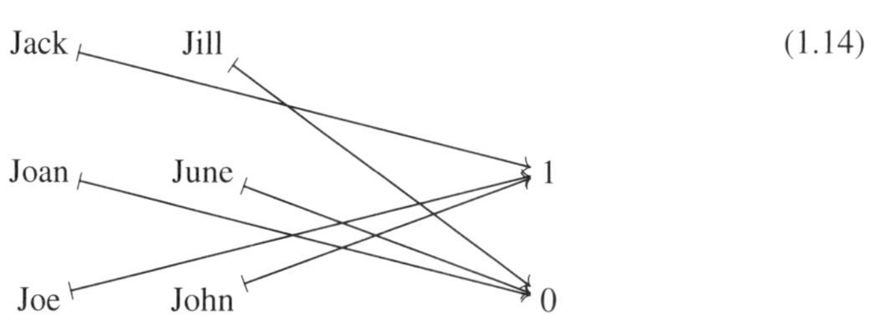

(1.14)

□

Exercise 1.4.18. For the sets of numbers, we know that $\mathbf{N} \subsetneq \mathbf{Z} \subsetneq \mathbf{Q} \subsetneq \mathbf{R} \subsetneq \mathbf{C}$. Give the characteristic function for each of these proper subsets. ■

A characteristic function assigns the elements of S to one of two possible values. There might be a need to assign one of many values to every element of S. For example, function $S \longrightarrow \{a,b,c,d\}$ assigns to every element of S one of these letters, which can stand for different ideas. In general, the function

$$S \longrightarrow \{1, 2, \ldots, n\} \tag{1.15}$$

assigns every element of S one of n numbers. We can also assign to every element of S an element of $[0,1]$, the real interval between 0 and 1. Such a function may correspond to assigning a probability to every element.

Example 1.4.19. In school, every student usually has an associated grade point average (GPA). This is written as a function $Students \longrightarrow [0,4]$. □

If S is a set, then there is a function called the **diagonal function** $\Delta\colon S \longrightarrow S \times S$, which takes every element to an ordered pair of the same element. In symbols, for s in S, we have

$$\Delta(s) = (s, s). \tag{1.16}$$

If $f\colon S \longrightarrow S'$ and $g\colon T \longrightarrow T'$ are functions, then there exists a function $f \times g\colon S \times T \longrightarrow S' \times T'$ that takes an ordered pair of elements and applies f to the first element and g to the second. In symbols, the function is defined for elements s of S and t of T as

$$(f \times g)((s,t)) = (f(s), g(t)) \in S' \times T'. \tag{1.17}$$

In a sense, this is a parallel process. Function f processes s, while function g processes t.

Exercise 1.4.20. Let $f\colon \mathbf{N} \longrightarrow \mathbf{R}$ be defined by $f(n) = \sqrt{n}$ and $g\colon \mathbf{R} \longrightarrow \mathbf{Z}$ be the ceiling function, denoted as $g(r) = \lceil r \rceil$. (The ceiling functions outputs the lowest integer that is greater than or equal to the input.) What is $(f \times g)((5, -5.1))$? ■

Definition 1.4.21. There are some special types of functions. We say that $f\colon S \longrightarrow T$ is

- **One-to-one** or **injective** if different elements in S go to different elements in T. That is, for all s and s' in S, if $s \neq s'$, then $f(s) \neq f(s')$. Another way to say this is that if $f(s) = f(s')$, then it must be that $s = s'$. This means that if the function takes elements to the same output, the elements must have started off equal.
- **Onto** or **surjective** if for every element t in T, there is an s in S such that $f(s) = t$.
- An **isomorphism**, a **one-to-one correspondence**, or a **bijection** if f is one-to-one and onto/surjective. That is, for every element s of S, there is a unique element t of T so $f(s) = t$; and for every element t of T, there is a unique element s of S so $f(s) = t$. When there are sets S and T with an isomorphism between them, we say that the sets are **isomorphic** and write this as $S \cong T$.

Isomorphism of sets is not the same idea as equality of sets. Consider a simple example: the set $\{x\}$ and the set $\{y\}$. Although these two sets have exactly the same number of elements, they are not equal. They are only isomorphic.

Exercise 1.4.22. Explain why two finite sets that have the same number of elements are isomorphic. ■

Exercise 1.4.23. Show that the Cartesian plane, $\mathbf{R} \times \mathbf{R}$, is isomorphic to the plane of complex numbers, $\mathbf{C}$. ■

One of the central ideas about sets is that, given sets S and T, we can form a set that consists of all functions from S to T. So while we looked at examples of particular functions, we will also be interested in the collection of *all* functions from one set to another. This collection will have interesting structure. We call this collection a **set of functions**, a **function set**, or a **Hom set**, and we denote it as $Hom(S,T)$ or T^S. (The notation $Hom(S,T)$ comes from the word "**hom**omorphism," which is a vestige of the algebraic origins of the idea. The notation T^S is similar to exponentiation because the function set has similar properties to exponentiation.)

Exercise 1.4.24. Write down the set of all the functions from the set $\{a,b,c\}$ to the set $\{0,1\}$. ■

We saw that every element in a set S can be described as function $\{*\} \longrightarrow S$. This correspondence between elements of S and functions from $\{*\}$ to S shows that

$$S \cong S^{\{*\}} = Hom(\{*\}, S). \tag{1.18}$$

Using characteristic functions, we saw in Definition 1.4.16 that there is a correspondence between subsets of S and functions from S to $\{0,1\}$. This correspondence can be stated as

$$\mathcal{P}(S) \cong \{0,1\}^S = Hom(S, \{0,1\}). \tag{1.19}$$

We will denote the set as $\{0,1\}$ as 2 and then write this as

$$\mathcal{P}(S) \cong 2^S = Hom(S, 2). \tag{1.20}$$

Example 1.4.25. Consider the simple binary addition operation $+\colon \mathbf{N}\times\mathbf{N} \longrightarrow \mathbf{N}$. Let us write this function with its inputs clearly marked as follows

$$(\ )+(\ )\colon \mathbf{N}\times\mathbf{N} \longrightarrow \mathbf{N}. \tag{1.21}$$

Now consider function $(\ )+5\colon \mathbf{N} \longrightarrow \mathbf{N}$. This is a function with only one input. We could also make another function of one variable $(\ )+7\colon \mathbf{N} \longrightarrow \mathbf{N}$. In fact, we can do this for any natural number. We can define a function that inputs a natural number and outputs a function from natural numbers to natural numbers. That is, there is a function $\Phi\colon \mathbf{N} \longrightarrow Hom(\mathbf{N},\mathbf{N})$, which is defined as $\Phi(n) = (\ )+n$. The information described by the $(\ )+(\ )$ function is the same as the information described by the Φ function.

Notice that what we said about + really applies to every function with two inputs. If $f\colon S\times T \longrightarrow U$ is a function from $S\times T$, then for every $t \in T$, there is a function $f(\ ,t)\colon S \longrightarrow U$. This shows that there is a function $f'\colon T \longrightarrow Hom(S,U)$. It is easy to see that the assignment described by any function f has the same information as the assignment described by function f'. That is, f and f' relay the same information. □

This example brings to light the following important theorem about sets.

Theorem 1.4.26. For sets S, T and U, there is an isomorphism:

$$Hom(S\times T, U) \cong Hom(T, Hom(S,U)) \qquad \text{or} \qquad U^{S\times T} \cong (U^S)^T \tag{1.22}$$

★

Proof. To show that these two sets are isomorphic, consider $f\colon S\times T \longrightarrow U$. From this function, let us define an $f'\colon T \longrightarrow Hom(S,U)$. For a t in T, we have function $f'(t)\colon S \longrightarrow U$, which is defined as follows: for s in S, let $f'(t)(s) = f(s,t)$. This function has the same information as f. Constructing f from f' is left to the reader. ♣

Let us count how many functions are between two finite sets. Consider S with $|S| = m$ and T with $|T| = n$, and a function $f\colon S \longrightarrow T$. For each element s in S, there are n possible values of $f(s)$ in T. For two elements in S, there are $n\cdot n$ possibilities of choices in T. In total, there are $n\cdot n\cdot\cdots\cdot n$ (m times) possible maps. So

$$|Hom(S,T)| = |T^S| = n^m = |T|^{|S|}. \tag{1.23}$$

Remark 1.4.27. For three sets S, T, and U, we have

$$\begin{aligned}
|Hom(S\times T,U)| &= |U^{S\times T}| && \text{by definition or notation}\\
&= |U|^{|S\times T|} && \text{by Equation (1.23)}\\
&= |U|^{|S|\cdot|T|} && \text{by Equation (1.6)}\\
&= (|U|^{|S|})^{|T|} && \text{by arithmetic}\\
&= |Hom(S,U)|^{|T|} && \text{by Equation (1.23)}\\
&= |Hom(T,Hom(S,U))| && \text{by definition.}
\end{aligned}$$

Notice that the rule about exponentiation usually learned as children, $m^{(n\cdot p)} = (m^n)^p$ is expanded to a rule about sets and functions. ♠

Operations on Functions

Often we are going to take two functions and perform an operation to get another function. Three such operations are composition, extension, and lifting.[1] Remarkably, many ideas about functions can be understood as operations in one of these three forms.

The simplest operation is **composition**. If there is a function $f\colon S \longrightarrow T$ and a function $g\colon T \longrightarrow U$, then the composite of them is the function $h = g\circ f\colon S \longrightarrow U$, which is defined on an s in S as $h(s) = g(f(s))$. We write these functions as

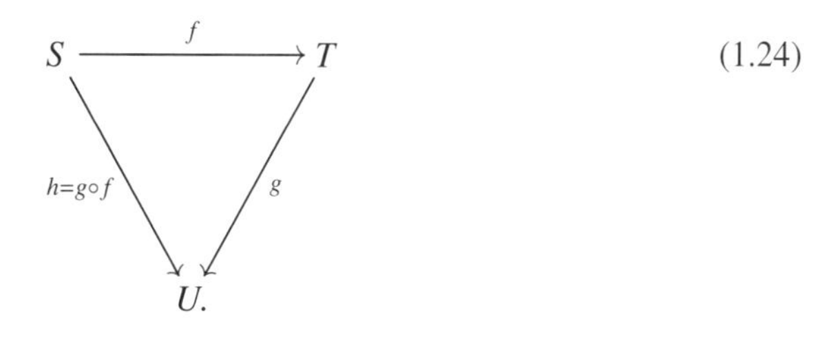

[1]Technically, however, extension and lifting are not operations. Composition is an operation because if you take two composable functions, you will get a unique composiiton function. In contrast, given two functions of a certain type, their extension and their lifiting are not necessarily unique. There might be many extension and many liftings for two functions. But we will use the word "operation" for these cases as well.

We say that f and g are **factors** of h or h **factors through** T. This diagram is called a **commutative diagram**. It means that if you start with any element s in S and you apply the functions f followed by g to go from S to U, you will get to the same element as applying function h to element s. In detail, for all s, we have that $g(f(s)) = h(s)$. The diagram is commutative because we can go around the triangle this way or that way. There will be many such diagrams in the coming pages. In all the cases, start from any set and follow all the paths of composible functions to another set, and you will get the same element. Throughout this text, unless otherwise stated, all diagrams are commutative.

Example 1.4.28. Consider the set of real numbers, **R**. Let a and b be real numbers. Consider function $(\ )\cdot a\colon \mathbf{R} \longrightarrow \mathbf{R}$, which takes any real number and multiplies it by a. There is also function $(\ )\cdot b\colon \mathbf{R} \longrightarrow \mathbf{R}$, which takes any real number and multiplies it by b. The composition of these two functions is function $(\ )\cdot(a\cdot b)\colon \mathbf{R} \longrightarrow \mathbf{R}$, which takes any real number and multiplies it by $a \cdot b$, as in the following commutative diagram:

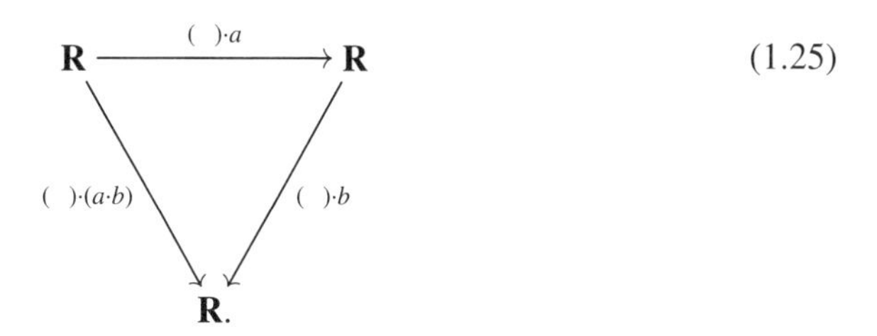

(1.25)

One can make similar compositions with other arithmetic operations. □

Exercise 1.4.29. Show that function composition is associative. That is, let $f\colon S \longrightarrow T$, $g\colon T \longrightarrow U$ and $h\colon U \longrightarrow V$, and show that $h \circ (g \circ f) = (h \circ g) \circ f$. ■

The fact that function composition is associative will be used many times throughout this text.

When dealing with the identity function, the input is the same as the output. This has an interesting consequence when dealing with composition. If you compose a function with an identity map, then you get the original function. In detail, for $f\colon S \longrightarrow T$, $id_S : S \longrightarrow S$, and $id_T : T \longrightarrow T$, we have

$$f \circ id_S = f \qquad \text{and} \qquad id_T \circ f = f. \tag{1.26}$$

We can see these equations as the following commutative diagram:

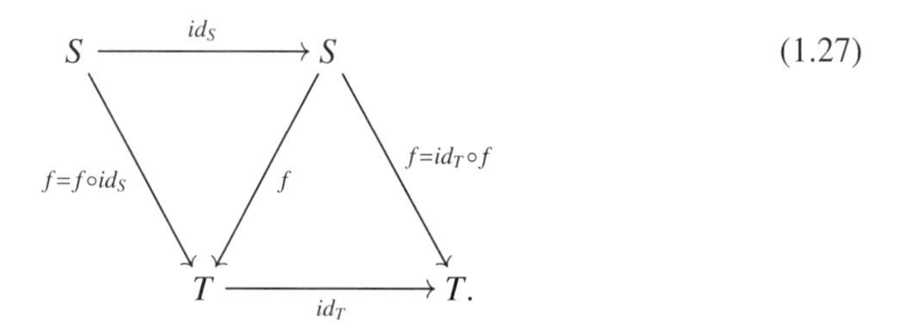

(1.27)

Example 1.4.30. Evaluation of a function can be seen as composition. Let $f: S \longrightarrow T$ be a function, and let an element be described by the function $g: \{*\} \longrightarrow S$. Then the value of f on the element that g chooses is the element that $f \circ g$ chooses, as in

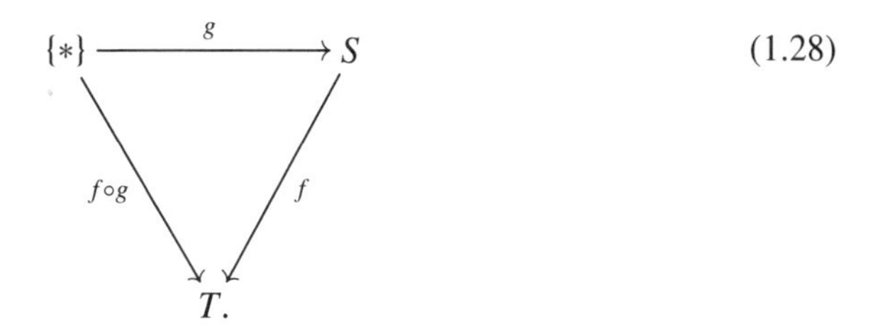

(1.28)

The function $f \circ g: \{*\} \longrightarrow T$ picks out the output of f. If g performs the assignment $* \mapsto s$, then $f \circ g$ performs the assignment $* \mapsto f(s)$. □

Example 1.4.31. If $f: S \longrightarrow T$ and A is a subset of S with inclusion function $inc: A \hookrightarrow S$, then the restriction of f to A is the function $f|: A \longrightarrow T$, which is given as the composition

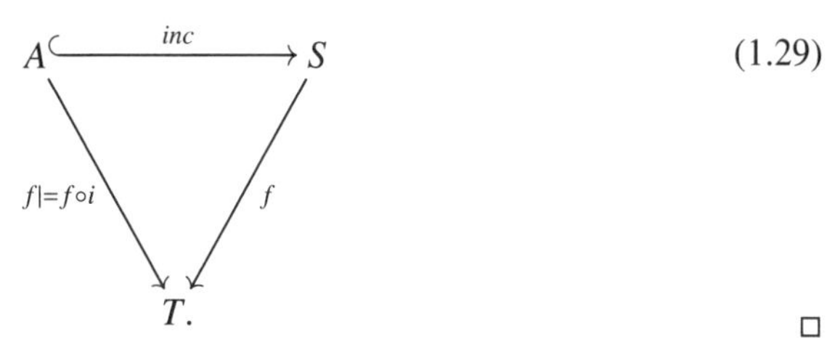

(1.29)

□

Theorem 1.4.32. The three properties of functions that we saw in Definition 1.4.21 can be described with function composition. For nonempty sets S and T, the function $f: S \longrightarrow T$ is

- **One-to-one** if and only if there exists $g: T \longrightarrow S$ such that $g \circ f = id_S$:

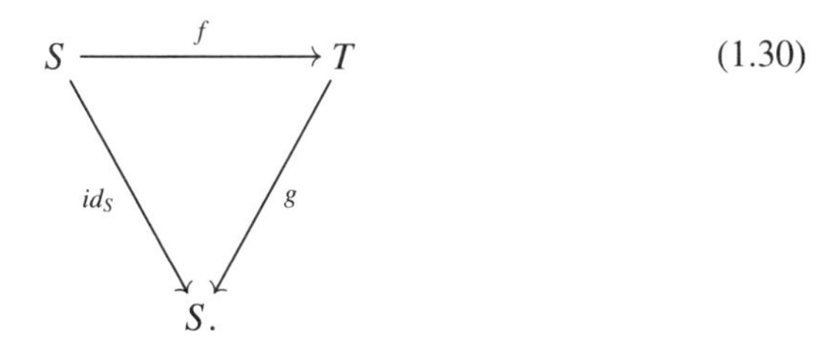

(1.30)

(Proof: The existence of g such that the diagram commutes implies that f is one-to-one. If $f(s) = f(s')$, then apply g to both sides of the equation and get $g(f(s)) = g(f(s'))$. But $g \circ f = id_S$ implies $s = s'$.

If f is one-to one, then there exists g such that the diagram commutes. Let $t \in T$. Assign $g(t)$ to be the unique s such that $f(s) = t$. If t is not an output of f, then it does not matter what value you give to $g(t)$.)

- **Onto** if and only if there exists $g\colon T \longrightarrow S$ such that $f \circ g = id_T$:

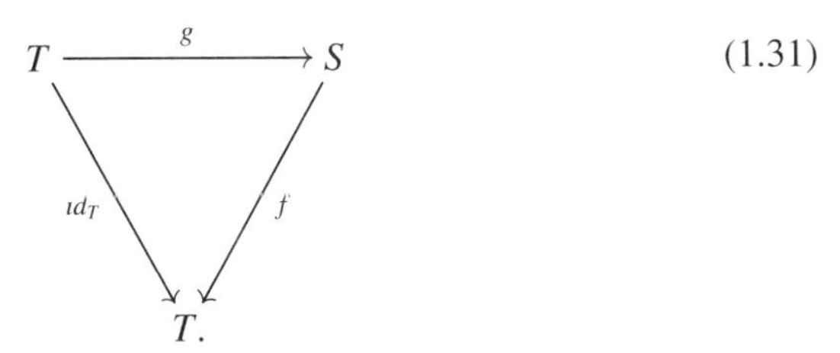

(1.31)

(Proof: The existence of g implies f is onto. Function f is onto because for any $t \in T$, function g has $g(t) = s$ for some $s \in S$. This s gives an input to f whose output is t; that is, $f(s) = f(g(t)) = id_T(t) = t$.

"Onto" implies the existence of g such that the diagram commutes. Let $g(t)$ equal any s such that $f(s) = t$. There must be one such t because f is onto. This proof assumes the axiom of choice, which we will meet later.)

- **Isomorphism** or **one-to-one correspondence** if and only if there exists $g\colon T \longrightarrow S$ such that $g \circ f = id_S$ and $f \circ g = id_T$. Or, putting the previous two triangles together, we have

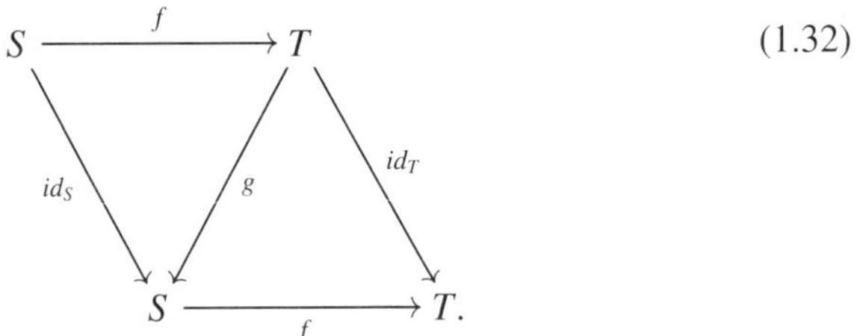

(1.32)

Another way to express this is the following diagram:

g∘f = id_S S f g T id_T = f∘g (1.33)

We will see variations of this diagram again and again. ★

A second operation of functions is an **extension**. In detail, if $f\colon R \longrightarrow T$ is a function and R is a subset of S with the inclusion function $inc\colon R \hookrightarrow S$, then an extension of f along inc is a function $\hat{f}\colon S \longrightarrow T$ such that the following commutes:

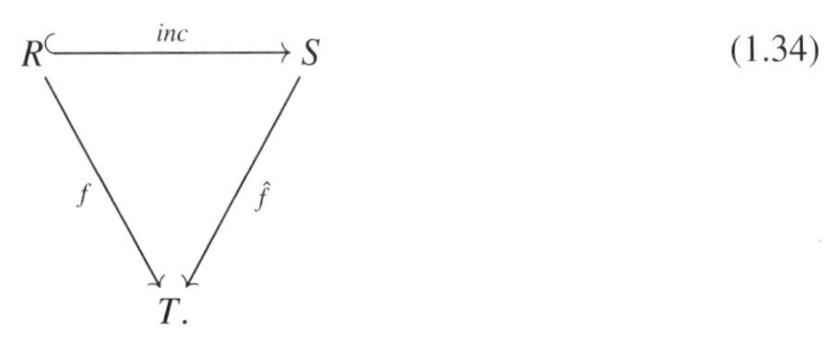

(1.34)

In English, $\hat{f}$ extends f to a larger domain.

Example 1.4.33. As a simple example, consider R to be a set of students and

$$f\colon R \longrightarrow \{A, B, C, D, F\} \tag{1.35}$$

assigns every student a grade. If some new students came into the class, the teacher would have to extend f to give grades to all the students (including the new ones) as $\hat{f}\colon S \longrightarrow \{A, B, C, D, F\}$. We want $\hat{f}$ to assign the same grades as f did for any of the original students. This is clear with the following commutative diagram:

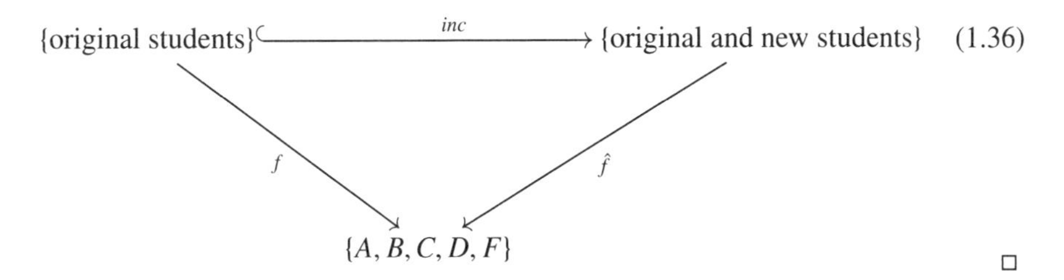

□

Example 1.4.34. Let $\{3, 5\}$ be a set of two real numbers. There is an obvious inclusion of the two real numbers into the set of all numbers $inc\colon \{3, 5\} \hookrightarrow \mathbf{R}$. Let $f\colon \{3, 5\} \longrightarrow \mathbf{R}$ be any function that picks two values. Then there exists a function $\hat{f}\colon \mathbf{R} \longrightarrow \mathbf{R}$ that extends f.

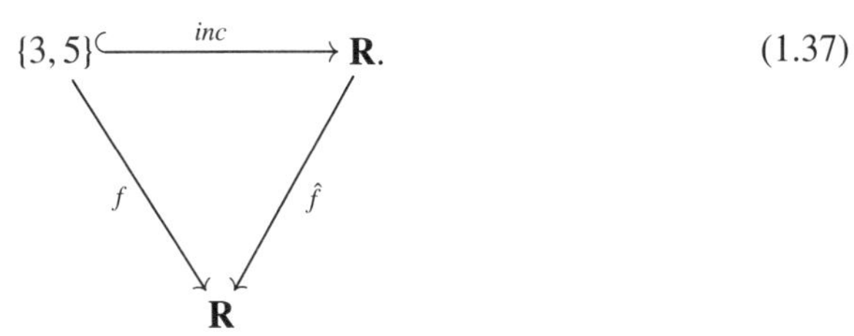

This extension is another way of describing the simple idea that, given any two points on the plane, there is a straight line that passes through both of them. This can be visualized as follows:

(1.38)

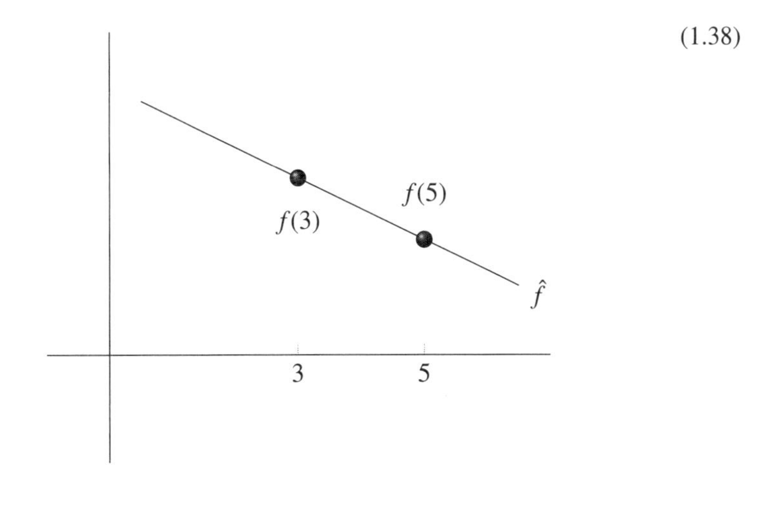

In detail, the extended line is given by the following formula:

$$\begin{aligned}\hat{f}(x) = mx + b = \frac{\Delta y}{\Delta x}x + b &= \frac{f(5) - f(3)}{5-3}x + \frac{5f(3) - 3f(5)}{5-3} \\ &= \frac{f(5) - f(3)}{2}x + \frac{5f(3) - 3f(5)}{2}. \end{aligned} \tag{1.39}$$

Be aware that this extension is one of many ways to extend f. There is nothing special about this extension other than it is usually taught in the first year of high school. □

Thinking of a straight line as a function, the previous example of an extension can be ... extended ...

Example 1.4.35. Let $\{x_0, x_1, x_2, \ldots, x_n\}$ be a set of $n+1$ different real numbers, and let $inc\colon \{x_0, x_1, x_2, \ldots, x_n\} \hookrightarrow \mathbf{R}$ be the inclusion function. Every $f\colon \{x_0, x_1, x_2, \ldots, x_n\} \longrightarrow \mathbf{R}$ has an extension called $\hat{f}\colon \mathbf{R} \longrightarrow \mathbf{R}$ along inc, which is a polynomial function of degree at most n:

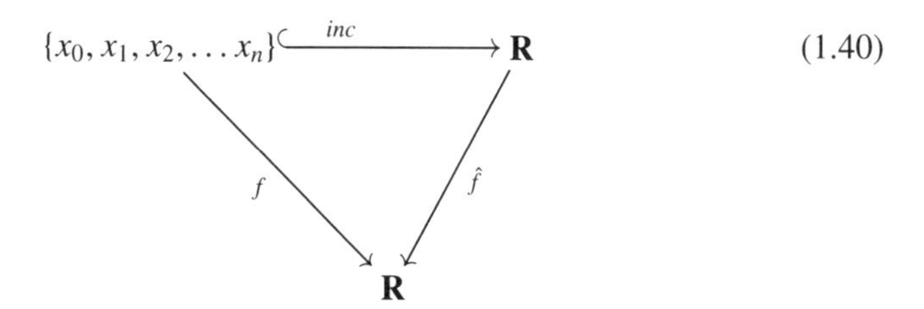

(1.40)

This can be visualized as

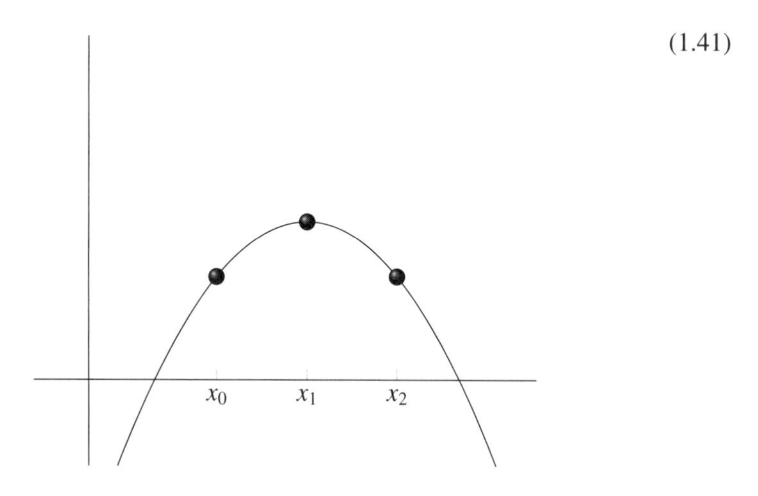

(1.41)

Function $\hat{f}$ is called the "Lagrange interpolating polynomial" of the points described by f. (We will not use this in this text.) □

While extensions are usually about inclusion functions, we can also use the setup of an extension for functions that are not inclusion functions.

Example 1.4.36. Function $f\colon S \longrightarrow T$ is a **constant function** if it outputs the same value for any input. That means there is some $t_0 \in T$ such that for all $s \in S$, we have $f(s) = t_0$. We can describe a constant function using the same notation of an extension, but without the inclusion function. In particular, f is a constant function if there is an "extension" $\hat{f}\colon \{*\} \longrightarrow T$ of f, as in the diagram

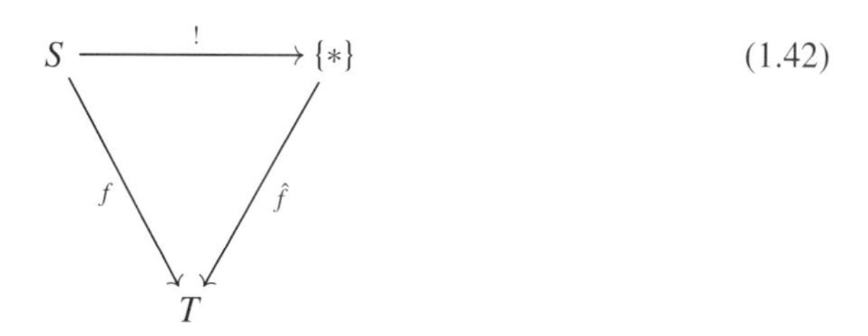

(1.42)

where function $!\colon S \longrightarrow \{*\}$ is the unique function that always outputs $*$, the only element that it can output. Another way of saying this is that f is a constant function if it can be written as a function that factors through $\{*\}$. □

Exercise 1.4.37. Show that if $id_S\colon S \longrightarrow S$ can be extended along the function $f\colon S \longrightarrow T$, then f is a one-to-one function. ■

The third operation of functions is a **lifting**. Consider an onto function $p\colon T \longrightarrow T'$. Let $f\colon S \longrightarrow T'$ be any function. A lifting of f along p is function $\hat{f}\colon S \longrightarrow T$, which makes the following triangle commute:

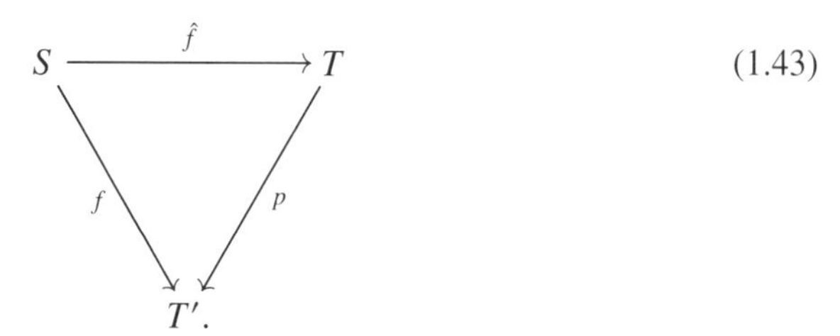

(1.43)

In a sense, we lift the map f from the target of p to the source of p. Example 1.4.38 presents the simplest example of lifting.

Example 1.4.38. Consider the following commutative diagram:

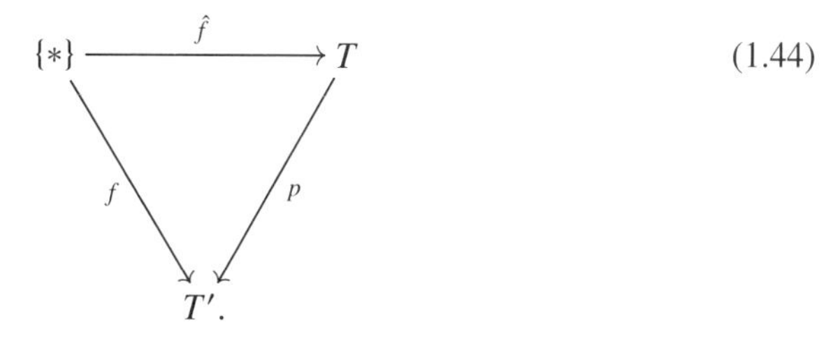

(1.44)

Here, a function $f\colon \{*\} \longrightarrow T'$ picks out an element of T'. A lifting of f is $\hat{f}\colon \{*\} \longrightarrow T$, which picks out an element of T. The fact that the diagram is commutative means that

p will take outputs of $\hat{f}$ to outputs of f. In other words, if f picked out $t_0 \in T'$, then $\hat{f}$ will pick an element in T that will map onto t_0 under p. There might be many liftings. The set of all possible elements that a lifting can pick is denoted as $p^{-1}(t_0) \subseteq T$, which is called the "preimage" of p. □

Example 1.4.39. Here is a cute example of a lifting from the world of politics. Let T be the set of 320 million American citizens, and let T' be the set of 50 states. Function p takes every citizen to the state that they live in. Let S be a set of three elements such as $\{a, b, c\}$. The function $f\colon S \longrightarrow T'$ chooses 3 states. A lifting of f along p is a function $\hat{f}\colon S \longrightarrow T$, which will choose three citizens. Each of the three will be from the states that f chooses. There are obviously many such liftings. □

Example 1.4.40. Let us build on the last example. Let T, T', and p be as in that example. Let S be the set $\{a, b, c\} \times T'$ (i.e., pairs of letters and states). The $f\colon S \longrightarrow T'$ function is defined as follows: $f(b, \text{New Jersey}) = \text{New Jersey}$ (i.e., f takes a letter and a state and outputs the same state). Notice that for each state, there are three elements in S that go to that state. For example,

$$f(a, \text{New Jersey}) = f(b, \text{New Jersey}) = f(c, \text{New Jersey}) = \text{New Jersey}.$$

A lifting of f along p is function $\hat{f}\colon S \longrightarrow T$, which will choose three citizens from each state. There are many such liftings. □

Exercise 1.4.41. Show that if $id_T : T \longrightarrow T$ can be lifted along the function $f\colon S \longrightarrow T$, then f is an onto function. ■

One can see these three operations—composition, extension, and lifting—as three sides of a triangle:

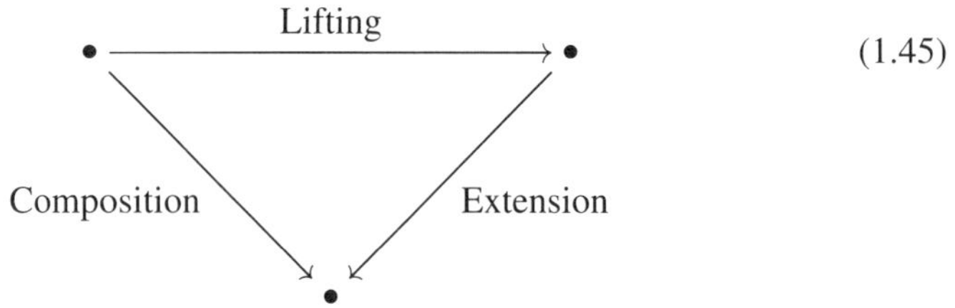

(1.45)

Each side uses the other two sides as the input to the operation. Composition will be used on every page of this book. We will see (especially in Section 9.2) that the extension and lifting operations are very important in many contexts besides sets and functions.

The rest of this mini-course will be concerned with equivalence relations, graphs, and groups. These three subjects are structures based on sets. The eager reader might want to skip these pages and go directly to the beginning of Chapter 2.

Equivalence Relations

We are not only interested in how a set is related to other sets. Sometimes the elements of a set are related to each other in interesting ways as well.

Definition 1.4.42. Let S be a set. A **relation** on S is a subset R of set $S \times S$. The ordered pair (s_1, s_2) in R means s_1 is related to s_2.

Example 1.4.43. Let S be the set of citizens of the United States. Consider the following relations on this set:

- R_1 consists of those (s, t) where s and t are cousins.
- R_2 consists of those (s, t) where s is the same age as or older than t.
- R_3 consists of those (s, t) where s and t live in the same state.
- R_4 consists of those (s, t) where s and t belong to the same political party. □

The three properties of a relation given in Definition 1.4.44 will characterize the notion of "sameness."

Definition 1.4.44. The relation $R \subseteq S \times S$ on a set S is

- **Reflexive** if every element is related to itself: for all s in S, (s, s) is in R.
- **Symmetric** whenever one element is related to another, then the other is related to the first: for all s and t in S, if (s, t) is in R, then (t, s) is in R.
- **Transitive** whenever s is related to t and t is related to u, then s is related to u: for all s, t and u in S, if (s, t) is in R and (t, u) is in R, then (s, u) is in R.

Example 1.4.45. Let us look at which properties are satisfied from the relations of Example 1.4.43:

- The cousin relation R_1 is not reflexive (no one is their own cousin); it is symmetric; but it is not transitive. So x can be a cousin to y through y's mother's side and y can be a cousin to z through y's father's side. In this case, x will, in general, not be a cousin to z.
- The older relation R_2 is reflexive (everyone is the same age as themselves), not symmetric (if x is older than y, then y is not older than or the same age as x), and it is transitive.
- The state relation R_3 is reflexive, symmetric, and transitive.
- The political party relation R_4 is reflexive, symmetric, and transitive. □

Many times, a set of elements can be split up or partitioned into different subsets where each subset will have all the elements that have a particular property. For example, the set of cars can be split up by color. So there will be a subset of blue cars, a subset of red cars, a subset of green cars, etc. The collection of all such subsets will form a set itself. Formally, this can be said as follows.

Definition 1.4.46. A relation on a set is an **equivalence relation** if it is reflexive, symmetric, and transitive. We write such relations as $\sim \subseteq S \times S$ and write $r \sim s$ for $(r, s) \in \sim$. With an equivalence relation on the set S, we can describe disjoint subsets of S called **equivalence classes**. If s is an element of S, then the equivalence class of s is the set of all elements that are related to it, as follows:

$$[s] = \{r \in S : r \sim s\} \tag{1.46}$$

That is, $[s]$ is the set of all elements that are the same as s. For a given set S and an equivalence relation $\sim$ on S, we form a **quotient set** denoted as $S/\sim$. The elements of $S/\sim$ are all the equivalence classes of elements in S. There is an obvious **quotient function** from S to $S/\sim$ that takes s to $[s]$.

Example 1.4.47. Let us examine the equivalence classes for the equivalence relations of Example 1.4.43:

- Each equivalence class for relation R_3 consists of all the residents of a particular state. The quotient set contains the 50 equivalence classes corresponding to the 50 states (here, we are ignoring nonstate entities like Guam and Washington, D.C.). The quotient function takes every citizen to the state in which they reside.
- Each equivalence class for relation R_4 consists of all the people belonging to a particular political party. The quotient set is a set whose elements correspond to political parties. The quotient function takes every citizen to the political party to which they belong (we are ignoring independents). □

Graphs

A directed graph is a common structure (based on sets) that has applications everywhere. Directed graphs also have many similarities to categories.

Definition 1.4.48. A **directed graph** $G = (V(G), A(G), src_G, trg_G)$ is both of the following:

- A set of vertices, $V(G)$
- A set of arrows, $A(G)$

Furthermore, the following are true:

- Every arrow has a source: there is a function $src_G : A(G) \longrightarrow V(G)$
- Every arrow has a target: there is a function $trg_G : A(G) \longrightarrow V(G)$.

If f is an element of $A(G)$ with $src_G(f) = x$ and $trg_G(f) = y$, we draw this arrow as

$$x \xrightarrow{\quad f \quad} y. \tag{1.47}$$

An example of a graph is shown in Figure 1.3.

Example 1.4.49. Graphs are everywhere:

- A street map can be thought of as a directed graph in which the vertices are street corners and there is an arrow from one corner to the other if there is a one-way

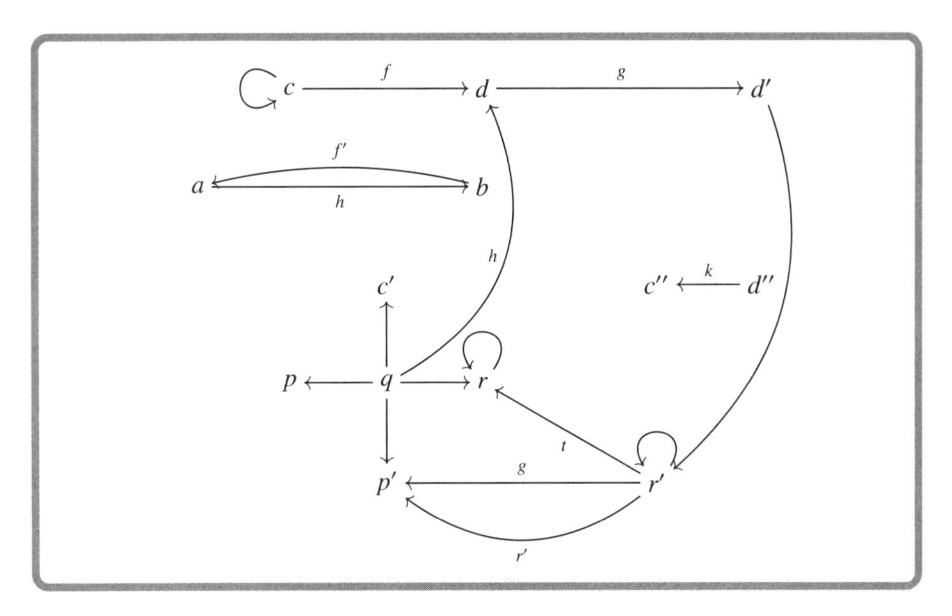

Figure 1.3. A directed graph.

street between them. When there is a two-way street, we might write it like this:

$$* \underset{\longleftarrow}{\overset{\longrightarrow}{}} * \qquad \text{or} \qquad * \longrightarrow *. \tag{1.48}$$

Such an arrow is called a "symmetric edge."

- An electrical circuit can be viewed as a directed graph. The vertices are the branching points, and the edges might have resistors, batteries, capacitors, diodes, and other elements. The arrows describe the direction of the flow of electricity.
- Computer networks can be seen as directed graphs in which the vertices are computers and there is an arrow from one computer to another if there is a way for the first computer to communicate with the second.
- The billions of web pages in the World Wide Web form a directed graph. The vertices are the web pages, and there is an arrow if there is a link from one web page to another.
- Facebook can be seen as a directed graph. Every personal Facebook account is a vertex, and there are arrows between two Facebook accounts if they are friends. Notice that if x is friends with y, then y must be friends with x. So all the arrows are symmetric edges and the graph is called "symmetric."
- All the people on Earth form a graph. The vertices are the people. There is an arrow from x to y if x knows y. (We are not being specific as to what it means to "know" someone.) There is an idea called "six degrees of separation," which says that in this graph, you never need to traverse more than six arrows to get from any person to any other person. We are all connected!

- The collection of all sets and functions form a giant graph. Described in detail, the vertices are all sets. The arrows are functions from one set to another. (This will be a motivating example of a category.) □

A graph homomorphism is a way of mapping one graph to another. This will be similar to what happens when we talk about mapping one category to another category. Basically, the vertices map to the vertices and the arrows map to the arrows, but we insist that they match up well in that mapping.

Definition 1.4.50. Let $G = (V(G), A(G), src_G, trg_G)$ and $G' = (V(G'), A(G'), src_{G'}, trg_{G'})$ be graphs. A **graph homomorphism** $H\colon G \longrightarrow G'$ consists of

- A function that assigns vertices to vertices, $H_V\colon V(G) \longrightarrow V(G')$
- A function that assigns arrows to arrows, $H_A\colon A(G) \longrightarrow A(G')$

These two maps must respect the source and target of each arrow. That means that both of the following are true:

- For all f in $A(G)$, $H_V(src_G(f)) = src_{G'}(H_A(f))$.
- For all f in $A(G)$, $H_V(trg_G(f)) = trg_{G'}(H_A(f))$.

Saying that these axioms are satisfied is the same as saying that the following two squares commute:

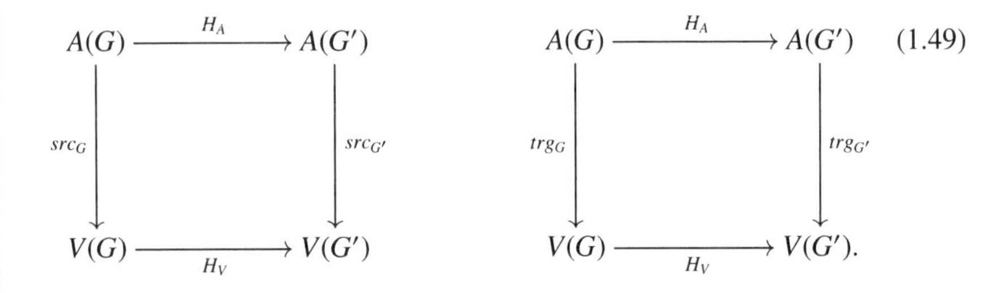

(1.49)

Another way to understand these requirements is to see what the maps H_V and H_A do to a single arrow f (i.e., $f \mapsto H_A(f)$. We use the wavy arrow so the reader can see what is going on easier.)

Graph G **Graph G′** (1.50)

$src_G(f) \overset{H_V}{\mapsto} H_V(src_G(f)) = src_{G'}(H_A(f))$

$f \overset{H_A}{\mapsto} H_A(f)$

$trg_G(f) \overset{H_V}{\mapsto} H_V(trg_G(f)) = trg_{G'}(H_A(f))$

Just as we can determine many properties of sets by examining functions, we can determine many properties of graphs by examining graph homomorphisms from simple graphs.

Example 1.4.51. Here are some properties of graphs that are determined by graph homomorphisms.

- A vertex of graph G can be described by a graph homomorphism from the one-vertex graph $(*)$ (no arrows) as follows $H\colon * \longrightarrow G$.
- A directed edge of a graph can be determined by a graph homomorphism from the graph $* \longrightarrow *$ to G.
- A triangle in graph G can be determined by a graph homomorphism from the graph:

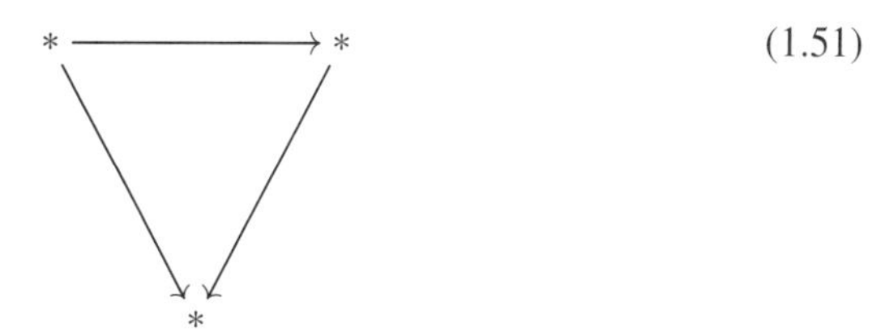

  to graph G.
- A path of length n in graph G can be determined by a graph homomorphism from the "snake" graph:

$$* \longrightarrow * \longrightarrow * \longrightarrow \cdots \longrightarrow * \tag{1.52}$$

  of length n to G. □

Exercise 1.4.52. Prove the following fact about graph G. If there is *not* a graph homomorphism that is surjective on vertices from G to the graph

$$\overset{f_a}{\circlearrowleft} \quad \overset{f_b}{\circlearrowleft}$$
$$a \qquad b \tag{1.53}$$

then G is weakly connected (for any two vertices, there is a sequence of edges in either direction between them). ■

Exercise 1.4.53. Use graph homomorphisms to determine different types of paths in a graph:

- How do you describe a simple path in a graph (a simple path is a path that does not have repeated vertices)?
- What about a cycle of length n (a cycle is a path that starts and ends at the same vertex)?
- Do the same for a simple cycle of length n (a simple cycle is a cycle in which the only repeating vertex is the starting point which is the ending point). ■

Exercise 1.4.54. Show that the composite of graph homomorphisms is a graph homomorphism. Also, show that the composition is an associative operation. ■

Exercise 1.4.55. Define the **identity graph homomorphism**, I_G, for any graph G. Show that if $H\colon G \longrightarrow G'$ is a graph homomorphism, then $H \circ I_G = H$ and $I_{G'} \circ H = H$. ■

Groups

Another important structure that is based on sets and related to categories is a group. It is nice to see the definition of a group from a function perspective.

First, here is a discussion of operations. We all know what we mean by operations on numbers. If you take numbers x and y, you can perform the addition operation, $x + y$. You can also perform other operations, like $x - y$ or $y \cdot x$. All of these are examples of binary operations. Operations are really just functions. For a given set, S, a **binary operation** is a function $f\colon S \times S \longrightarrow S$. For two elements s and s', we write the value of f as $f(s, s')$. A **unary operation** is a function that takes one element of S and outputs one element of S (i.e., $f\colon S \longrightarrow S$). An example of a unary operation is the inverse operation that takes x and returns x^{-1}. There are also **ternary operations**, $f\colon S \times S \times S \longrightarrow S$, and n**-ary operations**,

$$f\colon \underbrace{S \times S \times \cdots \times S}_{n \text{ times}} \longrightarrow S \tag{1.54}$$

for all natural numbers n. If $n = 0$, then we write the 0-ary product as the set with one element $\{*\}$ and a **0-ary operation** is written as $f\colon \{*\} \longrightarrow S$, which basically picks out an element of S. Such an operation describes an element that does not change (i.e., a constant).

Let us put this all together and give the formal definition of a group.

Definition 1.4.56. A **group** $(G, \star, e, (\ )^{-1})$ is set G with the following three operations:

- A binary operation: A function $\star\colon G \times G \longrightarrow G$.
- An identity: There is a special element e in G called the identity of the group. This can be stated in a functional way: there is a 0-ary operation $u: \{*\} \longrightarrow G$, where $u(*) = e$.
- An inverse operation: A unary operation $(\ )^{-1}\colon G \longrightarrow G$

These three operations satisfy the following axioms:

- The binary operation is associative: for all x, y and z, we have $(x \star y) \star z = x \star (y \star z)$.
- The identity acts like a unit of the binary operation (e.g., when you multiply a number by 1, the result does not change—that is, $1 \cdot n = n$, so 1 is a "unit"): for all x, $x \star e = x = e \star x$.
- Applying the binary operation to an element with its inverse gives the identity: for all x in G, $x \star x^{-1} = e = x^{-1} \star x$.

Example 1.4.57. Here are some examples of groups:

- The additive integers: $(\mathbf{Z}, +, 0, -(\ ))$. Addition and negation are the usual operations.
- The additive real numbers: $(\mathbf{R}, +, 0, -(\ ))$. Addition and negation are the usual operations.
- The multiplicative positive reals: $(\mathbf{R}^+, \cdot, 1, (\ )^{-1})$ where $\mathbf{R}^+$ are the positive real numbers, the operation $\cdot$ is multiplication, and the function $(\ )^{-1}$ takes r to $\frac{1}{r}$.
- Clock arithmetic: $(\{0, 1, 2, 3, \ldots, 11\}, +, 0, -)$, where addition and subtraction is going around the clock. Here, 0 is the unit because when you add 0 to any number, you get back to the original number. Notice that we could have used another number besides 11. In fact, any nonnegative integer would have worked.
- The **trivial group**: $(\{0\}, +, 0, -)$. This is the world's smallest group. It has only one element, and the operations work as expected. □

Parts of the three axioms of a group can be seen as commutative diagrams in Figure 1.4. Around each of the commutative diagrams are maps showing the values of the functions on elements.

Exercise 1.4.58. The second diagram in Figure 1.4 shows the $x = x \star e$ axiom. Give a commutative diagram for the $x = e \star x$ axiom. ■

Exercise 1.4.59. The third diagram in Figure 1.4 shows the $e = x \star x^{-1}$ axiom. Give the commutative diagram for the $e = x^{-1} \star x$ axiom. ■

> **Important Categorical Idea 1.4.60. Descriptions Using Morphisms.** Many times, even when we have a nice, clear definition or description of a mathematical structure in terms of elements, we still want a description in terms of functions or morphisms. The reason that a description using functions or morphisms is important is that once we have it, we can use it in many different categories. Whereas a description in terms of elements is good in only one context, a description in terms of functions or morphisms can be used in many different categories and contexts. For example, we will see this morphism definition of a group arise in other contexts besides sets and functions. ○

Just as a function is a way of mapping one set to another and a graph homomorphism is a way of mapping one graph to another, a group homomorphism is a way of mapping one group to another.

Definition 1.4.61. Let $(G, \star, e, (\ )^{-1})$ and $(G', \star', e', (\ )'^{-1})$ be groups. A **group homomorphism** $f\colon (G, \star, e, (\ )^{-1}) \longrightarrow (G', \star', e', (\ )'^{-1})$ is a function $f\colon G \longrightarrow G'$ that satisfies the following two axioms:

- The function respects the group operation: for all $x, y \in G$, $f(x \star y) = f(x) \star' f(y)$.
- The function respects the unit: $f(e) = e'$.

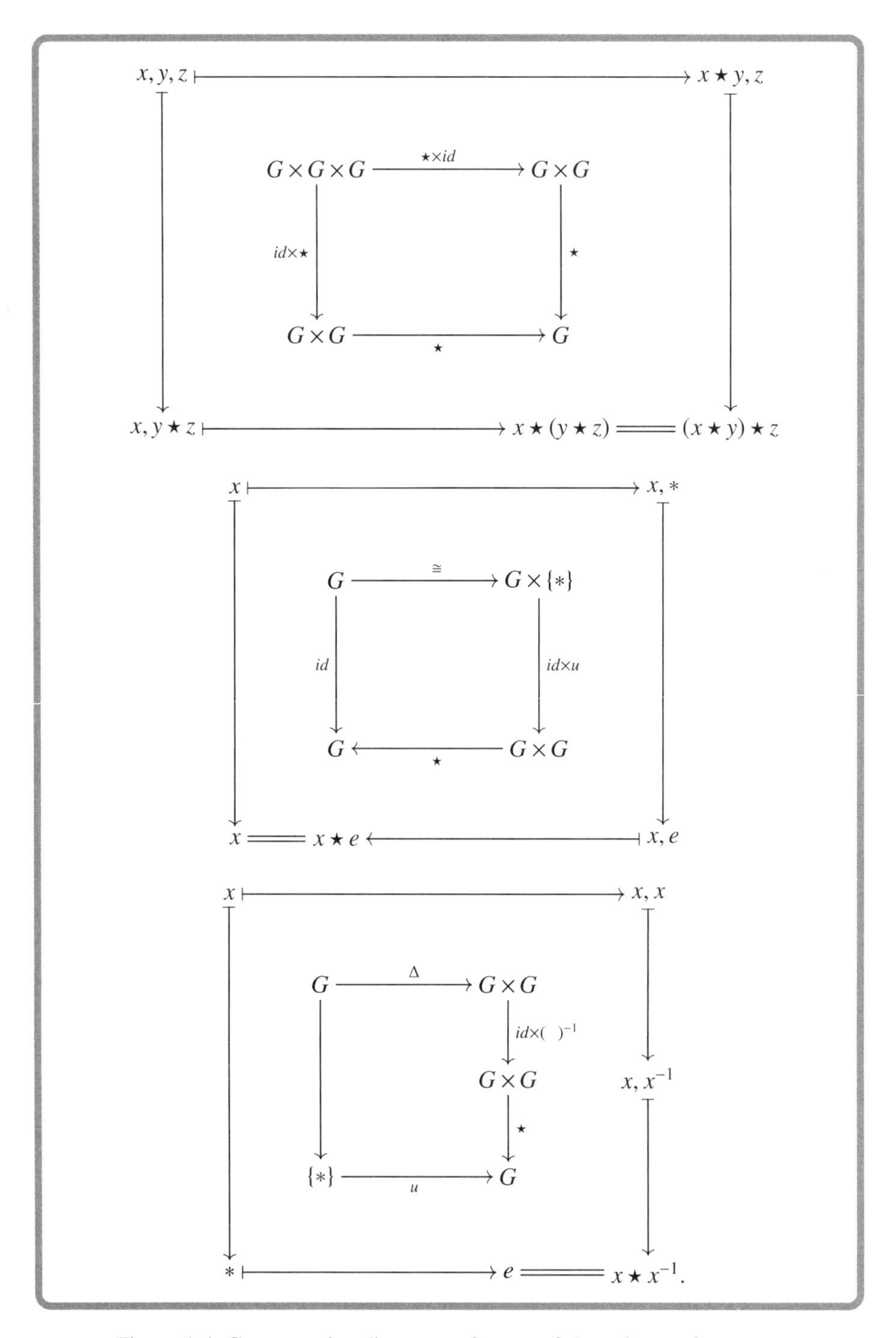

Figure 1.4. Commutative diagrams of some of the axioms of a group.

We can write these two requirements as the following two commutative diagrams:

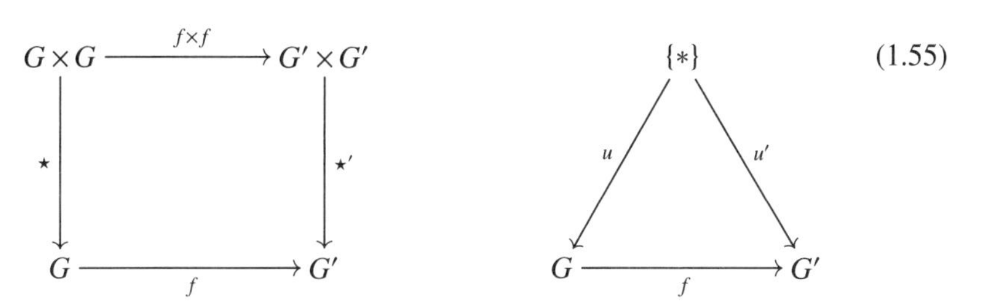

(1.55)

Technical Point 1.4.62. We did not insist that f respect inverses. Do not worry about that. This statement is true without saying it because it is a consequence of the other two axioms. First, notice that in any group, x^{-1} is the unique inverse of element x. To see this, imagine that x has two inverses, y and y'. Consider the following sequence of equalities:

$$\begin{aligned} y &= y \star e && \text{by the unit axiom} \\ &= y \star (x \star y') && \text{because } y' \text{ is the inverse of } x \\ &= (y \star x) \star y' && \text{by the associativity axiom} \\ &= e \star y' && \text{because } y \text{ is the inverse of } x \\ &= y' && \text{by the unit axiom.} \end{aligned}$$

This shows that $y = y'$. Now let us use this fact to show that inverses are preserved by group homomorphisms. First, consider

$$e' = f(e) = f(x \star x^{-1}) = f(x) \star' f(x^{-1}). \quad (1.56)$$

This shows that the inverse of $f(x)$ is $f(x^{-1})$. Since inverses are unique, we proved $f(x)^{-1} = f(x^{-1})$. ♡

Example 1.4.63. Here are some examples of group homomorphisms:

- There is always a unique group homomorphism from any group to the trivial group, where every element of the group goes to 0 of the trivial group.
- There is always a unique group homomorphism from the trivial group to any group in which the 0 of the trivial group goes to the identity of the group.
- There is an inclusion group homomorphism $inc\colon \mathbf{Z} \longrightarrow \mathbf{R}$.
- There is a group homomorphism $\mathbf{Z} \longrightarrow \{0, 1, 2, 3, \ldots, 11\}$ that takes every whole number x and sends it to the remainder when x is divided by 12.
- Let b be some positive real number called the "base." There is a exponential function

$$b^{(\ )}\colon (\mathbf{R}, +, 0, -) \longrightarrow (\mathbf{R}^{+}, \cdot, 1, (\ )^{-1}) \quad (1.57)$$

that takes a real number r and sends it to b^r. The two requirements to be a group homomorphism turn out to mean that $b^{r+r'} = b^r \cdot b^{r'}$ ($b^{(\ )}$ takes addition to multiplication) and $b^0 = 1$.

- There is a logarithm function (which is the inverse of the exponential function):

$$Log_b\colon (\mathbf{R}^+, \cdot, 1, (\ )^{-1}) \longrightarrow (\mathbf{R}, +, 0, -). \tag{1.58}$$

Function Log_b takes a positive real number r to $Log_b(r)$. The requirements to be a group homomorphism are the well-known facts that $Log_b(r \cdot r') = Log_b(r) + Log_b(r')$ (Log_b takes multiplication to addition) and $Log_b(1) = 0$. □

Exercise 1.4.64. Show that the composite of group homomorphisms is a group homomorphism. Also, show that the composition is an associative operation. ■

Exercise 1.4.65. Define the **identity group homomorphism**, id_G, for every group $(G, \star, 0, (\ )^{-1})$. Show that if $f\colon (G, \star, e, (\ )^{-1}) \longrightarrow (G', \star', e', (\ )'^{-1})$ is a group homomorphism, then $f \circ id_G = f$ and $id_{G'} \circ f = f$. ■

Suggestions for Further Study

Most of the material found in this section can be found in any discrete mathematics or finite mathematics textbook (e.g., [226, 225]). This material can also be found in many precalculus textbooks.

The importance of looking at functions between sets is central to all of category theory. This is also stressed by two books coauthored by F. William Lawvere, one of the leaders of category theory. Together with Robert Rosebrugh, Lawvere wrote *Sets for Mathematics* [165], and together with Stephen H. Schanuel, he wrote *Conceptual Mathematics* [166].

The novice can find basic group theory in any introduction to modern algebra or abstract algebra (e.g., [15, 79]).

The idea that most of the operations on functions can be seen as compositions, extensions, and liftings (as in Diagram (1.45)) was taken from [265], where much of category theory is built from these operations. We will see more of extensions and liftings in Section 9.2.

If you really want to learn more about categorical thinking, roll up your sleeves and let us get to the rest of this book!

2

Categories

A good stack of examples, as large as possible, is indispensable for a thorough understanding of any concept, and when I want to learn something new, I make it my first job to build one.

Paul Halmos
[100], page 63

With the ideas of set and functions in hand, we move on to the world of categories. Section 2.1 begins with a formal definition of categories. It then proceeds to list a giant stack of examples of categories from all over. In Section 2.2, we discuss some simple properties of morphisms in categories. We elaborate on some simple categories related to a category in Section 2.3. The chapter ends with Section 2.4, which is a mini-course that teaches the basics of linear algebra. The study of linear algebra is essentially an in-depth exploration of the category of vector spaces.

2.1 Basic Definitions and Examples

Before formally defining a category, let us summarize what we saw in Section 1.4 concerning sets and functions. The collection of sets and functions form a category. By carefully examining this collection, we will see what is needed in the definition of a category.

Example 2.1.1. Consider the collection of all sets. There are functions between sets. If f is a function from set S to set T, then we write it as $f\colon S \longrightarrow T$. We call S the **domain** of f and T the **codomain** of f. Certain functions can be composed: for $f\colon S \longrightarrow T$ and $g\colon T \longrightarrow U$, there exists a function $g \circ f\colon S \longrightarrow U$, which is defined for s in S as $(g \circ f)(s) = g(f(s))$. This composition operation is associative, which means that for $f\colon S \longrightarrow T$, $g\colon T \longrightarrow U$, and $h\colon U \longrightarrow V$, both ways of associating the functions $h \circ (g \circ f)$ and $(h \circ g) \circ f$ are equal to the function described as follows:

$$s \mapsto f(s) \mapsto g(f(s)) \mapsto h(g(f(s))). \tag{2.1}$$

That is, $h \circ (g \circ f) = (h \circ g) \circ f$, and on s of S, this function has the value $h(g(f(s)))$. For every set S, there is a function $id_S : S \longrightarrow S$, which is called the **identity function** and is defined for s in S as $id_S(s) = s$. These identity functions have the following properties: for all $f\colon S \longrightarrow T$, it is true that $f \circ id_S = f$ and $id_T \circ f = f$. The collection of sets and functions form a category called $\mathbb{Set}$. This category is easy to understand, and we use it to hone our ideas about many structures of category theory. □

Now we give the formal definition of a category.

Definition 2.1.2. A **category** $\mathbb{A}$ is a collection of **objects** $Ob(\mathbb{A})$ and a collection of **morphisms** $Mor(\mathbb{A})$ which has the following structure:

- Every morphism has an object associated to it called its *domain*: there is a function $dom_{\mathbb{A}} : Mor(\mathbb{A}) \longrightarrow Ob(\mathbb{A})$.
- Every morphism has an object associated to it called its *codomain*: there is a function $cod_{\mathbb{A}} : Mor(\mathbb{A}) \longrightarrow Ob(\mathbb{A})$. We write

$$f\colon a \longrightarrow b \qquad \text{or} \qquad a \xrightarrow{\quad f \quad} b \tag{2.2}$$

  to express the fact that $dom_{\mathbb{A}}(f) = a$ and $cod_{\mathbb{A}}(f) = b$.
- Adjoining morphisms can be composed: if $f\colon a \longrightarrow b$ and $g\colon b \longrightarrow c$, then there is an associated morphism $g \circ f : a \longrightarrow c$. We can write these morphisms as

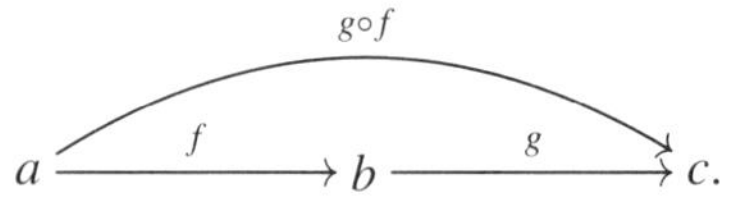

$$a \xrightarrow{\quad f \quad} b \xrightarrow{\quad g \quad} c. \quad (\text{with } g \circ f : a \to c \text{ arcing over}) \tag{2.3}$$

- Every object has an identity morphism: there is a function $ident_{\mathbb{A}} : Ob(\mathbb{A}) \longrightarrow Mor(\mathbb{A})$. We denote the identity of a as $id_a : a \longrightarrow a$, or

$$\overset{id_a}{\circlearrowright} a. \tag{2.4}$$

This structure must satisfy the following two axioms:

- Composition is associative: given $f\colon a \longrightarrow b$, $g\colon b \longrightarrow c$, and $h\colon c \longrightarrow d$, the two ways of composing these maps are equal:

$$h \circ (g \circ f) = (h \circ g) \circ f; \tag{2.5}$$

  that is, they are the same map from a to d.
- Composition with the identity does not change the morphism: for any $f\colon a \longrightarrow b$, the composition with id_a is f (i.e., $f \circ id_a = f$), and composition with id_b is also f (i.e., $id_b \circ f = f$).

Basic Examples

We will go through many examples here. One might feel overwhelmed by all the examples. You do not have to "get it" on the first reading. The point is that all the examples have the same feel. There are objects, there are morphisms, and they have to satisfy certain properties. Press on!

Example 2.1.3. Let us mention three examples of categories that we already saw in this text. Although we did not call them "categories," the text and exercises showed that they each have the structure of a category:

- Sets and functions form the category $\mathbb{S}\mathbf{et}$.
- Directed graphs and graph homomorphisms give us $\mathbb{G}\mathbf{raph}$.
- Groups and group homomorphisms make up $\mathbb{G}\mathbf{roup}$. □

The definition of a category is a mouthful that has many parts to it. There are several important comments concerning this definition:

- We called the elements of $Mor(\mathbb{A})$ the "morphisms" of the category. We will also interchangeably use the words **maps** and **arrows** to refer to them.
- $Ob(\mathbb{A})$ and $Mor(\mathbb{A})$ are called "collections" rather than the more set-theoretical "sets" or "classes." The reason for this is because we do not want to get bogged down in the language of set theory. If you know the language of set theory, then realize that sometimes our objects and morphisms will be sets and sometimes proper classes (classes that are not sets). Often, we will not specify which and just use the word "collection."
- It is important to notice that if we have the morphisms $f\colon a \longrightarrow b$ and $g\colon b \longrightarrow c$, then we write the composition as $g \circ f$ rather than $f \circ g$. We do this because in many categories, the morphisms will be types of functions. When we apply the composition of functions, it looks like $g(f(\;))$, which is notationally closer to $g \circ f$ than $f \circ g$. As we get more and more used to the language, we will write gf rather than $g \circ f$.
- Another way of seeing the definition of a category is to discuss collections of morphisms. In Chapter 1, we saw that for sets S and T, we can look at the collection of all set functions from S to T. Now we look at the set of all morphisms between two objects. For objects a and b in category $\mathbb{A}$, there is a collection of *all* the morphisms from a to b, which we write as $Hom_{\mathbb{A}}(a,b)$. The name "*Hom*" comes from the word "homomorphism," which is a vestige of the algebraic background of category theory. We call these collections **Hom sets**, even though the collections might not be sets. Composition in the category in terms of the Hom sets becomes the operation

$$\circ\colon Hom_{\mathbb{A}}(b,c) \times Hom_{\mathbb{A}}(a,b) \longrightarrow Hom_{\mathbb{A}}(a,c) \qquad (2.6)$$
$$(g,f) \longmapsto g \circ f.$$

This means that we take an $f\colon a \longrightarrow b$ and a $g\colon b \longrightarrow c$ and return $g \circ f\colon a \longrightarrow c$.

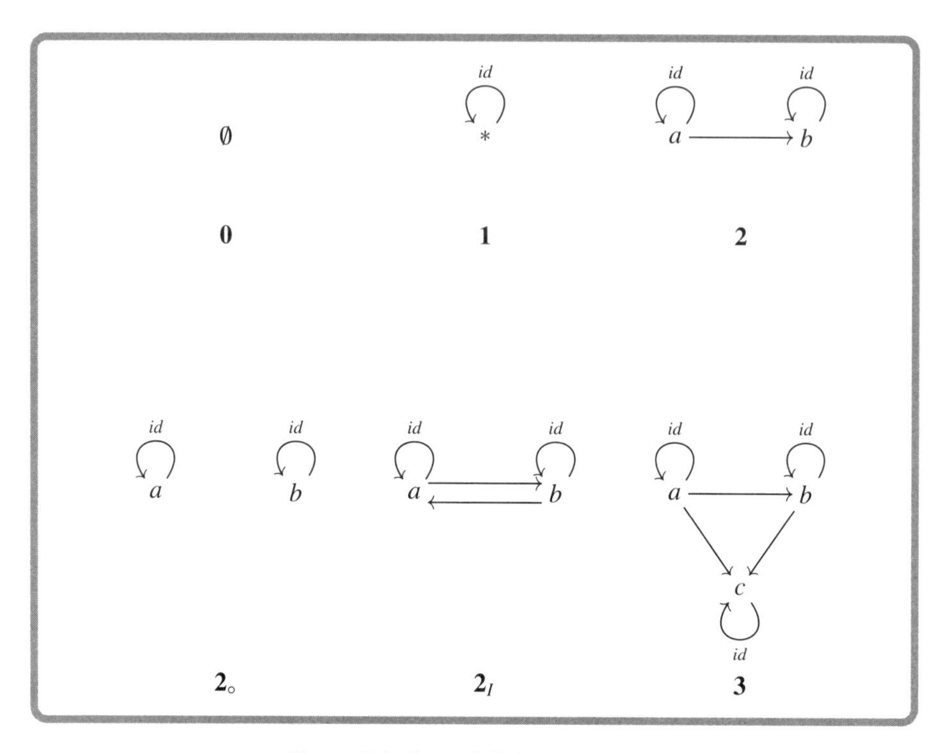

Figure 2.1. Several finite categories.

> We refer to mappings between Hom sets as "functions" even though the Hom sets might be a proper class. The fact that every element a of $\mathbb{A}$ has an identity element means that there is a special morphism in $Hom_{\mathbb{A}}(a, a)$ that satisfies the properties stated here. As time goes on, and it is obvious what category we mean, we will drop the subscript and write $Hom(a, b)$.

The categories $\mathbb{Set}$, $\mathbb{Graph}$, and $\mathbb{Group}$ have collections of objects and morphisms that are infinite. Let us, however, look at some examples of categories with finite collections of objects and morphisms.

Example 2.1.4. These **finite categories** are depicted in Figure 2.1. In detail,

- **0**, the empty category, has no objects and no morphisms. All the axioms of being a category are trivially true.
- **1** has one object and the single identity morphism on that object.
- **2** has two objects and three morphisms. Two of the morphisms are identity morphisms on the two objects, and the third morphism goes from one object to the other.
- $\mathbf{2}_\circ$ has the two objects and the two identity maps but does not have the nonidentity morphism.
- $\mathbf{2}_I$ is like **2**, but there are two nonidentity morphisms, and their compositions are the identity morphisms. In total, it has two objects and four morphisms.

- **3** is a category with three objects, three identity morphisms, and three nonidentity morphisms. The nonidentity morphisms form a commutative triangle.

Although these categories may seem trivial, they will be very useful. They provide easy examples to explore concepts and will play important roles as we explore category theory. □

Example 2.1.5. Not only does the collection of all sets form a category, but each individual set can also be thought of as having the structure of a category. Let S be a set (it can be finite, infinite, or even a proper class, if you understand the jargon of set theory). We form the category $d(S)$, where the objects are the elements of S and the only morphisms are identity morphisms. The composition operation can compose identity maps only with themselves. We call a category with only identity morphisms a **discrete category**. For example, the set $S = \{a, b, c, d\}$ becomes the following category:

$$\begin{array}{cc} \overset{id_a}{\circlearrowleft} & \overset{id_b}{\circlearrowleft} \\ a & b \\ \overset{id_c}{\circlearrowleft} & \overset{id_d}{\circlearrowleft} \\ c & d. \end{array} \tag{2.7}$$

□

Let us go through some more examples of categories.

Example from Computers

Here is an example of a category for someone who appreciates computer science.

Example 2.1.6. The category of computable functions $\mathbb{CompFunc}$ is central to computer science. A function is **computable** if there is a computer program that can tell a computer how to describe the function. That means, there is a computer program (written in some programming language), and if $f(x) = y$, then when x is entered into the computer as input, the program will output y. Computable functions take certain forms of data as input and return certain forms of data as output. The kind of input and output is called a **type**. A type is a class of data. Computers deal with types like *Nat* (natural numbers), *Int* (integers), *Real*, *Bool* (Boolean), and *String*, etc. The objects of $\mathbb{CompFunc}$ are sequences (or products) of types (for example, $Int \times Bool \times Bool \times Real$). Given two sequences of types, a morphism of this category will be a computable function from the first sequence of types to the second sequence of types. A typical computable function might look like

$$f\colon Int \times String \times Bool \longrightarrow Bool \times Real \times Real \times Nat. \tag{2.8}$$

The composition of two computable functions is easily seen to be a computable function itself (the program for the first program can be "composed" or "tagged onto" the program for the second function in order to form a program for the composition function). Just like functions, composition of computable functions are associative. For every list of types, there exists a useless computable function that accepts data of the appropriate type and outputs the same data without changing it. Such functions serve as the identity morphisms in this category. Composition with the identity functions does not change the function. (Notice that the name of this category comes from the morphisms, not the objects of the category.)

While we will mention this category several times, the book *Theoretical Computer Science for the Working Category Theorist* [274] is dedicated to this category. □

Example from Logic

Example 2.1.7. The category $\mathbb{P}\mathbb{r}\mathbb{o}\mathbb{p}$ is about propositional logic. The objects of the category are propositional statements, which are statements that are either true or false. Statements can be combined with logical operations like "and" (or "conjunction") $\wedge$, "or" (or "disjunction") $\vee$, "implication" $\Rightarrow$, "biconditional" (or "bi-implication") $\Leftrightarrow$, and "negation" (or "not") $\neg$. There is a single morphism from proposition P to proposition Q if and only if P logically implies Q (sometimes the word "entails" is used instead of "implies.") For example, there are arrows $P \wedge Q \longrightarrow P$, $P \wedge Q \longrightarrow Q$, and $P \longrightarrow P \vee Q$. The composition in the category exists because if P implies Q, i.e., $(P \longrightarrow Q)$ and Q implies R, i.e., $(Q \longrightarrow R)$, then it is obvious that P implies R, i.e., $(P \longrightarrow R)$. Associativity follows from the fact that there is at most one morphism between any two objects. The identities in the category come from the fact that for every propositional statement P, it is tautologically true that P implies P $(P \longrightarrow P)$. This category is different from the other infinite categories that we have already seen because between any two objects in the category, there is either a single morphism or no morphism at all. The fact that there is at most one morphism between objects means that for any objects P and Q, the set $Hom_{\mathbb{P}\mathbb{r}\mathbb{o}\mathbb{p}}(P, Q)$ is either the empty set or a one-object set. We will examine more of the structure of this category in Section 4.8. □

Examples from Mathematics

Let us move on to many more examples. If at any point while going through these examples you get lost, skip to the next one. Return to the skipped examples later.

Since category theory started as a branch of mathematics, there are many examples found in mathematics. We begin with algebraic structures. These are sets with operations that satisfy certain axioms. The morphisms between these algebraic structures are usually set functions that respect the operations.

We already saw the definition of a group. A group is just one type of algebraic structure. Here is a list of some of the algebraic structures we will encounter in this book. One can see the way these different algebraic structures are related to each other with the Venn diagram A.1 in Appendix A. There are concrete examples of such structures in Example 2.1.10.

Definition 2.1.8.

- A **magma** $(M, \star)$ is a set M with a binary operation (an operation with two inputs) $\star\colon M \times M \longrightarrow M$. This operation is called the "multiplication" of the magma. It is not assumed that this operation satisfies any axiom.
- A **semigroup** $(M, \star)$ is a magma whose binary operation is associative, that is, for all x, y, and z in M, $x \star (y \star z) = (x \star y) \star z$.
- A **monoid** $(M, \star, e)$ is a semigroup whose binary operation has an identity element; that is, there is an element e such that for all x in M, $x \star e = x = e \star x$.
- A **commutative monoid** $(M, \star, e)$ is a monoid whose binary operation is commutative. That is, the multiplication satisfies the axiom that for all x and y in M, $x \star y = y \star x$.
- A **group** $(M, \star, e, -(\ ))$ is a monoid that has an inverse operation. That is, there is a function $-(\ )\colon M \longrightarrow M$ such that for all x in M, $x \star -x = e = -x \star x$.
- A **commutative group** or an **abelian group** $(M, \star, e, -)$ is a group whose binary operation is commutative. That is, the multiplication satisfies the axiom that for all x and y in M, $x \star y = y \star x$. Another way to think about it is as a commutative monoid with an inverse operation.
- A **ring** $(M, \star, e, -, \odot, u)$ is an abelian group with another binary associative operation $\odot\colon M \times M \longrightarrow M$ and another identity element u (i.e., for all x in M, we have $x \odot u = x = u \odot x$; another way to say that is $(M, \odot, u)$ forms a monoid), and for which the new operation distributes over the old operation. That means that for all x, y, and z in M,

$$x \odot (y \star z) = (x \odot y) \star (x \odot z) \qquad \text{and} \qquad (y \star z) \odot x = (y \odot x) \star (z \odot x). \tag{2.9}$$

- A **field** $(M, \star, e, -, \odot, u, (\ )^{-1})$ is a ring with a partial inverse for the second binary operation. This means that there is an operation $(\ )^{-1}: M \longrightarrow M$, which is defined for all x in M except the identity element e. The inverse operation satisfies the axiom: for all $x \neq e$, $x \odot x^{-1} = u = x^{-1} \odot x$.

Monoids will play a major role in this text, so it pays to spell out all the details of their definition and to state them with commutative diagrams.

Definition 2.1.9. A **monoid** is a triple $(M, \star, e)$ where

- M is a set of elements.
- $\star\colon M \times M \longrightarrow M$ is a set function, i.e., a binary operation.
- e is an element of M; that is, there is a set function that picks out e in M, $v\colon \{*\} \longrightarrow M$.

These ingredients must satisfy the following requirements: $\star$ is associative, and e must behave like a unit; that is, the commutativity of the following two diagrams:

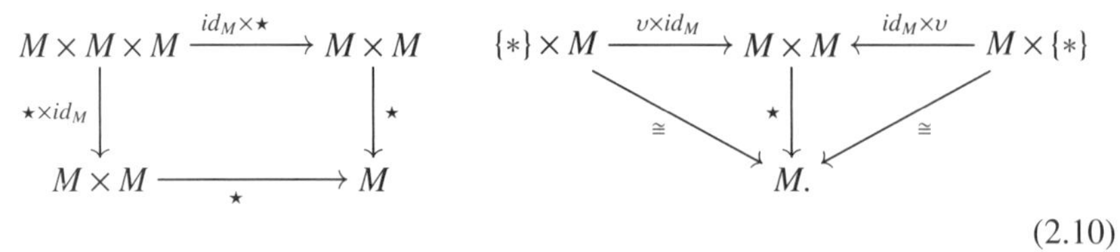

(2.10)

Some care must be given when understanding these diagrams. The horizontal map $id_M \times *$ is a product of the map $id_M \colon M \longrightarrow M$ and the map $\star \colon M \times M \longrightarrow M$. Using Equation (1.17), we take the product of these maps to get the required map. This means that we perform the operation id_M and the operation $\star$ in parallel. This gives us the operation $id_M \times \star$. Similarly, the map $\upsilon \times id_M$ is a product of two simpler maps. It is imperative that you understand the way that these two diagrams work before going further. The rest of the book will have diagrams like these.

Example 2.1.10. Figure 2.2 on page 40 has many examples of algebraic structures using the major number systems, $\mathbf{N}, \mathbf{Z}, \mathbf{Q}, \mathbf{R}, \mathbf{C}$, and the major operations, $+, -, \cdot, (\ )^{-1}$. We will also include examples from the positive rational numbers $\mathbf{Q}^+$ and the positive real numbers $\mathbf{R}^+$.

Magmas:				
$(\mathbf{N}, +)$	$(\mathbf{Z}, +)$	$(\mathbf{Q}, +)$	$(\mathbf{R}, +)$	$(,+)$
$(\mathbf{N}, \cdot)$	$(\mathbf{Z}, \cdot)$	$(\mathbf{Q}, \cdot)$	$(\mathbf{R}, \cdot)$	$(,\cdot)$
Semigroups:				
$(\mathbf{N}, +)$	$(\mathbf{Z}, +)$	$(\mathbf{Q}, +)$	$(\mathbf{R}, +)$	$(,+)$
$(\mathbf{N}, \cdot)$	$(\mathbf{Z}, \cdot)$	$(\mathbf{Q}, \cdot)$	$(\mathbf{R}, \cdot)$	$(,\cdot)$
Monoids:				
$(\mathbf{N}, +, 0)$	$(\mathbf{Z}, +, 0)$	$(\mathbf{Q}, +, 0)$	$(\mathbf{R}, +, 0)$	$(,+, 0)$
$(\mathbf{N}, \cdot, 1)$	$(\mathbf{Z}, \cdot, 1)$	$(\mathbf{Q}, \cdot, 1)$	$(\mathbf{R}, \cdot, 1)$	$(,\cdot, 1)$
Groups:				
	$(\mathbf{Z}, +, 0, -)$	$(\mathbf{Q}, +, 0, -)$	$(\mathbf{R}, +, 0, -)$	$(,+, 0, -)$
		$(\mathbf{Q}^+, \cdot, 1, (\ )^{-1})$	$(\mathbf{R}^+, \cdot, 1, (\ )^{-1})$	
Abelian group:				
	$(\mathbf{Z}, +, 0, -)$	$(\mathbf{Q}, +, 0, -)$	$(\mathbf{R}, +, 0, -)$	$(,+, 0, -)$
		$(\mathbf{Q}^+, \cdot, 1, (\ )^{-1})$	$(\mathbf{R}^+, \cdot, 1, (\ )^{-1})$	
Rings:				
	$(\mathbf{Z}, +, 0, -, \cdot, 1)$	$(\mathbf{Q}, +, 0, -, \cdot, 1)$	$(\mathbf{R}, +, 0, -, \cdot, 1)$	$(,+, 0, -, \cdot, 1)$
Fields:				
		$(\mathbf{Q}, +, 0, -, \cdot, 1, ()^{-1})$	$(\mathbf{R}, +, 0, -, \cdot, 1, ()^{-1})$	$(,+, 0, -, \cdot, 1, ()^{-1})$.

Figure 2.2. Number systems as examples of algebraic structures.

There are a few comments about these examples:

- Notice that all our examples of magmas are the same as our examples of semigroups. That is because we are dealing with the usual number operations, which are associative. We will see sets and operations that are magmas but not semigroups. (For the reader who is a member of the Illuminati, the octonions is an example of a number system that is not associative and hence is a magma but not a semigroup.)
- Notice that the examples for abelian groups are the same as groups. That is because, with these number systems, all the operations are commutative. (Examples of structures that are groups but not abelian groups are quaternions and the group of 2 by 2 invertible matrices.)
- Throughout this text, we will sometimes have operations that are similar to addition. In that case, we will write the unit of the operation as 0 and the inverse of that operation as $-(\ )$. We will also have operations that are similar to multiplication. In that case, we will write the unit as 1 and the inverse as $(\ )^{-1}$. Things get hairy, though, when we are talking about rings, which have both types of operations. □

Example 2.1.11. Here are some other common examples of algebraic structures. Some of these have nothing to do with numbers:

- The set $\mathbf{Z}$ with $(\ )-(\ )$ is only a magma because subtraction is not associative.
- The set $\mathbf{Q}$ with $(\ )^{-1}$ is not even a magma because for the number $0 \in \mathbf{Q}$, the inverse operation is not defined.
- Take the set of rooted binary trees. These are trees where each node has 0, 1, or 2 children, and there is only one special node, which has no parent, called a "root." This set has a binary operation that takes two rooted trees and connects each root to another root. This operation is described in Figure 2.3. The operation is not associative. This set and operation form a magma.
- Let Σ be a finite set, thought of as an alphabet of symbols. The set of all nonempty strings with symbols in Σ (sequences of letters, words) is denoted by Σ^+. This set has a concatenation operation $\bullet$, which is associative. The set and operation form a semigroup $(\Sigma^+, \bullet)$.

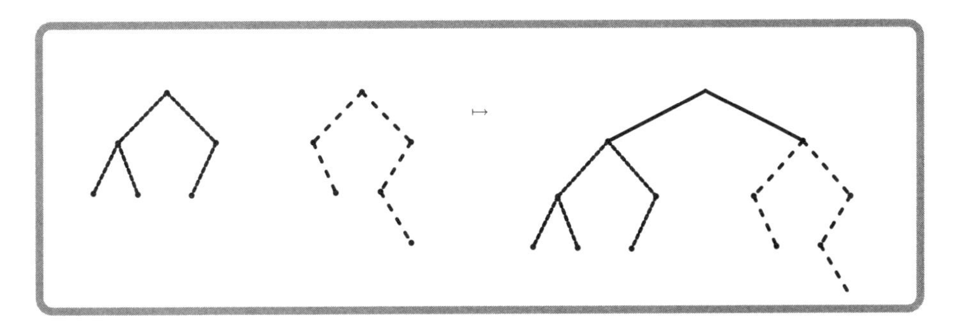

Figure 2.3. An operation on binary trees.

- The set of all strings in Σ, including the empty word $\emptyset$, is denoted as Σ^*. The empty word acts as a unit to the concatenation operation. That is, any word concatenated with the empty word is itself. This forms a monoid $(\Sigma^*, \bullet, \emptyset)$. □

An example of a monoid that will arise over and over again is the following.

Example 2.1.12. Let $\mathbb{A}$ be any category and a be any object in $\mathbb{A}$, and then consider all the morphisms that start and end at a, such as $Hom_{\mathbb{A}}(a,a)$. A morphism that starts and ends at the same object is called an **endomorphism**. We write this collection as $End(a)$. This collection forms a monoid. Given $f\colon a \longrightarrow a$ and $g\colon a \longrightarrow a$, we can multiply them as $f \circ g$. The multiplication is associative because the composition in $\mathbb{A}$ is associative. The unit is the identity morphism id_a because for all f, we have $f \circ id_a = f = id_a \circ f$. Putting this all together, we have the monoid $(End(a), \circ, id_a)$, which we call the **endomorphism monoid** of a in $\mathbb{A}$. □

Following Important Categorical Idea 1.4.9, we must discuss morphisms between each of these algebraic structures. For each type of algebraic structure, there is a notion of a **homomorphism** from one structure to another structure of the same type. If M and M' have the same type of structure, then a homomorphism $f\colon M \longrightarrow M'$ is a set function from M to M' that "respects" or "preserves" all the operations. For example, if $(M, \star, e, -, \odot, u)$ and $(M', \star', e', -', \odot', u')$ are both rings, then a homomorphism of rings is a set function $f\colon M \longrightarrow M'$ that satisfies the following axioms:

- f must respect the multiplication operations: for all x, y in M, $f(x \star y) = f(x) \star' f(y)$.
- f must respect the identity elements: $f(e) = e'$.
- f must respect the inverse operations: for all x in M, $f(-x) = -'(f(x))$.
- f must respect the other binary operations: for all x, y in M, $f(x \odot y) = f(x) \odot' f(y)$.
- f must respect the other identity elements: $f(u) = u'$.

(For some structures, a few of these rules are automatically satisfied if the others are satisfied. See [79], [15], or any book on abstract algebra for details.)

Because of the importance of monoids in this book, let us work out the details of being a monoid homomorphism in terms of diagrams.

Definition 2.1.13. Let $(M, \star, \upsilon)$ and $(M', \star', \upsilon')$ be monoids and $f\colon M \longrightarrow M'$ be a set function. Then f is a **monoid homomorphism** if it respects the multiplications and the units. This means that these two diagrams commute:

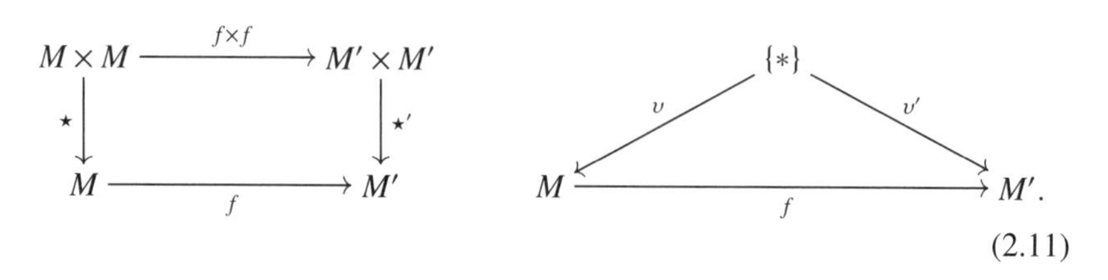

(2.11)

Theorem 2.1.14. The composition of homomorphisms is a homomorphism and the composition operation is associative. ★

Proof. Let M, M', and M'' be some type of structure with binary operations $\star, \star', \star''$, respectively. If $f\colon M \longrightarrow M'$ and $g\colon M' \longrightarrow M''$ are homomorphisms, then we have

$$\begin{aligned}
(g \circ f)(x \star y) &= g(f(x \star y)) && \text{by definition of composition}\\
&= g(f(x) \star' f(y)) && \text{because } f \text{ is a homomorphism}\\
&= g(f(x)) \star'' g(f(y)) && \text{because } g \text{ is a homomorphism}\\
&= (g \circ f)(x) \star'' (g \circ f)(y) && \text{by definition of composition.}
\end{aligned}$$

Hence $g \circ f\colon M \longrightarrow M''$ is a homomorphism. This composition is associative because they are basically set functions. ♣

Exercise 2.1.15. Show that for every algebraic structure, the identity functions satisfy the requirements of being homomorphisms. Show that the identity functions act as a unit for the composition operation. ■

From Theorem 2.1.14 and Exercise 2.1.15, we can see that each of the algebraic structures defined in Definition 2.1.8 (and many more), together with their homomorphisms, give an example of a category.

Example 2.1.16. For these algebraic structures, the collection of all the structures and their homomorphisms form a category. This gives us the categories $\mathbb{Magma}$, $\mathbb{SemiGp}$, $\mathbb{Monoid}$, $\mathbb{ComMonoid}$, $\mathbb{Group}$, $\mathbb{AbGp}$, $\mathbb{Ring}$, and $\mathbb{Field}$. The relationships between these categories will be examined in Chapter 4. □

Example 2.1.17. Not only does the entire collection of monoids form a category, but each individual monoid can be seen as a special type of category. If $(M, \star, e)$ is a monoid, then there is a category $A(M, \star, e)$, or just $A(M)$, whose morphisms are the elements of the monoid. The category consists of a single object $*$, and the morphisms in $A(M)$ are the elements of M which come and go from $*$. Such a category is called a **one-object category** or a **single-object category**. We might visualize this as in the following two examples:

b * a
c

t y
*
z ...

(2.12)

The $\cdots$ in the right diagram indicates that there might be many such arrows. Composition of morphisms are given by the monoid multiplication. That is, if there are elements of the monoid $m\colon * \longrightarrow *$ and $m'\colon * \longrightarrow *$, then their composition is $m' \star m\colon * \longrightarrow *$. The identity element of the monoid, e, becomes the identity morphism id_* in the category $A(M)$. □

The fact that a monoid is a category and the collection of all monoids forms a category can lead to confusion. When we speak of a monoid, are we thinking of it as an object in $\mathbb{Monoid}$ or as an individual category? We must specify the context.

Notice that the same trick will not work for semigroups and magmas. If we try to make a semigroup into a single-object category with the morphisms being the elements of the semigroup, it will fail simply because semigroups do not have units. Of course, it will fail for a general magma because the multiplication need not be associative, while the composition of morphisms in a category is associative.

We close our list of mathematical examples of categories with the notion of a topological space. This structure is one of the most important structures in modern mathematics and physics.

Definition 2.1.18. A **topological space** (T, τ) is a set T (the elements here are called "points") and τ a set of subsets of T that are called **open sets**. The subsets τ satisfies the following requirements:

- The empty set $\emptyset$ and the entire set T are in τ.
- The set τ is closed under finite intersection: given a finite set of subsets in τ, their intersection is in τ.
- The set τ is closed under arbitrary union: given any set of subsets in τ, their union is in τ.

The collection τ is called the **topology of** T.

The intuition is that the open sets determine which elements can be distinguished by maps in the category. If t and t' are in the same open set, then it is going to be hard to distinguish them. For typical spaces, an open set around a point contains all the points near it. There are many such open sets. Figure 2.4 is a typical space with points t_0 and t_1 and open sets around those points.

Example 2.1.19. For every set of points T, there are two extreme examples of topologies on T. If τ is the set of all subsets of T (i.e., the power set of T), then T is said to have the **discrete topology**. If τ only consists of the empty set $\emptyset$ and the entire set T, then all three requirements are satisfied and T is said to have the **indiscrete topology** or the **trivial topology**. The discrete topology has the most open sets possible, and the indiscrete topology has the least open sets possible. The other topologies fall somewhere between these two extremes. □

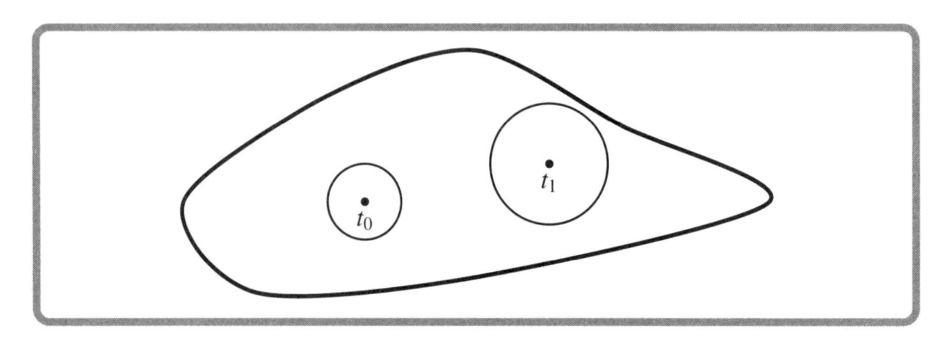

Figure 2.4. A space with two points and two open sets that contain those points.

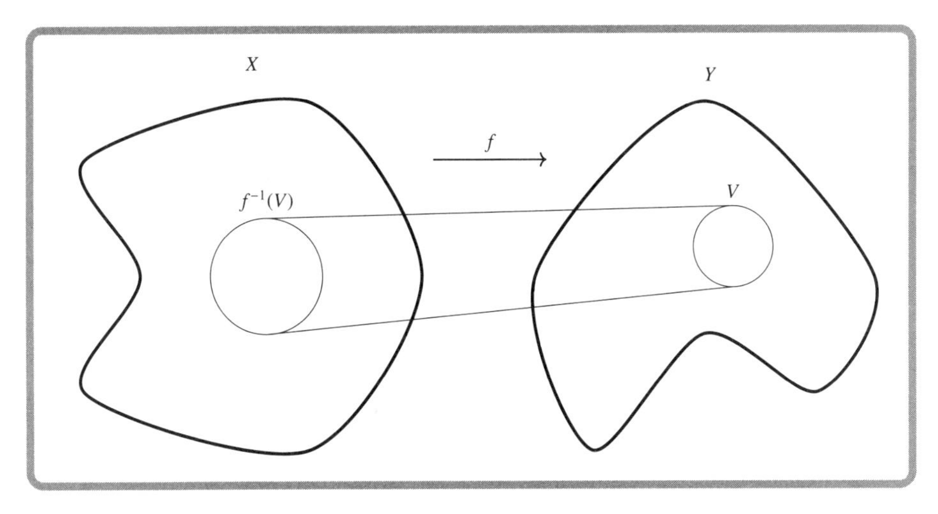

Figure 2.5. A continuous map. The preimage of an open set is an open set.

The definition of a topological space goes hand in hand with the definition of a map between topological spaces (see Important Categorical Idea 1.4.9). Such maps are called "continuous maps."

Definition 2.1.20. Given topological spaces (X, τ) and (Y, σ), a **continuous map** f from (X, τ) to (Y, σ), written as $f\colon (X, \tau) \longrightarrow (Y, \sigma)$ is a set function $f\colon X \longrightarrow Y$ that satisfies the following requirement:

$$\begin{aligned}&\text{for every open set } V \in \sigma,\\ &\qquad \text{the preimage } f^{-1}(V) = \{x \in X : f(x) \in V\} \subset X \text{ is an open set in } \tau. \qquad (2.13)\end{aligned}$$

Notice that the requirement, in a sense, goes "backward." We do not require open sets to go to open sets. Rather, we require that open sets come from open sets. Figure 2.5 shows a way of visualizing a continuous map and its requirement.

The following is a theorem about the extreme topologies.

Theorem 2.1.21. Consider topological spaces (X, τ) and (Y, σ), if one or the other of these is true:

- σ is the indiscrete topology, or
- τ is the discrete topology,

then any function $f\colon X \longrightarrow Y$ is a continuous map. ★

Proof.

- If σ is the indiscrete topology, its only two open sets are $\emptyset$ and Y. Then $f^{-1}(\emptyset) = \emptyset \subset X$ and, since f takes every element of X to some element in Y, we have $f^{-1}(Y) = X$, which are both in τ.

- If τ is the discrete topology, then no matter what open set is in σ, the preimage of it is in τ because τ has every subset of X. ♣

Exercise 2.1.22. Show that the composition of two continuous maps is a continuous map. ■

Exercise 2.1.23. Show that the composition operation is associative. ■

Exercise 2.1.24. Show that the identity function $id_T : (T, \tau) \longrightarrow (T, \tau)$ is a continuous map and acts like a unit to the composition operation. ■

Summing up Exercises 2.1.22, 2.1.23, and 2.1.24 gives the following category of topological spaces.

Example 2.1.25. The collection of all topological spaces and continuous maps form the category $\mathbb{T}\mathbb{o}\mathbb{p}$. □

Examples from Physics

Our physics examples are mostly mathematical structures that are used by physicists to describe physical phenomena. We will discuss vector spaces, manifolds, and matrices. Physicists use vector spaces to describe directions and lengths. They use manifolds to describe phenomena with enough structure to perform calculus-like operations. Matrices are arrays of numbers that store multidimensional information.

We will define and work with vector spaces in the mini-course in Section 2.4. Suffice it to say the following.

Example 2.1.26. For every field **K**, the category of **K**-vector spaces and linear transformations form a category called $\mathbf{K}\mathbb{V}\mathbb{e}\mathbb{c}\mathbb{t}$. We will mostly be concerned with $\mathbf{R}\mathbb{V}\mathbb{e}\mathbb{c}\mathbb{t}$ and $\mathbf{C}\mathbb{V}\mathbb{e}\mathbb{c}\mathbb{t}$. □

Physicists (as well as mathematicians) work with manifolds. These are "nice" topological spaces.

Definition 2.1.27. A smooth n-dimensional **manifold** or a n-**manifold** is a topological space that has the property that at every point, there is a surrounding open set that "looks like $\mathbf{R}^n$." We say it is "locally $\mathbf{R}^n$."

Rather than getting into the nitty-gritty details of the definition, let us look at many examples. Consider the surface of planet Earth. We all know that the Earth is a sphere (with some flattening at the poles). However, when we look around us, Earth looks like a flat plane (i.e., it looks like $\mathbf{R}^2$). Therefore the surface of the Earth is a two-dimensional manifold. For that matter, the surface of every sphere is a 2-manifold.

Let us go through some examples. A point is a zero-dimensional manifold. The empty set is an n-manifold for all n because it is true that every point in the empty set is locally $\mathbf{R}^n$. There are essentially two types of one-dimensional manifolds: an open line and a closed line, as in Figure 2.6. It is important that the ends of the curved line be "open" (as in $(0, 1) \subseteq \mathbf{R}$, as opposed to $[0, 1] \subseteq \mathbf{R}$, which is "closed.") Any point on the

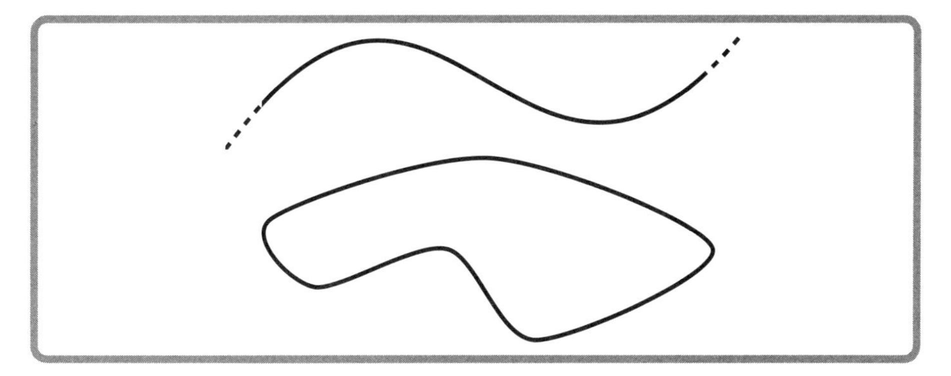

Figure 2.6. Two one-dimensional manifolds: an open line and a closed circle.

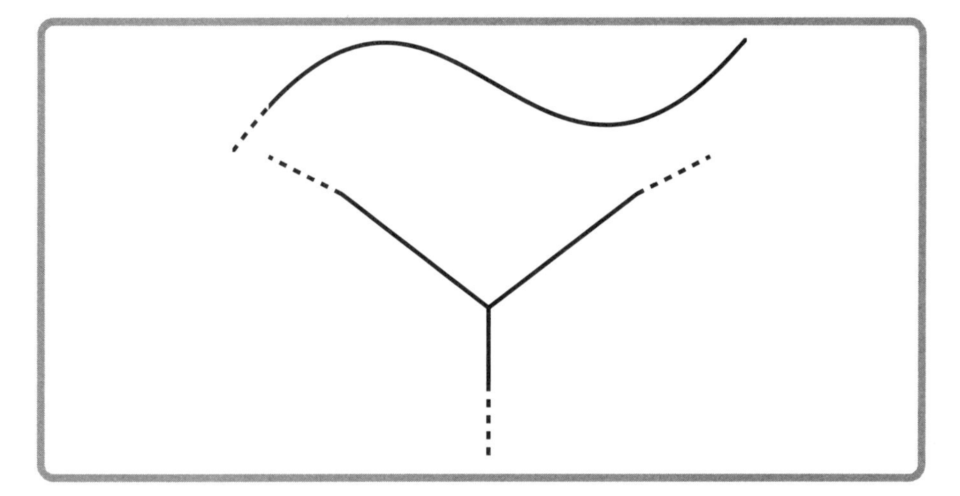

Figure 2.7. Two topological spaces that are not manifolds.

line, regardless how close the point is to the end, has a small neighborhood that looks like an "open" part of the real line.

In contrast, Figure 2.7 has two topological spaces that are *not* one-dimensional manifolds. The curved line on the left is not a 1-manifold because the end point at the right is not locally like the real line. The "Y" on the right is not a manifold because the one point in the middle has three lines coming out of it, and every point in **R** has two lines coming out of it. At that central point, it does not look like $\mathbf{R}^1$.

What about two-dimensional manifolds? Figure 2.8 has several 2-manifolds: (i) is a sphere, (ii) is a doughnut with one hole, (iii) is a doughnut with two holes, and (iv) is a doughnut with three holes. We can go on and discuss a doughnut with n holes. It is a theorem that these are all the finite (the technical term is **compact**) 2-manifolds. The number of holes is called the **genus** of the 2-manifold. The sphere has genus 0. At the moment, we do not need to look at 3-manifolds or higher.

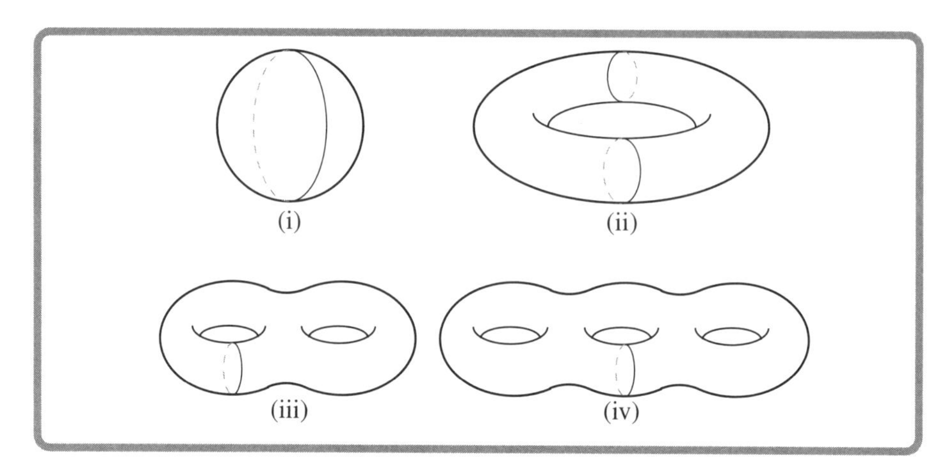

Figure 2.8. Several 2-manifolds.

Let us talk about maps between manifolds.

Definition 2.1.28. A map between manifolds is called a **smooth map** if it is a continuous map of topological spaces and it respects all the local $\mathbf{R}^n$ structure. (Again, we ignore the technical details of the definition because we will not be using them.)

Example 2.1.29. The n-dimensional manifolds and smooth maps form a category called n-$\mathbb{M}\mathrm{anif}$. All the different dimensional manifolds and all the possible smooth maps can be combined to form a category called $\mathbb{M}\mathrm{anif}$. □

Another category that is going to arise many times in physics and in many parts of this text is the category of matrices.

Example 2.1.30. The set of all matrices with real number entries forms a category called $\mathbf{R}\mathbb{M}\mathrm{at}$. The objects are the natural numbers, and the set of all morphisms from m to n is the set of n by m matrices with real number entries. To avoid confusion, we will write the matrix $A\colon m \longrightarrow n$ as A_{nm}. The composition in $\mathbf{R}\mathbb{M}\mathrm{at}$ is simply matrix multiplication. In detail, if A_{nm} is a m by n matrix and B_{pn} is a n by p matrix, then the matrix multiplication $B_{pn} \cdot A_{nm}$ is the composition of the morphisms as depicted here:

$$m \xrightarrow[A_{nm}]{} n \xrightarrow[B_{pn}]{} p. \qquad \text{(with } B_{pn}\cdot A_{nm}\colon m \longrightarrow p\text{)} \tag{2.14}$$

For every n, we say that there is a unique "empty" matrix in $Hom_{\mathbf{R}\mathbb{M}\mathrm{at}}(0, n)$, and similarly in $Hom_{\mathbf{R}\mathbb{M}\mathrm{at}}(n, 0)$. Like the elements in an empty set, such matrices do not exist. However, we need to say that they are there, for bookkeeping reasons. The associativity of composition follows from the fact that matrix multiplication is associative. The identity morphism $id_m\colon m \longrightarrow m$ will be to the identity m by m

matrix (1 down the diagonal and 0 everywhere else.) One can make similar definitions for matrices with entries in other number systems. We will deal with the categories $\mathbf{R}\mathbb{M}\mathrm{at}$, $\mathbf{Z}\mathbb{M}\mathrm{at}$, $\mathbf{Q}\mathbb{M}\mathrm{at}$, and $\mathbf{C}\mathbb{M}\mathrm{at}$, which will play major roles in the coming pages. We will also deal with Boolean matrices (entries that are either 0 or 1) as $Bool\mathbb{M}\mathrm{at}$. In general, if $\mathbf{K}$ is a field, then $\mathbf{K}\mathbb{M}\mathrm{at}$ is the category of matrices with entries in $\mathbf{K}$. □

Example 2.1.31. Consider the following example of multiplying matrices with entries in $\mathbf{N}$:

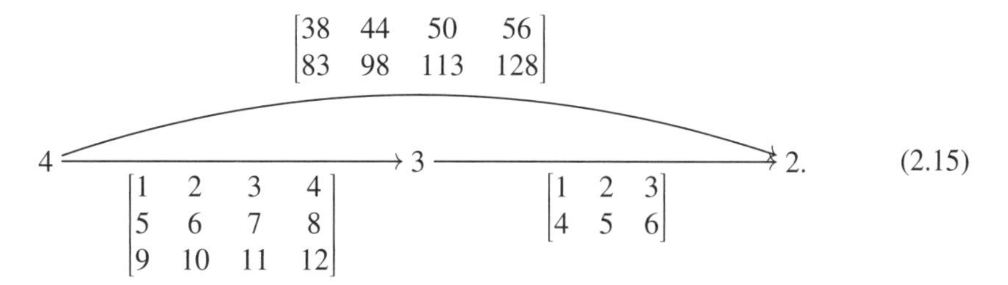

$\begin{bmatrix} 38 & 44 & 50 & 56 \\ 83 & 98 & 113 & 128 \end{bmatrix}$
4
3
2.
$\begin{bmatrix} 1 & 2 & 3 & 4 \\ 5 & 6 & 7 & 8 \\ 9 & 10 & 11 & 12 \end{bmatrix}$
$\begin{bmatrix} 1 & 2 & 3 \\ 4 & 5 & 6 \end{bmatrix}$

(2.15)

□

More Examples

Here are some more examples of categories that are related to computers and logic.

We met the category $\mathbb{S}\mathrm{et}$ of sets and functions between sets. There is another category with sets as objects that contains more morphisms. Relations are generalizations of functions that describe connections between sets.

Definition 2.1.32. A **relation** R from set S to set T, written as $R\colon S \not\longrightarrow T$, is a subset of $S \times T$. That is, $R \subseteq S \times T$.

Example 2.1.33. If S is a set of customers of a company and T is the set of products that a company sells, then relation R can be the relation that matches customers to products they purchased. We can do something very similar with students in college and classes that a student is registered to take. □

A function is a special type of relation.

Example 2.1.34. Every function $f\colon S \longrightarrow T$ can be seen as a relation $\hat{f}\colon S \not\longrightarrow T$:

$$\hat{f} = \{(s,t) : f(s) = t\} \subseteq S \times T. \tag{2.16}$$

Relations are more general than functions because (i) some elements in S might not be related to any element in T, and, furthermore, in a relation (ii) some elements in S might be related to more than one element in T. □

Relations can be composed. Given a relation $R\colon S \not\longrightarrow T$ and $Q\colon T \not\longrightarrow U$, the composition is written as $Q \circ R\colon S \not\longrightarrow U$ and is defined as

$$Q \circ R = \{(s,u) : \text{there exists a } t \in T \text{ such that } (s,t) \in R \text{ and } (t,u) \in Q\}. \tag{2.17}$$

Example 2.1.35. Consider the following three sets: $S = \{a, b, c, d, e\}$, $T = \{w, x, y, z\}$, and $U = \{1, 2, 3, 4\}$. Let $R\colon S \not\longrightarrow T$ and $Q\colon T \not\longrightarrow U$ be defined as $R = \{(a, x), (a, y), (c, x), (d, z)\}$ and $Q = \{(w, 4), (x, 2), (y, 2)\}$. Then $Q \circ R = \{(a, 2), (c, 2)\}$. □

Exercise 2.1.36. Show that the composition of relations is associative. ■

Exercise 2.1.37. Show that the composition of functions is a special case of the composition of relations. ■

Definition 2.1.38. For every set S, there is a special relation $id_S \subseteq S \times S$, called the **identity relation** and defined as $id_S = \{(s, s) : s \in S\}$.

Exercise 2.1.39. Show that for every relation $R\colon S \not\longrightarrow T$, composition with the identity relations act as they should. That is, $R \circ id_S = R$ and $id_T \circ R = R$. ■

Using Exercises 2.1.36 and 2.1.39, we can form the category of sets and relations.

Example 2.1.40. The category $\mathbb{Rel}$ of relations has sets as objects and relations between sets as morphisms. □

Definition 2.1.41. Some relations have special properties. Here are several possible properties of relations. The relation $R\colon S \not\longrightarrow T$ is as follows:

- **Total-valued** if for every $s \in S$, there is at least one element $t \in T$ such that $(s, t) \in R$
- **Single-valued** if for every $s \in S$, there is at most one $t \in T$ such that $(s, t) \in R$. In other words, if $(s, t) \in R$ and $(s, t') \in R$, then $t = t'$
- **One-to-one** if for every $t \in T$, there is at most one $s \in S$ such that $(s, t) \in R$. In other words, if $(s, t) \in R$ and $(s', t) \in R$, then $s = s'$
- **Onto** if for every $t \in T$, there is an $s \in S$ such that $(s, t) \in R$

While, in general, a relation is from one set to another set, it is also possible to have a relation from a set to itself. Such a relation tells how elements in a set are related to themselves. There are also special properties of relations from a set to itself. The relation $R\colon S \not\longrightarrow S$ is as follows:

- **Reflexive** if for every $s \in S$, we have $(s, s) \in R$
- **Symmetric** if for every $s, t \in S$, $(s, t) \in R$ implies $(t, s) \in R$
- **Transitive** if for every $s, t, u \in S$, $(s, t) \in R$ and $(t, u) \in R$ implies $(s, u) \in R$
- **Antisymmetric** if for every $s, t \in S$, $(s, t) \in R$ and $(t, s) \in R$ implies $s = t$
- **Total** if for every $s, t \in S$, either $(s, t) \in R$ or $(t, s) \in R$

Definition 2.1.42. For every relation $R\colon S \not\longrightarrow T$, there is a related **inverse relation** $R^{-1}\colon T \not\longrightarrow S$ that is defined as

$$R^{-1} = \{(t, s) : (s, t) \in R\} \subseteq T \times S. \tag{2.18}$$

Example 2.1.43. Let $S = \{a, b, c, d, e\}$ and $T = \{w, x, y, z\}$ be sets and $R\colon S \not\longrightarrow T$ be defined as $R = \{(a, x), (a, y), (c, x), (d, z)\}$, then $R^{-1}\colon T \not\longrightarrow S$ is $\{(x, a), (y, a), (x, c), (z, d)\}$. □

Similar to Theorem 1.4.32 on page 17 where we saw that properties of functions can be understood in terms of the composition of functions, here, the properties of relations can be understood in terms of the composition of relations.

Theorem 2.1.44. Relation $R\colon S \not\longrightarrow T$ has the property of being

- Total-valued if $id_S \subseteq R^{-1} \circ R$
- Single-valued if $R \circ R^{-1} \subseteq id_T$
- One-to-one if $R^{-1} \circ R \subseteq id_S$
- Onto if $id_T \subseteq R \circ R^{-1}$

There are similar methods to describe properties of relations from a set to itself. Relation $R\colon S \not\longrightarrow S$ has the property of being

- Reflexive if $id_S \subseteq R$
- Symmetric if $R = R^{-1}$
- Transitive if $R \circ R \subseteq R$
- Antisymmetric if $R \cap R^{-1} \subseteq id_S$
- Total if $R \cup R^{-1} = S \times S$ ★

Proof. We will prove that a relation is total-valued if and only if $id_S \subseteq R^{-1} \circ R$ and leave the rest as an exercise.

R is total-valued $\Longleftrightarrow$ for every $s \in S$, there is a $t \in T$ such that $(s, t) \in R$.
$\Longleftrightarrow$ for every $s \in S$, there is a $t \in T$ such that $(t, s) \in R^{-1}$.
$\Longleftrightarrow$ for every $s \in S$, there is a $t \in T$ such that $(s, t) \in R$ and $(t, s) \in R^{-1}$
$\Longleftrightarrow$ for every $s \in S$, we have $(s, s) \in R^{-1} \circ R$.
$\Longleftrightarrow$ $id_S \subseteq R^{-1} \circ R$. ♣

Exercise 2.1.45. Prove the other seven parts of Theorem 2.1.44. ■

Definition 2.1.46. We can use the properties given in Definition 2.1.41 to construct many definitions about arbitrary relations between sets. See Venn diagram A.2 in Appendix A.

- A **partial function** is a relation that is single-valued.
- A **function** is a partial function that is total-valued (i.e., a relation that is total-valued and single-valued).
- A **one-to-one function** is a function that is one-to-one as a relation.
- An **onto function** is a function that is onto.
- An **isomorphism** is a function that is one-to-one and onto.

We can also make definitions about relations between a set and itself. See Venn diagram A.3 in Appendix A.

- A **preorder** is a relation from a set to itself that is reflexive and transitive.
- A **partial order** is a preorder that is anti-symmetric, that is, it is a relation that is reflexive, transitive, and anti-symmetric.
- A **total order** is a partial order that is total, that is, it is a relation that is reflexive, transitive, anti-symmetric and total.
- An **equivalence relation** is a relation of a set to itself that is reflexive, symmetric, and transitive.

(In the center of the Venn diagram is a relation written as $*id$. It is called the "identity relation." Every element in the set is related only to itself. This is the only relation that is both symmetric and anti-symmetric. It is also an isomorphism, a preorder, and a partial order.)

These definitions are very important and need to be memorized. The notions of preorder and partial order will appear throughout the rest of the book.

Partial functions compose like relations. In detail, if $f\colon S \longrightarrow T$ and $g\colon T \longrightarrow U$, then $g \circ f\colon S \longrightarrow U$ is a partial function and on input s, it will have a the value $g(f(s))$ if s is defined for f and g is defined for $f(s)$. The identity relation is also a partial function.

Example 2.1.47. The category $\mathbb{P}\mathrm{ar}$ of partial functions has sets as objects and partial functions between the sets as morphisms. Composition and identity functions are the same as in $\mathbb{R}\mathrm{el}$. □

In summary, the categories, $\mathbb{R}\mathrm{el}$, $\mathbb{P}\mathrm{ar}$, and $\mathbb{S}\mathrm{et}$ all have sets as objects, all have the same composition, and all have the same identity morphisms. $\mathbb{R}\mathrm{el}$ is the biggest of the three and $\mathbb{S}\mathrm{et}$ is the smallest. (Since all three categories have sets as objects, it is disingenuous for $\mathbb{S}\mathrm{et}$ to have that name. The category $\mathbb{S}\mathrm{et}$ should really be called $\mathbb{F}\mathrm{un}$.)

Let us focus on partial orders and restate the definition.

Definition 2.1.48. A **partial order** $(P, \leq)$ is a set P and a relation $\leq \subseteq P \times P$ that satisfies the following requirements:

- $\leq$ is reflexive: for all $p \in P$, $p \leq p$.
- $\leq$ is transitive: for all p, q and r in P, if $p \leq q$ and $q \leq r$, then $p \leq r$.
- $\leq$ is antisymmetric: for all p and q in P, if $p \leq q$ and $q \leq p$, then $p = q$.

(Our use of the symbol $\leq$ is not to be confused with the standard use of the same symbol with numbers.)

As typical category theorists, right after defining a structure, we are interested in defining the morphisms between structures of this type.

Definition 2.1.49. Let $(P, \leq)$ and $(Q, \leq)$ be two partial orders. An **order-preserving function** $f\colon (P, \leq) \longrightarrow (Q, \leq)$ is a function from P to Q that satisfies the following axiom: for all p and p' in P, if $p \leq p'$, then $f(p) \leq f(p')$.

Exercise 2.1.50. Show that the composition of two order-preserving functions is order preserving. Furthermore, show that the composition is associative. ■

Exercise 2.1.51. Show that the identity function is order preserving and that the identity order-preserving function acts like a unit to the composition. ■

Putting this all together, we get another example of a category.

Example 2.1.52. The category $\mathbb{PO}$ consists of all partial orders and order-preserving maps. □

Example 2.1.53. Not only does the collection of all partial orders form a category, but each individual partial order forms a category. Let $(P, \leq)$ be a partial order. Consider the category $B(P, \leq)$ or $B(P)$, whose objects are the elements of P and a single morphism $p \longrightarrow q$ if and only if $p \leq q$. The transitivity of $\leq$ assures us that the composition works. The fact that there is at most one morphism between any two objects shows that the composition is associative. The reflexivity of $\leq$ corresponds to the identity maps. We sometimes abuse notation and write P for the category. □

Example 2.1.54. Everything we said about partial orders is true about preorders. Hence, there is a category of all preorders and order-preserving maps that we call $\mathbb{PreO}$. Furthermore, every preorder in $(P, \leq)$ forms a category. □

Important Categorical Idea 2.1.55. Contexts Are Central. Certain structures have arisen several times in different contexts. Consider the rational numbers **Q**, the real numbers **R**, and the complex numbers **C**. They are as follows (to name a few examples):

- Sets of numbers and hence objects of $\mathbb{Set}$.
- Algebraic objects (objects of the categories $\mathbb{Magma}$, $\mathbb{SemiGp}$, $\mathbb{Monoid}$, $\mathbb{Group}$, $\mathbb{AbGp}$, $\mathbb{Ring}$, and $\mathbb{Field}$). **R** is also a real vector space, while **C** is a real vector space and a complex vector space.
- As monoids, they each are also single-object categories $A(\mathbf{Q}), A(\mathbf{R})$, and $A()$.
- They are topological objects, i.e., they can be given topologies and hence are objects in $\mathbb{Top}$. **R** is also a one-dimensional manifold while **C** is a two-dimensional real manifold.
- They are objects in the categories $\mathbb{Par}$ and $\mathbb{Rel}$.
- They are partial orders, and hence objects in $\mathbb{PO}$ (For the cognoscenti, there are many partial orders that one can put on **C**. There is even a total order that one can put on **C**. But there is no total order that will respect the multiplication.)
- Each one is a partial order category: $B(\mathbf{Q})$, $B(\mathbf{R})$, and $B()$.

With a little thought, we can probably extend this list.

This fact that one concept can be seen in many contexts is sometimes confusing to the beginner. However, with time, one gets the hang of it. This possible confusion forces us to specify the context of the concept that we are discussing.

Let us stress this point with an example. While a group theorist looks at the real numbers one way, a topologist looks at the real numbers in another way, and a physicist and a computer scientist look at the real numbers in their own different ways, the category theorist must look at the real numbers in *all* possible ways. ○

Example 2.1.56. Let S be a set. Then $\mathcal{P}(S)$ is the power set of S (the set of all subsets of S). $\mathcal{P}(S)$ has the structure of a partial order. The order relation is $\subseteq$. That is, there is a morphism from one subset of S to another if the first is a subset of the second. The set $\mathcal{P}(S)$ is a partial order (and Boolean algebra, which we will meet in Section 6.4) and hence a category. □

We have seen that every monoid, every set, and every partial order forms a category. What about a graph? A graph does not necessarily have the structure of a category. A graph might not have identity morphisms, and a graph by itself does not have a composition operation. So while the category of all graphs and graph homomorphisms forms the category $\mathbb{Graph}$, each individual graph need not form a category.

Now we move on to some logic examples.

Example 2.1.57. The category of logical circuits is denoted as $\mathbb{Circuit}$. The objects are the natural numbers. The morphisms from m to n are all logical circuits with m inputs and n outputs. Figure 2.9 is a typical circuit, which has six inputs and two outputs (i.e., an element of $Hom_{\mathbb{Circuit}}(6, 2)$).

Let us just remember some simple logical gates: the AND, OR, NOT, NAND, and NOR gates. The AND, OR, XOR, and NAND gates are all elements of $Hom(2, 1)$, and the NOT gate is in $Hom(1, 1)$.

The composition of two circuits is as follows. Let us draw a circuit with m input wires and n output wires as

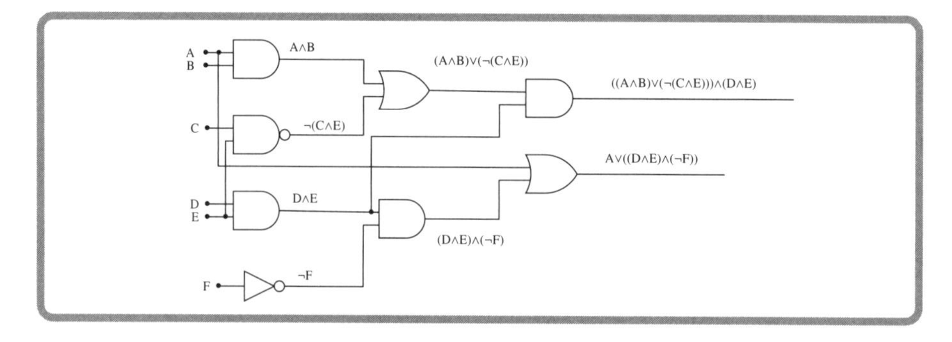

Figure 2.9. A logical circuit with six inputs and two outputs.

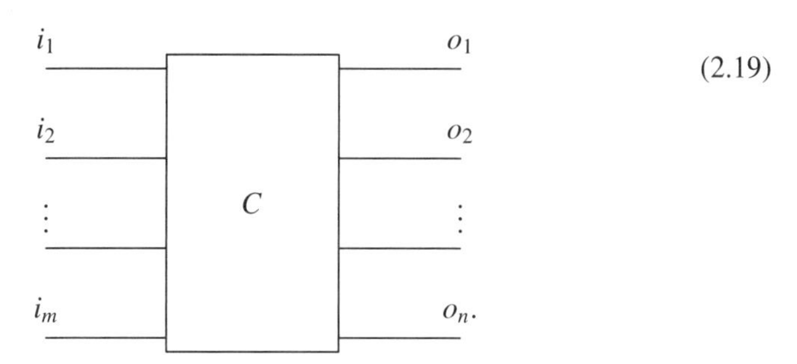

(2.19)

Now, given two circuits, C with m input wires and n output wires, and C' with n input wires and k output wires, we can compose them to form $C' \circ C$ as

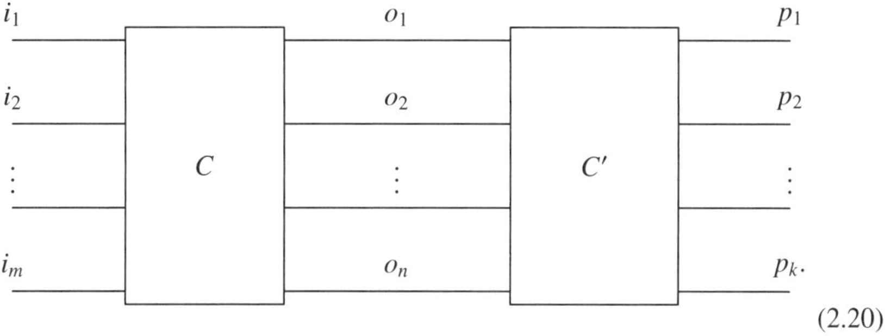

(2.20)

which has m input wires and k output wires.

The composition is obviously associative. For each n, the identity circuit is simply n straight wires that do nothing. For example, Id_5 looks like this:

i_1 o_1
i_2 o_2
i_3 o_3
i_4 o_4
i_5 o_5

(2.21)

□

Technical Point 2.1.58. There is a problem with the example of circuits. The technical issue is with the identity morphisms. We drew the identity morphisms as small wires. The problem is that when we compose these small wires with any circuit, we do not get back the original circuit. We get the original circuit with an extra wire attached to it, which is not the same circuit. As small as the identity wires are, when composed with another circuit, it does not form the same circuit as the original. While the composed circuit performs the same task as the original circuit, it is not the same circuit.

To solve this problem, we must modify the definition of the morphism in the category. Rather than say that the morphisms correspond to circuits, we must say that the morphisms correspond to equivalence classes of circuits. Two circuits are deemed equivalent if they are almost the same. That means they have the same connections, but we disregard how long the wires are. In that case, if we compose the class of identity wires with the class of a circuit, we get the same class as the circuit.

Another possible solution is to admit that the way that $\mathbb{C}\mathrm{ircuit}$ was originally defined is not really an example of a category. It is some type of souped-up graph with identity morphisms, but the identity morphisms do not act as they should. In fact, it forms something called a "bicategory,"' which we meet later in this book. For now, we use the equivalence classes definition.

Realize that the problem arises from the fact that circuits are physical objects as opposed to mathematical, idealized objects. The definition of a category was originally formulated for idealized objects, not physical objects. We will have to weaken the definition of a category to deal with physical objects. We will see this over and over again in later chapters. ♡

Let us close our first list of examples with the notion of proof.

Example 2.1.59. For this example, we will have to fix some type of logical system (we are being intentionally vague). For each such logical system, there will be a category called $\mathbb{P}\mathrm{roof}$. The objects of the category will be formal and exact logical statements, just as in $\mathbb{P}\mathrm{rop}$. A morphism from A to B will be a formal and exact proof that assumes that A is true and concludes that B is true. If there are formal proofs $f\colon A \longrightarrow B$ and $g\colon B \longrightarrow C$, then by concatenating the two proofs, there is a formal proof $g \circ f\colon A \longrightarrow C$. The proof with just the statement A goes from A to A, which corresponds to the identity on A. Notice that this category has the same objects as $\mathbb{P}\mathrm{rop}$, but in contrast to $\mathbb{P}\mathrm{rop}$, where there is at most one morphism between any two objects, here in $\mathbb{P}\mathrm{roof}$, there might be more than one proof between logical statements, and hence more than one morphism between two objects. □

Now that we finished our first batch of examples, we need to reflect on some general ideas about the size of categories. Set theorists make a distinction between collections called "classes" and collections called "sets." A class is a very "large" collection of entities, while a set is a somewhat "small" collection of entities that are part of a class. Mathematicians usually restrict themselves to work with sets rather than classes. One talks about sets with algebraic structure or sets with topological structure. One does not hear about classes with such structures. This text will not worry much about these issues. However, definitions are in order.

Definition 2.1.60. A category is called **small** if the objects and morphisms of the category are sets and not classes. If either the objects or the morphisms are a class, the category is called **large**. A category is called **locally small** if each of the Hom sets are sets and not classes.

Let us discuss the sizes of some categories that we have seen so far. Categories that are small are partial orders, preorders, and groups or monoids as single-object categories. Most of the rest of the categories that we dealt with are locally small but not small. For example, the category of sets is locally small but not small because the objects, which are the collection of all sets, are a class and not a set. Notice that even the subcollection of one-element sets is a class and not a set. The categories $\mathbb{Graph}$ and $\mathbb{Group}$ each have a proper class of objects, but their Hom sets are proper sets. These categories are locally small. For the same reason, $\mathbb{Top}$ and $\mathbb{Manif}$ are locally small.

Before closing this section, it is worth noting that there is another definition of a category, which is equivalent to Definition 2.1.2. "Equivalent" means that these two definitions describe the same structures. The main difference is that while the previous definition uses a collection of objects and a collection of morphisms, this definition only uses a collection of morphisms (i.e., it is a "single-sorted theory"). Objects are not mentioned in this definition, but they are essentially associated with special types of morphisms called "identity morphisms." The domain and codomain of every morphism are such identity morphisms. In other words, we can view function f as

$$dom(f) \circlearrowright \xrightarrow{\quad f \quad} \circlearrowleft cod(f) \,. \tag{2.22}$$

We did not start with this definition, as it is slightly less concrete and might discourage the novice. We mention it now because this definition is actually simpler (single-sorted rather than two-sorted) and it will be useful when talking about higher-dimensional categories (in particular, see Definition 9.4.7 in Section 9.4.) The definition is also important because it stresses the counterintuitive idea that it is the morphisms—not the objects—that are important in a category. The first-time reader can feel free to skip this definition.

Definition 2.1.61. A **category** $\mathbb{A}$ is a collection of morphisms $Mor(\mathbb{A})$ with the following structure:

- Every morphism has a domain morphism: there is a function $dom\colon Mor(\mathbb{A}) \longrightarrow Mor(\mathbb{A})$.
- Every morphism has a codomain morphism: there is a function $cod\colon Mor(\mathbb{A}) \longrightarrow Mor(\mathbb{A})$.
- There is a composition operation $\circ$: if f and g satisfy $cod_{\mathbb{A}}(f) = dom_{\mathbb{A}}(g)$, then there is an associated morphism, $g \circ f$.

These operations must satisfy the following axioms:

- The domain of a composite is the domain of the first: $dom(g \circ f) = dom(f)$.
- The codomain of a composite is the codomain of the second: $cod(g \circ f) = cod(g)$.
- The composition is associative: $h \circ (g \circ f) = (h \circ g) \circ f$.
- The domain morphism acts like an identity: $f \circ dom(f) = f$.
- The codomain morphism acts like an identity: $cod(f) \circ f = f$.
- The domain morphisms come and go from themselves: $dom(dom(f)) = dom(f) = cod(dom(f))$.
- The codomain morphisms come and go from themselves: $dom(cod(f)) = cod(f) = cod(cod(f))$.

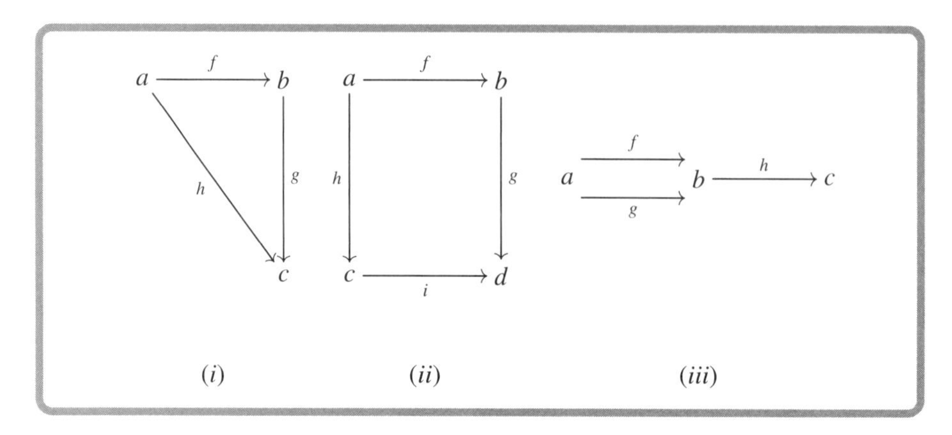

Figure 2.10. Three examples of commutative diagrams.

Exercise 2.1.62. Give an exact description of the finite categories **2**, $\mathbf{2}_I$, and **3** using the language of Definition 2.1.61. ■

Exercise 2.1.63. Show that every category that is described by Definition 2.1.2 can be described by Definition 2.1.61. ■

Exercise 2.1.64. Show that every category that is described by Definition 2.1.61 can be described by Definition 2.1.2. ■

Remark 2.1.65. It is interesting to look at the category of matrices $\mathbf{K}\mathbb{Mat}$ from the point of view of this definition where there are no objects. Usually, one thinks of the objects in $\mathbf{K}\mathbb{Mat}$ as the natural numbers. But the natural numbers are just used for bookkeeping and do not serve an essential role. With respect to Definition 2.1.61, the category only consists of the matrices and the information as to which matrices can be composed. ♠

2.2 Basic Properties

When dealing with categories, we show a part of a category and ask that those parts commute.

Definition 2.2.1. A **diagram** is a part of a category that has objects and morphisms between those objects. We say that a diagram **commutes** or is a **commutative diagram** if any two paths from the same starting object to the same finishing object actually describe the same morphism in the category.

For example, consider the commutative diagrams in Figure 2.10: (i) is a commutative triangle, and it says that $g \circ f = h$. (ii) is a commutative square, and it says that $g \circ f = i \circ h$. Saying that (iii) commutes means that $h \circ f = h \circ g$.

As another example, we can write the requirement that morphism composition has to be associative and that composition with the identity morphisms does not change the morphism, as shown in the following two commutative diagrams:

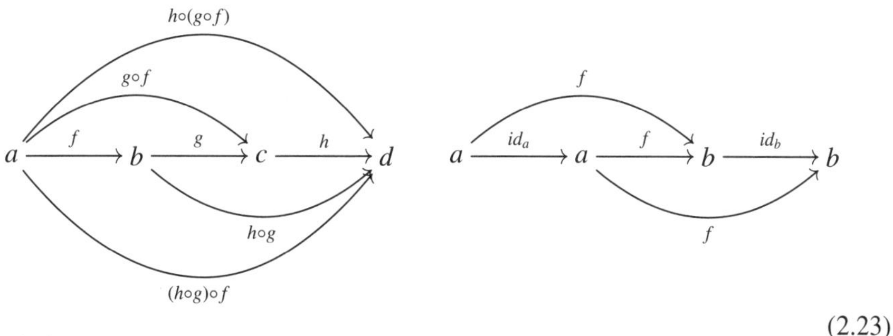

(2.23)

Within a category, there are special types of morphisms.

Definition 2.2.2. A morphism $f\colon a \longrightarrow b$ is a **monomorphism** or **monic** if, for all objects c and for all $g\colon c \longrightarrow a$ and $h\colon c \longrightarrow a$ with the relationship

$$c \overset{g}{\underset{h}{\rightrightarrows}} a \xrightarrow{\quad f \quad} b\,, \tag{2.24}$$

the following rule is satisfied:

$$\text{If } f \circ g = f \circ h, \text{ then } g = h. \tag{2.25}$$

Another way to say that f is monic is to say that f is **left cancelable**. That is, if f is on the left of two sides of an equation, we can cancel it.

Example 2.2.3. In $\mathbb{Set}$, a map $f\colon S \longrightarrow T$ is monic if and only if it is an injection (one-to-one). Assume that f is monic. Consider the one-object set $\{*\}$. Let $g\colon \{*\} \longrightarrow S$ and $h\colon \{*\} \longrightarrow S$ each pick out an element of S. The statement that f is monic amounts to

$$\text{If } f(g(*)) = f(h(*)), \text{ then } g(*) = h(*). \tag{2.26}$$

Thinking of $g(*)$ and $h(*)$ as elements in S, this is exactly the requirement that f is an injection.

To go the other way, remember that we saw in Theorem 1.4.32 that f is injective if and only if there is a f' such that $f' \circ f = id_a$. So if $f \circ g = f \circ h$, then we can compose both sides with f' to get $f' \circ f \circ g = f' \circ f \circ h$, and hence $g = h$. □

Exercise 2.2.4. Show that the composition of two monics is monic. ■

Exercise 2.2.5. Call a morphism $f\colon a \longrightarrow b$ a **split monomorphism** if there is a $f'\colon b \longrightarrow a$ such that $f' \circ f = id_a$. Show that every split monomorphism is a monomorphism. ■

Definition 2.2.6. A morphism $f: a \longrightarrow b$ is an **epimorphism** or **epic** if, for all objects c and for all $g: b \longrightarrow c$ and $h: b \longrightarrow c$ with the relationship

$$a \xrightarrow{f} b \underset{h}{\overset{g}{\rightrightarrows}} c, \tag{2.27}$$

the following rule is satisfied:

$$\text{If } g \circ f = h \circ f, \text{ then } g = h. \tag{2.28}$$

Another way to say that f is epic is to say that f is **right cancelable**.

Exercise 2.2.7. Show that in the category $\mathbb{S}\mathbf{et}$, a morphism is epic if and only if it is onto (surjective). ■

Exercise 2.2.8. Show that the composition of two epics is epic. ■

Exercise 2.2.9. Call a morphism $f: a \longrightarrow b$ a **split epimorphism** if there is a $f': b \longrightarrow a$ such that $f \circ f' = id_b$. Show that every split epimorphism is a epimorphism. ■

Notice that although the usual definitions of injective and surjective are given in terms of elements, the definitions of epic and monic are given here in terms of morphisms in a category. This is of fundamental importance in category theory.

Definition 2.2.10. A morphism $f: a \longrightarrow b$ is an **isomorphism** if there is a morphism $g: b \longrightarrow a$ such that $g \circ f = id_a$ and $f \circ g = id_b$. The morphisms f and g are called **inverses** of each other. In such a situation, we say that a and b are **isomorphic**, and we write this as $a \cong b$.

Exercise 2.2.11. Show that if a morphism has an inverse, it is unique. ■

Exercise 2.2.12. Show that any identity morphism is a isomorphism. ■

Exercise 2.2.13. Show that the composition of two isomorphisms is an isomorphism. ■

Exercise 2.2.14. Show that in any category, the relation of being isomorphic is an equivalence relation. ■

Theorem 2.2.15. If a morphism is an isomorphism, then it is monic and epic. ★

Proof. Simply use the inverse to show that the map is left cancelable and right cancelable. ♣

Remark 2.2.16. The converse is not true in every category. In the category of $\mathbb{S}\mathbf{et}$, if a morphism is epic and monic, then it is an isomorphism. However, within $\mathbb{R}\mathbf{ing}$, the category of rings, the inclusion $\mathbf{Z} \hookrightarrow \mathbf{Q}$ is epic and monic but not an isomorphism. Note that in the category of $\mathbb{Set}$, this inclusion is monic but not epic. (We can see this as an addendum to Important Categorical Idea 2.1.55. Not only do we have to see various objects within the context of the category that they are in, we also have to look at morphisms within the context of the category that they are in.) ♠

There are two special types of categories with only isomorphisms.

Definition 2.2.17. A **groupoid** is a category where all the morphisms are isomorphisms. A **group** is a special type of groupoid that has only one object.

Similar to what we saw in Example 2.1.17 about monoids, we have the following.

Example 2.2.18. Not only does the entire collection of groups form a category, but each individual group can be seen as a special type of category. If G is a group, then $C(G)$ is a category with one object and the morphisms from the single object to itself are the elements of G. The identity in the group becomes the identity of the category. The composition of the morphisms is the group multiplication. The associativity of the group multiplication becomes the associativity of morphism composition. The fact that every element of the group has an inverse is another way of saying that every morphism in the category is an isomorphism. Yet another way to describe a group is to say that a group is a monoid (a one-object category) where every morphism is invertible. □

Let us move on and talk about special types of objects in a category. There are certain objects that can be described by the way they relate to the morphisms in the category.

Definition 2.2.19. An object i in a category $\mathbb{A}$ is called an **initial object** if there is a unique morphism from i to every object in $\mathbb{A}$ (including i). An object t is called a **terminal object** if there is a unique morphism from every object (including t) to t. An object z is called a **zero object** if it is both an initial object and a terminal object.

When we want to stress that a map $f\colon a \longrightarrow b$ uniquely satisfies a property, we will write it as $a \xrightarrow[\exists!]{f} b$ or $f\colon a \xrightarrow{!} b$.

Example 2.2.20. The empty set $\emptyset$ is the initial object in the category of $\mathbb{Set}$ because for any set S, there is a unique function $f\colon \emptyset \longrightarrow S$ or $\emptyset \xrightarrow[\exists!]{f} S$. Any single element set $\{*\}$, $\{x\}$, or $\{\text{Wanda}\}$ is a terminal object in the category $\mathbb{Set}$ because for any set S, there is a unique function $f\colon S \longrightarrow \{*\}$. The category $\mathbb{Set}$ does not have a zero object. Notice that there is only one initial object in $\mathbb{Set}$, namely the empty set. In contrast, there is a whole class of terminal objects in $\mathbb{Set}$. While all the terminal objects are isomorphic, they are not equal to each other. □

Let us consider the algebraic categories. The empty set and the trivial operations on the empty set satisfy all the requirements of being a magma and a semi-group. The empty set is the initial object in the categories $\mathbb{Magma}$ and $\mathbb{SemiGp}$. In contrast, all the other algebraic categories that we looked at have at least one constant, and hence, the empty set does not satisfy the requirements.

Example 2.2.21. In $\mathbb{Prop}$, the initial object is the proposition that is always false, which is denoted $\bot$. It is the initial object because if you start with a falsehood, any proposition is a consequence. The terminal object is the proposition that is always true, denoted as $\top$ because it is a consequence of any proposition. (There is usually no zero object unless the logical system is inconsistent and $\top = \bot$. In that case, the logical system is totally worthless.) The category $\mathbb{Proof}$ does not have an initial or terminal object because, in general, there is more than one way to prove an implication. □

Example 2.2.22. Let $(P, \leq)$ be a partial order and $A(P, \leq)$ be its associated category. The initial object of that category is the bottom element. The bottom element is below every other element. This is not to be confused with an "atom," which is an element that has only the bottom element below it. Think of an atom as the smallest element (above nothing), and other elements are made of atoms. There might be many atoms, but there is at most one bottom. The terminal object is the top element. If P is nontrivial, then there is no zero object. □

Notice we say "an" initial object, not "the" initial object. The reason for this is that there might be more than one initial object. However, if there is more than one, then they are related in an interesting way.

Theorem 2.2.23. There is a unique isomorphism between any two initial objects. ★

Proof. Let i and i' be objects in $\mathbb{A}$, and assume that they are each an initial object. From the fact that i is initial and i' is an object in $\mathbb{A}$, there is a unique morphism, $f\colon i \longrightarrow i'$. From the fact that i' is initial and i is an object in $\mathbb{A}$, there is a unique morphism, $g\colon i' \longrightarrow i$. From the fact that i is an initial object and i is also an object, there is a unique morphism from i to i, which we know is the $id_i\colon i \longrightarrow i$. Since there is no other morphism from i to i, it must be that $g \circ f = id_i$. From the fact that i' is an initial object, $id_{i'}\colon i' \longrightarrow i'$ is the unique morphism from i' to i'. This means that $f \circ g = id_{i'}$. This is summarized in the following commutative diagram:

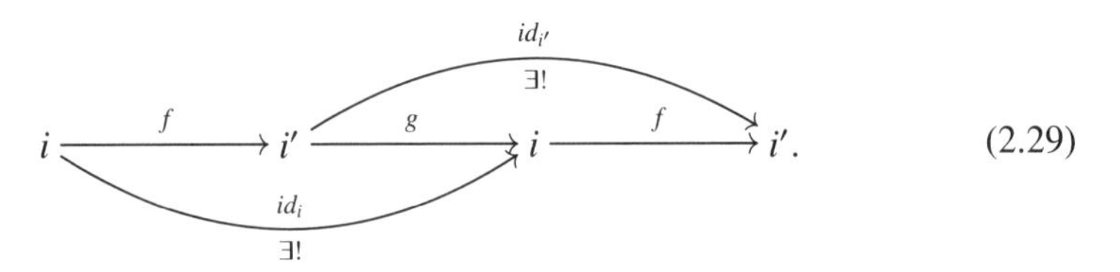

(2.29)

This shows that $g = f^{-1}$ and $f = g^{-1}$. Hence f and g are isomorphisms. This isomorphism $i \longrightarrow i'$ is unique because another isomorphism would entail a violation of i being initial, which means there is exactly one morphism (or isomorphism) from i to i'. ♣

This type of theorem—which shows how unique an object or a morphism is—arises frequently in category theory. We say that the initial object is "unique up to a unique isomorphism."

Important Categorical Idea 2.2.24. The Uniqueness of Morphisms. We will be very concerned with how unique an entity is. Sometimes we will describe an entity by giving its requirements, and there will be only one entity that satisfies the requirements. (For example, in the category of sets, there is only one empty set.) Sometimes many entities will satisfy the requirements and all those entities are isomorphic to each other. (For example, in the category of sets, there are many sets with three objects, such as $\{a,b,c\}$, $\{x,y,z\}$, etc. All these sets are isomorphic to each other; however, there are six possible isomorphisms between any two such sets.) There is an intermediate level: there are some requirements that are satisfied by many different entities, and all these entities are isomorphic to each other, but there is a *unique* isomorphism between them. (For example, in the category of sets, there are many one-element sets, and they are all isomorphic to each other with a unique isomorphism between them.)

We can summarize this hierarchy as follows:

- Unique
- Unique up to a unique isomorphism
- Unique up to an isomorphism

Technically, it is correct to use the word "the" only for an object characterized as unique. The word "a" or "an" is used when there are many possible objects that satisfy the requirement.

There will be more to say about this hierarchy in later chapters, especially when we talk about coherence theory in Chapter 5 and higher category theory in Section 9.4. ○

Exercise 2.2.25. Prove that there is a unique isomorphism between any two terminal objects. ■

Let us look at some more examples of initial objects, terminal objects, and zero objects in a category.

Example 2.2.26.

- $\mathbb{Top}, \mathbb{Manif}$: The empty set is the initial object and the single point is a terminal object. There are no zero objects.
- $\mathbb{Magma}, \mathbb{SemiGp}$: The empty set is the initial object. The single-element structure is the terminal object. There are no zero objects.
- $\mathbb{Monoid}, \mathbb{Group}, \mathbb{AbGp}, \mathbf{K}\mathbb{Vect}$: The single-element structure is the zero object.
- $\mathbb{Ring}$: There is a one-object ring where $0 = 1$, and this ring is terminal in the category. The ring of $\mathbf{Z}$ is initial in the category.
- $\mathbb{Field}$: The single element set is not an object in this category because we need two constants (0 and 1). However, this two-element field is neither initial nor terminal.

- $\mathbf{K}\mathbb{Mat}$: 0 is the zero object.
- $\mathbb{PO}, \mathbb{PreO}$: The empty set is the initial object. A one-element set is a terminal object. □

2.3 Related Categories

Just as there is an obvious notion of a subset of a set, so there is an obvious notion of a subcategory.

Definition 2.3.1. Category $\mathbb{A}$ is a **subcategory** of category $\mathbb{B}$ (or another way to say this is that category $\mathbb{B}$ is a **supercategory** of $\mathbb{A}$) if

- The objects of $\mathbb{A}$ are part of the objects of $\mathbb{B}$: $Ob(\mathbb{A}) \subseteq Ob(\mathbb{B})$.
- The morphisms of $\mathbb{A}$ are part of the morphisms of $\mathbb{B}$: $Mor(\mathbb{A}) \subseteq Mor(\mathbb{B})$.
- The composition operation for $\mathbb{A}$ is the same as the composition operation in $\mathbb{B}$ but is restricted to the elements of $\mathbb{A}$.
- The identity morphisms of $\mathbb{A}$ are the same as the identity morphisms of $\mathbb{B}$.

Example 2.3.2. The category $\mathbb{FinSet}$ of all finite sets and set functions between them is a subcategory of $\mathbb{Set}$. □

Example 2.3.3. The category $\mathbf{N}\mathbb{Mat}$ of matrices with natural number entries is a subcategory of $\mathbf{Z}\mathbb{Mat}$ in which the entries are integers. Notice that the objects of both of these two categories are the set of natural numbers. Since

$$\mathbf{N} \subseteq \mathbf{Z} \subseteq \mathbf{Q} \subseteq \mathbf{R} \subseteq, \tag{2.30}$$

we have that $\mathbf{Z}\mathbb{Mat}$ is a subcategory of $\mathbf{Q}\mathbb{Mat}$, which in turn is a subcategory of $\mathbf{R}\mathbb{Mat}$, which in turn is a subcategory of $\mathbf{C}\mathbb{Mat}$. □

Example 2.3.4. Consider the category $\mathbb{NANDCircuit}$. The objects are the natural numbers, and the set of morphisms from m to n is the set of all logical circuits made of NAND gates that have m input wires and n output wires. This category is a subcategory of $\mathbb{Circuit}$. Notice that $\mathbb{NANDCircuit}$ and $\mathbb{Circuit}$ have the same objects. It is also well known that every logical circuit with different types of gates can mimicked by a logical circuit with only NAND gates. This means that every morphism in $\mathbb{Circuit}$ has at least one corresponding morphism in $\mathbb{NANDCircuit}$. □

Example 2.3.5. The category $\mathbb{AbGp}$ is a subcategory of $\mathbb{Group}$. The category of $\mathbb{SemiGp}$ is a subcategory of $\mathbb{Magma}$. There are many other such subcategories when dealing with algebraic structures. □

Example 2.3.6. Since every partial order (reflexive, transitive, and anti-symmetric) is a preorder (reflexive and symmetric), and the morphisms between partial orders and preorders are order-preserving maps, the category of partial orders $\mathbb{PO}$ is a subcategory of $\mathbb{PreO}$. □

Example 2.3.7. Every category has a subcategory that is a groupoid. Just take all the morphisms that are isomorphisms. Since every identity morphism is an isomorphism, the groupoid subcategory has the same objects as the category. □

Given a set and an equivalence relation on the set, one can construct a quotient set. Similarly, given a category and a souped-up equivalence relation, there is a notion of a quotient category. The souped-up equivalence relation is a relation that respects the composition operation in the category.

Definition 2.3.8. Let $\mathbb{A}$ be a category. A **congruence relation** or **congruence** on $\mathbb{A}$ is an equivalence relation on the collection of morphisms of $\mathbb{A}$ that respects the composition. In detail, congruence $\sim$ is an equivalence relation on each of the Hom sets of the category that satisfies the following requirement: let $f, f' : a \longrightarrow b$ and $g, g' : b \longrightarrow c$; then the rule

$$\text{if } f \sim f' \text{ and } g \sim g', \text{ then } (g \circ f) \sim (g' \circ f') \tag{2.31}$$

is satisfied. Notice that a morphism can be equivalent only to other morphisms with the same source and target.

Technical Point 2.3.9. An equivalence relation and a congruence relation are only supposed to be on a set, not a proper class. We ignore this restriction and put relations on any collection. While some set theorists and mathematicians will be concerned with this, it will not negatively affect us. We beg the reader for forgiveness on this point. ♡

With the notion of a congruence, we go on to define a quotient category.

Definition 2.3.10. Let $\mathbb{A}$ be a category and $\sim$ be a congruence relation on $\mathbb{A}$; then a **quotient category** $\mathbb{A}/_{\sim}$ is a category constructed as follows: The objects of $\mathbb{A}/_{\sim}$ are the same as $\mathbb{A}$, and the morphisms of $\mathbb{A}/_{\sim}$ are the quotient collection of morphisms of $\mathbb{A}$ under the congruence relation $\sim$. This means that for any objects a and b, we have

$$Hom_{\mathbb{A}/_{\sim}}(a,b) = Hom_{\mathbb{A}}(a,b)/_{\sim}. \tag{2.32}$$

The composition operation in $\mathbb{A}/_{\sim}$ follows from the composition operation of $\mathbb{A}$. In detail, $[g] \circ [f] = [g \circ f]$.

Usually in a category, there are a lot of objects that are isomorphic but not equal to each other. There are special categories where there are no isomorphisms between different objects.

Definition 2.3.11. A category is **skeletal** if any two isomorphic objects are equal.

Every category $\mathbb{A}$ has an associated skeletal category $sk(\mathbb{A})$. The objects of $sk(\mathbb{A})$ are a collection of objects of $\mathbb{A}$ where each object of $\mathbb{A}$ is isomorphic to some object in $sk(\mathbb{A})$. In other words, every object in the original category has an isomorphic representation in the skeletal category. The skeletal category is a subcategory of the original category. (This construction demands the axiom of choice.)

Example 2.3.12 goes back to the spirit of the philosopher Gottlob Frege. He defined a natural number as the equivalence class of all finite sets with that number of elements.

Example 2.3.12. Consider the category of $\mathbb{FinSet}$. A skeletal category for $\mathbb{FinSet}$ is $Nat\mathbb{Set}$. The objects are the empty set and the sets $\{1\}, \{1,2\}, \{1,2,3\}, \ldots$ Notice that none of these objects is isomorphic to any other. Every finite set is isomorphic to one of these sets. The morphisms in $Nat\mathbb{Set}$ are all functions between these sets. □

Example 2.3.13. Let $(P, \leq)$ be a preorder category. A skeletal category of $(P, \leq)$ is a partial order category. In detail, we form a partial order category $(P_0, \leq)$, where the objects of P_0 are representatives of isomorphism classes of P. Two representatives p and p' have the relation $p \leq_0 p'$ if $p \leq p'$. □

Every category has an opposite category.

Definition 2.3.14. For category $\mathbb{A}$, the **opposite category** $\mathbb{A}^{op}$ is a category with the same objects as $\mathbb{A}$, but with the domain and codomain of each arrow of $\mathbb{A}$ reversed. Every morphism $a \longrightarrow b$ in $\mathbb{A}^{op}$ corresponds to a morphism $b \longrightarrow a$ in $\mathbb{A}$ or

$$Hom_{\mathbb{A}^{op}}(a,b) = Hom_{\mathbb{A}}(b,a). \tag{2.33}$$

The composition in $\mathbb{A}^{op}$ is given by the composition in $\mathbb{A}$. In detail, if

$$a \xrightarrow{\quad f \quad} b \xrightarrow{\quad g \quad} c \tag{2.34}$$

are two composible morphisms in $\mathbb{A}^{op}$, then there are two composible morphisms in $\mathbb{A}$:

$$a \xleftarrow{\quad f' \quad} b \xleftarrow{\quad g' \quad} c. \tag{2.35}$$

The composition $g \circ f$ in $\mathbb{A}^{op}$ will correspond to the composition $f' \circ g'$ in $\mathbb{A}$.

Example 2.3.15. $\mathbb{Rel}^{op}$ is almost the same as $\mathbb{Rel}$. (We will formulate what we mean by "almost the same" in Section 4.1.) The objects are all sets. The relation $R \subseteq S \times T$ is associated to the inverse relation $R^{-1} \subseteq T \times S$. Notice that this statement is not true for $\mathbb{Set}$ and $\mathbb{Par}$ because not every function or partial function has an inverse. □

Example 2.3.16. If $(P, \leq)$ is a preorder category, then the opposite category has the relation $\leq^{op}$, which is defined as $p \leq^{op} p'$ if and only if $p' \leq p$. This essentially turns the arrows around. □

Example 2.3.17. The category $\mathbf{K}\mathbb{Mat}$ has natural numbers as objects and a map $A\colon m \longrightarrow n$ is a n by m matrix with entries in $\mathbf{K}$. The category $\mathbf{K}\mathbb{Mat}^{op}$ has natural numbers as objects and a map $A\colon m \longrightarrow n$ is a m by n matrix with entries in $\mathbf{K}$. Every matrix $A\colon m \longrightarrow n$ in $\mathbb{K}\mathbb{M}\mathrm{at}$ will correspond to the transpose matrix $A^T\colon n \longrightarrow m$ in $\mathbb{K}\mathbb{M}\mathrm{at}^{op}$. (We will formally meet this operation on page 71.) □

Exercise 2.3.18. Show that $f\colon a \longrightarrow b$ in $\mathbb{A}$ is monic if and only if the corresponding $g\colon b \longrightarrow a$ in $\mathbb{A}^{op}$ is epic. ■

Exercise 2.3.19. Show that $\mathbb{A}$ has an initial object if and only if $\mathbb{A}^{op}$ has a terminal object. ■

There is an important operation on categories. Just as we can take two sets and form their Cartesian product, so we can take two categories and form their Cartesian product.

Definition 2.3.20. Given categories $\mathbb{A}$ and $\mathbb{B}$, there is a category that is called the **Cartesian product** of $\mathbb{A}$ and $\mathbb{B}$, written as $\mathbb{A} \times \mathbb{B}$. The objects are pairs of objects (a, b), where a is an object of $\mathbb{A}$ and b is an object of $\mathbb{B}$. Morphisms in the category are pairs of morphisms $(f, g)\colon (a, b) \longrightarrow (a', b')$, where $f\colon a \longrightarrow a'$ is in $\mathbb{A}$ and $g\colon b \longrightarrow b'$ is in $\mathbb{B}$. Composition is given "component-wise"; that is, the composition of $(f, g)\colon (a, b) \longrightarrow (a', b')$ and $(f', g')\colon (a', b') \longrightarrow (a'', b'')$ is

$$(f', g') \circ (f, g) = (f' \circ f, g' \circ g)\colon (a, b) \longrightarrow (a'', b''). \tag{2.36}$$

(This identity is an instance of the "interchange law," which will be discussed in Important Categorical Idea 3.1.18.) We leave it to the reader to prove associativity of composition. The unit for (a, b) is (id_a, id_b). The fact that this morphism works as an identity is obvious.

Exercise 2.3.21. Show that for any category $\mathbb{A}$, the category $\mathbb{A} \times \mathbf{1}$ is almost the same as $\mathbb{A}$. (We will formalize what we mean by "almost the same" in Section 4.1.) ■

Exercise 2.3.22. Draw the category $\mathbf{3} \times \mathbf{2}_I$. ■

Suggestions for Further Study

The categories and structures in this chapter are mostly pretty standard and can be found in any standard textbook on category theory (e.g., [180, 40, 32, 33, 108, 171, 222, 21]).

2.4 Mini-Course: Basic Linear Algebra

We have learned only the bare basics of category theory so far. The beauty and magic of category theory are apparent only when we relate one category to another, which we will do in Chapter 4. Nevertheless, one can still see a lot of nice category theoretic ideas within a single category.

Linear algebra is the study of directions, straight lines, and flat spaces. The mathematical structures that describe such notions are vector spaces. Functions that go from one vector space to another are called "linear transformations." The subject of linear algebra are vector spaces and linear transformations. The collection of vector spaces and linear transformations forms the category $\mathbb{Vect}$. This mini-course will explore the category of $\mathbb{Vect}$ with special emphasis on the morphisms.

We are going to focus on complex vector spaces, $\mathbf{C}\mathbb{Vect}$, which are vector spaces associated with complex numbers. We require such vector spaces later in this book, when we talk about quantum mechanics and many of the mini-courses.

A small disclaimer: Linear algebra is an important field associated with beautiful geometric ideas. There is no way we can show you anything more than a bird's-eye view of the field. We hope that this mini-course will whet your appetite to learn more linear algebra.

The Objects: Vector Spaces

The objects of $\mathbf{C}\mathbb{Vect}$ are complex vector spaces.

Definition 2.4.1. Consider $\mathbf{C}$, the field of complex numbers. A **complex vector space** is an abelian group $(V, +, 0, -)$. Reminder: V is a set, $+$ is an associative, commutative binary operation with identity 0 and inverse operation $-$) with a function $\cdot\colon \mathbf{C} \times V \longrightarrow V$, which is called **scalar multiplication** (or a **C-action**) on V. (A scalar is an element of a field or simply a number.) The elements of V will be called "vectors" and will correspond to the various directions. The $+$ and $-$ operations allow us to add and subtract directions. The scalar multiplication operation permits us to lengthen or shorten the directions. The $\cdot$ function must satisfy the following axioms:

- The scalar multiplication respects the addition in the abelian group: for all c in $\mathbf{C}$, and for all v, v' in V,

$$c \cdot (v + v') = (c \cdot v) + (c \cdot v'). \tag{2.37}$$

- The scalar multiplication respects the addition in the field: for all c, c' in $\mathbf{C}$, and for all v in V,

$$(c + c') \cdot v = (c \cdot v) + (c' \cdot v). \tag{2.38}$$

- The scalar multiplication respects the multiplication in the field: for all c, c' in $\mathbf{C}$, and for all v in V,

$$(cc') \cdot v = c \cdot (c' \cdot v). \tag{2.39}$$

- The scalar multiplication respects the unit of the field: for 1 in $\mathbf{C}$, and for all v in V,

$$1 \cdot v = v. \tag{2.40}$$

Example 2.4.2. There are many examples of complex vector spaces:

- For any positive integers m and n, the set of m by n matrices with elements in the field $\mathbf{C}$ has the structure of a vector space. We denote this complex vector space as $\mathbf{C}^{m \times n}$. The addition in this vector space is simply the addition of matrices. The negation operation multiplies every entry by -1. The zero vector is the m by n matrix whose entries are all 0. Given a matrix M and a scalar $c \in \mathbf{C}$, the scalar multiplication $c \cdot M$ is the matrix M with all its entries multiplied by c. (Another way of saying this is that for every m and n, the Hom set, $Hom_{\mathbb{CMat}}(m,n)$, is a complex vector space.)
- In particular, if $n = 1$, then $\mathbf{C}^{m\times 1} = \mathbf{C}^m$ is the set of all m-element column vectors. This vector space shares the same operations as $\mathbf{C}^{m\times n}$. Similarly, if $m = 1$, then $\mathbf{C}^{1\times n}$ is the set of n-element row vectors. These examples are the core of elementary linear algebra.
- In particular, if $m = 1$ and $n = 1$, then $\mathbf{C}^1 = \mathbf{C}$, the complex number, forms a complex vector space.
- The world's smallest complex vector space is the set with just the number 0. This is called the **trivial vector space**, which we denote as 0. The operations are $0 + 0 = 0$ and $c \cdot 0 = 0$.
- The collection, $Poly_{\mathbf{C}}(m)$, of complex polynomials in one variable where the highest degree is m or less forms a complex vector space. The addition is the regular addition of polynomials. The negation of a polynomial is the polynomial whose every term is negated. The 0 polynomial is $0 + 0x + 0x^2 + \cdots + 0x^m$. The scalar multiplication multiplies every term by a complex number; that is,

$$c \cdot (a_0 + a_1 z^1 + a_2 z^2 + \cdots + a_n z^n) = ca_0 + ca_1 z^1 + ca_2 z^2 + \cdots + ca_n z^n. \quad (2.41)$$

- The set, $Poly_{\mathbf{C}}$, of all polynomials with coefficients in the complex numbers forms a complex vector space. The operations are the same as in $Poly_{\mathbf{C}}(m)$.
- The set of all functions from $\mathbf{N}$ to $\mathbf{C}$, denoted as $Func(\mathbf{N}, \mathbf{C})$ or $\mathbf{C}^{\mathbf{N}}$, forms a complex vector space. The addition of two functions $f\colon \mathbf{N} \longrightarrow \mathbf{C}$ and $g\colon \mathbf{N} \longrightarrow \mathbf{C}$ is the function $f + g\colon \mathbf{N} \longrightarrow \mathbf{C}$, defined as $(f+g)(n) = f(n) + g(n)$. The negation of $f\colon \mathbf{N} \longrightarrow \mathbf{C}$ is $-f\colon \mathbf{N} \longrightarrow \mathbf{C}$, defined as $(-f)(n) = -(f(n))$. The zero function is the function that takes all natural numbers to 0. The scalar multiplication of $c \in \mathbf{C}$ with $f\colon \mathbf{N} \longrightarrow \mathbf{C}$ is $(c \cdot f)\colon \mathbf{N} \longrightarrow \mathbf{C}$ defined as $(c \cdot f)(n) = c(f(n))$.
- The set of all functions from the complex numbers to the complex numbers forms a complex vector space denoted as $Func(\mathbf{C}, \mathbf{C})$ or $\mathbf{C}^{\mathbf{C}}$. The operations are defined analogously to the operations of $Func(\mathbf{N}, \mathbf{C})$. □

Exercise 2.4.3. Let S be any set. Show that the set of all functions from S to $\mathbf{C}$ form a complex vector space. ■

This exercise—which shows that $Hom_{\mathbb{Set}}(S, \mathbf{C})$ is like a complex vector space just as $\mathbf{C}$ is a complex vector space—highlights an important idea that arises many times in all of category theory.

Important Categorical Idea 2.4.4. Hom Sets Inheriting Properties. In category $\mathbb{A}$, if a and b are objects in $\mathbb{A}$, then $Hom_{\mathbb{A}}(a,b)$ will inherit many of the properties of b. This simple fact arises over and over again. ○

Just as we can look at subsets of a set, and subcategories of a category, we can look at subvector spaces of a vector space.

Definition 2.4.5. A vector space W is a **subvector space** or **linear subspace** of vector space V if W is a subset of V and the following are true:

- W is closed under addition: for all w, w' in W, $w+w'$ is in W.
- W is closed under scalar multiplication: for all c in $\mathbf{C}$ and w in W, $c \cdot w$ is in W.
- W has the zero vector: 0 is in W.

Example 2.4.6. Here are some examples of linear subspaces:

- The trivial vector space, 0, is a linear subspace of every vector space.
- It is not hard to see that $\mathbf{C}$ is a linear subspace of $\mathbf{C}^m$ for any $m > 1$ (identify $c \in \mathbf{C}$ as $\begin{bmatrix} c \\ 0 \\ 0 \\ \vdots \\ 0 \end{bmatrix}$.) Similarly, $\mathbf{C}^m$ is a linear subspace of $\mathbf{C}^{m \times n}$ (identify $\begin{bmatrix} c_1 \\ c_2 \\ c_3 \\ \vdots \\ c_m \end{bmatrix}$ as $\begin{bmatrix} c_1 & 0 & 0 & \cdots & 0 \\ c_2 & 0 & 0 & \cdots & 0 \\ c_3 & 0 & 0 & \cdots & 0 \\ \vdots & \vdots & \vdots & \ddots & \vdots \\ c_m & 0 & 0 & \cdots & 0 \end{bmatrix}$.)
- For every m and n with $m \leq n$, $Poly_{\mathbf{C}}(m)$ is a linear subspace of $Poly_{\mathbf{C}}(n)$. $Poly_{\mathbf{C}}(m)$ is a linear subspace of $Poly_{\mathbf{C}}$.
- Consider the subset of all matrices in $\mathbf{C}^{m \times m}$ that contain only zeros, with a possible exception along the diagonal. This is a linear subspace of $\mathbf{C}^{m \times m}$.
- Consider the subset $Func_{Fin}(\mathbf{N}, \mathbf{C})$ of all functions $f \colon \mathbf{N} \longrightarrow \mathbf{C}$, whose values are all 0 except for a finite subset of $\mathbf{N}$. Such functions form a linear subspace of $Func(\mathbf{N}, \mathbf{C})$. □

The Morphisms: Linear Transformations

As we have seen (Important Categorical Idea 1.4.9), in category theory, the morphisms are as important or even more important than the objects.

Definition 2.4.7. Let V and W be vector spaces. A **linear map** or **linear transformation** T from V to W, written $T\colon V \longrightarrow W$ is a set function $T\colon V \longrightarrow W$ such that the following is true:

- T respects the addition operation: for all $v, v' \in V$,

$$T(v+v') = T(v) + T(v'). \tag{2.42}$$

  (Notice that the two $+$'s are operations in two different vector spaces. The one on the left is an operation in V, and the one on the right is an operation in W.)
- T respects the scalar multiplication: for all $c \in \mathbf{C}$ and $v \in V$,

$$T(c \cdot v) = c \cdot T(v). \tag{2.43}$$

  (Notice the two $\cdot$'s are in two different operations in two different vector spaces. The one on the left is an operation in V, and the one on the right is an operation in W.)

These two requirements are equivalent to the single requirement.

- For all $v, v' \in V$ and $c, c' \in \mathbf{C}$,

$$T((c \cdot v) + (c' \cdot v')) = (c \cdot T(v)) + (c' \cdot T(v')). \tag{2.44}$$

Exercise 2.4.8. Show that the single requirement of Equation (2.44) is equivalent to the requirements of Equations (2.42) and (2.43). ■

Example 2.4.9. A few examples of linear maps follow:

- If W is a linear subspace of V, then the inclusion map is a linear transformation.
- For vector spaces V and W, there is a simple linear transformation, called the **null map** or **zero map**, $N\colon V \longrightarrow W$, which takes every element of V to the 0 in W. That is, $N(v) = 0$.
- If A is an $m \times n$ matrix with entries in $\mathbf{C}$, then there is a linear map $T_A\colon \mathbf{C}^n \longrightarrow \mathbf{C}^m$ defined for $V \in \mathbf{C}^n$ as $T_A(V) = AV$. This is linear because
  - T_A respects the matrix addition:

$$T_A(V+V') = A(V+V') = AV + AV' = T_A(V) + T_A(V') \tag{2.45}$$

    and
  - T_A respects the scalar multiplication:

$$T_A(c \cdot V) = A(c \cdot V) = c \cdot (AV) = c \cdot T_A(V). \tag{2.46}$$

- There is a linear transformation from $\mathbf{C}^{m\times n}$ to $\mathbf{C}^{n\times m}$ that is called the **transpose** operation. This operation swaps the element in the i, j position with the element in the j, i position. We denote the transpose of the m by n matrix A as A^T. The definition is reduced to $A^T[i,j] = A[j,i]$. For example, the transpose operation works as follows:

$$\begin{bmatrix} a & b & c & d & e \\ f & g & h & i & j \\ k & l & m & n & o \end{bmatrix}^T = \begin{bmatrix} a & f & k \\ b & g & l \\ c & h & m \\ d & i & n \\ e & j & o \end{bmatrix}. \tag{2.47}$$

The transpose is a linear transformation because of the following:

– The transpose respects addition:

$$(A+B)^T[i,j] = (A+B)[j,i] = A[j,i] + B[j,i] = A^T[i,j] + B^T[i,j]. \tag{2.48}$$

– The transpose respects scalar multiplication:

$$(c \cdot A)^T[i,j] = (c \cdot A)[j,i] = c \cdot A[j,i] = c \cdot A^T[i,j]. \tag{2.49}$$

The transpose changes column vectors to row vectors and vice versa.

- For every vector space V, there is a unique linear transformation from V to the trivial vector space $V \longrightarrow 0$. Similarly, there is a unique linear map $0 \longrightarrow V$. This makes 0 a zero object in the category of vector spaces. □

Exercise 2.4.10. Consider the function $\pi_{1,3}\colon \mathbf{C}^3 \longrightarrow \mathbf{C}^2$, defined by $\pi_{1,3}(c_1, c_2, c_3) = (c_1, c_3)$. Show that $\pi_{1,3}$ is a linear map. ■

Theorem 2.4.11. Let $T\colon V \longrightarrow W$ be a linear map. Then T respects 0, that is, $T(0) = 0$. Also, T respects negation, that is, $T(-v) = -T(v)$. ★

Proof. To show that T respects 0, notice that

$$T(0) = T(0+0) = T(0) + T(0). \tag{2.50}$$

Now subtract $T(0)$ from both ends. This leaves $0 = T(0)$.

To show that T respects negation, notice that

$$0 = T(0) = T(v + (-v)) = T(v) + T(-v). \tag{2.51}$$

Now subtract $T(v)$ from both ends of the equation. This leaves $-T(v) = T(-v)$. ♣

Exercise 2.4.12. Show that the composition of two linear maps is a linear map. Moreover, show that this composition is associative. ■

Exercise 2.4.13. Show that for every vector space V, the identity function $I_V\colon V \longrightarrow V$, which is defined for $v \in V$ as $I_V(v) = v$, is a linear map. Show also that for $T\colon V \longrightarrow V'$, we have that $T \circ I_V = T = I_{V'} \circ T$. ■

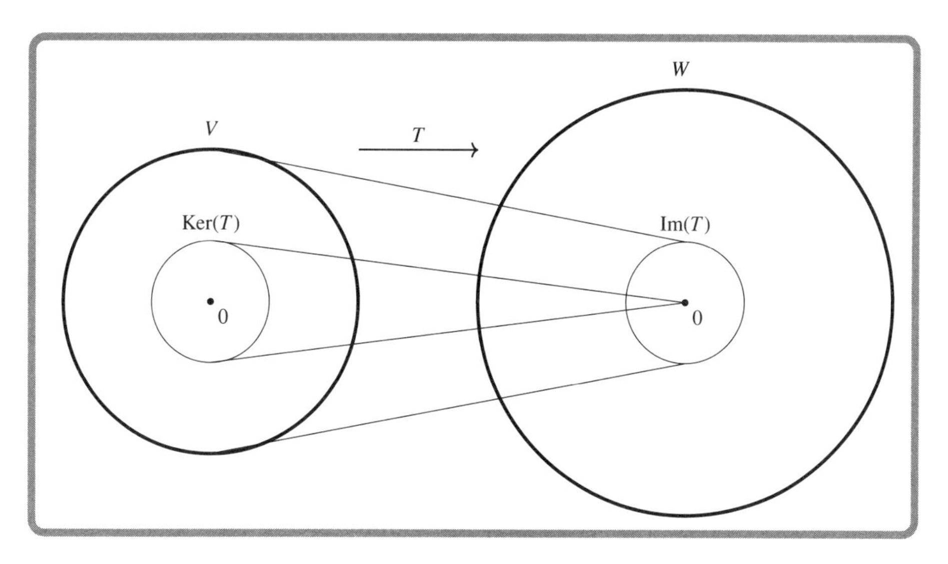

Figure 2.11. The kernel and image of a linear transformation.

Exercises 2.4.12 and 2.4.13 bring us to the main definition of this mini-course:

Definition 2.4.14. The collection of complex vector spaces and linear maps between them forms a category called $\mathbf{C}\mathbb{V}\mathrm{ect}$.

Remark 2.4.15. In the category $\mathbb{S}\mathrm{et}$, an element of set S is determined by a map from a terminal object to S (i.e., a map $\{*\} \longrightarrow S$). In the category $\mathbf{C}\mathbb{V}\mathrm{ect}$, a vector of a vector space V is determined by a linear map from $\mathbf{C}$ to V. (A map from the terminal object 0 in $\mathbf{C}\mathbb{V}\mathrm{ect}$ to V, i.e., $0 \longrightarrow V$ just picks out the 0 vector in V, which is not very helpful.) Any map $f\colon \mathbf{C} \longrightarrow V$ is totally determined by where it takes $1 \in \mathbf{C}$. In other words, f is defined by $f(1) \in \mathbf{C}$. For any $c \in \mathbf{C}$, the value $f(c) = f(c \cdot 1) = c \cdot f(1)$. In other words, $Hom_{\mathbf{C}\mathbb{V}\mathrm{ect}}(\mathbf{C}, V)$ is isomorphic to V. However, the output of f is not only $f(1)$. Rather, the output is all the scalar multiples of $f(1)$.

In particular, thinking of $\mathbf{C}$ as a complex vector space, we have that all linear maps $f\colon \mathbf{C} \longrightarrow \mathbf{C}$ are determined by the value $f(1)$. In other words, $Hom_{\mathbf{C}\mathbb{V}\mathrm{ect}}(\mathbf{C}, \mathbf{C})$ is isomorphic to $\mathbf{C}$. ♠

For any linear map $T\colon V \longrightarrow W$, there are a linear subspace of V, the kernel of T, and a linear subspace of W, the image of T, as in Figure 2.11.

Definition 2.4.16. For linear map $T\colon V \longrightarrow W$, the **kernel** of T, written as $Ker(T)$, is the linear subspace of V consisting of those vectors that go to the zero vector of W; that is,

$$Ker(T) = \{v \in V : T(v) = 0\}. \tag{2.52}$$

To see that $Ker(T)$ is a linear subspace of V, notice the following:

- $Ker(T)$ is closed under addition: if $T(v) = 0$ and $T(v') = 0$, then $T(v + v') = T(v) + T(v') = 0 + 0 = 0$.
- $Ker(T)$ is closed under scalar multiplication: if c is in $\mathbf{C}$ and $T(v) = 0$, then $T(c \cdot v) = c \cdot T(v) = c \cdot 0 = 0$.
- $Ker(T)$ contains the zero vector: $T(0) = 0$.

Notice that the composition map $Ker(T) \hookrightarrow V \xrightarrow{T} W$ is the null map.

Definition 2.4.17. For linear map $T\colon V \longrightarrow W$, the **image** of T, written $Im(T)$, is the linear subspace of W consisting of those vectors that are the output of T; that is,

$$Im(T) = \{w \in W : \text{there exists a } v \in V \text{ with } T(v) = w\}. \tag{2.53}$$

To see that the $Im(T)$ is a linear subspace of W, notice the following:

- $Im(T)$ is closed under addition: if $T(v) = w$ and $T(v') = w'$, then $w + w' = T(v) + T(v') = T(v + v')$.
- $Im(T)$ is closed under scalar multiplication: if c is in $\mathbf{C}$ and $T(v) = w$, then $c \cdot w = c \cdot T(v) = T(c \cdot v)$.
- $Im(T)$ contains the zero vector: $T(0) = 0$.

Theorem 2.4.18. For a linear map $T\colon V \longrightarrow W$, the following are equivalent: (i) T is monic, (ii) T is an injection, and (iii) $Ker(T) = \{0\}$. ★

Proof. The fact that monic is equivalent to injective is basically the same proof as in $\mathbb{Set}$ which can be seen in Example 2.2.3. Rather then use the one-object set $\{*\}$, use the one-dimensional vector space $\mathbf{C}$. If $T\colon V \longrightarrow W$ is an injection, then since $T(0) = 0$, it is the only vector that goes to 0. Hence, $Ker(T) = \{0\}$. For the other way, assume that $Ker(T) = \{0\}$ and $T(x) = T(y)$. Then $0 = T(x) - T(y) = T(x - y)$. This means that $x - y$ is in the kernel of T. That means $x - y = 0$, and hence, $x = y$. ♣

Theorem 2.4.19. A linear map $T\colon V \longrightarrow W$ is epic if and only if T is a surjection. ★

Proof. This is basically the same proof as in $\mathbb{Set}$, which can be seen in Exercise 2.2.7. Rather then use the one-object set $\{*\}$, use the one-dimensional vector space $\mathbf{C}$. ♣

There are two extremes for a linear map $T\colon V \longrightarrow W$. On the one hand, T can take all of V to $0 \in W$. That would make $Ker(T) = V$ and $Im(T) = \{0\}$. On the other hand, T can be injective with $Ker(T) = \{0\}$, and $Im(T)$ is isomorphic to a subspace of V. A typical T will lie somewhere between these two extremes.

Definition 2.4.20. A linear map is an **isomorphism** if it is has an inverse. If there is an isomorphism between two vector spaces, then the vector spaces are called **isomorphic**.

Theorem 2.4.21. A linear map is an **isomorphism** if and only if it is monic and epic. ★

Proof. If a linear map is an isomorphism, then use the inverse to show that it is left and right cancelable. If it is monic, then it is injective. If it is epic, then it is surjective. So there is a set theoretic inverse to the map. It remains to show that that inverse is linear. This follows from the fact that the map is linear and its inverse undoes what the map does. ♣

Exercise 2.4.22. Describe an isomorphism between $\mathbf{C}^{m+1}$ and $Poly_{\mathbf{C}}(m)$. ■

Exercise 2.4.23. Let S be any set with $|S| = m$. Show that $Func(S, \mathbf{C})$ is isomorphic to $\mathbf{C}^m$. ■

As we have said, we will stress morphisms in this mini-course. We will be able to prove many results with a tool called an "exact sequence." Knowing this construction will not only be good for this text, but exact sequences are used in many branches of higher mathematics, such as algebraic topology, homological algebra, and sheaf theory. (In fact, a main part of these fields is the study of how sequences *fail* to be exact sequences.)

Definition 2.4.24. An **exact sequence** is a diagram

$$V_1 \xrightarrow{T_1} V_2 \xrightarrow{T_2} V_3 \xrightarrow{T_3} \cdots \longrightarrow V_{n-1} \xrightarrow{T_{n-1}} V_n \cdots \tag{2.54}$$

of vector spaces and linear maps that satisfies $Im(T_i) = Ker(T_{i+1})$ for $i = 1, 2, \ldots, n-2, \ldots$. A **short exact sequence** is an exact sequence of the form

$$0 \longrightarrow V_1 \xrightarrow{T_1} V_2 \xrightarrow{T_2} V_3 \longrightarrow 0. \tag{2.55}$$

Notice that the composite of any two contiguous maps in an exact sequence is the null map. To see this, let T_i and T_{i+1} be two contiguous maps. For any v in the source of T_i, we have that $T_i(v) \in Im(T_i) = Ker(T_{i+1})$. Now apply T_{i+1} to that element, and we get that $T_{i+1}(T_i(v)) = 0$.

Example 2.4.25. Let us look at some examples of exact sequences:

- $0 \longrightarrow V \xrightarrow{T} W$ means that T is an injection.
- $V \xrightarrow{T} W \longrightarrow 0$ means that T is a surjection.
- $0 \longrightarrow V \xrightarrow{T} W \longrightarrow 0$ means that T is an isomorphism. □

Exercise 2.4.26. Show that $0 \longrightarrow V \longrightarrow 0$ is exact if and only if $V = 0$. ■

Bases and Dimension

Given a set of elements of V, we can use the addition and scalar multiplication operations to get other elements. (For convenience, we will drop the $\cdot$ as the scalar multiplication and write cx for $c \cdot x$.)

Definition 2.4.27. If $x_1, x_2, x_3, \ldots$ are vectors in a vector space[a] V and $c_1, c_2, c_3, \ldots$ are elements of $\mathbf{C}$, then

$$y = c_1 x_1 + c_2 x_2 + c_3 x_3 + \cdots \tag{2.56}$$

is a vector of V and is called a **linear combination** of $x_1, x_2, x_3, \ldots$

Definition 2.4.28. A **basis** $\mathcal{B} = \{b_1, b_2, b_3, \ldots\}$ for a vector space V is a set of vectors of V, such that every element of V can be written in exactly one way as a linear combination of elements of $\mathcal{B}$. That is, for every y in V, there is a unique sequence of scalars $c_1, c_2, c_3, \ldots$ such that

$$y = c_1 b_1 + c_2 b_2 + c_3 b_3 + \cdots \tag{2.57}$$

In a sense, we can say that the elements of the basis "generate" the entire vector space.

Exercise 2.4.29. Let V be a vector space with bases $\mathcal{B}$ and $\mathcal{B}'$. Show that $\mathcal{B}$ and $\mathcal{B}'$ have the same number of elements. ■

Definition 2.4.30. The **dimension** of a vector space is the cardinality of a base. If a vector space has a finite basis, then the dimension is written $dim(V)$.

Example 2.4.31. Let us go through all the examples given in Example 2.4.2 of vector spaces and describe particularly simple bases. Such simple bases are called **canonical bases**. We also determine the dimension of the vector space:

- For $\mathbf{C}^{m \times n}$, the canonical basis consists of the mn matrices, where each matrix has a single 1 as an entry and all the other entries are 0. In detail, the first three elements of the basis look like this:

$$\begin{bmatrix} 1 & 0 & 0 & \cdots & 0 \\ 0 & 0 & 0 & \cdots & 0 \\ 0 & 0 & 0 & \cdots & 0 \\ \vdots & \vdots & \vdots & \ddots & \vdots \\ 0 & 0 & 0 & \cdots & 0 \end{bmatrix}, \begin{bmatrix} 0 & 1 & 0 & \cdots & 0 \\ 0 & 0 & 0 & \cdots & 0 \\ 0 & 0 & 0 & \cdots & 0 \\ \vdots & \vdots & \vdots & \ddots & \vdots \\ 0 & 0 & 0 & \cdots & 0 \end{bmatrix}, \begin{bmatrix} 0 & 0 & 1 & \cdots & 0 \\ 0 & 0 & 0 & \cdots & 0 \\ 0 & 0 & 0 & \cdots & 0 \\ \vdots & \vdots & \vdots & \ddots & \vdots \\ 0 & 0 & 0 & \cdots & 0 \end{bmatrix} \tag{2.58}$$

  and the last one is this:

$$\begin{bmatrix} 0 & 0 & 0 & \cdots & 0 \\ 0 & 0 & 0 & \cdots & 0 \\ 0 & 0 & 0 & \cdots & 0 \\ \vdots & \vdots & \vdots & \ddots & \vdots \\ 0 & 0 & 0 & \cdots & 1 \end{bmatrix}. \tag{2.59}$$

[a] Our notation makes it seem as though the set is countably infinite. We use this notation for convenience. We allow for the set to be uncountably infinite.

The dimension is mn.

- For $\mathbf{C}^m$, a canonical basis is $\begin{bmatrix}1\\0\\0\\\vdots\\0\end{bmatrix}, \begin{bmatrix}0\\1\\0\\\vdots\\0\end{bmatrix}, \begin{bmatrix}0\\0\\1\\\vdots\\0\end{bmatrix}, \cdots, \begin{bmatrix}0\\0\\0\\\vdots\\1\end{bmatrix}$. The dimension is m.
- For $\mathbf{C}$, the canonical basis is 1 and the dimension is 1.
- The basis of the trivial vector space 0 is the empty set, which has no elements, and hence the dimension is 0.
- For $Poly_{\mathbf{C}}(m)$, the canonical basis is $1, z, z^2, z^3, \ldots, z^m$. The dimension is $m+1$.
- For $Poly_{\mathbf{C}}$, the canonical basis is $1, z, z^2, z^3, \ldots, z^m, \ldots$ The dimension is countably infinite.
- For $Func(\mathbf{N}, \mathbf{C})$, the canonical basis consists of a function for every $m \in \mathbf{N}$, $f_m \colon \mathbf{N} \longrightarrow \mathbf{C}$, which is defined as

$$f_m(n) = \begin{cases} 1 & : m = n \\ 0 & : m \neq n. \end{cases} \tag{2.60}$$

The dimension is countably infinite. Every element $f \in Func(\mathbf{N}, \mathbf{C})$ can be written as the sum

$$f = \sum_{i=0}^{\infty} f(i) f_i. \tag{2.61}$$

- For $Func(\mathbf{C}, \mathbf{C})$, the canonical basis consists of functions for every $z \in \mathbf{C}$, $f_z \colon \mathbf{C} \longrightarrow \mathbf{C}$, defined as

$$f_z(c) = \begin{cases} 1 & : z = c \\ 0 & : z \neq c. \end{cases} \tag{2.62}$$

The dimension is uncountably infinite. Every element $f \in Func(\mathbf{C}, \mathbf{C})$ can— analogous to Equation (2.61)— be written as the integral

$$f = \int_z f(z) f_z dz. \tag{2.63}$$

(Its only an analogy because these functions are not integrable.) □

Exercise 2.4.32. Let S be any set. Find a canonical basis for $Func(S, \mathbf{C})$. What is the dimension of $Func(S, \mathbf{C})$ in terms of S? ■

Theorem 2.4.33 will have many consequences.

Theorem 2.4.33. Fundamental theorem of linear algebra, or the **rank-nullity theorem**. Let V and W be finite dimensional vector spaces. For any linear map $T \colon V \longrightarrow W$, we have the following:

$$dim(Ker(T)) + dim(Im(T)) = dim(V). \tag{2.64}$$

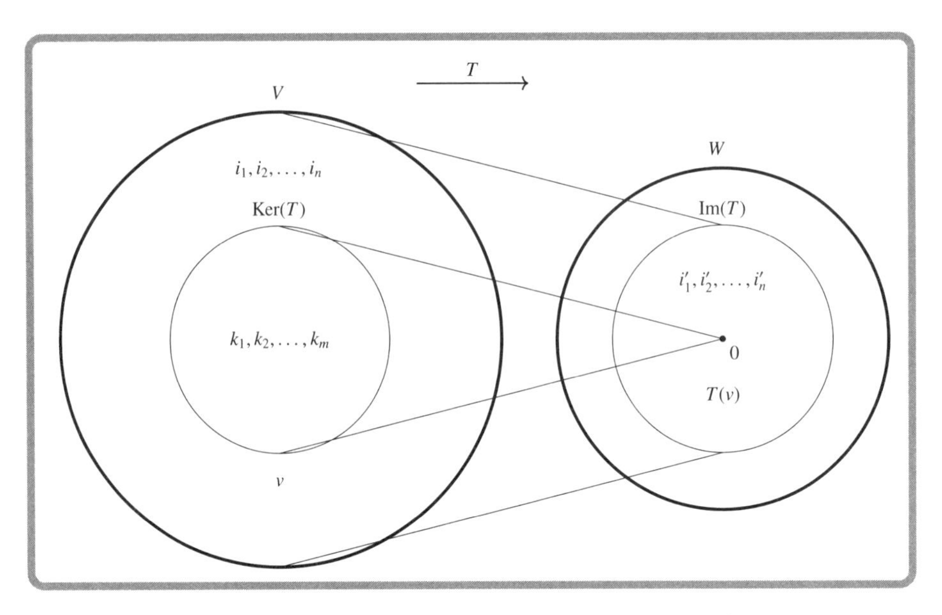

Figure 2.12. Bases for the kernel and image of a linear transformation.

Proof. Notice that $dim(W)$ does not enter into the equation. The reason for this is that any linear map $V \longrightarrow W$ can also be seen as a map $V \longrightarrow W'$, where W is a linear subspace of W'. So the target of the linear map can be arbitrarily big.

The proof is easier when keeping an eye on Figure 2.12. Let $dim(Ker(T)) = m$ and $\{k_1, k_2, \ldots, k_m\}$ be a basis for $Ker(T)$. Let $dim(Im(T)) = n$ and $\{i'_1, i'_2, \ldots, i'_n\}$ be a basis for $Im(T)$. Since the i's are in the image of T, for each i' there must be some elements of V that T takes to i'. Choose a value of $i \in V$ for each of the i'. Consider the set $\{i_1, i_2, \ldots, i_n\}$. We claim that $\{k_1, k_2, \ldots, k_m, i_1, i_2, \ldots, i_n\}$ is a basis for the entire vector space V. To show this, consider an arbitrary vector $v \in V$ and look at $T(v)$. Since $\{i'_1, i'_2, \ldots, i'_n\}$ is a basis of $Im(T)$, there are unique values of $c_1, c_2, \ldots, c_n$ such that

$$T(v) = c_1 i'_1 + c_2 i'_2 + \cdots + c_n i'_n. \tag{2.65}$$

Since each $i'_t = T(i_t)$, we can rewrite Equation (2.65) as

$$T(v) = c_1 T(i_1) + c_2 T(i_2) + \cdots + c_n T(i_n). \tag{2.66}$$

By linearity of T, this means

$$T(v) = T(c_1 i_1 + c_2 i_2 + \cdots + c_n i_n). \tag{2.67}$$

Bringing the right side to the left side, we have

$$T(v) - T(c_1 i_1 + c_2 i_2 + \cdots + c_n i_n) = 0 \tag{2.68}$$

or

$$T(v - (c_1 i_1 + c_2 i_2 + \cdots + c_n i_n)) = 0. \tag{2.69}$$

This means that $v-(c_1i_1+c_2i_2+\cdots+c_ni_n)$ is in the kernel of T, and since $\{k_1,k_2,\ldots,k_m\}$ is a basis for the kernel, there are unique scalars $d_1,d_2,\ldots,d_m$ such that

$$v-(c_1i_1+c_2i_2+\cdots+c_ni_n)=d_1k_1+d_2k_2+\cdots+d_mk_m. \tag{2.70}$$

That means that v can be written uniquely as

$$v=d_1k_1+d_2k_2+\cdots+d_mk_m+c_1i_1+c_2i_2+\cdots+c_ni_n. \tag{2.71}$$

So we have written an arbitrary element of V uniquely as a linear combination of $\{k_1,k_2,\ldots,k_m,i_1,i_2,\ldots,i_n\}$. We conclude that $dim(V)=m+n$. ♣

Theorem 2.4.34. Consider the exact sequence

$$0 \longrightarrow V_1 \xrightarrow{T_1} V_2 \xrightarrow{T_2} \cdots \longrightarrow V_{n-1} \xrightarrow{T_{n-1}} V_n \longrightarrow 0\,. \tag{2.72}$$

The dimensions of the vector spaces satisfy the following identity:

$$\sum_{i\text{ even}} dim(V_i)=\sum_{i\text{ odd}} dim(V_i). \tag{2.73}$$

Another way to say this is that

$$\sum_{i=1}^{n}(-1)^i dim(V_i)=0. \tag{2.74}$$

★

Proof. From Theorem 2.4.33, we know that for every i, $dim(V_i)=dim(Ker(T_i))+dim(Im(T_i))$. We therefore have

$$\sum_{i\text{ even}} dim(V_i)=\sum_{i\text{ even}} dim(Ker(T_i))+dim(Im(T_i)). \tag{2.75}$$

The fact that we are working with an exact sequence in which $Im(T_i)=Ker(T_{i+1})$ means that we can rewrite Equation (2.75) as

$$\begin{aligned}\sum_{i\text{ even}} dim(V_i) &= \sum_{i\text{ even}} dim(Ker(T_i))+dim(Ker(T_{i+1})) && \text{by Theorem 2.4.33}\\ &= \sum_{i\text{ odd}} dim(Ker(T_i))+dim(Ker(T_{i+1})) && \text{either } i \text{ or } i+1 \text{ is odd}\\ &= \sum_{i\text{ odd}} dim(V_i) && \text{by Theorem 2.4.33.}\end{aligned}$$

♣

Theorem 2.4.35. Let V and W be finite dimensional vector spaces. The vector spaces V and W are isomorphic if and only if $dim(V)=dim(W)$. This means that for each positive integer m, all the vector spaces of dimension m are isomorphic. We might choose the representative vector space of dimension m as $\mathbf{C}^m$. ★

Proof. Let $dim(V) = dim(W)$ with $\mathcal{B} = \{b_1, b_2, b_3, \ldots b_m\}$ a basis for V and $\mathcal{B}' = \{b'_1, b'_2, b'_3, \ldots b'_m\}$ a basis for W. We can form $T\colon V \longrightarrow W$ such that $T(b_i) = b'_i$ for all i. This T is an isomorphism. The other direction of the proof is simpler. If V is isomorphic to W, there must be an isomorphism (i.e., an exact sequence $0 \longrightarrow V \xrightarrow{T} W \longrightarrow 0$). The result then follows from a trivial instance of Equation (2.73). ♣

Operations on Vector Spaces

There are many ways of describing new vector spaces based on existing ones. We have already seen the notion of a subvector space and an inclusion linear map. Let us examine several other methods.

Let W be a vector space with subspaces V and V'. We define the **addition of subspaces** as

$$V + V' = \{v + v' : v \in V \text{ and } v' \in V'\}. \tag{2.76}$$

To see that this is, in fact, a subspace, notice that all the vectors are elements of W and the addition and scalar multiplication are both inherited from W. In particular, $V + V'$ is closed under addition. The only thing that remains to be shown is that 0 is in $V + V'$. This can easily be seen because $0 \in V$, $0 \in V'$, and $0 + 0 = 0$.

Let V and V' be vector spaces with $V \cap V' = 0$. If $\mathcal{B}$ is a basis of V and $\mathcal{B}'$ is a basis of V', then it is not hard to see that $\mathcal{B} \cup \mathcal{B}'$ is a basis of $V + V'$. And hence, the dimension of $dim(V + V') = dim(V) + dim(V')$.

There are obvious inclusion linear maps $inc_V\colon V \longrightarrow V + V'$ and $inc_{V'}\colon V' \longrightarrow V + V'$. These inclusion maps satisfy certain properties, which we will examine in Chapter 3. Suffice it to say (for now) that $V + V'$ is a coproduct of V and V'.

Let V be a vector space, and let W be a subvector space with inclusion map $inc\colon W \longrightarrow V$. We form the **quotient vector space**, V/W. Consider the following equivalence relation on the elements of V:

$$v \sim v' \text{ if and only if } v - v' \in W. \tag{2.77}$$

This is reflexive because $v - v = 0 \in W$. It is symmetric because if $v - v' \in W$, then $v' - v = -1 \cdot (v - v') \in W$. Transitivity comes from the fact that if $v - v' = w$ and $v' - v'' = w'$, then $v - v'' = w + w' \in W$. The objects of V/W will be the equivalence classes

$$[v] = v + W = \{v + w : w \in W\}. \tag{2.78}$$

Addition will be given as $[v] + [v'] = [v + v']$, and scalar multiplication is given as $c \cdot [v] = [c \cdot v]$. The zero object is $[w]$ for any $w \in W$.

Exercise 2.4.36. Show that these operations are well defined. This means that if $v \sim v'$ and $u \sim u'$, then $[v] + [u] = [v'] + [u']$ and $c \cdot [v] = c \cdot [v']$. ■

There is a quotient linear map $\pi\colon V \longrightarrow V/W$ that takes $v \in V$ to $[v]$. If $w \in W$, then $\pi(w) = [w] = [0]$, so $Ker(\pi) = W = Im(inc)$. We therefore have the exact sequence

$$0 \longrightarrow W \xrightarrow{inc} V \xrightarrow{\pi} V/W \longrightarrow 0, \tag{2.79}$$

and by Theorem 2.4.34, $dim(V) = dim(W) + dim(V/W)$ or $dim(V/W) = dim(V) - dim(W)$.

A nice example of a quotient space is the cokernel space, which is like a kernel, but with all the maps turned around.

Definition 2.4.37. Let $T\colon V \longrightarrow W$ be a linear map. The **cokernel** of T is the quotient space $W/Im(T)$, which we denote as $Cok(T)$. There is a map $\pi\colon W \longrightarrow Cok(T)$ and an exact sequence

$$V \xrightarrow{T} W \xrightarrow{\pi} Cok(T) \longrightarrow 0. \tag{2.80}$$

We saw that given sets S and T, the collection of all functions from S to T is a set T^S or $Hom_{\mathbb{Set}}(S, T)$. There is a similar idea for the collection of all linear maps from one vector space to another.

Theorem 2.4.38. For any vector spaces V and W, the collection of all linear maps from V to W, $Hom_{\mathbb{CVect}}(V, W)$, is a vector space and is called the **function space**. ★

Proof. The operations of the vector space are

- Addition: for $T\colon V \longrightarrow W$ and $T'\colon V \longrightarrow W$, the linear map $T + T'\colon V \longrightarrow W$ is defined for $v \in V$ as
$$(T + T')(v) = T(v) + T'(v), \tag{2.81}$$
where the addition on the right side is the addition operation of W.
We have to show that $(T + T')$ is a linear map:
  - Addition of maps preserves the addition of vectors:
$$(T + T')(v + v') = T(v + v') + T'(v + v') = T(v) + T(v') + T'(v) + T'(v') \tag{2.82}$$
$$= T(v) + T'(v) + T(v') + T'(v') = (T + T')(v) + (T + T')(v'), \tag{2.83}$$
and
  - Addition of maps preserves the scalar multiplication of vectors:
$$(T + T')(c \cdot v) = T(c \cdot v) + T'(c \cdot v) = c \cdot T(v) + c \cdot T'(v) = c \cdot (T + T')(v). \tag{2.84}$$

  The commutativity and associativity requirements for this addition essentially follow from the fact that they are satisfied in W:
- Negation: for $T\colon V \longrightarrow W$, the linear map $-T\colon V \longrightarrow W$ is defined for v in V as $(-T)(v) = -(T(v))$. We have to show that this map is linear:
- The negation of maps respects the addition of vectors:
$$(-T)(v + v') = -(T(v + v')) = -(T(v) + T(v')) = -(T(v)) - (T(v')) \tag{2.85}$$
$$= (-T)(v) + (-T)(v'), \tag{2.86}$$
and

- The negation of maps respects scalar multiplication:

$$(-T)(c \cdot v) = -(T(c \cdot v)) = -(c \cdot T(v)) = c \cdot -T(v). \tag{2.87}$$

- Zero: The unique null linear map $0\colon V \longrightarrow W$ is the map defined for v in V as $0(v) = 0$. ♣

Notice that much of the structure of $Hom_{\mathbb{CV}\mathrm{ect}}(V, W)$ was inherited from W. This is another instance of Important Categorical Idea 2.4.4.

When a category has the property that the Hom sets are also objects in the category, then it is called a **closed category**. We will meet this concept in depth in Chapter 7 and Section 9.1 of Chapter 9.

In terms of dimension, suffice it to say that for finite dimensional vector spaces,

$$dim(Hom_{\mathbb{V}\mathrm{ect}}(V, W)) = dim(V) \times dim(W). \tag{2.88}$$

Theorem 2.4.38 gives us many new examples of vector spaces. In particular, take W to be the complex number, $\mathbf{C}$, and then $Hom_{\mathbb{V}\mathrm{ect}}(V, \mathbf{C})$ is a complex vector space. We call this the **dual space** of V and write it as V^*. Remember that the objects of V^* are linear maps from V to $\mathbf{C}$. The operations are basically inherited from $\mathbf{C}$. Equation (2.88) tells us that $dim(V^*) = dim(V) \times 1 = dim(V)$.

Exercise 2.4.39. Let $T\colon W \longrightarrow W'$ be a linear map. Show that T induces a linear map $T_*\colon Hom_{\mathbb{V}\mathrm{ect}}(V, W) \longrightarrow Hom_{\mathbb{V}\mathrm{ect}}(V, W')$. (Hint: $T_*(F\colon V \longrightarrow W) = T \circ F$.) ■

Exercise 2.4.40. Let $T\colon V \longrightarrow V'$ be a linear map. Show that this induces a linear map $T^*\colon Hom_{\mathbb{V}\mathrm{ect}}(V', W) \longrightarrow Hom_{\mathbb{V}\mathrm{ect}}(V, W)$. (Hint: $T^*(F\colon V' \longrightarrow W) = F \circ T$.) ■

For a linear map $T\colon V \longrightarrow W$, there will be a linear map

$$W^* = Hom_{\mathbb{CV}\mathrm{ect}}(W, \mathbf{C}) \xrightarrow{\quad T^* \quad} Hom_{\mathbb{CV}\mathrm{ect}}(V, \mathbf{C}) = V^* \tag{2.89}$$

defined for the linear map $F\colon W \longrightarrow \mathbf{C}$ as $T^*(F) = F \circ T\colon V \longrightarrow W \longrightarrow \mathbf{C}$.

In Chapter 5, we will look at other operations for combining vector spaces. We will also continue our study of vector spaces in Section 5.6.

Suggestions for Further Study

There are many fine linear algebra textbooks that emphasize various aspects of the field, e.g., [197, 87, 154, 205]. The presentation in this chapter gained much from [198] and [23].

3

Structures within Categories

Thus we seem to have partially demonstrated that even in foundations,
not substance but invariant form is the carrier of the
relevant mathematical information.
F. William Lawvere
[160], page 1506

A category is a lot more than simply a collection of objects and morphisms between objects. Rather, there might be various relationships between certain objects and morphisms. These relationships can make certain objects have important properties. In this chapter, we describe many such relationships. We also describe operations that one can perform with the objects and morphisms in a category.

Section 3.1 describes the simple operations of products and coproducts. In Section 3.2, we discuss more general types of operations called "limits" and "colimits." Section 3.3 discusses slices and coslices of a category. These are methods of changing the morphisms of one category into the objects of another category. This chapter ends with Section 3.4, which is a mini-course that uses the simple idea of a product in a category to prove some of the most important theorems in mathematics and computer science. Thus the power of category theory is fully demonstrated by the end of the chapter!

3.1 Products and Coproducts

We begin with one of the simplest operations that one can perform with the objects of a category. Given objects a and b in a category, one can form their product as $a \times b$. The product means different things in different categories. While a product is one of the simplest types of structure, the ideas in this section arise over and over again in the rest of this book.

Before we come to the formal definition of what we mean by a product in an arbitrary category, let us carefully look at the category of sets and examine the idea of

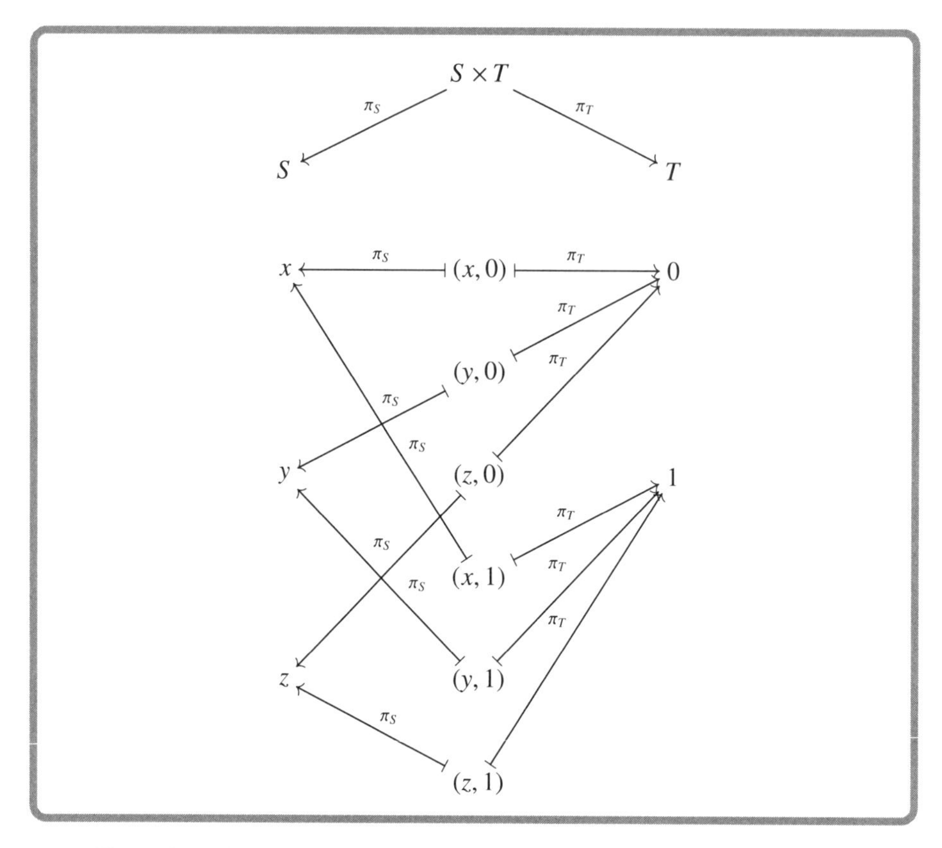

Figure 3.1. The Cartesian product of two sets and its projection functions.

a Cartesian product of sets. Let $S = \{x, y, z\}$ and $T = \{0, 1\}$. The Cartesian product is defined as the set of pairs of elements where the first element is from S and the second element is from T; that is,

$$S \times T = \{(x,0), (y,0), (z,0), (x,1), (y,1), (z,1)\}. \tag{3.1}$$

The set $S \times T$ contains the information of both S and T. There are functions $\pi_S : S \times T \longrightarrow S$ and $\pi_T : S \times T \longrightarrow T$ that "forgets" one of the elements of the pair. These functions are called **projection functions** and are defined on the elements as in Figure 3.1.

How does the Cartesian product of two sets and the projections relate to other sets? Let us look at three examples of other sets and morphisms. All three examples are depicted by the following diagram:

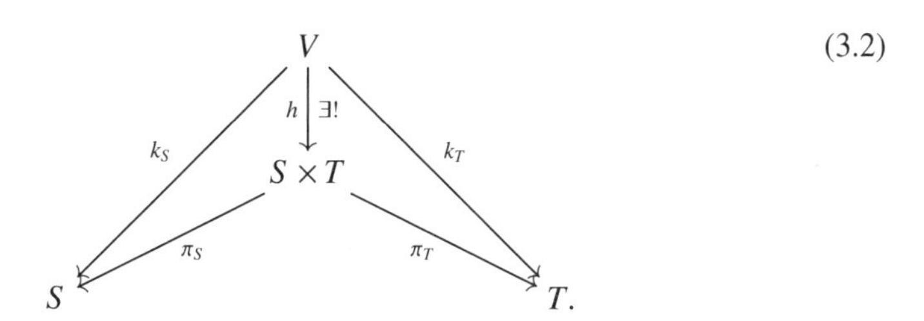

(3.2)

- Consider the set $V = \{(y,1),(z,0),(z,1)\}$. This set has some—but not all—pairs of elements from S and T. There are also projection maps $k_S : V \longrightarrow S$ and $k_T : V \longrightarrow T$, which have the same values as π_S and π_T on those pairs. While V feels like a product, it is not. V is a subset of $S \times T$ and there is an inclusion function $h: V \longrightarrow S \times T$ which makes Diagram (3.2) commute. The inclusion function is unique. From this vantage point, think of $S \times T$ as the set of pairs that is the largest or "best-fitting" set of pairs.
- Consider the set $V = \{a,b,c,d,e\}$ and two functions $k_S : V \longrightarrow S$ and $k_T : V \longrightarrow T$. For each element of V, these two functions pick an element in S and an element in T. For example, $k_S(c) = z$ and $k_T(c) = 0$. With k_S and k_T, one can make a new function $h: V \longrightarrow S \times T$ that uses the information of k_S and k_T. This new function associate c with $(z,0)$ in $S \times T$. That is, there is a unique implied function $h: V \longrightarrow S \times T$ that makes Diagram (3.2) commute.
- In the definition of $S \times T$, we used $(x,1)$ as the common notation for a pair of elements. While this is common, other notations could also be used. For example, there is $[x,1]$ or $\langle x,1\rangle$ or even $\{x,1,\{x\}\}$ (a set where the two elements are given and a third distinguished element describes which element is the first). If one was to use this last example as a pair of elements, then the set of all pairs of elements from S and T would look like

$$V = \{\{x,0,\{x\}\},\{y,0,\{y\}\},\{z,0,\{z\}\},\{x,1,\{x\}\},\{y,1,\{y\}\},\{z,1,\{z\}\}\}. \tag{3.3}$$

  Such a set V also has two projection functions $k_S : V \longrightarrow S$ and $k_T : V \longrightarrow T$. They are defined, for example, as $k_S(\{y,1,\{y\}\}) = y$. While $S \times T$ has the same number of elements as V, they are not the same sets. There is an obvious unique $h: V \longrightarrow S \times T$ that takes $\{x,1,\{x\}\}$ to $(x,1)$. In this case, h is a bijection (isomorphism).

The point of these three examples is that if there is a V and functions k_S and k_T, then there is a unique function h making the appropriate diagram commute. This property that $S \times T$ and its projection maps, π_S and π_T, have is called a **universal property**. It is a way of saying that $S \times T$ is the best set that has the information of S and T and maps to S and T. This property of the product is satisfied by *all* the objects and maps of the category that fit into the Diagram (3.2).

There is another way that one can look at the product. Given a set of real numbers, say

$$S = \left\{\frac{1}{2}, \frac{2}{3}, \frac{3}{4}, \ldots, \frac{n}{n+1}, \ldots\right\}, \tag{3.4}$$

one can ask for the **greatest lower bound** or the **infimum** of this set. This is a number X that has the following characteristics:

- A lower bound (i.e., a number, X, that is less than or equal to all the numbers in S)
- It is greatest of all the lower bounds (i.e., this number X is greater than (or equal to) all the lower bounds)

It is not hard to see that $X = \frac{1}{2}$ is the greatest lower bound of set S. Within the partial order of real numbers, a greatest lower bound corresponds to the fact that there is

- A morphism $X \longrightarrow s$ for all $s \in S$
- Any number Y that has a morphism $Y \longrightarrow s$ for all $s \in S$ will also have a morphism $Y \longrightarrow X$

Another way to say this is that X is

- A lower bound
- The "best-fitting" lower bound

A product is similar to a greatest lower bound, but it pertains to an arbitrary category, not just to a partial order category. A product is

- An object with projections.
- In addition, it is the best-fitting object with projections (i.e., if there is any other object with projections, then there is a unique morphism from the other object to the product).

With these examples in mind, let us make the following definition in an arbitrary category.

Definition 3.1.1. Let $\mathbb{A}$ be a category with objects a and b. A **product** of a and b is an object $a \times b$ with morphisms:

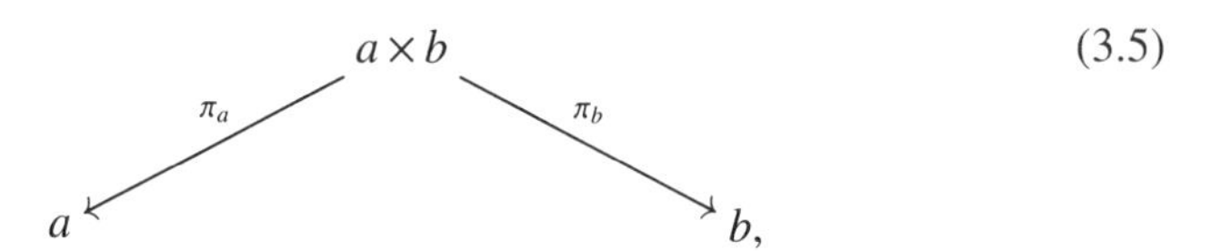

(3.5)

which satisfies the following universal property: for every object c and any two maps

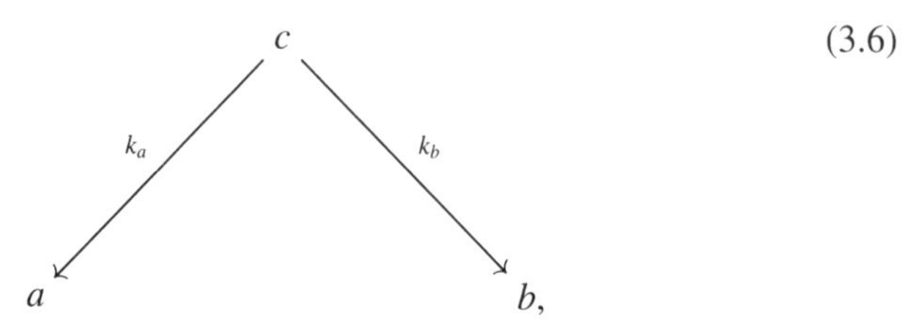

(3.6)

there is a unique map $h\colon c \longrightarrow a \times b$, which makes the following two triangles commute:

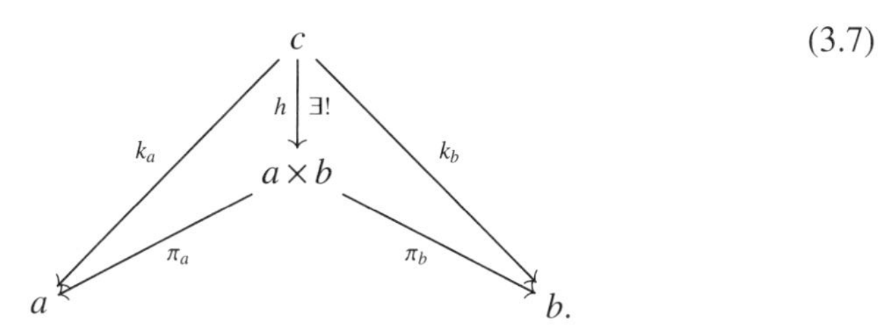

(3.7)

We call h the **induced** morphism of k_a and k_b, and we write it as $h = \langle k_a, k_b \rangle$. We write the h with the $\exists!$ to stress that h is a unique such map that satisfies the universal property.

Let us look at some examples.

Example 3.1.2. In $\mathbb{S}\mathrm{et}$, the product of sets S and T is the Cartesian product of sets. □

Example 3.1.3. In a partial order $(P, \leq)$, the product of two elements, p and q, is the **meet**. This is an element $p \wedge q$ such that

- $p \wedge q \leq p$ and $p \wedge q \leq q$.
- If there is any other element c such that $c \leq p$ and $c \leq q$, then $c \leq p \wedge q$.

Notice that $p \wedge q$ is exactly the greatest lower bound of p and q.

If $(P, \leq)$ is a total order, then $p \wedge q$ is the minimum of p and q. □

Example 3.1.4. In $\mathbb{C}\mathrm{omp}\mathbb{F}\mathrm{unc}$, the category of computable functions, the objects correspond to types. If T_1 and T_2 are types, then $T_1 \times T_2$ is a product type. Such a type consists of pairs of entities from type T_1 and T_2. There are projection functions that take pairs of elements and output one element. This means that there are computable functions $\pi_a\colon T_1 \times T_2 \longrightarrow T_1$ and $\pi_2\colon T_1 \times T_2 \longrightarrow T_2$. It is easy to see that they satisfy the property of being a product. □

Example 3.1.5. In the category of graphs, the product of two graphs G and H is a graph whose vertices are pairs of vertices from G and H, and there is an edge from (g, h) to (g', h') if there is an edge from g to g' in G and an edge from h to h' in H. It is not hard to see that the projection functions are graph homomorphisms. □

There are a few important ideas about the definition of a product. First, the property that the product satisfies is called a "universal property." It is the property that characterizes being a product of those two objects within the *whole* category. Also, notice that a category $\mathbb{A}$ with objects a and b might not have an object that satisfies this universal property. In that case, we say that "the product does not exist." Within a category where the product always exists, we say that the category "has binary products."

Be aware of what was done in this definition. We defined a special object in the category by not looking at what was in the object, but by examining the relationship of that object with all the objects in the category. We used morphisms to discuss the relationships of objects. This is in line with Important Categorical Idea 1.4.1.

Notice that we defined "a" product rather than "the" product. There might be more than one product. In other words, given a and b, there might be more than one object that satisfies the universal property of being a product. This is not a problem because the next theorem says that any two products of a and b are related in a special way.

Theorem 3.1.6. For objects a and b, assume that there is an object c and projection maps $\pi_a\colon c \longrightarrow a$ and $\pi_b\colon c \longrightarrow b$ that satisfy the universal property of making c into a product of a and b. Furthermore, assume that there also are an object c' and projection maps $\pi'_a\colon c' \longrightarrow a$ and $\pi'_b\colon c' \longrightarrow b$ that make c' into the product of a and b. Then there is a unique isomorphism from c to c' that commutes with the projections as follows:

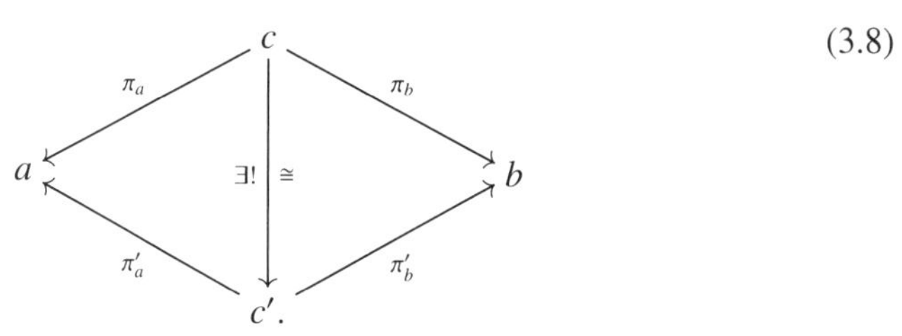

(3.8)

That is, the product is unique up to a unique isomorphism. ★

The proof is very similar to the proof to Theorem 2.2.23 that initial objects are unique up to a unique isomorphism. We leave it to the reader to work out the details.

Exercise 3.1.7. Prove Theorem 3.1.6. That is, show that when a product exists, it is unique up to a unique isomorphism. ■

From the definition of the product $a \times b$, for any object c and any pair of morphisms $k_a\colon c \longrightarrow a$ and $k_b\colon c \longrightarrow b$, there is an induced morphism $\langle k_a, k_b\rangle\colon c \longrightarrow a \times b$. Notice also that one can compose any morphism $h\colon c \longrightarrow a \times b$ with the projection morphisms to get two morphisms $\pi_a \circ h\colon c \longrightarrow a$ and $\pi_b \circ h\colon c \longrightarrow b$. We have just described two functions of Hom sets:

$$Hom_{\mathbb{A}}(c, a \times b) \underset{\langle\ ,\ \rangle}{\overset{\pi_a\circ(\ )\,,\ \pi_b\circ(\ )}{\rightleftarrows}} Hom_{\mathbb{A}}(c, a) \times Hom_{\mathbb{A}}(c, b). \qquad (3.9)$$

These two set functions are inverses of each other, and hence the two Hom sets are isomorphic. To see that these functions are inverses of each other, start with an $h\colon c \longrightarrow a \times b$ on the left side of Diagram (3.9). This goes to the morphisms $\pi_a \circ h$ and $\pi_b \circ h$ on the right side. Taking these two morphisms back to the left side gives us $\langle \pi_a \circ h, \pi_b \circ h\rangle$, as in

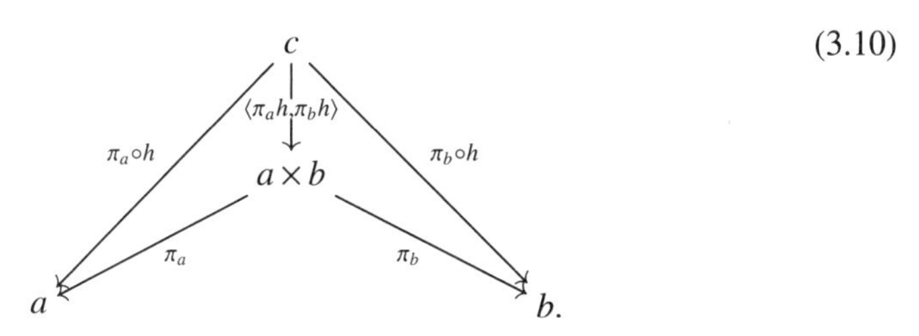

(3.10)

Since there is exactly one morphism that makes the two triangles commute, we get

$$h = \langle \pi_a \circ h, \pi_b \circ h \rangle. \tag{3.11}$$

On the other hand, if we start with two morphisms k_a and k_b on the right of Diagram (3.9), we get the morphism $\langle k_a, k_b \rangle$ on the left. Returning to the right gives us $\pi_a \circ \langle k_a, k_b \rangle$ and $\pi_b \circ \langle k_a, k_b \rangle$. From the two commutative triangles,

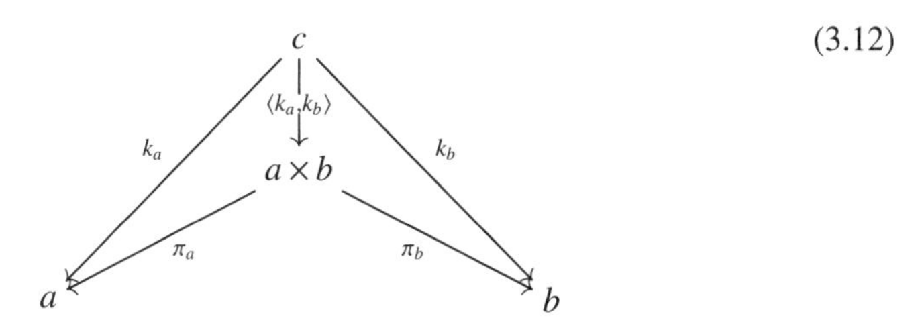

(3.12)

it is easy to see that

$$k_a = \pi_a \circ \langle k_a, k_b \rangle \qquad \text{and} \qquad k_b = \pi_b \circ \langle k_a, k_b \rangle, \tag{3.13}$$

thus showing that $\langle \ , \ \rangle$ and $\pi_a \circ (\ )$, $\pi_b \circ (\ )$ are inverse set functions.

This bijection of Hom sets can be taken as the definition of a product. Definition 3.1.8 puts it formally.

Definition 3.1.8. Let $\mathbb{A}$ be a category with objects a and b. A **product** of a and b is an object $a \times b$ such that there exists an isomorphism of Hom sets:

$$Hom_{\mathbb{A}}(c, a \times b) \cong Hom_{\mathbb{A}}(c, a) \times Hom_{\mathbb{A}}(c, b). \tag{3.14}$$

(Notice that the $\times$ in the left is a product of the category and the $\times$ on the right is the Cartesian product of Hom sets.) We need one more requirement–namely, this isomorphism should be "natural." We will see this requirement in Section 4.7.

Let us look at these Hom sets in terms of some of our examples.

Example 3.1.9. In the partial order $(P, \leq)$, the Hom sets are either the empty set or a set with one element:

$$Hom_P(p, q \wedge r) \cong Hom_P(p, q) \times Hom_P(p, r) \tag{3.15}$$

The left side is a one-element set if and only if each part of the right side is a one-element set. If either of the sets on the right is the empty set, then the set on the left is the empty set. □

Example 3.1.10. In $\mathbb{PO}$, let $(P, \leq)$ and $(Q, \leq)$ be partial orders. Then $(P \times Q, \sqsubseteq)$ is the partial order whose elements are ordered pairs of elements from P and Q and whose order is defined as follows:

$$(p, q) \sqsubseteq (p', q') \text{ if and only if } p \leq p' \text{ and } q \leq q'. \tag{3.16}$$

It is not hard to show that this is a partial order. This partial order is the product of $(P, \leq)$ and $(Q, \leq)$. □

Example 3.1.11. In $\mathbb{Top}$, if (X, τ) and (Y, σ) are topological spaces, then $(X \times Y, \delta)$ is their product, where $X \times Y$ is the Cartesian product of X and Y and δ is the **product topology** (see any introductory topology textbook). □

Example 3.1.12. In $\mathbb{Group}$, if $(G, \star, 1, (\ )^{-1})$ and $(G', \star', 1', [\ ]^{-1})$ are groups, then their product is $(G \times G', \cdot, (1, 1'), ((\ )^{-1}, [\ ]^{-1}))$. The elements are pairs of elements (x, x') where x is in G and x' is in G'. The multiplication $\cdot$ is given pointwise, i.e., on each component:

$$(x, x') \cdot (y, y') = (x \star y, x' \star' y'). \tag{3.17}$$

The unit of the group is $(1, 1')$. The inverse of (x, x') is $((x)^{-1}, [x']^{-1})$. We leave it to the reader to describe the projections and to show that the projections are group homomorphisms. □

In any category, consider the special case of taking a product of an object a with itself (i.e., $a \times a$). The induced morphism for $\langle id_a, id_a \rangle$, as in

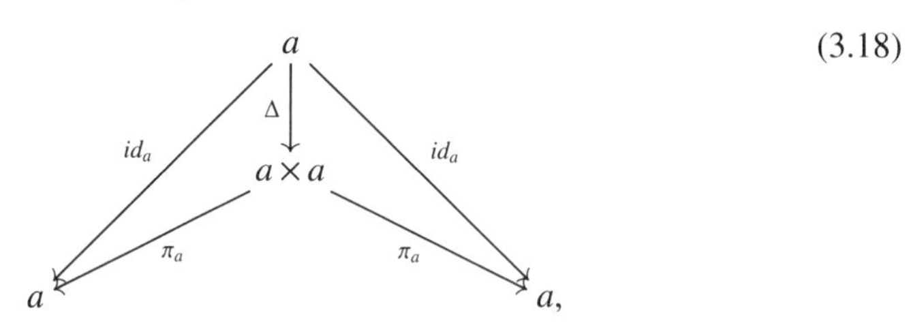

(3.18)

is denoted as Δ (Greek letter delta) and is called the **diagonal morphism**. This morphism has a very important role in the mini-course of self-reference at the end of this chapter (and will be noticeably missing when we talk about quantum mechanics and quantum computing).

Example 3.1.13. In $\mathbb{Set}$, the diagonal morphism $\Delta \colon S \longrightarrow S \times S$ takes each element s in S to (s, s) in $S \times S$. In $\mathbb{Group}$, the diagonal morphism $\Delta \colon G \longrightarrow G \times G$ is a group homomorphism defined as $\Delta(g) = (g, g)$. To see that this is a homomorphism, consider

$$\Delta(x \star y) = (x \star y, x \star y) = (x, x) \star' (y, y) = \Delta(x) \star' \Delta(y), \tag{3.19}$$

where $\star'$ is the multiplication in $G\times G$. In $\mathbb{Top}$, Δ works the same way that it works in $\mathbb{Set}$. □

In Equation (1.17) on page 13, we saw that within $\mathbb{Set}$, one can process parallel functions. There is a similar construction for products within any category. Let $a\times b$ be a product of a and b and $a'\times b'$ be a product of a' and b'. Consider morphisms $f\colon a\longrightarrow a'$ and $g\colon b\longrightarrow b'$, as in the following diagram:

$$\begin{array}{ccccc} a & \xleftarrow{\pi_a} & a\times b & \xrightarrow{\pi_b} & b \\ {\scriptstyle f}\downarrow & & \downarrow{\scriptstyle f\times g} & & \downarrow{\scriptstyle g} \\ a' & \xleftarrow[\pi'_a]{} & a'\times b' & \xrightarrow[\pi'_b]{} & b'. \end{array} \tag{3.20}$$

Since $f\circ\pi_a$ and $g\circ\pi_b$ satisfy the universal property of $a'\times b'$, there is an induced map $a\times b\longrightarrow a'\times b'$. This map is denoted as $f\times g$. Formally,

$$f\times g=\langle f\circ\pi_a, g\circ\pi_b\rangle. \tag{3.21}$$

There are two operations of morphisms: composition and taking the product. How do these two operations respect each other?

Theorem 3.1.14. Let $f\colon c\longrightarrow a$, $g\colon c\longrightarrow b$, and $h\colon d\longrightarrow c$, as in

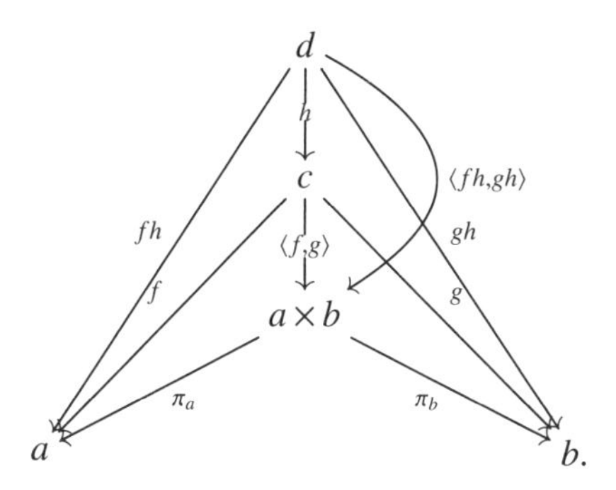

If $a\times b$ exists, then

$$\langle f\circ h, g\circ h\rangle=\langle f,g\rangle\circ h. \tag{3.22}$$

That is, the composition with h on the right distributes over the entries in the bracket. ★

Proof. Notice that all triangles in the diagram commute. The diagram shows the two morphisms and the theorem follows from the uniqueness promised by the universal property of the product. In other words, by uniqueness, the two morphisms are equal. ♣

There are many consequences of this theorem.

Theorem 3.1.15. Let $f\colon a \longrightarrow a'$, $g\colon b \longrightarrow b'$ and $h\colon c \longrightarrow a\times b$. Then we have

$$(f\times g)\circ h = \langle f\circ\pi_a, g\circ\pi_b\rangle \circ h = \langle f\circ\pi_a\circ h, g\circ\pi_b\circ h\rangle. \tag{3.23}$$

★

Proof. This follows from the definition $f\times g$ given in Equation (3.21) and Theorem 3.1.14. ♣

When we draw a diagram, the source and target of a projection map can easily be seen. In contrast, when we write an equation, it might be less obvious what the source and target of a projection map is. Notice that the following two maps are different:

$$\pi_a\colon a\times b \longrightarrow a \qquad \pi_a\colon a\times c \longrightarrow a. \tag{3.24}$$

When we need to be careful and make the distinction, we will differentiate the maps by placing their source as a superscript. The two maps here will be denoted as π_a^{ab} and π_a^{ac}, respectively.

The following properties of projection functions will be helpful.

Theorem 3.1.16. For any product $a\times b\times c$, we have

$$\pi_a^{ab}\pi_{ab}^{abc} = \pi_a^{abc}\colon a\times b\times c \longrightarrow a \tag{3.25}$$

and

$$\pi_{ab}^{abc} = \pi_a^{abc}\times\pi_b^{abc}\colon a\times b\times c \longrightarrow a\times b. \tag{3.26}$$

★

Proof. The first identity says that the following diagram commutes:

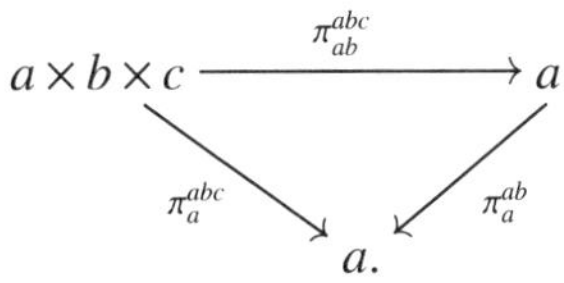

$$\begin{array}{ccc} a\times b\times c & \xrightarrow{\pi_{ab}^{abc}} & a\times b \\ & {\scriptstyle\pi_a^{abc}}\searrow \quad \swarrow{\scriptstyle\pi_a^{ab}} & \\ & a. & \end{array} \tag{3.27}$$

This commutes because $a\times b$ is a product of a and b and $a\times b\times c$ has projection maps to a and b. So by the universal property, there is a unique map from $a\times b\times c$ to $a\times b$ that makes the diagram commute.

The second identity comes from the commutativity of the following diagram:

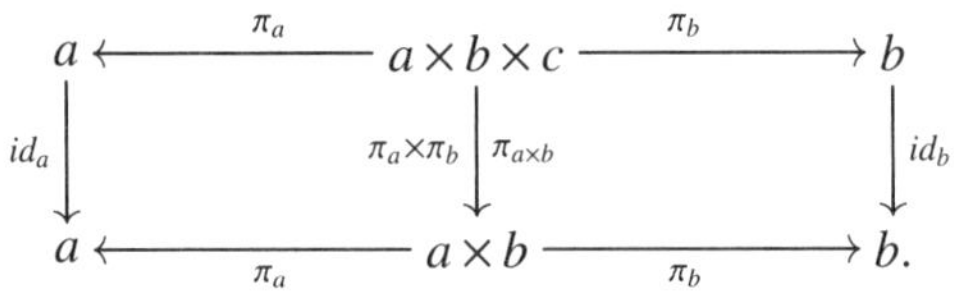

$$\begin{array}{ccccc} a & \xleftarrow{\pi_a} & a\times b\times c & \xrightarrow{\pi_b} & b \\ {\scriptstyle id_a}\downarrow & & {\scriptstyle\pi_a\times\pi_b}\downarrow\ \downarrow{\scriptstyle\pi_{a\times b}} & & \downarrow{\scriptstyle id_b} \\ a & \xleftarrow[\pi_a]{} & a\times b & \xrightarrow[\pi_b]{} & b. \end{array} \tag{3.28}$$

The induced map is $\pi_a\times\pi_b$. By the first identity, both rectangles commute with $\pi_{a\times b}$. The universal property of the product $a\times b$ ensures that there is only one such induced map, so they are equal. ♣

The following relationship between the $\times$ and $\circ$ operations will be very similar to other relationships between binary operations that we will find throughout this text.

Theorem 3.1.17. Let $f\colon a \longrightarrow a'$, $f'\colon a' \longrightarrow a''$, $g\colon b \longrightarrow b'$ and $g\colon b' \longrightarrow b''$. Assume that $a\times b, a'\times b'$, and $a''\times b''$ exist. Then we have the following equality, called the **interchange law**:

$$(f'\times g')\circ(f\times g) = (f'\circ f)\times(g'\circ g)\colon a\times b \longrightarrow a''\times b'', \tag{3.29}$$

as in

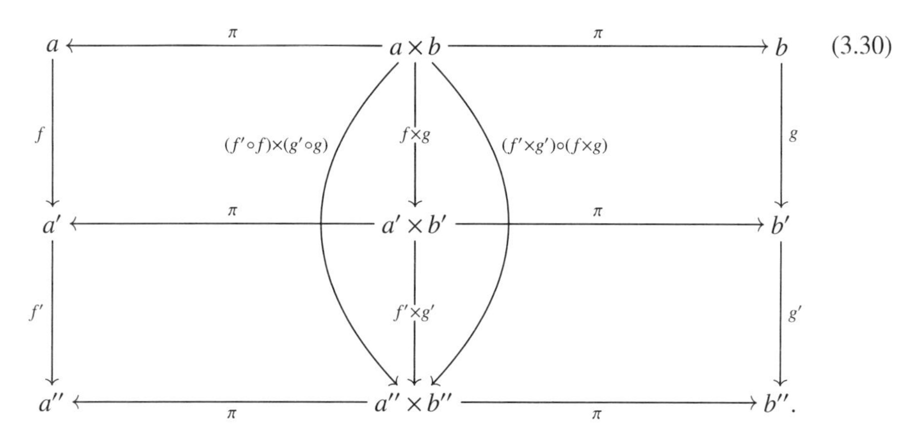

★

Proof.

$$\begin{aligned}(f'\times g')\circ(f\times g) &= \langle f'\pi, g'\pi\rangle\circ\langle f\pi, g\pi\rangle && \text{by definition of } \times\\ &= \langle f'\pi\circ\langle f\pi, g\pi\rangle, g'\pi\circ\langle f\pi, g\pi\rangle\rangle && \text{by Theorem 3.1.14}\\ &= \langle f'\circ f\pi, g'\circ g\pi\rangle && \text{by Equation (3.13)}\\ &= (f'\circ f)\times(g'\circ g) && \text{by definition of } \times.\end{aligned}$$

♣

Let us spend a few words meditating on the interchange law. Think of the functions f, f', g, and g' as processes. Consider the composition operation, $\circ$, as doing one process after another (sequential processes), and consider the product operation, $\times$, as doing two processes independently (parallel processes). The interchange law tells us how sequential and parallel process get along. On the one hand, we can think of first performing the parallel process $f\times g$ and then sequentially performing the parallel process $f'\times g'$. On the other hand, we can think of parallel-processing the compositions $f'\circ f$ and $g'\circ g$. The interchange law tells us that both ways of looking at these processes are correct.

Important Categorical Idea 3.1.18. The Interchange Law. In general, consider when there are two binary operations, say $\otimes$ and $\oplus$, operating on a collection. If the two operations satisfies the equality

$$(a \otimes b) \oplus (c \otimes d) = (a \oplus c) \otimes (b \oplus d), \tag{3.31}$$

then we call it an **interchange law** or **interchange rule**. It means that each operation respects the other operation (i.e., each operation is a homomorphism in terms of the other operation). This idea arises many times in category theory and will be at the heart of the theory of monoidal categories. ○

As an example of the interchange law, Theorem 3.1.17 describes how the operations of composition and product relate.

Example 3.1.19. The interchange law is of interest in the category $\mathbb{CompFunc}$ of computable functions. If $f\colon T_1 \longrightarrow T_2$ and $g\colon T_3 \longrightarrow T_4$, then one should think of $f \times g\colon T_1 \times T_3 \longrightarrow T_2 \times T_4$ as a type of parallel processing. The function $f \times g$ does each computation separately. In contrast, for composable maps f and f', one can think of $f' \circ f$ as sequential processing since one has to preform one process and then the other. □

We have seen the product of two objects. What about the product of three objects? (The following morphism will be very important when we talk about monoidal categories in Chapter 5.)

Theorem 3.1.20. Let a, b, and c be three objects in a category. If the products $a \times (b \times c)$ and $(a \times b) \times c$ exist, then there is a unique isomorphism $\alpha\colon a \times (b \times c) \longrightarrow (a \times b) \times c$ and an inverse $\alpha'\colon (a \times b) \times c \longrightarrow a \times (b \times c)$, such that α and α' each commutes with all the projection maps, as in this diagram:

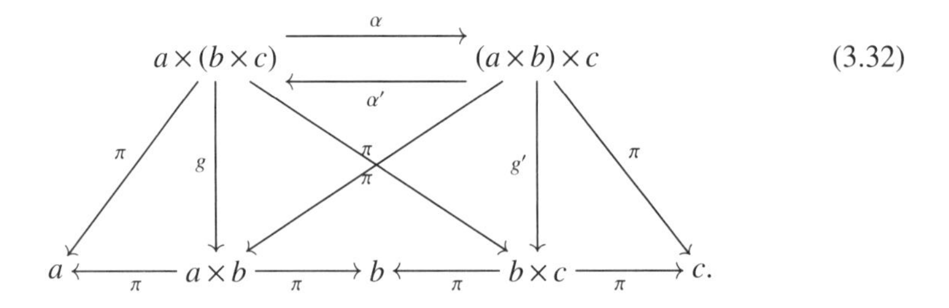

(3.32)

★

Proof. The projections are given in Diagram (3.32). However, they can be spaced out as follows:

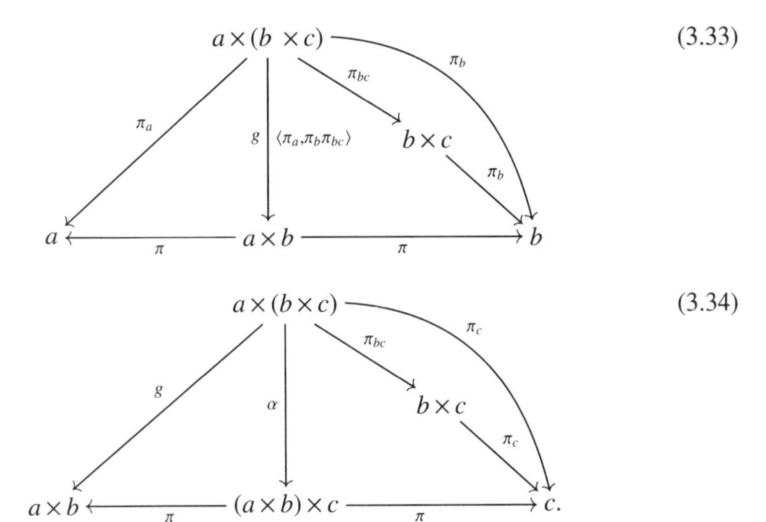

(3.33)

(3.34)

In detail, there are compositions of projection morphisms from $a \times (b \times c)$ to a and b, so there is a unique morphism

$$g = \langle \pi_a^{a(bc)}, \pi_b^{bc} \pi_{bc}^{a(bc)} \rangle = \langle \pi_a^{a(bc)}, \pi_b^{a(bc)} \rangle \colon a \times (b \times c) \longrightarrow a \times b. \tag{3.35}$$

The morphism g and the composition of projection morphisms from $a \times (b \times c)$ to c induce the unique morphism α. Formally,

$$\alpha = \langle g, \pi_c^{bc} \pi_{bc}^{a(bc)} \rangle = \langle \langle \pi_a^{a(bc)}, \pi_b^{a(bc)} \rangle, \pi_c^{a(bc)} \rangle \colon a \times (b \times c) \longrightarrow (a \times b) \times c. \tag{3.36}$$

The morphism α' goes in the opposite direction. Notice that the morphisms from $(a \times b) \times c$ to b and to c induce

$$g' = \langle \pi, \pi \rangle \colon (a \times b) \times c \longrightarrow b \times c. \tag{3.37}$$

The morphism g' and the morphism from $(a \times b) \times c$ to a induce the unique morphism α'. Formally stated,

$$\alpha' = \langle \pi_a^{(ab)c}, \langle \pi_b^{(ab)c}, \pi_c^{(ab)c} \rangle \rangle. \tag{3.38}$$

We leave it as an exercise to prove that α' is the inverse of α. ♣

Exercise 3.1.21. Formally show that α and α' are inverses of each other. ■

Example 3.1.22. In $\mathbb{Set}$, the associativity isomorphism for sets S, T, and U is $\alpha \colon S \times (T \times U) \longrightarrow (S \times T) \times U$, which is defined for $s \in S$, $t \in T$, and $u \in U$ as

$$\alpha((s, (t, u))) = ((s, t), u). \tag{3.39}$$

□

In a category with products, for every two objects a and b, what is the relationship between $a \times b$ and $b \times a$?

Theorem 3.1.23. In a category with products, for every two objects a and b, there is a **braid morphism** $br\colon a\times b \longrightarrow b\times a$, which is induced by projection maps as follows:

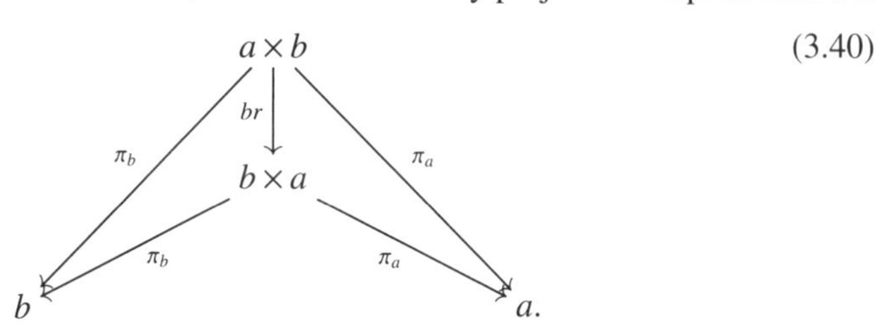

(3.40)

This means that

$$br = \langle \pi_b^{ab}, \pi_a^{ab} \rangle. \tag{3.41}$$

This map is an isomorphism that shows that $a\times b$ is isomorphic to $b\times a$. In general, they are not equal. (This will be a major topic in Chapters 6 and 7.) ★

Exercise 3.1.24. Describe br^{-1} and show that it is inverse to br. ■

Example 3.1.25. In $\mathbb{CompFunc}$, the objects correspond to types. If T_1 and T_2 are types, then there are maps going from $T_1\times T_2$ to $T_2\times T_1$ and the other way. □

Exercise 3.1.26. Let t be a terminal object in a category. Show that if $a\times t$ exists, then it is isomorphic to a. ■

One of the amazing aspects of category theory is that if you have some idea or construction and you turn all the arrows around to point in the opposite direction, then you have a related construction. The **coproduct** is what happens when you turn around the arrows of a product.

> **Important Categorical Idea 3.1.27. Duality.** Many properties and structures in categories come in pairs. For a given definition of a structure, one can make another **dual** definition with the arrows going in the opposite direction. This idea is called **duality** and happens very often. If a structure is called "X," then the dual structure is called "coX." ○

First, let us look at the coproduct in $\mathbb{Set}$ and see what we can learn about it. Let $S=\{x,y,z\}$ and $T=\{0,1\}$. The coproduct in $\mathbb{Set}$ is the disjoint union of these sets. In this example, the coproduct is just the union of both sets:

$$S \amalg T = \{x,y,z,0,1\}. \tag{3.42}$$

There are two inclusion functions, $inc_S : S \longrightarrow S\amalg T$ and $inc_T : T \longrightarrow S\amalg T$. (Note that these are in the opposite direction of the projection functions.) How does the disjoint union relate to other sets? Let us look at three examples of other sets and functions. Keep the following diagram in mind (notice that it is Diagram (3.2) with the arrows turned around).

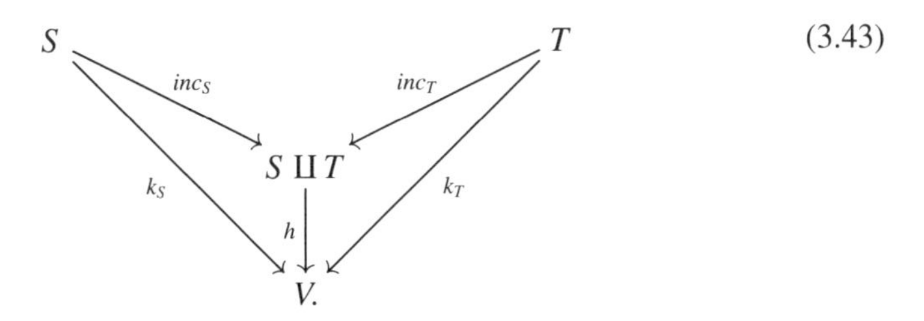

(3.43)

- Consider the set $V = \{r, w, x, y, z, a, 0, 1, 2\}$. The sets S and T are subsets of V, and there are inclusion functions $k_S : S \longrightarrow V$ and $k_T : T \longrightarrow V$. While V has a feel of a disjoint union, $S \amalg T$ is the real disjoint union because it contains nothing other than S and T. There is an inclusion function $h: S \amalg T \longrightarrow V$ that makes the triangles in Diagram 3.43 commute. From this vantage point, think of $S \amalg T$ as the smallest or best-fitting set that contains S and T. If there is any other set that contains S and T, then $S \amalg T$ is the best and includes into it.
- Consider the set $V = \{a, b, c, d, e\}$ and two functions $k_S : S \longrightarrow V$ and $k_T : T \longrightarrow V$. Let us say $k_S(x) = b$ and $k_T(0) = e$. There is a function $h: S \amalg T \longrightarrow V$, which unites the information of k_S and k_T. Such a function has $h(x) = b$ and $h(0) = e$. The function h defined like this ensures that the two triangles commute.
- Consider the set $V = \{x', y', z', 0', 1'\}$. While this is not the disjoint union of S and T, there still are obvious functions, $k_S : S \longrightarrow V$ and $k_T : T \longrightarrow V$, which take elements to their prime versions. There is an isomorphism of sets $h: S \amalg T \longrightarrow V$ that takes every element $h(x) = x'$.

There is another way that one can look at the coproduct. Given a set of numbers, say

$$S = \left\{\frac{1}{2}, \frac{2}{3}, \frac{3}{4}, \ldots, \frac{n}{n+1}, \ldots, 1\right\}, \tag{3.44}$$

one can ask for the **least upper bound** or the **supremum** of this set. This is a number X that is as follows:

- An upper bound (i.e., a number that is greater than or equal to all the numbers in S)
- Least of all the upper bounds (i.e., this number X is smaller or equal to all the upper bounds)

Within the partial order of real numbers, this corresponds to the fact that

- There is a morphism $s \longrightarrow X$ for all $s \in S$.
- Any number Y that has a morphism $s \longrightarrow Y$ for all $s \in S$ will also have a morphism $X \longrightarrow Y$.

Another way to say this is that X is a upper bound and is the best-fitting upper bound. It is not hard to see that $X = 1$ is the least upper bound of set S. A coproduct is similar to a least upper bound, but it pertains to an arbitrary category, not just to a partial order category.

With these examples in mind, let us define a coproduct in any category in Definition 3.1.28.

Definition 3.1.28. Let $\mathbb{A}$ be a category with objects a and b. A **coproduct** of a and b is an object $a+b$ (also written $a \amalg b$) with morphisms

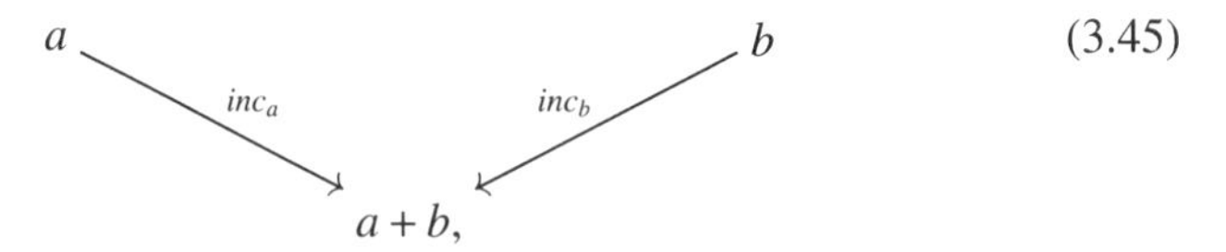

which satisfies the following universal property: if there is any other object c with two morphisms,

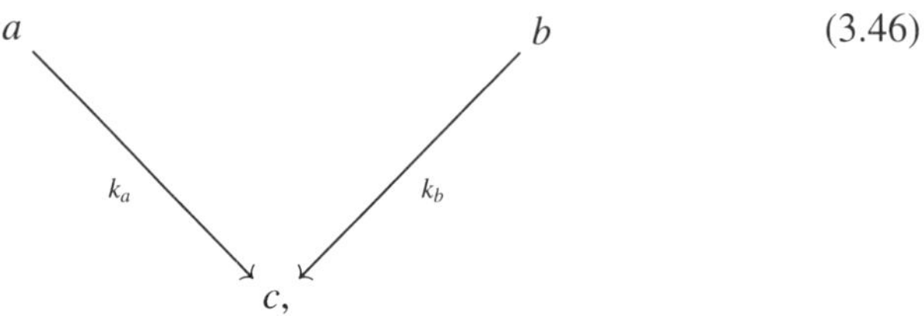

then there is a unique map $h\colon a+b \longrightarrow c$, which makes the following two triangles commute:

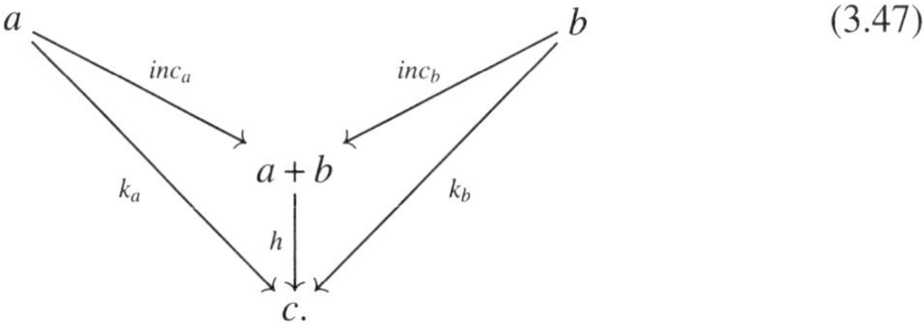

The morphisms $k_a\colon a \longrightarrow c$ and $k_b\colon b \longrightarrow c$ induce the morphism $a+b \longrightarrow c$. We will write this morphism as $[k_a, k_b]\colon a+b \longrightarrow c$. Notice that

$$[k_a, k_b] \circ inc_a = k_a \qquad \text{and} \qquad [k_a, k_b] \circ inc_b = k_b. \tag{3.48}$$

Example 3.1.29. As we showed just before the definition, in $\mathbb{Set}$, the coproduct is simply the disjoint union. □

Example 3.1.30. In computable functions $\mathbb{CompFunc}$, the coproduct of two types T_1 and T_2 is simply the disjoint union of T_1 and T_2. There are obvious inclusion functions. The universal property is also easy to see: if there is a computable function $f_1\colon T_1 \longrightarrow T'$ and a computable function $f_2\colon T_2 \longrightarrow T'$, then there is a computable function $h\colon T_1 \amalg T_2 \longrightarrow T'$. Function h depends on the input. If the input is of type T_1, then h executes f_1 on it. If the input is of type T_2, then h executes function f_2 on the input; that is,

$$h(x) = \begin{cases} f_1(x) & \text{if } x \text{ is of type } T_1 \\ f_2(x) & \text{if } x \text{ is of type } T_2\,. \end{cases} \tag{3.49}$$

□

Exercise 3.1.31. Show that coproducts are unique up to a unique isomorphism. ■

If a category has the property that for any two objects, a coproduct exists, then we say that the category "has binary coproducts."

Exercise 3.1.32. For the coproduct, show that the following two maps are inverse of each other:

$$Hom_{\mathbb{A}}(a+b,c) \underset{[\ ,\ ]}{\overset{(\ )\circ inc_a\ ,\ (\ )\circ inc_b}{\rightleftarrows}} Hom_{\mathbb{A}}(a,c)\times Hom_{\mathbb{A}}(b,c), \tag{3.50}$$

and hence the Hom sets are isomorphic. ■

Analogous with Definition 3.1.8, we have the equivalent definition of a coproduct given in Definition 3.1.33.

Definition 3.1.33. Let $\mathbb{A}$ be a category with objects a and b. A **coproduct** of a and b is an object $a+b$ such that the following two Hom sets are isomorphic:

$$Hom_{\mathbb{A}}(a+b,c)\cong Hom_{\mathbb{A}}(a,c)\times Hom_{\mathbb{A}}(b,c). \tag{3.51}$$

We will need one more requirement–namely, that the isomorphism should be "natural," which we will see in Section 4.6.

Example 3.1.34. Let $(P,\leq)$ and $(Q,\leq)$ be partial orders in $\mathbb{PO}$. Their coproduct is $(P+Q,\sqsubseteq)$, which is the partial order whose elements are the disjoint union of elements from P and Q and whose order is defined as follows:

$r\sqsubseteq r'$ if and only if r and r' are both in P and $r\leq r'$, <u>or</u> they are both in Q and $r\leq r'$. (3.52)

□

Exercise 3.1.35. Show that this partial order satisfies the universal property of being a coproduct. ■

Consider morphisms $f\colon a\longrightarrow a'$, $g\colon b\longrightarrow b'$ and the coproducts $a+b$ and $a'+b'$. There are morphisms from $inc'_{a'}\circ f\colon a\longrightarrow a'+b'$ and $inc'_{b'}\circ g\colon b\longrightarrow a'+b'$, so there is an induced morphism $a+b\longrightarrow a'+b'$, denoted as $f+g$, which makes the following two squares commute:

$$\begin{array}{ccccc} a & \xrightarrow{inc_a} & a+b & \xleftarrow{inc_b} & b \\ {\scriptstyle f}\downarrow & & \downarrow{\scriptstyle f+g} & & \downarrow{\scriptstyle g} \\ a' & \xrightarrow[inc_{a'}]{} & a'+b' & \xleftarrow[inc_{b'}]{} & b' \end{array}\ . \tag{3.53}$$

Keep in mind that

$$f+g=[inc_{a'}\circ f, inc_{b'}\circ g]. \tag{3.54}$$

Theorem 3.1.36 tells how the operations of composition and coproduct relate to each other.

Theorem 3.1.36. Let $f: a \longrightarrow c$, $g: b \longrightarrow c$, and $h: c \longrightarrow d$, as in

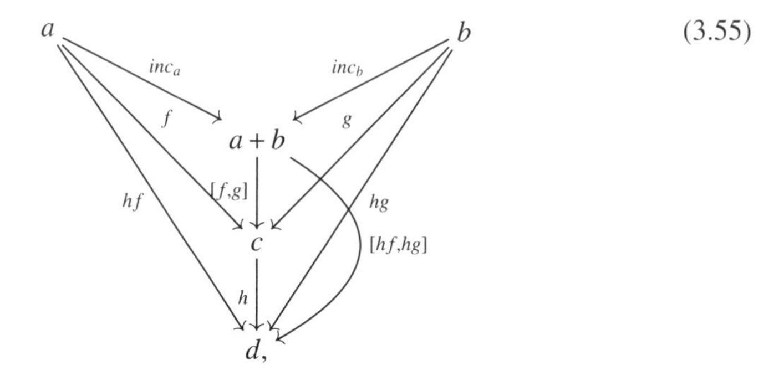

(3.55)

Then $[f, g]$ is a map from $a + b$ to c. This map can compose with $h: c \longrightarrow d$, producing the following important equation:

$$h \circ [f, g] = [hf, hg]. \tag{3.56}$$

★

Proof. The proof is the dual of the proof of Theorem 3.1.14. ♣

Exercise 3.1.37. Show that the dual of Theorem 3.1.20 is true. In other words, show that there is a unique isomorphism:

$$\alpha: a + (b + c) \longrightarrow (a + b) + c. \tag{3.57}$$

■

Example 3.1.38. In the category of topological spaces, $\mathbb{Top}$, the coproduct of two spaces, is their disjoint union. The coproduct in the category of manifolds, $\mathbb{Manif}$, is similarly defined. □

Analogous to the diagonal morphism and the braid map, there is a **codiagonal morphism**, ∇, and a **cobraid morphism**, *cobr*, for the coproduct. They are given by the following two diagrams:

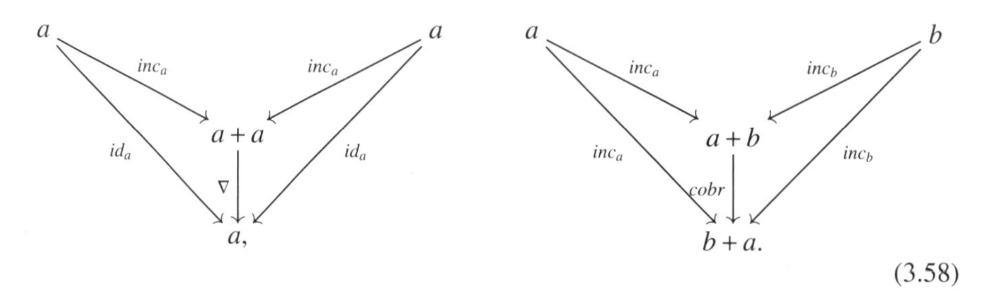

(3.58)

Exercise 3.1.39. Show that the cobraid morphism is an isomorphism. ■

How does the product relate to the coproduct? It is known that when dealing with numbers, we have the distributive law $x(y+z) = xy + xz$. In analogy, there is] Theorem 3.1.40.

Theorem 3.1.40. Let a, b, and c be objects in a category where binary products and coproducts exist. Then there is a unique morphism,

$$h\colon (a \times b) + (a \times c) \longrightarrow a \times (b + c), \tag{3.59}$$

which commutes with the appropriate projections and inclusions. ★

Proof. This can easily be seen in the following diagram of projections and inclusions:

(3.60)

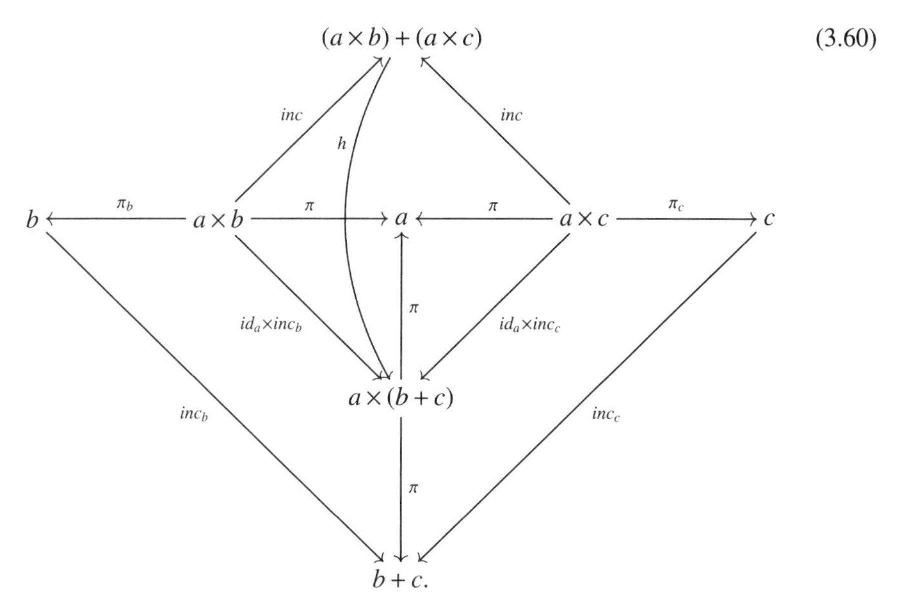

The morphism id_a and $inc_b\colon b \longrightarrow b+c$ induce $id_a \times inc_b$. Similarly, id_a and $inc_c\colon c \longrightarrow b+c$ induce $id_a \times inc_c$. The desired morphism is

$$h = [id_a \times inc_b, id_a \times inc_c] = [\langle \pi_a, inc_b \pi_b \rangle, \langle \pi_a, inc_c \pi_c \rangle]. \tag{3.61}$$

♣

While this morphism exists, it is not generally an isomorphism. In the special case where finite products and coproducts always exist and where this distributive morphism is an isomorphism (and another requirement), the category is called a **distributive category** (see, e.g., [51, 258]).

Exercise 3.1.41. Let S, T, and U be sets in $\mathbb{S}\mathrm{et}$. Show that there is an isomorphism

$$S \times (T + U) \longrightarrow (S \times T) + (S \times U), \tag{3.62}$$

which means that $\mathbb{S}\mathrm{et}$ is a distributive category. ■

3.2 Limits and Colimits

Limits and colimits are generalizations of products and coproducts. Before we move on to describe general limits, it is important to understand a product from a different perspective. One can think of a product of objects a and b in the category as **completing** the diagram or a **completion** of a diagram where a and b are points of a diagram:

$$a \qquad\qquad b \tag{3.63}$$

In other words, the product and the projection maps from the product to a and b are like a pot cover on a and b:

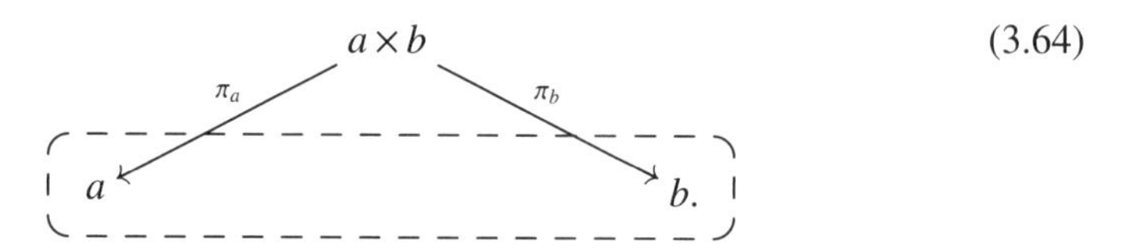

(3.64)

The universal property says that $a \times b$ is the best way to complete the diagram. That means that if there is any other completion c of a and b with maps $k_a\colon c \longrightarrow a$ and $k_b\colon c \longrightarrow b$, then there is a unique map $h\colon c \longrightarrow a \times b$ that makes all the triangles in the diagram

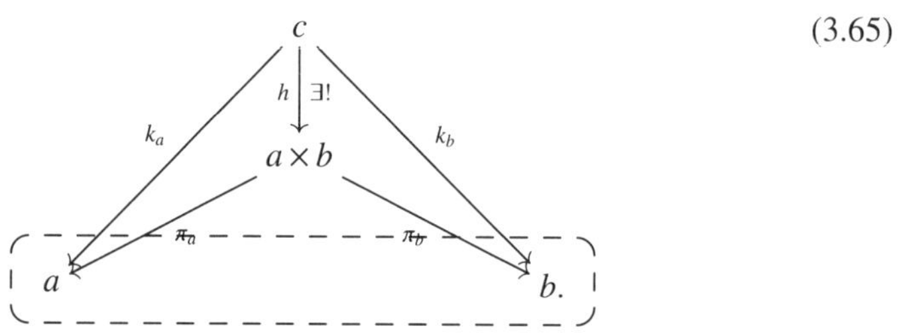

(3.65)

commute. In a similar way, we might say that $a \amalg b$ is the best way to complete Diagram (3.63) "from the bottom":

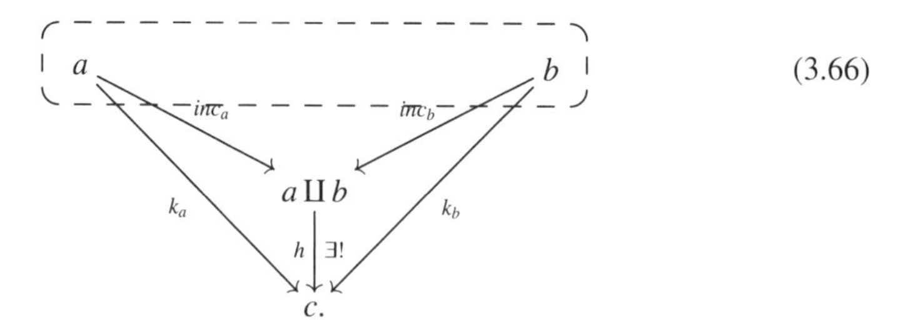

(3.66)

We say that we are **cocompleting** the diagram or this is a **cocompletion** of the diagram.

While binary products and coproducts are the completion of diagrams with two objects, limits and colimits are the completions and cocompletions of any type of diagram. For example, given a diagram with four objects a_1, a_2, a_3, and a_4, as in

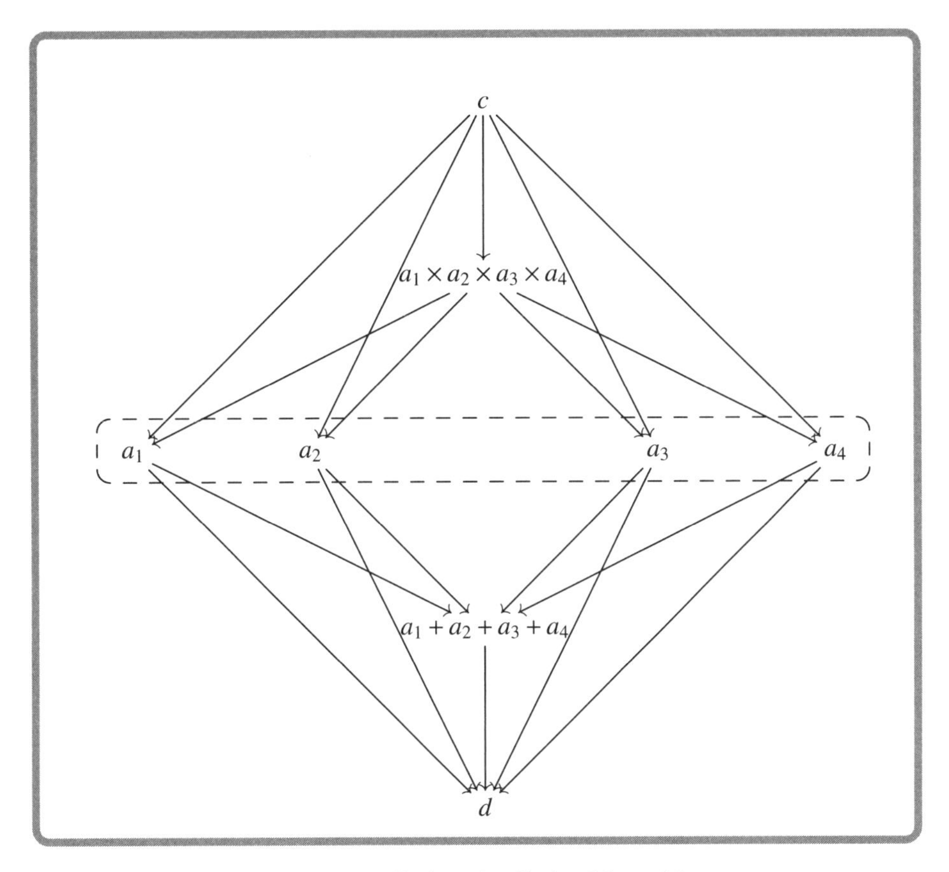

Figure 3.2. The limit and colimit of four objects.

Figure 3.2, we can talk about their product $a_1 \times a_2 \times a_3 \times a_4$ with the projection maps $\pi_1\colon a_1 \times a_2 \times a_3 \times a_4 \longrightarrow a_1$, $\pi_2\colon a_1 \times a_2 \times a_3 \times a_4 \longrightarrow a_2$, $\pi_3\colon a_1 \times a_2 \times a_3 \times a_4 \longrightarrow a_3$ and $\pi_4\colon a_1 \times a_2 \times a_3 \times a_4 \longrightarrow a_4$. This product satisfies the following universal property: if there are an object c and morphisms $g_1 : c \longrightarrow a_1$, $g_2 : c \longrightarrow a_2$, $g_3 : c \longrightarrow a_3$ and $g_4\colon c \longrightarrow a_4$, then there is a unique $h\colon c \longrightarrow a_1 \times a_2 \times a_3 \times a_4$ such that the expected diagrams commute. In a similar way, we can talk about the colimit of this diagram, which we denote as $a_1 + a_2 + a_3 + a_4$. In the same way, we can define the product and coproduct of any number of objects. In general, for a collection of objects $\{a_i\}$ where i is an index in a set S, then the product and coproduct are denoted as

$$\prod_{i \in S} a_i \qquad \text{and} \qquad \coprod_{i \in S} a_i. \tag{3.67}$$

Before we go on to more complicated diagrams, let us take a moment and think about completing simpler diagrams.

Exercise 3.2.1. Show that the limit of a diagram with just one object a is isomorphic to a. Show this is true for colimits as well. ■

Exercise 3.2.2. Show that a limit of the empty diagram is a terminal object of the category. Also, show that the colimit of the empty category is the initial object of the category. ■

Products are special types of limits and coproducts are special types of colimits. Untill now, the diagrams that we dealt with were discrete (i.e., without morphisms). Let's look at diagrams with some morphisms. Consider the following simple diagram:

$$a \xrightarrow{\ f\ } b. \tag{3.68}$$

The limit of this diagram will be an object c with two maps $\pi_a\colon c \longrightarrow a$ and $\pi_b\colon c \longrightarrow b$, such that the following triangle commutes:

(3.69)

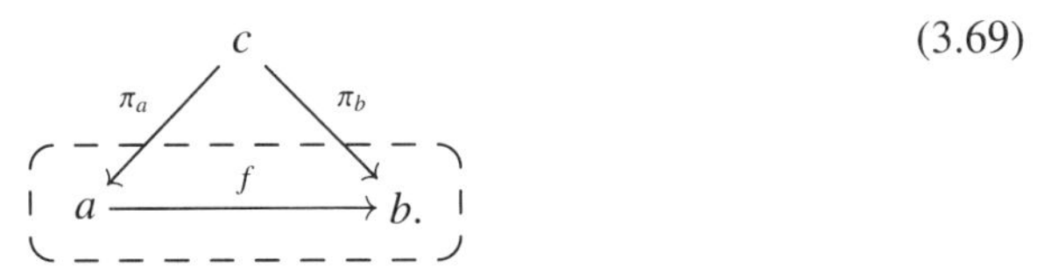

It satisfies the universal property that if there is any element d with maps $g_a\colon d \longrightarrow a$ and $g_b\colon d \longrightarrow b$ such that $f \circ g_a = g_b$ then there is a unique $h\colon d \longrightarrow c$ such that

(3.70)

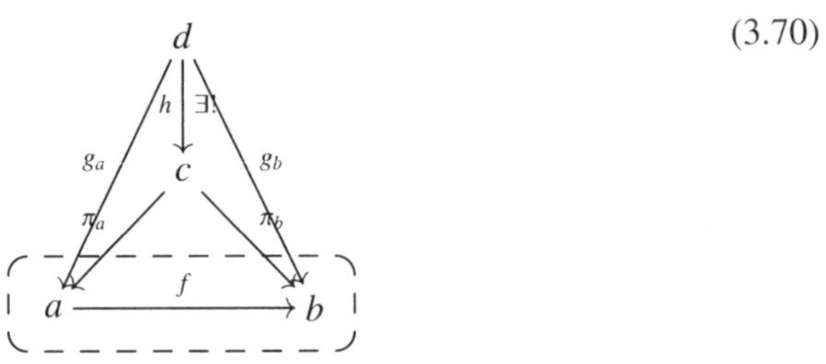

commutes. If you set $d = a$, $g_a = id_a$ and $g_b = f$, you will see that the limit c is isomorphic to a. In contrast, using the same trick, you can show that the colimit of this diagram is isomorphic to b.

Now we will examine the different types of limits and colimits in the category of sets.

Example 3.2.3. In $\mathbb{Set}$, given a set function $f\colon S \longrightarrow T$, the limit will be set R with two functions, $p_S : R \longrightarrow S$ and $p_T : R \longrightarrow T$, such that $f \circ p_S = p_T$. This set and these maps will satisfy a universal property. It is not hard to see that the set

$$R = \{(s, f(s)) \in S \times T\} \tag{3.71}$$

with the obvious projection functions satisfies the requirement of being a limit. This set is usually called the **graph of** f. □

Example 3.2.4. In $\mathbb{Set}$, given a set function $f\colon S \longrightarrow T$, the colimit will be set V with two functions, $inc_S : S \longrightarrow V$ and $inc_T : T \longrightarrow V$, such that $inc_T \circ f = inc_S$. This set and these maps will satisfy a universal property. The requirement is satisfied by the set

$$V = (S + T)/\sim, \tag{3.72}$$

where $\sim$ is the relation on the disjoint union that has $s \in S$ equivalent to $f(s) \in T$. The functions inc_S and inc_T are the obvious inclusion functions. □

Consider the diagram

$$a \underset{g}{\overset{f}{\rightrightarrows}} b. \tag{3.73}$$

A limit for this diagram will be an object c and two maps, $e\colon c \longrightarrow a$ and $j\colon c \longrightarrow b$, such that $f \circ e = j = g \circ e$, as in

$$\begin{array}{ccc} & c & \\ {}^{e}\swarrow & & \searrow^{j} \\ a & \underset{g}{\overset{f}{\rightrightarrows}} & b. \end{array} \tag{3.74}$$

We write the requirement as $f \circ e = g \circ e$ (without mentioning j) and then write the required commutative diagram as

$$c \xrightarrow{\;e\;} a \underset{g}{\overset{f}{\rightrightarrows}} b. \tag{3.75}$$

The universal property says that for any object d and map $k\colon d \longrightarrow a$ such that $f \circ k = g \circ k$, there is a unique map $h\colon d \longrightarrow c$ such that the triangle in the following diagram commutes:

$$\begin{array}{cccc} d & & & \\ {\scriptstyle h}\downarrow{\scriptstyle \exists!} & \searrow^{k} & & \\ c & \xrightarrow{\;e\;} & a & \underset{g}{\overset{f}{\rightrightarrows}} b. \end{array} \tag{3.76}$$

We call such a limit an **equalizer** of f and g.

Example 3.2.5. In $\mathbb{Set}$, an equalizer of set functions $f\colon S \longrightarrow T$ and $g\colon S \longrightarrow T$ is a set

$$R = \{s \in S : f(s) = g(s)\}. \tag{3.77}$$

There is an inclusion map $e\colon R \hookrightarrow S$. This set and inclusion map satisfy the universal property. □

A colimit of Diagram (3.73) is an object c and a map $p\colon b \longrightarrow c$ such that $p \circ f = p \circ g$. Furthermore, if there is any d and map $k\colon b \longrightarrow d$ such that $k \circ f = k \circ g$, then there is a unique $h\colon c \longrightarrow d$ such that the following diagram commutes:

$$\begin{array}{cccc} a \underset{g}{\overset{f}{\rightrightarrows}} & b & \xrightarrow{\;p\;} & c \\ & & \searrow_{k} & {\scriptstyle h}\downarrow{\scriptstyle \exists!} \\ & & & d. \end{array} \tag{3.78}$$

Such a colimit is called a **coequalizer**.

Example 3.2.6. In $\mathbb{S}\mathrm{et}$, a coequalizer for functions $f\colon S \longrightarrow T$ and $g\colon S \longrightarrow T$ is the set $V = T/\sim$, where $\sim$ is the equivalence relation that is generated by $f(s) \sim g(s)$ for all $s \in S$. The map $p\colon T \longrightarrow V$ takes every element $t \in T$ to the equivalence class that it belongs to in V (i.e., $t \mapsto [t]$). □

Let us move on to more complicated diagrams. We will no longer make a dashed line around the diagrams that we are completing. The limit of the diagram

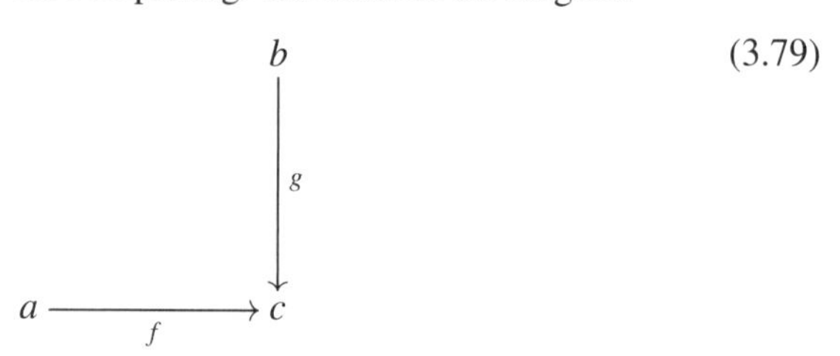

(3.79)

is called a **pullback**. It consists of an object $a \times_c b$ (which depends on the objects a, b, c, and the maps between them) and three maps, $\pi_a\colon a \times_c b \longrightarrow a$, $\pi_c\colon a \times_c b \longrightarrow c$, and $\pi_b\colon a \times_c b \longrightarrow b$, such that $f \circ \pi_a = \pi_c = g \circ \pi_b$. This object and these maps must be the best-fitting completion of this diagram. This means that they satisfy the following universal property: for any object e and morphisms $k_a\colon e \longrightarrow a$, $k_c\colon e \longrightarrow c$, and $k_b\colon e \longrightarrow b$ such that $f \circ k_a = k_c = g \circ k_b$, then there is a unique $h\colon e \longrightarrow a \times_c b$ that makes the following diagram commute:

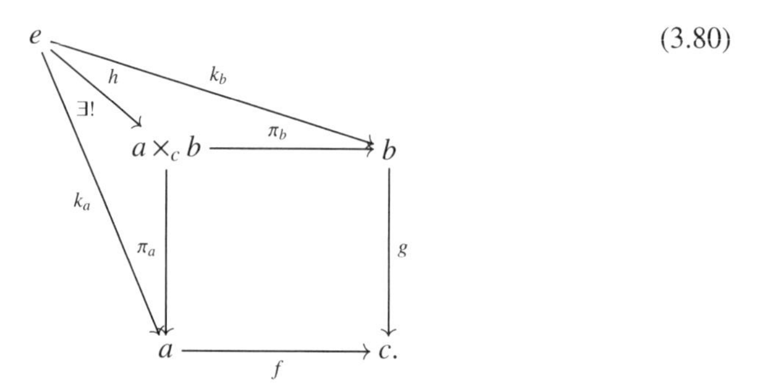

(3.80)

(There is no purpose in displaying π_c and k_c here.)

Notice that a product is a special type of pullback. The product is a pullback where target c is a terminal object of the category. In that case, there is a unique morphism from every element, and hence, the square always commutes. In addition, note that an equalizer is also a special type of pullback. It is the case where $b = a$ (i.e., the two maps of the pullback diagram have the same source).

Example 3.2.7. In $\mathbb{S}\mathrm{et}$, a pullback is sometimes called a **fiber product of sets**. Let $f\colon S \longrightarrow T$ and $g\colon R \longrightarrow T$; then the pullback is the set

$$P = \{(s, r) \in S \times R : f(s) = g(r)\}. \tag{3.81}$$

There are projection functions $\pi_S : P \longrightarrow S$ and $\pi_R\colon P \longrightarrow R$, which satisfy the universal properties.

A special case of a fiber product is when $g\colon R \hookrightarrow T$ is an inclusion of a subset. We then have the pullback diagram

$$\begin{array}{ccc} \{s \in S : f(s) \in R\} & \longrightarrow & R \\ \Big\downarrow & & \Big\downarrow g \\ S & \xrightarrow[f]{} & T. \end{array} \tag{3.82}$$

The fiber product of f and g is isomorphic to the set

$$f^{-1}(R) = \{s \in S : f(s) \in R\}, \tag{3.83}$$

which is called the **preimage** of f for the subset T. This is an example of an instance of a general theorem that the pullback of a monic is also a monic. □

The colimit of the diagram

$$\begin{array}{ccc} a & \xrightarrow{f} & b \\ {\scriptstyle g}\Big\downarrow & & \\ c & & \end{array} \tag{3.84}$$

is called a **pushout** of f and g. The colimit consists of an object $c +_a b$ and maps $h_a\colon a \longrightarrow c +_a b$, $h_b\colon b \longrightarrow c +_a b$, and $h_c\colon c \longrightarrow c +_a b$ such that $h_b \circ f = h_a = h_c \circ g$. This $c +_a b$ must be the best-fitting cocompletion of the diagram. This translates that for any object e and any maps $k_b\colon b \longrightarrow e$, $k_a\colon a \longrightarrow e$, and $k_c\colon c \longrightarrow e$, such that $k_b \circ f = k_a = k_c \circ g$, there is a unique $h\colon c +_a b \longrightarrow e$ that makes the following diagram commute:

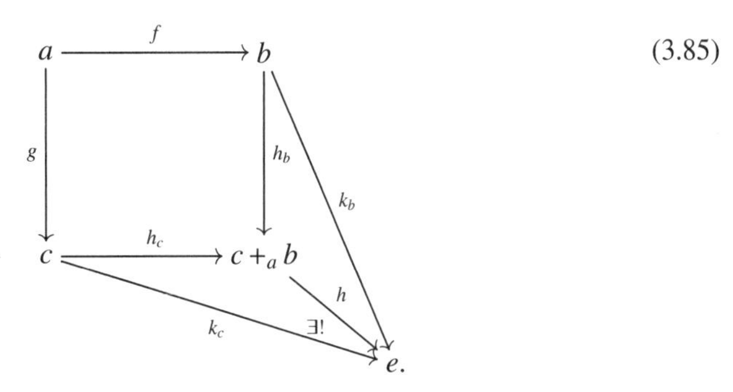

(3.85)

(There is no purpose to displaying h_a and k_a here.)

Example 3.2.8. In $\mathbb{Set}$, a pushout of the maps $f\colon S \longrightarrow R$ and $g\colon S \longrightarrow T$ is a set $P = (R + T)/\sim$, where $\sim$ is the equivalence relation on the set $R + T$ generated by the relation for all $s \in S$, $f(s) \sim g(s)$. There are obvious inclusions of R and T into P. □

Notice that a coproduct is a special type of pushout. It is the case where a is an initial object of the category, and hence, the square always commutes. Similarly, a coequalizer is a special type of pushout where $a = c$.

Now that we have seen many examples of limits and colimits, let us formally define a limit and colimit of any diagram in a category in Definition 3.2.9.

Definition 3.2.9. For an arbitrary diagram D in a category, a **limit** is an object of the category, denoted as $\varprojlim D$, and morphisms from that object to every object in D such that the appropriate diagrams commute. The object and projections must satisfy the universal properties outlined above. A **colimit** of a diagram D in the category will be an object, denoted as $\varinjlim D$, with morphisms from every object in the diagram to the colimit that makes the appropriate commutative diagram. The object and morphisms must satisfy the universal properties outlined above. In some older texts, a limit is sometimes also called an **inverse limit** or a **projective limit**, while a colimit is called a **direct limit** or an **inductive limit**. Since there are maps from the limit to all the objects of the diagram, and there are maps from all the objects of the diagram to the colimit, we might write this as

$$\varprojlim D \longrightarrow D \longrightarrow \varinjlim D. \tag{3.86}$$

There is a way of getting a limit by using a product and then taking an equalizer. First, let us look at a simple case. Let us revisit Diagram (3.69), where we took a limit of the single arrow $f\colon a \longrightarrow b$. Consider the object $a \times b$, which is the product of the two objects in the diagram. There are two projection maps, as in the following:

$$\begin{array}{ccc} & a \times b & \\ {}^{\pi_a}\swarrow & & \searrow^{\pi_b} \\ a & \xrightarrow[f]{} & b. \end{array} \tag{3.87}$$

Notice that there are two maps from $a \times b$ to b: π_b and $f \circ \pi_a$. We can now take the equalizer of these two maps as follows:

$$c \xrightarrow{\ e\ } a \times b \underset{\pi_b}{\overset{f\circ\pi_a}{\rightrightarrows}} b. \tag{3.88}$$

Setting $p_a = \pi_a \circ e$ and $p_b = \pi_b \circ e = f \circ \pi_a \circ e$, as in

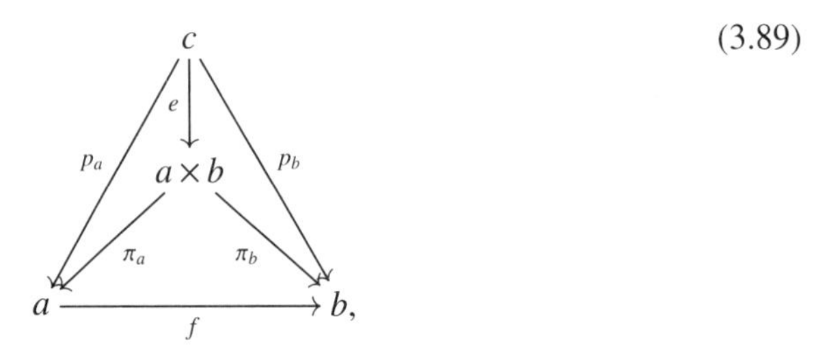

(3.89)

it is easy to see that c not only is the equalizer of the two maps, but it is also the limit of the diagram $f: a \longrightarrow b$.

We can generalize from a single arrow to all finite diagrams. any diagram with a finite number of objects and morphisms.

Theorem 3.2.10. Limits of finite diagrams can be obtained using finite products and equalizers. ★

Proof. Let D be a diagram in the category. We create the product of all the objects in D (i.e., the following object):

$$\prod_{a \in D} a. \tag{3.90}$$

There will be projection maps from this product to every object in the diagram. Let us form the product of all the objects that are the codomain of maps in the diagram (i.e., the following object):

$$\prod_{(f: a \longrightarrow b) \in D} b. \tag{3.91}$$

Now form the equalizers

$$c \xrightarrow{\quad e \quad} \prod_{a \in D} a \overset{s}{\underset{t}{\rightrightarrows}} \prod_{(f: a \longrightarrow b) \in D} b, \tag{3.92}$$

where both of the following occur:

- s is induced by maps going to b via projection maps.
- t is induced by maps that go to b via $f: a \longrightarrow b$.

Similar to Diagram (3.89) and the discussion that follows this, the equalizer c will be a limit. ♣

Theorem 3.2.11. Limits of finite diagrams can be obtained using pullbacks and a terminal object. ★

Proof. If there is a terminal object and pullbacks, we can create all finite products. Similarly, equalizers can be seen as a type of pullback. ♣

There is, of course, a dual theorem.

Theorem 3.2.12. Colimits of finite diagrams can be obtained using coproducts and coequalizers. Equivalently, all finite colimits can be obtained using pushouts and an initial object. ★

Exercise 3.2.13. Prove Theorem 3.2.12 ■

We will look at limits and colimits again from a more advanced perspective in Section 4.6.

3.3 Slices and Coslices

There are certain constructions that make the morphisms of one category into the objects of another category.

Definition 3.3.1. Given a category $\mathbb{A}$ and an object a of that category, the **slice category**, $\mathbb{A}/a$ (which is read "$\mathbb{A}$ over a"), is a category whose objects are pairs $(b, f\colon b \longrightarrow a)$, where b is an object of $\mathbb{A}$ and f is a morphism of $\mathbb{A}$, whose target is a. The morphisms of $\mathbb{A}/a$ from $(b, f\colon b \longrightarrow a)$ to $(b', f'\colon b' \longrightarrow a)$ are morphisms $g\colon b \longrightarrow b'$ of $\mathbb{A}$, which make the following triangle commute:

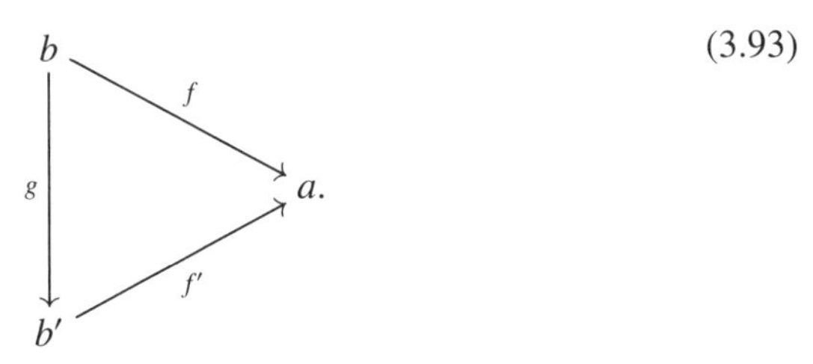

(3.93)

The composition of morphisms comes from the composition in $\mathbb{A}$, and the identity morphism for the object $(b, f\colon b \longrightarrow a)$ is id_b.

Example 3.3.2. Here are some examples of slice categories:

- For $\mathbf{R}$, the real numbers thought of as an element of $\mathbb{S}\mathrm{et}$, the category of $\mathbb{S}\mathrm{et}/\mathbf{R}$ is the collection of all $\mathbf{R}$-valued functions.
- For a partial order category $A(P, \leq)$ and $p \in P$, the category $A(P, \leq)/p$ is the partial order category of all elements below p, denoted as $p\downarrow$. □

Exercise 3.3.3. Show that $(a, id_a\colon a \longrightarrow a)$ is the terminal object in $\mathbb{A}/a$. ■

There is the dual notion of a slice category, expressed in Definition 3.3.4.

Definition 3.3.4. Given a category $\mathbb{A}$ and an object a of that category, the **coslice category**, $a/\mathbb{A}$ (read as "$\mathbb{A}$ under a"), is a category whose objects are pairs $(b, f\colon a \longrightarrow b)$, where b is an object of $\mathbb{A}$ and f is a morphism of $\mathbb{A}$, whose source is a. The morphisms of $a/\mathbb{A}$ from $(b, f\colon a \longrightarrow b)$ to $(b', f'\colon a \longrightarrow b')$ are morphisms $g\colon b \longrightarrow b'$ of $\mathbb{A}$, which make the following triangle commute:

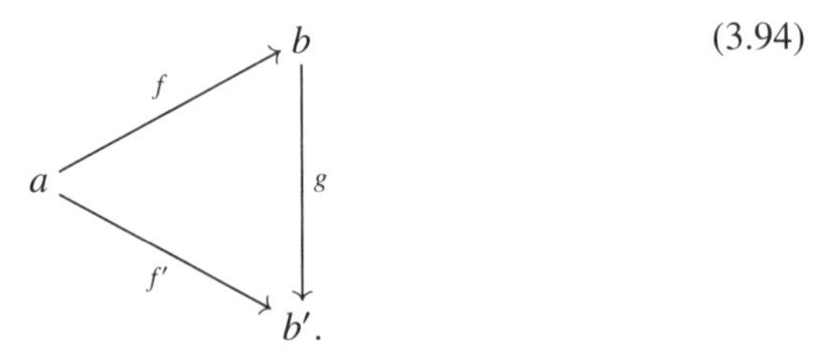

(3.94)

The composition of morphisms comes from the composition in $\mathbb{A}$, and the identity morphism for the object $(b, f\colon b \longrightarrow a)$ is id_b.

Example 3.3.5. Here are some examples of a coslice category:

- Consider the one-element set $\{*\}$. The category $\{*\}/\mathbb{Set}$ has objects that are sets with a function that picks out a distinguished element of the set. So an object in the category is effectively a pair (S, s_0), where S is a set and s_0 is a distinct element of that set. The morphisms from (S, s_0) to (T, t_0) are set functions that preserve the distinguished element. That is, $f: S \longrightarrow T$, such that $f(s_0) = t_0$. This is the category of **pointed sets**.
- Similarly, there is the category $\{*\}/\mathbb{Top}$, the category of **pointed topological spaces**, in which the objects are topological spaces with a distinguished object and morphisms are continious maps that preserve the distinguished point.
- For a partial order category $A(P, \leq)$ and p, an element in the partial order, the category $p/A(P, \leq)$ is $p\uparrow$, the partial order of the elements above p. □

Exercise 3.3.6. Show that $id_a: a \longrightarrow a$ is the initial object in $a/\mathbb{A}$. ■

Suggestions for Further Study

All the categorical structures in this chapter can be found in the standard textbooks on category theory (e.g., [180, 208, 40, 32, 33, 108, 171, 222, 21]). The main theorems showing that limits and colimits of certain types can be constructed from particular types of products, coproducts, pushouts, pullbacks, and other elements can be found in Section V.2 of [180], Section 1.9 of [208], Section 5.4 of [21], Section 2.8 of [40], Section 1.7 of [32], and Section 3.4 of [222].

3.4 Mini-Course: Self-Referential Paradoxes

With the simple idea of a product in a category, we know enough category theory to describe some of the most profound and influential theorems in mathematics and computer science of the past 150 years. In the next few pages, we will encounter Georg Cantor's theorem, which shows there are different types of infinity; Bertrand Russell's paradox, which proves that simple set theory is inconsistent; Kurt Gödel's famous incompleteness theorems, which demonstrates a limitation of the notion of proof; Alan Turing's realization that there are some problems that can never be solved by a computer; and much more.

What is truly amazing is that all these diverse and important theorems are consequences of a single simple theorem of category theory. This demonstrates the true power of category theory. What is still more shocking is that the central idea of this simple theorem goes back some 2,500 years ago to a conundrum about language called the Epimenides paradox (see page 136). This conundrum shows that language can talk about itself (i.e., it has self-reference). In the coming pages, we will see that not only language, but also many systems, have the ability to have objects in the system refer to other objects within the system, and even to themselves. This is the core of self-reference. It is shown that sets, language, logic, computers, and many other systems have the ability of self-reference. With self-reference, one goes on to form **self-referential paradoxes**, which are contradictions that come through an object using self-reference to negate itself. This is the explanation of all the important theorems discussed here.

Rather than presenting the various instances of the single categorical theorem in chronological order, we present the easier, more accessible instances first. After seeing the first few, we state and prove the categorical theorem and then continue with other instances.

Before we leap into all the examples, there is one technical definition that we have to describe. Let S be a set and $2 = \{0, 1\}$ be a set with two values that will correspond to true and false. We saw that a function $g: S \longrightarrow 2$ is a characteristic function and describes the subset of S that g takes to 1. Now consider a set function $f: S \times S \longrightarrow 2$. The function f accepts two elements of S and outputs either 0 or 1. For any element s_0 of S, consider the function f where the second input is always s_0. We say that s_0 is "hardwired into the function." This gives us a function

$$f(\ , s_0): S \longrightarrow 2 \tag{3.95}$$

with only one input. Since this function goes from S to 2, it is also a characteristic function and describes the subset of S. The subset is

$$\{s \in S : f(s, s_0) = 1\}. \tag{3.96}$$

For different f and different elements of S, there will be different characteristic functions that describe different subsets. We now ask a simple question: given $g: S \longrightarrow 2$ and $f: S \times S \longrightarrow 2$, is there an s_0 in S such that g characterizes the same subset as $f(\ , s_0)$? To restate, for a given g and f, is there an $s_0 \in S$ such that $g(\ ) = f(\ , s_0)$? If such an s_0 exists, then we say that g can be **represented** by f or g is **representable** by f.

The Barber Paradox

Bertrand Russell was not only a great logician, mathematician, and philosopher, but also a great expositor. To explain some of the central ideas of self-referential systems to a general audience, he supposedly conjured up the **barber paradox**. Imagine an isolated village on top of a mountain in the Austrian alps where it is difficult for villagers to leave and for itinerant barbers to come to the village. This village has exactly one barber, and there is a strict rule that is enforced:

> A villager cuts his own hair if and only if he does not go to the single barber.

This makes sense. After all, if the villager will cut his own hair, why should he go to the barber? On the other hand, if the villager goes to the barber, he will not need to cut his own hair. This works out very well for all the villagers except for one: the barber. Who cuts the barber's hair? If the barber cuts his own hair, then he is violating the village ordinance by cutting his own hair and having his hair cut by the barber. If he goes to the barber, then he is also cutting his own hair. This is illegal! What is an honest, law-abiding barber to do?

Now we formalize the problem. Let the set $Vill$ consist of all the villagers in the village. Let us also remember our set $2 = \{0, 1\}$. The function $f\colon Vill \times Vill \longrightarrow 2$ describes who cuts whose hair in the village. It is defined for villagers v and v' as

$$f(v, v') = \begin{cases} 1 & : \text{if the hair of } v \text{ is cut by } v' \\ 0 & : \text{if the hair of } v \text{ is not cut by } v'. \end{cases} \tag{3.97}$$

We can now express the village ordinance as saying that for all v,

$$f(v, v) = 1 \quad \text{if and only if} \quad f(v, \text{barber}) = 0. \tag{3.98}$$

The problem arises because this is true for all v, including $v = \text{barber}$. In this case, we get

$$f(\text{barber}, \text{barber}) = 1 \quad \text{if and only if} \quad f(\text{barber}, \text{barber}) = 0. \tag{3.99}$$

This is clearly a contradiction and cannot be true.

Let us be more categorical. Because $\mathbb{Set}$ has products, there is the diagonal function $\Delta\colon Vill \longrightarrow Vill \times Vill$, which is defined as $\Delta(v) = (v, v)$. This function is at the core of self-reference. It helps us see what f says when you evaluate an element v with itself (i.e., $f(v, v)$ and $f(\text{barber}, \text{barber})$). There is also a negation function $NOT\colon 2 \longrightarrow 2$, defined as $NOT(0) = 1$ and $NOT(1) = 0$, which swaps true and false. This function will be used to negate properties. Composing f with Δ and NOT gives us $g\colon Vill \longrightarrow 2$, as in the following commutative diagram:

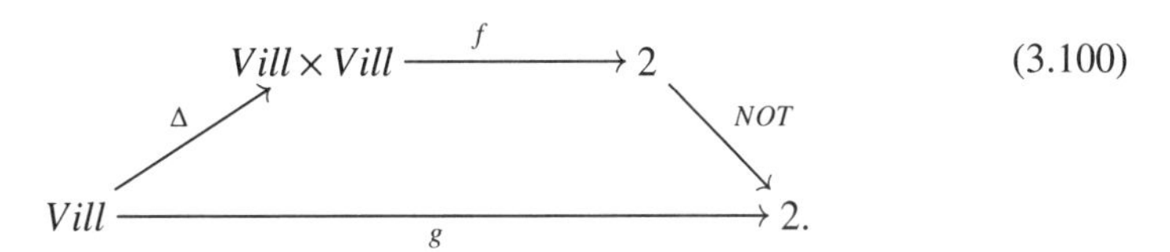

(3.100)

That is,

$$g = NOT \circ f \circ \Delta. \tag{3.101}$$

For villager v, $g(v) = NOT(f(\Delta(v))) = NOT(f(v, v))$. So $g(v) = 1$ if and only if $NOT(f(v, v)) = 1$ if and only if $f(v, v) = 0$ if and only if the hair of v is not cut by v. In other words, $g(v) = 1$ if and only if v does not cut his own hair. In terms of self-reference, the function g is the characteristic function of the subset of villagers who do not cut their own hair.

We now ask the simple question: can g be represented by f? In other words, is there a villager v_0 such that $g(\ ) = f(\ , v_0)$? The function g describes all those villagers who do not cut their own hair. It stands to reason that the barber is the villager who can represent g. After all, $f(\ \ , \text{barber})$ describes all the villagers who get their hair cut by the barber. We are asking if $g(\ )$ is the same function as $f(\ , \text{barber})$, and whether both functions characterize the same subset of villagers. Another way to pose this question is as follows: Is the set of villagers who do not cut their own hair the same as the set of villagers who get their hair cut by the barber? The answer is no. While it is true that for most $v \in Vill$,

$$g(v) = f(v, \text{barber}), \tag{3.102}$$

	f	v_1	v_2	v_3	v_4	v_5	$\cdots$	v_n
		Cutter						
Cuttee	v_1	1	0	0	0	0	$\cdots$	0
	v_2	0	0	0	1	0	$\cdots$	0
	v_3	0	0	1	0	0	$\cdots$	0
	v_4	0	0	0	?	0	$\cdots$	0
	v_5	0	0	0	1	0	$\cdots$	0
	$\vdots$	$\vdots$			$\vdots$		$\ddots$	$\vdots$
	v_n	0	0	0	0	0	$\cdots$	1

Figure 3.3. The function f as a matrix. Notice the diagonal is the opposite of the barber, column v_4.

it is not true for $v =$ barber. If it were true for the barber, then we would have

$$g(\text{barber}) = f(\text{barber}, \text{barber}). \tag{3.103}$$

But the definition of g is given as $g(\text{barber}) = NOT(f(\text{barber}, \text{barber}))$. We conclude that g is not represented by $f(\quad, \text{barber})$; in fact, it is not represented by any $f(\quad, v_0)$. That is, the set of villagers who do not cut their own hair cannot be the same as the set of villagers who get their hair cut by anyone.

It will be helpful to describe this problem in matrix form. Let us consider the set $Vill$ as $\{v_1, v_2, v_3, \ldots, v_n\}$. We can then describe the function $f\colon Vill \times Vill \longrightarrow 2$ as a matrix in Figure 3.3. Let us say that the barber is v_4. Notice that every row has exactly one 1 (every villager gets their haircut in only one place): either along the diagonal (the villager cuts their own hair) or in the v_4 column (the villager goes to the barber). Since it can only be one or the other, the numbers along the diagonal $1, 0, 1, ?, 0, \ldots, 1$ are almost the exact opposite of the numbers along the v_4 column $0, 1, 0, ?, 1, \ldots, 0$. This is a restatement of the rule of the village. There is only one problem: What is in the (v_4, v_4) position? We put a question mark in the matrix because that entry cannot be the opposite of itself. This way of seeing the problem will arise over and over again. Here, we can see why these paradoxes are related to proofs called **diagonal arguments**.

What is the resolution to this paradox? There are many attempts to solve this paradox, but they are not very successful. For example, the barber resigns as barber before cutting his own hair. (But that means that there is no barber in the village at all.) Or the wife of the barber cuts the barber's hair. (But that means that there are two barbers in the town.) Or the barber is bald. Or the barber is a long-haired hippie. Or the rule is ignored while the barber cuts his own hair, etc. All these are saying the same thing: the village with this contrived rule cannot exist. Because if the village with this rule existed, there would be a contradiction. There are no contradictions in the physical world. The only way that the world can be free of contradictions is if this proposed village with this strict rule does not exist.

Russell's Paradox

Bertrand Russell described the barber paradox to help explain a deeper, more important problem that he formulated, called **Russell's paradox**. This paradox concerns sets that are considered the foundation of much of mathematics.

As is known, sets contain elements. The elements can be anything. In particular, an element in a set can be a set itself. Here are a few sets to consider. $A = \{x, y, z\}$ simply has three elements. $B = \{s, t, \{x, y\}\}$ has the set $\{x, y\}$ as an element. $C = \{s, t, u, A\}$ contains set A as an element. It is not strange to have sets as elements of sets. The set of all classes given in a university can be thought of as containing elements where each element is the set of students in a class. For a set of certain people, one can imagine every person as the set of their cells. Consider set $D = \{x, y, D\}$. This set contains itself. A set containing itself is not so strange. Here are three examples of sets that contain themselves:

- The set of all ideas discussed in this book
- The set that contains all the sets that have more than three objects
- The set of abstract ideas

If you do not like sets that contain themselves, you might want to consider R, which is the collection of all sets that *do not* contain themselves. Formally,

$$R = \{\text{set } S : S \text{ does not contain } S\} = \{S : S \notin S\}. \tag{3.104}$$

Going back to our simple examples of sets, we have $A \in R$, $B \in R$, and $C \in R$, while $D \notin R$. Now ask yourself the simple question: Does R contain itself? In symbols, we ask if $R \in R$? Let us consider the possible answers. If $R \in R$, then since R fails to satisfy the requirements of being a member of R, we get that $R \notin R$. In contrast, if $R \notin R$, then since R satisfies the requirement of belonging to R, we have that $R \in R$. This is a contradiction.

Let us formulate this. There is a collection of all sets called *Set*. There is also a two-place function $f\colon Set \times Set \longrightarrow 2$, which describes which sets are elements of which sets:

$$f(S, S') = \begin{cases} 1 & : \text{if } S \in S' \\ 0 & : \text{if } S \notin S'. \end{cases} \tag{3.105}$$

We can use this f to describe all sets that do not contain themselves as follows. Consider g, formed by composing the following maps:

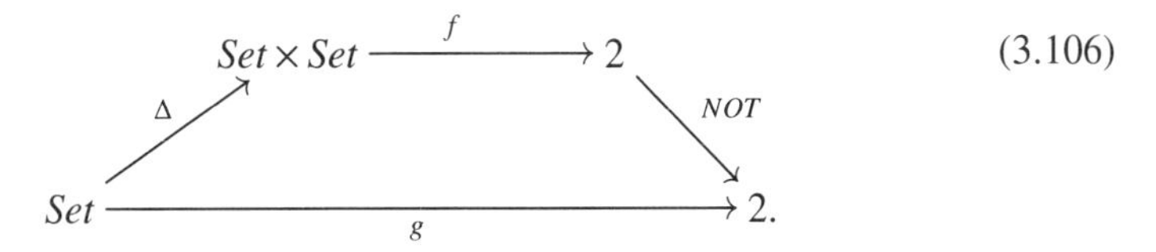

(3.106)

The value of $g(S)$ is defined to be $NOT(f(S, S))$. This means $g(S) = 1$ if and only if $f(S, S) = 0$. In terms of self-reference, g is the characteristic function of those sets that do not contain themselves.

Now we ask the simple question: Is there a set R such that $g(\ )$ is represented by f as $f(\ , R)$? That is, we want a set R such that

	f	S_1	S_2	S_3	S_4	S_5	$\cdots$
		Subset					
Element	S_1	NOT(1)=0	0	0	0	1	$\cdots$
	S_2	0	NOT(0)=1	0	1	0	$\cdots$
	S_3	0	0	NOT(1)=0	0	0	$\cdots$
	S_4	1	0	0	NOT(1)=0	0	$\cdots$
	S_5	0	1	0	1	NOT(0)=1	$\cdots$
	$\vdots$	$\vdots$			$\vdots$		$\ddots$

Figure 3.4. Function f as a matrix. Function g is the changed diagonal, and it is different for every column.

$$g(S) = 1 \quad \text{if and only if} \quad f(S, R) = 1 \tag{3.107}$$

and

$$g(S) = 0 \quad \text{if and only if} \quad f(S, R) = 0. \tag{3.108}$$

This means that R contains only the sets that do not contain themselves. The problem is that if such a set R exists, then we can ask about $g(R)$ (i.e., is $R \in R$). On the one hand, $g(R)$ is defined as $NOT(f(R, R))$; and on the other hand, if f represents g with R, then $g(R) = f(R, R)$. That is,

$$f(R, R) = g(R) = NOT(f(R, R)). \tag{3.109}$$

This is a contradiction.

Let us look at Russell's paradox from a matrix/array point of view. Let us consider the infinite collection *Set* as $\{S_1, S_2, S_3, \ldots\}$. We can then describe the function $f\colon Set \times Set \longrightarrow 2$ as the matrix in Figure 3.4. Notice that the diagonal is different from every column of the array. This is a way of saying that the diagonal (which is g) cannot be represented by any column of the array.

Before we move to the next instance, let us consider how to deal with this paradox. The only way to avoid this contradiction is to accept that function g cannot be represented by any element of *Set*. This means that the collection of all sets that do not contain themselves does not form a set (i.e., this collection is not an element of *Set*). While such a collection seems to be a well-defined notion, we have shown that if we say that this collection is an element of *Set*, then there is a contradiction. Mathematicians went on and used this to make a distinction between a "set" and a "class." They declared that classes are collections that are not sets. This distinction plays major roles in logic and higher mathematics.

Heterological Paradox

Now for a linguistic paradox. The **heterological paradox**, also called **Grelling's paradox** after Kurt Grelling, who first formulated it, concerns adjectives (words that modify nouns). Consider several adjectives and ask if they describe themselves; that is, if the adjective has the property of the adjective. "English" is English. In contrast, "French" is not French ("Français" is Français). "German" is not German ("Deutsch" is Deutsch). The word "abbreviated" is not abbreviated, whereas "unabbreviated" is unabbreviated; and "hyphenated" is not hyphenated, etc. We see that some adjectives describe themselves and some adjectives do not describe themselves. Call all adjectives that describe themselves "autological." In contrast, call all adjectives that do not describe themselves "heterological." We can start making a table of many adjectives.

autological	**heterological**
English	non-English
Français	French
Deutsch	German
noun	verb
unhyphenated	hyphenated
unabbreviated	abbreviated
polysyllabic	monosyllabic
⋮	⋮

It seems that we can split all adjectives into these two groups. Is that true? Let us ask a simple question. Is "heterological" autological or heterological? That is, does the adjective "heterologoical" belong on the left side or the right side of the table? Let us go through the two possibilities.

- If "heterological" is autological, then it is not heterological and it does not describe itself. This means that it is heterological.
- If "heterological" is heterological, then the adjective does describe itself and that makes it autological, not heterological.

We have a contradiction.

Let us formulate this paradox categorically. There is a set Adj of adjectives and a function $f\colon Adj \times Adj \longrightarrow 2$, which is defined for adjectives a and a' as follows:

$$f(a,a') = \begin{cases} 1 & : \text{if } a \text{ is described by } a' \\ 0 & : \text{if } a \text{ is not described by } a'. \end{cases} \tag{3.110}$$

Use f to formulate g as the composition of the following three maps:

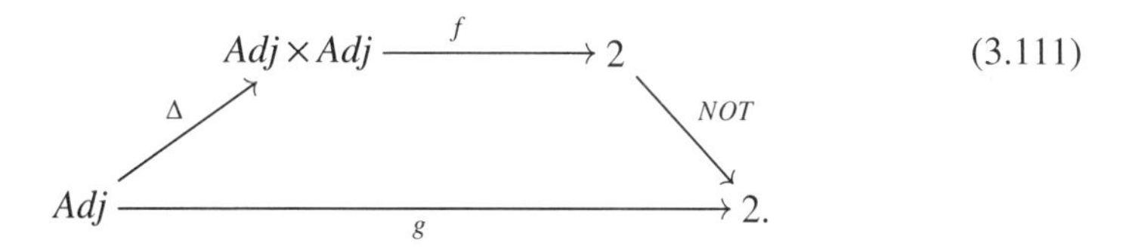

In terms of self-reference, the function g is the characteristic function of those adjectives that do not describe themselves. Can g be represented by some element in Adj? Is there some adjective, say "heterological," that can be used in f to represent g? That is, is it true that $g(\ ) = f(\ ,\text{"heterological"})$? We are asking if the subset of adjectives that do not describe themselves can be described by the word "heterological." If this was true, we would have that for all adjectives A,

$$g(A) = f(A, \text{“heterological”}). \tag{3.112}$$

But this cannot be true because then, we would have that

$$g(\text{“heterological”}) = f(\text{“heterological”}, \text{“heterological”}). \tag{3.113}$$

That would create a contradiction because by the definition of g, we have

$$g(\text{“heterological”}) = NOT(f(\text{“heterological”}, \text{“heterological”})). \tag{3.114}$$

The only conclusion we can come to is that $g(\ )$ cannot be represented by f. That is, the set of all adjectives that do not describe themselves cannot be represented by "heterological." However, that is exactly the definition of "heterological"!

The hetrological paradox can be described with an array similar to Figure 3.4, where $Adj = \{A_1, A_2, A_3, \ldots\}$. Again, we would have a changed diagonal that would be different from every column in the array.

How do we avoid this little paradox? There are two usual ways of resolving this paradox:

(i) Many philosophers say that the word "heterological" cannot exist. After all, we just showed that it is not always well defined. We cannot determine if a certain adjective ("hetrological") is heterological or not.

(ii) Another more obvious solution is just to ignore the problem. Human language is inexact and full of contradictions. Every time we use an oxymoron, we are stating a contradiction. Every time we ask for another piece of cake while lamenting the fact that we cannot lose weight, we are stating a contradiction. We can safely ignore the fact that heterological is not well defined for only one adjective.

A Philosophical Interlude on Paradoxes

The word "paradox" has many different definitions. For a logician, a paradox is a process where an assumption is made, and through valid reasoning, a contradiction is derived. We can visualize this as

$$\text{Assumption} \Longrightarrow \text{Contradiction}.$$

The logician then concludes that since the reasoning was valid and the contradiction cannot happen, it must be that the assumption was wrong. This is very similar to what mathematicians call "proof by contradiction" and philosophers call "*reductio ad absurdum*." In a sense, a paradox is a method of showing that the assumption is not part of rational thought. If one accepts the assumption, one will come to a contradiction, which is dangerous to rational thought.

We have so far seen the same pattern of proof in three areas: (i) villagers, (ii) sets, and (iii) adjectives. In all three of these paradoxes, the assumption is that the g function

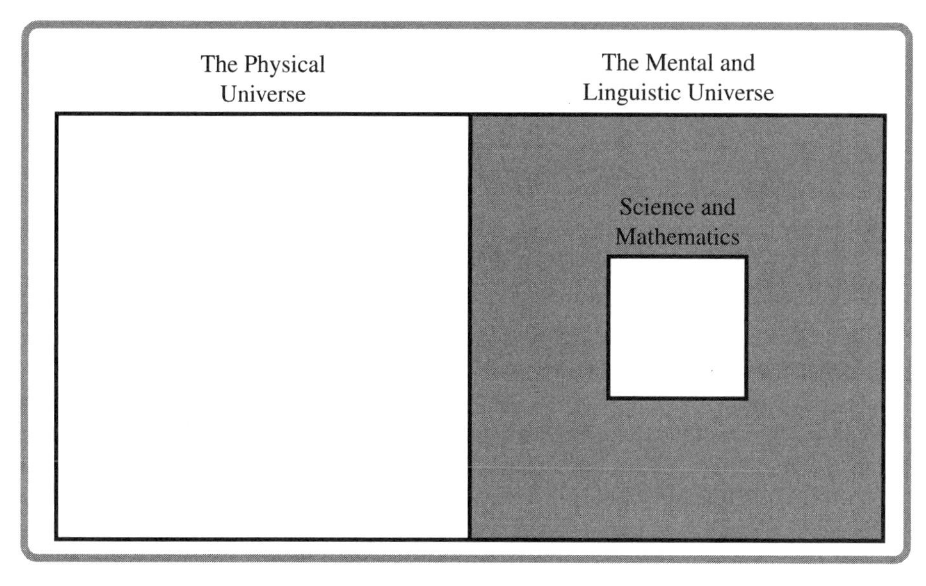

Figure 3.5. Contradictions can occur in gray areas. White areas cannot have contradictions.

can be represented by the f function. A contradiction is then derived, and we conclude that g is not represented by f. These three examples highlight three different realms where the alleged contradictions might be found (see Figure 3.5). Let us examine these carefully.

- **The Physical Universe**. A village with a particular rule is part of the physical universe. The physical universe does not have any contradictions. Facts and properties simply are, and no object can have two opposing properties. Whenever we come to such contradictions, we have no choice but to conclude that the assumption was wrong.
- **The Mental and Linguistic Universe**. In contrast to the physical universe, the human mind and human language—that the mind uses to express itself—are full of contradictions. We are not perfect machines. We have a lot of contradictory parts and desires. An oxymoron is a small contradiction that we all use in our speech. We all have conflicting thoughts in our head, and these thoughts are expressed in our speech. So when an assumption brings us to a contradiction in our thought or language, we do not need to take it very seriously. If an adjective is in two opposite classifications, it does not really concern us. In such a case, we cannot go back to our assumption and say that it is wrong. The entire paradox can be ignored.
- **Science and Mathematics**. There are, however, parts of human thought and language where we cannot tolerate contradictions: science and mathematics. These areas of exact thought are what we use to discuss the physical world (and more). If science and mathematics are to discuss / describe / model / predict the contradiction-free physical universe, then we better make sure that no

contradictions occur there. We first saw this in the early years of elementary school, when our teachers proclaimed that we are not permitted to divide by zero. Why not? We can divide by any other number. Why not zero? If we were permitted to divide by zero, an easy contradiction could be derived. Since mathematics and science cannot have contradictions, young fledglings are not permitted to divide by zero. To summarize, science and mathematics are products of the human mind and language, which we do not permit to have contradictions. If an assumption leads us to a contradiction in science or mathematics, then we must abandon the assumption.

What is gained by looking at self-referential paradoxes from the categorical point of view? Many have felt that these instances of self-referential paradoxes have a similar pattern (witness Bertrand Russell supposedly inventing the barber paradox to illustrate Russell's paradox). However, no one has ever formalized this feeling. The major advance that category theory has to offer the subject is to actually show that all these different self-referential paradoxes are really instances of a single categorical theorem. F. William Lawvere described a simple formalism that showed many of the major self-referential paradoxes and more. This shows that the logic of self-referential paradoxes is inherent in many systems. It also shows the unifying power of category theory.

There is, however, another positive aspect of our categorical formalism. Lawvere showed us how to have an exact mathematical description of the paradoxes while avoiding messy statements about what exists and what does not exist. In the categorical setting, the barber paradox does not say that a village with a rule does not exist. Rather, it says that a certain function cannot be represented by any element in the collection *Vill*. With Russell's paradox, a category theorist does not say that a certain collection does not form a set. Rather, she talks about representing a function g by f with an element in *Set*. Similarly, with the heterological paradox, we avoid the silly analysis as to whether a word exists or not. Rather, we say that a function cannot be represented by any element of a collection *Adj*. Although in our presentation of the paradoxes, we mention the way that philosophers have thought about these issues, in our categorical discussion, we successfully avoid metaphysical gobbledygook. For this alone, we should be appreciative of the categorical formalism.

Cantor's Inequalities

Let us continue our list of instances of self-referential paradoxes. At the end of the nineteenth century, Georg Cantor proved some important theorems about the sizes of sets. He first showed that every set is smaller than its powerset (set of subsets). That is, every set S is smaller than the set $\mathcal{P}(S)$. A more categorical way of saying this is that for any set S, there cannot be a surjection $h\colon S \longrightarrow \mathcal{P}(S)$. Yet another way of saying this is that for every purported surjection $h\colon S \longrightarrow \mathcal{P}(S)$, there will be some subset of S that will not be in the image of h. One can think of this as a proof by contradiction: we are going to assume (wrongly) that there is such a surjection h and derive a contradiction (because we will find something that is not in the image of h). Since this is formal mathematics, no such contradiction can exist, and hence our assumption that such a surjection exists must be false.

Given such an $h\colon S \longrightarrow \mathcal{P}(S)$, let us define $f_h\colon S \times S \longrightarrow 2$ for $s, s' \in S$ as follows:

$$f_h(s, s') = \begin{cases} 1 & : s \in h(s') \\ 0 & : s \notin h(s'). \end{cases} \tag{3.115}$$

Use f_h to construct g_h as follows:

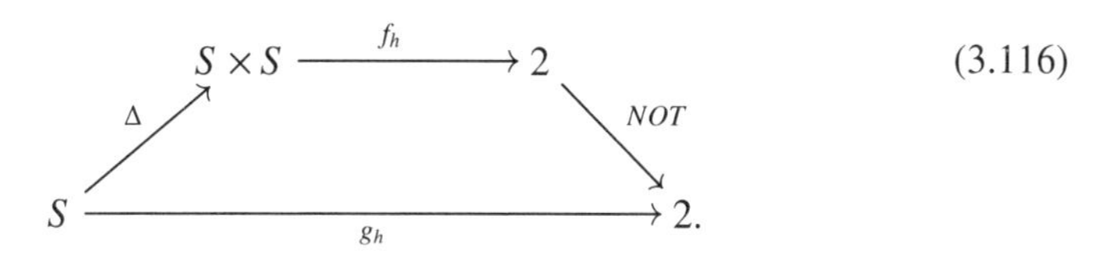

(3.116)

Function g_h is the characteristic function of the subset $C_h \subseteq S$, where each element s does not belong to $h(s)$; that is,

$$C_h = \{s \in S : s \notin h(s)\} \subseteq S. \tag{3.117}$$

Notice that f_h, g_h, and C_h depend on h. If we change h, we will get different functions and sets.

We claim that the subset C_h of S is not in the image of h (i.e., C_h is a "witness" or a "certificate" that h is not surjective). If C_h was in the image of h, there would be some $s_0 \in S$ such that $h(s_0) = C_h$. In that case, g_h would be represented by f_h with s_0. That is, for all $s \in S$,

$$g_h(s) = f_h(s, s_0), \tag{3.118}$$

but this would also be true for $s_0 \in S$, which would mean that

$$g_h(s_0) = f_h(s_0, s_0). \tag{3.119}$$

However, by the definition of g_h, we have that

$$g_h(s_0) = NOT(f_h(s_0, s_0)). \tag{3.120}$$

Since this cannot be, our assumption that $h(s_0) = C_h$ is wrong, and there is a subset of S that is not in the image of h.

Let us look at Cantor's inequality from a matrix/array point of view. We write the collection S as $\{s_1, s_2, s_3, \ldots\}$. Function $f_h\colon S \times S \longrightarrow 2$ can be described as the matrix in Figure 3.6. Notice that the diagonal is different with every column of the array. This is a way of saying that the diagonal (which is g_h) cannot be represented by any column of the array.

This is part of mathematics, and the only resolution for this paradox is to accept the fact that no such surjective h exists and $|S| < |\mathcal{P}(S)|$. Notice that this applies to any set. For finite S, this is obvious since $|S| = n$ implies $|\mathcal{P}(S)| = 2^n$. However, this is true for infinite S as well. What this shows is that $\mathcal{P}(S)$ is a different level of infinity than S. One can iterate this process and get $\mathcal{P}(S)$, $\mathcal{P}(\mathcal{P}(S))$, $\mathcal{P}(\mathcal{P}(\mathcal{P}(S)))$ ... each of which is at a different level of infinity.

	f_h	Subset of S $h(s_1)$	$h(s_2)$	$h(s_3)$	$h(s_4)$	$h(s_5)$	$\cdots$
Elements of S	s_1	NOT(1)=0	0	0	0	1	$\cdots$
	s_2	0	NOT(0)=1	0	1	0	$\cdots$
	s_3	0	0	NOT(1)=0	0	0	$\cdots$
	s_4	1	0	0	NOT(1)=0	0	$\cdots$
	s_5	0	1	0	1	NOT(0)=1	$\cdots$
	$\vdots$	$\vdots$			$\vdots$		$\ddots$

Figure 3.6. Function f_h as a matrix. Function g_h represents C_h and is the changed diagonal. It is different for every column.

Related to the Cantor's inequality discussed here is the theorem that the set of natural numbers $\mathbf{N} = \{0, 1, 2, 3, \ldots\}$ is smaller than the interval of all real numbers between 0 and 1 (i.e., $(0, 1) \subseteq \mathbf{R}$). The infinity that corresponds to the natural numbers is called **countable infinity**, while any larger infinity—such as the interval of real numbers— is called **uncountable infinity**.

This proof will be slightly different than the previous examples that we have seen. We include it because it has features that are closer to the upcoming general theorem. Rather than working with the set $2 = \{0, 1\}$, this proof works with the set $10 = \{0, 1, 2, 3, \ldots, 9\}$. Also, rather than working with the function $NOT\colon 2 \longrightarrow 2$, which swaps both elements of 2, we now work with the function $\alpha\colon 10 \longrightarrow 10$, which is defined as follows:

$$\alpha(0) = 1, \quad \alpha(1) = 2, \quad \alpha(2) = 3, \quad \ldots, \quad \alpha(8) = 9, \quad \alpha(9) = 0; \tag{3.121}$$

that is, $\alpha(n) = n + 1 \ \ Mod \ \ 10$. The most important feature of α is that every output is different than its input. There are many such functions from 10 to 10. We choose this one "although we could have chosen any of these for our purposes."

The proof that $\mathbf{N}$ is smaller than $(0, 1)$ works by showing every function $h\colon \mathbf{N} \longrightarrow (0, 1)$ defines a real number in $(0, 1)$ that is not in the image of h. We can think of this again as a proof by contradiction. We assume (wrongly) that there is a surjection $h\colon \mathbf{N} \longrightarrow (0, 1)$ and come to a contradiction, which proves that no such h can possibly exist.

With such an h, we can define a function $f_h\colon \mathbf{N} \times \mathbf{N} \longrightarrow 10$ that describes the decimal expansions of the real numbers that h describes. For $m, n \in \mathbf{N}$,

$$f_h(m, n) = \text{the } m\text{th digit of } h(n). \tag{3.122}$$

This means that f_h gives every output digit of the purported function h. Figure 3.7 will help explain f_h. The natural numbers on the left tell you the position. The function h assigns to every natural number on the top a real number below it. The numbers on the left are the first inputs to f_h, and the numbers on the top are the second inputs to f_h.

With such an f_h, one can go on to describe function g_h, with the—by now familiar—construction

		The numbers $h(n)$							
	f_h	0	1	2	3	4	5	6	⋯
		0	0	0	0	0	0	0	⋯
		.	.	.	.	.	.	.	⋯
Position	0	0	0	7	2	7	7	4	⋯
	1	0	1	2	2	7	6	7	⋯
	2	0	3	0	3	0	0	0	⋯
	3	0	6	2	0	1	2	0	⋯
	4	0	1	0	2	3	1	3	⋯
	5	0	1	0	3	0	1	5	⋯
	⋮	⋮			⋮			⋮	⋱

Figure 3.7. Function $f_h\colon \mathbf{N}\times\mathbf{N} \longrightarrow 10$, which describes a purported surjection h from $\mathbf{N}$ to $(0, 1)$.

		The number $h(n)$							
	f_h	0	1	2	3	4	5	6	⋯
		0	0	0	0	0	0	0	⋯
		.	.	.	.	.	.	.	⋯
Position	0	$\alpha(0)=1$	0	7	2	7	7	4	⋯
	1	0	$\alpha(1)=2$	2	2	7	6	7	⋯
	2	0	3	$\alpha(0)=1$	3	0	0	0	⋯
	3	0	6	2	$\alpha(0)=1$	1	2	0	⋯
	4	0	1	0	2	$\alpha(3)=4$	1	3	⋯
	5	0	1	0	3	0	$\alpha(1)=2$	5	⋯
	⋮	⋮			⋮			⋮	⋱

Figure 3.8. The changed diagonal, g_h, of a purported surjection h from $\mathbf{N}$ to $(0, 1)$.

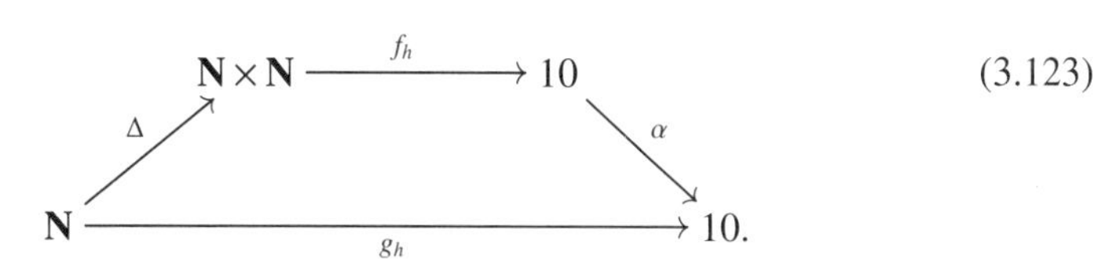

(3.123)

Function g_h also depends on h. Figure 3.8 will help explain the function g_h. It is the same as Figure 3.7, but with the elements along the diagonal changed by α. That is, the nth digit of the nth number is changed. The changed numbers are the outputs to function g_h. Thinking of the outputs of g_h as the digits of a real number, we are describing a real number between 0 and 1. We call this number G_h. In our case,

$$G_h = 0.121142\ldots \tag{3.124}$$

The claim is that g_h is not represented by f_h. This means that the number represented by g_h will not be the number represented by $f_h(\ \ , n_0)$ for any n_0. Another way to say this is that the number G_h will not be expressed in any column in the scheme described by Figure 3.7. This is obviously true because G_h was formed to be different from the first column because the number in the first position is different. It is different from the second column because it was formed to be different in the second position. It is different from the third column because it was formed to be different at the third digit, etc. In terms of self-reference, G_h is a real number that says

"This number is not the number in column n because the nth digit is different from the nth column's nth digit."

or

"This number is not in the image of h."

Conclusion: G_h is not on our list, and hence h is not surjective.

Let us show the end of the proof formally. If there were some n_0 that represented g_h, then for all m,

$$g_h(m) = f_h(m, n_0) \tag{3.125}$$

(i.e., G_h is the same as column n_o). But if this were true for all m, then it would be true for n_0 as well (that is, it would be true by every digit, including the one on the diagonal.) But that says that

$$g_h(n_0) = f_h(n_0, n_0). \tag{3.126}$$

However, g_h was defined for n_0 as

$$g_h(n_0) = \alpha(f_h(n_0, n_0)). \tag{3.127}$$

We conclude that no such n_0 exists, and g_h describes a number in (0, 1) that is not in the image of h. That is, h cannot be surjective and the set **N** is smaller than the set (0, 1).

The Main Theorem about Self-Referential Paradoxes

We have seen the same idea over and over in many different contexts. All these examples are instances of a single theorem of category theory. This again shows the unifying power and versatility of category theory.

We should note that all our examples up to now have been about sets and set maps. However, we will see that there are instances of the same phenomena in other categories. It pays to describe the theorem in its most general setting.

We begin with some needed preliminaries.

Definition 3.4.1. First, here is a simple definition in $\mathbb{S}\mathrm{et}$. Consider set Y and a set function $\alpha\colon Y \longrightarrow Y$. We call $s_0 \in Y$ a **fixed point** of α if $\alpha(s_0) = s_0$. That is, the output is the same as the input (i.e., the input is fixed by the function). We write the element s_0 by talking about a function $p\colon \{*\} \longrightarrow Y$ such that $p(*) = s_0$. Remember that $\{*\}$ is the terminal object in $\mathbb{S}\mathrm{et}$ and helps pick out elements of sets. Saying that s_0 is a fixed point of α amounts to saying that $\alpha \circ p = p$; that is, the following diagram commutes:

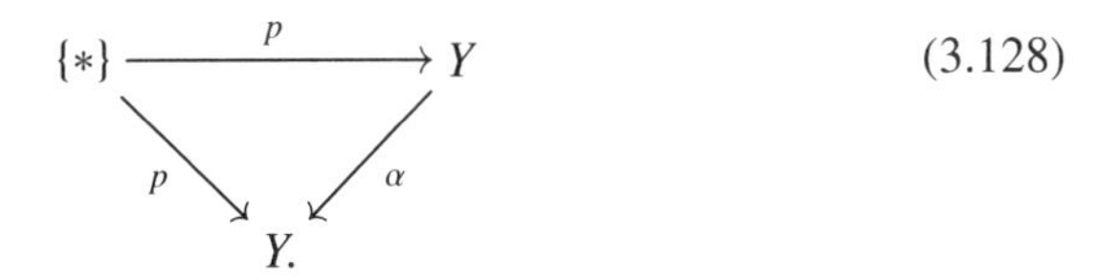

(3.128)

Let us generalize this to any category $\mathbb{A}$ with a terminal object 1. Let y be an object in $\mathbb{A}$ and $\alpha\colon y \longrightarrow y$ be a morphism in $\mathbb{A}$. Then we say that $p\colon 1 \longrightarrow y$ is a fixed point of α if $\alpha \circ p = p$.

Definition 3.4.2. Let us generalize from $\mathbb{S}\mathrm{et}$ to all categories the definition of one function representing another function. Remember that if $f\colon S \times S \longrightarrow Y$ is a set function and s_0 is an element in S, then $f(\ \ , s_0)\colon S \longrightarrow Y$ is a function of one input. We say that $g\colon S \longrightarrow Y$ is **represented** by f if there is an s_0 in S such that $g(\ ) = f(\ \ , s_0)$. In other words, for every x in S, $g(x) = f(x, s_0)$. What does it mean for $g\colon S \longrightarrow Y$ to *not* be represented by f? It means that for all s in S, $g(\ ) \neq f(\ \ , s)$. In detail, for all $s \in S$, there is some x in S such that $g(x) \neq f(x, s)$.

Let us generalize representability to any category $\mathbb{A}$ with a terminal object 1 and binary products. We need the isomorphism $i\colon a \longrightarrow a \times 1$. Let $f\colon a \times a \longrightarrow y$ and $g\colon a \longrightarrow y$ be morphisms in $\mathbb{A}$. Then g is representable by f if there is a morphism, $p\colon 1 \longrightarrow a$, such that

$$g = f \circ (id_a \times p) \circ i\colon a \longrightarrow y. \tag{3.129}$$

We can see this as

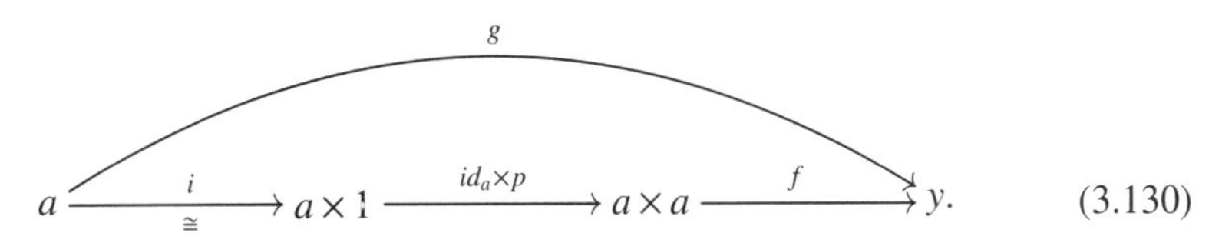

(3.130)

Function g is *not* representable by f if for all $p\colon 1 \longrightarrow a$, $g \neq f \circ (id_a \times p) \circ i$.

Now we will look at the main theorem as given by Lawvere (1969) [162].

Theorem 3.4.3. Cantor's theorem. Let $\mathbb{A}$ be a category with a terminal object and binary products. Let y be an object in the category and $\alpha\colon y \longrightarrow y$ be a morphism in the category. If α does not have a fixed point, then for all objects a and for all $f\colon a \times a \longrightarrow y$, there is a $g\colon a \longrightarrow y$ such that g is not representable by f. ★

Proof. Let $\alpha\colon y \longrightarrow y$ not have a fixed point. Then for any a and for any $f\colon a \times a \longrightarrow y$, we can compose f with Δ and α to form g as follows:

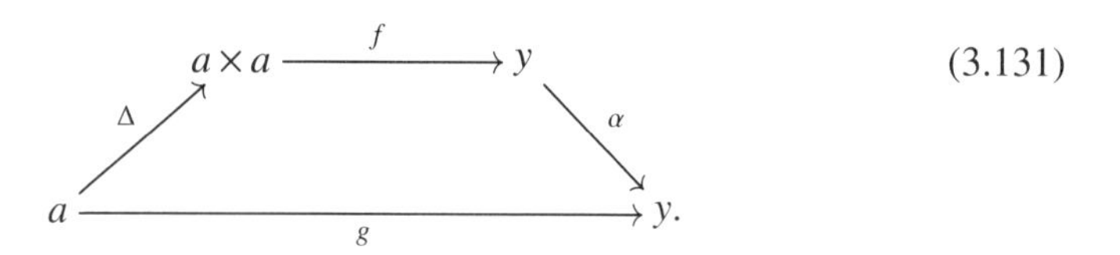

(3.131)

We claim that g is not representable by f. Assume (wrongly) that g is represented by f with $p\colon 1 \longrightarrow a$. That would mean that $g = f \circ (id_a \times p) \circ i$; that is, for all $q\colon 1 \longrightarrow a$,

$$g \circ q = f \circ (id_a \times p) \circ i \circ q. \tag{3.132}$$

That gives us

$$\begin{aligned}
g \circ p &= f \circ (id_a \times p) \circ i \circ p && \text{by substituting } q = p \\
&= f \circ \langle id_a \circ \pi_a, p \circ \pi_1 \rangle \circ i \circ p && \text{by definition of } \times \\
&= f \circ \langle id_a \circ \pi_a \circ i \circ p, p \circ \pi_1 \circ i \circ p \rangle && \text{by Theorem 3.1.14} \\
&= f \circ \langle p, p \circ \pi_1 \circ i \circ p \rangle && \text{because } \pi_a \circ i = id_a \\
&= f \circ \langle p, p \rangle && \text{bec. 1 is terminal and } \pi_1 i p = id_1 \\
&= f \circ \langle id_a, id_a \rangle \circ p && \text{by Theorem 3.1.14} \\
&= f \circ \Delta \circ p && \text{by definition of } \Delta.
\end{aligned}$$

But by the definition of g, we have

$$g \circ p = \alpha \circ f \circ \Delta \circ p. \tag{3.133}$$

These two equations imply that α has a fixed point:

$$(f \circ \Delta \circ p) = g \circ p = \alpha \circ (f \circ \Delta \circ p). \tag{3.134}$$

The fixed point for α is the map $(f \circ \Delta \circ p)$. Since α is assumed not to have any fixed point, the assumption that g is representable is wrong. ♣

Turing's Halting Problem

In the early 1930s, long before engineers actually created computers, Alan Turing, the "father of computer science," showed what computers *cannot* do. Loosely speaking, researchers[1] proved that no program can decide whether any program will go into an infinite loop. From this inexact statement, one already can see the self-reference: programs deciding the properties of programs.

Let us state a more exact version of Turing's theorem. First, here are some preliminaries. Programs come in many different forms. Here, we are concerned with programs that accept only a single natural number as input. For every such program, there is

[1]Most writers attribute the halting problem to Alan Turing. This is historically not true. He was the first to prove that certain problems were not decidable by machines, but he did not mention or prove the undecidability of the halting problem. This was originally done by Martin Davis. See more about this in Cristian Calude's book [49].

	Halt	**Program** 0	1	2	3	4	5	⋯
Input	0	0	1	0	0	0	1	⋯
	1	1	1	1	1	1	1	⋯
	2	0	1	0	0	0	0	⋯
	3	0	1	0	0	1	0	⋯
	4	0	0	1	1	1	1	⋯
	5	1	0	1	0	1	1	⋯
	⋮			⋮		⋮		⋱

Figure 3.9. A purported halting function.

a unique natural number that describes that program. To see this, realize that all computer programs are stored as a binary string. Every binary string can be seen as a natural number. We call the program associated with the natural number n "program n." This idea—that programs that act on numbers can be represented by numbers—shows that programs can be self-referential.

Programs that accept a single number can halt or they can go into an infinite loop. The **halting problem** asks for a program to accept a program and a number and determine if that program with that number will halt or go into an infinite loop. To be more exact, the halting problem asks for two numbers: (i) a number of a program that accepts a single number and (ii) an input to that program. Turing's theorem says that no such program can possibly exist. This is not a limitation of modern technology or of our current ability. Rather, this is an inherent limitation of computation.

The proof is, once again, a proof by contradiction. Assume (wrongly) that there is a program that accepts a program number and an input and can determine if that program will halt or go into an infinite loop when that number is entered into that program. Formally, such a program will describe a total computable function (i.e., a morphism in $\mathbb{CompFunc}$). The function named $Halt\colon \mathbf{N}\times\mathbf{N} \longrightarrow Bool$ defined on natural numbers $m, n \in \mathbf{N}$ is

$$Halt(m,n) = \begin{cases} 1 & : \text{if input } m \text{ into program } n \text{ halts} \\ 0 & : \text{if input } m \text{ into program } n \text{ goes into an infinite loop.} \end{cases} \tag{3.135}$$

This function can be illustrated by Figure 3.9, where the natural numbers on the top are the number of the programs and the natural numbers on the left are the input numbers. The numbers in the chart tell the value of *Halt*.

It is not hard to see that the function $\Delta\colon \mathbf{N} \longrightarrow \mathbf{N}\times\mathbf{N}$, defined as $\Delta(n) = (n,n)$, is a computable function. That is, one can program a computer to accept a single number and output the pair of the same numbers. Consider the partial NOT function $ParNOT\colon Bool \longrightarrow Bool$, defined as follows:

$$ParNOT(n) = \begin{cases} 1 & : \text{if } n = 0 \\ \uparrow & : \text{if } n = 1 \end{cases}, \tag{3.136}$$

	Program						
Halt	0	1	2	3	4	5	$\cdots$
Input 0	$\alpha(0)=1$	1	0	0	0	1	$\cdots$
1	1	$\alpha(1)=\uparrow$	1	1	1	1	$\cdots$
2	0	1	$\alpha(0)=1$	0	0	0	$\cdots$
3	0	1	0	$\alpha(0)=1$	1	0	$\cdots$
4	0	0	1	1	$\alpha(1)=\uparrow$	1	$\cdots$
5	1	0	1	0	1	$\alpha(1)=\uparrow$	$\cdots$
$\vdots$			$\vdots$		$\vdots$		$\ddots$

Figure 3.10. The changed diagonal of the purported halting function, where $\alpha = ParNOT$.

where $\uparrow$ means that it will go into an infinite loop. *ParNOT* is also a computable function. Since *Halt* is assumed to be computable, and the functions Δ and *ParNOT* are computable, then their composition as follows is also a computable function:

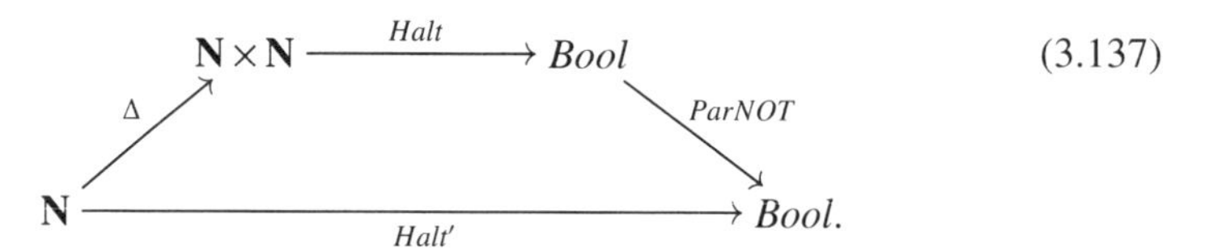

(3.137)

It is important to stress that this diagram is in $\mathbb{CompFunc}$, not in $\mathbb{Set}$. The new computable function, *Halt′*, accepts a number n as input and does the opposite of what the program n on input n does. That is, if the program n on input n halts, then $Halt'(n)$ will go into an infinite loop. Otherwise, if the program n on input n goes into an infinite loop, then $Halt'(n)$ will halt. In terms of self-reference, *Halt′* describes a program that does the opposite of what program n is supposed to do on input n. We can see the way that *Halt′* is defined in Figure 3.10.

Since *Halt′* is a computable function, the program for this computable function must have a number and be somewhere on our list of computable functions. However, it is not. *Halt′* was formed to be different from every column in the chart. What is wrong? We know that Δ and *ParNOT* are computable. We assumed that *Halt* was computable. It must be that our assumption about the *Halt* function was wrong, and *Halt* is not computable.

Let us formally show that *Halt′* is different from every column in the chart. Imagine that *Halt′* is computable and the number of *Halt′* is n_0. This means that *Halt′* is the n_0 column of our chart. Another way to say this is that *Halt′* is representable by $Halt(\ \ , n_0)$; that is, for all n,

$$Halt'(n) = Halt(n, n_0). \tag{3.138}$$

Now let us ask about $Halt'(n_0)$. We get

$$Halt'(n_0) = Halt(n_0, n_0). \tag{3.139}$$

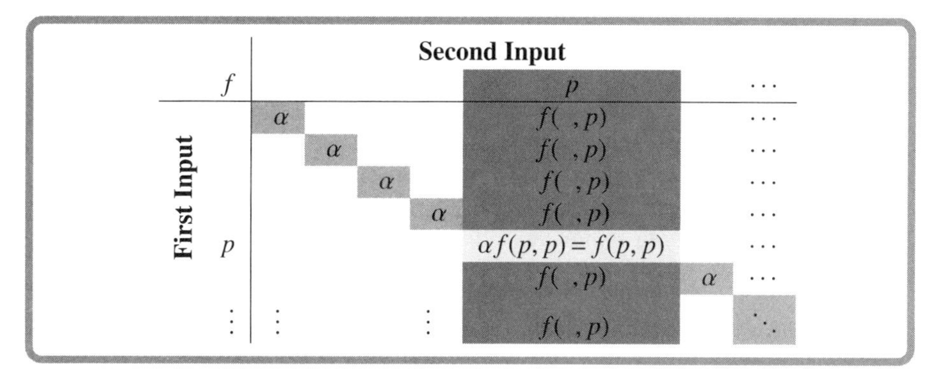

Figure 3.11. A highlighted column and diagonal. Their crossing point is a fixed point.

But we defined $Halt'(n_0)$ to be

$$Halt'(n_0) = ParNOT(Halt(n_0, n_0)), \tag{3.140}$$

so we have a contradiction.

In a sense, we can say that the computational task that $Halt'$ (and, in particular, $Halt'(n_0)$) performs is

> "If you ask whether this program will halt or go into an infinite loop, then this program will give the wrong answer."

Since computers cannot give the wrong answer, $Halt'$ cannot exist, so $Halt$ cannot exist either.

The Contrapositive of the Main Theorem about Self-Referential Paradoxes

Cantor's theorem is very important, and we need to express it in many equivalent ways. Let us consider the contrapositive of Cantor's theorem. In less technical terms, the matrix form of Cantor's theorem says:

- If α does not have a fixed point, then the diagonal—which is changed by α—is different from every column in the matrix.

The contrapositive then says:

- If α forms a changed diagonal that is the same as some column, then the point where the diagonal and the column meet will be a fixed point of α.

The intuition for the contrapositive can be viewed in Figure 3.11. Function g uses α and forms the changed diagonal of the matrix. The fact that g is representable by some column, say p, means that they are two ways of talking about the same thing: as a diagonal and as a column. At the crossing point, there is a fixed point where $\alpha(f(p,p)) = f(p,p)$.

With this intuition in hand, let us state the contrapositive.

Theorem 3.4.4. Let $\mathbb{A}$ be a category with a terminal object and binary products. Let y be an object in the category and $\alpha\colon y \longrightarrow y$ be a morphism in the category. If there is an object a and a morphism $f\colon a \times a \longrightarrow y$, such that $g = \alpha \circ f \circ \Delta$ is representable by f, then α has a fixed point. ★

Proof. Let $\mathbb{A}$, a, and f be as described in Theorem 3.4.4. Let $g = \alpha \circ f \circ \Delta$ be representable by $p\colon 1 \longrightarrow a$; that is, $g = f \circ (id_a \times p) \circ i$ (where i is the isomorphism $a \longrightarrow a \times 1$). The following commutative diagram is helpful:

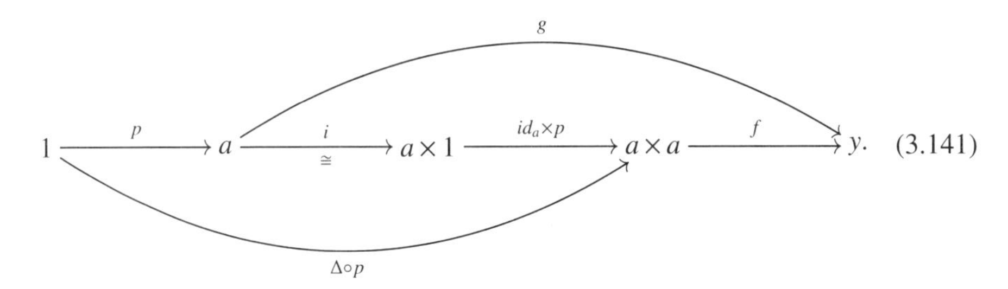

$$1 \xrightarrow{p} a \xrightarrow[\cong]{i} a \times 1 \xrightarrow{id_a \times p} a \times a \xrightarrow{f} y. \tag{3.141}$$

Since g is represented by f and defined as $g = \alpha \circ f \circ \Delta$, we have both equalities:

$$f \circ (id_a \times p) \circ i = g = \alpha \circ f \circ \Delta. \tag{3.142}$$

Precomposing all three maps by p (i.e., plugging p into the equations) gives us

$$f \circ (id_a \times p) \circ i \circ p = g \circ p = \alpha \circ (f \circ \Delta \circ p). \tag{3.143}$$

The left side shortens to

$$(f \circ \Delta \circ p) = g \circ p = \alpha \circ (f \circ \Delta \circ p). \tag{3.144}$$

And here, we see that $f \circ \Delta \circ p$ is a fixed point of α. ♣

Fixed Points in Logic

Now let us apply the contrapositive of Cantor's theorem to find fixed points in logic. Warning: this part will not be easy to understand and might require several readings. Some familiarity with basic logic will be beneficial. Remember Brooklyn College's motto: *Nil sine magno labore* ("Nothing without great effort").

First, here's some elementary logic. We are working in a system that can handle basic arithmetic. We will deal with logical formulas that accept at most one value, which is a number. A logical formula that accepts one value will be called a *predicate* and will be written as $\mathcal{A}(x)$, $\mathcal{B}(x)$, $C(x)$, etc. A logical formula that accepts no value (and hence is true or false) will be called a *sentence* and will be written as A, B, C, etc. Rather than considering the set of all predicates and the set of all sentences, we will be interested in equivalence classes of these sets. We take the sets of formulas and make an equivalence relation: two formulas are equivalent if they are provably logically equivalent. This means that there is a logical proof from one to the other.

When such a proof exists, we write $\mathcal{A}(x) \equiv \mathcal{B}(x)$ for predicates. We can generate a similar equivalence relations for sentences. When two sentences are related, we write $A \equiv B$. We will call the equivalence classes of predicates $Lind_1$ for the **Lindenbaum classes** of predicates. The equivalence classes of sentences will be denoted as $Lind_0$.

Since logical formulas are made of a finite string of symbols (like programs), all logical formulas can be encoded as natural numbers. We will write the natural number of a logical formula $\mathcal{A}(x)$ as $\ulcorner \mathcal{A}(x) \urcorner$ and the natural number of a sentence A as $\ulcorner A \urcorner$. By assigning a number to each formula, formulas can be made inputs to formulas. That means that logical formulas about numbers can then *evaluate* logical formulas about numbers. It is these numbers that will help logical formulas about numbers be self-referential.

Now we are going to get fixed points of logical predicates.

Theorem 3.4.5. For every logical predicate that takes a number $\mathcal{E}(x)$, there is a way of constructing a **fixed point**, which is a logical sentence C such that

$$\mathcal{E}(\ulcorner C \urcorner) \equiv C. \tag{3.145}$$

The process that goes from a $\mathcal{E}(x)$ to C is called a **fixed-point machine**. In a sense, C is a logical sentence that says,

"This logical statement has property $\mathcal{E}$." ★

With this fixed-point machine, we will find some of the most fascinating aspects of logic. On the first reading, you might want to skip the proof and get to where we apply the fixed-point machine to get Gödel's Incompleteness Theorem.

Proof. The proof uses the contrapositive of the main theorem about self-referential paradoxes. The category that we are working in is $\mathbb{Set}$. The a of the theorem is the set of equivalence classes $Lind_1$. The y of the theorem is the set of equivalence classes $Lind_0$. There is a function $f\colon Lind_1 \times Lind_1 \longrightarrow Lind_0$, defined as follows:

$$f(\mathcal{A}(x), \mathcal{B}(x)) = \mathcal{B}(\ulcorner \mathcal{A}(x) \urcorner). \tag{3.146}$$

That is, f applies the number of one predicate to another predicate. The function f determines[2] the truth value of the predicate $\mathcal{B}$ on the number corresponding to the $\mathcal{A}(x)$. Notice that at the end, we get out a sentence (with no inputs).

Let us talk about the function $\Delta\colon Lind_1 \longrightarrow Lind_1 \times Lind_1$. This function performs the following operation: $\mathcal{A} \longmapsto (\mathcal{A}, \mathcal{A})$.

The α of the theorem depends on some predicate $\mathcal{E}$, so we write it as $\alpha_{\mathcal{E}}$. Function $\alpha_{\mathcal{E}}$ applies the predicate to the number of a sentence A. It is a function $\alpha_{\mathcal{E}}\colon Lind_0 \longrightarrow Lind_0$, which is defined for sentence A as

$$\alpha_{\mathcal{E}}(A) = \mathcal{E}(\ulcorner A \urcorner). \tag{3.147}$$

---

[2]It remains to be shown that if $\mathcal{A}(x) \equiv \mathcal{A}'(x)$ and $\mathcal{B}(x) \equiv \mathcal{B}'(x)$, then

$$f(\mathcal{A}(x), \mathcal{B}(x)) = \mathcal{B}(\ulcorner \mathcal{A}(x) \urcorner) \equiv \mathcal{B}'(\ulcorner \mathcal{A}'(x) \urcorner) = f(\mathcal{A}'(x), \mathcal{B}'(x)).$$

This is dealt with in different ways in logic textbooks.

We have all the ingredients of the theorem, so let's use it. For every $\mathcal{A}(x)$, we can compose f with Δ and $\alpha_{\mathcal{E}}$ as follows:

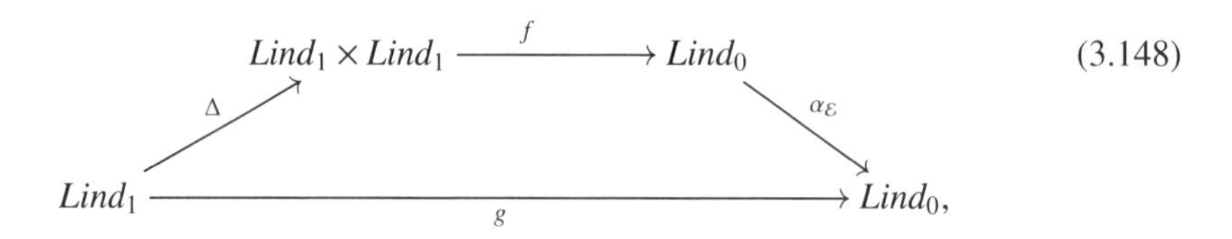

(3.148)

Function g is defined for $\mathcal{A}(x)$ as

$$\begin{aligned} g(\mathcal{A}(x)) &= (\alpha_{\mathcal{E}} \circ f \circ \Delta)(\mathcal{A}(x)) && \text{by definition of } g \\ &= \alpha_{\mathcal{E}}(f(\mathcal{A}(x), \mathcal{A}(x))) && \text{by definition of } \Delta \\ &= \alpha_{\mathcal{E}}(\mathcal{A}(\ulcorner \mathcal{A}(x) \urcorner)) && \text{by definition of } f \\ &= \mathcal{E}(\ulcorner \mathcal{A}(\ulcorner \mathcal{A}(x) \urcorner) \urcorner) && \text{by definition of } \alpha_{\mathcal{E}}. \end{aligned}$$

As always, g is the diagonal.

For the contrapositive of the theorem to be relevant, we must show that g is representable (i.e., that the diagonal is the same as some column). This means that we have to find a mathematical statement $\mathcal{G}(x)$ such that it describes the statements on the diagonal. That is, for any predicate $\mathcal{A}(x)$, we have $g(\mathcal{A}(x)) = f(\mathcal{A}(x), \mathcal{G}(x))$. To form $\mathcal{G}(x)$, we have to ensure that (i) it talks about itself (as all predicates on the diagonal do) and (ii) we replace it with $\mathcal{E}(x)$ (because this is a changed diagonal). To deal with (i), there is a helpful "doubling" function (or "diagonal" function), $D\colon \mathbf{N} \longrightarrow \mathbf{N}$, which is defined for all predicates $\mathcal{A}(x)$ as

$$D(\ulcorner \mathcal{A}(x) \urcorner) = \ulcorner \mathcal{A}(\ulcorner \mathcal{A}(x) \urcorner) \urcorner; \tag{3.149}$$

that is, D takes a number of a predicate and outputs the number of the sentence applied to its own number. Such a function can easily be computed by a program. This will be very important for describing predicates on the diagonal. To deal with (ii), consider

$$\mathcal{G}(x) = \mathcal{E}(D(x)). \tag{3.150}$$

This is perfect to deal with the changed diagonal.

Function g is representable by $\mathcal{G}(x) = \mathcal{E}(D(x))$ as follows:

$$\begin{aligned} g(\mathcal{A}(x)) &= \mathcal{E}(\ulcorner \mathcal{A}(\ulcorner \mathcal{A}(x) \urcorner) \urcorner) && \text{from the definition of } g \\ &= \mathcal{E}(D(\ulcorner \mathcal{A}(x) \urcorner)) && \text{by definition of } D \\ &= \mathcal{G}(\ulcorner \mathcal{A}(x) \urcorner) && \text{by definition of } \mathcal{G}(x) \\ &= f(\mathcal{A}(x), \mathcal{G}(x)) && \text{by definition of } f. \end{aligned}$$

Since this function is representable, there is a fixed point of $\alpha_{\mathcal{E}}$. This is exactly where the diagonal meets the column $\mathcal{G}(x)$. Explicitly, the fixed point is $\mathcal{G}(\ulcorner \mathcal{G}(x) \urcorner)$. This is an easy consequence of g's definition and the fact that g is representable.

				Second Input			
	f	$\mathcal{A}_0(x)$	$\mathcal{A}_1(x)$	$\mathcal{A}_2(x)$	$\cdots$	$\mathcal{A}_p(x)$	$\cdots$
First Input	$\mathcal{A}_0(x)$	$\mathcal{E}(\mathcal{A}_0(\ulcorner\mathcal{A}_0(x)\urcorner)$	$\mathcal{A}_1(\ulcorner\mathcal{A}_0(x)\urcorner)$	$\mathcal{A}_2(\ulcorner\mathcal{A}_0(x)\urcorner)$	$\cdots$	$\mathcal{A}_p(\ulcorner\mathcal{A}_0(x)\urcorner)$	$\cdots$
	$\mathcal{A}_1(x)$	$\mathcal{A}_0(\ulcorner\mathcal{A}_1(x)\urcorner)$	$\mathcal{E}(\mathcal{A}_1(\ulcorner\mathcal{A}_1(x)\urcorner)$	$\mathcal{A}_2(\ulcorner\mathcal{A}_1(x)\urcorner)$	$\cdots$	$\mathcal{A}_p(\ulcorner\mathcal{A}_1(x)\urcorner)$	$\cdots$
	$\mathcal{A}_2(x)$	$\mathcal{A}_0(\ulcorner\mathcal{A}_2(x)\urcorner)$	$\mathcal{A}_1(\ulcorner\mathcal{A}_2(x)\urcorner)$	$\mathcal{E}(\mathcal{A}_2(\ulcorner\mathcal{A}_2(x)\urcorner)$	$\cdots$	$\mathcal{A}_p(\ulcorner\mathcal{A}_2(x)\urcorner)$	$\cdots$
	$\vdots$	$\vdots$	$\vdots$	$\vdots$	$\ddots$	$\vdots$	$\cdots$
	$\mathcal{A}_p(x)$	$\mathcal{A}_0(\ulcorner\mathcal{A}_p(x)\urcorner)$	$\mathcal{A}_1(\ulcorner\mathcal{A}_p(x)\urcorner)$	$\mathcal{A}_2(\ulcorner\mathcal{A}_p(x)\urcorner)$	$\cdots$	$\mathcal{E}(\mathcal{A}_p(\ulcorner\mathcal{A}_p(x)\urcorner)$	$\cdots$
	$\vdots$	$\vdots$		$\vdots$		$\vdots$	$\ddots$

Figure 3.12. The chart of function f and the changed diagonal g. Column p is also highlighted. Since the diagonal will be the same as the column, they will meet at the crossing point, which will be the fixed point $\mathcal{E}(\mathcal{A}_p(\ulcorner\mathcal{A}_p(x)\urcorner)) = \mathcal{A}_p(\ulcorner\mathcal{A}_p(x)\urcorner)$.

In detail:

$$\begin{aligned}
\mathcal{E}(\mathcal{G}(\ulcorner\mathcal{G}(x)\urcorner) &= \alpha_{\mathcal{E}}(\mathcal{G}(\ulcorner\mathcal{G}(x)\urcorner)) && \text{by definition of } \alpha_{\mathcal{E}}\\
&= \alpha_{\mathcal{E}}(f(\mathcal{G}(x), \mathcal{G}(x))) && \text{by definition of } f\\
&= g(\mathcal{G}(x)) && \text{by definition of } g\\
&= f(\mathcal{G}(x), \mathcal{G}(x)) && \text{by the representability of } g\\
&= \mathcal{G}(\ulcorner\mathcal{G}(x)\urcorner) && \text{by definition of } f.
\end{aligned}$$

We can understand this better by looking at the chart in Figure 3.12, which is similar to the charts in Figure 3.8 and Figure 3.10. Assume that all the objects of $Lind_1$ are listed as

$$\mathcal{A}_0(x), \mathcal{A}_1(x), \mathcal{A}_2(x), \mathcal{A}_3(x), \mathcal{A}_4(x), \ldots. \tag{3.151}$$

Then function f is described by the chart. Function g is defined by changing the diagonal. The new line formed just by giving the element of $Lind_1$ is $\mathcal{G}(x) = \mathcal{E}(D(x))$. Since g is representable, there must be some line that has this property. This gives us the fixed point. ♣

Gödel's Incompleteness Theorem

Now let us use this fixed-point machine for some interesting predicates $\mathcal{E}(x)$ and get self-referential statements. First, consider some ideas about logic. Not only can we assign a unique number to every predicate and to every sentence, we can also assign a unique number to every statement in logic. This follows from the fact that statements are sequences of symbols. Every symbol can be given a unique number and every sequence of numbers can be given a unique number as well. Similarly, we can assign a unique natural number to every proof. After all, a proof is a sequence of statements. The numbers assigned to statements and proof are called the **Gödel numbers** of those statements and proofs.

Let $\text{Prov}(x, y)$ be the two-place predicate that is true when "y is the Gödel number of a proof of a statement whose Gödel number is x." Now we form the statement

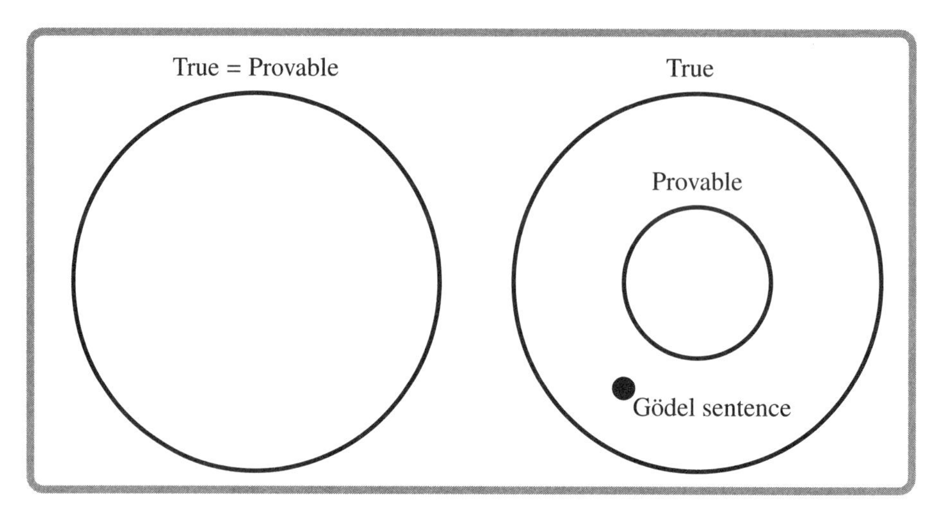

Figure 3.13. On the left: what was believed before Gödel's Theorem. On the right: what we know after Gödel's Theorem.

$$\mathcal{E}(x) = (\forall y)\neg \mathrm{Prov}(y, x). \tag{3.152}$$

This is true for x when no number y is a proof of statement x (i.e., this is true for statement x when no proof of x exists). With the fixed-point machine, we find a statement G (standing for Gödel) such that

$$G \equiv \mathcal{E}(\ulcorner G \urcorner) = (\forall y)\neg \mathrm{Prov}(y, \ulcorner G \urcorner). \tag{3.153}$$

G is a logical statement that essentially says,

"This statement is not provable for any proof y,"

or, succinctly,

"This statement is unprovable."

Let us assume that the logical system is sound (i.e., you cannot prove false statements). If G was false, then there would be a proof of G, and hence, there would be a proof of a false statement. In that case, the system is not sound. On the other hand, if G is true, then it essentially says that G is true but unprovable.[3]

This is one of the most important theorems in twentieth-century mathematics, and it is worthwhile to spend a few minutes meditating on its significance. Before Gödel came along, it was believed that any statement that was true was also provable. After Gödel, we can see that the set of provable statements is a proper subset of the set of statements that are true. There are statements that are true that cannot be proved within the particular system. Figure 3.13 depicts the distinction.

[3]We are ignoring technical details like ω-consistency and Rosser's theorem. To learn more details, see Section 3.5 of [195] and Chapter 17 of [39].

Tarski's Theorem

Alfred Tarski's theorem shows that a logical system cannot tell which of its predicates are true. Assume (wrongly) that there is some logical predicate $\mathcal{T}(x)$ that accepts a number and tells if the statement of that number is true. This formula will be true when "x is the Gödel number of a true statement in the theory." We can then use $\mathcal{T}(x)$ to form the statement

$$\mathcal{E}(x) = \neg\mathcal{T}(x). \tag{3.154}$$

This says that $\mathcal{E}(x)$ is true when $\mathcal{T}(x)$ is false. Now place $\mathcal{E}(x)$ into the fixed-point machine. We will get a statement C such that

$$C \equiv \mathcal{E}(\ulcorner C \urcorner) = \neg\mathcal{T}(\ulcorner C \urcorner). \tag{3.155}$$

The logical sentence C essentially says,

"This statement is false."

It is a logical version of the liar paradox (see page 136.) The logical sentence C is true if and only if it is false. A logical system cannot be consistent and have such a statement. The only assumption that we made was that $\mathcal{T}(x)$ can be formulated in the language of the system. It follows that this assumption is false. So while Gödel showed us that certain logical systems have limitations with respect to the notion of provability, Tarski showed us that those logical systems cannot deal with their own truthfulness.

Parikh Sentences

Rohit Parikh used the fixed-point machine to formulate some fascinating sentences that express properties about the length of proofs. Consider the two-place predicate Prflen(m, x), which is true if "there is a proof of length m (in symbols) of a statement whose Gödel number is x." A computer can actually decide if this is true or false because there are only a finite number of proofs of length m. Now consider the logical formula

$$\mathcal{E}_n(x) = \neg(\exists m < n \quad \text{Prflen}(m, x)). \tag{3.156}$$

The statement $\mathcal{E}_n(x)$ is true if the formula with Gödel number x does not have a proof whose length is less than n. With the fixed-point machine, we find a sentence C_n that satisfies

$$C_n \equiv \mathcal{E}_n(\ulcorner C_n \urcorner) = \neg(\exists m < n \quad \text{Prflen}(m, \ulcorner C_n \urcorner)). \tag{3.157}$$

The logical sentence C_n essentially says,

"This statement does not have a proof of length less than n."

So long as the logical system is sound, C_n will be true and will not have a proof of length less than n. This is interesting in itself, however, Parikh went further. He showed that although C_n does not have a short proof (you can make n as large as you want), there actually is a short proof of the fact that C_n is provable. With the predicate

$\mathcal{P}(x) = \exists y \text{Prov}(y, x)$, we will show that although there is no short proof of C_n, there is a short proof of $\mathcal{P}(\ulcorner C_n \urcorner)$. The short proof is basically a formalization of the following short argument:

1. If C_n does not have any proof, then C_n is true (i.e., $\neg\mathcal{P}(\ulcorner C_n \urcorner) \longrightarrow C_n$).
2. If C_n is true, we can check all proofs of length less then n and prove C_n (i.e., $C_n \longrightarrow \mathcal{P}(\ulcorner C_n \urcorner)$).
3. From arguments 1 and 2, we have shown that if C_n does not have a proof, then we can prove C_n (i.e., $\neg\mathcal{P}(\ulcorner C_n \urcorner) \longrightarrow \mathcal{P}(\ulcorner C_n \urcorner)$). (This is not a contradiction. It just shows that $\neg\mathcal{P}(\ulcorner C_n \urcorner)$ cannot be true.)
4. We conclude that there is a proof of C_n (i.e., $\mathcal{P}(\ulcorner C_n \urcorner)$).

Parikh went even further. He iterated $\mathcal{P}$ such that $\mathcal{P}^t(A) = \mathcal{P}(\ulcorner \mathcal{P}^{t-1}(A) \urcorner)$ and for every C, there is the following sequence:

$$\mathcal{P}^0(\ulcorner C \urcorner) = C, \quad \mathcal{P}^1(\ulcorner C \urcorner), \quad \mathcal{P}^2(\ulcorner C \urcorner), \quad \ldots, \quad \mathcal{P}^k(\ulcorner C \urcorner), \ldots. \tag{3.158}$$

Parikh then showed that for every k, there is a formula C_n^k that does not have a short proof, nor does the fact that it is provable have a short proof, nor does the fact that it is provable that it is provable ... have a short proof. However, eventually it will have a short proof. Expressed in symbols: There is no short proof of $\mathcal{P}^t(\ulcorner C_n^k \urcorner)$ for $t < k$, but there is a short proof of $\mathcal{P}^k(\ulcorner C_n^k \urcorner)$. This is just one of the many gems found in Parikh's papers.

Epimenides and the Liar

Before we close this mini-course, it pays to look at two famous paradoxes that are not exactly instances of Cantor's theorem but are close enough that they are easy to describe. The two examples are (i) the Epimenides paradox (which is related to the liar's paradox) and (ii) the time travel paradox.

Chronologically, the granddaddy of all the self-referential paradoxes is the **Epimenides paradox**. Epimenides (sixth or seventh century BC), a philosopher from Crete who lived in the sixth or seventh century BC, was a curmudgeon who did not like his neighbors in Crete. He is quoted as saying, "All Cretans are liars." The problem is that Epimenides himself is a Cretan. He is talking about himself and his statement. If his statement is true, then this very utterance is also a lie and hence is not true. On the other hand, if what he is saying is false, then he is not a liar and what he said is true. This seems to be a contradiction.

On deeper analysis, one sees that Epimenides's utterance actually is not a contradiction. Even if it is true that all Cretans are liars, that does not mean that every sentence that every Cretan ever made is false. A liar is someone who lied once, not necessarily someone who lies all the time. We have all lied, and hence we are all liars! So the statement could be true and it will not negatively affect his own statement. On the other hand, if the statement is false, that implies that there is at least one pious Cretan who always tells the truth all the time. Presumably, Epimenides thinks that he is that righteous truth-teller. The statement could be true or false and it will not be a contradiction either way.

Even though the paradox of Epimenides is flawed, there are other similar types of sentences that are paradoxical. Look at the following sentences:

- I always lie.
- This sentence is false.
- **The only sentence that is in boldface on this page is not true.**

These are all declarative statements that are true if and only if they are false—hence, they are contradictions. They are simple examples of paradoxical statements. There are many such statements and they are all instances of the **Liar Paradox**.

Let's back up a little and see what it going on here. Consider the statement "This sentence is false." It is an English sentence that refers to itself. Usually, a declarative sentence refers to some object. For our purpose, language gets interesting because a sentence has the ability to refer to itself.

Here are true sentences that refer to themselves:

- This sentence has five words.
- *This sentence is in italics.*
- This is an example of a sentence that shows no originality.

There are also false sentences that refer to themselves:

- This sentence has six words.
- This sentence is in italics.
- This is an example of a sentence that shows originality.
- The world will little note, nor long remember what we say here... (from Abraham Lincoln's Gettysburg Address—arguably one of the most famous speeches ever made)

The liar paradoxes are about sentences that go further. They are not false statements about themselves. Rather they are sentences that negate themselves.

Let's look at the self-reference of language in a formal way. There is a set of English sentences that we will call *Sent*. We can describe a function $f\colon Sent \times Sent \longrightarrow 2$. Function f is defined for sentences s and s' as

$$f(s,s') = \begin{cases} 1 & : s \text{ is negated by } s' \\ 0 & : s \text{ is not negated by } s'. \end{cases} \tag{3.159}$$

We are interested in sentences that do not negate other sentences but negate themselves. To deal with such sentences, we are going to compose f with function Δ and NOT as follows:

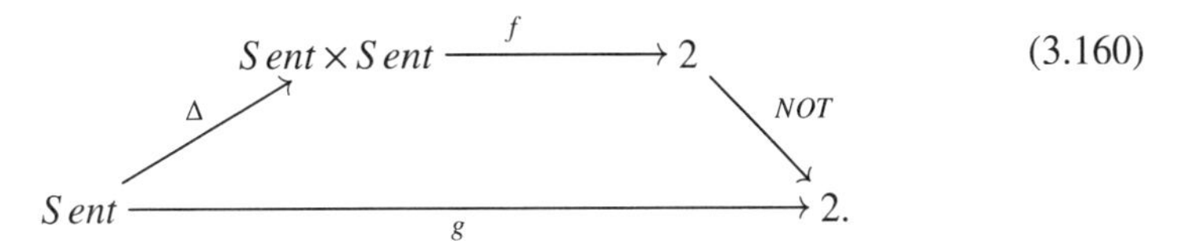

(3.160)

The value $g(S)$ is defined as $NOT(f(S,S))$. Function g is the characteristic function of the subset of sentences that negate themselves. Up to now, we have been mimicking the setup of Cantor's theorem. However, this is where the resemblance stops. It is not clear what we would mean by talking about g being representable by f. What would it mean for sentence S to represent a subset of sentences? Nevertheless, sentences that negate themselves are declarative sentences that are contradictions.

Time Travel Paradoxes

A significant amount of science fiction is about time travel paradoxes. If time travel were possible, a time traveler might go back in time and shoot his childless grandfather, guaranteeing that the time traveler was never born. Homicidal behavior is not necessary to achieve such paradoxical results, either. The time traveler might just make sure that his parents never meet (In Back to the Future, Marty McFly dealt with similar issues.), or he might simply go back in time and make sure that he does not enter the time machine. These actions would imply a contradiction and hence cannot happen. The time traveler should not shoot his own grandfather (moral reasons notwithstanding) because if he shoots his own grandfather, he will not exist and will not be able to travel back in time to shoot his own grandfather. So by performing an action, he is guaranteeing that the action cannot be performed. The event of shooting your own grandfather is self-referential. Usually, one event affects other events, but here an event affects itself. Since the physical universe does not permit contradictions, we must deny the assumption that time travel exists.

By now, it is easy to formalize this paradox. There is a collection, *Events*, of all physical events. There is also a function $f\colon Events \times Events \longrightarrow 2$, which is defined for two events e and e' as

$$f(e,e') = \begin{cases} 1 & : \text{if } e \text{ is negated by } e' \\ 0 & : \text{if } e \text{ is not negated by } e'. \end{cases} \tag{3.161}$$

Some events negate other events and some events do not. If e and e' are not in each other's space-time cone, then they will not affect each other[4]. We now move on to describe the subset of all physical events that negate themselves in the usual way:

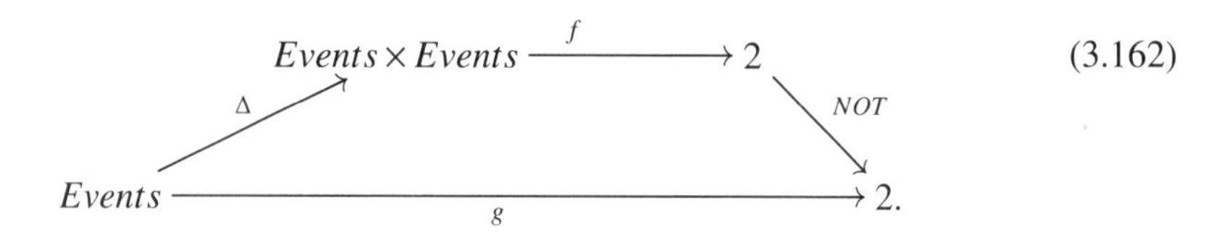

[4]We will look at exceptions to this rule when we discuss entanglement in our study of quantum mechanics in Section 10.2 of Chapter 10.

Function g is the characteristic function of those events that negate themselves. Such events cannot exist. Up to now, the pattern has been the same with Cantor's theorem. However, we do not go on to talk about representing g by f. What would it mean for $f(\ \ , e)$ to represent a subset of events? While time travel paradoxes do not necessarily fit into Cantor's theorem, there is a sense of self-reference here and the events described by g cannot possibly exist.

Is there a resolution to the time travel paradox? Since events are part of the physical world, the obvious resolution is that time travel is not a possibility. However, we need not be so drastic. We can be a little subtle. Einstein's theory of relativity tells us that time travel is not possible in the usual way that we think of the universe. However, Einstein's friend and neighbor, Kurt Gödel, wrote an interesting paper on relativity theory that constructed a model of the universe in which time travel would be possible. In this "Gödel universe," it would be very hard, but not impossible, for time travel to be a reality. Gödel, the greatest logician of the past 1,000 years, was aware of the logical problems of time travel. A writer, Rudy Rucker, tells of an interview with Gödel in which Rucker asks about the time-travel paradoxes [228]. Gödel responded by saying, "Time-travel is possible, but no person will ever manage to kill his past self . . . The *a priori* is greatly neglected. Logic is very powerful." That means that the universe simply will not allow you to kill your past self. Just as the barber paradox shows that certain villages with strict rules cannot exist, the physical universe will not allow you to perform an action that will cause a contradiction. This leads us to even more mind-blowing questions. What would happen if someone took a gun back in time to shoot an earlier version of himself? How will the universe stop him? Does he not have the free will to perform the dastardly deed? Will the gun fail to shoot? If the bullet fires and is properly aimed, will the bullet stop short of his body? It is indeed bewildering to live in a world with self-reference.

Suggestions for Further Study

The idea that category theory can describe all these diverse self-referential phenomena was brilliantly formulated by F. William Lawvere in a short paper published in 1969 [162]. Lawvere and Schanuel discuss these ideas in Session 29 of their book [166]. Lawvere and Rosebrugh also deal with these concepts in Section 7.3 of their book [165]. In 2003, I published a paper [270] whose goal was to make these ideas understandable for novices to category theory. The paper contains nineteen instances of Cantor's theorem and its contrapositive obtained from logic, mathematics, and theoretical computer science.

A larger discussion of all these self-referential paradoxes for the general reader can be found in [272]. See Chapter 10 of that book, and in particular page 344, for a unified listing of many self referential paradoxes.

The philosophical distinction between the physical world, the linguistic world, and the language of mathematics and science, (depicted in Figure 3.5) was first stated in [273] and [275].

Classical discussions about some of the instances of the theorem can be found here:

- Turing's halting problem: Section 5.3 of [195], Section 4.2 of [241], Chapter V of [185], Section 4.1 of [39] and Section 6.2 of my [272].
- Gödel's incompleteness theorem: Section 3.5 of [195], Section 6.2 of [241], Chapter VII of [185], Chapter 17 of [39], and Section 9.4 of my book [272].
- Tarski's theorem: Section 3.6 of [195] and Section 17.1 of [39].
- Parikh sentences: Parikh's fascinating article [203].

4

Relationships between Categories

... comme Grothendieck nous l'a appris, les objets d'une catégorie ne jouent pas un grand rôle, ce sont les morphismes qui sont essentiels.
... as Grothendieck taught us, the objects of a category do not play a great role, it is the morphisms that are essential.
Jean-Pierre Serre
[234], page 335.

Category theory is about relating different categories. We will see that there are many possible relationships between categories. In this chapter, we formally introduce functors between categories and natural transformations between functors. We then move on to employ these structures to relate categories in a myriad of ways.

Section 4.1 introduces the many types of functors between categories. In Section 4.2, we meet natural transformations between functors. A special relationship between categories called "equivalence" is discussed in Section 4.3. Another relationship between categories, called "adjunction," is dealt with in Section 4.4. Section 4.5 discusses operations on categories and functors. With all these tools in hand, we revisit the notions of limits and colimits in Section 4.6. The famous Yoneda lemma is introduced in Section 4.7. The chapter ends with Section 4.8, which is a mini-course that uses these categorical structures to describe logical systems.

4.1 Functors

While categories are interesting by themselves, the true power of category theory is seeing the way that categories are related to each other. Just as a function is the main way of expressing a relationship between sets, a functor is going to be the main way of showing a relationship between categories. A functor assigns to an object of one category an object of another category and similar with morphisms. The assignment respects domains and codomains. We will show how to connect seemingly disparate areas using functors.

Remark 4.1.1. For those who know the language of graph theory and group theory, the concept of a functor is not so strange. If you are unfamiliar with these areas, skip this paragraph. If you think of a category as a souped-up, directed graph, then a functor is a souped-up, directed graph homomorphism. That is, it takes objects of one category to objects of another category. It also takes morphisms from one category to the morphisms of another category. We further require that the sources and targets are matched up as they are in graphs (see page 26). However, a functor must also respect the composition of morphisms and the identity morphisms of categories. A functor is also a generalization of a group homomorphism from one group to another. Just as a group homomorphism respects the multiplication operation, so too a functor respects the composition of morphisms. ♠

Definition 4.1.2. Given two categories $\mathbb{A}$ and $\mathbb{B}$, a **functor** F from $\mathbb{A}$ to $\mathbb{B}$, written as $F\colon \mathbb{A} \longrightarrow \mathbb{B}$, is a rule that assigns to every object a of $\mathbb{A}$ an object $F(a)$ of $\mathbb{B}$, and assigns to every morphism $f : a \longrightarrow a'$ in $\mathbb{A}$ a morphism $F(f)\colon F(a) \longrightarrow F(a')$ in $\mathbb{B}$. These assignments must satisfy the following requirements:

- Functors respect the compositions of morphisms: for $f\colon a \longrightarrow a'$ and $f'\colon a' \longrightarrow a''$ in $\mathbb{A}$, we require that $F(f' \circ_{\mathbb{A}} f) = F(f') \circ_{\mathbb{B}} F(f)$, where the $\circ_{\mathbb{A}}$ on the left is the composition in $\mathbb{A}$, while the $\circ_{\mathbb{B}}$ on the right is the composition in $\mathbb{B}$. (We omit such subscripts when they are clear from the context.)
- Functors respect identity morphisms; that is, they take identity morphisms in one category to identity morphisms in the second category: for all a in $\mathbb{A}$, we require $F(id_a) = id_{F(a)}$ where id_a is in $\mathbb{A}$ while $id_{F(a)}$ is in $\mathbb{B}$.

A functor $F\colon \mathbb{A} \longrightarrow \mathbb{B}$ can be thought of as a pair of functions $F_0\colon Ob(\mathbb{A}) \longrightarrow Ob(\mathbb{B})$ and $F_1\colon Mor(\mathbb{A}) \longrightarrow Mor(\mathbb{B})$. The fact that $f : a \longrightarrow a'$ in $\mathbb{A}$ must go to $F(f)\colon F(a) \longrightarrow F(a')$ in $\mathbb{B}$ means that the assignments respect the domain and codomain functions, as in the commuting of the following two diagrams:

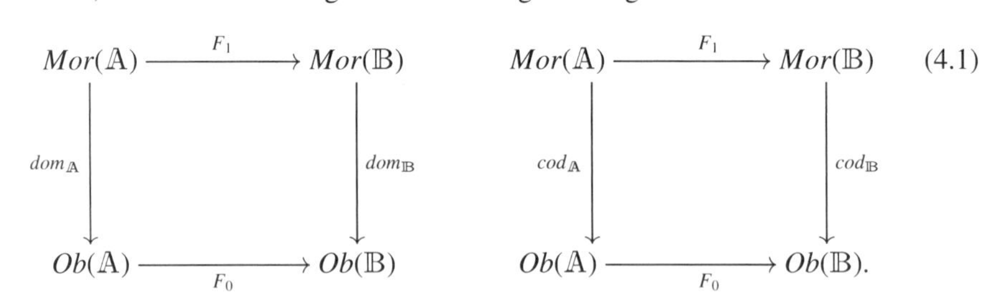

(4.1)

These two commuting squares are similar to Diagram (1.49) where we met the definition of a graph homomorphism.

A functor $F\colon \mathbb{A} \longrightarrow \mathbb{B}$ takes a morphism $a \longrightarrow a'$ in $\mathbb{A}$ to a morphism $F(a) \longrightarrow F(a')$ in $\mathbb{B}$, which means that for any a, a' in $\mathbb{A}$ there is a function of Hom sets:

$$Hom_{\mathbb{A}}(a, a') \longrightarrow Hom_{\mathbb{B}}(F(a), F(a')). \tag{4.2}$$

These functions of Hom sets will have a central importance in the coming pages.

We begin with some simple examples of functors.

Example 4.1.3. Let $\mathbb{A}$ be any category, then there is an **identity functor** $Id_{\mathbb{A}} : \mathbb{A} \longrightarrow \mathbb{A}$, which is defined for object a as $Id_{\mathbb{A}}(a) = a$ and defined similarly for morphisms in $\mathbb{A}$. □

Example 4.1.4. Consider the natural numbers **N** as a partial order category. There is a functor $(\ ) + 5 \colon \mathbf{N} \longrightarrow \mathbf{N}$, which accepts a number m and adds 5 to it (i.e., $m + 5$). This is a functor because it respects morphisms (if $m \leq m'$, then $m + 5 \leq m' + 5$) and it respects the identity $(id_m) + 5 = id_{m+5}$. □

Example 4.1.5. There is a functor $\mathcal{P} \colon \mathbb{S}\mathrm{et} \longrightarrow \mathbb{S}\mathrm{et}$ that takes a set S to its power set $\mathcal{P}(S)$. A set function $f \colon S \longrightarrow S'$ will go to the set function $\mathcal{P}(f) \colon \mathcal{P}(S) \longrightarrow \mathcal{P}(S')$, which is defined for $X \subseteq S$ as its image under f. In other words,

$$\mathcal{P}(f)(X) = f(X) = \{f(x) \in S' \mid x \in S\} \subseteq S'. \tag{4.3}$$

The function $\mathcal{P}(f)$ is also denoted f_* and is called the **direct image** of f. The requirements for being a functor are easily seen to be satisfied. This functor is called the **direct image functor**. □

Any functor from a category to itself is called an **endofunctor**. Examples 4.13, 4.14, and 4.15 are about endofunctors.

Example 4.1.6. Consider the real number, **R**, and the integer, **Z**, each thought of as a partial order. There is a functor $Floor \colon \mathbf{R} \longrightarrow \mathbf{Z}$ that takes any real number r to $Floor(r)$, the greatest integer that is less than or equal to r. For example, $Floor(3.7563) = 3$ and $Floor(-5.87) = -6$. The assignment $Floor$ is a functor because if $r \leq r'$, then $Floor(r) \leq Floor(r')$. The floor functor is also denoted as $Floor(r) = \lfloor r \rfloor$. There is also a ceiling functor, $Ceil \colon \mathbf{R} \longrightarrow \mathbf{Z}$, which takes any real number, r, to the least integer that is greater or equal to r. The ceiling functor is denoted as $Ceil(r) = \lceil r \rceil$. □

Example 4.1.7. There is a functor $D \colon \mathbb{S}\mathrm{et} \longrightarrow \mathbb{G}\mathrm{raph}$ that takes every set S to the discrete graph $D(S)$. This is a graph with only objects and no morphisms. Given a set function $f \colon S \longrightarrow S'$, there is a similar graph homomorphism $D(S) \longrightarrow D(S')$ that takes objects of $D(S)$ to objects of $D(S')$ as described by f. The requirements of being a functor are easily seen to be satisfied. □

Example 4.1.8. There is a **forgetful functor** $U \colon \mathbb{G}\mathrm{roup} \longrightarrow \mathbb{S}\mathrm{et}$ from the category of groups to the category of sets that "forgets" the group structure. Remember that a group is a set G with extra structure, that is, $(G, \star, e, (\ )^{-1})$. The functor U takes a group and forgets the rest of the structure, that is, it takes a group to its underlying set. This means that U performs the following operation: $(G, \star, e, (\ )^{-1}) \mapsto G$. A group homomorphism is a set function that respects all the structure. The functor U will take a group homomorphism to its underlying set function, that is, a group homomorphism $f \colon G \longrightarrow G'$ will go to the set function $U(F) \colon U(G) \longrightarrow U(G')$. It is not hard to see that the requirements for U being a functor are satisfied.

There is a similar functor from the category of monoids $U\colon \mathbb{Monoid} \longrightarrow \mathbb{Set}$ that forgets the monoid structure. □

The following two functors will be of importance in the coming sections.

Example 4.1.9. For every set B, there is a functor $L_B\colon \mathbb{Set} \longrightarrow \mathbb{Set}$ that is defined on set A as $L_B(A) = A \times B$. Morphism $f\colon A \longrightarrow A'$ goes to $L_B(f) = f \times id_B\colon A \times B \longrightarrow A' \times B$. For every set B, there is also a functor $R_B\colon \mathbb{Set} \longrightarrow \mathbb{Set}$ that is defined on the input set C as $R_B(C) = Hom_{\mathbb{Set}}(B, C)$. For $f\colon C \longrightarrow C'$, we define $R_B(f)\colon Hom_{\mathbb{Set}}(B, C) \longrightarrow Hom_{\mathbb{Set}}(B, C')$ on input $g\colon B \longrightarrow C$ as $R_B(f)(g) = f \circ g\colon B \longrightarrow C \longrightarrow C'$, which is in $Hom_{\mathbb{Set}}(B, C')$. Functor R_B is called a **representable functor**. It is represented by B. □

Example 4.1.10. There is a functor $F\colon \mathbb{Set} \longrightarrow \mathbb{Monoid}$ that takes every set S to the **free monoid** $F(S) = S^*$. This monoid consists of the set of all strings of elements in S. In other words, think of S as a set of letters or an alphabet; then $F(S)$ is the set of words that can be made from those letters. Given two words $X = s_1 s_2 \cdots s_m$ and $Y = s'_1 s'_2 \cdots s'_n$, the multiplication is the concatenation of the two strings $X \bullet Y = s_1 s_2 \cdots s_m s'_1 s'_2 \cdots s'_n$. The unit of the monoid is the empty word $\emptyset$, which has no letters. It is easy to see that $X \bullet \emptyset = X = \emptyset \bullet X$. For a set function $f\colon S \longrightarrow T$, the value of $F(f)\colon F(S) \longrightarrow F(T)$ takes a word in S such as $s_1 s_2 \cdots s_m$ to $f(s_1) f(s_2) \cdots f(s_m)$, which is a word in T. This functor is important in mathematics and is called the **free monoid functor**. It is also important in computer science, where it is called the **list functor**. □

There are many properties of functors. With sets, we ask if a function is one-to-one or onto. With categories, we ask if a functor is one-to-one or onto with respect to objects and with respect to morphisms.

Definition 4.1.11. There are properties of functors that depend on the way it assigns objects from one category to the other. A functor $F\colon \mathbb{A} \longrightarrow \mathbb{B}$ can be **injective on objects**, **surjective on objects**, or **bijective on objects**. There are also properties of functors that depend on morphisms. We call a functor **full** if, for all a and a', the map $Hom_{\mathbb{A}}(a, a') \longrightarrow Hom_{\mathbb{B}}(F(a), F(a'))$ is onto. We call a functor **faithful** if for all a and a', the map $Hom_{\mathbb{A}}(a, a') \longrightarrow Hom_{\mathbb{B}}(F(a), F(a'))$ is one-to-one. When a functor is both full and faithful, we call it **fully faithful**.

Let us discuss various families or types of functors.

If $\mathbb{A}$ is a subcategory of $\mathbb{B}$, then there is an **inclusion functor** from $\mathbb{A}$ to $\mathbb{B}$ that takes every object in $\mathbb{A}$ into the same object in $\mathbb{B}$. We denote such a functor as $\mathbb{A} \hookrightarrow \mathbb{B}$. Inclusion functors are injective on objects and faithful. We also say that such functors are the **identity on objects**; that is, the functor takes an object a to a.

Example 4.1.12. Examples of inclusion functors abound:

- There is an inclusion of finite sets into sets, $\mathbb{FinSet} \hookrightarrow \mathbb{Set}$.
- Abelian groups include into all groups, $\mathbb{AbGp} \hookrightarrow \mathbb{Group}$.
- As we saw in Section 2.1, there are three related categories that have sets as objects. The category $\mathbb{Set}$ has set functions as morphisms, $\mathbb{Par}$ has partial

functions as morphisms, and $\mathbb{R}\mathrm{el}$ has relations as morphisms. There are inclusion functors

$$\mathbb{S}\mathrm{et} \hookrightarrow \mathbb{P}\mathrm{ar} \hookrightarrow \mathbb{R}\mathrm{el}. \tag{4.4}$$

Notice that all three of these categories have the same objects, namely sets, and both inclusion functors are the identity on objects.

- If you think of the sets of numbers $\mathbf{N}, \mathbf{Z}, \mathbf{Q}, \mathbf{R}$, and $\mathbf{C}$ with addition as monoids (one-object categories), then there are inclusion functors:

$$\mathbf{N} \hookrightarrow \mathbf{Z} \hookrightarrow \mathbf{Q} \hookrightarrow \mathbf{R} \hookrightarrow \mathbf{C}. \tag{4.5}$$

All these functors take the single object to the single object (hence they are technically bijective on objects) and are faithful on the set of morphisms.

- If you think of the numbers $\mathbf{N}, \mathbf{Z}, \mathbf{Q}, \mathbf{R}$, and $\mathbf{C}$ as partial orders (in fact, all except the usual way of thinking about $\mathbf{C}$ are total orders), then these inclusion functors are injective on objects and full and faithful on morphisms.
- Remember that for a field $\mathbf{K}$, the category of $\mathbf{K}\mathbb{M}\mathrm{at}$ has the natural numbers as objects and for $m, n \in \mathbf{N}$, $Hom_{\mathbf{K}\mathbb{M}\mathrm{at}}(m, n)$ is the set of n by m matrices with entries in $\mathbf{K}$. By examining the elements in the matrices and noticing that these numbers are subsets of each other, we have the following inclusion functors:

$$\mathbf{N}\mathbb{M}\mathrm{at} \hookrightarrow \mathbf{Z}\mathbb{M}\mathrm{at} \hookrightarrow \mathbf{Q}\mathbb{M}\mathrm{at} \hookrightarrow \mathbf{R}\mathbb{M}\mathrm{at} \hookrightarrow \mathbf{C}\mathbb{M}\mathrm{at}. \tag{4.6}$$

Notice that all these categories have the natural numbers as objects and all the functors are the identity on objects.

- There is an inclusion of the category of partial orders into the category of preorders. Since both categories have order-preserving maps as their morphisms, the inclusion $\mathbb{P}\mathbb{O} \hookrightarrow \mathbb{P}\mathrm{re}\mathbb{O}$ is full and faithful.
- There is an inclusion for every n of n-$\mathbb{M}\mathrm{anif} \hookrightarrow \mathbb{M}\mathrm{anif}$. □

Exercise 4.1.13. Explain why there is no inclusion of monoids from *Bool* into $\mathbf{N}$. Similarly explain there is no inclusion from $Bool\mathbb{M}\mathrm{at}$ into $\mathbf{N}\mathbb{M}\mathrm{at}$. ■

Definition 4.1.14. A special type of inclusion functor is an **embedding**. This is a functor that is not only an inclusion (injective on objects and faithful) but also full. This means that between any two objects in the subcategory, the Hom sets are isomorphic. A subcategory with such an embedding is called a full subcategory.

Example 4.1.15. The following are embeddings: $\mathbb{F}\mathrm{in}\mathbb{S}\mathrm{et} \hookrightarrow \mathbb{S}\mathrm{et}$, $\mathbb{A}\mathrm{b}\mathbb{G}\mathrm{p} \hookrightarrow \mathbb{G}\mathrm{roup}$, $\mathbb{P}\mathbb{O} \hookrightarrow \mathbb{P}\mathrm{re}\mathbb{O}$, and n-$\mathbb{M}\mathrm{anif} \hookrightarrow \mathbb{M}\mathrm{anif}$. □

Exercise 4.1.16. Show that the other examples in Example 4.1.12 are inclusions but not embeddings. ■

We saw in Example 4.1.8 a functor that forgets the group structure of a group. There are many functors that are called **forgetful functors**, which forget or disregard part of the structure of an object in a category. Such functors are usually denoted by the letter U, which might stand for "underlying." Since two different morphisms will go to two different underlying morphisms, forgetful functors are faithful.

Here are some examples of forgetful functors:

Example 4.1.17.

- Similar to the forgetful functor $U\colon \mathbb{Group} \longrightarrow \mathbb{Set}$, there are forgetful functors from categories like $\mathbb{Magma}$, $\mathbb{Monoid}$, $\mathbb{Ring}$, $\mathbb{Field}$, etc., to $\mathbb{Set}$.
- We do not have to forget all the structure. For example, given a ring $(M, \star, e, -, \odot, u)$, we can forget its second operation and second unit to get an abelian group $(M, \star, e, -)$. This gives us a functor $U\colon \mathbb{Ring} \longrightarrow \mathbb{AbGp}$. We can further forget its inverse operation and get a monoid $(M, \star, e)$. This is a functor $U\colon \mathbb{Ring} \longrightarrow \mathbb{Monoid}$. Rather than forgetting the second operation, second unit, and inverse operation, we can forget its first operation, first unit, and inverse operation and get a monoid: $(M, \odot, u)$. This is a *different* functor $U'\colon \mathbb{Ring} \longrightarrow \mathbb{Monoid}$. There are many more examples along this path.
- There is a functor $\mathbb{Top} \longrightarrow \mathbb{Set}$ that forgets the open set structure, $U((T, \tau)) = T$. There is also a forgetful functor that takes a manifold (which is a topological space with more structure) and forgets the manifold structure. The output is a topological space. This gives us a functor $\mathbb{Manif} \longrightarrow \mathbb{Top}$. Furthermore, we can forget the manifold structure "all the way down" to $\mathbb{Set}$ and get a forgetful functor $\mathbb{Manif} \longrightarrow \mathbb{Set}$.
- Given a complex scalar multiplication on a vector space, one can forget the action of the imaginary numbers and get a real scalar multiplication on the vector space. In other words, the scalar multiplication $\cdot\colon \mathbf{R} \times V \longrightarrow V$ is simply the restriction of $\cdot\colon \mathbf{C} \times V \longrightarrow V$ to the real numbers. This entails a forgetful functor $\mathbf{C}\mathbb{Vect} \longrightarrow \mathbf{R}\mathbb{Vect}$.
- There is a forgetful functor from the slice category $\mathbb{A}/a$ to $\mathbb{A}$ that sends object $f\colon b \longrightarrow a$ to b. A morphism $g\colon b \longrightarrow b'$ in $\mathbb{A}/a$ will simply go to $g\colon b \longrightarrow b'$ in $\mathbb{A}$. There are similar forgetful functors from coslice catgeories. □

In contrast to an inclusion functor, a forgetful functor is usually not injective on objects. For example, the forgetful functor from $\mathbb{Top}$ to $\mathbb{Set}$ is not injective on objects because there might be many different topologies that one can put on a single set. Similarly, the forgetful functor from graphs to sets is not injective on objects because there are many different graphs for a set of vertices.

Remark 4.1.18. This brings to light a general thought about the relationship between structures. Consider two types of structures: A-structures and B-structures. There are a lot of possible different relationships between the two structures. Let us highlight two possible relationships:

- There can be an inclusion functor from the category of A-structure to the category of B-structure. This is the case when the A-structures are special types of B-structures that satisfy more requirements. Another way to say this is that

A-structures have more requirements than B-structures. For example, an abelian group is a special type of group that satisfies commutativity; a finite set is a special type of set that is finite; or a n-dimensional manifold is a special type of manifold.

- There can be a forgetful functor from the category of A-structures to the category of B-structures. This is the case when the A-structures have more operations than the B-structures. Another way to say this is that A-structures have more structure than B-structures. For example, a ring has more operations than a group, and hence, there is a forgetful functor from rings to groups. Another example: a partial order has more structure than a set, so there is a forgetful functor from partial orders to sets. Yet another example: a topological space has more structure than a set, so there is a forgetful functor from topological spaces to sets.

We can see this clearly with all the algebraic structures described in Definition 2.1.8 on page 39. We write $\hookrightarrow$ for an inclusion functor and $\xrightarrow{U}$ for a forgetful functor.

$$\begin{array}{ccccccccc} \mathbb{Field} & \xrightarrow{U} & \mathbb{Ring} & \xrightarrow{U} & \mathbb{Group} & \xrightarrow{U} & \mathbb{Monoid} & \xrightarrow{U} & \mathbb{SemiGp} \\ & & & & \uparrow & & & & \uparrow \\ & & & & \mathbb{AbGp} & & & & \mathbb{Magma} \end{array} \tag{4.7}$$

We also highlight these different relationships in the Venn diagrams in Appendix A. Inclusion relationships are described with regular lines, while forgetful functors are described with zigzag lines. ♠

Just as functions can be composed, functors can be composed. For $F: \mathbb{A} \longrightarrow \mathbb{B}$ and $G: \mathbb{B} \longrightarrow \mathbb{C}$, the composition $G \circ F: \mathbb{A} \longrightarrow \mathbb{C}$ is defined as $(G \circ F)(a) = G(F(a))$. The composition is defined for morphism $f: a \longrightarrow a'$ as $(G \circ F)(f) = G(F(f)): G(F(a)) \longrightarrow G(F(a'))$. If there are two composable morphisms f and f' in $\mathbb{A}$, then we will have expressions like $(G \circ F)(f' \circ f)$. The $\circ$ on the left denotes composition of functors, and the $\circ$ on the right denotes composition in $\mathbb{A}$.

Exercise 4.1.19. Show that the composition of two functors is a functor. ■

Exercise 4.1.20. Show that composition of functors is associative. ■

Exercise 4.1.21. Show that the composition of a functor with appropriate identity functors is the original functor. That is, for functor $F: \mathbb{A} \longrightarrow \mathbb{B}$, we have $F \circ Id_{\mathbb{A}} = F = Id_{\mathbb{B}} \circ F$. ■

We are ready to move up a level. Up to now we have dealt with two structures and a functor between them. Now we will talk about *all* structures and *all* the functors between them.

Example 4.1.22. The composition of functors and the identity functors bring to light one of the most important examples of a category: $\mathbb{CAT}$ is the category of all categories and functors between them. We are mostly interested in a subcategory $\mathbb{C}\mathrm{at}$, which consists of all locally small categories and the functors between them. □

Exercise 4.1.23. Show that **0** is the initial object in $\mathbb{C}\mathrm{at}$ and **1** is a terminal object in $\mathbb{C}\mathrm{at}$. ■

Example 4.1.24. In this text, we have already seen many functors that take values in $\mathbb{C}\mathrm{at}$. When we wrote about them, we did not have the language to describe them as functors. Now we do.

- Every set can be thought of as a discrete category. In Example 2.1.5 on page 37, we saw that for every set S, there is a discrete category $d(S)$. This is a functor $d\colon \mathbb{S}\mathrm{et} \longrightarrow \mathbb{C}\mathrm{at}$. The objects of $d(S)$ are the elements of S and the only morphisms are identity morphisms. If $f\colon S \longrightarrow S'$, then there is a corresponding functor $d(f)\colon d(S) \longrightarrow d(S')$ where $d(f)(s) = f(s)$.
- Every monoid is a one-object category. In Example 2.1.17 on page 43, we saw that for every monoid M, there is a one-object category $A(M)$ whose morphisms are the elements of M and whose composition is the monoid multiplication. This is a functor $A\colon \mathbb{M}\mathrm{onoid} \longrightarrow \mathbb{C}\mathrm{at}$. If $f\colon M \longrightarrow M'$ is a monoid homomorphism, then $f(m \star m') = f(m) \star' f(m')$. This requirement shows that the obvious map $A(f)\colon A(M) \longrightarrow A(M')$ that takes the single object to the single object and respects the composition is a functor.
- Every partial order is a category. In Example 2.1.53 on page 53, we saw that for every partial order $(P, \leq)$, there is a category $B(P, \leq)$, or just $B(P)$, whose objects are the objects of P, and there is a single morphism from p to p' in $B(P)$ if and only if $p \leq p'$. This is a functor $B\colon \mathbb{PO} \longrightarrow \mathbb{C}\mathrm{at}$. An order-preserving $f\colon (P, \leq) \longrightarrow (P', \leq')$ induces a functor $B(f)\colon B(P, \leq) \longrightarrow B(P', \leq')$. Whatever we said about partial orders is also true for preorders. This means that there is a functor called $\mathbb{P}\mathrm{re}\mathbb{O} \longrightarrow \mathbb{C}\mathrm{at}$.
- Every group is a one-object category where all the morphisms are isomorphisms. In Example 2.2.18 on page 61, we saw that every group can be thought of as a one-object category where the elements of the group become invertible morphisms in the category. The composition in the category is group multiplication. The inverse of the morphism is the inverse of the related element. This describes a functor called $C\colon \mathbb{G}\mathrm{roup} \longrightarrow \mathbb{C}\mathrm{at}$. A group homomorphism $f\colon G \longrightarrow G'$ become a functor $C(f)\colon C(G) \longrightarrow C(G')$.
- The power set of a set is a partial order category. In Example 2.1.56 on page 54, we saw that for every set S, $\mathcal{P}(S)$ has the structure of a partial order. This gives us a functor $\mathbb{S}\mathrm{et} \longrightarrow \mathbb{PO}$, and we saw that there is a functor $B\colon \mathbb{PO} \longrightarrow \mathbb{C}\mathrm{at}$. Composing these two functors gives us $\mathcal{P}\colon \mathbb{S}\mathrm{et} \longrightarrow \mathbb{C}\mathrm{at}$. Every set function $f\colon S \longrightarrow S'$ induces the direct image functor $\mathcal{P}(f)\colon \mathcal{P}(S) \longrightarrow \mathcal{P}(S')$ that we met in Example 4.1.5.
- The opposite operation takes a category to its opposite category. In Definition 2.3.14 on page 66, we saw that for a category $\mathbb{A}$, there is a category $\mathbb{A}^{op}$. This is a functor $(\ )^{op}\colon \mathbb{C}\mathrm{at} \longrightarrow \mathbb{C}\mathrm{at}$. It also takes a functor $F\colon \mathbb{A} \longrightarrow \mathbb{B}$ to the functor $F^{op}\colon \mathbb{A}^{op} \longrightarrow \mathbb{B}^{op}$ (which is defined in the obvious way). □

Exercise 4.1.25. Show that $(\quad)^{op} \circ (\quad)^{op} = Id_{\mathbb{C}\mathrm{at}}$ (i.e., for any category $\mathbb{A}$, $((\mathbb{A}^{op})^{op}) = \mathbb{A}$). ■

When dealing with sets, we have a method of saying that two sets are essentially the same. On page 18, we called a set function $f\colon S \longrightarrow T$ an isomorphism if there is a $g\colon T \longrightarrow S$ such that for all s in S, $g(f(s)) = s$, and for all t in T, $f(g(t)) = t$. Another way to say that is $g \circ f = Id_S$ and $f \circ g = Id_T$. Set S is isomorphic to set T if there is an ismorphism between them, and we write it as $S \cong T$.

Now we extend this idea to categories.

Definition 4.1.26. Let $\mathbb{A}$ and $\mathbb{B}$ be categories. A functor $F\colon \mathbb{A} \longrightarrow \mathbb{B}$ is an **isomorphism** if there is a functor $G\colon \mathbb{B} \longrightarrow \mathbb{A}$ called the **inverse** of F such that $G \circ F = Id_{\mathbb{A}}$ and $Id_{\mathbb{B}} = F \circ G$ (we will soon explain why we are writing it this way as opposed to the equivalent $F \circ G = Id_{\mathbb{B}}$). Similar to Diagram (1.33), we can express this as

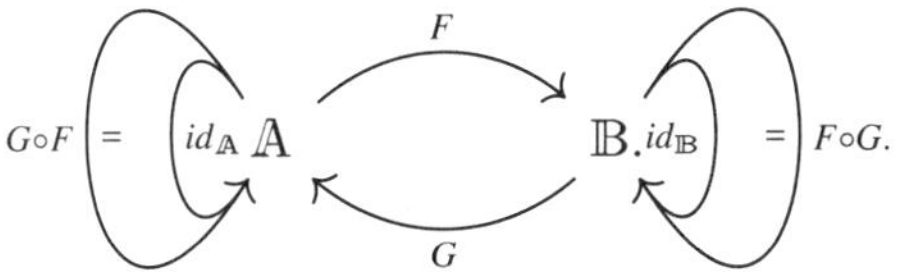

$$G\circ F = id_{\mathbb{A}} \circlearrowright \mathbb{A} \underset{G}{\overset{F}{\rightleftarrows}} \mathbb{B} \circlearrowleft id_{\mathbb{B}} = F\circ G. \qquad (4.8)$$

Category $\mathbb{A}$ is **isomorphic** to category $\mathbb{B}$ if there is an isomorphism between them, and which we write as $\mathbb{A} \cong \mathbb{B}$.

Exercise 4.1.27. Show that categories $\mathbb{A}$ and $\mathbb{B}$ are isomorphic if and only if there is a functor $F\colon \mathbb{A} \longrightarrow \mathbb{B}$ that is bijective on objects, full, and faithful. ■

Essentially, an isomorphism of categories means that the categories have exactly the same structure. The two functors essentially rename the objects and morphisms of the categories.

Example 4.1.28. Let us examine some simple examples of isomorphism of categories:

- An identity functor is an isomorphism.
- We saw in Exercise 2.3.21 on page 67 that for any category $\mathbb{A}$, the product $\mathbb{A} \times \mathbf{1}$ is essentially the same as $\mathbb{A}$. Specifically, the projection functor $\pi_{\mathbb{A}}\colon \mathbb{A} \times \mathbf{1} \longrightarrow A$ is an isomorphism.
- $\mathbb{R}\mathrm{el}^{op} \cong \mathbb{R}\mathrm{el}$. The isomorphism functor takes the set S to S and the relation $R \subseteq S \times T$ to the inverse relation $R^{-1} \subseteq T \times S$ (see Definition 2.1.42.) Notice that this works for $\mathbb{R}\mathrm{el}$, but it does not work with $\mathbb{S}\mathrm{et}$ or $\mathbb{P}\mathrm{ar}$.
- $\mathbf{K}\mathbb{M}\mathrm{at}^{op} \cong \mathbf{K}\mathbb{M}\mathrm{at}$. The isomorphism functor $(\quad)^T\colon \mathbf{K}\mathbb{M}\mathrm{at}^{op} \longrightarrow \mathbf{K}\mathbb{M}\mathrm{at}$ is the identity on the natural numbers and takes $A\colon m \longrightarrow n$ to $A^T\colon n \longrightarrow m$. The inverse of the isomorphism functor is itself, that is, $(A^T)^T = A$. □

Important Categorical Idea 4.1.29. Weakening Structures. The notion of an isomorphism of category is a legitimate idea, but it is a very strong requirement. The necessity that $G \circ F$ is equal to $Id_{\mathbb{A}}$ and $F \circ G$ is equal to $Id_{\mathbb{B}}$ ensures that there are not a lot of interesting, nontrivial examples of isomorphism of categories. The main problem is that insisting on equality is too strong. The weaker we make this requirement, the more phenomena we can model. We will weaken these requirements in Sections 4.3 and 4.4 and find many interesting examples. In general, the weaker the assumption, the more phenomena we can encapsulate. How weak can we go? This topic is central to higher category theory, which we will meet in the coming pages. ○

Example 4.1.30. Let $\mathbb{A}$ be a locally small category and $a \in \mathbb{A}$. We know that for every $b \in \mathbb{A}$, there is a set $Hom_{\mathbb{A}}(a, b)$. This brings to light the functor

$$Hom_{\mathbb{A}}(a, \ ): \mathbb{A} \longrightarrow \mathbb{S}\mathrm{et}. \tag{4.9}$$

For every object $b \in \mathbb{A}$, there is a set $Hom_{\mathbb{A}}(a, b)$; and for every $f: b \longrightarrow b'$ in $\mathbb{A}$, there is a morphism of sets:

$$Hom_{\mathbb{A}}(a, f): Hom_{\mathbb{A}}(a, b) \longrightarrow Hom_{\mathbb{A}}(a, b'), \tag{4.10}$$

which takes $g: a \longrightarrow b$ to $f \circ g: a \longrightarrow b \longrightarrow b'$ in $Hom_{\mathbb{A}}(a, b')$. That is, $Hom_{\mathbb{A}}(a, f)(g) = f \circ g$. Since this functor is induced by f, we sometimes denote $Hom_{\mathbb{A}}(a, f)$ as f_*.

Let us show that $Hom_{\mathbb{A}}(a, \ )$ satisfies the requirement of being a functor:

- $Hom_{\mathbb{A}}(a, \ )$ preserves composition: for $f: b \longrightarrow b'$ and $f': b' \longrightarrow b''$, then

$$Hom_{\mathbb{A}}(a, f' \circ f)(g) = f' \circ f \circ g = (Hom_{\mathbb{A}}(a, f') \circ Hom_{\mathbb{A}}(a, f))(g). \tag{4.11}$$

- $Hom_{\mathbb{A}}(a, \ )$ preserves the identity morphisms: for $id_b: b \longrightarrow b$,

$$Hom_{\mathbb{A}}(a, id_b)(g) = id_b \circ g = g = Id_{Hom(a,b)}(g). \tag{4.12}$$

We will see later that functors of the form $Hom_{\mathbb{A}}(a, \ )$ are **representable functors**. Here, a represents the functor. □

Functors actually come in two flavors: **covariant functors** and **contravariant functors**. Although we did not have the name yet, all the functors that we have seen up to now are covariant funtors. The word "covariant" means that the functor varies in the same way that the source category varies. In other words, if there is a morphism from a to a' in the source category, then the covariant functor F takes that morphism from $F(a)$ to $F(a')$ in the target category. In stark contrast, a contravariant functor would go "contra" (against or opposite) the way that the source category varies. In other words, if there is a morphism from a to a' in the source category, then the contravariant functor F takes that morphism from $F(a')$ to $F(a)$ in the target category. Definition 4.1.31 formalizes this.

Definition 4.1.31. Given two categories $\mathbb{A}$ and $\mathbb{B}$, a **contravariant functor** F from $\mathbb{A}$ to $\mathbb{B}$, written as $F\colon \mathbb{A} \longrightarrow \mathbb{B}$, is a rule that assigns to every object a of $\mathbb{A}$ an object $F(a)$ of $\mathbb{B}$, and assigns to every morphism $f: a \longrightarrow a'$ in $\mathbb{A}$ a morphism $F(f)\colon F(a') \longrightarrow F(a)$ in $\mathbb{B}$ (notice the direction). These assignments must satisfy the following two requirements:

- Functors reverse the composition of morphisms: for $f\colon a \longrightarrow a'$ and $f'\colon a' \longrightarrow a''$ in $\mathbb{A}$, we require that $F(f' \circ f) = F(f) \circ F(f')$, where the $\circ$ on the left is the composition in $\mathbb{A}$ while the $\circ$ on the right is the composition in $\mathbb{B}$.
- Functors respect identity morphisms: for all a in $\mathbb{A}$, we require $F(id_a) = id_{F(a)}$, where id_a, is in $\mathbb{A}$ while $id_{F(a)}$ is in $\mathbb{B}$.

Example 4.1.32. In Example 4.1.30, we saw that $Hom_{\mathbb{A}}(a, \ )$ is a covariant functor. In contrast, $Hom_{\mathbb{A}}(\ , a)$ is a contravariant functor. In detail, for every locally small category $\mathbb{A}$ and every object $a \in \mathbb{A}$, there is a contravariant functor $Hom_{\mathbb{A}}(\ , a)\colon \mathbb{A} \longrightarrow \mathbb{Set}$. For object $b \in \mathbb{A}$, there is a set $Hom_{\mathbb{A}}(b, a)$, and for $f\colon b \longrightarrow b'$, there is a set function

$$Hom_{\mathbb{A}}(f, a)\colon Hom_{\mathbb{A}}(b', a) \longrightarrow Hom_{\mathbb{A}}(b, a), \tag{4.13}$$

which takes $g\colon b' \longrightarrow a$ in $Hom_{\mathbb{A}}(b', a)$ to $g \circ f\colon b \longrightarrow b' \longrightarrow a$ in $Hom_{\mathbb{A}}(b, a)$. Since this map is induced by f and it is contravariant, we sometimes denote $Hom_{\mathbb{A}}(f, a)$ as f^*. Bear in mind the contrast between the covariant f_* and the contravvariant f^*.

Let us show that $Hom_{\mathbb{A}}(\ , a)$ satisfies the requirement of being a contravariant functor:

- $Hom_{\mathbb{A}}(\ , a)$ reverses composition: for $f\colon b \longrightarrow b'$ and $f'\colon b' \longrightarrow b''$,

$$Hom_{\mathbb{A}}(f' \circ f, a)(g) = g \circ f' \circ f = (Hom_{\mathbb{A}}(f, a) \circ Hom(f', a))(g). \tag{4.14}$$

- $Hom_{\mathbb{A}}(\ , a)$ preserves the identity morphisms: for $id_b\colon b \longrightarrow b$,

$$Hom_{\mathbb{A}}(id_b, a)(g) = g \circ id_b = g = id_{Hom(b,a)}(g). \tag{4.15}$$

We will see later that functors of the form $Hom_{\mathbb{A}}(\ , a)$ are also called **representable functors**. Here, a represents the functor. □

Let us now apply this to a concept from linear algebra.

Example 4.1.33. In Exercise 2.4.3 on page 69 we saw that for every set S, the set $Func(S, \mathbf{C})$ has the structure of a complex vector space. If $f\colon S \longrightarrow S'$ is a set function, then $Func(f, \mathbf{C})\colon Func(S', \mathbf{C}) \longrightarrow Func(S, \mathbf{C})$ is a linear map, and hence,

$$Func(\ , \mathbf{C})\colon \mathbb{Set} \longrightarrow \mathbf{C}\mathbb{Vect} \tag{4.16}$$

is a contravariant functor. □

Example 4.1.34. In Example 4.1.5, we saw the covariant direct image functor from $\mathbb{Set}$ to $\mathbb{Set}$ that takes set S to $\mathcal{P}(S)$. There is a contravariant version of this. The functor $\mathcal{P}' \colon \mathbb{Set} \longrightarrow \mathbb{Set}$ performs the same action on the objects (i.e., $\mathcal{P}'(S)$ is the power set of S) but is contravariant. For set function $f \colon S \longrightarrow T$, the functor is defined for $Y \subseteq T$ as

$$\mathcal{P}'(f)(Y) = f^{-1}(Y) = \{x \in S \mid f(x) \in Y\} \subseteq S. \tag{4.17}$$

This functor can be visualized as

$$\begin{array}{ccc} f^{-1}(Y) & \longrightarrow & Y \\ \big\downarrow & & \big\downarrow \\ S & \xrightarrow[f]{} & T. \end{array} \tag{4.18}$$

This functor is called the **preimage functor** or the **inverse image functor**. □

There is really no reason to use the nomenclature of contravariant functors for the simple reason that every contravariant functor $F \colon \mathbb{A} \longrightarrow B$ has a related covariant functor $F' \colon \mathbb{A}^{op} \longrightarrow \mathbb{B}$ that performs the same action. Remember that in $\mathbb{A}^{op}$, the arrows are all turned around, and the composition is reversed as in Definition 2.3.14 on page 66. Thus the contravariant functors in the previous examples can be written as covariant functors:

$$Func(\ \ , \mathbf{C}) \colon \mathbb{Set}^{op} \longrightarrow \mathbf{C}\mathbb{Vect} \qquad \text{and} \qquad \mathcal{P}' \colon \mathbb{Set}^{op} \longrightarrow \mathbb{Set}. \tag{4.19}$$

Exercise 4.1.35. Explain why a functor $F \colon \mathbb{A}^{op} \longrightarrow \mathbb{B}$ has a related functor $F' \colon \mathbb{A} \longrightarrow \mathbb{B}^{op}$ that performs the same action. ■

Exercise 4.1.36. Show that $f \colon b \longrightarrow b'$ is an epimorphism if and only if for all a, $Hom(f,a) \colon Hom(b',a) \longrightarrow Hom(b,a)$ is monomorphism. ■

Exercise 4.1.37. Show that $f \colon b \longrightarrow b'$ is a split monomorphism if and only if for all a, $Hom(f,a) \colon Hom(b',a) \longrightarrow Hom(b,a)$ is an epimorphism. ■

Definition 4.1.38. A split monomorphism $f \colon a \longrightarrow b$ in a category $\mathbb{A}$ is also called an **absolute monomorphism** because for any covariant functor $F \colon \mathbb{A} \longrightarrow \mathbb{B}$, we have that $F(f)$ is a split monomorphism. This follows from the fact that if f' is the partial inverse of f, then $F(f')$ is the partial inverse of $F(f)$. We can make a similar statement about an **absolute epimorphism**.

Exercise 4.1.39. Show that contravariant functors take absolute monomorphisms to absolute epimorphisms and absolute epimorphisms to absolute monomorphisms. ■

Functors might have more than one input.

Definition 4.1.40. Given categories $\mathbb{A}$, $\mathbb{B}$, and $\mathbb{C}$, a **bifunctor** $F \colon \mathbb{A} \times \mathbb{B} \longrightarrow \mathbb{C}$ is simply a functor from the product of $\mathbb{A}$ and $\mathbb{B}$ to $\mathbb{C}$. In particular, F is a rule that assigns to every object a of $\mathbb{A}$ and b of $\mathbb{B}$ an object $F(a,b)$ of $\mathbb{C}$, and assigns to every morphism $f: a \longrightarrow a'$ in $\mathbb{A}$ and morphism $g: b \longrightarrow b'$ in $\mathbb{B}$ a morphism $F(f,g): F(a,b) \longrightarrow F(a',b')$ in $\mathbb{C}$. These assignments must satisfy the following two requirements:

- Functors respect the compositions of morphisms: for $f: a \longrightarrow a'$, $f': a' \longrightarrow a''$, $g: b \longrightarrow b'$ and $g: b' \longrightarrow b''$, we require that

$$F(f' \circ f, g' \circ g) = F(f', g') \circ F(f, g), \tag{4.20}$$

  where the $\circ$ on the right is the composition in $\mathbb{C}$.
- Functors respect identity morphisms: for all a in $\mathbb{A}$ and b in $\mathbb{B}$, we require $F(id_{(a,b)}) = id_{F(a,b)}$, where $id_{(a,b)}$ is in $\mathbb{A} \times \mathbb{B}$ while $id_{F(a,b)}$ is in $\mathbb{C}$.

Technical Point 4.1.41. Many times, rather than writing the name of the bifunctor before the input, like $G(a, b)$ we write the name of the bifunctor as an operation between the input, for example, $a \Box b$. If we write the bifunctor between the inputs, then Equation (4.20) becomes

$$(f' \circ f) \Box (g' \circ g) = (f' \Box g') \circ (f \Box g). \tag{4.21}$$

This is another instance of an interchange law that we saw in Important Categorical Idea 3.1.17 on page 93 with morphism composition as one operation and the bifunctor as the other operation. ♡

Definition 4.1.40 can be generalized to a **multifunctor** that takes inputs from a finite number of categories:

$$F: \mathbb{A}_1 \times \mathbb{A}_2 \times \cdots \times \mathbb{A}_n \longrightarrow \mathbb{B}. \tag{4.22}$$

The functor might be contravariant in some of the inputs. In such a case, we can look at the opposite category of those inputs and consider only covariant functors. For example, the functor

$$Hom_{\mathbb{A}}(\ ,\ ): \mathbb{A}^{op} \times \mathbb{A} \longrightarrow \mathbb{S}\mathrm{et} \tag{4.23}$$

is a bifunctor where the first input works contravariantly and the second input works covariantly.

Example 4.1.42. Let $\mathbb{A}$ be any category with binary products. We can think of the binary product as a functor *Prod* that takes two objects of $\mathbb{A}$ as inputs and outputs their product. That is, there is a functor $Prod: \mathbb{A} \times \mathbb{A} \longrightarrow \mathbb{A}$ such that $Prod(a, a') = a \times a'$. In particular, the product of two categories can be described in this way. That is, there is a functor $Prod: \mathbb{C}\mathrm{at} \times \mathbb{C}\mathrm{at} \longrightarrow \mathbb{C}\mathrm{at}$. □

Technical Point 4.1.43. There is a slight problem that we have to worry about with the above example. A product of two elements in a category is defined up to a unique isomorphism. In general, there is no unique product (or coproduct) of two elements. So which product should the functor choose? Sometimes we will just assume that the functor chooses one. Sometimes we will simply ignore the question because whatever choice we make will work. ♡

There are a few functors related to the product of two categories.

Example 4.1.44. For any categories $\mathbb{A}$ and $\mathbb{B}$, there are **projection functors** $\pi_{\mathbb{A}}: \mathbb{A}\times\mathbb{B}\longrightarrow\mathbb{A}$ and $\pi_{\mathbb{B}}: \mathbb{A}\times\mathbb{B}\longrightarrow\mathbb{B}$. The functor $\pi_{\mathbb{A}}$ is defined on objects as $\pi_{\mathbb{A}}(a,b)=a$, and on morphisms as $\pi_{\mathbb{A}}(f,g)=f$. The other projection is defined analogously. For any categories $\mathbb{A}$ and $\mathbb{B}$, there is a **braid functor** $br_{\mathbb{A},\mathbb{B}}: \mathbb{A}\times\mathbb{B}\longrightarrow\mathbb{B}\times\mathbb{A}$ that takes the object (a,b) to (b,a) and the morphism (f,g) to (g,f). For any category $\mathbb{A}$, there is a **diagonal functor** $\Delta_{\mathbb{A}}: \mathbb{A}\longrightarrow\mathbb{A}\times\mathbb{A}$ that takes a to (a,a) and is defined similarly for morphisms. □

Exercise 4.1.45. Show that the braid functor is an isomorphism by finding its inverse. ■

Example 4.1.46. Let $\mathbb{A}$ be a category and $\sim$ be a congruence on $\mathbb{A}$. As we saw in Definition 2.3.10, we can form the quotient category $\mathbb{A}/_{\sim}$. There is a **quotient functor** $\pi\colon \mathbb{A}\longrightarrow\mathbb{A}/_{\sim}$ that is defined as $\pi(a)=a$ and takes every $f\colon a\longrightarrow a'$ to $[f\colon a\longrightarrow a']$, its equivalence class. The functor is the identity on objects—full, but generally not faithful. □

The rest of this section will just be examples of functors. Some of these functors will play prominent roles in the rest of this text.

Example 4.1.47. Let $(M,\star,e)$ be a monoid. An M**-set** is a set S with a way that the elements of M "act on" or change the elements of S. The action is a function $\cdot\colon M\times S\longrightarrow S$, which must satisfy the following two commutative diagrams:

(4.24)

(Reminder: the map $u\colon\{*\}\longrightarrow M$ is the set function that chooses the unit e of the monoid.) In detail, this means the following requirements are satisfied:

- The action must respect the monoid multiplication. This means that if m and m' are elements of M, then the action of $m\star m'$ is the same as first acting with m' and then acting with m. That is, for all m, m' in M; and for all $s\in S$, we have that

$$(m\star m')\cdot s = m\cdot(m'\cdot s), \tag{4.25}$$

where $\star$ is the monoid multiplication.
- The action of the identity of the monoid does not make any changes. That is, for e, the identity of the monoid, and for all $s\in S$,

$$e\cdot s = s. \tag{4.26}$$

One can view an M-set as a functor. Think of M as a one-object category $A(M)$. Then any functor $F: A(M) \longrightarrow \mathbb{S}\mathrm{et}$ is an M-set. The functor F takes the single object $*$ to a set (say S). For every $m \in M$, which corresponds to a morphism $m: * \longrightarrow *$ in the one-object category, the image of m under F describes the action of m, that is, $F(m): S \longrightarrow S$. This is simply an instance of the following isomorphism:

$$Hom(M, Hom(S,S)) \cong Hom(M \times S, S). \tag{4.27}$$

□

Exercise 4.1.48. Show that a functor $A(M) \longrightarrow \mathbb{S}\mathrm{et}$ ensures that Diagrams (4.24) are commutative. ■

Example 4.1.49. There is a functor $O: \mathbb{T}\mathrm{op}^{op} \longrightarrow \mathbb{PO}$, which takes a topological space and outputs the partial order of all the open sets. Letting (T, τ) be a topological space, we write $O(T)$ (rather than $O((T, \tau))$) for the partial order of all the open sets. If U and U' are open sets of T, then $U \leq U'$ if and only if $U \subseteq U'$. The reason for $\mathbb{T}\mathrm{op}^{op}$ and not $\mathbb{T}\mathrm{op}$ is that $f: T \longrightarrow T'$ is a continuous map of topological spaces if f^{-1} takes open sets of T' to open sets of T. This translates into the fact that $O(f): O(T') \longrightarrow O(T)$. It turns out that many properties of a topological space can be determined by simply examining the partial order of open sets. Sometimes this area is humorously called "Pointless Topology" [124]. This is also a beginning of topos theory, which we will see in Section 9.5. □

Example 4.1.50. Let us connect the world of relations and matrices. Consider the natural number n as the set $\bar{n} = \{1, 2, \ldots, n\}$. In this scenario, zero is the empty set. Define the full subcategory $Nat\mathbb{R}\mathrm{el}$ of $\mathbb{R}\mathrm{el}$ to be a category whose objects are these finite sets of natural numbers. In detail, the morphisms from $\bar{m} = \{1, \ldots, m\}$ to $\bar{n} = \{1, \ldots, n\}$ are relations from the first set to the second set. There is a functor $F: Nat\mathbb{R}\mathrm{el} \longrightarrow Bool\mathbb{M}\mathrm{at}$, where $Bool\mathbb{M}\mathrm{at}$ is the category of matrices with Boolean entries. Functor F is bijective on objects (i.e., $F(\bar{m}) = m$). For $R \subseteq \{1, 2, \ldots, m\} \times \{1, 2, \ldots, n\}$, we define matrix $F(R)$ as $(F(R))[i, j] = 1$ if and only if $(j, i) \in R$. □

Exercise 4.1.51. Show that F in Example 4.1.50 is a functor. ■

Exercise 4.1.52. Show that F in Example 4.1.50 is an isomorphism of categories. ■

Example 4.1.53. Similar to $Nat\mathbb{R}\mathrm{el}$, there are also $Nat\mathbb{P}\mathrm{ar}$ (objects are the sets of natural numbers and morphisms are partial functions) and $Nat\mathbb{S}\mathrm{et}$ (objects are sets of natural numbers and morphisms are all functions). As in Diagram (4.4), there is

$$Nat\mathbb{S}\mathrm{et} \hookrightarrow Nat\mathbb{P}\mathrm{ar} \hookrightarrow Nat\mathbb{R}\mathrm{el}, \tag{4.28}$$

and each of these inclusions is the identity on objects and are faithful but not full. There are restriction functors to functor F in Example 4.1.50 as follows:

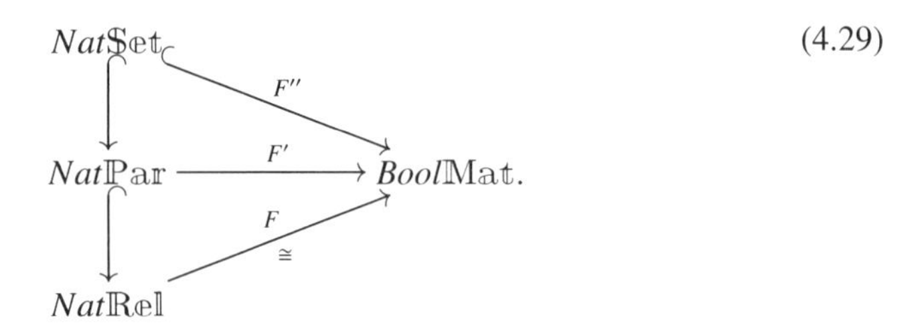

Let us characterize the image of each of these functors:

- The image of F will be any Boolean matrix.
- The image of F' will have any Boolean matrix where there is no more than a single 1 on each row.
- The image of F'' will have exactly one 1 on each row. □

Exercise 4.1.54. There is a subcategory $Nat\mathbb{I}\mathrm{so} \hookrightarrow Nat\mathbb{S}\mathrm{et}$ whose objects are all sets of natural numbers and whose morphisms are only isomorphisms. There is a restriction of F'' to this subcategory. Characterize the image of this restriction. ■

Example 4.1.55. Consider the category $\mathbf{K}\mathbb{M}\mathrm{at}$ of all matrices with entries in $\mathbf{K}$. Let $\mathbf{K}\mathbb{FDV}\mathrm{ect}$ be the full subcategory of $\mathbf{K}\mathbb{V}\mathrm{ect}$ consisting of all finite-dimensional $\mathbf{K}$ vector spaces. There is a functor $F\colon \mathbf{K}\mathbb{M}\mathrm{at} \longrightarrow \mathbf{K}\mathbb{FDV}\mathrm{ect}$ that takes m to the vector space $\mathbf{K}^m$. The functor takes the morphism $A\colon m \longrightarrow n$ (an n by m matrix) to the linear transformation T_A, which is defined for $B \in \mathbf{K}^m$ as $T_A(B) = AB \in \mathbf{K}^n$. The F is a functor because if $A\colon m \longrightarrow n$ and $A'\colon n \longrightarrow p$, then $T_A\colon \mathbf{K}^m \longrightarrow \mathbf{K}^n$, $T_{A'}\colon \mathbf{K}^n \longrightarrow \mathbf{K}^p$,

$$F(A' \circ A) = T_{(A' \cdot A)} = T_{A'} \circ T_A = F(A') \circ F(A), \tag{4.30}$$

and $F(id_m) = T_{Id_{\mathbf{K}^m}} = Id_{\mathbf{K}^m}$. □

Let us consider some examples from logic and computer science.

Example 4.1.56. There is an interesting functor from the category $\mathbb{P}\mathrm{roof}$ of proofs to the category $\mathbb{P}\mathrm{rop}$ of propositions, $Q\colon \mathbb{P}\mathrm{roof} \longrightarrow \mathbb{P}\mathrm{rop}$. The objects of both categories are all propositions of a certain logical system, and Q is the identity on objects. The functor Q takes a proof to the implication that it proves. In particular, if A and B are two propositions and $f\colon A \longrightarrow B$ is a proof that A implies B (there might be many), then $Q(f)\colon Q(A) \longrightarrow Q(B)$. If Q is full (i.e., every implication has a proof in the system), then we say that the logical system is **complete**. In contrast, if Q is not full (i.e., there is an implication that does not have a proof), then we say that the logical system is **incomplete**.

Things get even more interesting when we deal with two logical systems where one system is a subsystem of the other. Then we will have two such functors and the following diagram:

$$\begin{array}{ccc} \mathbb{P}\mathrm{roof}' & \hookrightarrow & \mathbb{P}\mathrm{roof} \\ {\scriptstyle Q'}\downarrow & & \downarrow{\scriptstyle Q} \\ \mathbb{P}\mathrm{rop}' & \hookrightarrow & \mathbb{P}\mathrm{rop}. \end{array} \tag{4.31}$$

Q' can describe a complete logical system and Q can describe a larger incomplete system. (For example, the right system might be Peano arithmetic, which we know from Gödel's incompleteness theorem, is incomplete. At the same time, the left system is a smaller complete system such as Presburger arithmetic.) On the other hand, Q' might be an incomplete logical system, and we can add more logical axioms to form the complete logical system on the right. □

The next three functors relate matrices, Boolean functions, and logical circuits.

Example 4.1.57. There is a functor $FuncDesc\colon \mathbb{Circuit} \longrightarrow Bool\mathbb{Func}$ that describes logical circuits as Boolean functions. Remember from Example 2.1.57 on page 54 that $\mathbb{Circuit}$ is the category of logical circuits. The objects are the natural numbers and the morphisms are logical circuits. Notice that a circuit with m input wires and n output wires has 2^m possible inputs and 2^n possible outputs. Consider the category $Bool\mathbb{Func}$, whose objects are natural numbers and whose morphisms from m to n are the total computable functions from $\{0,1\}^m$ to $\{0,1\}^n$. Composition in $Bool\mathbb{Func}$ is simple function composition. The functor takes every circuit to the Boolean function that it performs. The functoriality means that the composition of circuits gives you the composition of the Boolean functions that they perform. This means

$$FuncDesc(C' \circ C) = FuncDesc(C') \circ FuncDesc(C). \tag{4.32}$$

This functor is the identity on objects and is full. But it is not faithful because there are many logical circuits that perform the same function. □

Example 4.1.58. There is a functor $MatrixDesc\colon \mathbb{Circuit} \longrightarrow Bool\mathbb{Mat}^{op}$ that describes the operations of a logical circuit with a Boolean matrix. Recall that $Bool\mathbb{Mat}$ is the category of Boolean matrices. The objects are the natural numbers and the morphisms are matrices with entries that are either 0 or 1. The functor is defined as follows. On object m of $\mathbb{Circuit}$, $MatrixDesc(m) = 2^m$. Circuit C in $\mathbb{Circuit}$ with m inputs and n outputs will go to the 2^m by 2^n matrix $MatrixDesc(C)$, whose i, j entry is 1 if and only if the jth binary input to C outputs the ith output. We can see how this functor works with some examples of simple logical gates. The inputs are on the top of the matrices and the output is on the left:

$$MatrixDesc(\text{NOT}) = \begin{matrix} & \begin{matrix} 0 & 1 \end{matrix} \\ \begin{matrix} 0 \\ 1 \end{matrix} & \begin{bmatrix} 0 & 1 \\ 1 & 0 \end{bmatrix} \end{matrix} \qquad MatrixDesc(\text{OR}) = \begin{matrix} & \begin{matrix} 00 & 01 & 10 & 11 \end{matrix} \\ \begin{matrix} 0 \\ 1 \end{matrix} & \begin{bmatrix} 1 & 0 & 0 & 0 \\ 0 & 1 & 1 & 1 \end{bmatrix} \end{matrix} \tag{4.33}$$

$$MatrixDesc(\text{AND}) = \begin{matrix} & \begin{matrix} 00 & 01 & 10 & 11 \end{matrix} \\ \begin{matrix} 0 \\ 1 \end{matrix} & \begin{bmatrix} 1 & 1 & 1 & 0 \\ 0 & 0 & 0 & 1 \end{bmatrix} \end{matrix} \tag{4.34}$$

$$MatrixDesc(\text{NAND}) = \begin{matrix} & \begin{matrix} 00 & 01 & 10 & 11 \end{matrix} \\ \begin{matrix} 0 \\ 1 \end{matrix} & \begin{bmatrix} 0 & 0 & 0 & 1 \\ 1 & 1 & 1 & 0 \end{bmatrix}. \end{matrix} \tag{4.35}$$

The functor is contravariant because the composition of circuits goes to matrix multiplication in opposite order, that is, for $C\colon m \longrightarrow n$ and $C'\colon n \longrightarrow p$, we have

$$MatrixDesc(C' \circ C) = MatrixDesc(C) \cdot MatrixDesc(C'). \tag{4.36}$$

This functor is injective on objects and full, but not faithful. It is full because every such matrix can be described by a circuit. It is not faithful because there can be many circuits that can describe the same Boolean matrix. A variation of this functor will play a major role in our mini-course on Quantum Computing in Section 10.3. □

Exercise 4.1.59. Show that *MatrixDesc* satisfies the requirements of being a functor. ■

Example 4.1.60. There is a functor $FuncEval\colon Bool\mathbb{F}\mathbb{u}\mathbb{n}\mathbb{c} \longrightarrow Bool\mathbb{M}\mathbb{a}\mathbb{t}^{op}$ that describes Boolean functions as Boolean matrices. On objects, *FuncEval* is like *MatrixDesc* and takes m to 2^m. The functor *FuncEval* takes a Boolean function $f\colon \{0,1\}^m \longrightarrow \{0,1\}^n$ to the 2^m by 2^n matrix $FuncEval(f)$, whose i, j entry is 1 if and only if the jth binary input to f outputs the ith output. This functor is contravariant for the same reason as the *MatrixDesc* is. This functor is injective (but not surjective) on objects, full and faithful. □

Examples 4.158, 4.1.59, and 4.1.60 have to be looked at together. They express a triangle of functors as follows:

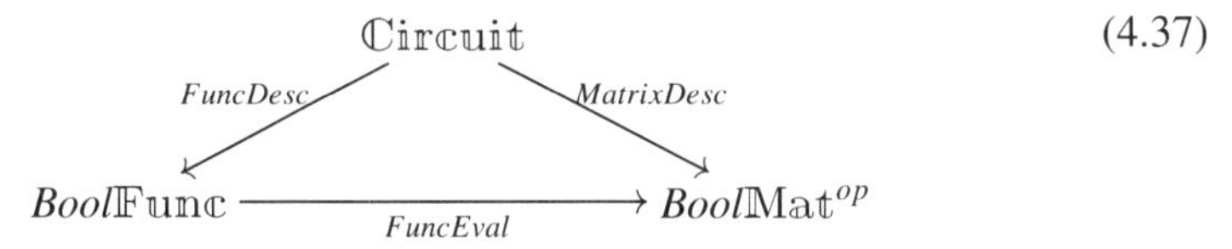

(4.37)

It is not hard to see that the triangle commutes.

Example 4.1.61. Remember from Example 2.3.4 the category $\mathbb{NANDCircuit}$. The objects are the natural numbers, and the set of morphisms from m to n is the set of all logical circuits made of NAND gates that have m input wires and n output wires. This category is a subcategory of $\mathbb{Circuit}$. Notice that $\mathbb{NANDCircuit}$ and $\mathbb{Circuit}$ have the same objects. There is an obvious inclusion functor $inc\colon \mathbb{NANDCircuit} \hookrightarrow \mathbb{Circuit}$, which is the identity on the objects. Every circuit in $\mathbb{Circuit}$ is equivalent to a circuit with only NAND gates and fanout operations. This means that there is a functor $F\colon \mathbb{Circuit} \longrightarrow \mathbb{NANDCircuit}$ that is the identity on objects and takes every circuit to an equivalent NAND circuit as in the following diagram:

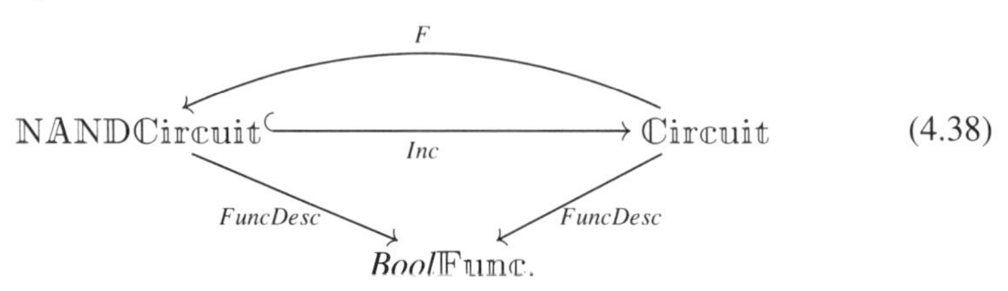

(4.38)

The two triangles commute, and $F \circ Inc = Id$. In general, $Inc \circ F \neq Id$. □

We will see many more examples of functors in the pages to come.

4.2 Natural Transformations

Functors are just the beginning of the story. Just as functors relate categories, natural transformations relate functors. A functor goes from a category to a category, while a natural transformation goes from a functor to a functor. Intuitively, if you think of functors $F\colon \mathbb{A} \longrightarrow \mathbb{B}$ and $G\colon \mathbb{A} \longrightarrow \mathbb{B}$ as ways of providing images of $\mathbb{A}$ in $\mathbb{B}$, then a natural transformation from F to G is a way of going from the image of F to the image of G. One can visualize a natural transformation as Diagram 4.1.

Definition 4.2.1. Let $F\colon \mathbb{A} \longrightarrow \mathbb{B}$ and $G\colon \mathbb{A} \longrightarrow \mathbb{B}$ be functors. A **natural transformation** α from F to G, written as $\alpha\colon F \Longrightarrow G$ or

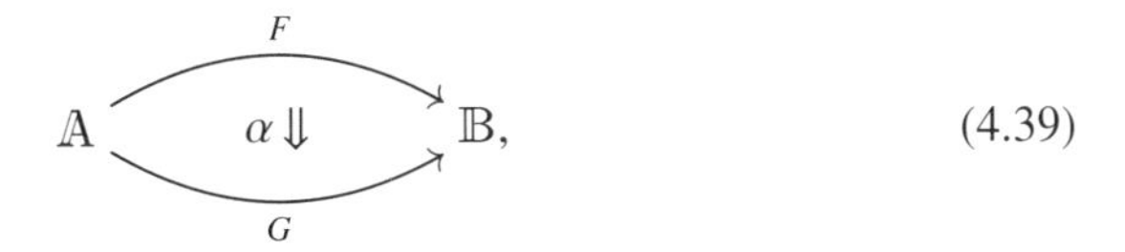

(4.39)

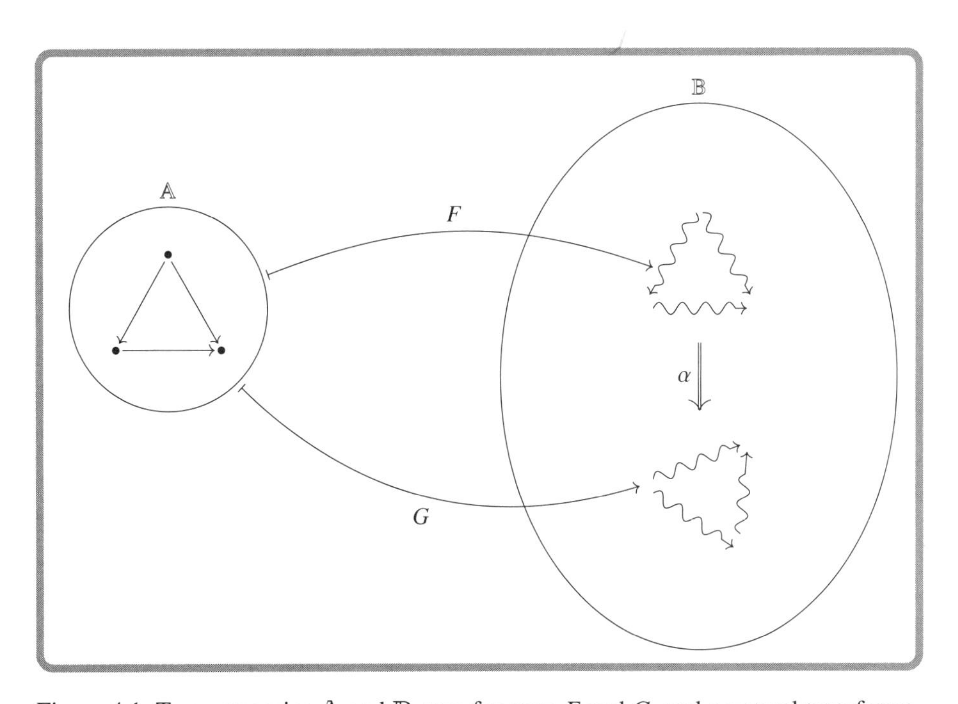

Figure 4.1. Two categories $\mathbb{A}$ and $\mathbb{B}$, two functors F and G, and a natural transformation α that takes the image of F to the image of G.

is a rule that assigns to every object a in $\mathbb{A}$ a morphism $\alpha_a\colon F(a) \longrightarrow G(a)$, called the **component of** α **at** a. This assignment must further satisfy the following **naturality condition**: for every morphism $f\colon a \longrightarrow a'$ in $\mathbb{A}$, the square

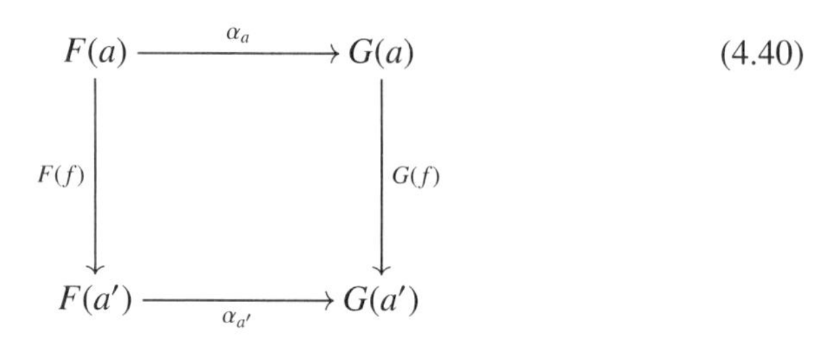

(4.40)

commutes.

Notice that a natural transformation is written with a $\Longrightarrow$, but each of its components is written with a $\longrightarrow$.

The functors F and G can be described as in Diagram (4.1). We can add α to those diagrams as shown in

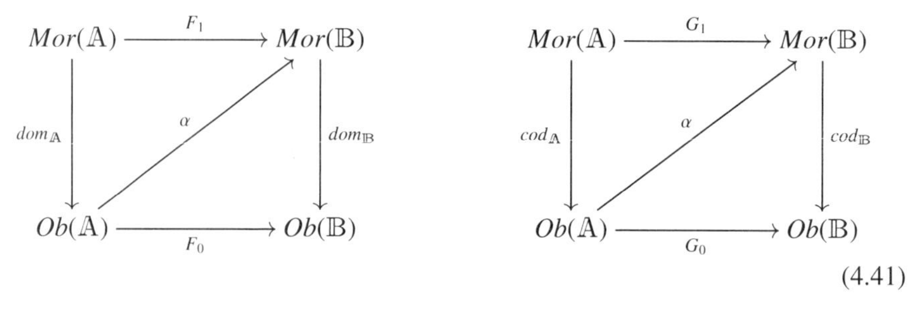

(4.41)

and insist that the lower triangles commute. Let us explain the squares. The diagonal map takes an object a to $\alpha_a\colon F(a) \longrightarrow G(a)$. The lower triangle in the left square says that α_a goes from $F(a)$, while the lower triangle in the right square says that α_a ends in $G(a)$. Notice that the top triangle do not, in general, commute.

First, here are some examples.

Example 4.2.2. Just as every category has a unique identity functor, every functor has a unique **identity natural transformation**. Such a natural transformation does not change the functor. Formally, for every $F\colon \mathbb{A} \longrightarrow \mathbb{B}$, there is a natural transformation $\iota_F\colon F \Longrightarrow F$ (ι is the Greek letter iota) where each component is $(\iota_F)_a = id_{F(a)}\colon F(a) \longrightarrow F(a)$. □

Example 4.2.3. In computer science, there is a **list functor** $List\colon \mathbb{Set} \longrightarrow \mathbb{Set}$, which takes a set of elements to the set of all lists or sequences of the elements of the set. (In Example 4.1.10, this functor was also called the "free monoid functor.") For $S = \{a, b, c\}$, we have $List(S) = \{\emptyset, a, b, c, aa, ab, ac, ba, bb, bc, ca, cb, cc, aaa, \ldots\}$. For a set map $f\colon S \longrightarrow T$, $List$ takes every element in the list to the value of the function. For example, if $f\colon \{a, b, c\} \longrightarrow \{x, y\}$ where $a \mapsto y$, $b \mapsto x$, and $c \mapsto y$, then the function

List(f) will take the word *accbab* to the word *yyyxyx*. There are three natural transformations associated with this functor:

- *Reverse*: *List*( ) $\Longrightarrow$ *List*( ). This natural transformation takes a word to the reversed word. For example, $Reverse_S(accbab) = babcca$.
- *Unit*: $Id_{\mathbb{S}\mathrm{et}}(\ ) \Longrightarrow$ *List*( ). This natural transformation takes an element of the original set (alphabet) to the word of that single letter. For example, $Unit_S(b) = b$.
- *Flatten*: *List*(*List*( )) $\Longrightarrow$ *List*( ). This natural transformation takes a list of lists and makes it into a list of words. For example, for the set $S = \{a, b, c\}$, an element of *List*(*List*(S)) is $aba, bbbbc, caaba, bca$. This element will go to the element $ababbbbccaababca$.

One needs to show that all these are, in fact, natural. We leave this to the reader. □

Example 4.2.4. Given two M sets $\cdot\colon M \times S \longrightarrow S$ and $\cdot'\colon M \times S' \longrightarrow S'$, an **$M$-set homomorphism** is a set function $f\colon S \longrightarrow S'$ that respects the action. This means that for all $m \in M$ and $s \in S$, f satisfies $f(m \cdot s) = m \cdot' f(s)$. In terms of diagrams, this means that the square

$$\begin{array}{ccc} M \times S & \xrightarrow{id_M \times f} & M \times S' \\ {\scriptstyle \cdot}\downarrow & & \downarrow{\scriptstyle \cdot'} \\ S & \xrightarrow[f]{} & S' \end{array} \tag{4.42}$$

commutes. If $F\colon M \longrightarrow \mathbb{S}\mathrm{et}$ and $G\colon M \longrightarrow \mathbb{S}\mathrm{et}$ describe M-sets, then a natural transformation from F to G is the same thing as an M-set homomorphism. □

There are two types of composition of natural transformations: vertical composition and horizontal composition. **Vertical composition**, $\circ_V$, takes a natural transformation $\alpha\colon F \Longrightarrow G$ and a natural transformation $\beta\colon G \Longrightarrow H$ and gives a natural transformation $\beta \circ_V \alpha\colon F \Longrightarrow H$. This can be visualized as in Figure 4.2 or as follows:

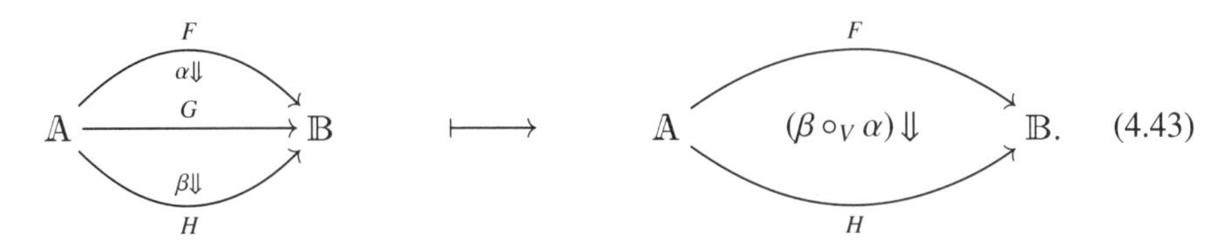

The component of $\beta \circ_V \alpha$ on element a of $\mathbb{A}$ is defined as

$$(\beta \circ_V \alpha)_a = \beta_a \circ \alpha_a\colon F(a) \longrightarrow G(a) \longrightarrow H(a), \tag{4.44}$$

which is natural because each of the following squares are commutative, and hence the whole diagram is commutative:We will formally meet functor categories

$$\begin{array}{ccccc} F(a) & \xrightarrow{\alpha_a} & G(a) & \xrightarrow{\beta_a} & H(a) \\ {\scriptstyle F(f)}\downarrow & & \downarrow{\scriptstyle G(f)} & & \downarrow{\scriptstyle H(f)} \\ F(a') & \xrightarrow[\alpha_{a'}]{} & G(a') & \xrightarrow[\beta_{a'}]{} & H(a'). \end{array} \tag{4.45}$$

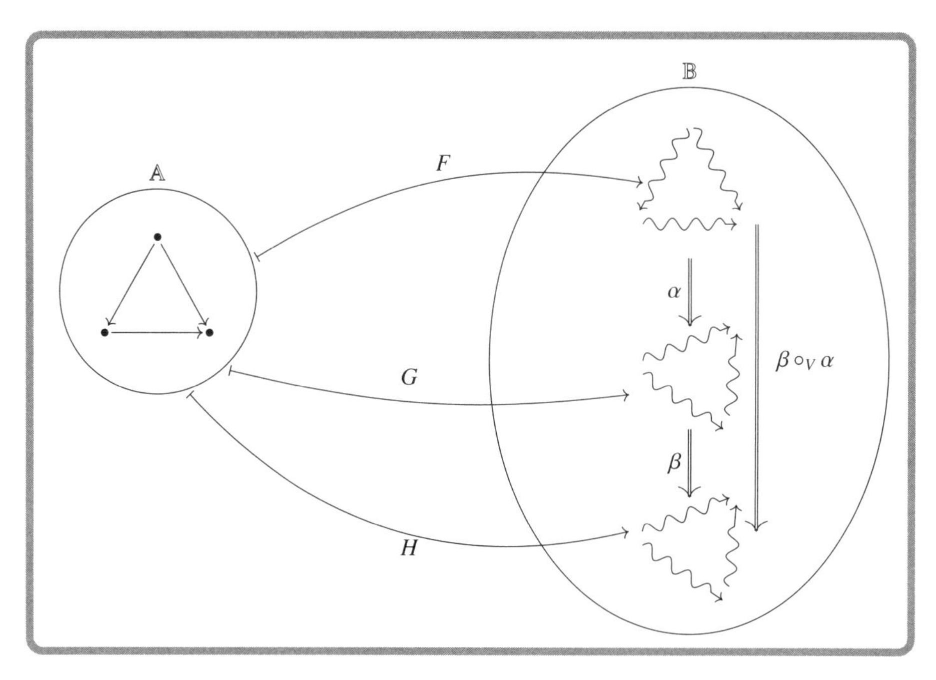

Figure 4.2. Vertical composition of natural transformations.

Exercise 4.2.5. Show that vertical composition is associative. ■

Exercise 4.2.6. Show that the identity natural transformation is a unit for vertical composition. ■

Remark 4.2.7. The definition of vertical composition with Exercises 4.2.5 and 4.2.6 show that the collection of functors from $\mathbb{A}$ to $\mathbb{B}$ and natural transformations between such functors form a category. We call such a structure a **functor category** and denote it as $\mathbb{B}^{\mathbb{A}} = Hom_{\mathbb{C}\mathrm{at}}(\mathbb{A}, \mathbb{B})$. We will formally meet functor categories in Section 4.5. ♠

With the notion of vertical composition of natural transformation, we can talk about isomorphic natural transformations.

Definition 4.2.8. A natural transformation $\alpha\colon F \Longrightarrow G$ is a **natural isomorphism** if there exists a $\beta\colon G \Longrightarrow F$ such that $\beta \circ_V \alpha = \iota_F$ and $\alpha \circ_V \beta = \iota_G$. In such a case, F and G are called **isomorphic functors**.

It can easily be seen that a natural transformation α is a natural isomorphism if and only if every one of its components $\alpha_a\colon F(a) \longrightarrow G(a)$ is an isomorphism.

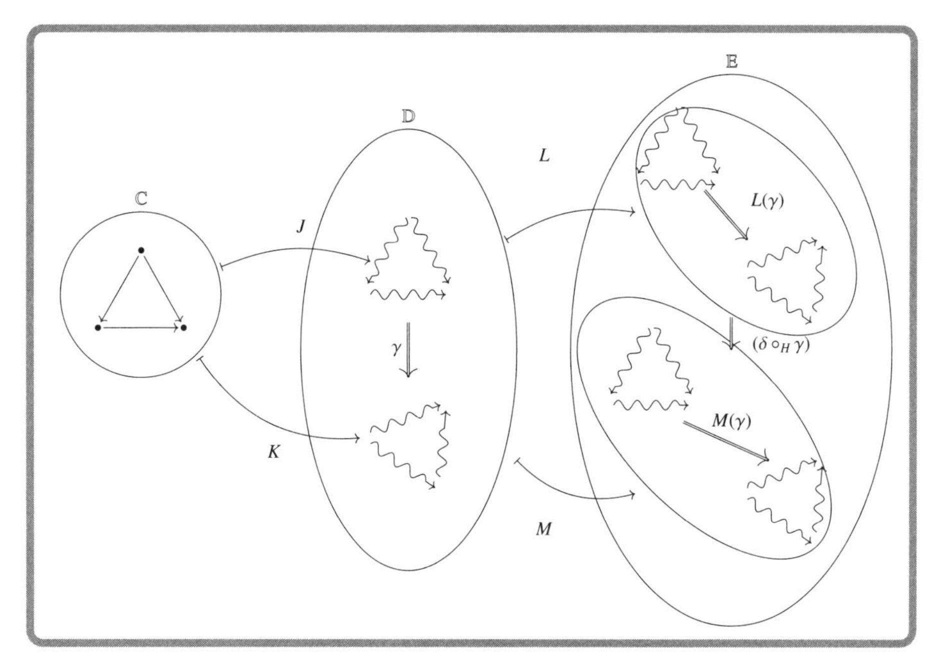

Figure 4.3. Horizontal composition of natural transformations.

Horizontal composition, $\circ_H$, can be visualized by Figure 4.3 and the following diagram:

$$\mathbb{C} \underset{K}{\overset{J}{\Downarrow \gamma}} \mathbb{D} \underset{M}{\overset{L}{\Downarrow \delta}} \mathbb{E} \quad \longmapsto \quad \mathbb{C} \underset{M \circ K}{\overset{L \circ J}{\Downarrow (\delta \circ_H \gamma)}} \mathbb{E}, \tag{4.46}$$

where the $c \in \mathbb{C}$ component of $\delta \circ_H \gamma$ is defined as

$$(\delta \circ_H \gamma)_c = \delta_{K(c)} \circ L(\gamma_c)\colon L(J(c)) \longrightarrow L(K(c)) \longrightarrow M(K(c)). \tag{4.47}$$

This is equivalent to defining it as

$$(\delta \circ_H \gamma)_c = M(\gamma_c) \circ \delta_{J(c)}\colon L(J(c)) \longrightarrow M(J(c)) \longrightarrow M(K(c)) \tag{4.48}$$

because the following square in $\mathbb{E}$ commutes out of the naturality of δ:

$$\begin{array}{ccc} LJ(c) & \xrightarrow{\delta_{J(c)}} & MJ(c) \\ {\scriptstyle L(\gamma_c)}\downarrow & \searrow^{(\delta \circ_H \gamma)_c} & \downarrow{\scriptstyle M(\gamma_c)} \\ LK(c) & \xrightarrow[\delta_{K(c)}]{} & MK(c). \end{array} \tag{4.49}$$

The mapping $\delta \circ_H \gamma$ is natural because for all $f\colon c \longrightarrow c'$ in $\mathbb{C}$, there is

$$\begin{array}{ccccc} LJ(c) & \xrightarrow{\delta_{Jc}} & MJ(c) & \xrightarrow{M(\gamma_c)} & MK(c) \\ {\scriptstyle LJ(f)}\downarrow & & \downarrow{\scriptstyle MJ(f)} & & \downarrow{\scriptstyle MK(f)} \\ LJ(c') & \xrightarrow{\delta_{Jc'}} & MJ(c') & \xrightarrow{M(\gamma_{c'})} & MK(c'). \end{array} \tag{4.50}$$

The left square commutes because of the naturality of δ. The right square commutes because of the naturality of γ and the functoriality of M.

Exercise 4.2.9. Show that horizontal composition is associative. ■

When there is a natural transformation of the following form:

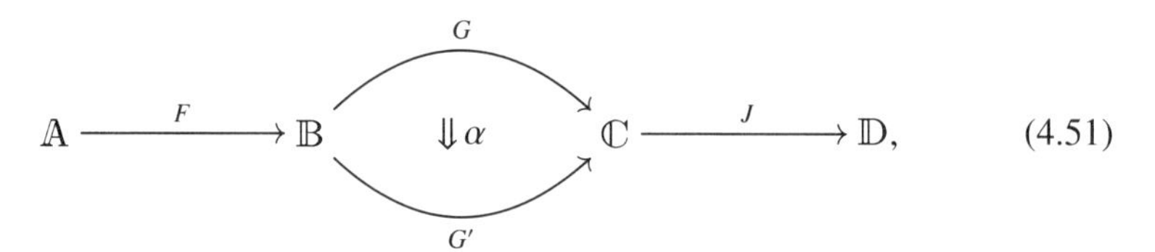

(4.51)

then we consider the unwritten identity natural transformations ι_F and ι_J. We denote the horizontal composition of $\alpha \circ_H \iota_F$ as αF and $\iota_J \circ_H \alpha$ as $J\alpha$. These compositions of functors and natural transformations are called **whiskering**.

The vertical and horizontal compositions relate as follows:

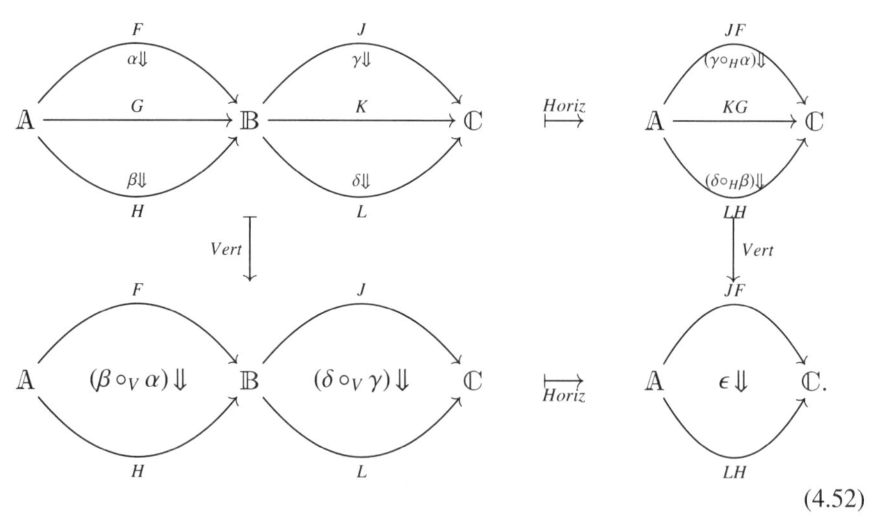

(4.52)

The vertical maps are describing vertical composition of natural transformations, while the horizontal maps are horizontal compositions. The diagram shows two ways to go from the four natural transformations to one natural transformation. Using the definitions of the compositions gives the same natural transformation. This means that the $\epsilon \Downarrow$ in the bottom-right corner is

$$(\delta \circ_V \gamma) \circ_H (\beta \circ_V \alpha) = (\delta \circ_H \beta) \circ_V (\gamma \circ_H \alpha). \tag{4.53}$$

The fact that these two compositions form the same natural transformation is another instance of the interchange law, (see Important Categorical Idea 3.1.18). Here, the two operations are vertical and horizontal compositions of natural transformations.

Exercise 4.2.10. Use the definitions of $\circ_V$ and $\circ_H$ to prove Equation 4.53. ■

Remark 4.2.11. While $\mathbb{CAT}$ and $\mathbb{C}\mathrm{at}$ each contain categories and functors, they are not the complete picture. Natural transformations are a third level of structure. When all three levels are gathered, one has the structure of a **2-category**. Such structures contain objects (also called **0-cells**), morphisms between objects (also called **1-cells**), and **2-cells** between morphisms. We denote a 2-category as a category with a line above it, as in $\overline{\mathbb{A}}$. In particular, the 2-category versions of $\mathbb{CAT}$ and $\mathbb{C}\mathrm{at}$ are denoted as $\overline{\mathbb{CAT}}$ and $\overline{\mathbb{C}\mathrm{at}}$, respectively. We will see more about 2-categories throughout the rest of this text. They will be formally defined and discussed in terms of higher category theory in Section 9.4. ♠

The rest of this section contains examples of natural transformations.

Example 4.2.12. This example of a natural isomorphism comes from the first few pages of Eilenberg and Mac Lane's premiere paper introducing category theory [68]. It is a motivating example to show the importance of the naturality condition.

There is a functor $(\ )^*\colon \mathbf{K}\mathbb{FDV}\mathrm{ect} \longrightarrow \mathbf{K}\mathbb{FDV}\mathrm{ect}^{op}$ that takes a vector space V to its **dual vector space** $V^* = Hom_{\mathbf{K}\mathbb{FDV}\mathrm{ect}}(V, \mathbf{K})$. We saw the dual vector space at the end of Section 2.4. Let V be a finite-dimensional $\mathbf{K}$-vector space with basis $\mathcal{E} = \{e_1, e_2, \ldots, e_n\}$. Then V^* is a set of functions that has the structure of a finite-dimensional vector space and has basis $\mathcal{F} = \{f_1, f_2, \ldots, f_n\}$, where each $f_i\colon V \longrightarrow \mathbf{K}$ is defined as $f_i(e_j) = 1$ if $i = j$ and $f_i(e_j) = 0$ otherwise. For an arbitrary $v \in V$ where $v = k_1e_1 + k_2e_2 + \cdots + k_ne_n$, we have $f_i(v) = k_i$.

There is a perfectly legitimate isomorphism of vector spaces $\varphi_{V,\mathcal{E},\mathcal{F}}\colon V \longrightarrow V^*$, which depends on $\mathcal{E}$ and $\mathcal{F}$. It is defined as

$$\varphi_{V,\mathcal{E},\mathcal{F}}(k_1e_1 + k_2e_2 + \cdots + k_ne_n) = k_1f_1 + k_2f_2 + \cdots + k_nf_n. \tag{4.54}$$

The inverse is obviously

$$\varphi^{-1}_{V,\mathcal{E},\mathcal{F}}(k_1f_1 + k_2f_2 + \cdots + k_nf_n) = k_1e_1 + k_2e_2 + \cdots + k_ne_n. \tag{4.55}$$

The fact that these isomorphisms depend on a basis has an "unnatural" feeling to it. When two vector spaces are isomorphic, why should the isomorphism depend on how the elements are expressed? It seems more—dare we say—natural for there to be an isomorphism that is independent of how the elements are expressed.

There is an isomorphism that does not depend on how it is presented. Compose the functor $(\ )^*\colon \mathbf{K}\mathbb{FDV}\mathrm{ect} \longrightarrow \mathbf{K}\mathbb{FDV}\mathrm{ect}^{op}$ with itself to get a covariant functor $(\ )^{**}\colon \mathbf{K}\mathbb{FDV}\mathrm{ect} \longrightarrow (\mathbf{K}\mathbb{FDV}\mathrm{ect}^{op})^{op} = \mathbf{K}\mathbb{FDV}\mathrm{ect}$, which takes every vector space V to its "double dual" V^{**}. The elements of

$$V^{**} = Hom_{\mathbf{K}\mathbb{FDV}\mathrm{ect}}(Hom_{\mathbf{K}\mathbb{FDV}\mathrm{ect}}(V, \mathbf{K}), \mathbf{K}) \tag{4.56}$$

are functions $\psi\colon V^* \longrightarrow \mathbf{K}$.

There is an isomorphism of vector spaces $\theta_V\colon V \longrightarrow V^{**}$ that is defined on an element $v \in V$ to be the function that always evaluates on the element v. That is,

$$\theta_V(v) = ev[v], \tag{4.57}$$

where the function $ev[v]\colon V^* \longrightarrow \mathbf{K}$ is defined for any linear map $f\colon V \longrightarrow \mathbf{K}$ as

$$ev[v](f) = f(v). \tag{4.58}$$

This means that $ev[v]$ simply evaluates at v. The linear transformation θ_V is an isomorphism because it is injective, and since $dim(V) = dim(V^*) = dim(V^{**})$, we know that it is an isomorphism. (Interestingly, one can describe an inverse $\theta_V^{-1}\colon V^{**} \longrightarrow V$, but only if you specify a basis. In other words, the isomorphism is natural, but the proof of the fact that it is an isomorphism is not natural.) These θ_V are components of a natural transformation:

$$\mathbf{K}\mathbb{FDV}\mathbb{ect} \underset{(\ )^{**}}{\overset{Id}{\Downarrow \theta}} \mathbf{K}\mathbb{FDV}\mathbb{ect}. \tag{4.59}$$

The fact that it is natural means that for all linear transformations $T\colon V \longrightarrow W$, we have

$$\begin{array}{ccc} V & \xrightarrow{\theta_V} & V^{**} \\ {\scriptstyle T}\downarrow & & \downarrow{\scriptstyle T^{**}} \\ W & \xrightarrow[\theta_W]{} & W^{**}. \end{array} \tag{4.60}$$

Stated in words, we have shown that although a vector space, V, is isomorphic with V^*, the isomorphism depends on the basis. In contrast, V is naturally isomorphic to V^{**}. This means that the isomorphism does not depend on the way the elements are described. This naturality is one of the most important ideas in category theory. □

Example 4.2.13. Let $\mathbb{A}$ be a category with finite products. For objects a and b, there is a functor $Hom_{\mathbb{A}}(\ , a \times b)\colon \mathbb{A} \longrightarrow \mathbb{S}\mathbb{et}$ that takes c to $Hom_{\mathbb{A}}(c, a \times b)$. There is another functor, $Hom_{\mathbb{A}}(\ , a) \times Hom_{\mathbb{A}}(\ , b)\colon \mathbb{A} \longrightarrow \mathbb{S}\mathbb{et}$, which takes c to the set $Hom_{\mathbb{A}}(c, a) \times Hom_{\mathbb{A}}(c, b)$. The rule that assigns to every pair of maps f and g the induced map $\langle f, g \rangle$ is a natural isomorphism. Formally,

$$\langle\ ,\ \rangle\colon Hom_{\mathbb{A}}(\ , a) \times Hom_{\mathbb{A}}(\ , b) \Longrightarrow Hom_{\mathbb{A}}(\ , a \times b). \tag{4.61}$$

"Naturality" here means that if there is a function $f\colon c' \longrightarrow c$ in $\mathbb{A}$, then the following diagram commutes:

$$\begin{array}{ccc} Hom_{\mathbb{A}}(c,a)\times Hom_{\mathbb{A}}(c,b) & \xrightarrow[\cong]{\langle\ ,\ \rangle_c} & Hom_{\mathbb{A}}(c,a\times b) \\ {\scriptstyle f^*\times f^*}\downarrow & & \downarrow{\scriptstyle f^*} \\ Hom_{\mathbb{A}}(c',a)\times Hom_{\mathbb{A}}(c',b) & \xrightarrow[\cong]{\langle\ ,\ \rangle_{c'}} & Hom_{\mathbb{A}}(c',a\times b). \end{array} \tag{4.62}$$

The fact that this natural transformation is a natural isomorphism follows from our discussion on page 88. □

Example 4.2.14. The category $\mathbb{Group}$ has groups as objects and homomorphisms between groups as morphisms. There are no nontrivial morphisms between homomorphisms, which means that $\mathbb{Group}$ does not have 2-cells. However, if we look at groups as one-object categories within the 2-category $\overline{\mathbb{Cat}}$, and homomorphisms as functors between such one-object categories, then there are natural transformations between such functors. Let G and G' be groups thought of as one-object categories, and let $F, H\colon G \longrightarrow G'$ be functors. Then consider a natural transformation $\alpha\colon F \Longrightarrow H$, as in

$$G \underset{H}{\overset{F}{\underset{\alpha\Downarrow}{\rightrightarrows}}} G'. \tag{4.63}$$

Since there is only one object $*$ in G, there is only one component of α (namely, $\alpha_*\colon F(*) \longrightarrow H(*)$). The morphism α_* is an element of G'. The naturality condition amounts to the commutativity of

$$\begin{array}{ccc} F(*) & \xrightarrow{\alpha_*} & H(*) \\ {\scriptstyle F(g)}\downarrow & & \downarrow{\scriptstyle H(g)} \\ F(*) & \xrightarrow[\alpha_*]{} & H(*) \end{array} \tag{4.64}$$

for all morphisms g in G. This means that a natural transformation is an element α_* in G' such that for all g in G, we have that $\alpha_* F(g) = H(g)\alpha_*$ or $F(g) = (\alpha_*)^{-1} H(g)\alpha_*$.

Notice that in the same way, we can talk about a natural transformation between two monoid homomorphisms, as in

$$M \underset{H}{\overset{F}{\underset{\alpha\Downarrow}{\rightrightarrows}}} M'. \tag{4.65}$$

In that case, we have $\alpha_* F(m) = H(m)\alpha_*$. In general, we cannot write $F(m) = (\alpha_*)^{-1} H(m)\alpha_*$ because α_* need not be invertible. However, there might be an invertible element in M' and we can insist that our natural transformation uses this invertible element. We will see variations of this idea in Section 7.4. □

4.3 Equivalences

We saw (in Important Categorical Idea 4.1.29) that the notion of isomorphism of categories is very strong and hence does not arise often in a nontrivial way. However, if we weaken the notion of isomorphism, we get the notion of equivalence of categories, which arises often. If $F\colon \mathbb{A} \longrightarrow \mathbb{B}$ is an isomorphism, then for every b in $\mathbb{B}$, there is a unique a in $\mathbb{B}$ such that $F(a) = b$. This basically says that the structure of $\mathbb{A}$ and $\mathbb{B}$ are exactly the same. Here, we weaken this so that for every b in $\mathbb{B}$, there is some a in $\mathbb{A}$ so that $F(a)$ is not necessarily equal to b but is *isomorphic* to b. This is the essence of an equivalence of categories.

Definition 4.3.1. Categories $\mathbb{A}$ and $\mathbb{B}$ have an **equivalence** between them if there are functors $F\colon \mathbb{A} \longrightarrow \mathbb{B}$ and $G\colon \mathbb{B} \longrightarrow \mathbb{A}$ such that $G \circ F \cong Id_{\mathbb{A}}$ and $Id_{\mathbb{B}} \cong F \circ G$. Similar to Diagrams (1.33) and (4.8), we can express this as

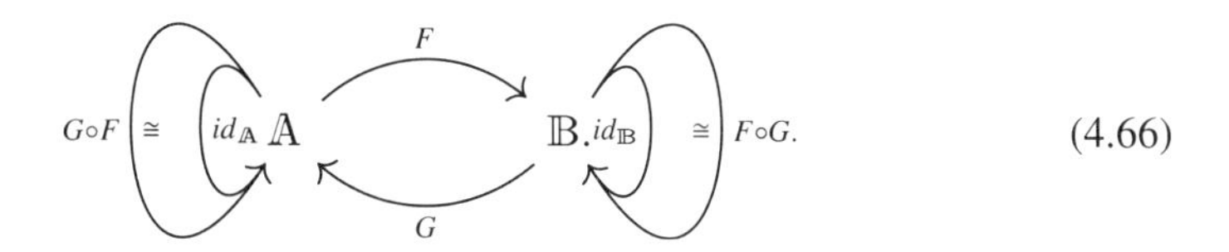

(4.66)

Functors F and G are called **quasi-inverses** of each other, and we say that the two categories are **equivalent categories**. We denote an equivalence as $\mathbb{A} \simeq \mathbb{B}$. Unpacking the definition shows that, for every a in $\mathbb{A}$, there is a b in $\mathbb{B}$, such that $G(b)$ is isomorphic to a, and for every b in $\mathbb{B}$, there is an a in $\mathbb{A}$, such that $F(a)$ is isomorphic to b.

Let us describe another way of discussing the equivalence of categories. From our discussion, there arises a definition of a special type of functor from Definition 4.3.2.

Definition 4.3.2. We say that a functor $F\colon \mathbb{A} \longrightarrow \mathbb{B}$ is **essentially surjective** if for all b in $\mathbb{B}$, there is an a in $\mathbb{A}$ such that $F(a)$ is isomorphic to b.

Theorem 4.3.3. Categories $\mathbb{A}$ and $\mathbb{B}$ are equivalent if and only if there is a functor $F\colon \mathbb{A} \longrightarrow \mathbb{B}$ that is (i) full, (ii) faithful, and (iii) essentially surjective. ★

Proof. ($\Longleftarrow$) Let $F\colon \mathbb{A} \longrightarrow \mathbb{B}$ be full, faithful, and essentially surjective. We describe G, a quasi-inverse to F. For b in $\mathbb{B}$, we let $G(b)$ be an a such that $F(a) \cong b$, which definitely exists because F is essentially surjective. (For the cognoscenti, this assumes the axiom of choice.) This will mean that there is an isomorphism $\beta_b\colon FG(b) \longrightarrow b$. Given $h\colon b \longrightarrow b'$ in $\mathbb{B}$, we can form

$$FGb \xrightarrow[\cong]{\beta} b \xrightarrow{h} b' \xrightarrow[\cong]{\beta^{-1}} FGb'. \tag{4.67}$$

Since the source and the target of this morphism is in the image of F and F is full and faithful, there is a unique $\hat{h}\colon Gb \longrightarrow Gb'$ such that $F(\hat{h})$ is this morphism. Set

$G(h) = \hat{h}$, which defines G and shows that β is natural. Since $GF(a)$ can equal a and the isomorphism can be the identity, we can see that the $\alpha\colon GF \Longrightarrow Id_{\mathbb{A}}$ is natural.

($\Longrightarrow$) Let $\alpha\colon G \circ F \longrightarrow Id_{\mathbb{A}}$ and $\beta\colon F \circ G \longrightarrow Id_{\mathbb{B}}$ be two natural isomorphisms. The functor F is as follows:

- Essentially surjective. For every b, there is a $G(b)$ and an isomorphism $\beta_b\colon FG(b) \longrightarrow b$.
- Faithful. If f and f' are morphisms in $\mathbb{A}$ and $F(f) = F(f')$, then by composing with G, we have $GF(f) = GF(f')$. By further composing with $\alpha_{a'}$ and α_a^{-1} as in the following diagram:

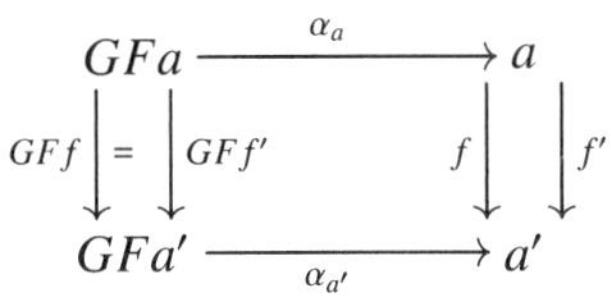

$$\begin{array}{ccc} GFa & \xrightarrow{\alpha_a} & a \\ {\scriptstyle GFf}\downarrow \; = \; \downarrow{\scriptstyle GFf'} & & {\scriptstyle f}\downarrow \; \downarrow{\scriptstyle f'} \\ GFa' & \xrightarrow[\alpha_{a'}]{} & a' \end{array} \tag{4.68}$$

We can see that

$$\alpha_{a'}(GFf)\alpha_a^{-1} = \alpha_{a'}(GFf')\alpha_a^{-1}. \tag{4.69}$$

The naturality of α implies that $f = f'$. We can make a similar argument and show that G is also faithful.
- Full. If $g\colon F(a) \longrightarrow F(a')$, then $G(g)\colon GF(a) \longrightarrow GF(a')$. Let $f\colon a \longrightarrow a'$ be defined as $\alpha_{a'} \circ G(g) \circ \alpha_a^{-1}$. Also, consider $GF(f)\colon GF(a) \longrightarrow GF(a')$. We have the commutativity of the following two squares:

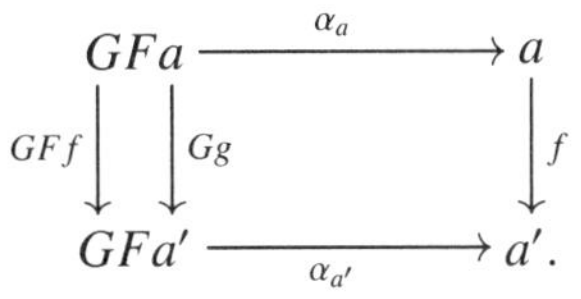

$$\begin{array}{ccc} GFa & \xrightarrow{\alpha_a} & a \\ {\scriptstyle GFf}\downarrow \; \downarrow{\scriptstyle Gg} & & \downarrow{\scriptstyle f} \\ GFa' & \xrightarrow[\alpha_{a'}]{} & a'. \end{array} \tag{4.70}$$

One square commutes out of the definition of f and another square commutes out of the naturality of α. From this, we get $GF(f) = G(g)$. Use the fact that G is faithful, and we get that $F(f) = g$. Thus, F is full. ♣

For category $\mathbb{A}$, the skeletal category $sk(\mathbb{A})$ includes into $\mathbb{A}$. But we can say more.

Theorem 4.3.4. Every category is equivalent to its skeletal category. Furthermore, if two categories are equivalent, then their skeletal categories are isomorphic; that is,

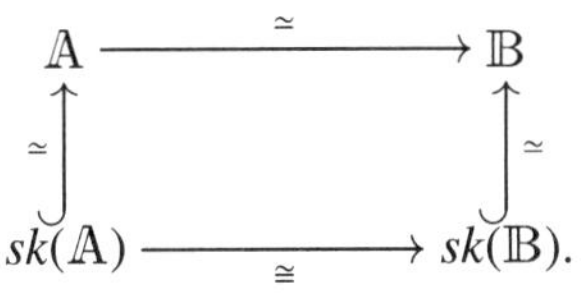

$$\begin{array}{ccc} \mathbb{A} & \xrightarrow{\simeq} & \mathbb{B} \\ {\scriptstyle \simeq}\uparrow & & \uparrow{\scriptstyle \simeq} \\ sk(\mathbb{A}) & \xrightarrow[\cong]{} & sk(\mathbb{B}). \end{array} \tag{4.71}$$

In particular, any two skeletal categories of a category are isomorphic. ★

Proof. Let $F\colon \mathbb{A} \longrightarrow \mathbb{B}$ be an equivalence of categories, and let a' be an element of $sk(\mathbb{A})$. Functor F takes a' to b in $\mathbb{B}$. There is an element b' in $sk(\mathbb{B})$ that is isomorphic to b. (The axiom of choice was used here.) We make a functor $F'\colon sk(\mathbb{A}) \longrightarrow sk(\mathbb{B})$ that takes a' to b'. For each a, there is exactly one such b, and vice versa. There is a similar discussion for morphisms. Hence, this functor is an isomorphism. ♣

Let us look at some examples of equivalence of categories and skeletal categories.

Example 4.3.5. The world's simplest example of an equivalence is the relationship between the one-object category $\mathbf{1}$ and category $\mathbf{2}_I$, which have two objects and a single isomorphism between them. We can view it as

$$* \qquad \simeq \qquad a \xrightarrow{\quad\cong\quad} b. \tag{4.72}$$

In detail, there is a unique functor $!\colon \mathbf{2}_I \longrightarrow \mathbf{1}$, and there is a functor $L\colon \mathbf{1} \longrightarrow \mathbf{2}_I$ such that $L(*) = a$. It is obvious that $L\circ ! = id_{\mathbf{1}}$ and $! \circ L \cong Id_{\mathbf{2}_I}$. Another way to say this is that L is essentially surjective. Thus, we have shown that $sk(\mathbf{2}_I) = \mathbf{1}$. Notice that $\mathbf{2}_I$ is a groupoid. We can extend this example to all groupoids that have a unique isomorphism between any two objects. Such groupoids are called **contractable**. They are connected groupoids where all diagrams commute. It is easy to see that any contractable groupoid is equivalent to $\mathbf{1}$. □

Example 4.3.6. Let $sk(\mathbb{FinSet}) = Nat\mathbb{Set}$. The skeletal category of $\mathbb{FinSet}$ is $Nat\mathbb{Set}$, which we met in Example 4.1.53 on page 155. The objects of $Nat\mathbb{Set}$ are the sets of natural numbers, and the morphisms are all functions between them. There is an inclusion $inc\colon Nat\mathbb{Set} \hookrightarrow \mathbb{FinSet}$ that is full and faithful. The functor inc is essentially surjective because for every finite set S with $|S| = n$, there is an isomorphism between S and set $\bar{n}$. Thus, it is an equivalence. Since no two elements of $Nat\mathbb{Set}$ are isomorphic, it is a skeletal category. □

Example 4.3.7. We saw in Example 2.3.13, that for every preorder $(P, \leq)$, there is a partial order $(P_0, \leq_0)$ that is its skeletal category. There is an inclusion $inc\colon (P_0, \leq_0) \hookrightarrow (P, \leq)$ that is full, faithful, and essentially surjective (i.e., every $p \in P$ is in some isomorphism class represented by an element in P_0). The quasi-inverse of this map, $\pi\colon (P, \leq) \longrightarrow (P_0, \leq_0)$, is also of interest. This function takes every element to its isomorphism representative. Notice that $\pi \circ inc = id_{(P_0, \leq_0)}$, but $inc \circ \pi$ need not equal $id_{(P,\leq)}$. □

Example 4.3.8. Related to the previous example, is the idea that if two partial order categories $(P, \leq)$ and $(P', \leq')$ are equivalent, then they are isomorphic. This is because the only isomorphisms in either category are identities. □

Example 4.3.9. The following equivalence is a very useful way of thinking of partial functions. Let $f\colon S \longrightarrow T$ be a partial function. Now consider the total function

$\hat{f}: (S + \{*\}) \longrightarrow (T + \{*\})$, defined as follows:

$$\hat{f}(x) = \begin{cases} f(x) & : \text{if } f(x) \text{ is defined} \\ * & : \text{if } f(x) \text{ is undefined, or if } x = *. \end{cases} \tag{4.73}$$

Since $\hat{f}$ takes the $*$ in one set to the $*$ in the other set, the function is a map of pointed sets (i.e., $*/\mathbb{S}\mathrm{et}$). This describes a functor from $\mathbb{P}\mathrm{ar}$ to $*/\mathbb{S}\mathrm{et}$ that takes S to $S + \{*\}$ and takes f to $\hat{f}$. This functor is full and faithful. The functor going the other way is simpler. Take a function of pointed $g: (S, s) \longrightarrow (T, t)$ and form the partial function $g': S \longrightarrow T$, which is not defined when $g(x) = t$. Putting these two functors together gives us the equivalence of categories $\mathbb{P}\mathrm{ar} \simeq */\mathbb{S}\mathrm{et}$. □

Example 4.3.10. In Example 4.1.55, we saw that there is a functor $F: \mathbf{K}\mathbb{M}\mathrm{at} \longrightarrow \mathbf{K}\mathbb{FDV}\mathrm{ect}$ that is defined on a natural number m as $F(m) = \mathbf{K}^m$. The functor takes the morphism $A: m \longrightarrow n$ (an n by m matrix) to linear transformation T_A, where T_A is defined for $B \in \mathbf{K}^m$ is $T_A(B) = AB$. The functor is full and faithful. The fact that it is essentially surjective follows from the fact that every finite-dimensional vector space of dimension n is isomorphic to $\mathbf{K}^n$. Hence, we have shown that $\mathbf{K}\mathbb{M}\mathrm{at} \simeq \mathbf{K}\mathbb{FDV}\mathrm{ect}$. The quasi-inverse of F takes any m-dimensional vector space to m and any linear transformation to the matrix that induces it. Notice that the image of the functor F is a skeletal category of $\mathbf{K}\mathbb{FDV}\mathrm{ect}$. □

Why are equivalences important? When two categories are equivalent, they essentially have the same structure and the functors respect that structure. This means that if $F: \mathbb{A} \longrightarrow \mathbb{B}$ is an equivalence, and a is an initial object in $\mathbb{A}$, then $F(a)$ is an initial object of $\mathbb{B}$. Even more, if $F(a)$ is the initial object of $\mathbb{B}$, then a is the initial object of $\mathbb{A}$. We will see that this is true for most of the limits and colimits that we met in Chapter 3. We will give a more formal statement of the power of equivalence at the end of Section 4.6.

4.4 Adjunctions

If we weaken the notion of an equivalence of categories, we come to the notion of an adjunction of categories. This weakening of requirements explains why the concept of an adjunction of categories arises very often. Whenever we weaken a notion, the new notion will be applicable in more instances (see Important Categorical Idea 4.1.29). This ubiquity gives adjunctions the status of being one of the most important ideas in all of category theory.

The importance of this notion impels us to give four equivalent definitions of an adjunction between two categories. Let $\mathbb{A}$ and $\mathbb{B}$ be categories and $L: \mathbb{A} \longrightarrow \mathbb{B}$ and $R: \mathbb{B} \longrightarrow \mathbb{A}$ be functors. If any of the four definitions are satisfied, then all four are satisfied, and we say the following:

- L and R form an **adjunction** between $\mathbb{A}$ and $\mathbb{B}$.
- L is a **left adjoint** of R.
- R is a **right adjoint** of L.

We denote such an adjunction as $R \vdash L$ or $L \dashv R$, where the dash always points to the left adjoint. We also denote this as

$$\mathbb{A} \underset{R}{\overset{L}{\underset{\perp}{\rightleftarrows}}} \mathbb{B}. \tag{4.74}$$

The first definition is the easiest way to see an adjunction as a generalization of equivalence.

Definition 4.4.1. (I) There are natural transformations (that need not be natural isomorphisms, as in an equivalence of categories) $\eta\colon Id_{\mathbb{A}} \Longrightarrow R \circ L$, called the **unit**, and $\varepsilon\colon L \circ R \Longrightarrow Id_{\mathbb{B}}$, called the **counit**. Similar to Diagrams (1.33), (4.8), and (4.66), we can express this as

$$R\circ L \Leftarrow id_{\mathbb{A}} \ \mathbb{A} \underset{R}{\overset{L}{\rightleftarrows}} \mathbb{B}.\ id_{\mathbb{B}} \Leftarrow L\circ R. \tag{4.75}$$

The unit and counit must satisfy the following **triangle identities**:

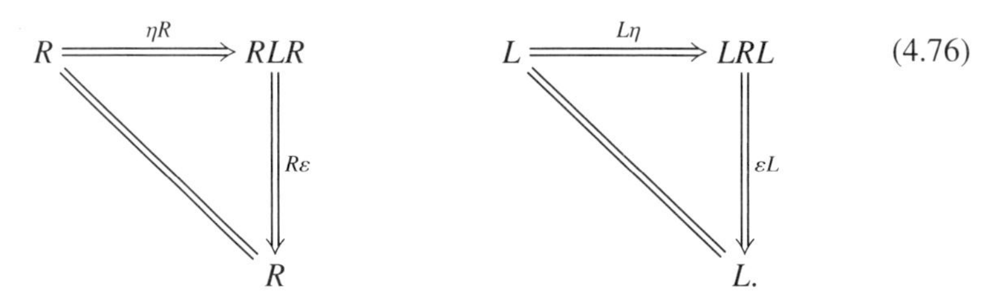

$$R \xrightarrow{\eta R} RLR \xrightarrow{R\varepsilon} R \qquad L \xrightarrow{L\eta} LRL \xrightarrow{\varepsilon L} L. \tag{4.76}$$

The equal signs mean that the composition of the natural transformations gives the original functor.

The second definition stresses the maps between the objects of the two categories.

Definition 4.4.2. (II) There is an adjunction when there is a certain relationship between the images of the functors that go from category to category. In detail, for all a in $\mathbb{A}$ and b in $\mathbb{B}$, there is a natural isomorphism of sets:

$$Hom_{\mathbb{B}}(L(a), b) \xrightarrow{\Phi_{a,b}} Hom_{\mathbb{A}}(a, R(b)). \tag{4.77}$$

Another way to say this is that there is a natural isomorphism between the functors $Hom_{\mathbb{B}}(L(\ \ ),\ \ )$ and $Hom_{\mathbb{A}}(\ \ , R(\ \ ))$, where both functors take the form

$\mathbb{A}^{op} \times \mathbb{B} \longrightarrow \mathbb{S}\mathrm{et}$. We can write the fact that these functions are in correspondence with each other as follows:

$$\begin{array}{ccc} \mathbb{A} & & a \longrightarrow R(b) \\ L \downarrow \uparrow R & & \\ \mathbb{B} & & L(a) \longrightarrow b. \end{array} \tag{4.78}$$

Because L occurs on the left of the isomorphism, it is called a "left adjoint." In contrast, R occurs on the right and is called a "right adjoint."

The fact that the isomorphism is natural (or satisfies the naturality condition) means that for every $f\colon a' \longrightarrow a$ in $\mathbb{A}$ and $g\colon b \longrightarrow b'$ in $\mathbb{B}$, the following two squares commute:

$$\begin{array}{ccc} Hom_{\mathbb{B}}(L(a), b) & \xrightarrow{\Phi_{a,b}} & Hom_{\mathbb{A}}(a, R(b)) \\ \downarrow {\scriptstyle Hom_{\mathbb{B}}(L(f),b)} & & \downarrow {\scriptstyle Hom_{\mathbb{A}}(f,R(b))} \\ Hom_{\mathbb{B}}(L(a'), b) & \xrightarrow[\Phi_{a',b}]{} & Hom_{\mathbb{A}}(a', R(b)) \end{array} \tag{4.79}$$

$$\begin{array}{ccc} Hom_{\mathbb{B}}(L(a), b) & \xrightarrow{\Phi_{a,b}} & Hom_{\mathbb{A}}(a, R(b)) \\ \downarrow {\scriptstyle Hom_{\mathbb{B}}(L(a),g)} & & \downarrow {\scriptstyle Hom_{\mathbb{A}}(a,R(g))} \\ Hom_{\mathbb{B}}(L(a), b') & \xrightarrow[\Phi_{a,b'}]{} & Hom_{\mathbb{A}}(a, R(b')). \end{array} \tag{4.80}$$

For the sake of clarity, it pays to look at how the maps in Diagram 4.79 work. Remember that the functor L is covariant. However, the functor $Hom_{\mathbb{B}}(L(\ \ ), b)$ is contravariant, which means that it takes $f\colon a' \longrightarrow a$ to $Hom_{\mathbb{B}}(L(f), b)\colon Hom_{\mathbb{B}}(L(a), b) \longrightarrow Hom_{\mathbb{B}}(L(a'), b')$. In detail, the left vertical set map takes a morphism $h\colon L(a) \longrightarrow b$ to

$$L(a') \xrightarrow{L(f)} L(a) \xrightarrow{h} b. \tag{4.81}$$

The right set map takes $h'\colon a \longrightarrow R(b)$ to

$$a' \xrightarrow{f} a \xrightarrow{h'} R(b). \tag{4.82}$$

The third definition deals with the universal property of the round-trip $R \circ L$.

Definition 4.4.3. (III) There is a universal property that describes the relationship of a starting object to the ending object of this round-trip process. Specifically, there exists a natural transformation $\eta\colon Id_{\mathbb{A}} \Longrightarrow R \circ L$ called the **unit**, which satisfies the following universal property: for any morphism in $\mathbb{A}$ of the form $f\colon a \longrightarrow R(b)$, there is a unique morphism in $\mathbb{B}$ of the form $f'\colon L(a) \longrightarrow b$ such that the following triangle commutes:

$$\begin{array}{ccc} a & \xrightarrow{\eta_a} & R(L(a)) \\ & \searrow^{f} & \downarrow R(f') \\ & & R(b) \end{array} \qquad \begin{array}{c} L(a) \\ \downarrow f' \\ b. \end{array} \tag{4.83}$$

The fourth definition deals with universal property of the round-trip $L \circ R$. This definition is very similar to the previous one.

Definition 4.4.4. (IV) There exists a natural transformation $\varepsilon\colon L \circ R \Longrightarrow Id_{\mathbb{B}}$ called the **counit**, which satisfies the following universal property: for any morphism in $\mathbb{B}$ of the form $g\colon L(a) \longrightarrow b$, there is a unique morphism in $\mathbb{A}$ of the form $g'\colon a \longrightarrow R(b)$, such that the following triangle commutes:

$$\begin{array}{c} R(b) \\ \uparrow g' \\ a \end{array} \qquad \begin{array}{ccc} L(R(b)) & \xrightarrow{\varepsilon_b} & b \\ \uparrow L(g') & \nearrow_{g} & \\ L(a). & & \end{array} \tag{4.84}$$

Theorem 4.4.5. All four definitions of adjunctions are equivalent. ★

Proof. The first-time reader should skip these technical details and go directly to the examples.

(I) implies (II). Given η and ε, we define Φ using R and η. In detail, for $h\colon L(a) \longrightarrow b$, we define

$$\Phi_{a,b}(h) = R(h) \circ \eta_a\colon a \longrightarrow RL(a) \longrightarrow R(b). \tag{4.85}$$

Its inverse Φ^{-1} is defined using ε and L. In detail, for $h'\colon a \longrightarrow R(b)$ as

$$\Phi^{-1}_{a,b}(h') = \varepsilon_b \circ L(h')\colon L(a) \longrightarrow LR(b) \longrightarrow b. \tag{4.86}$$

The naturality of Φ and Φ^{-1} comes from the naturality of η and ε. These two maps are inverse because for a given $h\colon L(a) \longrightarrow b$, we have

$$\begin{aligned} \Phi^{-1}(\Phi(h)) &= \Phi^{-1}(\, a \xrightarrow{\eta_a} RL(a) \xrightarrow{R(h)} R(b)\,) && \text{by definition of } \Phi \\ &= L(a) \xrightarrow{L(\eta_a)} LRL(a) \xrightarrow{LR(h)} LR(b) \xrightarrow{\varepsilon_b} b && \text{by definition of } \Phi^{-1} \end{aligned}$$

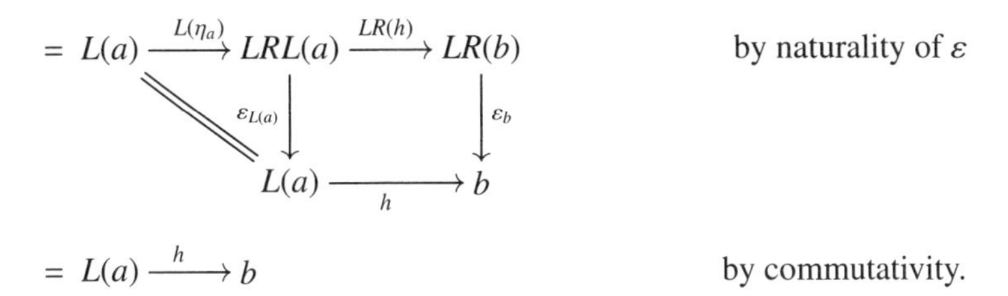

by naturality of ε

$$= L(a) \xrightarrow{h} b$$ by commutativity.

Furthermore, given $h' \colon a \longrightarrow R(b)$,

$$\Phi(\Phi^{-1}(h)) = \Phi(\ L(a) \xrightarrow{L(h')} LR(b) \xrightarrow{\varepsilon_b} R(b)\ )$$ by definition of Φ^{-1}

$$= a \xrightarrow{\eta_a} RL(a) \xrightarrow{RL(h')} RLR(b) \xrightarrow{R\varepsilon_b} R(b)$$ by definition of Φ

$$\begin{array}{ccccc} = RL(a) & \xrightarrow{RL(h')} & RLR(b) & \xrightarrow{R\varepsilon_b} & R(b) \\ \uparrow{\scriptstyle \eta_a} & & \uparrow{\scriptstyle \eta_{R(b)}} & \diagup\!\!\diagup & \\ a & \xrightarrow[h']{} & R(b) & & \end{array}$$ by naturality of η

$$= a \xrightarrow{h'} R(b)$$ by commutativity.

(II) implies (III). Setting $b = L(a)$ in Diagram (4.77) gives us

$$Hom_{\mathbb{B}}(L(a), L(a)) \xrightarrow{\Phi_{a,L(a)}} Hom_{\mathbb{A}}(a, R(L(a))). \tag{4.87}$$

On the left side, there is id_{La}. Take $\Phi_{a,La}(id_{La})$ as η_a. The naturality of η comes from the naturality of Φ. To show the universal property, notice that for any $f \colon a \longrightarrow R(b)$, there will correspond a unique $f' = \Phi^{-1}_{a,b}(f)$. The triangle commutes because $R(f') \circ \eta_a = R(\Phi^{-1}_{a,b}(f)) \circ \Phi_{a,La}(id_{La}) = f$. This last equality can be seen from the commutativity of the following diagram:

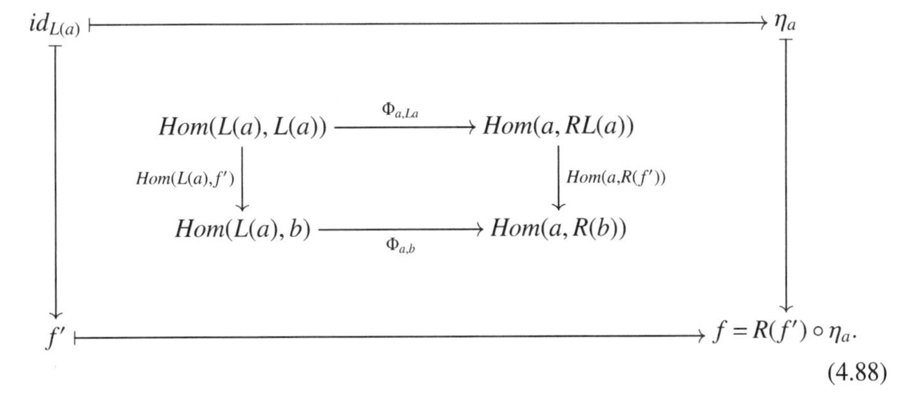

(4.88)

(III) implies (I). We have a unit. To find the counit at b, look at $\eta_{R(b)} \colon R(b) \longrightarrow RLR(b)$. Now look at the map induced by the identity $id_{R(b)}$. Set the counit ε_b to be the induced

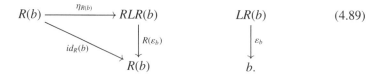

map, as in the following diagram:

$$\begin{array}{ccc} R(b) & \xrightarrow{\eta_{R(b)}} & RLR(b) \\ & {\scriptstyle id_R(b)}\searrow & \Big\downarrow{\scriptstyle R(\varepsilon_b)} \\ & & R(b) \end{array} \qquad\qquad \begin{array}{c} LR(b) \\ \Big\downarrow{\scriptstyle \varepsilon_b} \\ b. \end{array} \tag{4.89}$$

The commutativity of this diagram is the left triangle identity of Definition (I). We leave it to the reader to prove the naturality of ε and to prove the right triangle identity. (See, e.g., page 100 of [40].)

We leave the proof that Definition (IV) is equivalent to the other definitions for the reader. ♣

Exercise 4.4.6. Show that Definition (IV) is equivalent to the other definitions. ■

Technical Point 4.4.7. There is a little redundancy in these definitions. Once a functor and a natural transformation is given, the other functor and natural transformation is partially free. (For the cognoscenti, this assumes the axiom of choice.) We state it with the redundancy for simplicity's sake. ♡

Now let's look at the examples.

Example 4.4.8. We begin with one of the world's simplest examples of an adjunction. We generalize Example 4.3.5, where we showed that $\mathbf{1} \simeq \mathbf{2}_I$ to demonstrate that $\mathbf{1}$ is adjoint to $\mathbf{2} = a \longrightarrow b$. In detail, there is a unique functor $!\colon \mathbf{2} \longrightarrow \mathbf{1}$ and there is a functor $L\colon \mathbf{1} \longrightarrow \mathbf{2}$ such that $L(*) = a$. It is obvious that $L\circ ! = id_{\mathbf{1}}$ and

$$Hom_{\mathbf{1}}(*, !(b)) \cong Hom_{\mathbf{2}}(L(*), b). \tag{4.90}$$

The set on the right has only one element. Similarly, the functor ! has right adjoint $R\colon \mathbf{1} \longrightarrow \mathbf{2}$, such that $R(*) = b$. This can be seen as

$$Hom_{\mathbf{1}}(!(b), *) \cong Hom_{\mathbf{2}}(a, R(*)). \tag{4.91}$$

Both cases are summarized as

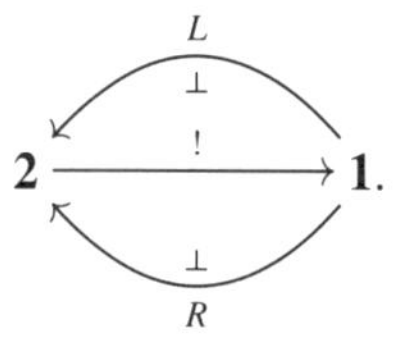

(4.92)

Notice that a is the initial object of $\mathbf{2}$, while b is its terminal object. □

Example 4.4.9. Consider the real numbers $\mathbf{R}$ and the integers $\mathbf{Z}$ as partial order categories. There is an inclusion function $inc\colon \mathbf{Z} \longrightarrow \mathbf{R}$. This inclusion function has a left adjoint. Let us call the left adjoint $L\colon \mathbf{R} \longrightarrow \mathbf{Z}$ and see if we can figure out what it is.

The definition of the adjunction says that for all n in $\mathbf{Z}$ and for all r in $\mathbf{R}$, we have

$$Hom_{\mathbf{Z}}(L(r), n) \cong Hom_{\mathbf{R}}(r, inc(n)). \tag{4.93}$$

Notice that $inc(n)$ is just n in the real numbers and both of these categories are partial orders, so the Hom sets are either the empty set or a one-element set. This means that the isomorphism can be interpreted as an if and only if statement:

$$L(r) \leq n \text{ if and only if } r \leq inc(n) = n. \tag{4.94}$$

The right side is exactly true when r is less than or equal to n. Consider the number 7.27. The following statements about it are true:

$$7.27 \nleq 5, \quad 7.27 \nleq 6, \quad 7.27 \nleq 7, \quad 7.27 \leq 8, \quad 7.27 \leq 9. \tag{4.95}$$

This forces the left side to be true:

$$L(7.27) \nleq 5, \quad L(7.27) \nleq 6, \quad L(7.27) \nleq 7, \quad L(7.27) \leq 8, \quad L(7.27) \leq 9 \tag{4.96}$$

In other words, $L(7.27) = 8$. That is, the left adjoint L is the functor that for input r outputs the smallest integer that is larger than or equal to r. That is the ceiling function, which shows us that $inc \vdash \lceil \ \rceil$. The unit $Id_R \Longrightarrow inc \circ \lceil \ \rceil$ is the arrow showing that every real number r is less than or equal to its ceiling (i.e., $r \leq \lceil r \rceil$). The counit of this adjunction is the identity. That is, for every natural number n, it is a fact that $\lceil n \rceil \leq n$. □

Exercise 4.4.10. Show that the right adjoint of the inclusion function $inc\colon \mathbf{Z} \longrightarrow \mathbf{R}$ is the floor function $\lfloor \ \rfloor\colon \mathbf{R} \longrightarrow \mathbf{Z}$. Describe the unit and counit. ■

Example 4.4.9 and Exercise 4.4.10 can be summarized as

(4.97)

These adjunctions are between partial order categories. Such adjunctions have a special name, given in Definition 4.4.11.

Definition 4.4.11. An adjunction between two preordered or partially ordered categories is called a **Galois connection**. Definition (II) of an adjunction in the case of preorder or partially ordered categories reduces to

$$L(a) \leq b \qquad \text{if and only if} \qquad a \leq R(b). \tag{4.98}$$

Example 4.4.12. In Example 4.1.9, we saw that for every set B, there are two functors from $\mathbb{Set}$ to $\mathbb{Set}$: $L_B(A) = A \times B$ and $R_B(C) = Hom_{\mathbb{Set}}(B, C)$. The functor L_B is left adjoint to R_B. Using Definition (II) of adjoint functors amounts to showing that

$$Hom_{\mathbb{Set}}(L_B(A), C) \cong Hom_{\mathbb{Set}}(A, R_B(C)), \tag{4.99}$$

which translates into

$$Hom_{\mathbb{Set}}(A \times B, C) \cong Hom_{\mathbb{Set}}(A, Hom_{\mathbb{Set}}(B, C)). \tag{4.100}$$

We already saw that these two sets are isomorphic in Theorem 1.4.26.

The counit of this adjunction will be very important. The counit is a map $\varepsilon\colon L_B \circ R_B \Longrightarrow Id_{\mathbb{Set}}$. On set C, this turns out to be

$$\varepsilon_C\colon Hom_{\mathbb{Set}}(B, C) \times B \longrightarrow C. \tag{4.101}$$

This morphism takes a set function $f\colon B \longrightarrow C$ and $b \in B$, and outputs $f(b)$. This function is called the **evaluation function**. The universal property of the counit says that for every set D and every function $g\colon D \times B \longrightarrow C$, there is a unique $g'\colon D \longrightarrow Hom_{\mathbb{Set}}(B, C)$ such that

$$\begin{array}{ccc} Hom_{\mathbb{Set}}(B, C) \times B & \xrightarrow{\varepsilon_C} & C \\ {\scriptstyle g' \times id_B} \uparrow & \nearrow_{g} & \\ D \times B. & & \end{array} \tag{4.102}$$

This is true because for a g, we set g' to be the function that takes d to $f_d\colon B \longrightarrow C$, where f_d is defined by $f_d(b) = g(d, b)$. This satisfies the commutative triangle because for any $(d, b) \in D \times B$, we have that $g(d, b) = \varepsilon_C(g'(d), b) = \varepsilon_C(f_d, b)$. This says that the function set, $Hom_{\mathbb{Set}}(A, C)$, is the "best-fitting" set to deal with all evaluations.

The unit is a little less familiar and is called the **co-evaluation function.** The unit is a natural transformation $\eta\colon Id_{\mathbb{Set}} \Longrightarrow R_B \circ L_B$. Its component on set A is $\eta_A\colon A \longrightarrow Hom_{\mathbb{Set}}(B, A \times B)$, which takes $a \in A$ and outputs $f_a\colon B \longrightarrow A \times B$. The function f_a is defined as $b \mapsto (a, b)$. We leave the universal property of the co-evaluation function for the reader. □

A generalization of Example 4.4.8 is given in Example 4.4.13.

Example 4.4.13. For every category $\mathbb{A}$, there is a unique functor $!\colon \mathbb{A} \longrightarrow \mathbf{1}$. The right adjoint, R, of this functor would satisfy the following requirement:

$$Hom_{\mathbb{A}}(a, R(*)) \cong Hom_{\mathbf{1}}(!(a), *) = \{id_*\}. \tag{4.103}$$

This means that R takes the single object of $\mathbf{1}$ to the object $R(*)$ of $\mathbb{A}$ that has exactly one morphism from any object a of $\mathbb{A}$ to that object. The functor R picks out a terminal object of $\mathbb{A}$ if it exists. The adjoint will exist if and only if the terminal object exists. The unit of the adjunction is the unique map from the object to the terminal object. The counit is always the identity. Similarly, the left adjoint of ! picks out an initial object of $\mathbb{A}$ if it exists. We can summarize these two adjunctions with

□

Example 4.4.14. In Example 4.1.42, we saw that any category $\mathbb{A}$ with products has a functor $Prod\colon \mathbb{A}\times\mathbb{A} \longrightarrow \mathbb{A}$ that gives the product of objects and morphisms. In Example 4.1.44, we saw that for every category $\mathbb{A}$, there is a diagonal functor $\Delta\colon \mathbb{A} \longrightarrow \mathbb{A}\times\mathbb{A}$ that takes objects and morphisms to their double. These two functors are adjoint to each other: $\Delta \dashv Prod$. In terms of Definition II, this amounts to

$$Hom_{\mathbb{A}\times\mathbb{A}}(\Delta(c),(a,b)) \cong Hom_{\mathbb{A}}(c, Prod(a,b)). \tag{4.105}$$

This can be written as

$$Hom_{\mathbb{A}\times\mathbb{A}}((c,c),(a,b)) \cong Hom_{\mathbb{A}}(c, a\times b), \tag{4.106}$$

which is almost the same as Equation 3.9 on page 88. There is a similar co-product functor $coprod\colon \mathbb{A}\times\mathbb{A} \longrightarrow \mathbb{A}$. This functor is left adjoint to Δ. We can summarize this as

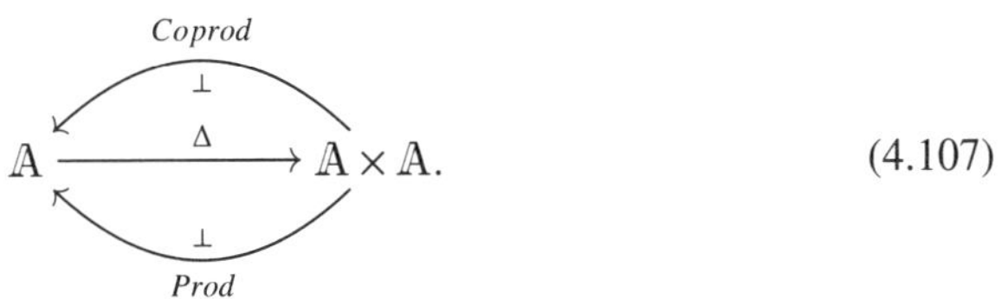

(4.107)

□

The adjunction given in Example 4.4.15 is a paradigm for many examples of **free-forgetful adjunctions**.

Example 4.4.15. There is a forgetful functor $U\colon \mathbb{M}\mathrm{onoid} \longrightarrow \mathbb{S}\mathrm{et}$ that takes every monoid to its underlying set. In Example 4.1.10, we saw the free monoid functor $F\colon \mathbb{S}\mathrm{et} \longrightarrow \mathbb{M}\mathrm{onoid}$, which takes every set S to S^*. We show that F is left adjoint to U. This means that for all sets S and for all monoids M, there is the following natural isomorphism:

$$Hom_{\mathbb{M}\mathrm{onoid}}(F(S), M) \cong Hom_{\mathbb{S}\mathrm{et}}(S, U(M)). \tag{4.108}$$

Given any set map $f\colon S \longrightarrow U(M)$, let $f'\colon F(S) \longrightarrow M$ be defined as follows: for an element $w = s_1 s_2 s_3 \cdots s_n$ of the free monoid, $f'(w) = f'(s_1 s_2 s_3 \cdots s_n) = f(s_1)f(s_2)f(s_3)\cdots f(s_n)$. For a monoid homomorphism $g\colon F(S) \longrightarrow M$, set the corresponding $g'\colon S \longrightarrow U(M)$ to be defined by $g'(s) = g(s)$.

The unit of this adjunction at set S is the set function $\eta_S\colon S \longrightarrow UF(S)$, which includes every letter as the one-element word.

It will be beneficial for us to examine the free monoid on one object, such as $*$. The monoid will consist of $*, **, ***, \ldots$. There will also be the empty set as the unit. This monoid is isomorphic to the monoid of natural numbers $(\mathbf{N}, +, 0)$. The universal property of the unit η can be expressed with the diagram

$$\begin{array}{ccc} \{*\} & \xrightarrow{\eta_{\{*\}}} & U(F(\{*\})) = U(\mathbf{N}) \\ & {}_{f}\searrow & \downarrow U(f') \\ & & U(M) \end{array} \qquad\qquad \begin{array}{c} F(\{*\}) = \mathbf{N} \\ \downarrow f' \\ M. \end{array} \tag{4.109}$$

This says that for every set function $f\colon \{*\} \longrightarrow U(M)$, which is a function that picks out an element m of M, there is a monoid homomorphism $f'\colon \mathbf{N} \longrightarrow M$ such that the above triangle commutes. The output of the function f' is $m, mm, mmm, \ldots$. Let us restate this in a way that will be useful. The free monoid on one object will have the property that for every monoid M and every element m in M, there is a unique morphism from the free monoid on one object to M that takes $*$ to m.

Let us summarize the properties of $F(\{*\})$ in three ways:

- The monoid $F(\{*\})$ is the free monoid on one generator.
- For every object m in M, there is a unique morphism $f\colon F(*) \longrightarrow M$ such that $f(*) = m$.
- When we substitute $S = \{*\}$ in Equation (4.108), we get

$$Hom_{\mathbb{Monoid}}(F(\{*\}), M) \cong Hom_{\mathbb{Set}}(\{*\}, U(M)) \cong U(M). \tag{4.110}$$

This means that there is an isomorphism $Hom_{\mathbb{Monoid}}(F(\{*\}), M) \cong U(M)$, where $U(M)$ is the set of elements of M.

When we discuss the free functor of an adjunction, we are describing another way of talking about the universal property of a structure. The ideas discussed in this example will be fundamental in the coming chapters when we discuss the free element of other structures. This will be extremely important in coherence theory. □

Example 4.4.16 will be the paradigm of the next few examples and exercises. The details are worked out for the $\mathbb{Cat}$ example because that is our primary interest.

Example 4.4.16. Consider the categories $\mathbb{Cat}$ and $\mathbb{Set}$. There is a forgetful functor $U\colon \mathbb{Cat} \longrightarrow \mathbb{Set}$[1] that takes a category $\mathbb{A}$ to the set of objects of $\mathbb{A}$ and forgets the morphism and the rest of the structure of the category. It also takes a functor and outputs its underlying set function on objects.

Functor U has a left adjoint $d\colon \mathbb{Set} \longrightarrow \mathbb{Cat}$, which takes any set to its discrete category and any set function to its discrete functor. The name "discrete" is a vestige of topological language. The adjunction is

$$Hom_{\mathbb{Cat}}(d(S), \mathbb{A}) \cong Hom_{\mathbb{Set}}(S, U(\mathbb{A})). \tag{4.111}$$

The right side of this isomorphism consists of *all* set functions from S to $U(\mathbb{A})$. Imagine a set function $f\colon S \longrightarrow U(\mathbb{A})$. This will correspond to the functor $\hat{f}\colon d(S) \longrightarrow \mathbb{A}$. In other words, $\hat{f}$ is the same function as f if we ignore the maps in $\mathbb{A}$.

The U functor also has a right adjoint, c, which stands for "continuous." This functor corresponds to the "indiscrete" topology. The functor takes set S and forms the category that has the elements of S as the objects and exactly one morphism between any two objects. The adjunction is

$$Hom_{\mathbb{Cat}}(\mathbb{A}, c(S)) \cong Hom_{\mathbb{Set}}(U(\mathbb{A}), S). \tag{4.112}$$

[1] There is a size problem here. We defined $\mathbb{Cat}$ to have all locally small categories, and now we are relating them to $\mathbb{Set}$, which consists of sets (and not classes). There are at least two possible solutions to this problem: (i) Define a category with only small categories. Such a category would match up well with $\mathbb{Set}$. Or (ii) define a category with sets and classes. This would match up well with $\mathbb{Cat}$. Either solution would solve the problem. We omit these solutions because we appreciate the simplicity of $\mathbb{Cat}$ and $\mathbb{Set}$.

A function $f\colon U(\mathbb{A}) \longrightarrow S$ does not have to worry about the arrows in $\mathbb{A}$ because there are no arrows in $U(\mathbb{A})$. Such a function will correspond to a functor $\hat{f}\colon \mathbb{A} \longrightarrow c(S)$ because for all a and a' in $\mathbb{A}$, there will always be exactly one arrow $\hat{f}(a) \longrightarrow \hat{f}(a')$ in $c(S)$ where all arrows $a \longrightarrow a'$ can go to.

But the story is not over. The left adjoint d has a left adjoint. There is a functor π_0 (again, a vestige of topological language) that takes category $\mathbb{A}$ to $\pi_0(\mathbb{A})$, the set of components of $\mathbb{A}$. In detail, if there is a morphism from a to a' in $\mathbb{A}$, then these two elements are in the same component. Let us examine the following adjunction:

$$Hom_{\mathbb{Cat}}(\mathbb{A}, d(S)) \cong Hom_{\mathbb{Set}}(\pi_0(\mathbb{A}), S). \tag{4.113}$$

Since there are no nonidenity morphisms in category $d(S)$, a functor $F\colon \mathbb{A} \longrightarrow d(S)$ will take a morphism $f\colon a \longrightarrow a'$ in $\mathbb{A}$ to an identity morphism in $d(S)$. This will correspond to a set function $\hat{F}\colon \pi_0(\mathbb{A}) \longrightarrow S$.

We can summarize all the functors in this example as

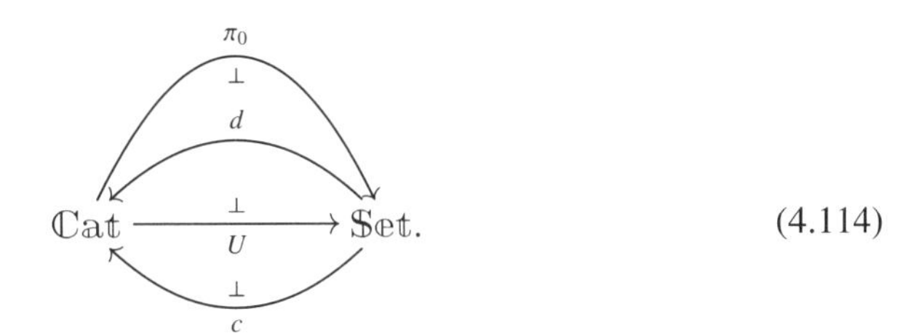

Now let us list the units and counits of these adjunctions:

- For the $U \vdash d$ adjunction, the unit $\eta_S : S \longrightarrow U(d(S))$ takes set S to the same set. The counit $\varepsilon_{\mathbb{A}} : d(U(\mathbb{A})) \longrightarrow \mathbb{A}$ is the inclusion of a category $\mathbb{A}$ stripped of its arrows into the original category $\mathbb{A}$.
- For the $c \vdash U$ adjunction, the unit $\eta_{\mathbb{A}} : \mathbb{A} \longrightarrow c(U(\mathbb{A}))$ is an identity-on-objects functor from category $\mathbb{A}$ onto the category with the same objects but with exactly one morphism between any two objects. The counit $\varepsilon_S : U(c(S)) \longrightarrow S$ takes a set S to the same set.
- For the $d \vdash \pi_0$ adjunction, the unit $\eta_{\mathbb{A}} : \mathbb{A} \longrightarrow d(\pi_0(\mathbb{A}))$ takes category $\mathbb{A}$ to the discrete category of its components. The counit $\varepsilon_S : \pi_0(d(S)) \longrightarrow S$ on a set S is the identity on the set because the components of a discrete category are exactly the same as the elements of the set. □

Exercise 4.4.17. Graphs are simpler than categories. Use Example 4.4.16 to define the functors and show that the following adjunctions hold between graphs and sets:

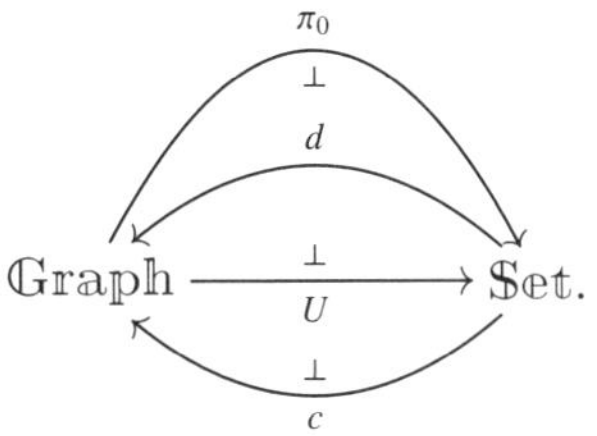

$$\mathbb{Graph} \underset{U}{\longrightarrow} \mathbb{Set}. \qquad \pi_0 \dashv d \dashv U \dashv c \tag{4.115}$$

(Hint: The functors and proofs are almost exactly the same as in Example 4.4.16.) ■

Example 4.4.18. Similar to the above example and exercise, there is the following adjunctions between topological spaces and sets:

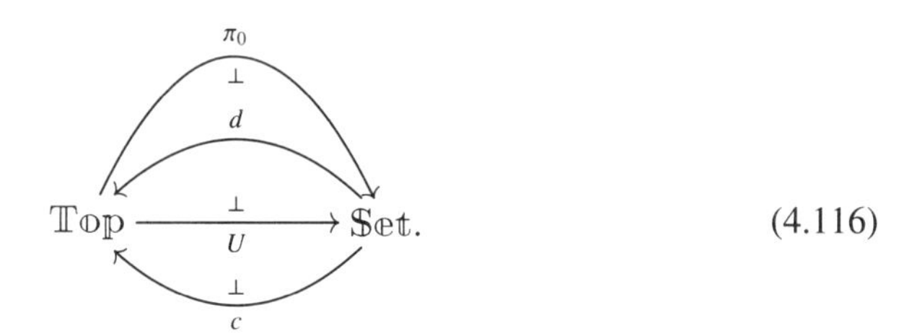

The U functor takes a topological space and forgets the topological structure. The output of U is simply the underlying set. The functors d and c take a set to the topological space with the discrete and indiscrete (continuous) topology, respectively (see Example 2.1.19.) The fact that they satisfy the universal properties is exactly the contents of Theorem 2.1.21. The functor π_0 takes a topological space and outputs the set of connected components of that space. □

Example 4.4.19. There is a forgetful functor $U\colon \mathbb{C}\mathrm{at} \longrightarrow \mathbb{G}\mathrm{raph}$. This functor has a left adjoint $L\colon \mathbb{G}\mathrm{raph} \longrightarrow \mathbb{C}\mathrm{at}$. The functor L takes graph G to a category $L(G)$, which will be called the "free category over G." Such a category will have the same objects as G but with more edges added. To make a graph into a category, an identity has to be added for each object, and compositions have to be added for every composable pair of morphisms. In detail, $L(G)$ is a category with the same objects as G and the morphisms are the set of paths in G. Another way to say this is that the morphisms are all composible strings of morphisms in the graph. We might envision them as

$$x_1 \longrightarrow x_2 \longrightarrow x_3 \longrightarrow \cdots \longrightarrow x_n. \qquad (4.117)$$

Included are paths of length zero, which correspond to identity morphisms. Composition is simply concatenating two such strings of arrows. Composition with paths of length zero gives the original path, thus ensuring that the paths of length zero are the identities. The composition of paths is an associative operation. Hence, $L(G)$ is a category.

To get a feel for this functor, let us examine this functor on two graphs:

- Category $L(*)$, where $*$ is the one-object graph with no arrows. The only morphism added is the path of length 0 (i.e., the identity). This category will be **1**, the one-object category with one identity.
- In stark contrast, consider graph $*'$, which consists of one object and one arrow from the single object to itself:

$$\circlearrowright \; *. \qquad (4.118)$$

Then $L(*')$ will consist of all the possible compositions of that one morphism and will look like

(4.119)

> There is one morphism for every natural number. This one-object category is the monoid of natural numbers.

This L functor is very similar to the free monoid functor of Example 4.1.10. □

Example 4.4.20. In Example 4.3.7, we met functors between a partial order and a preorder. These functors can be boosted to the category of all partial orders and all preorders. There is an inclusion $Inc\colon \mathbb{PO} \hookrightarrow \mathbb{P}\mathbb{r}\mathbb{e}\mathbb{O}$. What about going the other way? Let $(P, \leq)$ be a preorder. There is a relation $\approx$ on the objects of P as follows: $p \approx p'$ if and only if $p \leq p'$ and $p' \leq p$. This is clearly an equivalence relation. We can form a partial order $(P/\approx, \sqsubseteq)$ whose objects are equivalence classes of objects of P and $[p] \sqsubseteq [p']$ if and only if $p \leq p'$. This defines a functor $\Pi\colon \mathbb{P}\mathbb{r}\mathbb{e}\mathbb{O} \longrightarrow \mathbb{PO}$. Notice that $\Pi \circ Inc = Id_{\mathbb{PO}}$, but $Inc \circ \Pi$, is not equal or isomorphic to $Id_{\mathbb{P}\mathbb{r}\mathbb{e}\mathbb{O}}$. In fact, the map $Id_{\mathbb{P}\mathbb{r}\mathbb{e}\mathbb{O}} \Longrightarrow Inc \circ \Pi$ is the unit of an adjunction with $Inc \vdash \Pi$. One can see the adjunction by noticing that any order-preserving map of partial orders $f\colon \Pi(P) \longrightarrow P'$ has a related order-preserving map of preorders $\widehat{f}\colon P \longrightarrow Inc(P')$, and vice versa. The map $\widehat{f}$ will take isomorphic elements of P to the element that f took them to. □

Here is an interesting example about prime numbers.

Example 4.4.21. Consider the functor $L\colon \mathbf{N} \longrightarrow \mathbf{N}$, which is defined by $L(n) = P_n$, the nth prime number. So

$$L(0) = 0, L(1) = 2, L(2) = 3, L(3) = 5, L(4) = 7, \ldots \tag{4.120}$$

This functor has a right adjoint $R\colon \mathbf{N} \longrightarrow \mathbf{N}$ defined as $R(n) = \pi(n)$, the number of primes that are less than or equal to n. So

$$R(0) = 0, R(1) = 0, R(2) = 1, R(3) = 2, R(4) = 2, \ldots \tag{4.121}$$

The adjunction says that for all integers m and n,

$$P_m \leq n \text{ if and only if } m \leq \pi(n). \tag{4.122}$$

The unit is actually an equality: $m = \pi(P_m)$. However, the counit is generally the inequality $P_{\pi(n)} \leq n$. □

Let us prove some properties about adjunctions. First, we would like to put the notion of an adjunction in the context of equivalence of categories. We pointed out that an adjunction is a weakening of the notion of an equivalence. That is, every equivalence is a special type of adjunction where the unit and the counit are isomorphisms. But this is not really true. After all, our definition of an equivalence never required the two functors to satisfy the triangle identities (Diagram (4.76)). We can rectify the situation

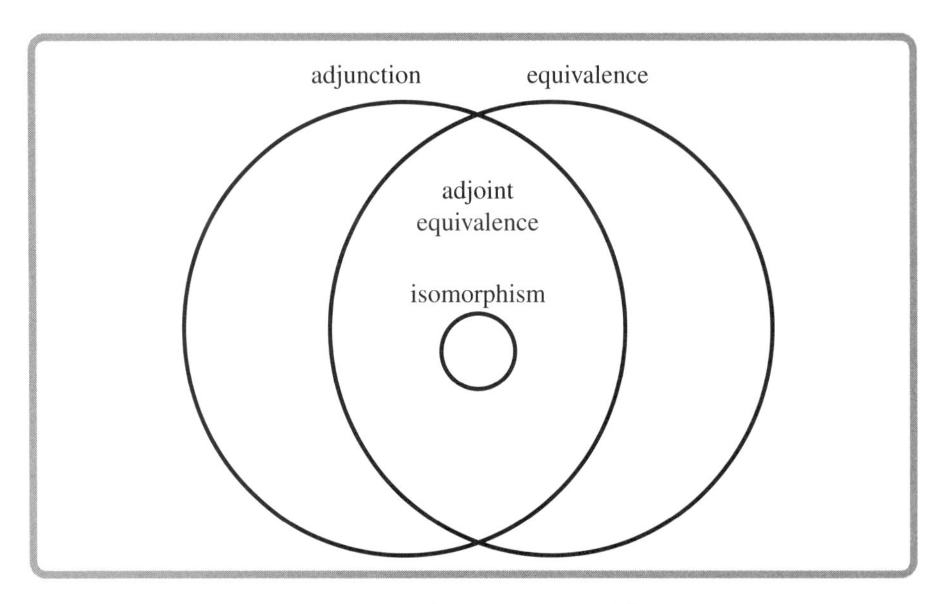

Figure 4.4. Adjunctions, equivalences, and adjoint equivalences.

by calling an equivalence that also satisfies the triangle identities (and hence all four definitions of an adjunction) an **adjoint equivalence**. This can be visualized as the Venn diagram in Figure 4.4. Theorem 4.4.22 relates the functors in the right circle with the functors in the intersection.

Theorem 4.4.22. Every equivalence can be turned into an adjoint equivalence. ★

Proof. Given an equivalence $L\colon \mathbb{A} \longrightarrow \mathbb{B}$ with quasi-inverse $R\colon \mathbb{B} \longrightarrow \mathbb{A}$, and natural isomorphisms $\eta\colon Id_{\mathbb{A}} \Longrightarrow RL$ and $\varepsilon\colon LR \Longrightarrow Id_{\mathbb{B}}$, we shall find an isomorphic equivalence that satisfies Definition (I) of being an adjunction. The new adjunction will have the same L, R, and η but a new $\varepsilon'\colon LR \Longrightarrow Id_{\mathbb{B}}$.

Start with the natural isomorphisms η and ε. We aim at getting the first of the triangle identities (Diagram (4.76)). By composition, ηR and $R\varepsilon$ are also natural isomorphisms. Their composition

$$R \xRightarrow{\ \eta R\ } RLR \xRightarrow{\ R\varepsilon\ } R \tag{4.123}$$

is also a natural isomophism, which we call φ. Thus, we get the commutative diagram

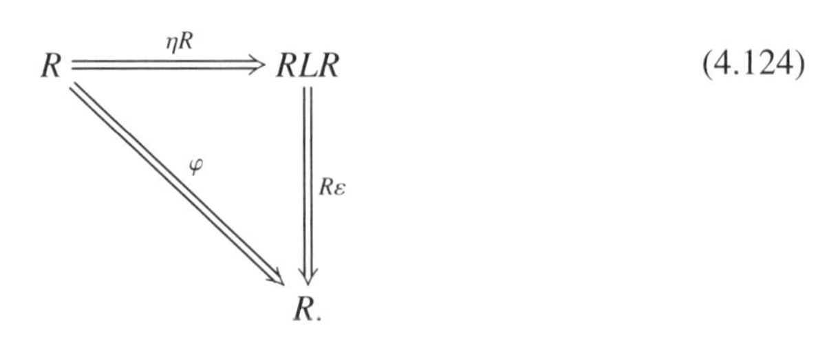

The problem is that φ need not be the identity. In fact, we can take it to be a measure of how far $R\varepsilon \circ_V \eta R$ is from being an identity. There is also an interest in the inverse of $\varphi, \varphi^{-1}\colon R \Longrightarrow R$.

Our aim is to "nudge" ε by φ to get a triangle identity that works. Set the new $\varepsilon' = \varepsilon \circ L(\varphi^{-1})$. Apply R to this to get $R(\varepsilon') = R(\varepsilon) \circ R(L(\varphi^{-1}))$. We can now extend the previous commutative diagram as

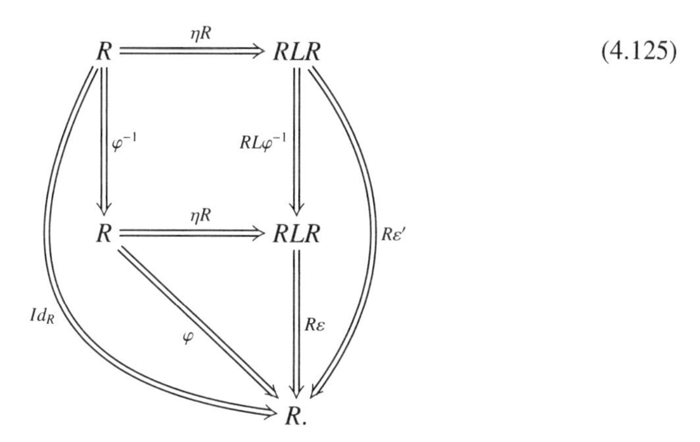

(4.125)

The top square commutes out of the naturality of η, and the entire diagram is the first triangle identity with the nudged ε. What about the second triangle identity? Consider the following commutative diagram:

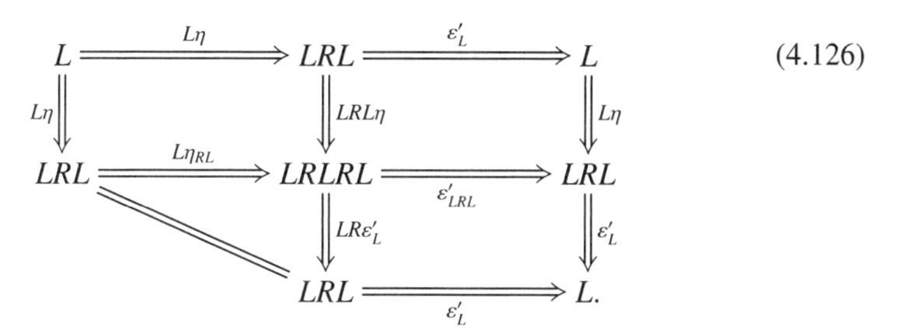

(4.126)

The top-left square commutes out of the naturality of η. The two right squares commute out of the naturality of ε'. The lower-left triangle is the L applied to the first of the triangle identities. Since all the parts commute, the entire diagram commutes, which means that

$$(\varepsilon'_L \circ L\eta) \circ (\varepsilon'_L \circ L\eta) = (\varepsilon'_L \circ L\eta). \tag{4.127}$$

(A map f such that $f \circ f = f$ is called **idempotent**.) Composing both sides of the equation by $(\varepsilon'_L \circ L\eta)^{-1}$ gives us that $(\varepsilon'_L \circ L\eta) = \iota_L$. This is exactly the second triangle identity. ♣

There is another connection between adjunctions and equivalences. Embedded within every adjunction sits an equivalence of categories.

Theorem 4.4.23. For every adjunction $L\colon \mathbb{A} \longrightarrow \mathbb{B}$ and $R\colon \mathbb{B} \longrightarrow \mathbb{A}$ with $R \vdash L$, there are subcategories $(\mathbb{A}) \hookrightarrow \mathbb{A}$ and $(\mathbb{B}) \hookrightarrow \mathbb{B}$ such that L and R, which are restricted to these subcategories, form an equivalence of categories. We can combine this idea with Diagram (4.71) to get the following three layers:

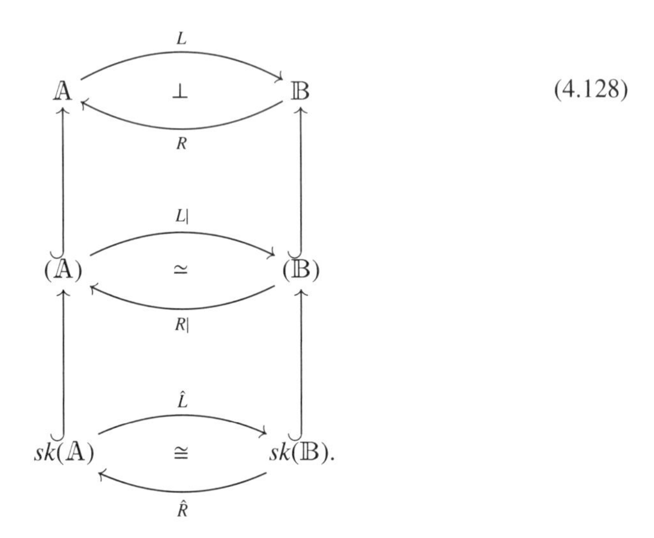

(4.128)

In the top level, the unit and the counit are morphisms; in the middle level, the unit and the counit are isomorphisms; and on the bottom level, the unit and the counit are identity morphisms. ★

Proof. Simply let ($\mathbb{A}$) be the full subcategory of A consisting of objects where the unit is an isomorphism, and let ($\mathbb{B}$) be the full subcategory of $\mathbb{B}$ consisting of the objects where the counit is an isomorphism. For the skeletal category level, we have to use a modification of L and R because those functors might not take skeletal objects of one category to skeletal objects of the other category. This can easily be done. ♣

Theorem 4.4.24. The composition of right adjoints is a right adjoint. Similarly, the composition of left adjoints is a left adjoint. ★

Proof. Consider the following two right adjoints:

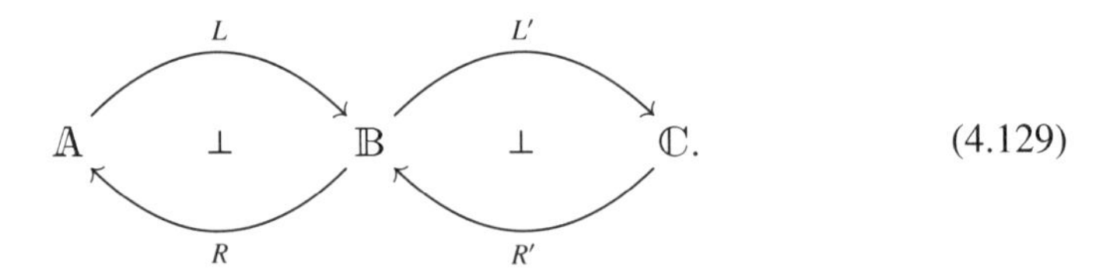

(4.129)

Then we have

$$Hom_{\mathbb{C}}(R'R(a), c) \cong Hom_{\mathbb{B}}(R(a), L'(c)) \cong Hom_{\mathbb{A}}(a, LL'(c)), \tag{4.130}$$

where the left natural isomorphism follows from the $R' \vdash L'$ adjunction and the right natural isomorphism follows from the $R \vdash L$ adjunction. The proof for left adjoints is very similar. ♣

Theorem 4.4.25. A right adjoint to a functor is unique up to a unique isomorphism. That is, if $R \vdash L$ and $R' \vdash L$, then R is isomorphic to R' by a unique isomorphism. There is a similar dual statement about the uniquness of left adjoints. ★

Proof. From the adjunctions, we have

$$Hom(a, R(b)) \cong Hom(L(a), b) \cong Hom(a, R'(b)). \tag{4.131}$$

Setting $a = R(b)$, this becomes

$$Hom(R(b), R(b)) \cong Hom(L(R(b)), b) \cong Hom(R(b), R'(b)). \tag{4.132}$$

The first Hom set contains $id_{R(b)}$. The morphism in the third Hom set that corresponds to this is the component τ_b of a natural transformation $\tau\colon R \Longrightarrow R'$. Similarly, setting $a = R'(b)$ in the first line gives

$$Hom(R'(b), R(b)) \cong Hom(LR'(b), b) \cong Hom(R'(b), R'(b)). \tag{4.133}$$

The third Hom set contains $id_{R'(b)}$. The morphism that corresponds to this in the first Hom set is the b component to τ^{-1}. The naturality of τ and τ^{-1} follows from the naturality of the isomorphisms of these Hom sets. To see that τ is the inverse of τ^{-1}, consider the following commutative diagram, taken from the relevant parts of these isomorphisms:

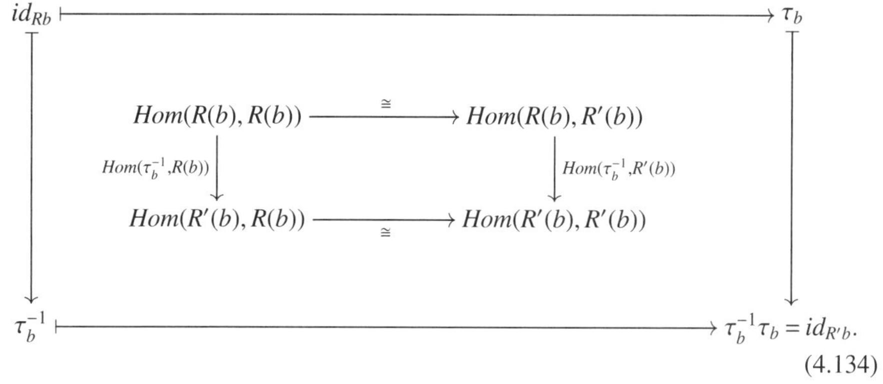

(4.134)

There is a similar square to show that $\tau_b\tau_b^{-1} = id_{Rb}$.

The uniqueness of τ follows from the fact that these sets are isomorphic and any such isomorphism has to go to the identity morphism. ♣

Exercise 4.4.26. Show that if $L\colon \mathbb{A} \longrightarrow \mathbb{B}$ is left adjoint to $R\colon \mathbb{B} \longrightarrow \mathbb{A}$, then $L^{op}\colon \mathbb{A}^{op} \longrightarrow \mathbb{B}^{op}$ is right adjoint to $R^{op}\colon \mathbb{B}^{op} \longrightarrow \mathbb{A}^{op}$. ■

Definition 4.4.27 occurs often.

Definition 4.4.27. Let $\mathbb{A}$ be a full subcategory of $\mathbb{B}$; then $\mathbb{A}$ is called a **reflective subcategory** if the inclusion has a left adjoint:

$$\mathbb{A} \underset{\perp}{\rightleftarrows} \mathbb{B}. \tag{4.135}$$

Dually, a full subcategory is called a **coreflective subcategory** if the inclusion has a right adjoint.

We close this section with an important adjunction from linear algebra.

Example 4.4.28. There is a forgetful functor $U\colon \mathbf{K}\mathbb{V}\mathbb{ect} \longrightarrow \mathbb{S}\mathbb{et}$ that takes a complex vector space and outputs its underlying set. The functor has a left adjoint $F\colon \mathbb{S}\mathbb{et} \longrightarrow \mathbf{K}\mathbb{V}\mathbb{ect}$ called the **free vector space functor**, which takes a set and outputs the vector space whose basis is the elements of the set. For example, if the set $S = \{s_1, s_2, \ldots, s_n\}$ is a finite set, then the elements of $F(S)$ look like this:

$$k_1 s_1 + k_2 s_2 + \cdots + k_n s_n, \tag{4.136}$$

where k_i are elements of $\mathbf{K}$. The addition of such elements combines like elements. Scalar multiplication is done as

$$k \cdot (k_1 s_1 + k_2 s_2 + \cdots + k_n s_n) = kk_1 s_1 + kk_2 s_2 + \cdots + kk_n s_n. \tag{4.137}$$

The unit of the adjunction at set S is $\eta\colon S \longrightarrow U(F(S))$, which takes element s to $1s$. This is called "insertion of generators." It is easy to see the universal property of the unit, which we leave to the reader. Functor F is essentially surjective because every vector space is a free vector space. (It is also faithful, but it is not full.)

The restriction of F to finite sets outputs finite-dimensional vector spaces. Notice that when one forgets the vector space structure of a finite-dimensional complex vector space, one does not necessarily get a finite set. This means that there is no forgetful functor from $\mathbf{C}\mathbb{FD}\mathbb{V}\mathbb{ect}$ to $\mathbb{F}\mathbb{in}\mathbb{S}\mathbb{et}$.[2] This all can be summarized as

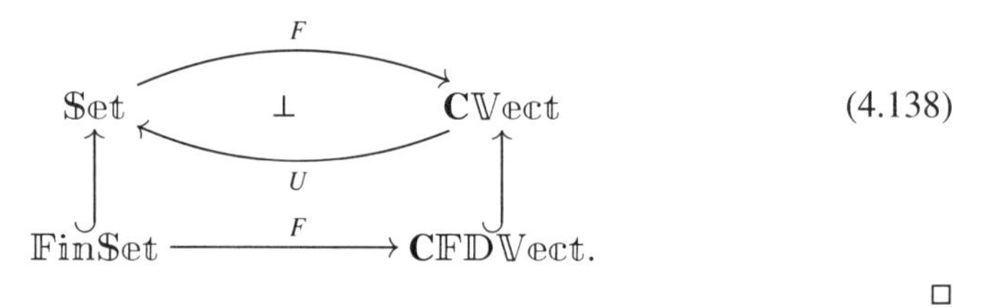

(4.138)

□

4.5 Exponentiation and Comma Categories

In this section, we will meet various operations on categories and functors. We start by discussing functor categories (which we first met in Remark 4.2.7). We will also meet comma categories, which are ways of making the morphisms of one category into the objects of another category.

We have seen that given categories $\mathbb{A}$ and $\mathbb{B}$, we can form category $\mathbb{A} \times \mathbb{B}$. Here, we will take two categories and form their functor category. This is analogous to taking two sets, S and T, and forming the set of functions $T^S = Hom_{\mathbb{S}\mathbb{et}}(S, T)$.

[2] If we use a finite field instead of the complex numbers, such a forgetful functor does exist and there will be an adjunction with the free functor. I am indebted to Mirco Mannucci for pointing this out to me.

Definition 4.5.1. Given categories $\mathbb{A}$ and $\mathbb{B}$, there exists the **functor category**, written as $\mathbb{B}^{\mathbb{A}}$ or $Hom_{\mathbb{C}\mathrm{at}}(\mathbb{A},\mathbb{B})$, whose objects are all functors from category $\mathbb{A}$ to category $\mathbb{B}$, and whose morphisms are all natural transformations between those functors.

The fact that for any two objects in $\mathbb{C}\mathrm{at}$, $\mathbb{A}$ and $\mathbb{B}$, the Hom set, $Hom_{\mathbb{C}\mathrm{at}}(\mathbb{A},\mathbb{B})$, has the structure of a category means that $\mathbb{C}\mathrm{at}$ is a closed category. We will see more about this in Sections 7.2 and 9.1.

Let us look at some simple examples.

Example 4.5.2.

- If $\mathbb{A}$ is $\mathbf{1}$, the category with a single object and a single identity morphism, then the objects of $\mathbb{B}^{\mathbf{1}}$ are functors that pick out a single object of $\mathbb{B}$. The natural transformations essentially pick out morphisms of $\mathbb{B}$. Analogous to Equation (1.18), we have $\mathbb{B}^{\mathbf{1}} \cong \mathbb{B}$.
- If $\mathbb{A}$ is category $\mathbf{2}_\circ$, the discrete category with two objects and no nonidentity morphisms, then the functor $\mathbf{2}_\circ \longrightarrow \mathbb{B}$ picks out two objects of $\mathbb{B}$. The natural transformations are pairs of morphisms in $\mathbb{B}$. This means that

$$\mathbb{B}^{\mathbf{2}_\circ} \cong \mathbb{B} \times \mathbb{B}. \tag{4.139}$$

- If $\mathbb{A} = \mathbf{2}$, the category with two objects and a morphism between the two objects, then the functor $\mathbf{2} \longrightarrow \mathbb{B}$ picks out a morphism in $\mathbb{B}$. Consider a functor that picks out the morphism $f\colon b_1 \longrightarrow b_2$ and a functor that picks out the morphism $f'\colon b_3 \longrightarrow b_4$. Then a natural transformation from the first functor to the second functor amounts to a commutative diagram:

$$\begin{array}{ccc} b_1 & \xrightarrow{g} & b_3 \\ {\scriptstyle f}\downarrow & & \downarrow{\scriptstyle f'} \\ b_2 & \xrightarrow[g']{} & b_4. \end{array} \tag{4.140}$$

  Composition corresponds to horizontal composition of natural transformations and can be seen as pasting one box on top of the other. This functor category is called the **arrow category** of $\mathbb{B}$.
- Let $\mathbb{A}$ be the category

$$* \rightrightarrows *. \tag{4.141}$$

  Since names of objects and morphisms do not really matter, we might view this category as

$$A \underset{trg}{\overset{src}{\rightrightarrows}} V, \tag{4.142}$$

  where A stands for arrows, V stands for vertices, *src* stands for source, and *trg* stands for target. Then the objects in the functor category $\mathbb{S}\mathrm{et}^{\mathbb{A}}$ are functors $F\colon \mathbb{A} \longrightarrow \mathbb{S}\mathrm{et}$ that pick out two sets, $F(A)$ and $F(V)$, and two set morphisms

$F(src)\colon F(A) \longrightarrow F(V)$ and $F(trg)\colon F(A) \longrightarrow F(V)$. This is nothing more than a directed graph. Natural transformations are exactly directed graph homomorphisms that make Diagram (1.49) commutes. To summarize,

$$\mathbb{S}\mathrm{et}^{\ * \rightrightarrows *} \cong \mathbb{G}\mathrm{raph}. \tag{4.143}$$

- We can take the previous item and go further. Consider the category

$$\mathbb{A} = A \underset{trg}{\overset{src}{\rightrightarrows}} V \longrightarrow L_V. \tag{4.144}$$

Then the category $\mathbb{S}\mathrm{et}^{\mathbb{A}}$ is a directed graph with an added function from the vertices of the graph to a set of labels. This will give the category of graphs with labeled vertices. We also have directed graphs with labeled arrows and labeled vertices:

$$\mathbb{A} = L_A \longleftarrow A \underset{trg}{\overset{src}{\rightrightarrows}} V \longrightarrow L_V. \tag{4.145}$$

- We already saw in Example 4.1.47 that if $\mathbb{A}$ is a monoid, M, which is thought of as a one-object category, then the functor $F\colon M \longrightarrow \mathbb{S}\mathrm{et}$ is going to pick out a set S, and for every morphism $m\colon * \longrightarrow *$, there will be a set function $F(m)\colon S \longrightarrow S$. We also saw in Example 4.2.4, that natural transformations are homomorphisms of M-sets. Thus $\mathbb{S}\mathrm{et}^M$ is the category of M-sets. Similarly, if G is a group, thought of as a one-object category, then $\mathbb{S}\mathrm{et}^G$ is the category of G-sets.
- In particular, when the monoid M is the natural number $\mathbf{N}$, then $\mathbb{S}\mathrm{et}^{\mathbf{N}}$ is the category of $\mathbf{N}$-sets. An object in this category is a set S with maps

$$S \xrightarrow{F_0} S \xrightarrow{F_1} S \xrightarrow{F_2} \cdots. \tag{4.146}$$

We can think of these diagrams as describing how systems change in discrete time. The element $s \in S$ in time t follows the map to become a member of S in time $t+1$. The objects in this category are called **dynamical systems** or **discrete time dynamical systems**. A small example of such a dynamical system is a set $S = \{a, b, c, d, e, f, g, h, i, j\}$ and a map $F_0\colon S \longrightarrow S$, described as

$s \in S$	a	b	c	d	e	f	g	h	i	j
$F_0(s)$	j	h	c	f	e	g	i	b	d	e

- We can also talk about the monoid $\mathbf{R}$ of real numbers. In this case, $\mathbb{S}\mathrm{et}^{\mathbf{R}}$ becomes **continuous-time dynamical systems**. It is harder to draw a diagram of such a system. There is a set S, and for every $r \in \mathbf{R}$, there is a set function $t_r\colon S \longrightarrow S$. Many physical systems can be described by such dynamical systems.

- When $\mathbb{A} = \mathbf{0}$, the empty category, then for any category $\mathbb{B}$, we have $\mathbb{B}^{\mathbf{0}} = \mathbf{1}$ because there is exactly one functor from the empty category to any other category.
- To what extent can any locally small category $\mathbb{A}$ be seen as an functor category? We will see in Section 4.7 that every small category $\mathbb{A}$ can be embedded in a functor category $\mathbb{S}\mathrm{et}^{\mathbb{A}^{op}}$. □

Important Categorical Idea 4.5.3. Exponentiation Is Central. Exponentiation is very important throughout category theory. When we examine $\mathbb{B}^{\mathbb{A}}$, we might think of $\mathbb{A}$ as a diagram or a shape and think of $\mathbb{B}$ as a context where the diagrams take place. Other ways to think of $\mathbb{A}$ is as the "ideal model," "syntax," or "cookie cutter," and every functor $F\colon \mathbb{A} \longrightarrow \mathbb{B}$ describes a "semantic model" of the ideal in $\mathbb{B}$ or the "cookie" in $\mathbb{B}$. Sometimes we will look at special types of functors $F\colon \mathbb{A} \longrightarrow \mathbb{B}$ that preserve some of the structures of $\mathbb{A}$ and $\mathbb{B}$ (e.g., product-preserving functors, colimit-preserving functors, or monoidal-preserving functors, etc.).

Another fundamental idea is the relationship between $\mathbb{A}$ (the ideal) and $\mathbb{B}^{\mathbb{A}}$ (the collection of models of the ideal). It is a particularly important to look at the case when $\mathbb{B} = \mathbb{S}\mathrm{et}$. We will examine the relationship between $\mathbb{A}$ and $\mathbb{S}\mathrm{et}^{\mathbb{A}^{op}}$ when we talk of the Yoneda lemma in Section 4.7. ○

How do functor categories relate to each other? We saw that for any category $\mathbb{A}$, there are covariant and contravariant functors $Hom_{\mathbb{A}}(a, \ )$ and $Hom_{\mathbb{A}}(\ , a)$, respectively. Let us spell out the details for the case when $\mathbb{A} = \mathbb{C}\mathrm{at}$. For any category $\mathbb{C}$, the functor $F\colon \mathbb{B} \longrightarrow \mathbb{B}'$ induces a functor:

$$F_* = Hom_{\mathbb{C}\mathrm{at}}(\mathbb{C}, F)\colon \mathbb{B}^{\mathbb{C}} \longrightarrow \mathbb{B}'^{\mathbb{C}}, \tag{4.147}$$

which is defined as

$$H\colon \mathbb{C} \longrightarrow \mathbb{B} \qquad \mapsto \qquad F \circ H\colon \mathbb{C} \longrightarrow \mathbb{B} \longrightarrow \mathbb{B}'. \tag{4.148}$$

For any category $\mathbb{C}$, the functor $G\colon \mathbb{A} \longrightarrow \mathbb{A}'$, induces a functor:

$$G^* = Hom_{\mathbb{C}\mathrm{at}}(G, \mathbb{C})\colon \mathbb{C}^{\mathbb{A}'} \longrightarrow \mathbb{C}^{\mathbb{A}}, \tag{4.149}$$

which is defined as

$$H\colon \mathbb{A}' \longrightarrow \mathbb{C} \qquad \mapsto \qquad H \circ G\colon \mathbb{A} \longrightarrow \mathbb{A}' \longrightarrow \mathbb{C}. \tag{4.150}$$

Example 4.5.4. Here are some examples of induced functors.

- Let $\mathbb{A} = \mathbf{2}_\circ$. Consider the functor $H\colon \mathbf{1} \longrightarrow \mathbf{2}_\circ$, which takes the single object in $\mathbf{1}$ to the first object in $\mathbf{2}_\circ$. The functor H^* takes a functor $F\colon \mathbf{2}_\circ \longrightarrow \mathbb{B}$ to a functor $F \circ H\colon \mathbf{1} \longrightarrow \mathbf{2}_\circ \longrightarrow \mathbb{B}$, which outputs the first object that F chose. This is essentially the projection functor $\pi_1\colon \mathbb{B} \times \mathbb{B} \longrightarrow \mathbb{B}$. The functor $H'\colon \mathbf{1} \longrightarrow \mathbf{2}_\circ$, which chooses the second object, induces the other projection function (i.e., $H'^* = \pi_2$).

- Let $\mathbb{A} = A \underset{trg}{\overset{src}{\rightrightarrows}} V$. Consider the functor $J\colon \mathbf{1} \longrightarrow \mathbb{A}$, which takes the single object in $\mathbf{1}$ to the V in $\mathbb{A}$. Any functor $F\colon \mathbb{A} \longrightarrow \mathbb{S}\mathrm{et}$ is a graph. The composition with J gives a functor $F \circ J\colon \mathbf{1} \longrightarrow \mathbb{A} \longrightarrow \mathbb{S}\mathrm{et}$, which picks out the set of vertices of the graph. So J^* is the forgetful functor from the category $\mathbb{G}\mathrm{raph}$ to $\mathbb{S}\mathrm{et}$ that gives the underlying set of vertices. There is a similar functor that gives the underlying set of arrows.
- Let $K\colon \mathbf{1} \longrightarrow M$ be a functor that takes the single object in the one-object category $\mathbf{1}$ to the single object in the one-object category of the monoid M. The functor $K^*\colon \mathbb{S}\mathrm{et}^M \longrightarrow \mathbb{S}\mathrm{et}^{\mathbf{1}}$ takes an M–set to the underlying set without the action (i.e., K^* forgets the action of the M-set).
- Let $inc\colon \mathbf{N} \hookrightarrow \mathbf{R}$ be the inclusion of the one-object monoid of natural numbers to the one-object monoid of real numbers. The induced functor $inc^*\colon \mathbb{S}\mathrm{et}^{\mathbf{R}} \longrightarrow \mathbb{S}\mathrm{et}^{\mathbf{N}}$ takes a continuous time dynamical system $\mathbf{R} \longrightarrow \mathbb{S}\mathrm{et}$ to the discrete time dynamical system $\mathbf{N} \longrightarrow \mathbf{R} \longrightarrow \mathbb{S}\mathrm{et}$. The discrete system picks out those parts of the continuous system that correspond to whole numbers. One way to think of this is that there is some continuous dynamical system, but an experimenter looks at the system at separate time clicks and records the observations. Much of the physical sciences is done this way.
- Let $\mathbb{A} = A \underset{trg}{\overset{src}{\rightrightarrows}} V$, and $\mathbf{N}$ be the one-object natural numbers. Consider the functor $F\colon \mathbb{A} \longrightarrow \mathbf{N}$ that takes the objects A and V to the $*$ and $F(src) = id_*$ and $F(trg) = 1\colon * \longrightarrow *$. Notice we are not interested in the entire category of natural numbers. We are only interested in the first nontrivial element of the monoid. The induced functor is

$$F^*\colon \mathbb{S}\mathrm{et}^{\mathbf{N}} \longrightarrow \mathbb{S}\mathrm{et}^{\mathbb{A}} = \mathbb{G}\mathrm{raph}, \tag{4.151}$$

which takes a discrete time dynamical system to the graph that has arrows showing how the system changes at the first-time click. In detail, the source of every arrow will be where the element of the dynamical system starts, and the target of the arrow will be where the element goes after one-time click. For example, the discrete time dynamical system of Example 4.5.2 becomes the graph

(4.152)

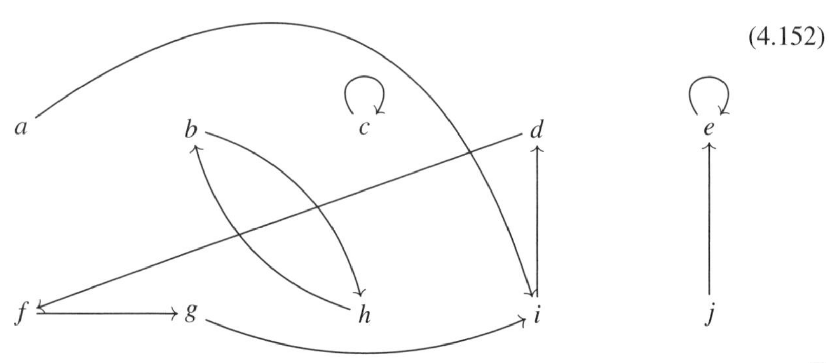

□

There is another example that will be extremely important for the next section and hence worth spending some time spelling out all the ingredients. For every category $\mathbb{A}$, there is a unique functor $!\colon \mathbb{A} \longrightarrow \mathbf{1}$. This functor induces a functor, which we denote as Δ:

$$\Delta = !^*\colon \mathbb{B}^{\mathbf{1}} \longrightarrow \mathbb{B}^{\mathbb{A}}. \tag{4.153}$$

In detail, Δ is defined on objects as

$$F\colon \mathbf{1} \longrightarrow \mathbb{B} \qquad \mapsto \qquad \Delta(F) = F \circ !\colon \mathbb{A} \longrightarrow \mathbf{1} \longrightarrow \mathbb{B}. \tag{4.154}$$

That is, if $F(*) = b$ (where $*$ is the single object in $\mathbf{1}$), then $\Delta(F) = F \circ !\colon \mathbb{A} \longrightarrow \mathbf{1} \longrightarrow \mathbb{B}$ is going to take every object of $\mathbb{A}$ to b. A morphism $f\colon a \longrightarrow a'$ in $\mathbb{A}$ is going to go to the identity morphism $id_b\colon b \longrightarrow b$. For example, if the category is $\mathbb{A} = \mathbf{2}_\circ$ then any functor $F\colon \mathbf{1} \longrightarrow \mathbb{B}$ that picks out an element b will go to the functor $\Delta(F) = F \circ !\colon \mathbf{2}_\circ \longrightarrow \mathbf{1} \longrightarrow \mathbb{B}$. This will send each element of $\mathbf{2}_\circ$ to b. This is exactly the diagonal morphism $\Delta(b)$.

Now let us generalize from $2_\circ$ to an arbitrary category $\mathbb{A}$. If category $\mathbb{A}$ is

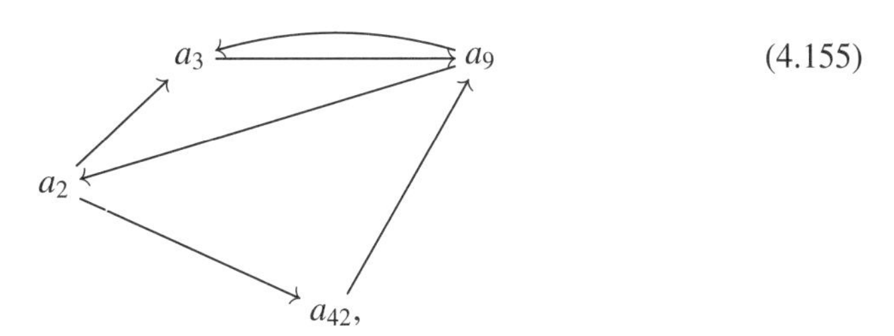

(4.155)

then one can imagine the image of $\Delta(b)$ as

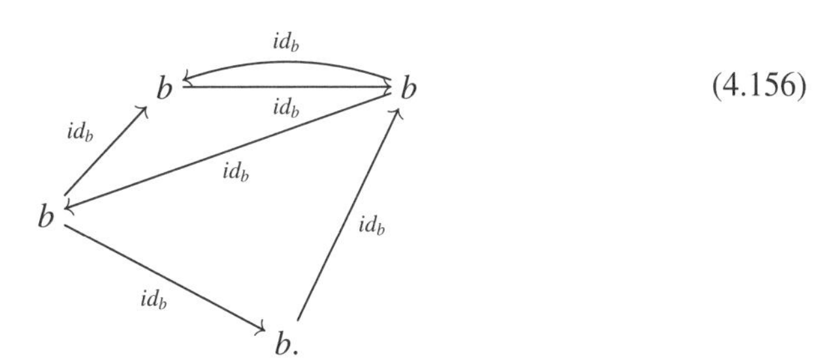

(4.156)

Of course, the image of the functor is simply one object and the identity map; however, we can think of the image as it appears here. Although we are overloading the word, we call Δ the **diagonal functor**. It is similar to the diagonal functor that we saw earlier, where $\Delta\colon \mathbb{A} \longrightarrow \mathbb{A} \times \mathbb{A}$. The old one is defined as $\Delta(a) = (a, a)$. This means that in each position of the output, the value is a. The functor $\Delta\colon \mathbb{B} \longrightarrow \mathbb{B}^{\mathbb{A}}$ also outputs the same value.

Now for constructions that make the morphisms of one category into the objects of another category.

Definition 4.5.5. Given two functors

$$\mathbb{A} \xrightarrow{\ F\ } \mathbb{C} \xleftarrow{\ G\ } \mathbb{B}, \tag{4.157}$$

we can form the **comma category** (F,G), sometimes also written as $(F \downarrow G)$. The objects of this category are triples (a,f,b), where a is an object of $\mathbb{A}$, b is an object $\mathbb{B}$, and $f\colon F(a) \longrightarrow G(b)$ is a morphism in $\mathbb{C}$. A morphism from (a,f,b) to (a',f',b') in (F,G) consists of a pair of morphisms (g,h) where $g\colon a \longrightarrow a'$ in $\mathbb{A}$ and $h\colon b \longrightarrow b'$ in $\mathbb{B}$ such that the following square commutes:

$$\begin{array}{ccc} F(a) & \xrightarrow{\ f\ } & G(b) \\ {\scriptstyle F(g)}\downarrow & & \downarrow{\scriptstyle G(h)} \\ F(a') & \xrightarrow[\ f'\ ]{} & G(b') \end{array} \quad . \tag{4.158}$$

Composition of morphisms come from the fact that two commuting squares placed one on top of the other also commute. The identity morphisms are obvious.

Some examples of comma categories are familiar already.

Example 4.5.6.

- If $\mathbb{A} = \mathbf{1}$ and $F\colon \mathbf{1} \longrightarrow \mathbb{C}$ picks out the object c_0 and $\mathbb{B} = \mathbb{C}$ with $G = Id_{\mathbb{C}}$, as in

$$\mathbf{1} \xrightarrow{\ F\ } \mathbb{C} \xleftarrow{\ Id_{\mathbb{C}}\ } \mathbb{C}, \tag{4.159}$$

  then the comma category (F,G) is the coslice category $c_0/\mathbb{C}$.
- If $\mathbb{B} = \mathbf{1}$ and $G\colon \mathbf{1} \longrightarrow \mathbb{C}$ picks out the object c_0 and $\mathbb{A} = \mathbb{C}$ with $F = Id_{\mathbb{C}}$, as in

$$\mathbb{C} \xrightarrow{\ Id_{\mathbb{C}}\ } \mathbb{C} \xleftarrow{\ G\ } \mathbf{1}, \tag{4.160}$$

  then the comma category (F,G) is the slice category $\mathbb{C}/c_0$.
- If $\mathbb{A} = \mathbb{B} = \mathbb{C}$ and $F = G = Id_{\mathbb{C}}$, as in

$$\mathbb{C} \xrightarrow{\ Id_{\mathbb{C}}\ } \mathbb{C} \xleftarrow{\ Id_{\mathbb{C}}\ } \mathbb{C}, \tag{4.161}$$

  then the comma category $(Id_{\mathbb{C}}, Id_{\mathbb{C}})$ is nothing more than the arrow category $\mathbb{C}^{\longrightarrow}$, which we saw on page 189. □

There are forgetful functors $U_1\colon (F,G) \longrightarrow \mathbb{A}$ and $U_2\colon (F,G) \longrightarrow \mathbb{B}$, which are defined on objects as follows: functor U_1 takes (a,f,b) to a and U_2 takes (a,f,b) to b.

This means that for slice categories $a/\mathbb{A}$, there is a forgetful functor to $\mathbb{A}$. There are similar forgetful functors for coslice categories.

How do comma categories relate to each other? First, let us look at slice categories. Consider slice categories $\mathbb{C}/c$ and $\mathbb{C}/c'$. A morphism $g\colon c \longrightarrow c'$ in $\mathbb{C}$ will induce the functor $\mathbb{C}/c \longrightarrow \mathbb{C}/c'$, which takes object $f\colon a \longrightarrow c$ of $\mathbb{C}/c$ to $g \circ f\colon a \longrightarrow c \longrightarrow c'$ in $\mathbb{C}/c'$. Morphisms in $\mathbb{C}/c$ can be dealt with in the same manner. There are similar statements about coslice categories.

Theorem 4.5.7 shows how general comma categories relate to each other.

Theorem 4.5.7. Given functors

$$\mathbb{A} \xrightarrow{\quad F \quad} \mathbb{C} \xleftarrow{\quad G \quad} \mathbb{B} \tag{4.162}$$

and

$$\mathbb{A} \xrightarrow{\quad F' \quad} \mathbb{C} \xleftarrow{\quad G' \quad} \mathbb{B}, \tag{4.163}$$

there are comma categories (F,G) and (F',G'). Any functor $H\colon \mathbb{C} \longrightarrow \mathbb{C}'$ where the following two triangles commute:

$$\begin{array}{ccc} \mathbb{A} & \xrightarrow{\quad F \quad} & \mathbb{C} \\ {\scriptstyle F'}\big\downarrow & \swarrow{\scriptstyle H} & \big\uparrow{\scriptstyle G} \\ \mathbb{C}' & \xleftarrow[\quad G' \quad]{} & \mathbb{B} \end{array} \tag{4.164}$$

induces a functor $\widehat{H}\colon (F,G) \longrightarrow (F',G')$ that is defined as

$$(a, f\colon Fa \longrightarrow Gb, b) \qquad \mapsto \qquad (a, Hf\colon HFa \longrightarrow HGb, b) = (a, Hf\colon F'a \longrightarrow G'b, b). \tag{4.165}$$

On morphisms, $\widehat{H}$ takes (f,g) to (f,g); that is, the Hom sets are isomorphic.

Furthermore, if there are adjoint functors $R\colon \mathbb{C} \longrightarrow \mathbb{C}'$ and $L\colon \mathbb{C}' \longrightarrow \mathbb{C}$ with $R \vdash L$ and each functor makes the corresponding triangle commute:

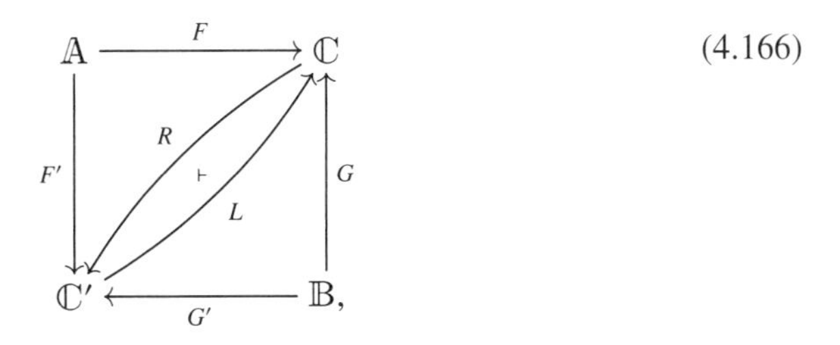

then their induced functors are also adjoint:

$$(F,G) \underset{\widehat{L}}{\overset{\widehat{R}}{\rightleftarrows}} (F',G'). \qquad \top \tag{4.167}$$

★

We leave the proof to the reader.

4.6 Limits and Colimits Revisited

With the knowledge of natural transformations and adjoint functors, we can see limits and colimits in a new light. Let $\mathbb{D}$ be a category that we employ as a diagram in an exponent. We call it a **diagram category** or a **shape category**. For example, category $\mathbb{D}$ might look like

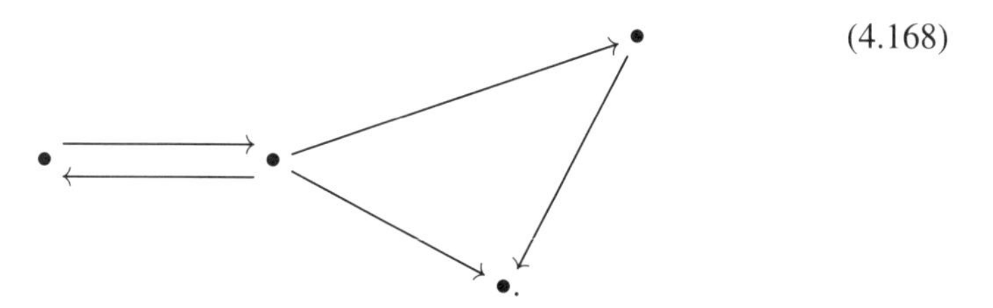

(4.168)

For category $\mathbb{B}$, the category $\mathbb{B}^{\mathbb{D}}$ is the collection of all functors from $\mathbb{D}$ to $\mathbb{B}$ and natural transformations between them. For example, the image of a typical functor $F\colon \mathbb{D} \longrightarrow \mathbb{B}$ might look like this:

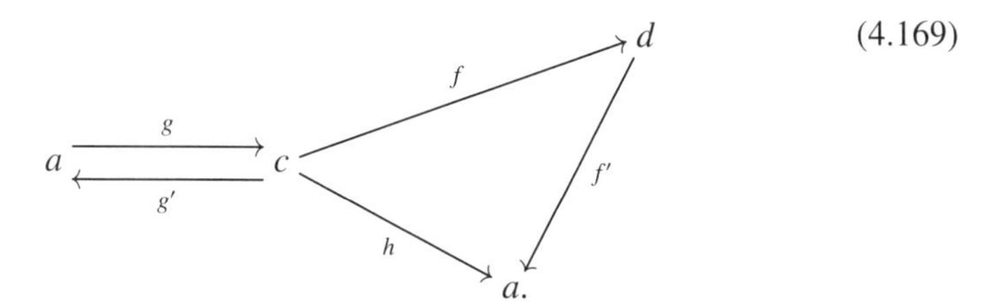

(4.169)

As we saw in Section 4.5, there is a diagonal functor $\Delta\colon \mathbb{B} \longrightarrow \mathbb{B}^{\mathbb{D}}$, which takes every $b \in \mathbb{B}$ to the diagram with all the objects being b and all the morphisms being id_b. For Diagram (4.168), $\Delta(b)$ looks like this:

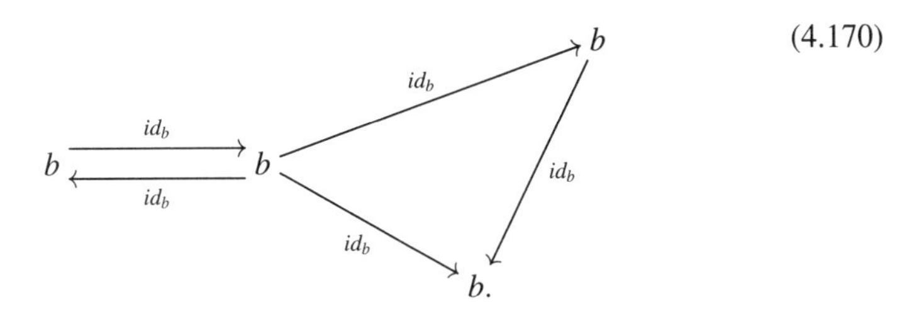

(4.170)

A natural transformation from the functor $\Delta(b)$ to a functor $F\colon \mathbb{D} \longrightarrow \mathbb{B}$ (i.e., a morphism in the category $\mathbb{B}^{\mathbb{D}}$) is a commutative diagram that looks like this:

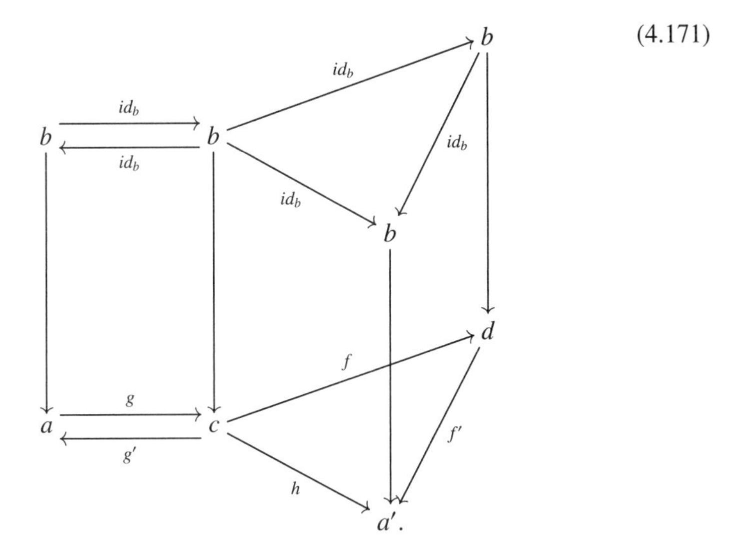

(4.171)

The vertical morphisms are the components of the natural transformation. We can shorten this to

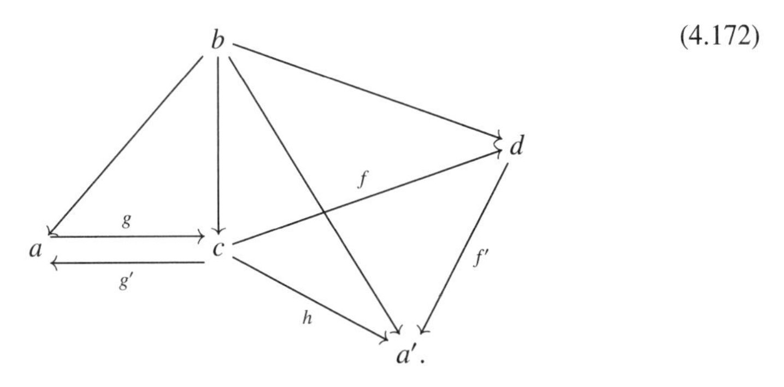

(4.172)

Definition 4.6.1. A **cone over** $F\colon \mathbb{D} \longrightarrow \mathbb{B}$ **with base** b is a natural transformation in $\mathbb{B}^{\mathbb{D}}$ from $\Delta(b)$ to F. Consider when there are two cones over F, $\Delta(b) \Longrightarrow F$

and $\Delta(b') \Longrightarrow F$. A map from the cone with base b to the cone with base b' is a map $b \longrightarrow b'$ such that all the expected diagrams commute:

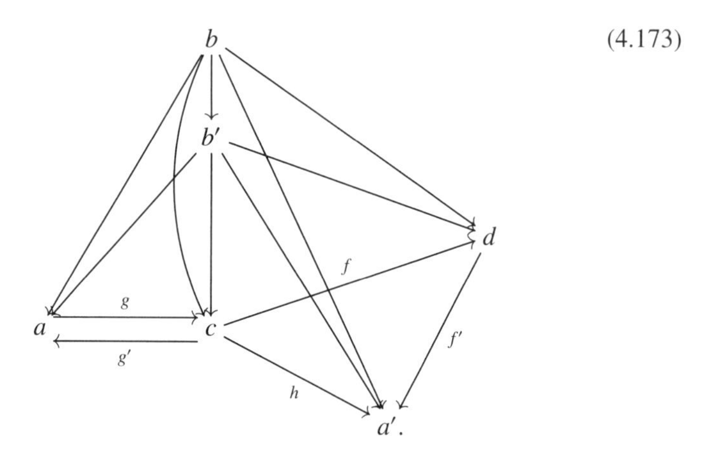

(4.173)

With this concept of a morphism between cones over F, we define the category of cones over $F\colon \mathbb{D} \longrightarrow \mathbb{B}$, which we denote as $\mathbb{Cone}(F)$.

There is a dual notion of a **cocone over** F **with base** b, which is a natural transformation in $\mathbb{B}^{\mathbb{D}}$ from $F\colon \mathbb{D} \longrightarrow \mathbb{B}$ to $\Delta(b)$. There is an obvious definition of a morphism between cocones and the category of cocones over F, denoted as $\mathbb{Cocone}(F)$.

A formal way of constructing the category of cones for a functor F is to consider the comma category of the following two functors:

$$\mathbb{B} \xrightarrow{\ \ \Delta\ \ } \mathbb{B}^{\mathbb{D}} \xleftarrow{\ \ Const_F\ \ } \mathbf{1}, \tag{4.174}$$

where the right functor chooses functor F. Similarly, the category of cocones of F is the comma category of

$$\mathbf{1} \xrightarrow{\ \ Const_F\ \ } \mathbb{B}^{\mathbb{D}} \xleftarrow{\ \ \Delta\ \ } \mathbb{B}\,. \tag{4.175}$$

With the category of cones, we can talk about the best-fitting cone.

Definition 4.6.2. The **limit** of a diagram $F\colon \mathbb{D} \longrightarrow \mathbb{B}$ is the terminal object in the category $\mathbb{Cone}(F)$ of cones over F. That is, it is the cone over F with the property that every cone has a unique map to it. In detail, the limit is an object $Lim(F)$ of $\mathbb{B}$ and a morphism $Lim(F) \longrightarrow F(d)$ for every d in $\mathbb{D}$. The obvious compositions of morphisms commute.

The **colimit** of a diagram $F\colon \mathbb{D} \longrightarrow \mathbb{B}$ is the initial object in the category $\mathbb{Cocone}(F)$ of cocones over F. That is, it is a cocone over F with the property that there is a cocone map from it to every cocone over F. Again, the colimit is an object $Colim(F)$ of $\mathbb{B}$ and maps $F(d) \longrightarrow Colim(F)$ for every d in $\mathbb{D}$. The obvious compositions of morphisms commute.

From the vantage point of looking at limits as terminal objects in the category of cones, one easily sees that a limit has the same uniqueness up to a unique isomorphism as the terminal object of a category.

Let us elaborate the details of this definition. A cone is a natural transformation $\Delta(b) \Longrightarrow F$. The limit of F is an object $Lim(F)$ in category $\mathbb{B}$. A cone of F with the limit as a base of the cone is a natural transform $\Delta(Lim(F)) \Longrightarrow F$. Saying that the limit is the terminal cone means that for every cone $\Delta(b) \Longrightarrow F$, there is a unique morphism $b \longrightarrow Lim(F)$. In terms of Hom sets, this become a statement about adjoint functors:

$$Hom_{\mathbb{B}^{\mathbb{D}}}(\Delta(b), F) \cong Hom_{\mathbb{B}}(b, LimF). \tag{4.176}$$

There is a similar analysis of coloimits, and we get

$$Hom_{\mathbb{B}^{\mathbb{D}}}(F, \Delta(b)) \cong Hom_{\mathbb{B}}(ColimF, b). \tag{4.177}$$

Both of these adjoint functors can be encapsulated as

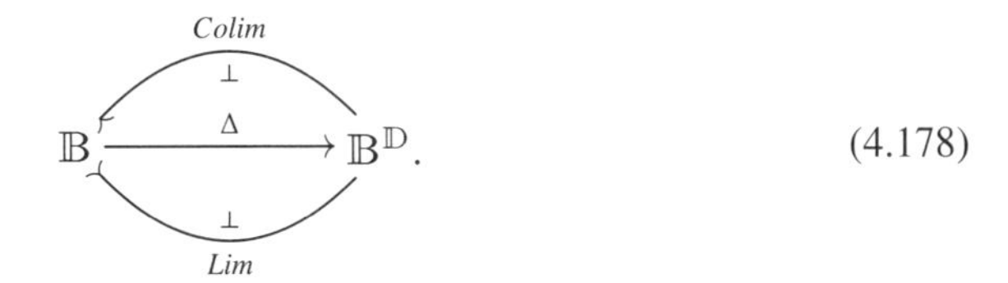

This is a generalization of Diagram (4.107) that summarizes the many examples of limits and colimits that we saw in Sections 3.1 and 3.2.

Let us examine the properties of limits and colimits. First, a definition:

Definition 4.6.3. Let $G: \mathbb{B} \longrightarrow \mathbb{C}$ be a functor. Notice that for any diagram $F: \mathbb{D} \longrightarrow \mathbb{B}$, there is also an induced diagram $G \circ F: \mathbb{D} \longrightarrow \mathbb{C}$. We say that G **preserves limits** if for all $F: \mathbb{D} \longrightarrow \mathbb{B}$, G takes the limit of F to the limit of $G \circ F$. Expressed in symbols, $G(Lim(F)) = Lim(G \circ F)$. We say that G **reflects limits** if for all $F: \mathbb{D} \longrightarrow \mathbb{B}$ and any cone $\lambda: \Delta(b) \Longrightarrow F$, $G(\lambda)$ is a limit of $G \circ F$ implies λ is a limit of F.

There are similar definitions about preserving and reflecting colimits.

Theorem 4.6.4. Right adjoints preserve limits and left adjoints preserve colimits. ★

Proof. Figure 4.5 will be helpful. Let $R: \mathbb{B} \longrightarrow \mathbb{C}$ be a right adjoint to $L: \mathbb{C} \longrightarrow \mathbb{B}$ and $F: \mathbb{D} \longrightarrow \mathbb{B}$. The diagram of F is at the bottom left of the figure. Assume that b is the limit of F in $\mathbb{B}$. Apply R to the bottom left to get the diagram on the bottom right. Our aim is to show that $R(b)$ is the limit of the diagram on the bottom right. By functoriality, there is a map from $R(b)$ to every element of the diagram. Furthermore, for any c in $\mathbb{C}$ with maps $c \longrightarrow R(x)$, where x is any element of the diagram on the bottom right, there is, by adjointness, a map $L(c) \longrightarrow x$ on the left. Since b is the limit on the left, by the universal property of the limit, there is a unique map $L(b) \longrightarrow c$ making all the diagrams commute. By adjointness again, there is a unique map $c \longrightarrow R(b)$ on the right ensuring that $R(b)$ is the limit. ♣

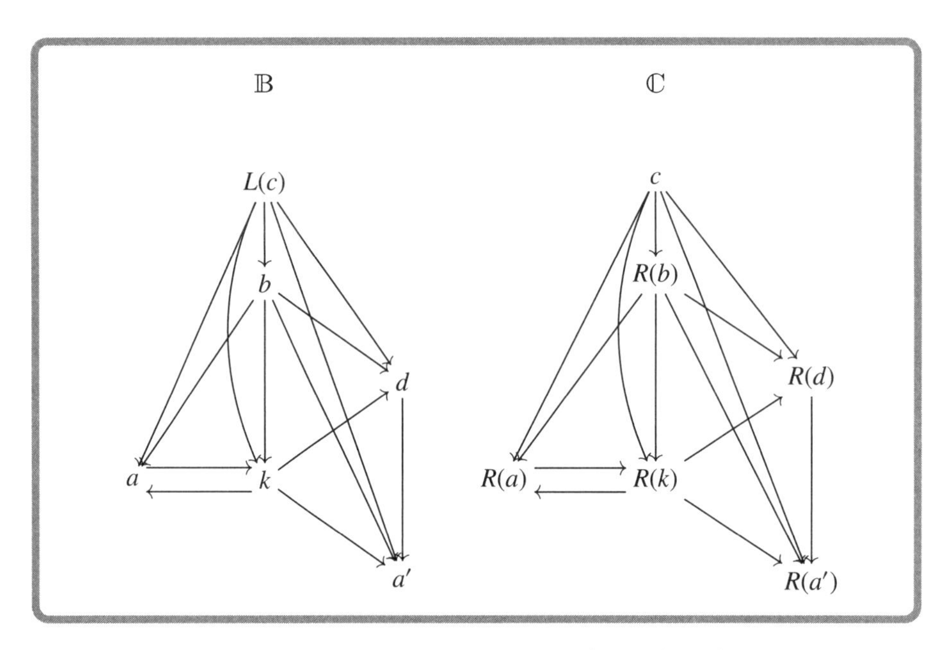

Figure 4.5. A cone and its image under a right adjoint functor.

Exercise 4.6.5. Show that left adjoints preserve colimits. ■

Definition 4.6.6. A category **has finite limits** or is **finitely complete** if it has limits of finite diagrams. A category **has limits** or is **complete** if it has all limits of small diagrams (the collection of objects and maps are sets). We similarly define a category that **has finite colimits**, is **finitely cocomplete**, or is **cocomplete**.

Theorem 4.6.7. A category is complete if and only if it has all products and equalizers. A category is cocomplete if and only if it has all coproducts and coequalizers. ★

Proof. This is a generalization of Theorem 3.2.10. ♣

Remark 4.6.8. It is easy to see that an equivalence of categories is both a right and a left adjoint. Such functors preserve and reflect limits and colimits. However, there is something very strange about equalizers and coequalizers. Consider an equivalence of categories $F\colon \mathbb{A} \longrightarrow \mathbb{B}$. Imagine that it works as follows:

$$\mathbb{A} \xrightarrow{\quad F \quad} \mathbb{B} \tag{4.179}$$

$$a \xrightarrow{f} b \xrightarrow[\cong]{p} c \overset{g}{\underset{h}{\rightrightarrows}} d \qquad\qquad F(a) \xrightarrow{F(f)} \begin{matrix} F(b) \\ = \\ F(c) \end{matrix} \overset{F(g)}{\underset{F(h)}{\rightrightarrows}} F(d).$$

In category $\mathbb{A}$, there is an isomorphism $p\colon b \longrightarrow c$. The functor F can have $F(b)$ and $F(c)$ be the same object of $\mathbb{B}$ and $F(p) = id_{F(b)} = id_{F(c)}$. Furthermore, it could be that in category $\mathbb{B}$, the morphism $F(f) = F(pf)$ is the equalizer of $F(g)$ and $F(h)$. So we have that F is an equivalence of categories, the image $F(f) = F(pf)$ is an equalizer of $F(g)$ and $F(h)$, and pf is an equalizer of g and h, but f is not an equalizer of g and h. (The equivalence still reflects equalizers because in our definition of reflecting limits, we assumed that the cone exists in $\mathbb{A}$ already. Here, we are talking of just equalizers in the target and not those formed in the source.) ♠

Important Categorical Idea 4.6.9. The Many Ways of Describing Structures. We have already seen three related ways of describing categorical structures: (i) universal properties, (ii) limits and colimits, and (iii) adjoint functors. Each of these is important in its own right, and we will continue to talk about each of them. However, it is important to realize that they are three ways of talking about the same thing. With each of these, you can describe the other two. We saw this explicitly in Definition 4.6.2, where we learned that limits and colimits are really objects with universal properties (terminal and initial objects) in categories of cones and cocones. These objects are chosen by left and right adjoints. We also explicitly saw this in Definitions (III) and (IV) of adjoint functors, where the units and counits have universal properties. It is important to see them individually and as reflections of each other. In Section 9.2, we will see that Kan extensions are yet another equivalent way of describing categorical structures. ○

4.7 The Yoneda Lemma

As we saw in Important Categorical Idea 4.5.3, the relationship between category $\mathbb{A}$ and category $\mathbb{Set}^{\mathbb{A}^{op}}$ is fundamental. In this section, we will elaborate on this point. The Yoneda lemma, first put forward by Nobuo Yoneda, arises everywhere and has the reputation of being one of the most fundamental theorems in category theory.

This section describes two ideas. First, the Yoneda Embedding Theorem shows that category $\mathbb{A}$ embeds or "sits nicely" inside $\mathbb{Set}^{\mathbb{A}^{op}}$. To do this, we will identify the objects of $\mathbb{A}$ with certain objects of $\mathbb{Set}^{\mathbb{A}^{op}}$. Second, the Yoneda lemma shows that every object in $\mathbb{Set}^{\mathbb{A}^{op}}$ is determined by the way that it interacts with the objects from $\mathbb{A}$. The implications of these two ideas will be elaborated at the end of this section.

Let us begin with $\mathbb{Set}^{\mathbb{A}^{op}}$. The objects of this category are functors $F\colon \mathbb{A}^{op} \longrightarrow \mathbb{Set}$ (we will see in Section 9.5 of Chapter 9 that such a functor is called a **presheaf**) and the morphisms are natural transformations of such functors. In Example 4.1.32, we saw special functors of the type $Hom_{\mathbb{A}}(\ \ , a)\colon \mathbb{A}^{op} \longrightarrow \mathbb{Set}$, where a is an object in $\mathbb{A}$. Any functor that is isomorphic to such a functor is called a **representable functor**. In a sense, these functors are represented by the objects in $\mathbb{A}$.

There is a functor $y_{\mathbb{A}}\colon \mathbb{A} \longrightarrow \mathbb{Set}^{\mathbb{A}^{op}}$ called the **Yoneda embedding functor**, that takes every object $a \in \mathbb{A}$ to the functor it represents, that is, $y_{\mathbb{A}}(a) = Hom_{\mathbb{A}}(\ \ , a)$. The functor $y_{\mathbb{A}}$ is defined on morphisms as follows: for every $f\colon a \longrightarrow a'$ in $\mathbb{A}$, there is a natural transformation

$$y_{\mathbb{A}}(f)\colon Hom_{\mathbb{A}}(\ ,a) \Longrightarrow Hom_{\mathbb{A}}(\ ,a'). \tag{4.180}$$

For object b in $\mathbb{A}$, the b component is

$$y_{\mathbb{A}}(f)_b\colon Hom_{\mathbb{A}}(b,a) \longrightarrow Hom_{\mathbb{A}}(b,a') \tag{4.181}$$

and is defined as

$$k\colon b \longrightarrow a \qquad \longmapsto \qquad f \circ k\colon b \longrightarrow a \longrightarrow a'. \tag{4.182}$$

One can see that $y_{\mathbb{A}}(f)$ is natural by showing that the following square is commutative for any $g\colon b' \longrightarrow b$:

$$\begin{array}{ccc} Hom_{\mathbb{A}}(b,a) & \xrightarrow{\ y_{\mathbb{A}}(f)_b\ } & Hom_{\mathbb{A}}(b,a') \\ {\scriptstyle Hom_{\mathbb{A}}(g,a)}\downarrow & & \downarrow{\scriptstyle Hom_{\mathbb{A}}(g,a')} \\ Hom_{\mathbb{A}}(b',a) & \xrightarrow[\ y_{\mathbb{A}}(f)_{b'}\ ]{} & Hom_{\mathbb{A}}(b',a'). \end{array} \tag{4.183}$$

For a given $h\colon b \longrightarrow a$ in the upper-left corner, both paths around the square go to $f \circ h \circ g\colon b' \longrightarrow a'$ in the lower-right corner.

Exercise 4.7.1. Show that $y_{\mathbb{A}}$ respects composition. That is, show that for $f\colon a \longrightarrow a'$ and $f'\colon a' \longrightarrow a''$ in $\mathbb{A}$, we have $y_{\mathbb{A}}(f' \circ f) = y_{\mathbb{A}}(f') \circ_V y_{\mathbb{A}}(f)$. ■

Exercise 4.7.2. Show that $y_{\mathbb{A}}$ respects identity morphisms. That is, show $y_{\mathbb{A}}(id_a) = Id_{Hom(\ ,a)}$. ■

These two exercises show that $y_{\mathbb{A}}\colon \mathbb{A} \longrightarrow \mathbb{Set}^{\mathbb{A}^{op}}$ is indeed a functor.

Some examples of this functor are needed. Consider the following two categories:

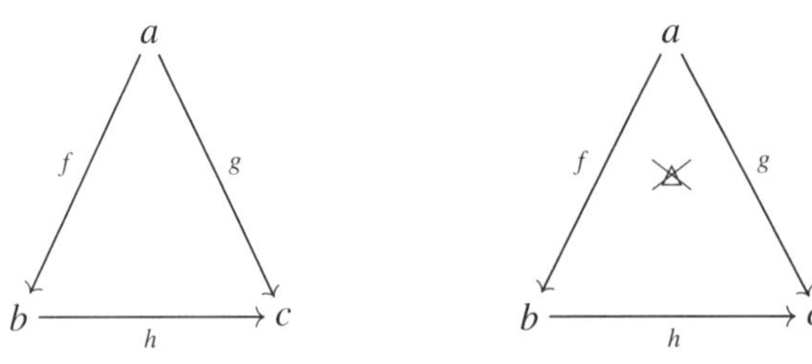

These two categories are almost exactly alike except that **3′** does not commute. We place the ⨯ to symbolize this failure of commutativity. Formally, this means that

$\mathbf{3}$	$\mathbf{3'}$
$Hom_{\mathbf{3}}(x,a) = \begin{cases} \{id_a\} & : \text{if } x = a \\ \emptyset & : \text{if } x = b \\ \emptyset & : \text{if } x = c \end{cases}$	$Hom_{\mathbf{3'}}(x,a) = \begin{cases} \{id_a\} & : \text{if } x = a \\ \emptyset & : \text{if } x = b \\ \emptyset & : \text{if } x = c \end{cases}$
$Hom_{\mathbf{3}}(x,b) = \begin{cases} \{f\} & : \text{if } x = a \\ \{id_b\} & : \text{if } x = b \\ \emptyset & : \text{if } x = c \end{cases}$	$Hom_{\mathbf{3'}}(x,b) = \begin{cases} \{f\} & : \text{if } x = a \\ \{id_b\} & : \text{if } x = b \\ \emptyset & : \text{if } x = c \end{cases}$
$Hom_{\mathbf{3}}(x,c) = \begin{cases} \{g = hf\} & : \text{if } x = a \\ \{h\} & : \text{if } x = b \\ \{id_c\} & : \text{if } x = c \end{cases}$	$Hom_{\mathbf{3'}}(x,c) = \begin{cases} \{g, hf\} & : \text{if } x = a \\ \{h\} & : \text{if } x = b \\ \{id_c\} & : \text{if } x = c \end{cases}$

Figure 4.6. The representable functors for **3** and **3′**.

$g = h \circ f$ in **3**, but $g \neq h \circ f$ in **3′**. Figure 4.6 lists the representable functors for these categories.

The morphisms of **3** and **3′** induce morphisms of representable functors. Let us examine the morphisms induced by $h\colon b \longrightarrow c$:

$$\begin{array}{cc}
\underline{\mathbf{3}} & \underline{\mathbf{3'}} \\
Hom_{\mathbf{3}}(\ ,b) \xrightarrow{y_{\mathbf{3}}(h)} Hom_{\mathbf{3}}(\ ,c) & Hom_{\mathbf{3'}}(\ ,b) \xrightarrow{y_{\mathbf{3'}}(h)} Hom_{\mathbf{3'}}(\ ,c) \\
f \xmapsto{y_{\mathbf{3}}(h)_a} g = hf & f \xmapsto{y_{\mathbf{3'}}(h)_a} hf \\
id_b \xmapsto{y_{\mathbf{3}}(h)_b} h & id_b \xmapsto{y_{\mathbf{3'}}(h)_b} h \\
\{\ \} \xmapsto{y_{\mathbf{3}}(h)_c} id_c & \{\ \} \xmapsto{y_{\mathbf{3'}}(h)_c} id_c.
\end{array} \tag{4.185}$$

The main point of these calculations is to show that all the information about categories **3** and **3′** are in the representable functors and in the induced morphisms of the representable functors. This is the central idea of the Yoneda Embedding Theorem: a locally small category can be totally described as representable functors. Thus, a version of the category **3** "sits inside" the category $\mathbb{S}\mathrm{et}^{\mathbf{3}^{op}}$, and a version of the category **3′** sits inside the category $\mathbb{S}\mathrm{et}^{\mathbf{3'}^{op}}$. This idea is true for any small category.

Theorem 4.7.3. The Yoneda Embedding Theorem For any locally small category $\mathbb{A}$, the covariant functor $y_{\mathbb{A}}\colon \mathbb{A} \longrightarrow \mathbb{S}\mathrm{et}^{\mathbb{A}^{op}}$ is injective on objects, full, and faithful. ★

Proof. The functor has the following characteristics:

- Injective on objects. Let $y_{\mathbb{A}}(a) = y_{\mathbb{A}}(a')$. This means that they are equal on every component, including a. Hence, $y_{\mathbb{A}}(a)_a = y_{\mathbb{A}}(a')_a$, which implies that $Hom(a,a) = Hom(a,a')$. The first set contains $id_a \colon a \longrightarrow a$. Since that morphism is in the second Hom set, a must equal a'.
- Full. Consider an arbitrary natural transformation $\beta \colon Hom(\ ,a) \Longrightarrow Hom(\ ,a')$. The naturality of β means that for any $k \colon b \longrightarrow a$, the following square commutes:

$$\begin{array}{ccc} Hom_{\mathbb{A}}(a,a) & \xrightarrow{\beta_a} & Hom_{\mathbb{A}}(a,a') \\ {\scriptstyle Hom(f,a)}\downarrow & & \downarrow{\scriptstyle Hom(f,a')} \\ Hom_{\mathbb{A}}(b,a) & \xrightarrow[\beta_b]{} & Hom_{\mathbb{A}}(b,a'). \end{array} \tag{4.186}$$

  In the upper-left corner is id_a. The top horizontal map takes id_a to map $f \colon a \longrightarrow a'$. The right vertical map takes f to $f \circ k$. The left vertical map takes id_a to $k \colon b \longrightarrow a$. For this square to commute, β_b must take k to $f \circ k$. But this is exactly the definition of $y_{\mathbb{A}}(f \colon a \longrightarrow a')$ as described in Diagram (4.182). This means that $\beta, = y_{\mathbb{A}}(f \colon a \longrightarrow a')$.
- Faithful. If $y_{\mathbb{A}}(f) = y_{\mathbb{A}}(f')$ then they are equal at the a component, that is, $y_{\mathbb{A}}(f)_a = y_{\mathbb{A}}(f')_a$. This means that they have the same value at id_a. Seeing as both are defined by Diagram (4.182), we have that $f = f \circ id_a = f' \circ id_a = f'$. ♣

It is important to keep in mind that $y_{\mathbb{A}}$ is a covariant functor, but each of its images is a contravariant functor.

A theorem that shows that a structure can be represented as another structure is usually called a **representation theorem**. Perhaps the Yoneda embedding theorem should be called the **Yoneda Representation Theorem**.

Exercise 4.7.4. Consider the **contravariant Yoneda embedding functor** $y^{\mathbb{A}} \colon \mathbb{A}^{op} \longrightarrow \mathbb{Set}^{\mathbb{A}}$ that takes a to $Hom_{\mathbb{A}}(a,\ \ )$ and $f \colon a \longrightarrow a'$ to $y^{\mathbb{A}}(f) = Hom_{\mathbb{A}}(a, f)$. Show that $y^{\mathbb{A}}(f)$ is natural. Also, show that $y^{\mathbb{A}}$ is injective on objects, full, and faithful. The contravariant Yoneda embedding functor says that $\mathbb{A}^{op}$ sits nicely inside $\mathbb{Set}^{\mathbb{A}}$. Keep in mind that $y^{\mathbb{A}}$ is contravariant, but each of its images is a covariant functor. ■

The main idea of the Yoneda lemma is not only that a small category embeds nicely inside its functor category, but other elements in its functor category are determined by their interactions with the embedded category. As an example, consider category **3**. We have the embedded representable functors, and for every $F \colon \mathbf{3}^{op} \longrightarrow \mathbb{Set}$, there are natural tranformations from the representable functors to F. We can envision such maps as the left diagram that follows:

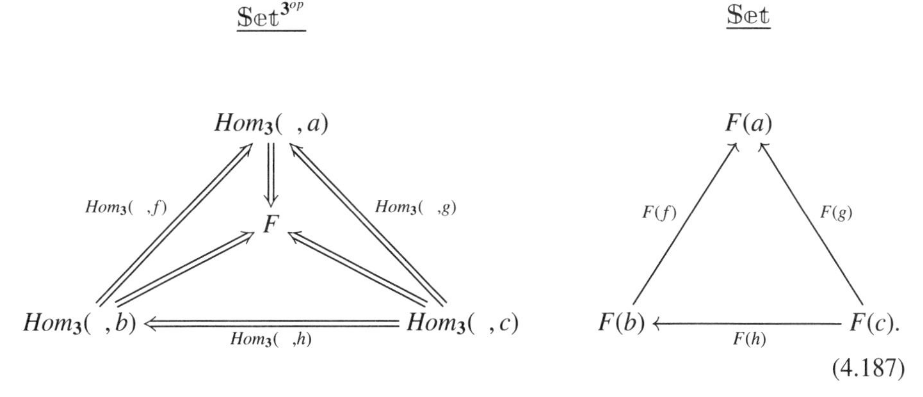

(4.187)

We will see that for the functor $F\colon \mathbb{A}^{op} \longrightarrow \mathbb{S}\mathrm{et}$, the maps from the representable functors to F determine the values of F shown on the right. In other words, the representable functors determine all the functors.

Theorem 4.7.5. The Yoneda lemma. For every locally small category $\mathbb{A}$, and for every $F\colon \mathbb{A}^{op} \longrightarrow \mathbb{S}\mathrm{et}$, the set of natural transformations from $Hom_{\mathbb{A}}(\ ,a)$ to F is naturally isomorphic with the set $F(a)$; that is,

$$Hom_{\mathbb{S}\mathrm{et}^{\mathbb{A}^{op}}}(Hom_{\mathbb{A}}(\ ,a), F) \quad \cong \quad F(a). \tag{4.188}$$

★

Proof. We describe an isomorphism $\Theta\colon Hom_{\mathbb{S}\mathrm{et}^{\mathbb{A}^{op}}}(Hom_{\mathbb{A}}(\ ,a), F) \longrightarrow F(a)$. The following diagram will be helpful:

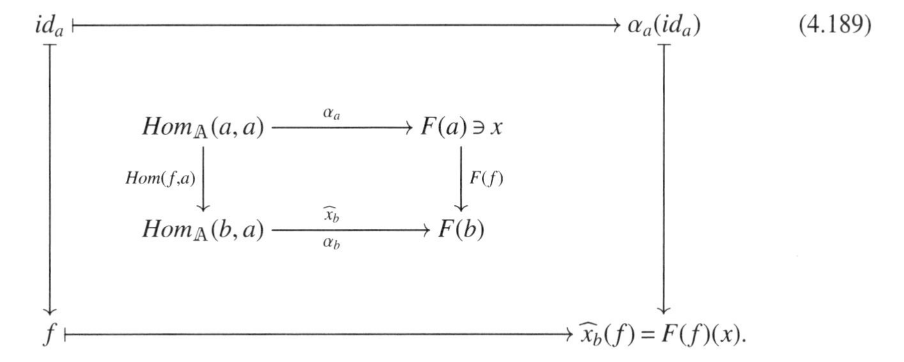

(4.189)

For a natural transformation $\alpha\colon Hom_{\mathbb{A}}(\ ,a) \Longrightarrow F$, we define $\Theta(\alpha) = \alpha_a(id_a)$. In detail, $\alpha_a : Hom_{\mathbb{A}}(a,a) \longrightarrow F(a)$ and $id_a \in Hom_{\mathbb{A}}(a,a)$, so $\alpha_a(id_a) \in F(a)$. The inverse of Θ is $\Theta^{-1}\colon F(a) \longrightarrow Hom_{\mathbb{S}\mathrm{et}^{\mathbb{A}^{op}}}(Hom_{\mathbb{A}}(\ ,a), F)$. For an element x in the set $F(a)$, we define $\Theta^{-1}(x) = \widehat{x}$, where $\widehat{x}\colon Hom_{\mathbb{A}}(\ ,a) \Longrightarrow F$. For $b \in \mathbb{A}$, the component $\widehat{x}_b\colon Hom_{\mathbb{A}}(b,a) \longrightarrow F(b)$ is defined for a map $f\colon b \longrightarrow a$ in $\mathbb{A}$ as $\widehat{x}_b(f) = F(f)(x)$. In detail, $F(f)\colon F(a) \longrightarrow F(b)$ and $x \in F(a)$, so $F(f)(x) \in F(b)$.

It remains to show that Θ and Θ^{-1} are inverses, which can be done as follows:

- $\Theta(\Theta^{-1}(x)) = \Theta(\widehat{x}) = \widehat{x}_a(id_a) = F(id_a)(x) = id_{F(a)}(x) = x$.
- $\Theta^{-1}(\Theta(\alpha)) = \Theta^{-1}(\alpha_a(id_a)) = \widehat{\alpha_a(id_a)}$. Let us see how this natural transformation is defined at component b. The morphism $\widehat{\alpha_a(id_a)}_b$ is defined for $f\colon b \longrightarrow a$ as $F(f)(\alpha_a(id_a))$. By the naturality of α, this is exactly α_b. Thus, we have shown that $\Theta^{-1}(\Theta(\alpha)) = \alpha$. ♣

Exercise 4.7.6. Show that the Yoneda lemma is natural in a and F. ■

Notice that the fact that $y_{\mathbb{A}}$ is full and faithful is basically a consequence of the Yoneda lemma. To see this, set F of the Yoneda lemma to $Hom_{\mathbb{A}}(\ \ , a')$. This gives us

$$Hom_{\mathbb{S}\mathrm{et}^{\mathbb{A}^{op}}}(Hom_{\mathbb{A}}(\ \ , a), Hom_{\mathbb{A}}(\ \ , a')) \quad \cong \quad Hom_{\mathbb{A}}(a, a'). \tag{4.190}$$

Example 4.7.7. What does the Yoneda embedding functor say about partial orders? Let P be a partial order. Then $Hom_P(p, p')$ is either a one-element set (if $p \leq p'$) or the empty set (if $p \not\leq p'$). The functor $Hom_p(\ \ , p)$ tells whether any element is less than or equal to p. Putting this all together, Equation (4.190) gives us the following obvious property of partial orders:

$$[\text{ For all } q \in P,\ \text{if } q \leq p \text{ then } q \leq p'] \qquad \text{if and only if} \qquad p \leq p'; \tag{4.191}$$

($\Longleftarrow$ is obvious, and $\Longrightarrow$ is true by setting $q = p$.) □

There is a dual to the Yoneda lemma:

Theorem 4.7.8. The **contravariant Yoneda lemma**. For every category $\mathbb{A}$, and for every $F\colon \mathbb{A} \longrightarrow \mathbb{S}\mathrm{et}$, the set of natural transformations from $Hom_{\mathbb{A}}(a,\ \ )$ to F is naturally isomorphic with the set $F(a)$; that is,

$$Hom_{\mathbb{S}\mathrm{et}^{\mathbb{A}}}(Hom_{\mathbb{A}}(a,\ \ ), F) \quad \cong \quad F(a). \tag{4.192}$$

★

Exercise 4.7.9. Prove the contravariant Yoneda lemma. ■

Why are the Yoneda Embedding Theorem and the Yoneda lemma so important? They formalize ideas that we met many times:

- Way back in Section 1.4, we saw that elements in a set, S, can be described by morphisms $\{*\} \longrightarrow S$.
- Similarly, to find triplets in S, one should look at morphisms $\{a, b, c\} \longrightarrow S$.
- We also saw that objects in a graph are determined by graph homomorphisms from trivial graphs.
- Certain types of paths in a graph are determined by certain types of graph homomorphisms to the graph.
- Vectors in a $\mathbf{K}$-vector space V are described with linear transformations $\mathbf{K} \longrightarrow V$.

- We will see later that paths in a topological space (or manifold) T are determined by maps $[0,1] \longrightarrow T$.

In all these examples, the properties of an object in a category are determined by the maps to that object. The maps probe the objects. Both the Yoneda Embedding Theorem and the Yoneda lemma show the full power of category theory by showing that the properties of an object are totally determined by the maps to the object. In other words, one should study the morphisms of a category to understand the structure of the objects in the category. This is the core of category theory.

There is another way to think about the Yoneda lemma. It says that every object in $\mathbb{Set}^{\mathbb{A}^{op}}$ can be written in a universal way as a bunch of maps from representable functors. Another way to say this is that an arbitrary element of $\mathbb{Set}^{\mathbb{A}^{op}}$ can be written as a colimit of representable functors. As we saw, the category $\mathbb{Set}$ is complete and cocomplete. The category $\mathbb{Set}^{\mathbb{A}^{op}}$ inherits this cocompleteness. The Yoneda lemma says that the best (i.e., smallest) category that cocompletes $\mathbb{A}$ is $\mathbb{Set}^{\mathbb{A}^{op}}$. Another way to say this is that if you freely add colimits to $\mathbb{A}$, you get a category equivalent to $\mathbb{Set}^{\mathbb{A}^{op}}$. This functor category is the cocompletion of $\mathbb{A}$. A consequence of this is that if $\mathbb{A}$ starts off cocomplete, then $\mathbb{A}$ is equivalent to $\mathbb{Set}^{\mathbb{A}^{op}}$.

Suggestions for Further Study

One can read more about these concepts in most category theory textbooks. See [188, 187, 150] for more on the history of these concepts. Unfortunately, I am not a historian; however, certain historical ideas have to be mentioned.

Categories, functors, and natural transformations were first formulated in Eilenberg and Mac Lane's paper in 1945 [68].

The notion of equivalence of categories was not mentioned until Section 1.2 of Alexander Grothendieck's important paper "Sur quelques points d'algèbre homologique" [95]. This was based on his 1955 lectures at the University of Kansas and is called the "Tohoku" paper. The paper was magnificently translated [98] as "Some aspects of homological algebra" by Marcia Barr with the help of Michael Barr as the mathematical reviser and TeX editor. There was a ten-year delay from the beginning of category theory until the notion of equivalence of categories. That is rather shocking since the notion is basically a homotopical notion and world-class algebraic topologists were working in this area. For more about the history of this topic, see Jean-Pierre Marquis's article [189], and Rick Jardine's essay [122].

Adjoint functors were not defined until 1958. The legend is that Daniel Kan came to the notion of adjoint functors when his thesis advisor, Sammy Eilenberg, was dismissive of one of Kan's ideas. Kan took this as a challenge and proved that he was right about the notion of adjoint functors. The paper [133] came out in 1958. The thirteen-year delay between the beginnings of category theory and the notion of adjoint functors is also mystifying.

Legend has it that Nobuo Yoneda told Saunders Mac Lane about the Yoneda's lemma result in Gare du Nord while they were both visiting Paris in the 1950s. Mac Lane was taken by these results and named it. The many ideas related to the Yoneda lemma are clearly explained in a three-part blog by Tai-Danae Bradley starting with [46], and in a paper by Tom Leinster [167].

4.8 Mini-Course: Basic Categorical Logic

Here, we describe how category theorists look at the structures of logic. The first part is concerned with propositional logic, which deals with true or false statements about properties of particular objects. The second part is concerned with predicate logic which deals with general properties. Logic will be met several times in this book (e.g., Sections 9.5 and 9.6), and knowledge of this mini-course will be assumed in those sections.

Propositional Logic

Logic is about **propositions**, which are statements that are true or false. Propositional logic is concerned with statements about properties of entities. For example, "Category theory is easy" is a true proposition and "Category theory is boring" is totally false. Our central focus here will be the category of propositions, or $\mathbb{Prop}$. Remember that the preorder category, $\mathbb{Prop}$, has propositions as objects, and there is a single morphism from proposition P to proposition Q if P implies (or entails) Q. For example, there is a morphism from the proposition "Joan studies category theory" to the proposition "Joan will be able to learn a lot of science." It is also true that "Joan will be able to learn a lot of science" implies that "Joan will be happy." Combining these two implications means that "Joan studies category theory" implies that "Joan will be happy." There are obviously many examples of such implications.

Logic is not only about propositions, but about operations on propositions and how propositions relate to each other. Let us look at several operations on propositions.

Conjunction. In $\mathbb{Prop}$, the **conjunction** or the **logical and** of the proposition "Jack will be able to learn a lot of physics" and the proposition "Jack will be able to learn a lot of mathematics" is the proposition "Jack will be able to learn a lot of physics and mathematics." This operation is the product in $\mathbb{Prop}$. Let us show this. First, the proposition "Jack will be able to learn a lot of physics and mathematics" implies that "Jack will be able to learn a lot of physics," and it also implies that "Jack will be able to learn a lot of mathematics." These two implications will correspond to projection maps in the category. Furthermore, this proposition also satisfies the universal property of being a product. Consider any proposition that implies the two propositions. For example, "Jack is studying category theory" implies that "Jack will be able to learn a lot of physics" and also that "Jack will be able to learn a lot of mathematics." From this, it is obvious that "Jack is studying category theory" implies "Jack will be able to learn a lot of physics and mathematics." This can all be summarized in Figure 4.7, which is exactly the form of Diagram (3.7) where the universal property of being a product is shown.

In symbols, the conjunction is denoted with $\wedge$. The conjunction of propositions P and Q is written as $P \wedge Q$. Let us use symbols to show that conjunction is the product in

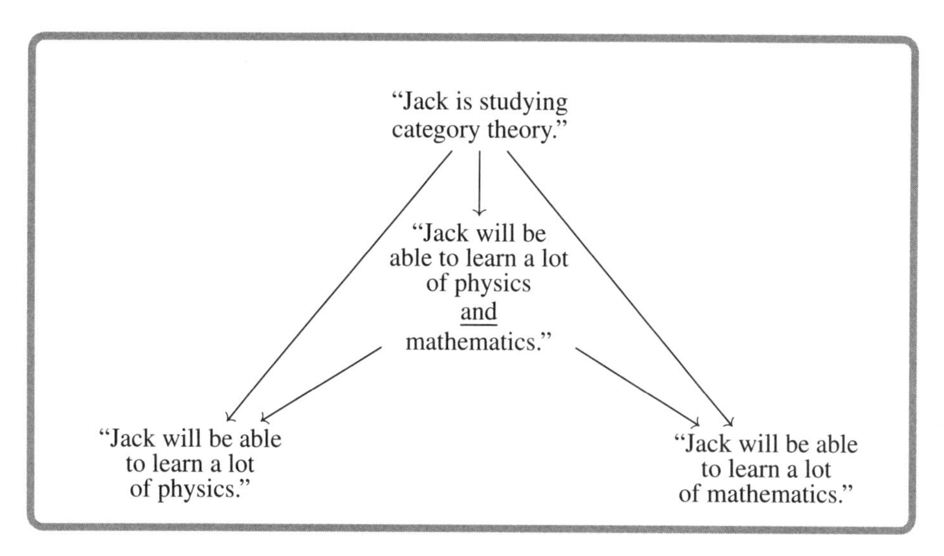

Figure 4.7. An example demonstrating the universal property of conjunction.

$\mathbb{Prop}$. It is obvious that from $P \wedge Q$, one can conclude P, and also from $P \wedge Q$, one can conclude Q. These two implications are the projection maps of a product. This means that for any two propositions P and Q, there are maps $P \wedge Q \longrightarrow P$ and $P \wedge Q \longrightarrow Q$. To show that $P \wedge Q$ satisfies the universal property of being a product, realize that if $S \longrightarrow P$ and $S \longrightarrow Q$, then we can conclude that $S \longrightarrow P \wedge Q$. Putting this all together, we have

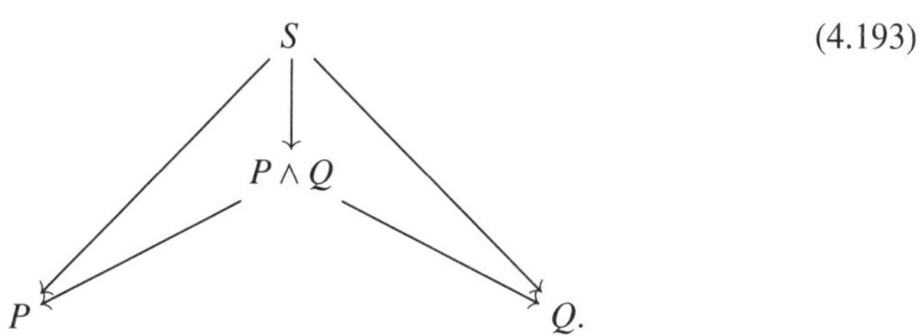

(4.193)

In terms of categories, the conjunction operation is a bifunctor:

$$\wedge\colon \mathbb{Prop} \times \mathbb{Prop} \longrightarrow \mathbb{Prop}. \quad (4.194)$$

The functoriality of this operation means that if $P \longrightarrow P'$ and $Q \longrightarrow Q'$, then $(P \wedge Q) \longrightarrow (P' \wedge Q')$. We can describe the universal property of the product (similar to Diagram (3.9)) as

$$Hom_{\mathbb{Prop}}(S, P \wedge Q) \cong Hom_{\mathbb{Prop}}(S, P) \times Hom_{\mathbb{Prop}}(S, Q). \quad (4.195)$$

The left side has a single arrow if both of the right Hom sets have a single element and neither of them is the empty set. Let us restate this using the product of Hom sets:

$$Hom_{\mathbb{Prop}}(S, P \wedge Q) \cong Hom_{\mathbb{Prop}^2}((S, S), (P, Q)). \quad (4.196)$$

Using the diagonal functor $\Delta\colon \mathbb{Prop} \longrightarrow \mathbb{Prop}^2$, we get

$$Hom_{\mathbb{Prop}}(S, P \wedge Q) \cong Hom_{\mathbb{Prop}^2}(\Delta(S), (P, Q)). \tag{4.197}$$

We have just proved that $\wedge\colon \mathbb{Prop}^2 \longrightarrow \mathbb{Prop}$ is right adjoint to $\Delta\colon \mathbb{Prop} \longrightarrow \mathbb{Prop}^2$; that is,

$$\mathbb{Prop} \underset{\wedge}{\overset{\Delta}{\rightleftarrows}} \mathbb{Prop}^2 \quad (\bot). \tag{4.198}$$

This is in accord with the notion of a product or any type of limit that we saw in Section 4.6.

Implication. Another operation on propositions is the **logical implication**. This takes two propositions and forms a conditional if-then statement out of them.

Some examples are in order.

Example 4.8.1. Here are some examples:

- Given the proposition "Jordan understands category theory" and the proposition "Jordan will see the world in a new light," we can form the proposition "If Jordan understands category theory, then he will see the world in a new light." This statement is a conditional statement that is true if the first part actually implies the second part. (This particular example is, in fact, true.)
- The proposition "It is raining" implies "There are clouds in the sky." This gives us the proposition "If it is raining, then there are clouds in the sky."
- The proposition "Drive more than 60 miles per hour" might imply "Getting a speeding ticket," so we can form the proposition "If you drive more than 60 miles per hour, then you will get a speeding ticket." Another way to say this proposition is "Driving more than 60 miles per hour is a sufficient condition to getting a speeding ticket."
- The proposition "If the moon is made of green cheese, then the sky is blue." is an implication. Notice the first part really has nothing to do with the second part. The implication is true because the sky is blue. □

In symbols, given propositions P and Q, we can form $P \Rightarrow Q$. This statement is true if P implies Q. It is important to keep in mind the distinction between $\Rightarrow$ and $\longrightarrow$. The arrow $\Rightarrow$ is an operation of two propositions and is a logical symbol. In contrast, the arrow $\longrightarrow$ is a categorical symbol that describes when there is an implication of propositions in $\mathbb{Prop}$. The two symbols are related. We will see in a few lines that $\Rightarrow$ is the internal Hom of $\mathbb{Prop}$. However, the symbols should not be confused.

In terms of categories, the $\Rightarrow$ operation can be seen as a bifunctor:

$$\Rightarrow\colon \mathbb{Prop}^{op} \times \mathbb{Prop} \longrightarrow \mathbb{Prop}. \tag{4.199}$$

It is easy to see that it is covariant on the second input: If $Q \longrightarrow Q'$, then $(P \Rightarrow Q) \longrightarrow (P \Rightarrow Q')$. It is contravariant in the first input: If $P \longrightarrow P'$, then $(P' \Rightarrow Q) \longrightarrow (P \Rightarrow Q)$.

The $\wedge$ and $\Rightarrow$ operations are related. Similar to what we saw about sets in Example 4.1.9, for every proposition Q, there is a functor $(\ )\wedge Q\colon \mathbb{Prop} \longrightarrow \mathbb{Prop}$ that is defined on proposition P as $P \wedge Q$. There is also a functor $Q \Rightarrow (\ )\colon \mathbb{Prop} \longrightarrow \mathbb{Prop}$, which is defined on proposition S as $Q \Rightarrow S$. Similar to the adjunction that we saw in Example 4.4.12 we have the following adunction:

Theorem 4.8.2. For every proposition Q, the functor $(\ ) \wedge Q$ is left adjoint to $Q \Rightarrow (\ )$; that is,

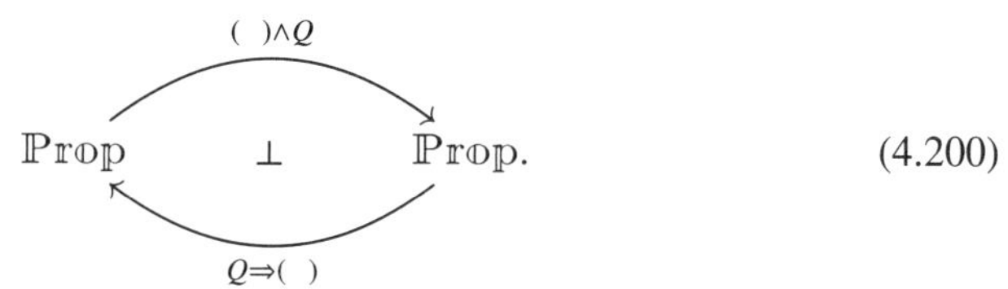

(4.200)

★

Proof. The adjunction means

$$Hom_{\mathbb{Prop}}(P \wedge Q, S) \cong Hom_{\mathbb{Prop}}(P, Q \Rightarrow S). \tag{4.201}$$

or

$$(P \wedge Q) \longrightarrow S \qquad \text{if and only if} \qquad P \longrightarrow (Q \Rightarrow S). \tag{4.202}$$

In English, if $P \wedge Q$ implies S, then just P itself implies that if Q is also true, then S is true. The other direction is similar. ♣

It pays to examine the unit and counit of this adjunction. The unit is $P \longrightarrow (Q \Rightarrow (P \wedge Q))$. This says that if P is true, then Q not only implies Q but also implies P. The counit is a little more famous. It says:

$$(Q \Rightarrow P) \wedge Q \longrightarrow P. \tag{4.203}$$

In English, the counit expresses the fact that if Q implies P, and Q is true, then P is also true. This rule is called **modus ponens**, which means the "way of pushing." We are pushing this implication forward.

Disjunction. Another operation on propositions is the disjunction operation, or the **logical or** operation. Given the proposition "Joan is good at category theory" and the proposition "2+2=4," we can form the proposition "Joan is good at category theory or 2+2=4." In symbols, given the propositions P and Q, we can form the proposition $P \vee Q$.

In terms of categories, the disjunction operation is a bifunctor:

$$\vee\colon \mathbb{Prop} \times \mathbb{Prop} \longrightarrow \mathbb{Prop}. \tag{4.204}$$

The functoriality means that if $P \longrightarrow P'$ and $Q \longrightarrow Q'$, then $(P \vee Q) \longrightarrow (P' \vee Q')$.

The disjunction operation is the coproduct in the category $\mathbb{Prop}$. Consider the propositions "Jill plays music" and "Jill studies category theory." The disjunction of these propositions is "Jill plays music or studies category theory." There are obvious

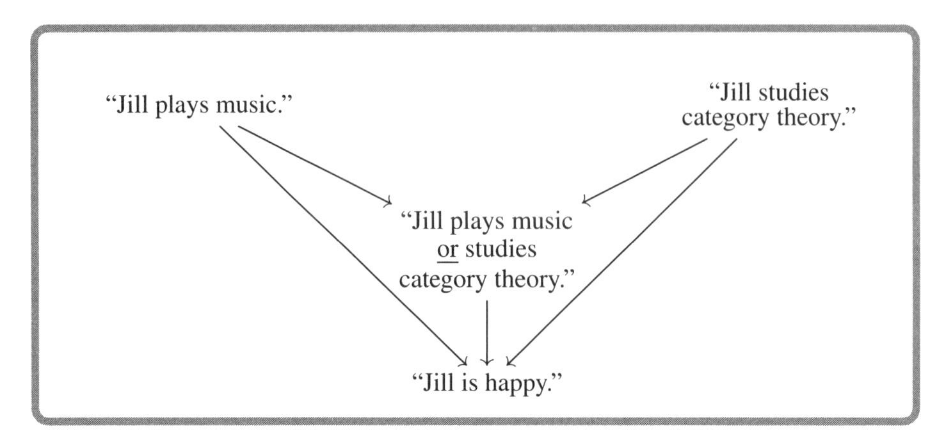

Figure 4.8. An example showing the universal property of $\vee$ as a coproduct.

implications: "Jill plays music" $\longrightarrow$ "Jill plays music or studies category theory" and "Jill studies category theory." $\longrightarrow$ "Jill plays music or studies category theory." To show that the disjunction is the coproduct, we must show that it satisfies the universal property. Consider the proposition "Jill is happy." There are obvious implications: "Jill plays music" $\longrightarrow$ "Jill is happy" and "Jill studies category theory" $\longrightarrow$ "Jill is happy." From this, we can see that "Jill plays music or studies category theory" $\longrightarrow$ "Jill is happy." This is summarized with Figure 4.8, which is similar to Diagram (3.47), where the universal property of being a coproduct was shown. We can describe the universal property of the coproduct similar to Diagram (3.51) as

$$Hom_{\mathbb{Prop}}(P \vee Q, S) \cong Hom_{\mathbb{Prop}}(P, S) \times Hom_{\mathbb{Prop}}(Q, S). \tag{4.205}$$

The left side has a single arrow if both of the right Hom sets have a single element and neither of them is the empty set. Let us restate this using the product of Hom sets:

$$Hom_{\mathbb{Prop}}(P \vee Q, S) \cong Hom_{\mathbb{Prop}^2}((P, Q), (S, S)). \tag{4.206}$$

Restated with the diagonal functor $\Delta \colon \mathbb{Prop} \longrightarrow \mathbb{Prop}^2$, we get

$$Hom_{\mathbb{Prop}}(P \vee Q, S) \cong Hom_{\mathbb{Prop}^2}((P, Q), \Delta(S)). \tag{4.207}$$

We have just proved that $\vee \colon \mathbb{Prop}^2 \longrightarrow \mathbb{Prop}$ is left adjoint to $\Delta \colon \mathbb{Prop} \longrightarrow \mathbb{Prop}^2$. This is in accord with the notion of a colimit that we saw in Section 4.6. The universal properties of $\wedge$ and $\vee$ can be summarized as

$$\mathbb{Prop} \xrightarrow{\Delta} \mathbb{Prop}^2 \quad (\vee \dashv \Delta \dashv \wedge,\ \text{with } \vee, \wedge \colon \mathbb{Prop}^2 \longrightarrow \mathbb{Prop}). \tag{4.208}$$

Negation. Another operation is the negation operation or the **logical not** operation operation. For the proposition "Jack studied category theory," the negation is "Jack did not study category theory." In symbols, for proposition P, we write the negation as $\neg P$. We can also define negation as the following: $\neg P = P \Rightarrow F$, where F is the proposition that is always false. With this definition, we view the negation operation as a contravariant functor:

$$\neg\colon \mathbb{Prop}^{op} \longrightarrow \mathbb{Prop}. \tag{4.209}$$

The functoriality of negation means that if $P \longrightarrow P'$, then $(\neg P') \longrightarrow (\neg P)$. This says that if an implication is true, so is its contrapositive. One can "internalize" this result, where it becomes

$$((P \Rightarrow P') \wedge (\neg P')) \longrightarrow (\neg P). \tag{4.210}$$

This rule is also one of the important rules in logic and is known as **modus tollens**, which is Latin for "the way of pulling." One is pulling the negation of the second proposition back to the first proposition.

Bi-implication. The final logical operation that we introduce is the bi-implication or **logical if and only if**. Two propositions bi-imply each other if each implies the other. We use the symbol $\Leftrightarrow$ for this operation. For example, "Today is Tuesday" $\Leftrightarrow$ "Tomorrow is Wednesday." In symbols, $P \Leftrightarrow Q$ is interpreted as $(P \Rightarrow Q) \wedge (Q \Rightarrow P)$.

With these operations in hand, let us look at some properties of logical systems. First, look at Definition 4.8.3.

Definition 4.8.3. Consider the composition of the negation operation with itself:

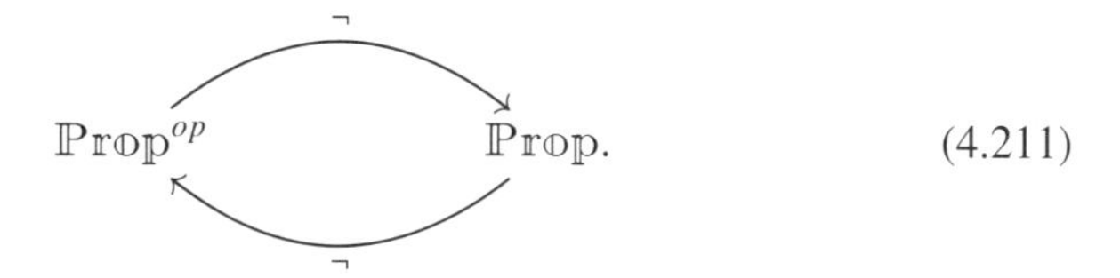

(4.211)

For logical systems called **intuitionistic**, these two functors form an adjunction. The unit of the adjunction says that for every P in $\mathbb{Prop}$, we have $P \longrightarrow \neg\neg P$. (The counit says that for all P, we have $\neg\neg P \longrightarrow P$ in $\mathbb{Prop}^{op}$, which is equivalent to the unit in $\mathbb{Prop}$.) There are special types of intuitionistic logical systems called **Boolean**, where the $\neg$ functors form an equivalence of categories. In a Boolean system, for proposition P, we have $\neg\neg P \Leftrightarrow P$.

Theorem 4.8.4. In a Boolean system, the following **DeMorgan's laws** hold:

$$(\neg P \vee \neg Q) \Leftrightarrow \neg(P \wedge Q) \qquad \text{and} \qquad (\neg P \wedge \neg Q) \Leftrightarrow \neg(P \vee Q). \tag{4.212}$$

Proof. Consider the following square within a Boolean system of logic:

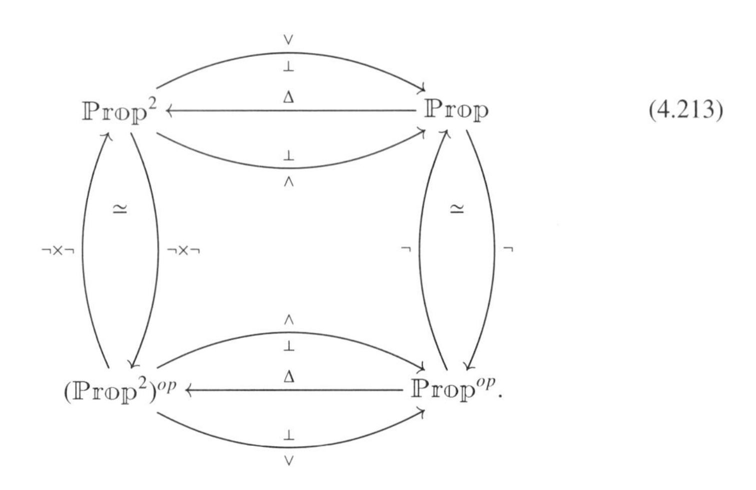

(4.213)

The vertical functors are equivalences, and hence right and left adjoints to each other. The functors on the bottom invert their adjointness because they are in the opposite category (see Exercise 4.4.26). We will go through this proof carefully because the rest of the proofs in this section are very similar. The two paths from the lower-right corner to the upper-left corner are equal, as can be seen here:

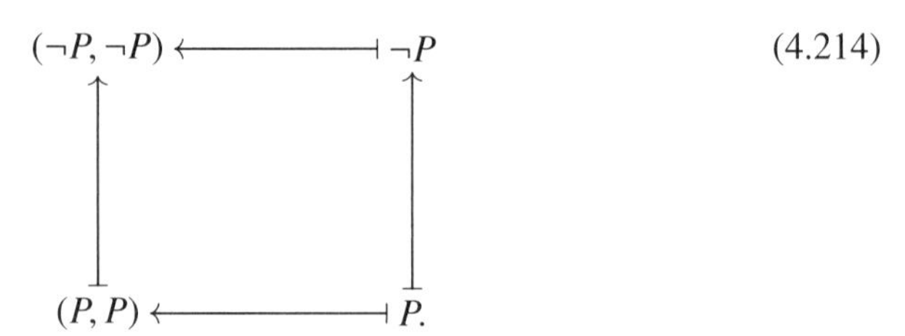

(4.214)

They both take proposition P in $\mathbb{Prop}^{op}$ to $(\neg P, \neg P)$ in $\mathbb{Prop}^2$. Since any two right adjoints to this map are isomorphic, the two right adjoint paths from the upper-left corner to the lower-right corner are isomorphic. In terms of elements, this is

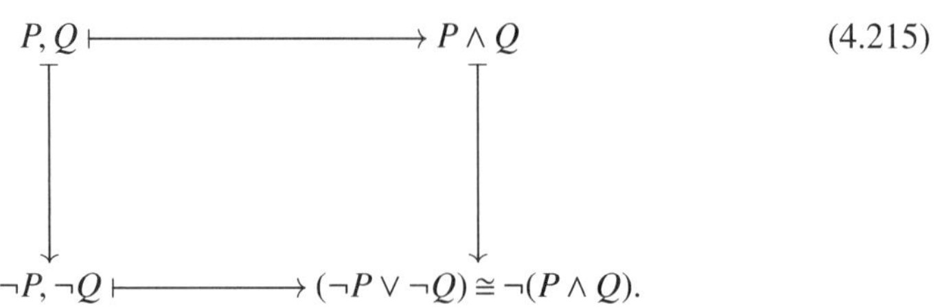

(4.215)

This entails the first DeMorgan's law. The left adjoints are also unique up to isomorphism, and hence the isomorphism of the two left adjoint maps give the second

DeMorgan's law. In detail, this is

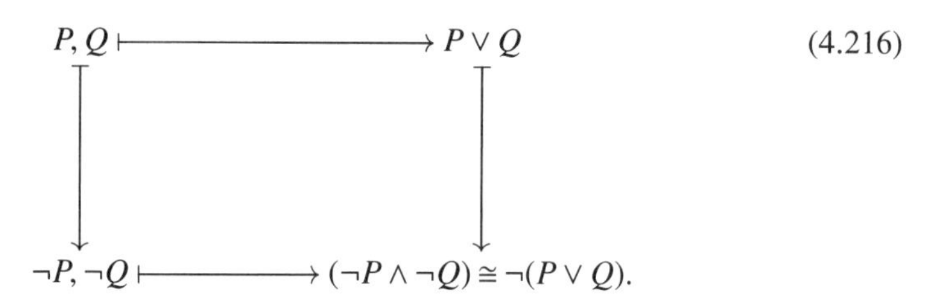

(4.216)

♣

Definition 4.8.5. While $\mathbb{Prop}$ is a preorder, the partial order associated with it is of importance as well. The skeletal category of $\mathbb{Prop}$ is a partial order, which we will call the **Lindenbaum category** and denote as $\mathbb{Lind}$. The objects correspond to equivalence classes of propositions. Two propositions P and Q are equivalent if $P \Leftrightarrow Q$. Keep in mind that $P \Leftrightarrow Q$ in $\mathbb{Prop}$ if and only if $P = Q$ in $\mathbb{Lind}$.

Predicate Logic

Logic gets much more interesting when one moves beyond dealing with the properties of particular entities and starts dealing with properties in general. This is done with **predicates.** For every property, there is a predicate that says whether entities have this property. For every predicate, there is a set of possible inputs to the predicate called the **domain of discourse** for the predicate.

Example 4.8.6. We begin with some fun examples:

- There is a predicate $H(x)$ that says whether someone is happy or not. So H(Bill) is true if Bill is happy and false if Bill is not. The domain of discourse for H is the set of all people.
- Another example of a predicate is $KCT(x)$, which is true if x knows category theory and false if x does not know category theory. The domain of discourse for KCT is also the set of all people.
- There are also predicates with more than one input. For example, $M(x, y)$ is the predicate that is true when x is the mother of y. The domain of discourse for M is the set of pairs of people.

Here are some more mathematical examples:

- $E(n)$ is true when n is an even natural number. So $E(1032)$ is true, but $E(777)$ is false. The domain of discourse for E is the set of natural numbers.
- $P(n)$ is true when n is a prime number. So $P(7)$ is true, but $P(57)$ (the "Grothendieck prime") is false. The domain of discourse for E is the set of natural numbers.
- $G(m, n)$ is true when m is greater than n. The domain of discourse for G is pairs of natural numbers.

- The predicate $A(x, n)$ is true if person x is n years old. The domain of discourse for this predicate is the set of pairs of people and natural numbers. □

One can think of a predicate as a function from the domain of discourse to the set {True, False}. A predicate describes a subset of the domain of discourse where the predicate is true; that is, the subset that contains those elements of the domain of discourse that have the property. For example, the predicate $H(x)$ describes the subset of people who are happy, and the predicate $P(n)$ describes the subset of natural numbers that are prime.

A proposition is about one particular entity and can be viewed as a predicate with no inputs. Another way to think about a proposition is that there is only one element in the domain of discourse and no need for a variable to describe that one element.

The same logical operations ($\wedge, \vee, \Rightarrow, \neg$, and $\Leftrightarrow$) that work for propositions work for predicates as well. The operations are interpreted in the same way. Some examples are in order:

Example 4.8.7.

- $KCT(x) \Rightarrow H(x)$ says that if someone knows category theory, then they are happy.
- $(M(x, y) \wedge KCT(y)) \Rightarrow H(x)$ says that the mother of a child who knows category theory is happy.
- $(A(x, m) \wedge A(y, n) \wedge M(x, y)) \Rightarrow G(m, n)$ says that a mother's age is greater than her child's age.
- $E(n) \Rightarrow \neg E(n+1)$ says that if a number is even, then its successor is not even.
- $(G(n, 2) \wedge P(n)) \Rightarrow \neg E(n)$ says that every prime that is more than 2 is not even. □

In addition to such logical operations, predicates have **quantifiers**. These are devices that tell the quantity "how much" or the quantity of elements of the domain of discourse. There are two main quantifiers:

- A **universal quantifier**, written as $\forall$. The predicate $\forall y Q(x, y)$ is true if and only if $Q(x, y)$ is true for all possible y in the domain of discourse.
- An **existential quantifier**. written as $\exists$. The predicate $\exists y Q(x, y)$ is true if and only if $Q(x, y)$ is true for some possible y in the domain of discourse.

Some examples are needed here:

Example 4.8.8.

- $\forall m \exists n G(n, m)$. This says that for all m, there is an n such that n is greater than m. In other words, the set of natural numbers is infinite.
- $\exists n(E(n) \wedge P(n))$. This says that there is a number that is even and is prime. This is true because 2 is a number that is both even and prime.
- $\forall n(P(n) \Rightarrow (\exists m G(m, n) \wedge P(m)))$. This says that for any number, if it is a prime number, then there is a larger prime number. In other words, the collection of prime numbers is infinite. This statement is true.

- $\forall n((P(n) \wedge P(n+2)) \Rightarrow (\exists m G(m,n) \wedge (P(m) \wedge P(m+2))))$. To understand this, one must understand the concept of twin primes. Two primes are twin if they are separated by one intermediate number. For example, 3 and 5 are twin primes. So are 11 and 13, as well as 599 and 601. This statement says that if n and $n+2$ are twin primes, then there is a larger m such that m and $m+2$ are twin primes. This means there are an infinite amount of twin primes. It is not known if this statement is true or false. □

Quantifiers work on predicates with many variables. Consider a set of variables $\{x_1, x_2, \ldots, x_n\}$. We can write this set as $\bar{x}$. Then a predicate with all these variables can be written as $P(\bar{x})$. A predicate $S(\bar{x}, y)$ has $n+1$ variables. One can then talk about $\forall y S(\bar{x}, y)$ and $\exists y S(\bar{x}, y)$. The predicate $\exists y \forall x \exists b B(a, b, c, d, x, y, z)$ uses seven variables. The variables y, x, and b are **bound** by quantifiers. The rest of the variables are not bound and are called **free**.

Let us look at predicate logic from the categorical point of view. Collections of predicates form preorder categories.

Definition 4.8.9. Let $\bar{x} = \{x_1, x_2, \ldots, x_n\}$ be a set of variables. Then the collection of all predicates that have any of these variables as free variables forms a category called $\mathbb{Pred}(\bar{x})$. There is a morphism $R(\bar{x}) \longrightarrow S(\bar{x})$ whenever R is true (for its variables) implies that S is true (for its variables). The identity morphisms come from the obvious fact that every predicate implies itself.

The category $\mathbb{Prop}$ of propositional statements is the category of $\mathbb{Pred}(\emptyset)$. If $\bar{x} \subseteq \bar{y}$ are two sets of variables, then a predicate $P(\bar{x})$ can also be seen as a predicate $P(\bar{y})$. This means that there is an inclusion of categories $\mathbb{Pred}(\bar{x}) \hookrightarrow \mathbb{Pred}(\bar{y})$.

The same way that logical operations on propositions can be viewed as functors on $\mathbb{Prop}$, logical operations on predicts can be seen as functors on $\mathbb{Pred}(\bar{x})$. These functors satisfy the same universal properties as they did with $\mathbb{Prop}$. This means that for $\bar{x}$, we have the following functors and adjunctions:

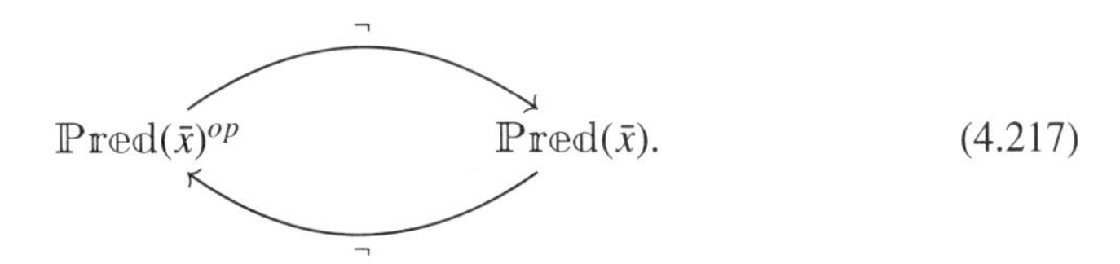

(4.217)

For any predicate $Q(\bar{x})$, there is the adjunction

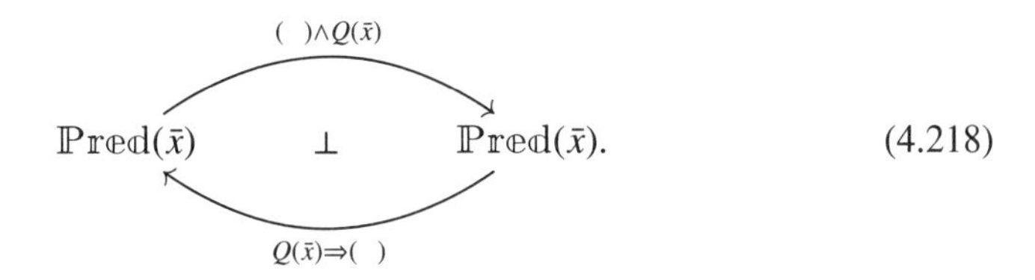

(4.218)

We also have

$$\mathbb{P}\mathrm{red}(\bar{x}) \begin{array}{c} \xleftarrow{\quad\vee\quad} \\ \perp \\ \xrightarrow{\quad\Delta\quad} \\ \perp \\ \xleftarrow[\quad\wedge\quad]{} \end{array} \mathbb{P}\mathrm{red}(\bar{x})^2. \tag{4.219}$$

Diagram (4.217) might be an adjunction or an equivalence depending on whether the logical system is intuitionistic or Boolean, respectively.

Within predicate logic, the quantifiers also become functors with universal properties between the predicate categories. In detail, if $Q(\bar{x}, y)$ is a predicate in $\mathbb{P}\mathrm{red}(\bar{x}, y)$, then $\forall y Q(\bar{x}, y)$ is a predicate in $\mathbb{P}\mathrm{red}(\bar{x})$. The same is true for the predicate $\exists y Q(\bar{x}, y)$. These two mappings are functors:

$$\forall y\colon \mathbb{P}\mathrm{red}(\bar{x}, y) \longrightarrow \mathbb{P}\mathrm{red}(\bar{x}) \qquad \text{and} \qquad \exists y\colon \mathbb{P}\mathrm{red}(\bar{x}, y) \longrightarrow \mathbb{P}\mathrm{red}(\bar{x}). \tag{4.220}$$

Functoriality comes from the fact that if $Q(\bar{x}, y) \longrightarrow R(\bar{x}, y)$ in $\mathbb{P}\mathrm{red}(\bar{x}, y)$, then $\forall y Q(\bar{x}, y) \longrightarrow \forall y R(\bar{x}, y)$ and $\exists y Q(\bar{x}, y) \longrightarrow \exists y R(\bar{x}, y)$ in $\mathbb{P}\mathrm{red}(\bar{x})$.

What universal properties do these functors satisfy? To answer this, we have to understand an earlier example of an adjunction. Let S and T be sets. As we saw, the power sets $\mathcal{P}(S)$ and $\mathcal{P}(T)$ form partial order categories. In Example 4.1.34 on page 151, we saw that for any set function $f\colon S \longrightarrow T$, there is an induced functor $f^{-1}\colon \mathcal{P}(T) \longrightarrow \mathcal{P}(S)$ called the "preimage" or the "inverse image functor" and is defined on subset $Y \subseteq T$ as

$$f^{-1}(Y) = \{x \ : \ f(x) = y \text{ for some } y \in Y\}. \tag{4.221}$$

(In the language of exponentiation that we saw in Section 4.5, $\mathcal{P}(S) = 2^S$, $\mathcal{P}(T) = 2^T$, and $f^{-1} = 2^f$.) The preimage functor has a left adjoint $\exists f\colon \mathcal{P}(S) \longrightarrow \mathcal{P}(T)$ called the **direct image functor** or the **image functor**, which we met in Example 4.1.5 and is defined on subset $X \subseteq S$ as

$$\exists f(X) = f(X) = \{f(x) \ : \ x \in X\} = \{y \ : \ \text{there are an } x \in X \text{ and } f(x) = y\}. \tag{4.222}$$

The adjunction can be visualized by examining Figure 4.9 and noticing that

$$\exists f(X) \subseteq Y \text{ if and only if } X \subseteq f^{-1}(Y). \tag{4.223}$$

The functor $f^{-1}\colon \mathcal{P}(T) \longrightarrow \mathcal{P}(S)$ also has a right adjoint $\forall f\colon \mathcal{P}(S) \longrightarrow \mathcal{P}(T)$, which we call the **coimage functor** and is defined for $X \subseteq S$ as

$$\forall f(X) = \{y \ : \ \text{for all } s \in S, \text{ if } f(s) = y, \text{ then } s \in X\}. \tag{4.224}$$

In other words, $\forall f(X)$ consists of those elements in T where all the preimages are only in X and nowhere else. Since $\exists f(X)$ consists of those elements in T where there is some preimage in X, you can see that for any $X \subseteq S$, we have

$$\forall f(X) \subseteq \exists f(X). \tag{4.225}$$

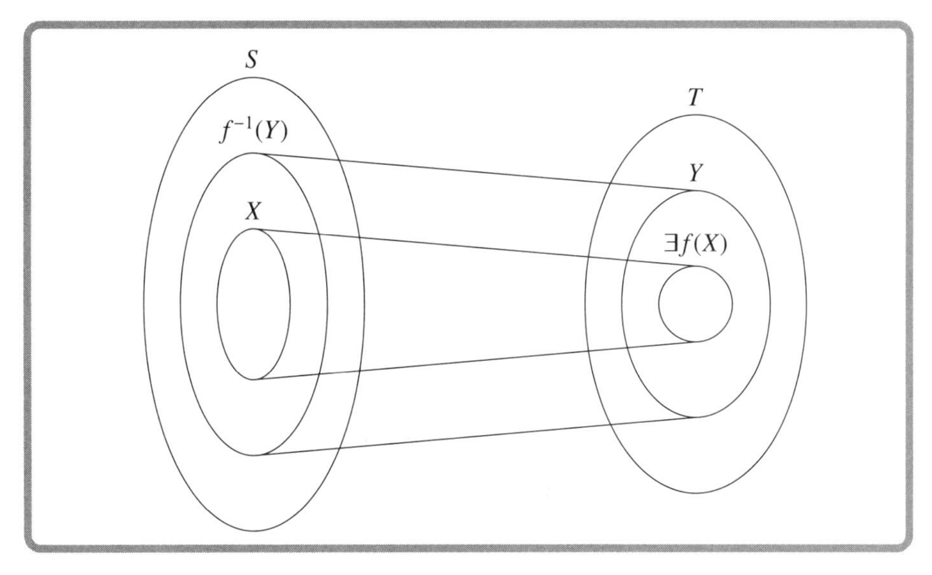

Figure 4.9. The preimage–direct image adjunction: $\exists f(X) \subseteq Y$ if and only if $X \subseteq f^{-1}(Y)$.

The fact that $\forall f$ is right adjoint to f^{-1} means that for $X \subseteq S$ and $Y \subseteq T$,

$$f^{-1}(Y) \subseteq X \text{ if and only if } Y \subseteq \forall f(X). \tag{4.226}$$

These two adjunctions can be summarized with

$$\mathcal{P}(T) \xrightarrow{f^{-1}} \mathcal{P}(S), \qquad \exists f \dashv f^{-1} \dashv \forall f \quad (\exists f, \forall f : \mathcal{P}(S) \longrightarrow \mathcal{P}(T)). \tag{4.227}$$

To get closer to our goal, let us look at a projection set function $\pi\colon S \times T \longrightarrow T$. This function induces three functors: π^{-1}, $\exists\pi$, and $\forall\pi$. The workings of the functors can be seen in Figure 4.10. The functions are defined for $X \subseteq S \times T$ and $Y \subseteq T$ as

- $\pi^{-1}(Y) = \{(s,t) \in S \times T \mid t \in Y\} \subseteq S \times T$.
- $\exists\pi(X) = \{t \in T \mid \text{there exists a } (s,t) \in X \text{ and } \pi(s,t) = t\} \subseteq T$.
- $\forall\pi(X) = \{t \in T \mid \text{for all } (s,t) \in X, \text{ we have } \pi(s,t) = t\} \subseteq T$.

These functors are adjoint as follows:

$$\mathcal{P}(T) \xrightarrow{\pi^{-1}} \mathcal{P}(S \times T), \qquad \exists\pi \dashv \pi^{-1} \dashv \forall\pi \quad (\exists\pi, \forall\pi : \mathcal{P}(S \times T) \longrightarrow \mathcal{P}(T)). \tag{4.228}$$

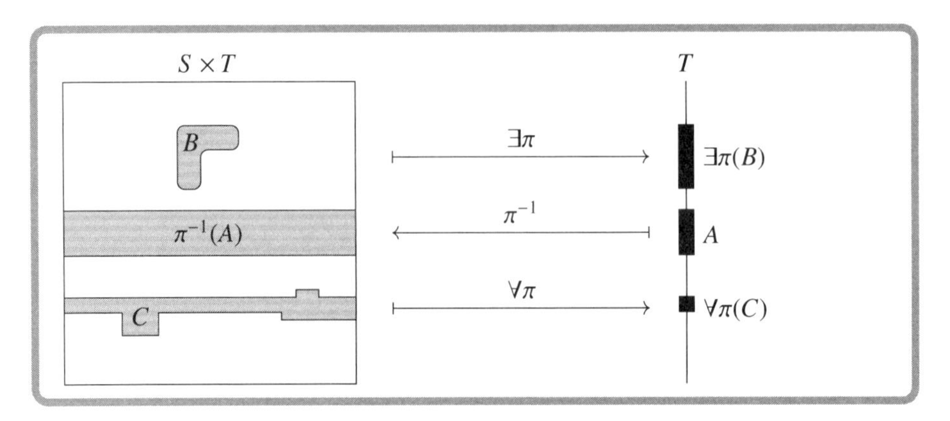

Figure 4.10. The preimage, direct image, and coimage functors for a projection function. Here, $\exists\pi$ projects B, π^{-1} takes A to the whole strip, and $\forall\pi$ takes C to the part that has the entire strip highlighted.

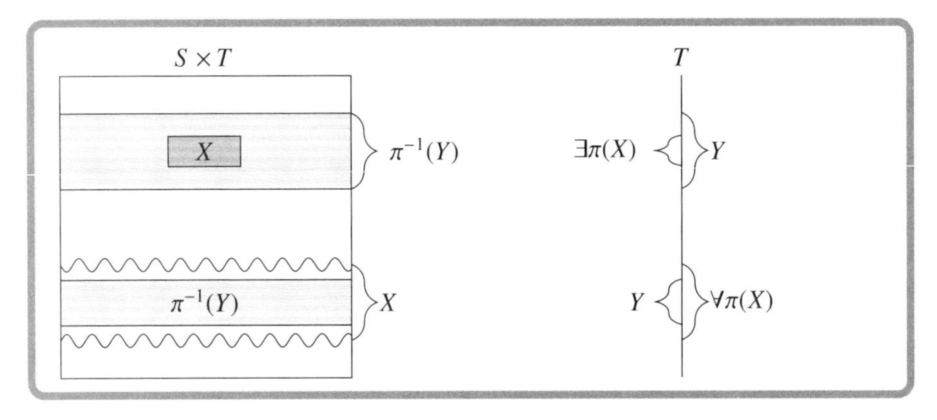

Figure 4.11. Top: The preimage–direct image adjunction of a projection function. Bottom: The preimage-coimage adjunction of a projection function.

The two adjunctions can be seen in Figure 4.11.and are written as the following statements:

$$X \subseteq \pi^{-1}(Y) \quad \text{if and only if} \quad \exists\pi(X) \subseteq Y. \tag{4.229}$$

$$\pi^{-1}(Y) \subseteq X \quad \text{if and only if} \quad Y \subseteq \forall\pi(X). \tag{4.230}$$

With these adjunctions in mind, we can talk about functors between the categories of predicates. For every two variables x and y, there is an inclusion functor $\Delta_y\colon \mathbb{Pred}(x) \hookrightarrow \mathbb{Pred}(x,y)$. It takes predicate $Q(x)$ to the predicate $Q(x,y)$, where the y variable is not used. We denote this inclusion functor Δ_y because it is like the diagonal functor in the sense that if $Q(x)$ is true, then $Q(x,y)$ is true for *all* possible values of y. This is reminiscent of the diagonal functor $\Delta\colon \mathbb{B} \longrightarrow \mathbb{B}^{\mathbb{A}}$, where $\Delta(b)$ is a functor whose output is b for *all* possible values of a in $\mathbb{A}$.

The predicate $Q(x,y)$ describes a subset of pairs of elements of the domain of discourse. The predicate $Q(x)$ describes a subset of elements of the projection of the domain of discourse. The Δ_y functor is exactly the same as the π^{-1} of the previous discussion. As such, the Δ_y also has two adjoints. Let us boost this up from a single variable x to a set of variables $\bar{x}$. Given a predicate $Q(\bar{x})$ in $\mathbb{Pred}(\bar{x})$ and a predicate $R(\bar{x},y)$ in $\mathbb{Pred}(\bar{x},y)$, we have the following adjunction, similar to Equations (4.229) and (4.230):

$$R(\bar{x},y)) \longrightarrow \Delta_y(Q(\bar{x})) \quad \text{if and only if} \quad \exists y R(\bar{x},y)) \longrightarrow Q(\bar{x}). \tag{4.231}$$

$$\Delta_y(Q(\bar{x})) \longrightarrow R(\bar{x},y)) \quad \text{if and only if} \quad Q(\bar{x}) \longrightarrow \forall y R(\bar{x},y)). \tag{4.232}$$

Hence, there are the following two adjunctions:

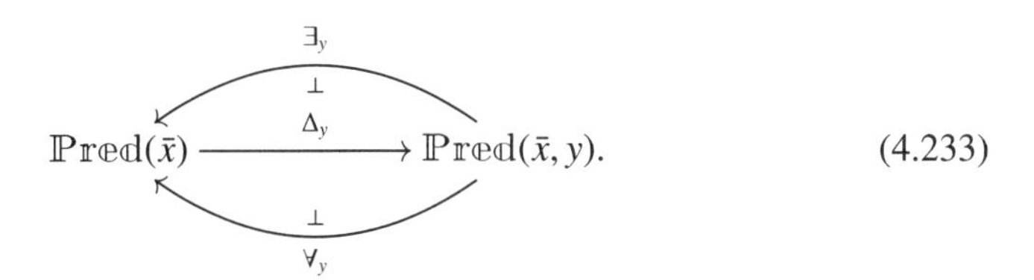

(4.233)

This diagram is similar to Diagram (4.219), which makes sense because the universal quantifier is a generalization of $\wedge$, and the existential quantifier is a generalizatio of $\vee$.

The units and counits of these adjunctions are interesting. Given predicate $Q(\bar{x})$ in $\mathbb{Pred}(\bar{x})$ and predicate $R(\bar{x},y)$ in $\mathbb{Pred}(\bar{x},y)$, we have the following units and counits:

- The unit of the $\exists_y$ and Δ_y adjunction is $R(\bar{x},y) \longrightarrow \Delta_y(\exists y R(\bar{x},y))$.
- The counit of the $\exists_y$ and Δ_y adjunction is $\exists y(\Delta_y(Q(\bar{x}))) \longrightarrow Q(\bar{x})$.
- The unit of the Δ_y and $\forall_y$ adjunction is $Q(\bar{x}) \longrightarrow \forall y(\Delta_y(Q(\bar{x})))$.
- The counit of the Δ_y and $\forall_y$ adjunction is $\Delta_y(\forall y R(\bar{x},y)) \longrightarrow R(\bar{x},y)$.

Exercise 4.8.10. Give the triangle identities for these adjunctions. ■

How do the quantifiers respect the other logical operations? We begin with the negation operation. There is the following generalization of Theorem 4.8.4:

Theorem 4.8.11. In a Boolean logical system, the following **generalized DeMorgan's laws** hold:

$$\neg\forall_y P(\bar{x},y) \Leftrightarrow \exists_y \neg P(\bar{x},y) \qquad \text{and} \qquad \neg\exists_y P(\bar{x},y) \Leftrightarrow \forall_y \neg P(\bar{x},y). \tag{4.234}$$

Proof. Consider the following square with a Boolean system of logic:

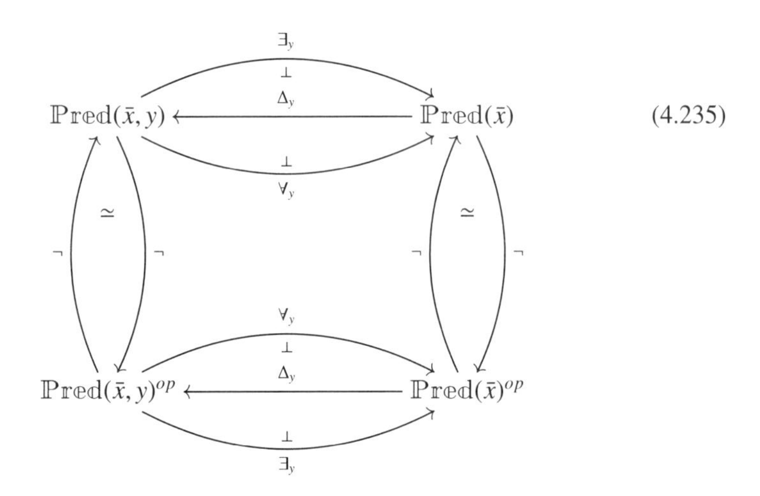

(4.235)

The proof follows along the same line as the proof of Theorem 4.8.4. ♣

What about the other logical operations?

Theorem 4.8.12. The existential quantifier (which is like an infinite disjunction) respects the disjunction. Similarly, the universal quantifier (which is like a conjunction) respects the conjunction. This amounts to the following:

$$\exists x P(x) \vee \exists x Q(x) \iff \exists x (P(x) \vee Q(x)). \tag{4.236}$$

$$\forall x P(x) \wedge \forall x Q(x) \iff \forall x (P(x) \wedge Q(x)). \tag{4.237}$$

★

Proof. Consider the following diagram:

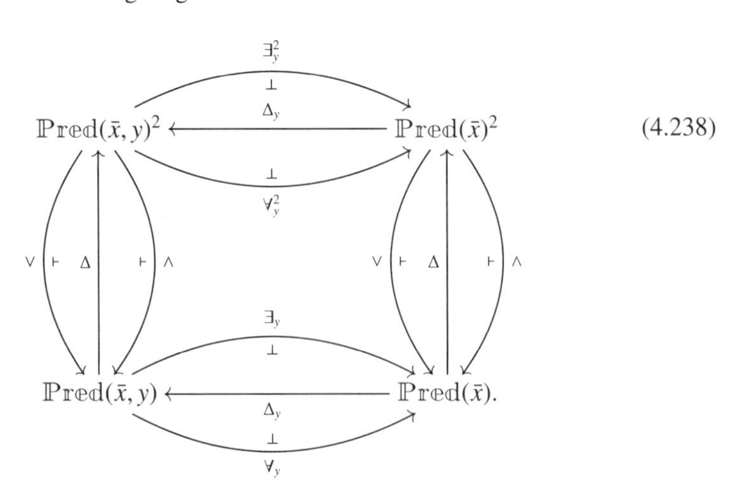

(4.238)

The two maps from the lower-right to the upper-left amount to $Q(\bar{x}) \mapsto (\Delta_y Q(\bar{x}), \Delta_y Q(\bar{x}))$. This means that the two compositions of two right adjoints are isomorphic. Similarly, the composition of two left adjoints are isomorphic. ♣

Exercise 4.8.13. Prove the following propositions:

$$\exists y \exists z P(\bar{x}, y, z) \iff \exists z \exists y P(\bar{x}, y, z). \tag{4.239}$$

$$\forall y \forall z P(\bar{x}, y, z) \iff \forall z \forall y P(\bar{x}, y, z). \tag{4.240}$$

■

Remark 4.8.14. It should be noticed that logic and sets are intimately related.

- There is a relationship between the conjunction operation, $\wedge$, and the intersection set operation, $\cap$. The intersection of two sets is the set of elements that are in one set and in the other set. In symbols, for sets S and T,

$$S \cap T = \{x \ : \ x \in S \wedge x \in T\}. \tag{4.241}$$

- There is a relationship between the disjunction operation, $\vee$, and the union set operation, $\cup$. The union of two sets is the set of elements that are in one set or the other set. In symbols, for sets S and T,

$$S \cup T = \{x \ : \ x \in S \vee x \in T\}. \tag{4.242}$$

- There is a relationship between the negation operation, $\neg$, and the complement set operation, $\{\ \}^c$. The complement of set S is the set of elements that are not in S. In symbols, for set S,

$$S^c = \{x \ : \ \neg(x \in S)\}. \tag{4.243}$$

We will explore this connection again when we talk about topos theory in Section 9.5. ♠

Before we close this mini-course, it pays to meditate on what was accomplished here. What was gained by presenting logic from a categorical point of view? Our goal here was not simply to show that logic can be done in the language of category theory. Rather, our goal was to demonstrate several important ideas:

- We showed that logic is united with the other fields that we met in these pages. The unity comes from the fact that logic employs the same tools of products, coproducts, functoriality, equivalences, and adjunction that are used in other areas.
- We showed that the various logical operations do not stand alone. Category theory shows that the operations are intimately connected to each other and can be defined in terms of each other with universal properties.
- We showed that many of the truths of propositional and predicate logic are simple consequences of the universal properties of the operations. A logical statement is not true because it seems true. Rather, the statement has to be true because of the way that operations are defined in terms of other operations.

Categorical logic is a wonderful modern contribution to the ancient field of logic.

Suggestions for Further Study

Classical logic texts where one can learn more about these ideas are Yuri Manin's [185], Elliott Mendelson's [195], and Herbert Enderton's [69]. There is also some nice logic in Kenneth A. Ross and Charles Wright's work [226].

The central idea of this mini-course—that logical operations and quantifiers arise from adjoint functors—was originally stated by F. W. Lawvere [164]. That paper is one of the most important papers in category theory. Of course, categorical logic is impossible without Joachim Lambek and Philip Scott's text [151]. Robert Goldblatt's writing [91] is also wonderful. Quantifiers as adjoint can be found in Section IV.5 of [180], Section 9.5 of [21], Section 7.4.4 of [77], and Chapter 15 of [91].

There are many excellent survey papers that introduce the field of categorical logic, such as [209, 78, 210, 148, 2]. A warning is in order: be aware that there is no uniformity as to what is in the area of categorical logic and what is not. The intersection of the content of these papers is very small. Jean-Pierre Marquis and Gonzalo E. Reyes have a fascinating paper [190] on the early history of categorical logic. Go and explore!

Over the course of this book, we will learn more categorical logic. In Section 9.5 we will learn some topos theory, and in Section 9.6 we will look at homotopy type theory.

5

Monoidal Categories

A mathematician is a person who can find analogies between theorems; a better mathematician is one who can see analogies between proofs and the best mathematician can notice analogies between theories. One can imagine that the ultimate mathematician is one who can see analogies between analogies.

Stefan Banach

Many times in these pages, we have seen that objects and morphisms in certain categories can be combined. The first time we met this idea was in Section 1.4, where we saw the formation of the Cartesian product in the category $\mathbb{Set}$. Given two sets S and T, we form $S \times T$. Furthermore, given two set maps $f\colon S \longrightarrow T$ and $g\colon S' \longrightarrow T'$, we form the map $f \times g\colon S \times S' \longrightarrow T \times T'$. We also saw that we can combine sets by taking their disjoint union. Other examples of combining include taking the product of two partial orders (Example 3.1.10) and two groups (Example 3.1.12), as discussed in Chapter 3. All these categories where one can combine objects and morphisms are examples of monoidal categories. The word "monoidal" comes from the fact that these categories with extra structure are reminiscent of monoids where elements are combined.

Monoidal categories come in many different varieties. The differences depend on what rules the combination of objects and morphisms follow. To get a feel for this, let us remember some basic arithmetic. There are operations $\otimes$ on numbers (such as $+$ or $\times$) that satisfy the associativity axiom, $a \otimes (b \otimes c) = (a \otimes b) \otimes c$. There are also operations (such as $-$ and $\div$) that fail this associativity axiom. There are similar notions about commutativity, where there are some operations $\otimes$ (such as $+$ or $\times$) that satisfy the commutativity axiom $a \otimes b = b \otimes a$ and some operations (such as $-$ and $\div$) that do not satisfy the commutativity axiom. The story with categories is even more varied. We will see in this chapter that for categories, there can be many possible relationships between $a \otimes (b \otimes c)$ and $(a \otimes b) \otimes c$. Similarly, there are many possible relationships between $a \otimes b$ and $b \otimes a$. The multiple possibilities make the theory of monoidal categories very rich.

The chapter goes from the simplest type of monoidal structure to the more complicated. Strict monoidal categories—which are similar to the monoids we already met—are in Section 5.1. In Section 5.2, we move on to categories with products (or coproducts) that we already met in Chapter 3. In Section 5.3, we meet the general notion of a monoidal category and also prove theorems about such categories. Section 5.4 is an introduction to coherence theory. String diagrams are introduced in Section 5.5. In the conclusion of this chapter, Section 5.6 presents a mini-course with some advanced topics in linear algebra. These topics are necessary for some later parts of this book.

5.1 Strict Monoidal Categories

We begin with the simplest possible type of a monoidal category.

Definition 5.1.1. A **strict monoidal category** $(\mathbb{A}, \otimes, I)$ is a category $\mathbb{A}$ with the following extra structure:

- A way of combining objects and morphisms: a bifunctor $\otimes\colon \mathbb{A} \times \mathbb{A} \longrightarrow \mathbb{A}$, called the **tensor product** or **monoidal product**
- An object I of $\mathbb{A}$, called the **unit**

This extra structure satisfies the following requirements:

- The bifunctor $\otimes$ is associative: for all objects a, b, and c of $\mathbb{A}$, $a \otimes (b \otimes c) = (a \otimes b) \otimes c$. Similarly, for all morphisms f, g, and h, $f \otimes (g \otimes h) = (f \otimes g) \otimes h$. We say that the tensor product is **strictly associative**.
- The I acts like a two-sided unit of $\otimes$: for all objects a of $\mathbb{A}$, $a \otimes I = a = I \otimes a$.

Example 5.1.2. $(\mathbf{N}, +, 0)$. A simple example of a strict monoidal category is the natural numbers with addition. The objects of the category are the natural numbers. The morphisms are only identities. The tensor product is addition and the unit is 0. The requirements are easily seen to be satisfied. It is also easy to see that for the same discrete category $\mathbf{N}$, the multiplication and unit 1 form a strict monoidal category $(\mathbf{N}, \cdot, 1)$. □

Example 5.1.3. $(\Sigma^*, \bullet, \emptyset)$. An example of a strict monoidal category that will be very important in the coming pages is the monoid of strings in an alphabet. Let Σ be an alphabet; that is, a finite set of symbols (or letters). The set of all strings (or words, or lists) of symbols in Σ, denoted as Σ^*, forms a monoidal category. The objects of the category are strings and the only morphisms are identity morphisms. The tensor product of the monoidal category is concatenation $\bullet$ (combining one string after another) . That is, given two strings, w and w', their tensor is simply their concatenation, $w \bullet w'$. This operation is associative. The empty string $\emptyset$ is the unit: $w \bullet \emptyset = w = \emptyset \bullet w$. As special instances of such strict monoidal categories, consider the strict monoidal categories of strings in one symbol, $(\{1\}^*, \bullet, \emptyset)$, and strings of two symbols $(\{0, 1\}^*, \bullet, \emptyset)$. □

Exercise 5.1.4. Show that when $\Sigma = \{1\}$ is a singleton set, then the strict monoidal category $(\{1\}^*, \bullet, \emptyset)$ is essentially the same as the strict monoidal category of the natural numbers $(\mathbf{N}, +, 0)$. (We will formalize what it means to be "essentially the same" in Chapter 6.) ■

The collections of objects in Examples 5.1.3 and 5.1.4 are all monoids. Now let us make a general statement about all monoids.

Example 5.1.5. $(M, \star, e)$. Any monoid can be thought of as a strict monoidal category. The category is $d(M)$, the discrete category of elements of M. The tensor product is the monoid multiplication $\star$, and the unit of the strict monoidal category is the unit of the monoid. □

Notice that category theorists can think of a monoid as a category in at least two different ways. On the one hand, they are one-object categories where the morphisms come and go to the single object and the composition in the category is the monoid multiplication. On the other hand, a monoid is a discrete category with a monoidal category structure where the tensor product is the monoid multiplication. We have to specify what we mean.

Example 5.1.6. $(P, \wedge, 1)$. Any partial order category $(P, \leq)$ that has products is a strict monoidal category. We call the product **meet** and write it as $\wedge$. For any three objects, p, q, and r, we have $p \wedge (q \wedge r) = (p \wedge q) \wedge r$, which means that the meet is associative. The unit is the terminal object, 1, which satisfies $p \wedge 1 = p$. (For any objects, p and q, we have $p \wedge q = q \wedge p$, which means that the meet is commutative.) Such a partial order is called a **bounded meet semilattice**. The same is true for a partial order category with coproducts where the operation is called **join** and denoted as $\vee$. The unit is the initial object, 0. In that case, we have the strict monoidal category $(P, \vee, 0)$. Such a partial order is called a **bounded join semilattice**.

A special case of partial order category is the following: $(\mathbf{2} = \{0, 1\}, \wedge, 1)$. The partial order with two elements such that $0 \leq 1$ is a strict monoidal category. For completeness, let us just give the monoidal structure: $0 \wedge 0 = 0, 0 \wedge 1 = 0, 1 \wedge 0 = 0$, and $1 \wedge 1 = 1$. From these, we can see the associativity and the fact that 1 is the unit. □

The next example will be very important in Section 6.2.

Example 5.1.7. $(\mathbb{A}^{\mathbb{A}}, \circ, Id_{\mathbb{A}})$. Let $\mathbb{A}$ be any category. Category $\mathbb{A}^{\mathbb{A}}$ consists of functors from $\mathbb{A}$ to $\mathbb{A}$ (endofunctors) and natural transformations between such functors. This category has a strict monoidal category structure. The tensor product of objects is the composition $\circ$. In detail, if $F\colon \mathbb{A} \longrightarrow \mathbb{A}$ and $G\colon \mathbb{A} \longrightarrow \mathbb{A}$, then the tensor product is $F \circ G$. Notice that both $F \circ G$ and $G \circ F$ exist but they need not be equal. The tensor of morphisms (natural transformations) $\alpha\colon F \Longrightarrow F'$ and $\beta\colon G \Longrightarrow G'$ is $\beta \circ \alpha = \beta \circ_H \alpha$, where $\circ_H$ is the horizontal composition of natural transformations. Vertical composition of natural transformations is the regular composition of maps in $\mathbb{A}^{\mathbb{A}}$. The unit of the monoidal structure is the identity functor $Id_{\mathbb{A}}$. □

Remark 5.1.8. Some foreshadowing is warranted before we proceed. The category of matrices and the category of finite-dimensional vector spaces are intimately related. In this chapter, we will show that the category of matrices has two monoidal structures, which we will denote as $\oplus$ and $\otimes$. Related to these are two monidal structures on the category of finite-dimensional vector spaces, which are denoted by the same symbols. Exactly how matrices and vector spaces are related and how all these different structure are united will be formalized in Chapter 6. Keep in mind the larger picture while going through the technical details. ♠

Example 5.1.9. ($\mathbf{K}\mathbb{M}\mathtt{at}, \oplus, 0$). The category $\mathbf{K}\mathbb{M}\mathtt{at}$ has a strict monoidal structure called the **direct sum**, which is denoted by $\oplus$. On objects, the monoidal structure is defined as $m \oplus n = m + n$. Remember that a morphism $A\colon n \longrightarrow m$ corresponds to an m by n matrix and is denoted as $A_{m,n}$. The m by n matrix with all zeros is denoted as $0_{m,n}$. The direct sum is defined as

$$A_{m,n} \oplus B_{m',n'} = \begin{bmatrix} A_{m,n} & 0_{m,n'} \\ 0_{m',n} & B_{m',n'} \end{bmatrix} \tag{5.1}$$

which is an $m + m'$ by $n + n'$ matrix.

Formally, the direct sum of matrices is the function

$$\oplus\colon \mathbb{C}^{m \times n} \times \mathbb{C}^{m' \times n'} \longrightarrow \mathbb{C}^{(m+m') \times (n+n')} \tag{5.2}$$

and is defined as

$$(A \oplus B)[j,k] = \begin{cases} A[j,k] & : \text{if } j \leq m \text{ and } k \leq n \\ B[j-m, k-n] & : \text{if } j > m \text{ and } k > n \\ 0 & : \text{otherwise.} \end{cases} \tag{5.3}$$

This operation preserves matrix multiplication (which is morphism composition in this category):

$$\begin{aligned}
(A_{m,n} \oplus B_{m',n'}) \cdot (A'_{n,p} \oplus B'_{n',p'}) &= \begin{bmatrix} A_{m,n} & 0_{m,n'} \\ 0_{m',n} & B_{m',n'} \end{bmatrix} \cdot \begin{bmatrix} A'_{n,p} & 0_{n,p'} \\ 0_{n',p} & B'_{n',p'} \end{bmatrix} \\
&= \begin{bmatrix} A_{m,n} \cdot A'_{n,p} & 0_{m,p'} \\ 0_{m',p} & B_{m',n'} \cdot B'_{n',p'} \end{bmatrix} \\
&= (A_{m,n} \cdot A'_{n,p}) \oplus (B_{m',n'} \cdot B_{n',p'}).
\end{aligned}$$

This equation is yet another instance of the ubiquitous interchange law (see Important Categorical Idea 3.1.18.)

The unit for $\oplus$ is 0 (i.e., $n \oplus 0 = n + 0 = n$). We are employing a morphism that represents a (nonexistent) zero-by-zero matrix with nothing in it.

One can see that $\oplus$ is strictly associative as follows:

$$\begin{aligned}
(A_{m,n} \oplus B_{m',n'}) \oplus C_{m'',n''} &= \begin{bmatrix} \begin{bmatrix} A_{m,n} & 0_{m,n'} \\ 0_{m',n} & B_{m',n'} \end{bmatrix} & 0_{m+m',n''} \\ 0_{m'',n+n'} & C_{m'',n''} \end{bmatrix} \\
&= \begin{bmatrix} A_{m,n} & 0_{m,n'} & 0_{m,n''} \\ 0_{m',n} & B_{m',n'} & 0_{m',n''} \\ 0_{m'',n} & 0_{m'',n'} & C_{m'',n''} \end{bmatrix} \\
&= \begin{bmatrix} A_{m,n} & 0_{m,n'+n''} \\ 0_{m'+m'',n} & \begin{bmatrix} B_{m',n'} & 0_{m',n''} \\ 0_{m'',n'} & C_{m'',n''} \end{bmatrix} \end{bmatrix} \\
&= A_{m,n} \oplus (B_{m',n'} \oplus C_{m'',n''}).
\end{aligned}$$

$$\begin{bmatrix} a_0 \\ a_1 \\ a_2 \\ a_3 \end{bmatrix} \otimes \begin{bmatrix} b_0 \\ b_1 \\ b_2 \end{bmatrix} = \begin{bmatrix} a_0 \cdot \begin{bmatrix} b_0 \\ b_1 \\ b_2 \end{bmatrix} \\ a_1 \cdot \begin{bmatrix} b_0 \\ b_1 \\ b_2 \end{bmatrix} \\ a_2 \cdot \begin{bmatrix} b_0 \\ b_1 \\ b_2 \end{bmatrix} \\ a_3 \cdot \begin{bmatrix} b_0 \\ b_1 \\ b_2 \end{bmatrix} \end{bmatrix} = \begin{bmatrix} a_0b_0 \\ a_0b_1 \\ a_0b_2 \\ a_1b_0 \\ a_1b_1 \\ a_1b_2 \\ a_2b_0 \\ a_2b_1 \\ a_2b_2 \\ a_3b_0 \\ a_3b_1 \\ a_3b_2 \end{bmatrix}.$$

Figure 5.1. The Kronecker product of two vectors. Every entry in the first vector is scalar multiplied with the second vector.

It pays to mention that a special case of this strict monoidal structure is $(Bool\mathbb{M}\mathbb{a}\mathbb{t}, \oplus, 0)$. This is the category of matrices with Boolean entries. □

There is another strict monoidal structure on $\mathbf{K}\mathbb{M}\mathbb{a}\mathbb{t}$.

Example 5.1.10. $(\mathbf{K}\mathbb{M}\mathbb{a}\mathbb{t}, \otimes, 1)$. On objects, $\otimes$ is defined as $m \otimes n = m \cdot n$. The tensor of $A_{m,n}$ with $B_{m',n'}$ is the **Kronecker product of matrices** $A_{m,n} \otimes B_{m',n'}$, which is defined as follows: every entry of $A_{m,n}$ is scalar multiplied with the matrix $B_{m',n'}$. Examples are needed: Figure 5.1 shows the Kronecker product of two vectors, and Figure 5.2 shows the Kronecker product of two matrices.

Formally, the tensor product of matrices is the function

$$\otimes : \mathbb{C}^{m \times n} \times \mathbb{C}^{m' \times n'} \longrightarrow \mathbb{C}^{(m \cdot m') \times (n \cdot n')} \tag{5.4}$$

and is defined as

$$(A \otimes B)[j,k] = A[\lfloor j/m' \rfloor, \lfloor k/n' \rfloor] \cdot B[j \ \ Mod \ \ m', k \ \ Mod \ \ n']. \tag{5.5}$$

The unit of the monoidal structure is the 1 and the unit morphism is the 1 by 1 identity matrix [1]. We leave it to the reader to prove that the Kronecker product of matrices respects matrix multiplication. □

Exercise 5.1.11. Prove that the Kronecker product of matrices respects matrix multiplication; that is,

$$(A_{m,n} \cdot A'_{n,p}) \otimes (B_{m',n'} \cdot B'_{n',p'}) = (A_{m,n} \otimes B_{m',n'}) \cdot (A'_{n,p} \otimes B'_{n',p'}). \tag{5.6}$$

(Yet another instance of the ubiquitous interchange law that we saw in Important Categorical Idea 3.1.18.) ■

For the matrices

$$A = \begin{bmatrix} a_{0,0} & a_{0,1} \\ a_{1,0} & a_{1,1} \end{bmatrix} \quad \text{and} \quad B = \begin{bmatrix} b_{0,0} & b_{0,1} & b_{0,2} \\ b_{1,0} & b_{1,1} & b_{1,2} \\ b_{2,0} & b_{2,1} & b_{2,2} \end{bmatrix},$$

the Kronecker product is

$$A \otimes B = \begin{bmatrix} a_{0,0} \cdot \begin{bmatrix} b_{0,0} & b_{0,1} & b_{0,2} \\ b_{1,0} & b_{1,1} & b_{1,2} \\ b_{2,0} & b_{2,1} & b_{2,2} \end{bmatrix} & a_{0,1} \cdot \begin{bmatrix} b_{0,0} & b_{0,1} & b_{0,2} \\ b_{1,0} & b_{1,1} & b_{1,2} \\ b_{2,0} & b_{2,1} & b_{2,2} \end{bmatrix} \\ a_{1,0} \cdot \begin{bmatrix} b_{0,0} & b_{0,1} & b_{0,2} \\ b_{1,0} & b_{1,1} & b_{1,2} \\ b_{2,0} & b_{2,1} & b_{2,2} \end{bmatrix} & a_{1,1} \cdot \begin{bmatrix} b_{0,0} & b_{0,1} & b_{0,2} \\ b_{1,0} & b_{1,1} & b_{1,2} \\ b_{2,0} & b_{2,1} & b_{2,2} \end{bmatrix} \end{bmatrix}$$

$$= \begin{bmatrix} a_{0,0} \cdot b_{0,0} & a_{0,0} \cdot b_{0,1} & a_{0,0} \cdot b_{0,2} & a_{0,1} \cdot b_{0,0} & a_{0,1} \cdot b_{0,1} & a_{0,1} \cdot b_{0,2} \\ a_{0,0} \cdot b_{1,0} & a_{0,0} \cdot b_{1,1} & a_{0,0} \cdot b_{1,2} & a_{0,1} \cdot b_{1,0} & a_{0,1} \cdot b_{1,1} & a_{0,1} \cdot b_{1,2} \\ a_{0,0} \cdot b_{2,0} & a_{0,0} \cdot b_{2,1} & a_{0,0} \cdot b_{2,2} & a_{0,1} \cdot b_{2,0} & a_{0,1} \cdot b_{2,1} & a_{0,1} \cdot b_{2,2} \\ a_{1,0} \cdot b_{0,0} & a_{1,0} \cdot b_{0,1} & a_{1,0} \cdot b_{0,2} & a_{1,1} \cdot b_{0,0} & a_{1,1} \cdot b_{0,1} & a_{1,1} \cdot b_{0,2} \\ a_{1,0} \cdot b_{1,0} & a_{1,0} \cdot b_{1,1} & a_{1,0} \cdot b_{1,2} & a_{1,1} \cdot b_{1,0} & a_{1,1} \cdot b_{1,1} & a_{1,1} \cdot b_{1,2} \\ a_{1,0} \cdot b_{2,0} & a_{1,0} \cdot b_{2,1} & a_{1,0} \cdot b_{2,2} & a_{1,1} \cdot b_{2,0} & a_{1,1} \cdot b_{2,1} & a_{1,1} \cdot b_{2,2} \end{bmatrix}.$$

Figure 5.2. The Kronecker product of matrices. Every entry in the first matrix is scalar multiplied with the second matrix.

Example 5.1.12. We saw in Example 5.1.2 that the discrete category $(\mathbf{N}, +, 0)$ is a strict monoidal category. If we look at $\mathbf{N}$ as a total order, then it is also a strict monoidal category. One must check that the bifunctor preserves addition; that is, if $m \leq m'$ and $n \leq n'$, then $m + n \leq m' + n'$. This not only works for addition but also for multiplication, which means that $(\mathbf{N}, \cdot, 1)$ is a strict monoidal category. These two strict monoidal operations are not independent of each other. Rather, the multiplication distributes over the addition.

Along the same lines, the totally ordered real numbers $\mathbf{R}$ has a strict monoidal category structure $(\mathbf{R}, +, 0)$. The nonnegative real numbers $\mathbf{R}^+$ has another strict monoidal category structure $(\mathbf{R}^+, \cdot, 1)$. These two monoidal structures are related, as we will show in Example 6.1.6 in Chapter 6. □

While we discuss many variations of strict monoidal categories in Chapter 7, there is one variation that arises with many of our examples, and it pays to describe it now.

Definition 5.1.13. A strict monoidal category $(\mathbb{A}, \otimes, I)$ is **strictly symmetric** if for all objects a and b in $\mathbb{A}$, we have $a \otimes b = b \otimes a$, and for all morphisms f, g, we have $f \otimes g = g \otimes f$.

Example 5.1.14. Of the examples of strict monoidal categories that we described, every commutative monoid can be thought of as a strictly symmetric, strict monoidal category. The strict monoidal categories $(\mathbf{N}, +, 0)$, $(\{1\}^*, \bullet, \emptyset)$, $(P, \wedge, 1)$, $(P, \vee, 0)$, and $(\mathbf{2}, \wedge, 1)$ are all strictly symmetric. In contrast, $(\{0, 1\}^*, \bullet, \emptyset)$ (where $0 \bullet 1 \neq 1 \bullet 0$) and both strict monoidal category structures on $\mathbb{K}\mathbb{M}\mathrm{at}$ are not strictly symmetric. □

Exercise 5.1.15. Show that both monoidal structures on $\mathbb{K}\mathbb{M}\mathrm{at}$ are not strictly symmetric by finding two 2-by-2 matrices, A and B, and showing that $A \oplus B \neq B \oplus A$ and $A \otimes B \neq B \otimes A$. ■

We conclude this section with a theorem about strict monoidal categories that are monoids.

Theorem 5.1.16. Let $(M, \star, e)$ be a monoid, thought of as a one-object category. If M has a strict monoidal category structure $(M, \square, I)$, then $\star = \square$, and the monoid is a commutative monoid. Another way to say this is that every one-object strict monoidal category is a commutative monoid. ★

Proof. First, we show that the unit of the monoid is the same as the unit of the monoidal category:

$$\begin{aligned}
e &= e \star e && e \text{ is a unit of the monoid.}\\
&= (I \square e) \star (e \square I) && I \text{ is a unit of the monoidal category.}\\
&= (I \star e) \square (e \star I) && \text{from bifunctoriality (i.e., Equation (4.20)).}\\
&= I \square I && e \text{ is a unit of the monoid.}\\
&= I && I \text{ is a unit of the monoidal category.}
\end{aligned}$$

By the bifunctoriality of $\square$, we have that

$$(m \star n) \square (m' \star n') = (m \square m') \star (n \square n'). \tag{5.7}$$

Setting $n = m' = e$ gives

$$(m \star e) \square (e \star n') = (m \square e) \star (e \square n'), \tag{5.8}$$

which reduces to

$$m \square n' = m \star n'. \tag{5.9}$$

That is, the two multiplications are the same. Setting $m = n' = e$ in Equation (5.7) gives us

$$(e \star n) \square (m' \star e) = (e \square m') \star (n \square e), \tag{5.10}$$

which reduces

$$n \square m' = m' \star n. \tag{5.11}$$

However, since $\square = \star$ as operations, we get that $n \star m' = m' \star n$. ♣

This theorem states that a set with two monoid structures that respect each other is a commutative monoid. The theorem is known as the **Eckmann–Hilton argument**.

5.2 Cartesian Categories

Many of the examples of combining objects and morphisms come from a category having a finite product structure. In general, such structures fail to be strictly associative. The simplest example is the category of sets with the Cartesian product. Given any sets S and T, there is a product $S \times T$. In general, for any three sets S, T and U, the product is not associative; that is,

$$S \times (T \times U) \neq (S \times T) \times U. \tag{5.12}$$

The set on the left contains elements of the form $\langle s, \langle t, u\rangle\rangle$, while the set on the right contains elements of the form $\langle\langle s, t\rangle, u\rangle$. Although these sets are not equal, there is an isomorphism from one set to the other.

Even though finite products in a general category fail to be strictly associative, we saw in Theorem 3.1.20 that any category $\mathbb{A}$ with finite products entails that for all a, b, and c, there is an ismorphism $a \times (b \times c) \longrightarrow (a \times b) \times c$. This isomorphism is actually a component of a natural isomorphism from the functor

$$(\ ) \times ((\ ) \times (\ )) \colon \mathbb{A}^3 \longrightarrow \mathbb{A} \tag{5.13}$$

to the functor

$$((\ ) \times (\ )) \times (\ ) \colon \mathbb{A}^3 \longrightarrow \mathbb{A}. \tag{5.14}$$

We also saw in Exercise 3.1.26 that if t is the terminal object in $\mathbb{A}$, then for every object a, there is an isomorphism $a \times t \longrightarrow a$. This isomorphism is a component of a natural isomorphism from the functor $(\ ) \times t \colon \mathbb{A} \longrightarrow \mathbb{A}$ to the identity functor $Id_{\mathbb{A}} \colon \mathbb{A} \longrightarrow \mathbb{A}$. Furthermore, Theorem 3.1.23 showed us that the product structure induces a braid map $br \colon a \times b \longrightarrow b \times a$, which is an isomorphism. This isomorphism is a component of the a natural isomorphism from the functor $(\ ) \times (\ )$ to the functor $((\ ) \times (\ )) \circ br$.

Let us formalize all of these notions about a category with a finite product.

Definition 5.2.1. A **Cartesian category** $(\mathbb{A}, \times, t)$ is a category with finite products (the terminal object, t, is the product over the empty diagram). The finite product structure induces the following natural isomorphisms:

- A way of reassociating the product: a **reassociator** natural isomorphism

$$\alpha \colon (\ ) \times ((\ ) \times (\ )) \Longrightarrow ((\ ) \times (\ )) \times (\ ). \tag{5.15}$$

  This means that for every a, b, and c, there is a component that is the isomorphism

$$\alpha_{a,b,c} \colon a \times (b \times c) \longrightarrow (a \times b) \times c. \tag{5.16}$$

- A way of eliminating the unit on the right: a **right unitor** natural isomorphism $\rho \colon (\ ) \times t \Longrightarrow Id_{\mathbb{A}}$. This means that for any a, there is a component that is an isomorphism: $\rho_a \colon a \times t \longrightarrow a$.
- A way of eliminating the unit on the left: a **left unitor** natural isomorphism $\lambda \colon t \times (\ ) \Longrightarrow Id_{\mathbb{A}}$. This means that for any a, there is a component that is an isomorphism: $\lambda_a \colon t \times a \longrightarrow a$.

- A way of reordering a product: a **braiding** natural isomorphism $\gamma\colon (\ )\times(\ )\Longrightarrow ((\ )\times(\ ))\circ br$. This means that for every a and b, there is a component that is an isomorphism: $\gamma_{a,b}\colon a\times b\longrightarrow b\times a$.

These natural isomorphisms all interact and satisfy more axioms. For pedagogical reasons, we will resist listing these axioms until we get to Definition 5.3.1 on page 237.

There are many examples of Cartesian categories, some of which we have already seen.

Example 5.2.2. $(\mathbb{Set}, \times, \{*\})$, $(\mathbb{Top}, \times, \{*\})$, and $(\mathbb{Manif}, \times, \{*\})$. We are familiar with the product structure in $\mathbb{Set}$. The product structures in $\mathbb{Top}$ and $\mathbb{Manif}$ are less familiar. One must show that the Cartesian product of topological spaces has a topological structure that must conform with the projection functions. In detail, one must show that if X and Y are topological spaces, then so is $X\times Y$, and the projections $\pi\colon X\times Y\longrightarrow X$ and $\pi\colon X\times Y\longrightarrow Y$ are continuous maps. One must also show that the product of two manifolds is locally like $\mathbf{R}^n$ and that the projection functions are smooth. In all three of these examples, the terminal object is the one-element set. We are omitting here many details that can easily be found in any topology textbook. □

Example 5.2.3. $(\mathbb{Cat}, \times, \mathbf{1})$ and $(\mathbb{Graph}, \times, \mathbf{1})$. We saw in Example 4.1.42 that $\mathbb{Cat}$ has a Cartesian category structure. We showed how to form the product of two categories. The unit is the terminal category that has a single object and a single identity morphism. The structure on the category of graphs is very similar. □

Example 5.2.4. The categories of most algebraic structures from Definition 2.1.8 have a Cartesian category structure. Example 3.1.12 showed how the product structure of groups is formed. In general, the product of two algebraic structures is the product of their underlying sets, and the operations are then performed pointwise (i.e., on each component).

There is a prominent exception within the algebraic structures mentioned in Definition 2.1.8. The category of fields does not have products. Given two fields, F_1 and F_2, one can form the set of ordered pairs $F_1\times F_2$. The pairs (a,b) can be added,

$$(a,b)+(a',b')=(a+a',b+b'), \tag{5.17}$$

and multiplied,

$$(a,b)\cdot(a',b')=(a\cdot a',b\cdot b'). \tag{5.18}$$

However, there are elements $(a,0)$, which are not zero, and yet they do not have inverses. This means that the pairs fail to be a field. Hence, the obvious product in $\mathbb{Field}$ does not form a Cartesian category structure. □

Example 5.2.5. $(\mathbf{K}\mathbb{FDVect}, \oplus, 0)$. The category of finite-dimensional $\mathbf{K}$ vector spaces and linear maps has a Cartesian category structure. There is a plethora of names for this operation: **direct sum**, **product**, and **Cartesian product**. This monoidal structure is similar to the monoidal structures described in Example 5.2.4. However, since

$\mathbb{KFDV}\mathbb{ect}$ will play such an important role in this text, we discuss it in detail here. If V and W are finite-dimensional vector spaces, then

$$V \oplus W = \{(v, w) : v \in V, w \in W\} \tag{5.19}$$

has a vector space structure. The vector space operations are defined as follows:

- Addition is pointwise (i.e., $(v, w) + (v', w') = (v + v', w + w')$).
- Scalar multiplication is pointwise (i.e., $c \cdot (v, w) = (c \cdot v, c \cdot w)$).
- The zero is $(0, 0)$.

The unit of this product is the trivial vector space 0. The details of showing that $(\mathbf{K}\mathbb{V}\mathbb{ect}, \oplus, 0)$ is a Cartesian category will be proven in the upcoming exercises. □

Exercise 5.2.6. Show that the product of two vector spaces satisfies all the axioms of being a vector space. ■

Exercise 5.2.7. Show that the obvious projection maps

$$V \xleftarrow{\pi_V} V \oplus W \xrightarrow{\pi_W} W \tag{5.20}$$

defined by $(v, w) \mapsto v$ and $(v, w) \mapsto w$ are linear maps, and they satisfy the universal property of being a product. ■

Exercise 5.2.8. Show that for three vector spaces V, W, and X, the vector space $V \oplus (W \oplus X)$ is isomorphic to $(V \oplus W) \oplus X$. ■

Exercise 5.2.9. Show that the trivial vector space 0 acts like a unit (i.e., show that $V \oplus 0$ and $0 \oplus V$ are isomorphic to V). ■

Exercise 5.2.10. Show that $V \oplus W$ is isomorphic to $W \oplus V$. ■

Here are two important theorems about the direct sum of vector spaces:

Theorem 5.2.11. For any finite-dimensional vector spaces V and W, $dim(V \oplus W) = dim(V) + dim(W)$. ★

Proof. If $\mathcal{B} = \{b_1, b_2, \ldots, b_m\}$ is a basis for V and $\mathcal{B}' = \{b'_1, b'_2, \ldots, b'_n\}$ is the basis for W, then

$$\{(b_1, 0), (b_2, 0), \ldots, (b_m, 0), (0, b'_1), (0, b'_2), \ldots, (0, b'_n)\} \tag{5.21}$$

is a basis for $V \oplus W$. ♣

Theorem 5.2.12. For any short exact sequence

$$0 \longrightarrow U \xrightarrow{T} V \xrightarrow{S} W \longrightarrow 0, \tag{5.22}$$

V is isomorphic to $U \oplus W$. ★

Proof. This is an easy consequence of Theorem 2.4.34 and Theorem 2.4.35. ♣

Example 5.2.13. Any preorder category $(P, \leq)$ that has finite products has a Cartesian category structure. The product is called a **meet** and denoted $\wedge$. The unit is the terminal 1, which satisfies the requirement that for all p, there is a unique isomorphism from $p \wedge 1$ to p. Notice that in contrast to Example 5.1.6, where we discussed partial orders, for preorders, $p \wedge (q \wedge r)$ need not be equal to $(p \wedge q) \wedge r$. In preorder categories, there also might be many terminal objects.

Special cases of such preorder categories are the categories of propositions and predicates that we saw in Section 4.8. We showed that $(\mathbb{Prop}, \wedge, True)$, and for all $\bar{x}$, $(\mathbb{Pred}(\bar{x}), \wedge, True)$, are Cartesian categories.

Notice that in contrast to $\mathbb{Prop}$, in general, the category $\mathbb{Proof}$ does not have a Cartesian category structure because $\wedge$ does not have universal properties. For example, there might be many proofs of the fact that $P \wedge Q$ implies P. □

Example 5.2.14. $(\mathbb{CompFunc}, \times, *)$. We saw in Exercise 3.1.19 that the category of computable functions with the multiplication $\times$ and the terminal type $*$ is a Cartesian category. □

Example 5.2.15. The direct sum in $\mathbb{KMat}$ has to be reexamined. Example 5.1.9 showed that $(\mathbb{KMat}, \oplus, 0)$ has a strict monoidal category structure. In fact, it is an unusual example of a Cartesian category where $\oplus$ is strictly associative. It pays to carefully show that the $\oplus$ monoidal structure satisfies the universal property of being a product in $\mathbb{KMat}$. The object $m \oplus n = m + n$ is the product of m and n, with the projection given as

$$m \xleftarrow{\begin{bmatrix} I_m & 0_{m,n} \end{bmatrix}} m + n \xrightarrow{\begin{bmatrix} 0_{n,m} & I_n \end{bmatrix}} n. \tag{5.23}$$

To see that these projections satisfy the universal properties, consider two morphisms (matrices) $A_{m,p}\colon p \longrightarrow m$ and $B_{n,p}\colon p \longrightarrow n$, as in

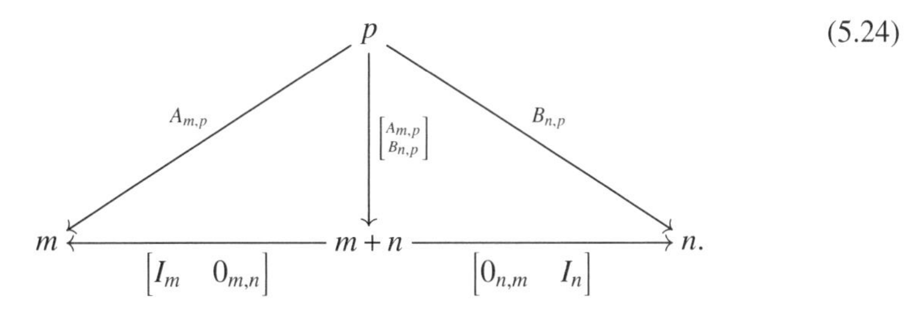

The $m + n$ by p matrix with $A_{m,p}$ on top of $B_{n,p}$ uniquely satisfies this property. The left triangle commutes because

$$\begin{bmatrix} I_m & 0_{m,n} \end{bmatrix} \cdot \begin{bmatrix} A_{m,p} \\ B_{n,p} \end{bmatrix} = A_{m,p}. \tag{5.25}$$

The right triangle is done similarly. □

From duality, we know that whatever we said about the product and the terminal object also applies to the coproduct and the initial object.

Definition 5.2.16. A **co-Cartesian category** or a **coproduct category** $(\mathbb{A}, +, i)$ or $(\mathbb{A}, \amalg, i)$ is a category with finite coproducts (the initial object, i, is the coproduct over the empty diagram.) The finite coproduct structure induces the following natural isomorphisms:

- A way of reassociating the coproduct: a **reassociator** natural isomorphism

$$\alpha\colon (\ ) + ((\ ) + (\ )) \Longrightarrow ((\ ) + (\ )) + (\ ). \tag{5.26}$$

- A way of eliminating the unit on the right: a **right unitor** natural isomorphism $\rho\colon (\ ) + i \Longrightarrow Id_{\mathbb{A}}$.
- A way of eliminating the unit on the left: a **left unitor** natural isomorphism $\lambda\colon i + (\ ) \Longrightarrow Id_{\mathbb{A}}$.
- A way of reordering a coproduct: a **braiding** or a **cobraiding** natural isomorphism $\gamma\colon (\ ) + (\ ) \Longrightarrow ((\ ) + (\ )) \circ br$.

These natural isomorphisms all interact and satisfy more axioms. We will state those axioms in Definition 5.3.1 on page 237.

There are many examples of co-Cartesian categories, some of which we have already seen.

Example 5.2.17. $(\mathbb{Set}, +, \emptyset)$, $(\mathbb{Top}, +, \emptyset)$, and $(\mathbb{Manif}, +, \emptyset)$. All three categories have coproduct structures. The empty set, the empty topological space, and the empty manifold are the initial objects in their respective categories. One must show that in the cases of $\mathbb{Top}$ and $\mathbb{Manif}$, the coproduct exists and the inclusion maps satisfy the universal properties and are continuous and smooth. □

Example 5.2.18. $(\mathbb{Cat}, +, \mathbf{0})$ and $(\mathbb{Graph}, +, \emptyset)$. The category $\mathbb{Cat}$ has a co-Cartesian category structure. The coproduct of two categories is their disjoint union. The unit is the empty category $\mathbf{0}$. The category of graphs has a similar structure. □

Example 5.2.19. $(Bool\mathbb{Func}, +, 0)$. The objects are the natural numbers and the morphisms from m to n are all the set functions from the set $\{0,1\}^m$ to $\{0,1\}^n$. The tensor product on objects is the addition of natural numbers. The morphisms is done as follows. Let $f\colon \{0,1\}^m \longrightarrow \{0,1\}^n$ and $g\colon \{0,1\}^{m'} \longrightarrow \{0,1\}^{n'}$. Then $f + g\colon \{0,1\}^{m+m'} \longrightarrow \{0,1\}^{n+n'}$ is defined by f for the first m digits of the input and by g for the last m' digits. □

Example 5.2.20. A preorder category $(P, \leq)$ with finite coproducts has a co-Cartesian category structure where the coproduct is called **join** and denoted as $\vee$. The initial object is denoted as 0 and for all p, there is the isomorphism $p \vee 0 \longrightarrow p$. In contrast to a partial order category, for an arbitrary preorder category, $p \vee (q \vee r)$ need not be equal to $(p \vee q) \vee r$. The categories that we saw in our mini-course in basic categorical logic in Chapter 4, $(\mathbb{Prop}, \vee, False)$ and, for all $\bar{x}$, $(\mathbb{Pred}(\bar{x}), \vee, False)$, are co-Cartesian categories. □

Exercise 5.2.21. In Example 5.1.9, we showed that $\mathbf{K}\mathbb{Mat}$ has a strict monoidal structure. In Example 5.2.15, we showed that it has a Cartesian monoidal structure. Show that the product in $\mathbf{K}\mathbb{Mat}$ also makes it a co-Cartesian category. ■

Exercise 5.2.22. We saw that $(\mathbf{K}\mathbb{Vect}, \oplus, 0)$ is a Cartesian category. In fact, it is also a co-Cartesian category. Show that the obvious inclusion maps

$$V \xrightarrow{inc_V} V \oplus W \xleftarrow{inc_W} W \tag{5.27}$$

defined by $v \mapsto (v, 0)$ and $w \mapsto (0, w)$ are linear maps and satisfy the universal properties of a coproduct. The $\oplus$ operation, which is both a product and a coproduct, is called a **biproduct**. ■

Notice that some categories, such as $\mathbb{Set}, \mathbb{Top}, \mathbb{Manif}, \mathbb{Prop}, \mathbb{Cat}, \mathbb{Par}$, $\mathbf{K}\mathbb{Mat}$, and $\mathbf{K}\mathbb{Vect}$, have both Cartesian and co-Cartesian category structures. We saw in Theorem 3.1.40 that there is a distributive morphism relating the two structures. In distributive categories, this morphism is an isomorphism. In some categories, like $\mathbf{K}\mathbb{FDVect}$ and $\mathbb{AbGp}$, the product and the coproduct are the same.

5.3 Monoidal Categories

The structures we have seen up to now, strict monoidal categories, Cartesian categories, and co-Cartesian categories, are special types of a more general structure that we will present here. By weakening the requirements, we get a notion that is more applicable (see Important Categorical Idea 4.1.29.)

Definition 5.3.1. A **monoidal category** $(\mathbb{A}, \otimes, I, \alpha, \lambda, \rho)$ has the following structure:

- A category $\mathbb{A}$
- A way of combining objects and morphisms: a bifunctor called the **tensor product** or the **monoidal product** $\otimes\colon \mathbb{A} \times \mathbb{A} \longrightarrow \mathbb{A}$
- A special object I of $\mathbb{A}$ called the **unit**
- A way of reassociating the monoidal product: a natural isomorphism called a **reassociator**:

$$\alpha\colon (\ ) \otimes ((\ ) \otimes (\ )) \Longrightarrow ((\ ) \otimes (\ )) \otimes (\ ). \tag{5.28}$$

That is, for every a, b, and c, the component is an isomorphism

$$\alpha_{a,b,c}\colon a \otimes (b \otimes c) \longrightarrow (a \otimes b) \otimes c. \tag{5.29}$$

- A way of eliminating the unit on the right: a natural isomorphism called a **right unitor**:

$$\rho\colon (\ )\otimes I \Longrightarrow Id_{\mathbb{A}}. \tag{5.30}$$

That is, for any a, the component is an isomorphism:

$$\rho_a\colon a\otimes I \longrightarrow a. \tag{5.31}$$

- A way of eliminating the unit on the left: a natural isomorphism **left unitor**:

$$\lambda\colon I\otimes(\ ) \Longrightarrow Id_{\mathbb{A}}. \tag{5.32}$$

That is, for any a, the component is an isomorphism:

$$\lambda_a\colon I\otimes a \longrightarrow a. \tag{5.33}$$

The following requirements must be satisfied:

- The reassociator must cohere with itself: for all objects a, b, c, and d, the following **pentagon coherence condition**, **Mac Lane's coherence condition**, **Stasheff's coherence condition**

$$\begin{array}{ccccc} a\otimes(b\otimes(c\otimes d)) & \xrightarrow{id_a\otimes\alpha_{b,c,d}} & a\otimes((b\otimes c)\otimes d) & \xrightarrow{\alpha_{a,b\otimes c,d}} & (a\otimes(b\otimes c))\otimes d \\ \downarrow{\scriptstyle \alpha_{a,b,c\otimes d}} & & & & \downarrow{\scriptstyle \alpha_{a,b,c}\otimes id_d} \\ (a\otimes b)\otimes(c\otimes d) & & \xrightarrow[\alpha_{a\otimes b,c,d}]{} & & ((a\otimes b)\otimes c)\otimes d \end{array} \tag{5.34}$$

commutes. (This is one of the most important diagrams in this book!)

- The reassociator must cohere with the right and left unitors: for all objects a and b of $\mathbb{A}$, the following **triangle coherence condition**

$$\begin{array}{ccc} a\otimes(I\otimes b) & \xrightarrow{\alpha_{a,I,b}} & (a\otimes I)\otimes b \\ & {\scriptstyle id_a\otimes\lambda_b}\searrow \quad \swarrow{\scriptstyle \rho_a\otimes id_b} & \\ & a\otimes b & \end{array} \tag{5.35}$$

commutes.

There is an in-depth discussion of the coherence conditions in Section 5.4.

While most of the variations of monoidal categories will be given in Chapter 7, there is one variation that deals with commutativity that arises so often that we provide it here.

Definition 5.3.2. A **symmetric monoidal category** $(\mathbb{A}, \otimes, I, \alpha, \lambda, \rho, \gamma)$ is a monoidal category $(\mathbb{A}, \otimes, I, \alpha, \lambda, \rho)$ with a natural isomorphism called a **braiding** that permutes two objects:

$$\gamma\colon \otimes \Longrightarrow \otimes \circ br. \tag{5.36}$$

That is, for objects a and b in $\mathbb{A}$, there is a natural isomorphism:

$$\gamma_{a,b}\colon a \otimes b \longrightarrow b \otimes a. \tag{5.37}$$

In addition to being a monoidal category, a symmetric monoidal category must satisfy the following conditions:

- The braiding must be its own inverse: the **symmetry coherence condition**

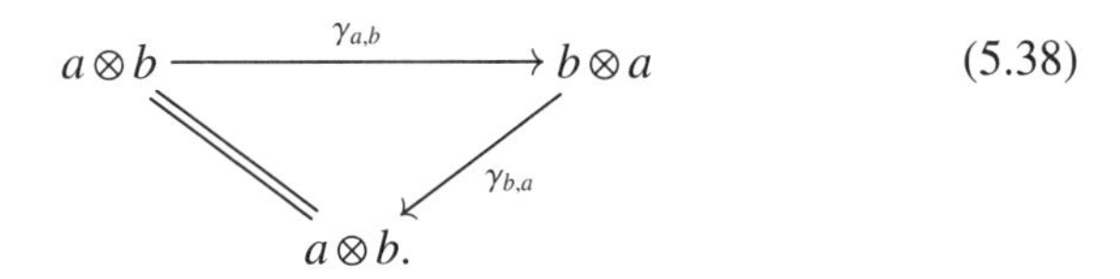

(5.38)

- The braiding must cohere with itself and the associator: the **hexagon coherence conditions**

$$\begin{array}{ccccc}
(a \otimes b) \otimes c & \xrightarrow{\alpha^{-1}_{a,b,c}} & a \otimes (b \otimes c) & \xrightarrow{\gamma_{a,b\otimes c}} & (b \otimes c) \otimes a \\
\downarrow{\scriptstyle \gamma_{a,b}\otimes Id_c} & & & & \downarrow{\scriptstyle \alpha^{-1}_{b,c,a}} \\
(b \otimes a) \otimes c & \xrightarrow[\alpha^{-1}_{b,a,c}]{} & b \otimes (a \otimes c) & \xrightarrow[id_b \otimes \gamma_{a,c}]{} & b \otimes (c \otimes a)
\end{array} \tag{5.39}$$

$$\begin{array}{ccccc}
a \otimes (b \otimes c) & \xrightarrow{\alpha_{a,b,c}} & (a \otimes b) \otimes c & \xrightarrow{\gamma_{a\otimes b,c}} & c \otimes (a \otimes b) \\
\downarrow{\scriptstyle id_a \otimes \gamma_{b,c}} & & & & \downarrow{\scriptstyle \alpha_{c,a,b}} \\
a \otimes (c \otimes b) & \xrightarrow[\alpha_{a,c,b}]{} & (a \otimes c) \otimes b & \xrightarrow[\gamma_{a,c} \otimes id_b]{} & (c \otimes a) \otimes b.
\end{array} \tag{5.40}$$

- The braiding must cohere with the left and right unitors: the **triangle symmetry coherence condition**

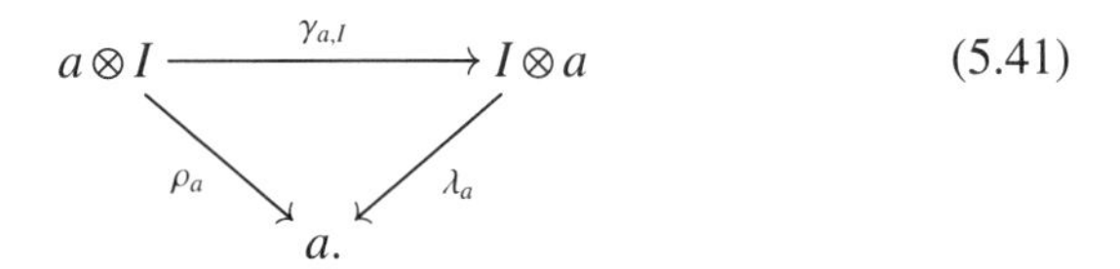

(5.41)

Technical Point 5.3.3. When Diagram (5.38) commutes, then Diagram (5.39) commuting implies that Diagram (5.40) commutes, and vice versa. The way to see this is to realize that the commuting of Diagram (5.38) shows that $\gamma^{-1}_{a,b} = \gamma_{b,a}$. Keeping in mind that $(f \circ g)^{-1} = g^{-1} \circ f^{-1}$, and if you change any diagram that commutes by replacing morphisms with their inverses, the resulting diagram will also commute. We mention the seemingly superfluous requirement because we will need both conditions in Section 7.1. ♡

A monoidal category can be strictly associative and still be a nonstrictly symmetric monoidal category.

Definition 5.3.4. A **strictly associative symmetric monoidal category** is a symmetric monoidal category where α, λ, and ρ are all identity maps. In this case, Diagrams (5.39) and (5.40) shorten to

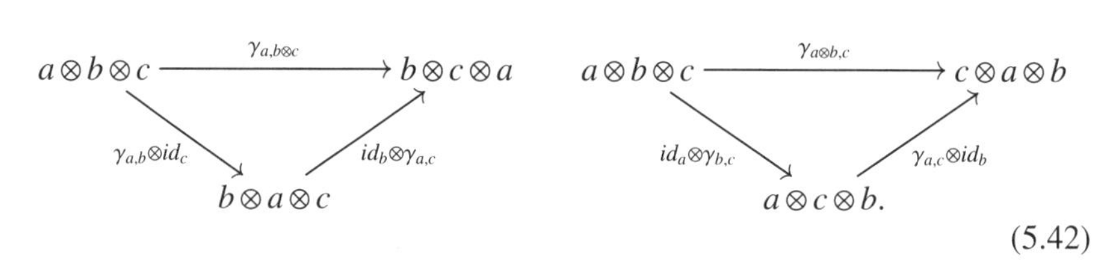

(5.42)

Let us now explore some examples of monoidal categories.

Example 5.3.5. Every strict monoidal category is a monoidal category where the natural isomorphisms α, ρ, and λ are identity-natural transformations. Obviously, all the coherence conditions are satisfied. □

Example 5.3.6. Every Cartesian category and co-Cartesian category is a symmetric monoidal category. We shall concentrate on Cartesian categories and leave co-Cartesian categories as an exercise. In Section 3.1 we showed that products induce reassociators (Theorem 3.1.20), unitors (Exercise 3.1.26), and braidings (Theorem 3.1.23). We are left to show that the induced isomorphisms satisfy all the coherence conditions for a monoidal category.

First, we show that the reassociator satisfies the pentagon coherence condition (Diagram (5.34)). It is possible to show this by looking at Figure 5.3, which has a pentagon of four letters (with the Cartesian product omitted). The figure has all the projections. It would be extremely painful (but possible) to "diagram chase" and show that the universal properties of the product ensure that the pentagon commutes.

A slightly less painful way of showing that the pentagon commutes is algebraically. (Feel free to skip this proof on the first reading.) First let us remember some basic properties of brackets and reassociators. These equations are central to the **calculus of projection functions**:

$$\alpha_{a,b,c} = \langle\langle \pi_a^{a(bc)}, \pi_b^{a(bc)}\rangle, \pi_c^{a(bc)}\rangle \tag{5.43}$$

$$f\times g = \langle f\circ\pi, g\circ\pi\rangle \tag{5.44}$$

$$\langle f,g\rangle\circ h = \langle f\circ h, g\circ h\rangle \tag{5.45}$$

$$\pi_a\circ\langle f,g\rangle = f \qquad \text{and} \qquad \pi_b\circ\langle f,g\rangle = g. \tag{5.46}$$

These last two equations needs some elaboration. As we saw in Theorem 3.1.16, every contiguous sequence of projections can be written as $\pi_x^{[\text{bracketing of } a,b,c,d,\ldots,z]}$. When such a projection is composed with a bracketing as $\langle\langle a,b\rangle, c\langle\cdots,z\rangle\rangle$, the projection will output the entry where the x is in the bracketing. Here are some examples:

$$\pi_b^{a(bc)}\circ\langle X\langle Z,Y\rangle\rangle = Z \qquad \pi_d^{a(b(cd))}\circ\langle W,\langle X,\langle T,G\rangle\rangle\rangle = G \qquad \pi_c^{cba}\circ\langle X,Y,Z\rangle = X. \tag{5.47}$$

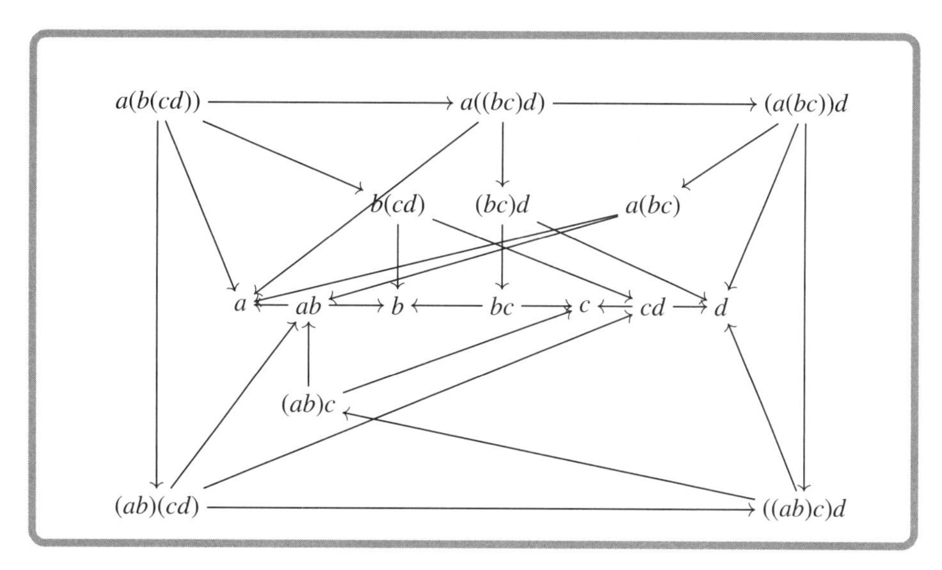

Figure 5.3. The pentagon coherence condition for a Cartesian category with all its projections.

In general,

$$\pi_x^{[\text{bracketing of } (((a,b,c,d,\ldots,z)))]} \circ \text{bracketing of } \langle\langle\langle a,b,c\cdots,z\rangle\rangle\rangle = x. \tag{5.48}$$

With these properties in our toolbox, we can prove the theorem. This will be proven by showing that whether one goes around the pentagon clockwise or counterclockwise, the result is the same map.

Clockwise:

$= (\alpha_{a,b,c} \times id_d) \circ \alpha_{a,bc,d} \circ (id_a \times \alpha_{b,c,d}),$

$= (\alpha_{a,b,c} \times id_d) \circ \alpha_{a,bc,d} \circ \langle \pi_a^{a(b(cd))}, \alpha_{b,c,d}\pi_{b(cd)}^{a(b(cd))}\rangle,$ Equation (5.44)

$= (\alpha_{a,b,c} \times id_d)\circ$

$\langle\langle \pi_a^{a((bc)d)}, \langle \pi_b^{a((bc)d)}, \pi_c^{a((bc)d)}\rangle\rangle, \pi_d^{a((bc)d)}\rangle \circ \langle \pi_a^{a(b(cd))}, \alpha_{b,c,d}\pi_{b(cd)}^{a(b(cd))}\rangle,$ Equation (5.43)

$= (\alpha_{a,b,c} \times id_d)\circ$

$\langle\langle \pi_a^{a((bc)d)} \circ \langle \pi_a^{a(b(cd))}, \alpha_{b,c,d}\pi_{b(cd)}^{a(b(cd))}\rangle,$

$\langle \pi_b^{a((bc)d)} \circ \langle \pi_a^{a(b(cd))}, \alpha_{b,c,d}\pi_{b(cd)}^{a(b(cd))}\rangle,$

$\pi_c^{a((bc)d)} \circ \langle \pi_a^{a(b(cd))}, \alpha_{b,c,d}\pi_{b(cd)}^{a(b(cd))}\rangle\rangle\rangle,$

$\pi_d^{a((bc)d)} \circ \langle \pi_a^{a(b(cd))}, \alpha_{b,c,d}\pi_{b(cd)}^{a(b(cd))}\rangle\rangle,$ Equation (5.45)

$= (\alpha_{a,b,c} \times id_d)\circ$

$\langle\langle \pi_a^{a((bc)d)} \circ \langle \pi_a^{a(b(cd))}, \langle\langle \pi_b^{a(b(cd))}, \pi_c^{a(b(cd))}\rangle \pi_d^{a(b(cd))}\rangle\rangle,$

$$\langle \pi_b^{a((bc)d)} \circ \langle \pi_a^{a(b(cd))}, \langle\langle \pi_b^{a(b(cd))}, \pi_c^{a(b(cd))} \rangle \pi_d^{a(b(cd))} \rangle\rangle,$$

$$\pi_c^{a((bc)d)} \circ \langle \pi_a^{a(b(cd))}, \langle\langle \pi_b^{a(b(cd))}, \pi_c^{a(b(cd))} \rangle \pi_d^{a(b(cd))} \rangle\rangle,$$

$$\pi_d^{a((bc)d)} \circ \langle \pi_a^{a(b(cd))}, \langle\langle \pi_b^{a(b(cd))}, \pi_c^{a(b(cd))} \rangle \pi_d^{a(b(cd))} \rangle\rangle, \qquad \text{Equation (5.43)}$$

$$= (\alpha_{a,b,c} \times id_d) \circ \langle\langle \pi_a^{a(b(cd))} \langle \pi_b^{a(b(cd))}, \pi_c^{a(b(cd))} \rangle\rangle, \pi_d^{a(b(cd))} \rangle, \qquad \text{Equation (5.48)}$$

$$= (\alpha_{a,b,c} \pi_{a(bc)}^{(a(bc))d}, \pi_d^{(a(bc))d}) \circ \langle\langle \pi_a^{a(b(cd))} \langle \pi_b^{a(b(cd))}, \pi_c^{a(b(cd))} \rangle\rangle, \pi_d^{a(b(cd))} \rangle, \qquad \text{Equation (5.44)}$$

$$= \langle\langle\langle \pi_a^{(a(bc))d}, \pi_b^{(a(bc))d} \rangle, \pi_c^{(a(bc))d} \rangle, \pi_d^{(a(bc))d} \rangle \circ$$

$$\langle\langle \pi_a^{a(b(cd))}, \langle \pi_b^{a(b(cd))}, \pi_c^{a(b(cd))} \rangle\rangle, \pi_d^{a(b(cd))} \rangle, \qquad \text{Equation (5.43)}$$

$$= \langle\langle\langle \pi_a^{(a(bc))d} \circ \langle\langle \pi_a^{a(b(cd))} \langle \pi_b^{a(b(cd))}, \pi_c^{a(b(cd))} \rangle\rangle, \pi_d^{a(b(cd))} \rangle,$$

$$\pi_b^{(a(bc))d} \circ \langle\langle \pi_a^{a(b(cd))} \langle \pi_b^{a(b(cd))}, \pi_c^{a(b(cd))} \rangle\rangle, \pi_d^{a(b(cd))} \rangle\rangle,$$

$$\pi_c^{(a(bc))d} \circ \langle\langle \pi_a^{a(b(cd))} \langle \pi_b^{a(b(cd))}, \pi_c^{a(b(cd))} \rangle\rangle, \pi_d^{a(b(cd))} \rangle\rangle,$$

$$\pi_d^{(a(bc))d} \circ \langle\langle \pi_a^{a(b(cd))} \langle \pi_b^{a(b(cd))}, \pi_c^{a(b(cd))} \rangle\rangle, \pi_d^{a(b(cd))} \rangle\rangle, \qquad \text{Equation (5.45)}$$

$$= \underline{\underline{\langle\langle\langle \pi_a^{a(b(cd))}, \pi_b^{a(b(cd))} \rangle, \pi_c^{a(b(cd))} \rangle, \pi_d^{a(b(cd))} \rangle.}} \qquad \text{Equation (5.48)}$$

We also have the following:

Counterclockwise:

$$= \alpha_{ab,c,d} \circ \alpha_{a,b,cd},$$

$$= \langle\langle\langle \pi_a^{(ab)(cd)} \pi_b^{(ab)(cd)} \rangle, \pi_c^{(ab)(cd)} \rangle, \pi_d^{(ab)(cd)} \rangle \circ \alpha_{a,b,cd}, \qquad \text{Equation (5.43)}$$

$$= \langle\langle\langle \pi_a^{(ab)(cd)} \circ \alpha_{a,b,cd}, \pi_b^{(ab)(cd)} \circ \alpha_{a,b,cd} \rangle, \pi_c^{(ab)(cd)} \circ$$

$$\alpha_{a,b,cd} \rangle, \pi_d^{(ab)(cd)} \circ \alpha_{a,b,cd} \rangle, \qquad \text{Equation (5.45)}$$

$$= \langle\langle\langle \pi_a^{(ab)(cd)} \circ \langle\langle \pi_a^{a(b(cd))}, \pi_b^{a(b(cd))} \rangle, \langle \pi_c^{a(b(cd))} \pi_d^{a(b(cd))} \rangle\rangle,$$

$$\pi_b^{(ab)(cd)} \circ \langle\langle \pi_a^{a(b(cd))}, \pi_b^{a(b(cd))} \rangle, \langle \pi_c^{a(b(cd))} \pi_d^{a(b(cd))} \rangle\rangle,$$

$$\pi_c^{(ab)(cd)} \circ \langle\langle \pi_a^{a(b(cd))}, \pi_b^{a(b(cd))} \rangle, \langle \pi_c^{a(b(cd))} \pi_d^{a(b(cd))} \rangle\rangle,$$

$$\pi_d^{(ab)(cd)} \circ \langle\langle \pi_a^{a(b(cd))}, \pi_b^{a(b(cd))} \rangle, \langle \pi_c^{a(b(cd))} \pi_d^{a(b(cd))} \rangle\rangle\rangle \qquad \text{Equation (5.43)}$$

$$= \underline{\underline{\langle\langle\langle \pi_a^{a(b(cd))}, \pi_b^{a(b(cd))} \rangle \pi_c^{a(b(cd))} \rangle, \pi_d^{a(b(cd))} \rangle.}} \qquad \text{Equation (5.48)}$$

The double underline is the description of the unique morphism from the beginning to the end of the pentagon.

The correct way to prove that the product satisfies the pentagon coherence condition was suggested to me by Todd Trimble. Consider the five functors of the form $\mathbb{A} \times \mathbb{A} \times \mathbb{A} \times \mathbb{A} \longrightarrow \mathbb{A}$, which represent the five ways of associating the four letters. All five of these functors are essentially right adjoints to versions of the diagonal map $\Delta\colon \mathbb{A} \longrightarrow \mathbb{A} \times \mathbb{A} \times \mathbb{A} \times \mathbb{A}$, as in Diagram (4.178). We saw in Theorem 4.4.25 that all right adjoints are unique *up to a unique isomorphism*. This means that there is a unique isomorphism between any two of the five functors. In particular, there is a unique isomorphism from the beginning to the end of the pentagon.

Next, we show that unitors of a Cartesian category satisfy the triangle coherence condition. Remember that $\rho_a = \pi_a^{aI} \colon a \times I \longrightarrow a$ and $\lambda_b = \pi_b^{Ib} \colon I \times b \longrightarrow b$. The triangle commutes because

$$\begin{aligned}
(\rho_a \times id_b) \circ \alpha_{a,I,b} &= \langle \pi_a^{(aI)b}, \pi_b^{(aI)b} \rangle \circ \alpha_{a,I,b} \rangle && \text{Equation (5.44)}\\
&= \langle \pi_a^{(aI)b}, \pi_b^{(aI)b} \rangle \circ \langle\langle \pi_a^{a(Ib)}, \pi_I^{a(Ib)} \rangle, \pi_b^{a(Ib)} \rangle && \text{Equation (5.43)}\\
&= \langle \pi_a^{(aI)b} \circ \langle\langle \pi_a^{a(Ib)}, \pi_I^{a(Ib)} \rangle, \pi_b^{a(Ib)} \rangle,\\
&\quad \pi_b^{(aI)b} \circ \langle\langle \pi_a^{a(Ib)}, \pi_I^{a(Ib)} \rangle, \pi_b^{a(Ib)} \rangle\rangle && \text{Equation (5.45)}\\
&= \langle \pi_a^{a(Ib)}, \pi_b^{Ib} \rangle, && \text{Equation (5.48)}\\
&= id_a \times \lambda_b. && \text{Equation (5.44)}
\end{aligned}$$

The fact that the triangle symmetry condition is satisfied was shown to be true in Exercise 3.1.24.

We have to show that the braiding of a Cartesian category satisfies the hexagon coherence conditions. There is nothing really to be gained by proving the full hexagon coherence condition since we already dealt with α and γ. Readers can do it on their own. In way of a hint, the map in the first hexagon from the starting object to the last object is

$$\langle \pi_b^{(ab)c}, \langle \pi_c^{(ab)c}, \pi_a^{(ab)c} \rangle\rangle, \tag{5.49}$$

and for the second hexagon is

$$\langle\langle \pi_c^{a(bc)}, \pi_a^{a(bc)} \rangle, \pi_b^{a(bc)} \rangle. \tag{5.50}$$

Rather than proving the hexagon, we show that the triangle on the left of Diagram (5.42) is satisfied:

$$\begin{aligned}
(id_b \times \gamma_{a,c}) \circ (\gamma_{a,b} \times id_c) &= \langle \pi_b^{bac}, \gamma_{a,c}\pi_{a,c}^{bac} \rangle \circ \langle \gamma_{a,b}\pi_{a,b}^{abc}, \pi_c^{abc} \rangle && \text{Equation (5.44)}\\
&= \langle \pi_b^{bac}, \pi_c^{bac}, \pi_a^{bac} \rangle \circ \langle \pi_b^{abc}, \pi_a^{abc}, \pi_c^{abc} \rangle && \text{by definition of } \gamma\\
&= \langle \pi_b^{bac} \circ \langle \pi_b^{abc}, \pi_a^{abc}, \pi_c^{abc} \rangle,\\
&\quad \pi_c^{bac} \circ \langle \pi_b^{abc}, \pi_a^{abc}, \pi_c^{abc} \rangle,\\
&\quad \pi_a^{bac} \circ \langle \pi_b^{abc}, \pi_a^{abc}, \pi_c^{abc} \rangle\rangle, && \text{Equation (5.45)}\\
&= \langle \pi_b^{abc}, \pi_c^{abc}, \pi_a^{abc} \rangle, && \text{Equation (5.48)}\\
&= \langle \pi_{bc}^{abc}, \pi_a^{abc} \rangle, && \text{Equation (5.48)}\\
&= \gamma_{a,bc}. && \text{by definition of } \gamma.
\end{aligned}$$

The final commutative diagram to show is the triangle symmetry condition (Diagram (5.41)) is satisfied:

$$\begin{aligned}
\lambda_a \circ \gamma_{a,I} &= \pi_a^{Ia} \circ \langle \pi_I^{aI}, \pi_a^{aI} \rangle && \text{by definition of } \lambda \text{ and } \gamma.\\
&= \pi_a^{aI} && \text{Equation 5.48.}\\
&= \rho_a && \text{by definition of } \rho.
\end{aligned}$$

□

Important Categorical Idea 5.3.7. Cartesian Versus Monoidal. Although Cartesian categories are a special type of monoidal category, there could be significant differences between a Cartesian category and a general monoidal category. The tensor in a Cartesian category satisfies universal properties. This means that there are projection functions that satisfy all the universal properties that we saw in Section 3.1. For example, in a Cartesian category, any object a has a diagonal map $\Delta\colon a \longrightarrow a \times a$. In a general monoidal category, such a morphism need not exist. Also, in a Cartesian category, the monoidal product is automatically symmetric, while in a general monoidal category, this is not necessarily true. This will have profound ramifications in the rest of this book. ○

Exercise 5.3.8. Prove that every co-Cartesian category is a symmetric monoidal category. ■

Example 5.3.9. There is a category called the **association category**, $\mathbb{A}\mathrm{ssoc}$, that is the paradigm of monoidal categories. The objects are the associations of letters. For example, the following are three associations:

$$\bullet(\bullet\bullet), \qquad (\bullet(\bullet\bullet))(\bullet(\bullet(\bullet\bullet))), \qquad \bullet(\bullet(((\bullet)\bullet))\bullet)\bullet. \tag{5.51}$$

Each $\bullet$ is a placeholder, and only the arrangement of the parentheses is important. Some of the associations will have the letter I in some of their positions, which will correspond to having a unit element in that position. For example:

$$\bullet(I\bullet), \qquad (\bullet(I\bullet))(I(\bullet(\bullet I))), \qquad \bullet(I(((\bullet\bullet)I))\bullet)\bullet. \tag{5.52}$$

The morphisms in this category are easy to describe: there is exactly one morphism between any two objects in the category. This makes the category $\mathbb{A}\mathrm{ssoc}$ a contractible groupoid and a preorder category. The monoidal structure is also easy to describe: given associations w and w', the tensor product is $w \otimes w' = w \bullet w'$. Given $f\colon w \longrightarrow w'$ and $g\colon w'' \longrightarrow w^3$, the tensor product $f \otimes g$ is the unique isomorphism $w \bullet w'' \longrightarrow w' \bullet w^3$. The unit of the monoidal structure is the association I. The uniqueness of the morphisms gives us the isomorphisms

$$\alpha_{w,w',w''}\colon w \bullet (w' \bullet w) \longrightarrow (w \bullet w') \bullet w'', \qquad \rho_w\colon w \bullet I \longrightarrow w, \qquad \lambda_w\colon I \bullet w \longrightarrow w. \tag{5.53}$$

The uniqueness of the morphisms also ensure that the pentagon and triangle coherence conditions are satisfied.

We will see in Section 6.2 how $(\mathbb{A}\mathrm{ssoc}, \otimes, I)$ is the paradigm monidal category. □

Example 5.3.10. There is a category called the **symmetry category**, $\mathbb{S}\mathrm{ym}$, that is the paradigm of symmetric monoidal categories. This category is built out of the symmetric groups, which we first define here. For each natural number n, there is a **symmetric group** on n elements or the nth symmetric group, written S_n, whose elements are ways of permuting the first n numbers. These groups are very important in all of mathematics and physics. A typical element of S_6 is

$$(1,2,3,4,5,6) \longmapsto (3,4,1,6,5,2). \tag{5.54}$$

This permutation takes 1 to 3, 2 to 4, 3 to 1, etc. Another typical element is the permutation

$$(1,2,3,4,5,6) \longmapsto (5,3,6,4,2,1). \tag{5.55}$$

These permutations can be composed by doing the first one and then the second one. The composition of these two permutations is the permutation that takes 1 to 3, which further goes to 6, etc. Here is the composition:

$$(1,2,3,4,5,6) \longmapsto (6,4,5,1,2,3). \tag{5.56}$$

There is an identity permutation which does not change anything; that is,

$$(1,2,3,4,5,6) \longmapsto (1,2,3,4,5,6). \tag{5.57}$$

Every permutation has an inverse. The inverse of the Line (5.54) permutation is

$$(1,2,3,4,5,6) \longmapsto (3,6,1,2,5,4). \tag{5.58}$$

A graphical way of describing these permutations is having a line from the top number to the number it goes to. The permutations in Lines (5.54) and (5.55) can be written as

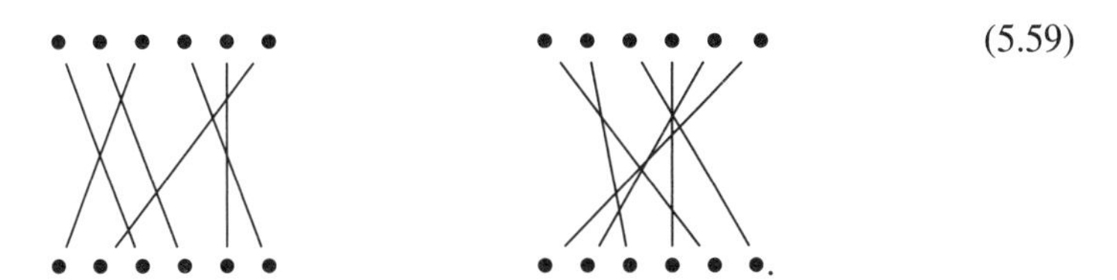

(5.59)

The composition of Diagrams (5.54) and (5.55) can be seen as

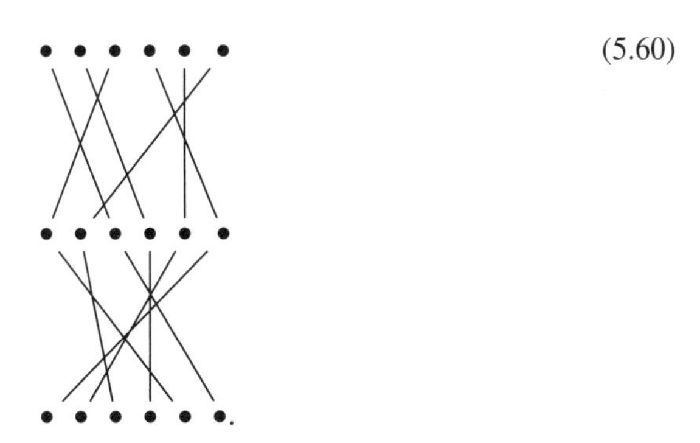

(5.60)

Follow the lines from top to bottom. This is a way of describing permutation (5.56).

The identity permutation in S_6 can be viewed as

(5.61)

This corresponds to not changing anything. The composition of any permutation with the identity is the original permutation.

The inverse of Line (5.54) is its diagram turned upside-down:

(5.62)

One can see that it is the inverse by looking at the composition of the permutation with its inverse:

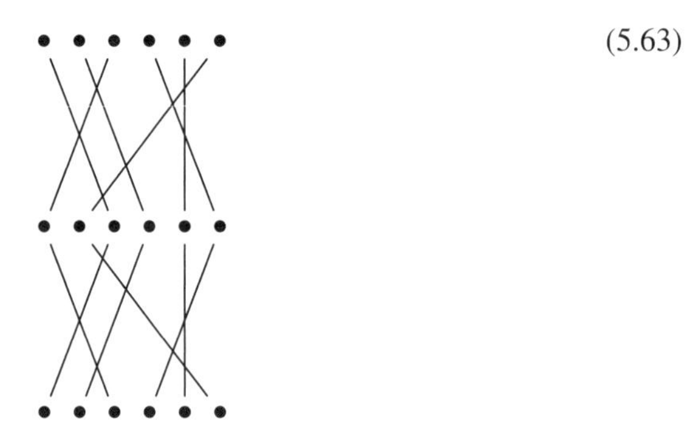

(5.63)

which is the identity.

Using these graphical pictures, one can see that every permutation of the group can be constructed as a combination of small permutations that only change a number and its neighbor. We name these small permutations as $s_1, s_2, \ldots, s_{n-1}$. For S_6, these small permutations graphically look like this:

$s_1 =$, $s_2 =$, $\cdots$, $s_5 =$. (5.64)

Here, s_i permutes the number i and $i+1$. Formally, this amounts to

$$(1, 2, 3, \ldots, i, i+1, \ldots, n) \longmapsto (1, 2, 3, \ldots, i+1, i, \ldots, n). \tag{5.65}$$

The fact that Line (5.54) can be written as the composition means that it is equal to $s_2 s_4 s_1 s_5 s_4 s_3 s_2$, as can be seen here:

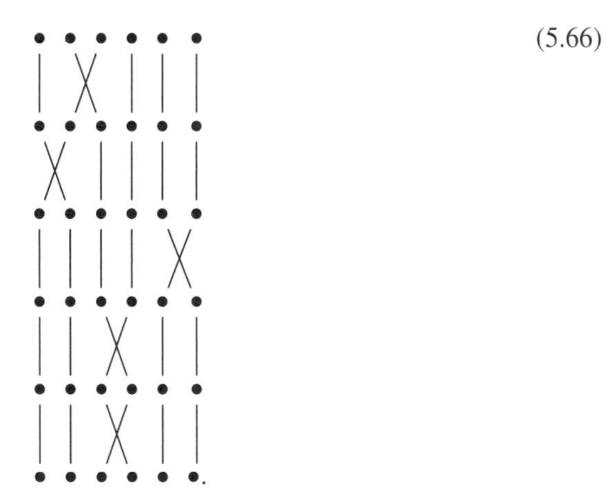

(5.66)

In a sense, the s_i are a basis for the group S_n. We call them **generator of the group**. Notice that there are other ways to express this permutation. For example, Equation (5.54) is also $s_2 s_4 s_1 s_5 s_4 s_3 s_2$.

For the group S_n, the generators $s_1, s_2, \ldots, s_{n-1}$ satisfy the following three equations:

- $s_i s_i = e$
- $s_i s_j = s_j s_i$ for $|i - j| > 1$
- $s_i s_{i+1} s_i = s_{i+1} s_i s_{i+1}$

Each of these equations can be understood graphically. The meaning of the first equation is that switching the two numbers and then switching them again is like doing nothing. (Another way of saying this is that every generator is its own inverse.) This is the content of the following diagram:

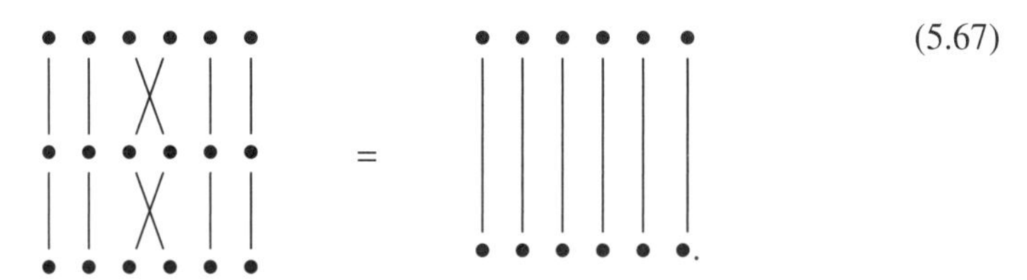

(5.67)

The meaning of the second equation says that when the numbers being switched are more than one number apart, then it does not matter in what order it is done. This corresponds to

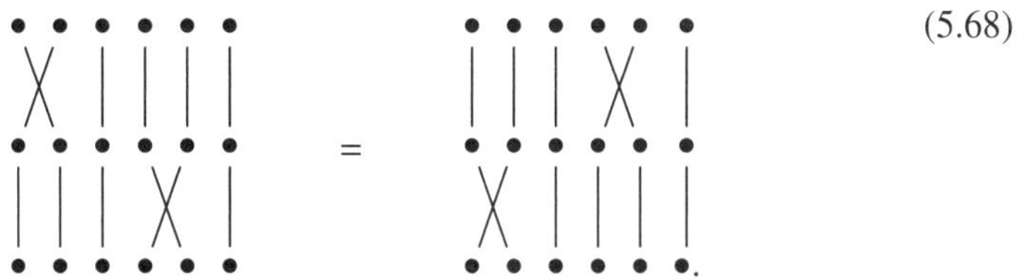

(5.68)

And finally, the third equation says that switching close numbers has the following property:

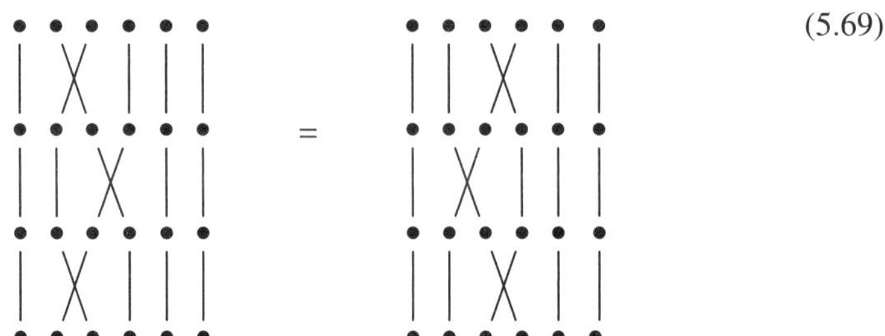

(5.69)

Both diagrams describe the permutation $(1,2,3,4,5,6) \longmapsto (1,4,3,2,5,6)$.

Now that we have these symmetry groups, let us gather them to form the category $\mathbb{S}\mathrm{ym}$. The objects are the natural numbers and the morphisms are given as follows:

$$Hom_{\mathbb{S}\mathrm{ym}}(m,n) = \begin{cases} S_m & : \text{if } m = n \\ \emptyset & : \text{if } m \neq n. \end{cases} \tag{5.70}$$

One can envision this category as follows:

$$\begin{array}{cccccc} S_0 & S_1 & S_2 & & S_n & \\ * & * & * & \cdots & * & \cdots \\ 0 & 1 & 2 & & n & \end{array} \tag{5.71}$$

This category has a symmetric monoidal category structure. The monoidal product $\oplus$ on objects is addition (i.e., $m \oplus n = m + n$). On the morphisms $f\colon m \longrightarrow m$ and $g\colon n \longrightarrow n$, go to the function $f \oplus g\colon m+n \longrightarrow m+n$, where this function acts on each of its parts. Formally,

$$(f \oplus g)(i) = \begin{cases} f(i) & : \text{if } i \leq m \\ g(i-m)+m & : \text{if } i > m. \end{cases} \tag{5.72}$$

We can visualize this as

$$\begin{array}{ccc} m & \oplus & n \\ {\scriptstyle f}\downarrow & & \downarrow{\scriptstyle g} \\ m & \oplus & n. \end{array} \tag{5.73}$$

Expressed in words, $\oplus$ does both permutations without interfering with each other. For example, the monoidal product of permutations in Lines (5.54) and (5.55) is

$$(1,2,3,4,5,6,7,8,9,10,11,12) \longmapsto (3,4,1,6,5,2,11,9,12,10,8,7). \tag{5.74}$$

This can be visualized by putting the diagrams side by side:

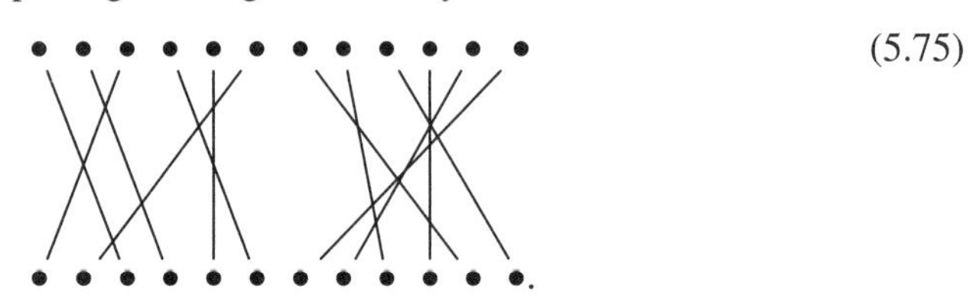

(5.75)

It is easy to see that $\oplus$ is strictly associative. The interesting part is the braiding. For every m and n, there is an element in S_{m+n}, written as $\gamma_{m,n}\colon m+n \longrightarrow n+m$, which is defined as

$$(1,2,\ldots m, m+1, m+2,\ldots m+n) \longmapsto (m+1, m+2, \ldots, m+n, 1, 2, \ldots, m). \quad (5.76)$$

Formally, this is

$$\gamma_{m,n}(p) = \begin{cases} p+m & : \text{if } p \leq m \\ p-m & : \text{if } p > m. \end{cases} \quad (5.77)$$

For example, if $m=4$ and $n=3$, then $\gamma_{4,3}$ looks like

(5.78)

We are left to show that this braiding satisfies the two triangle coherence conditions in Diagram (5.42). Given three integers m, n, and q, if we think of the m strands as one cable, the n strands as another cable, and the q strands as a third cable, then the two diagrams amount to

(5.79)

We will see in Section 6.2 how $(\mathbb{S}\mathrm{ym}, \oplus, \emptyset)$ is the paradigm symmetric monidal category. □

Exercise 5.3.11. Show that category $\mathbb{S}\mathrm{ym}$ is isomorphic to category $Nat\mathbb{I}\mathrm{so}$, which we met in Exercise 4.1.54. ■

Let us continue with our examples of monoidal categories.

Example 5.3.12. There is a monoidal category structure on $\mathbf{K}\mathbb{FDV}\mathrm{ect}$ where the monoidal product is the tensor product of vector spaces that corresponds to the Kronecker product of matrices. This operation on vector spaces will be very important for quantum theory and quantum computing. The tensor product operation takes two vector spaces, V and W, and forms $V \otimes W$. If $\mathcal{B} = \{b_1, b_2, \ldots, b_m\}$ is a basis for V and $\mathcal{B}' = \{b'_1, b'_2, \ldots, b'_n\}$ is a basis for W, then the basis for $V \otimes W$ will consist of vectors of the form

$$\{b \otimes b' : b \in \mathcal{B}, b' \in \mathcal{B}'\}. \quad (5.80)$$

A typical element of $V \otimes W$ will be a finite linear combination of these elements:

$$c_{1,1}(b_1 \otimes b'_1) + c_{1,2}(b_1 \otimes b'_2) + c_{1,3}(b_1 \otimes b'_3) + \cdots + c_{m,n}(b_m \otimes b'_n) = \sum_{i=1}^{m} \sum_{j=1}^{n} c_{i,j}(b_i \otimes b'_j). \tag{5.81}$$

The elements must satisfy a **bilinearity axiom**. This says that the tensor product respects the addition in the two vector spaces. This amounts to

$$(b + b') \otimes b'' = (b \otimes b'') + (b' \otimes b''), \tag{5.82}$$

and similarly,

$$b \otimes (b' + b'') = (b \otimes b') + (b \otimes b''). \tag{5.83}$$

The bifunctor $\otimes$ on the category $\mathbf{K}\mathbb{FDV}\mathbb{ect}$ is also defined on liner maps. If $T: V \longrightarrow W$ and $T': V' \longrightarrow W'$, then there is a linear map $T \otimes T': V \otimes V' \longrightarrow W \otimes W'$. This map is defined as

$$(T \otimes T')(v \otimes w) = T(v) \otimes T'(w). \tag{5.84}$$

The unit of the monoidal category structure is the one-dimensional vector space $\mathbf{K}$. The basis for this vector space is the set $\{1\}$. The basis for $V \otimes \mathbf{K}$ is $\{b_1 \otimes 1, b_2 \otimes 1, \ldots, b_m \otimes 1\}$.

The braiding is important. While $V \otimes W$ is not equal to $W \otimes V$, there is an isomorphism between the two. Basically, the isomorphism is induced by the map that takes the basis element $b \otimes b'$ to $b' \otimes b$. This braiding satisfies all the conditions of a symmetric monoidal category. □

In Theorem 5.2.11, we saw how the dimension of the direct product of vector spaces work. Theorem 5.3.13 is about the dimension of the tensor product of vector spaces.

Theorem 5.3.13. For finite dimensional vector spaces V and W, we have $dim(V \otimes W) = dim(V) \cdot dim(W)$. ★

Proof. A basis for the tensor product is the Cartesian product of the original basis; that is,

$$|\{b \otimes b' : b \in \mathcal{B}, b' \in \mathcal{B}'\}| = |\mathcal{B}| \cdot |\mathcal{B}'|. \tag{5.85}$$

♣

If the linear maps are given as matrices, then it is very easy to form their tensor product. In detail, if $T = T_A$ for some matrix A and $T' = T_B$ for some matrix B, then $T \otimes T' = T_{A \otimes B}$, where $A \otimes B$ is the Kronecker product of matrices.

Example 5.3.14. $(\mathbb{R}\mathbb{el}, \otimes, \{*\})$. The category $\mathbb{R}\mathbb{el}$ of sets and the relations between them form a symmetric monoidal category. The value of $\otimes$ on sets S and T is $S \times T$. For relation $Q: S \not\longrightarrow S'$ and $R: T \not\longrightarrow T'$, we form

$$Q \otimes R: S \times T \not\longrightarrow S' \times T', \tag{5.86}$$

which is defined as

$$((s,t),(s',t')) \in Q \otimes R \qquad \text{if and only if} \qquad (s,s') \in Q \text{ and } (t,t') \in R. \tag{5.87}$$

This is also written as

$$(s,t) \sim (s',t') \qquad \text{if and only if} \qquad s \sim s' \text{ and } t \sim t'. \tag{5.88}$$

Given three sets S, T, and U, the reassociator $\alpha_{S,T,U}\colon S \otimes (T \otimes U) \not\longrightarrow (S \otimes T) \otimes U$ is defined as

$$((s,(t,u)),((s,t),u)) \in (S \otimes (T \otimes U) \times (S \otimes T) \otimes U) \tag{5.89}$$

or

$$(s,(t,u)) \sim ((s,t),u). \tag{5.90}$$

This reassociator clearly satisfies the pentagon condition.

The unit object is the one-element set $\{*\}$ and the right unitor on the set S is defined as $(s,*) \sim s$. There is a similar relation for the left unitor.

The braiding for sets S and T is $\gamma_{S,T}\colon S \otimes T \longrightarrow T \otimes S$, which is defined as $(s,t) \sim (t,s)$. This braiding satisfies the symmetry condition and the hexagons.

It is worth noting that the monoidal product $\otimes$ is not the categorical product on $\mathbb{R}\mathrm{el}$. If it were a product, there would be projection maps and the following commutative diagram:

$$\begin{array}{ccccc} S & \xleftarrow{\pi} & S \otimes T & \xrightarrow{\pi} & T \\ \downarrow{\scriptstyle Q} & & \downarrow{\scriptstyle Q\otimes R} & & \downarrow{\scriptstyle R} \\ S' & \xleftarrow{\pi} & S' \otimes T' & \xrightarrow{\pi} & T'. \end{array} \tag{5.91}$$

The best guess for a projection map $\pi\colon S \otimes T \longrightarrow S$ is $(s,t) \sim s$ for all $s \in S$ and $t \in T$. It is similar for the other projections. Saying that the left side commutes means that if (i) $(s,t) \sim s$ and (ii)$(s,t) \sim (s',t')$, and (iii) $(s',t') \sim s'$, then $s \sim s'$. This simply need not be true for relations Q and R. □

Endomorphisms of the unit of a monoidal category will be important. First a theorem.

Theorem 5.3.15. For I, the unit of a monoidal category, we have $\rho_I = \lambda_I\colon I \otimes I \longrightarrow I$. ★

Proof. See Proposition 1.1 of [131], Lemma XI.2.3 of [134], or corollary 2.2.5 of [71]. ♣

Theorem 5.3.16. The endomorphisms of the unit of a monoidal category form a commutative monoid under composition. This means that for any pair $f\colon I \longrightarrow I$ and $g\colon I \longrightarrow I$, we have that $f \circ g = g \circ f\colon I \longrightarrow I$. Furthermore, we have that $f \otimes g = g \otimes f\colon I \otimes I \longrightarrow I \otimes I$. ★

Proof. We already know (from Example 2.1.12 in Chapter 2) that the endomorphisms of any object in any category form a monoid via composition. In particular, for the unit I of a monoidal category, $Hom(I,I) = End(I)$ is a monoid. Within a strict monoidal category, this monoid is commutative because of the Eckmann-Hilton argument given in Theorem 5.1.16. In detail, besides the monoid operation, which is composition, there is also the monoid operation from the strict monoidal operation $\otimes$. These two operations are equal and commutative.

However, for a general monoidal category, the argument is a tad more complicated. Start with two morphisms, $f\colon I \longrightarrow I$ and $g\colon I \longrightarrow I$. Notice that $f \otimes g$ and $g \otimes f$ are morphisms from $I \otimes I$ to $I \otimes I$, while $f \circ g$ and $g \circ f$ are morphisms from I to I. The map $\rho_I = \lambda_I$ is an isomorphism from $I \otimes I$ to I that connects them. The following diagram is helpful in understanding the proof:

$$f \otimes g \quad = \quad \begin{array}{c} I \xrightarrow{\quad\quad f \quad\quad} I \\ \otimes \xrightarrow{\rho_I} I \xrightarrow{f} I \xrightarrow{g} I \xrightarrow{\rho_I^{-1}} \otimes \\ I \xrightarrow[\quad\quad g \quad\quad]{} I \end{array} \quad = \quad \begin{array}{c} I \xrightarrow{\quad\quad g \quad\quad} I \\ \otimes \xrightarrow{\rho_I} I \xrightarrow{g} I \xrightarrow{f} I \xrightarrow{\rho_I^{-1}} \otimes \\ I \xrightarrow[\quad\quad f \quad\quad]{} I \end{array} \quad = \quad g \otimes f. \tag{5.92}$$

Use the bifunctoriality of $\otimes$ Equation (4.21) to get

$$\begin{aligned}(id_I \otimes g) \circ (f \otimes id_I) = (id_I \circ f) \otimes (g \circ id_I) = f \otimes g = (f \circ id_I) \otimes (id_I \circ g) \\ = (f \otimes id_I) \circ (id_I \otimes g)\end{aligned} \tag{5.93}$$

and

$$\begin{aligned}(id_I \otimes f) \circ (g \otimes id_I) = (id_I \circ g) \otimes (f \circ id_I) = g \otimes f = (g \circ id_I) \otimes (id_I \circ f) \\ = (g \otimes id_I) \circ (id_I \otimes f).\end{aligned} \tag{5.94}$$

The naturality of ρ and λ gives us

$$f \otimes id_I = \rho_I^{-1} \circ f \circ \rho_I \qquad \text{and} \qquad id_I \otimes g = \lambda_I^{-1} \circ g \circ \lambda_I. \tag{5.95}$$

From Theorem 5.3.15, we see that the right side of the previous line can be written as

$$id_I \otimes g = \rho_I^{-1} \circ g \circ \rho_I. \tag{5.96}$$

In this case, we have the following identities:

$$f \otimes g = (f \otimes id_I) \circ (id_I \otimes g) = \rho_I^{-1} \circ f \circ \rho_I \circ \rho_I^{-1} \circ g \circ \rho_I = \rho_I^{-1} \circ f \circ g \circ \rho_I. \tag{5.97}$$

$$g \otimes f = (id_I \otimes f) \circ (g \otimes id_I) = \rho_I^{-1} \circ f \circ \rho_I \circ \rho_I^{-1} \circ g \circ \rho_I = \rho_I^{-1} \circ f \circ g \circ \rho_I. \tag{5.98}$$

$$f \otimes g = (id_I \otimes g) \circ (f \otimes id_I) = \rho_I^{-1} \circ g \circ \rho_I \circ \rho_I^{-1} \circ f \circ \rho_I = \rho_I^{-1} \circ g \circ f \circ \rho_I. \tag{5.99}$$

$$g \otimes f = (g \otimes id_I) \circ (id_I \otimes f) = \rho_I^{-1} \circ g \circ \rho_I \circ \rho_I^{-1} \circ f \circ \rho_I = \rho_I^{-1} \circ g \circ f \circ \rho_I. \tag{5.100}$$

From the first two identities, we see that $f \otimes g = g \otimes f$. Combining this with the third identity gives us

$$f \otimes g = \rho_I^{-1} \circ (f \circ g) \circ \rho_I = \rho_I^{-1} \circ (g \circ f) \circ \rho_I = g \otimes f. \tag{5.101}$$

Since ρ is an isomorphism, we also get $f \circ g = g \circ f$.

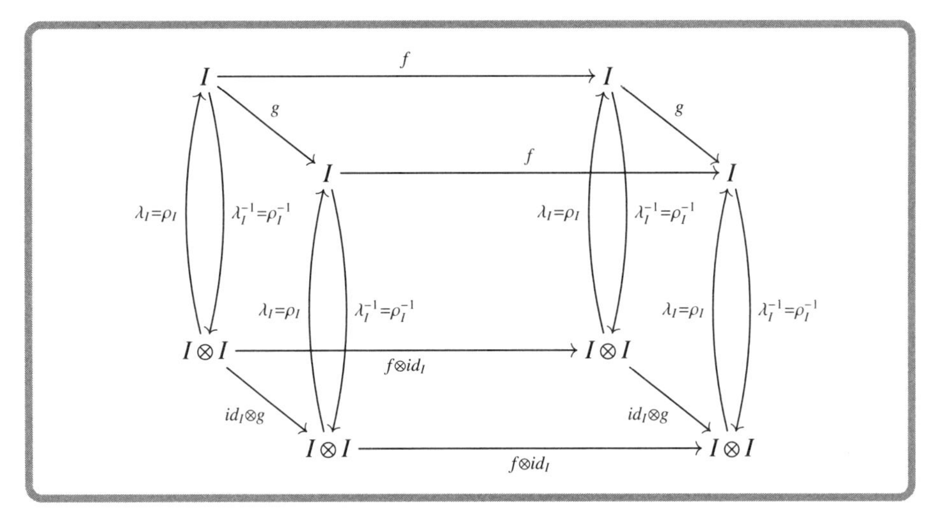

Figure 5.4. A cuboid proof of the commutativity of f and g.

There is a nifty way to see this proof that is worth meditation on.[1]Consider the three-dimensional cuboid in Figure 5.4. The bottom of the cuboid commutes because of the interchange law. The sides commute because of the naturality of $\lambda_I = \rho_I$. The vertical edges commute because the maps are isomorphisms. Therefore, the top rectangle commutes, which gives us the commutativity of f and g. ♣

Sometimes the endomorphisms of the unit have a lot more structure than just a commutative monoid.

Remark 5.3.17. It pays to remember what we saw in Remark 2.4.15. In the monoidal category of complex vector spaces $(\mathbb{CFDV}\mathtt{ect}, \otimes, \mathbf{C})$, the set of linear maps from $\mathbf{C}$ to $\mathbf{C}$ is—not only a commutative monoid but it is also—the field of complex numbers. By linearity, every map $f\colon \mathbf{C} \longrightarrow \mathbf{C}$ is totally determined by the value $f(1)$ because $f(c) = f(c\cdot 1) = c\cdot f(1)$. ♠

Example 5.3.18. $(\mathbb{C}\mathtt{ircuit}, \oplus, \emptyset)$. The category of logical circuits has a monoidal category structure. The tensor product is the disjoint union. That is, given two circuits, C with m input wires and n output wires, and C' with m' input wires and n' output wires, we can compose them to form $C \oplus C'$, as in Figure 5.5. which has $m+m'$ input wires and $n+n'$ output wires. The unit is the empty circuit with no input wires and no output wires. We will see in the coming pages that this category is *not* a symmetric monoidal category. □

When dealing with such a monoidal category, one must be concerned with a lot of baggage, like the reassociator, the units, and the coherence conditions. This is in sharp

[1] I found a slight variation of this proof online in a presentation "Categories and Quantum Informatics" by Chris Heunen.

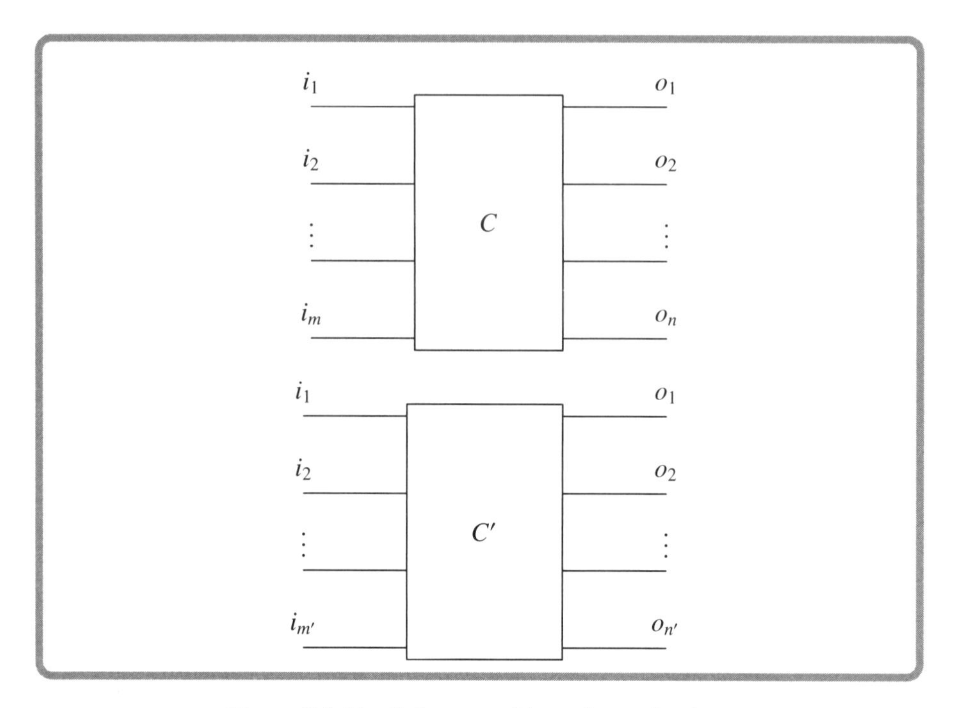

Figure 5.5. Parallel composition of two circuits.

contrast to a strict monoidal category, which does not have a lot of such baggage. We will see in the next chapter that monoidal categories have a special relationship with strict monoidal categories. This relationship will help us easily deal with all of these issues.

5.4 Coherence Theory

Let us spend a few minutes meditating on coherence conditions. What are they all about? To get an intuition about operations and their axioms, we return to basic arithmetic. Consider the real numbers and an operation $\otimes$, which satisfies the associativity axiom: $a \otimes (b \otimes c) = (a \otimes b) \otimes c$. The $+$ and $\times$ operations satisfy this requirement. If this single axiom is satisfied, we are assured that no matter how many numbers and no matter how they are associated (bracketed), there will be a unique final value of the expression. To reiterate, this means that for any numbers a, b, c, d, e, f, and g, and any two associations, their values will be equal. For example,

$$(((a \otimes b) \otimes (c \otimes d)) \otimes e) \otimes (f \otimes g) = (a \otimes (b \otimes c)) \otimes ((d \otimes (e \otimes f)) \otimes g). \tag{5.102}$$

In such a case, we might as well omit the parentheses and write $a \otimes b \otimes c \otimes d \otimes e \otimes f \otimes g$. In contrast, if the associativity axiom is not satisfied, as with the $-$ or $\div$ operation, then different associations can give different values.

Another example in basic arithmetic is commutativity. We ask if an operation $\otimes$ satisfies the commutativity axiom $a \otimes b = b \otimes a$. If the axiom is satisfied, we are assured that no matter how many numbers and no matter how they are ordered, there will be a unique final value. To reiterate, this means that for any numbers and for any two ordering of the numbers, their values are equal. For example,

$$c \otimes f \otimes a \otimes b \otimes g \otimes d \otimes e = g \otimes a \otimes b \otimes f \otimes c \otimes d \otimes e. \tag{5.103}$$

This implies that the order of the elements is irrelevant. In contrast, with operations that do not satisfy the commutativity axiom, such as $-$ and $\div$, different orders of the numbers give different values.

Now let us return to categories. If a bifunctor $\otimes$ satisfies a strict associativity axiom, $a \otimes (b \otimes c) = (a \otimes b) \otimes c$, or a strict commutativity axiom, $a \otimes b = b \otimes a$, then these categories will satisfy the same properties of basic arithmetic. However, the universe is not always so pretty, and many categories and bifunctors do not have strict associativity or strict commutativity.

What can we say about monoidal categories when they do not satisfy strict axioms? In a monoidal category, there are two functors, $(\ )\otimes((\ )\otimes(\ ))$ and $((\ )\otimes(\ ))\otimes(\ )$, with a natural isomorphism between them. In other words, there is the isomorphism $\alpha\colon a \otimes (b \otimes c) \longrightarrow (a \otimes b) \otimes c$. We now ask: What about functors that accept more letters and represent associations? How many natural isomorphisms are there between

$$(((a \otimes b) \otimes (c \otimes d)) \otimes e) \otimes (f \otimes g) \qquad \text{and} \qquad (a \otimes (b \otimes c))((d \otimes (e \otimes f)) \otimes g)? \tag{5.104}$$

For four letters, there are five ways of associating the letters, and the pentagon shows us that there are five isomorphisms between them. How many isomorphisms are there from $a \otimes (b \otimes (c \otimes d))$ to $((a \otimes b) \otimes c) \otimes d$? If we assume that the pentagon commutes, then there is exactly one such isomorphism. We can think of this geometrically as follows: Before we assume that the pentagon commutes, there is a ring of isomorphisms. Once we assume that the pentagon commutes, then think of the ring as a filled-in disk. In that case, there is a unique path (map) from any functor to any other.

What if one has more than four letters? To get a handle on this, let us look at the case where there are five letters. For five letters, there are fourteen ways of associating the letters. They are partially depicted in Figure 5.6. The shape forms a sphere. To see this shape more clearly, let us cut open the sphere and spread it out as in Figure 5.7. Notice that the two long curved arrows from the lower-left corner to the upper-right corner are the same map. This shape is made of 14 vertices and 21 arrows, which form three squares and six pentagons. The squares are all naturality squares that commute because α is a natural transformation. The pentagons are all instances of the Mac Lane pentagon condition. Here is the main point: if one assumes that α is a natural transformation, and that the pentagon commutes, then the entire diagram commutes. Rather than there being many morphisms from one vertex to another vertex, the entire diagram commutes and there is exactly one isomorphism made of α's between any two vertices. This is similar to what we saw with elementary arithmetic: if a single axiom is satisfied, then there is exactly one value.

What about six or more letters? It turns out that for any amount of letters, the figure is made of pentagons and squares. If you are working in a situation where there is coherence, then the pentagons commute. The squares commute from naturality. This means that with coherence and naturality, the whole shape commutes.

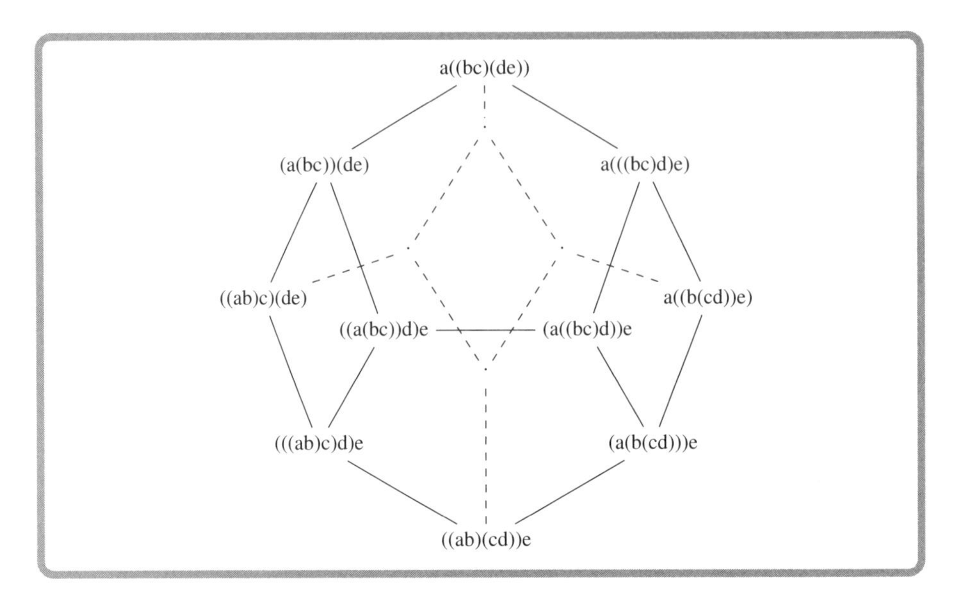

Figure 5.6. Some of the associations and reassociations for five letters.

In category theory, there is always a goal of having a unique isomorphism. We just proved that there is a unique isomorphism of a certain type in a monoidal category. Whenever we define a categorical structure, like an initial object or a product, we always prove that the object is "unique up to a unique isomorphism." What is this obsession with a *unique* isomorphism? Why can't there be more than one isomorphism between two objects in the category? Why do we make coherence conditions to ensure that there is a unique morphism between two vertices? As we saw in Important Categorical Idea 2.2.24, there is a hierarchy. When two structures are isomorphic, it shows that they have the same structure. The isomorphisms are ways of showing that they have the same structure. However, when there is a unique isomorphism, then there is a unique way of reordering so the two structures are the same.

A simple example is in order. Consider the three graphs

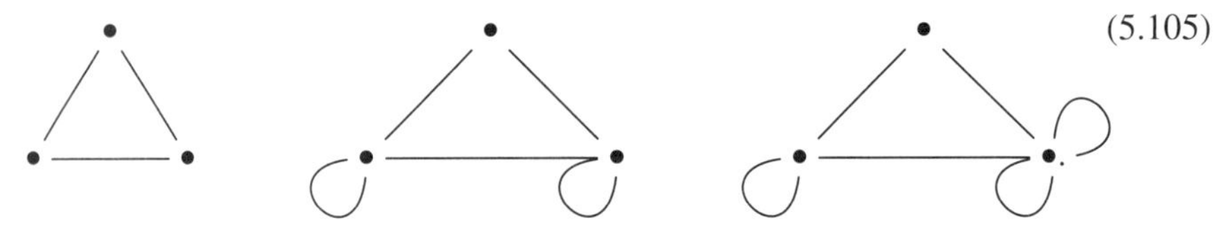
(5.105)

The graph on the left has three vertices and $3! = 6$ isomorphisms from it to itself. An isomorphism of the middle graph to itself must map the two bottom vertices to themselves or each other. There are only two maps that do this. In contrast, the graph on the right has exactly one isomorphism from it to itself (the trivial identity). This makes the definition of the graph on the right more unique since one cannot reshuffle the objects in a nontrivial way.

Mac Lane's pentagon condition is the beginning of a branch of category theory called **coherence theory**. In arithmetic and algebra, we study how different operations

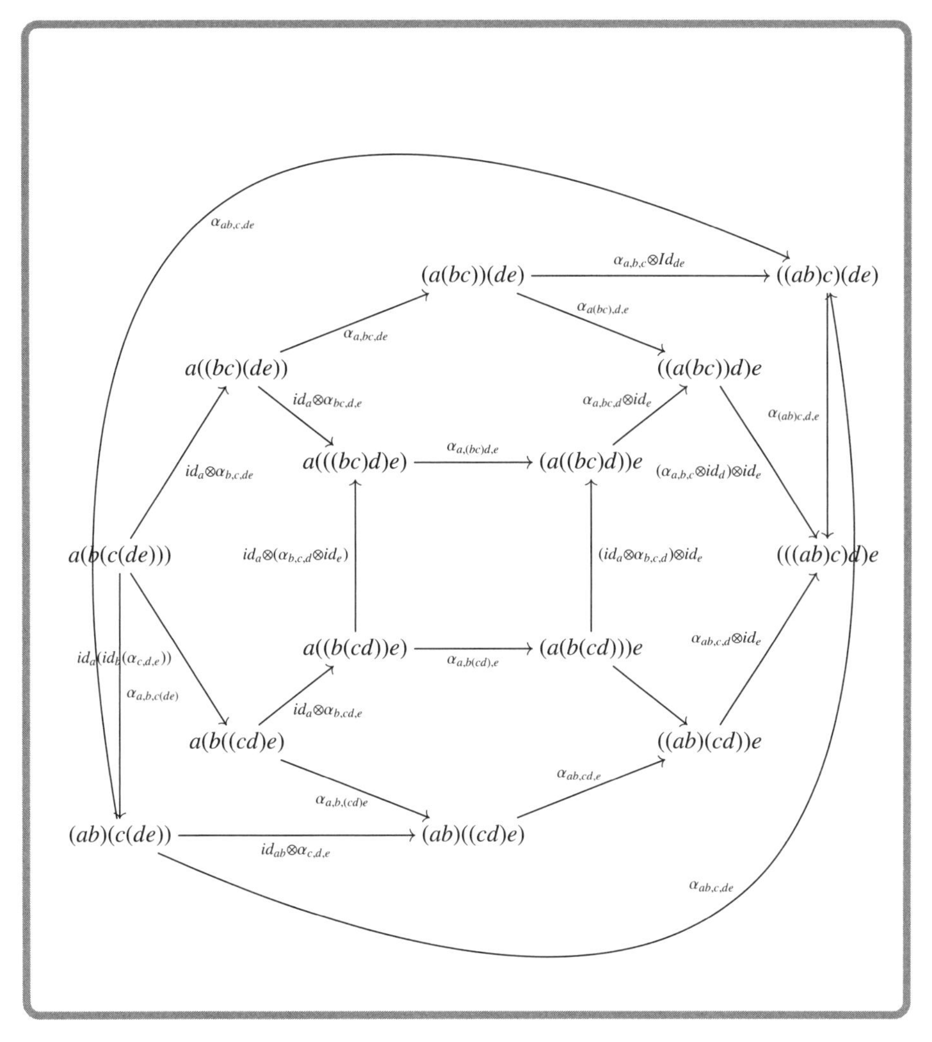

Figure 5.7. All the associations and reassociations for five letters, made clearer than in Figure 5.6.

relate to each other. For example, the rule $x \cdot (y+z) = (x \cdot y) + (x \cdot z)$ says that the multiplication operation distributes over the addition operation. In category theory, we go one level higher and study how different operations, which are represented by functors, relate to each other with natural transformations. With arithmetic and algebra, we are interested in determining if there is a single unique value. With coherence theory, we are concerned with determining the relationship between instances of the functors. Here, we study how these natural transformations "cohere" with themselves and with each other. In Mac Lane's pentagon condition, we see what happens when the reassociator coheres with itself. After we introduce functors between monoidal categories, in Section 6.2, we will describe certain coherence theorems that explain properties of monoidal categories, and the relationship of general monoidal categories to strict monoidal categories. Since coherence theory deals with the relationship of morphisms between operations, as opposed to relationships between operations, it is sometimes called **higher-dimensional algebra**. Later, when discussing higher category theory (Section 9.4), we will meet even higher coherence conditions.

To better understand coherence theory, we examine certain shapes or finite categories with the operations and the morphisms between them. The objects of the category correspond to the associations and the morphisms correspond to the reassociations. There is a sequence of such categories denoted as $A_1, A_2, \ldots, A_n, \ldots$ where each one is called an **associahedron** (which is similar to the word "polyhedron" but related to the word "association") and together, they are called **associahedra**. In detail, the objects correspond to associations written as functors and the morphisms correspond to reassociations written as natural isomorphisms. Since they are isomorphisms, the categories are in fact groupoids. Here are the associahedra:

- A_1: For one letter, there is exactly one object: a.
- A_2: For two letters there is exactly one object: $a \otimes b$.
- A_3: For three letters, there are two ways of associating them and an isomorphism between them:
$$a \otimes (b \otimes c) \longrightarrow (a \otimes b) \otimes c. \tag{5.106}$$
- A_4: For four letters there are five ways of associating them and five instances of isomorphism. This is depicted in the Mac Lane coherence condition shown in Diagram (5.34). Notice that the pentagon is the same shape as a two-dimensional circle.
- A_5: For five letters, we have Figures 5.6 and 5.7. Notice that this is a three-dimensional sphere.
- A_n: What about for six or more letters? The shapes get too complicated to draw, but mathematicians know a lot about them. For n letters, there are $n-1$ monoidal products between the letters and
$$C_{n-1} = \frac{\binom{2(n-1)}{n-1}}{n} = \frac{(2(n-1))!}{(n)!(n-1)!} \tag{5.107}$$
ways of associating or bracketing the letters. These numbers are called the **Catalan numbers**. The first few Catalan numbers are 1, 1, 2, 5, 14, 42, 132, 429, 1430, 4862, 16796, ... The fact that the number of bracketings is equal to the Catalan numbers is left for Theorem 5.4.1 at the end of this section. For n letters, the shape of the whole category is an $n-2$–dimensional sphere. The main point for all the associahedra is that they are made of commuting naturality squares and Mac Lane pentagons. If the pentagon commutes, then the whole diagram commutes and there is exactly one natural transformation between any two objects in the category.

It is important to notice that these associahedra are not subcategories of every monoidal category. In fact, in an arbitrary monoidal category, two different associations of objects might be the same object. The extreme instance where this happens is within a strict monoidal category where all the associations collapse to a single object. The associahedra are the extreme opposite. They are the descriptions of the associations that are all different.

What about commutativity? What shapes are formed if one looks at letters that can be permuted? Let us assume for a moment that we are dealing with strictly associative monoidal structure so we do not have to worry about parentheses. We will form categories (again groupoids) $P_1, P_3, P_3, \ldots$ each called a **permutohedron**; together, they are called the **permutahedra**, which describe the permutations of the monoidal product:

- P_1: For one letter, there is only one way of combining it: a.
- P_2: For two letters, there are two ways of combining them with a braiding between them: $ab \longrightarrow ba$.
- P_3: For three letters, there are $3! = 6$ ways of combining them. Some of the morphisms in P_3 look like this:

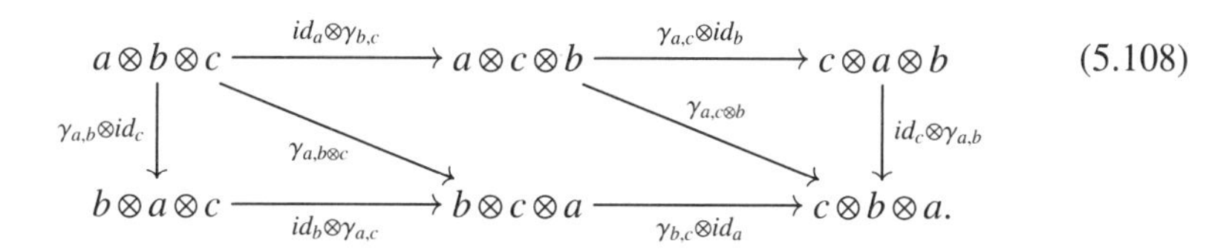

  The two triangles commute because they are instances of the left triangle in Diagram (5.42), which we assume commutes, and the middle quadrilateral commutes out of naturality. There are other instances of γ, but we are omitting them for readability.[2]
- P_n: For $n > 3$ letters, the categories get too complicated to draw. The shapes consist of $n!$ vertices. These shapes are made out of naturality squares and hexagons.

There are categories that take into account associativity and commutativity. We denote the shapes as $AP_1, AP_2, AP_3, \ldots,$ where each AP_n is called a **permuto-associahedron** and the collection is called the **permuto-associahedra**.

- AP_1: For one letter, there is only one vertex: a.
- AP_2: For two letters, there are two ways of ordering them with a braiding between them: $ab \longrightarrow ba$.

[2] A warning is in order. Some mistakenly say that Diagram (5.108) commutes because of the naturality of γ and the commutativity of *both* triangles in Diagram (5.42). This is false. The two triangles in Diagram (5.108) are instances of the left triangle in Diagram (5.42). There is another diagram:

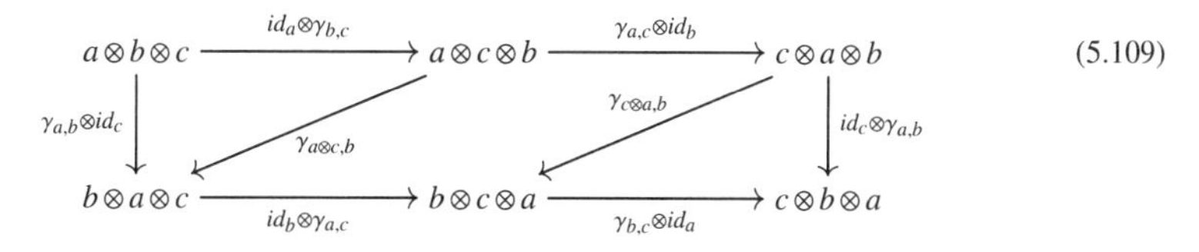

which commutes because of the naturality of γ and the commutativity of the right triangle in Diagram (5.42).

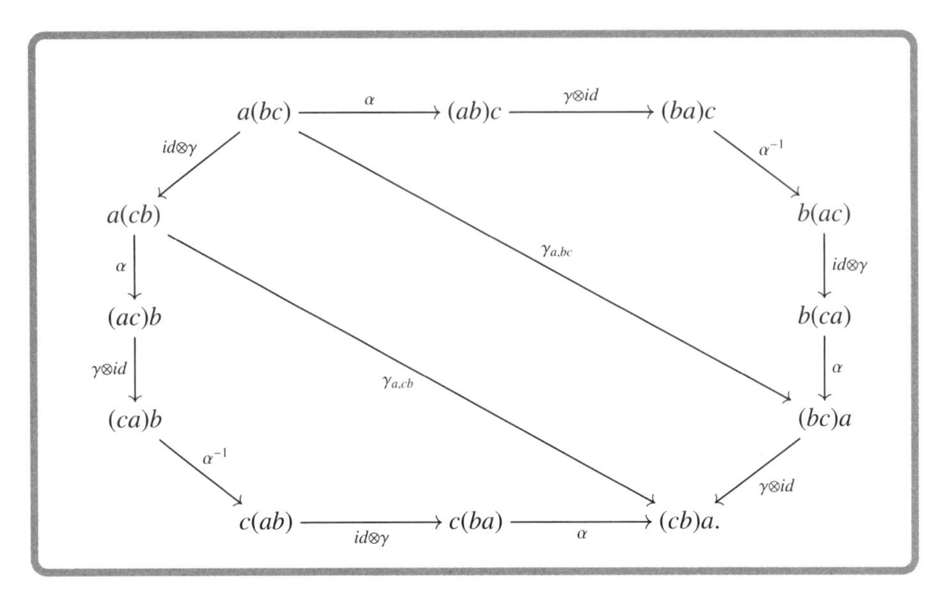

Figure 5.8. The permuto-associahedron for three letters.

- AP_3: For three letters, there are six permutations and for each permutation, there are two ways of associating. This gives twelve objects and is partially depicted in Figure 5.8. (Again, we omit instances of γ.) This diagram commutes because the middle rectangle is a naturality square. The other two hexagons are both instances of Diagram (5.39).
- AP_n: For $n > 3$ letters, the shapes get too big and complicated to draw. How many objects are there in AP_n? There are $n!$ ways of ordering them, and for each ordering, there are C_{n-1} bracketings. This gives us $\frac{(2n-2)!}{(n-1)!}$ objects. The first few are 1, 2, 12, 120, 1,680, 30,240, 665,280, and 17,297,280.

We will see in Section 8.2 that each of the collections of associahedra, permutahedra, and permuto-associahedra forms a categorical structure called an **operad**. In Section 6.2 we will encounter these ideas again and see more ramifications of the coherence conditions.

We close this section by relating the number of bracketings and the Catalan numbers.

Theorem 5.4.1. The number of legal bracketings of n letters is the $n-1$ Catalan number. ★

Proof. Figure 5.9 is helpful for understanding the proof. For n letters, there are $n-1$ multiplications, and for each multiplication, a left parenthesis and a right parenthesis are needed to make a complete association. When the set of left parentheses is the same size as the set of right parentheses, we call the association "balanced." In total, there are $2(n-1)$ positions for parentheses. One must choose $(n-1)$ of those to be left parentheses, while the rest are right parentheses. This explains $\binom{2(n-1)}{n-1}$ and is the size

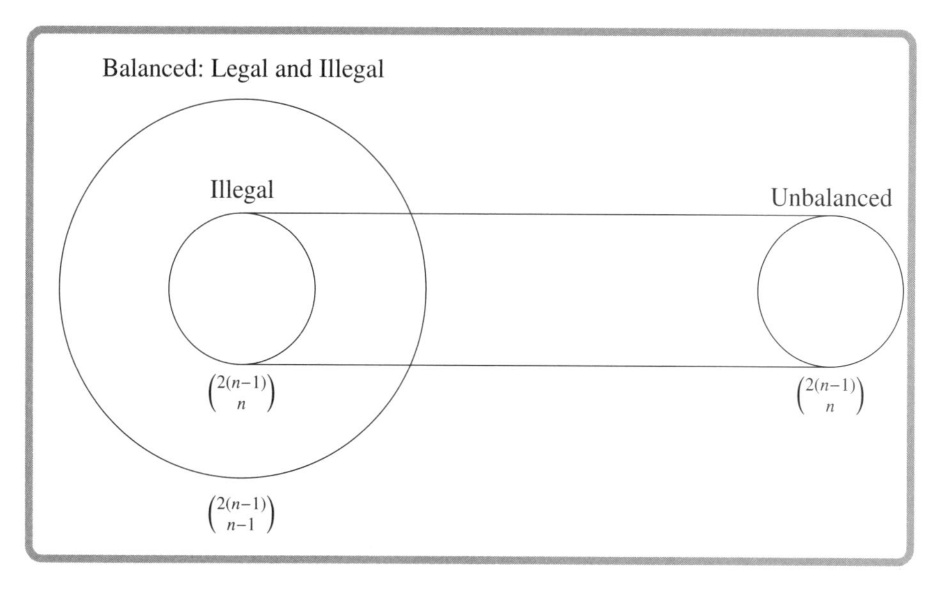

Figure 5.9. All balanced bracketings (legal and illegal) and an isomorphism with unbalanced bracketings.

of the large circle on the left of Figure 5.9. The large circle is the total number of legal and illegal balanced associations of letters. One must subtract from this the number of balanced illegal associations to get the number of legal associations:

$$\text{Total} - \text{Illegal} = \text{Legal}. \tag{5.110}$$

There is an isomorphism between the set of balanced illegal bracketings and the set of unbalenced illegal bracketings with n left parentheses and $n-2$ right parentheses. The isomorphism is as follows. Take a balanced illegal bracketing. The only reason why it is illegal is that at some point in reading the word, there will be more right parentheses than left parentheses. Call the first time this happens position t. For example, the word

$$(()))^t(())((()) $$

has 7 left and 7 right parentheses. From the left, all is well until we reach the $)^t$ in position 5. Notice that before that, there are 2 left and 2 right parentheses. Now swap all the left parentheses for right parentheses and vice versa in the first t positions. This gives us

$$))((({}^t(())((()).$$

We now have 8 left parentheses and 6 right parentheses. The inverse to this isomorphism is also simple. Given any bracketing with 8 left parentheses and 6 right parentheses, find the first position t where there are more left parentheses than right parentheses. Swapping all the parentheses through position t gives us an illegal balanced bracketing.

There are $\binom{2(n-1)}{n} = \binom{2(n-1)}{n-2}$ such unbalanced bracketings, and because of the isomorphism, there are $\binom{2(n-1)}{n}$ illegal balanced bracketings. Using Equation (5.110) to calculate the number of legal bracketings, we get the equality

$$\binom{2(n-1)}{n-1} - \binom{2(n-1)}{n} = C_{n-1}. \tag{5.111}$$

This is true because

$$\begin{aligned}
\frac{(2n-2)!}{(n-1)!(n-1)!} & \\
-\frac{(2n-2)!}{n!(n-2)!} & \\
=\frac{(2n-2)!n}{(n-1)!(n-1)!n} & \\
-\frac{(2n-2)!(n-1)}{n!(n-2)!(n-1)} & \qquad \text{to get common denominators} \\
=\frac{(2n-2)!n-(2n-2)!(n-1)}{n!(n-1)!} & \qquad \text{by combining fractions} \\
=\frac{(2n-2)!}{n!(n-1)!} & \qquad \text{by definition of factorial} \\
=C_{n-1} & \qquad \text{by definition.}
\end{aligned}$$

♣

5.5 String Diagrams

String diagrams are ways of describing the flow of morphisms within a category. We will introduce the concepts here and then build on them throughout the rest of this book. As the categorical constructions get more and more sophisticated and complicated, the string diagrams will start having all types of bells and whistles. We will meet many of these extra features in Chapter 7.

There is no consistency within the literature as to what direction the string diagrams should go:

- There are those inspired by physicists who have their diagrams go from bottom to top, similar to Feynman diagrams.
- Some authors have their diagrams go from top to bottom. This seems more natural, as we read from top to bottom. We will sometimes follow this convention when talking about braids and tangles, which we will explore in Chapter 7.
- Many researchers draw their diagrams from left to right. This saves space and is similar to the way that classical and quantum circuits are drawn.

While string diagrams are important and helpful, our presentation will not be done exclusively with string diagrams. My thesis adviser, Alex Heller, used to quip, "A clear sentence is worth a thousand pictures."

Since we began our journey, we have described an object in a category as node a and morphism from one node to another as arrow $a \xrightarrow{f} b$. String diagrams invert this convention. When discussing string diagrams, an object in a category corresponds to a line and the morphisms correspond to a box on the line that changes the line. So object a corresponds to

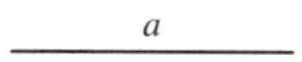

a (5.112)

and the morphism $f\colon a \longrightarrow b$ corresponds to

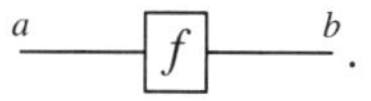

a f b. (5.113)

At times, we stress the direction or orientation of an line by making the line into an arrow:

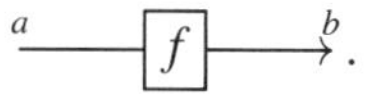

a f b. (5.114)

Composition of $f\colon a \longrightarrow b$ with $g\colon b \longrightarrow c$ is described as

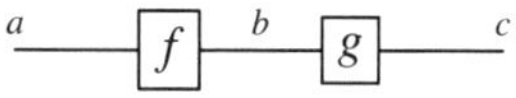

a f b g c or a f b g c. (5.115)

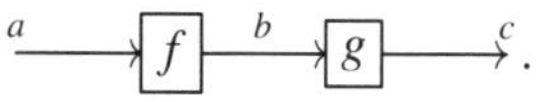

When there are arrows, one can describe a morphism going in the opposite direction; for example, $f'\colon b \longrightarrow a$ as the string diagram

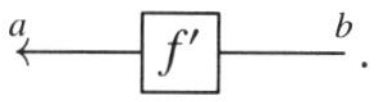

a f' b. (5.116)

For example, the dual of a linear transformation $T^*\colon W^* \longrightarrow V^*$ is described by

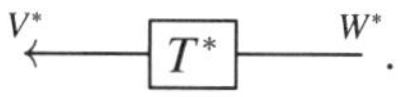

V^* T^* W^*. (5.117)

String diagrams get more interesting when dealing with monoidal categories. The morphism $f\colon a \otimes b \otimes c \longrightarrow x \otimes y$ is drawn as

(5.118)

When the domain of a morphism is the unit of a monoidal category, such as $f\colon I \longrightarrow x \otimes y \otimes z$, rather than draw it as

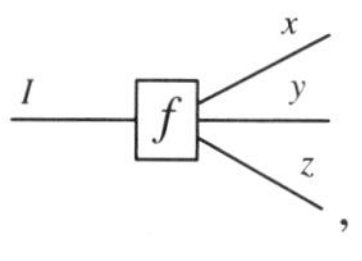

I f x y z, (5.119)

we draw it from the empty domain as

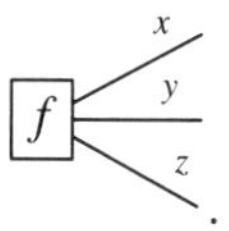

f x y z. (5.120)

Similarly, a morphism $f\colon a\otimes b \longrightarrow I$ is drawn as

(5.121)

Of course, $f\colon I \longrightarrow I$ will be written as

$$\boxed{f} \tag{5.122}$$

In the event where the name of the morphism is not important, we draw the morphisms $x\otimes y \longrightarrow I$ and $I \longrightarrow x\otimes y$ as

x y and x y. (5.123)

We are not only interested in string diagrams for the tensor product of objects. We form string diagrams for the tensor product of morphisms. The tensor product of $f\colon a \longrightarrow b$ and $g\colon c \longrightarrow d$ is

a f b. c g d (5.124)

The interchange law that we saw in Important Categorical Idea 3.1.17 on page 93 should be viewed with a string diagram. It says that four maps $f\colon a \longrightarrow a'$, $g\colon b \longrightarrow b'$, $f'\colon a' \longrightarrow a''$, and $g'\colon b' \longrightarrow b''$, expressed as

a f a′ f′ a″ b g b′ g′ b″, (5.125)

can be correctly viewed in two different ways:

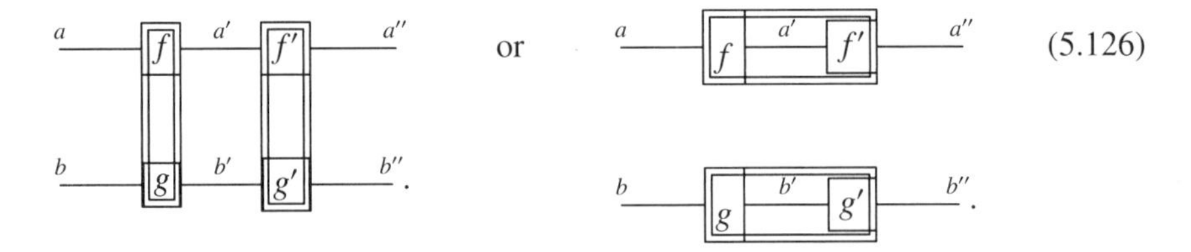

(5.126)

On the left is a sequential process of parallel processes, and on the right is a parallel process of sequential processes.

The triangle identities for an adjunction can be drawn as follows:

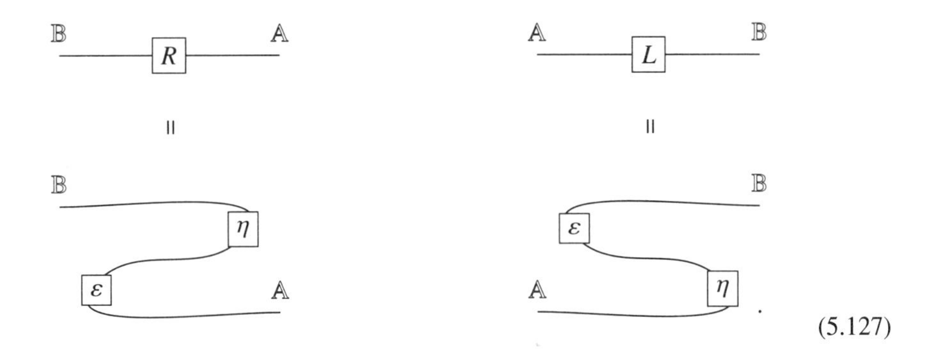

(5.127)

Theorem 5.3.16 is about that the endomorphisms of the unit of a monoidal category. The unit I and object $I \otimes I$ are depicted as an empty domain. The theorem shows that $\otimes$ and $\circ$ are equal and commutative. We can draw this as follows:

f g = f g = f g = g f = g f = g f.

(5.128)

We can use these string diagrams to visualize the coherence conditions of a monoidal category. Let us consider $a \otimes b$ as two strings that are close to each other. In contrast, $(a \otimes b) \otimes c$ has a and b as close strings and c as a further string. In this way, the reassociation α can be seen as a way of showing how strings change distances:

$\boldsymbol{a(bc)}$ $\boldsymbol{(ab)c}$

a a

b

b

c c.

(5.129)

The pentagon coherence condition can then be viewed as saying that the following two string diagrams are equal:

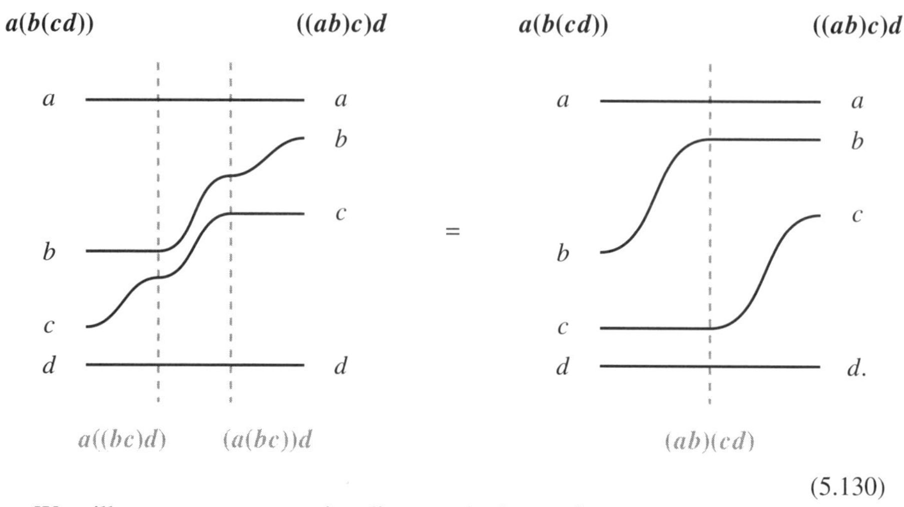

(5.130)

We will meet many more string diagrams in the coming pages.

Suggestions for Further Study

The definition of monoidal category goes back to Saunders Mac Lane [177] and Jean Bénabou [36]. The definition of symmetric monoidal category was established about the same time. For a beautiful early discussion of coherence theory, see Saunders Mac Lane [178].

Once the structure was known, it was widely generalized to other structures and applied. See, for example, [182, 145, 139, 146, 140, 144]. We will see some of the applications in the coming pages.

Variations of Theorem 5.3.16 can be found in [131], Proposition XI.2.4 of [134], and Proposition 2.2.10 of [71]. While the unit of a monoidal category seems like a fairly simple idea, there is a whole complicated web of ideas about coherence conditions for units. See, for example, [138, 229]. We presented a simple version in this chapter. For the whole story, see the introduction to [149].

For more on string diagrams, see the paper by John Baez and Aaron Lauda [28] and Peter Selinger's comprehensive paper [233].

This is just the beginning. The literature on coherence theory is huge, and we will learn some more of it as our tale unfolds.

5.6 Mini-Course: Advanced Linear Algebra

Here, we go further with our study of linear algebra. There are many advanced parts of linear algebra, but we focus on what we need. In particular, the material that we learn in this mini-course will be central for our mini-courses on basic quantum theory (Section 10.2) and quantum computing (Section 10.3), as well as for the rest of the text.

Let us summarize what we know from our mini-course on basic linear algebra (Section 2.4). The category $\mathbf{K}\mathbb{Vect}$ has $\mathbf{K}$ vector spaces as objects and linear maps as morphisms. We are mostly going to focus on the subcategory of finite-dimensional

K-vector spaces $\mathbf{K}\mathbb{FDV}\mathbb{ect}$. Earlier in this chapter, we saw that $\mathbf{K}\mathbb{FDV}\mathbb{ect}$ has two distinct monoidal category structures: the Cartesian category structure ($\mathbf{K}\mathbb{FDV}\mathbb{ect}$, $\oplus, 0$) called the "direct product," and the monoidal category structure ($\mathbf{K}\mathbb{FDV}\mathbb{ect}$, $\otimes, \mathbf{K}$) called the "tensor product." With every new notion, we will inquire how it respects these monoidal structures.

In this section, we will focus our ideas and only discuss vector spaces over the complex numbers, **C**, rather then over an arbitrary field **K**. So we will be looking at the category $\mathbf{C}\mathbb{V}\mathbb{ect}$ and $\mathbf{C}\mathbb{FDV}\mathbb{ect}$. We also look at complex vector spaces with more structure called "Hilbert spaces."

Hilbert Spaces

To arrive at the definition of a Hilbert space, we must ramp up our knowledge of complex numbers and complex matrices. Here are some operations on complex numbers and complex matrices:

Definition 5.6.1.

- If $c = a + bi$ is a complex number, then the **complex conjugate** of c is $\overline{c} = a - bi$. This defines a functor $\overline{(\ )}\colon \mathbf{C} \longrightarrow \mathbf{C}$. Notice the complex conjugate operation is idempotent: $\overline{\overline{c}} = c$ and that if c is a real number (i.e., $c = a + 0i$,) then $\overline{c} = c$, that is, the following commutes:

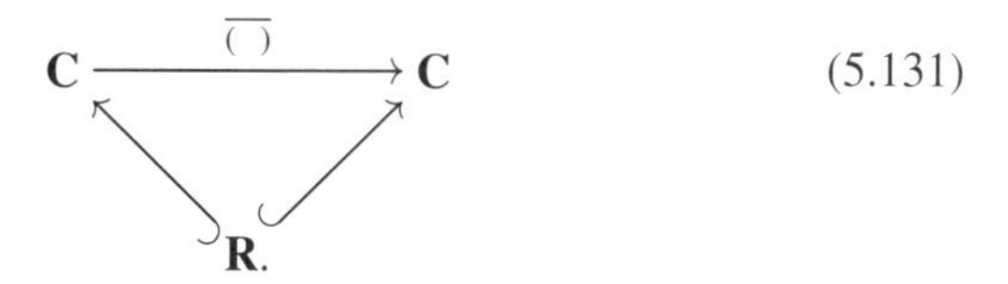

(5.131)

- If A is a matrix with complex entries, then we define the **conjugation** operation, which we denote as $\overline{A}$, to be the complex matrix whose every entry is the complex conjugate of the original. Formally,

$$\overline{A}[i, j] = \overline{A[i, j]}. \tag{5.132}$$

This is a functor $\overline{(\ )}\colon \mathbf{C}\mathbb{M}\mathbb{at} \longrightarrow \mathbf{C}\mathbb{M}\mathbb{at}$. Notice that conjugation is idempotent: $\overline{\overline{A}} = A$ and if A has only real entries, then $\overline{A} = A$, that is, the following commutes:

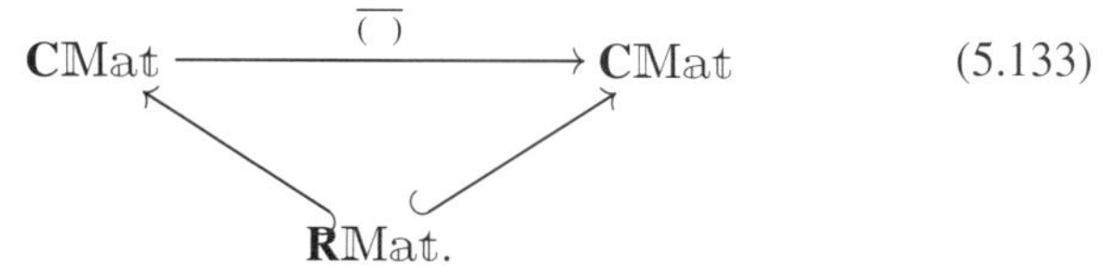

(5.133)

- If A is a matrix (with complex entries), then we define the **transpose** operation, denoted as A^T, to be the complex matrix whose entries are flipped across the main diagonal of the original matrix. Formally,

$$A^T[i,j] = A[j,i]. \tag{5.134}$$

This is a functor $(\ )^T : \mathbf{CMat} \longrightarrow \mathbf{CMat}^{op}$. Notice that the transpose operation is idempotent: $(A^T)^T = A$.

- We can combine the complex conjugation and the transpose to get the **adjoint** or **dagger** operation. If A is a matrix with complex entries, then $A^\dagger = \overline{A}^T = \overline{A^T}$. Formally,

$$A^\dagger[i,j] = \overline{A[j,i]}. \tag{5.135}$$

In terms of categories, the fact that conjugation and transpose are idempotent means that they are both "functors are isomorphisms." The $(\ )^\dagger : \mathbf{CMat} \longrightarrow \mathbf{CMat}^{op}$ functor is defined by composition as

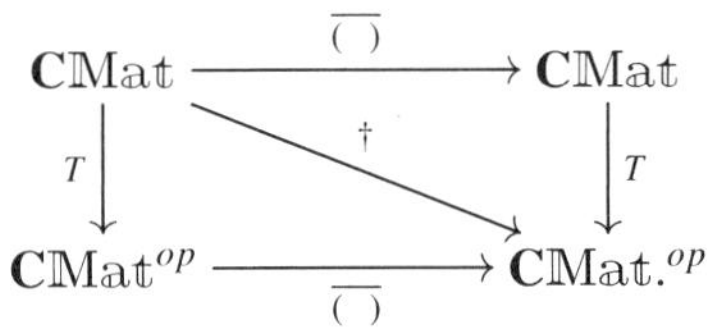

$$\begin{array}{ccc} \mathbf{CMat} & \xrightarrow{\overline{(\ )}} & \mathbf{CMat} \\ {\scriptstyle T}\downarrow & \searrow^{\dagger} & \downarrow{\scriptstyle T} \\ \mathbf{CMat}^{op} & \xrightarrow[\overline{(\ )}]{} & \mathbf{CMat}.^{op} \end{array} \tag{5.136}$$

Notice that dagger is idempotent: $(A^\dagger)^\dagger = A$ and hence an isomorphism. If A has only real entries, then $A^\dagger = A^T$. The following commutative diagram expresses the relationship of real and complex matrices with the $\dagger$ operation:

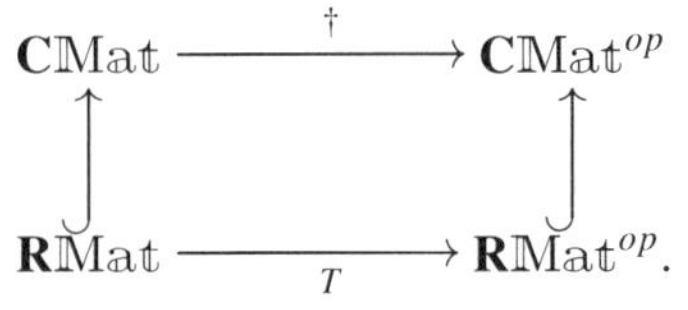

$$\begin{array}{ccc} \mathbf{CMat} & \xrightarrow{\dagger} & \mathbf{CMat}^{op} \\ \uparrow & & \uparrow \\ \mathbf{RMat} & \xrightarrow[T]{} & \mathbf{RMat}^{op}. \end{array} \tag{5.137}$$

- If A is an $n \times n$ matrix, then the **trace** of A is the sum of the diagonal elements. That is,

$$Tr(A) = \sum_{i=1}^{n} A[i,i]. \tag{5.138}$$

This is a functor $Tr\colon \mathbf{C}^{n \times n} \longrightarrow \mathbf{C}$. In particular, the trace of the identity matrix is equal to the dimension of the identity matrix, that is, $Tr(Id_n) = n$.

Exercise 5.6.2. Show that complex conjugation respects complex addition and multiplication; that is,

$$\overline{c_1 + c_2} = \overline{c_1} + \overline{c_2} \qquad \text{and} \qquad \overline{c_1 \cdot c_2} = \overline{c_1} \cdot \overline{c_2}. \tag{5.139}$$

■

Exercise 5.6.3. Prove the following properties about how conjugation, dagger, and trace respect the usual matrix operations:

Operation	Conjugation	Dagger	Trace
Matrix ad	$\overline{A+B}=\overline{A}+\overline{B}$	$(A+B)^\dagger = A^\dagger + B^\dagger$	$Tr(A+B) = TrA + TrB$
Scalar mlt	$\overline{c \cdot A} = \overline{c} \cdot \overline{A}$	$(c \cdot A)^\dagger = \overline{c} \cdot A^\dagger$	$Tr(c \cdot A) = c \cdot Tr(A)$
Matrix mlt	$\overline{A \cdot B} = \overline{A} \cdot \overline{B}$	$(A \cdot B)^\dagger = B^\dagger \cdot A^\dagger$	$Tr(A \cdot B) = Tr(B \cdot A)$
Direct sm	$\overline{A \oplus B} = \overline{A} \oplus \overline{B}$	$(A \oplus B)^\dagger = A^\dagger \oplus B^\dagger$	$Tr(A \oplus B) = TrA + TrB$
Kronecker	$\overline{A \otimes B} = \overline{A} \otimes \overline{B}$	$(A \otimes B)^\dagger = A^\dagger \otimes B^\dagger$	$Tr(A \otimes B) = TrA \cdot TrB$

■

Often, we need the ability to compare vectors in a complex vector space. That is, we want a function that accepts two vectors and outputs a complex number telling us how they relate. A Hilbert space is a complex vector space with a comparing function that satisfies certain properties.

Definition 5.6.4. An **inner product space** is a pair $(V, \langle\ ,\ \rangle)$ where

- V is a vector space.
- $\langle\ ,\ \rangle : V \times V \longrightarrow \mathbf{C}$ is a function called an **inner product** or a **Hermitian inner product**, which is used to compare vectors.

The inner product satisfies the following requirements:

- A vector measured with itself is nonnegative: for all v in V, $\langle v, v\rangle$ is a real number, $\langle v, v\rangle \geq 0$, and furthermore, $\langle v, v\rangle = 0$ if and only if $v = 0$.
- The inner product respects addition in each variable:

$$\langle v + v', w\rangle = \langle v, w\rangle + \langle v', w\rangle \qquad \text{and} \qquad \langle v, w + w'\rangle = \langle v, w\rangle + \langle v, w'\rangle. \tag{5.140}$$

- The inner product is linear with scalar multiplication in the first variable and **anti-linear** with the second variable:

$$\langle c \cdot v, w\rangle = c\langle v, w\rangle \qquad \text{and} \qquad \langle v, c \cdot w\rangle = \overline{c}\langle v, w\rangle. \tag{5.141}$$

- The inner product is not symmetric, but rather **skew-symmetric**:

$$\langle v, w\rangle = \overline{\langle w, v\rangle}. \tag{5.142}$$

Example 5.6.5. Some examples of complex inner product spaces include the following:

- $\mathbf{C}^n$: The inner product is given as $\langle v, w\rangle = v^\dagger w$.
- $\mathbf{C}^{m \times n}$: The inner product is given as $\langle A, B\rangle = Tr(A^\dagger B)$.
- $Func(\mathbf{N}, \mathbf{C})$: The inner product is given as $\langle f, g\rangle = \sum_{i=0}^{\infty} \overline{f(i)} g(i)$.
- $Func([a, b], \mathbf{C})$, where $[a, b] \subseteq \mathbf{R}$: The inner product is given as $\langle f, g\rangle = \int_a^b \overline{f(x)} g(x) dx$ (when the integral exists). □

We use the inner product to describe relationships between two vectors in a vector space. Two vectors v and w are called **orthogonal** if $\langle v, w\rangle = 0$.

The norm is a way of describing the length of a vector.

Definition 5.6.6. For a vector in a complex inner product space, the **norm** of a vector is $|v| = \sqrt{\langle v, v \rangle}$.

Example 5.6.7. For $\mathbf{C}^n$, if $v = [x_1, x_2, \ldots, x_n]^T$, then $|v| = \sqrt{x_1^2 + x_2^2 + \cdots + x_n^2}$. □

Exercise 5.6.8. Show that the norm satisfies the following properties: for all v and w in V and for $c \in \mathbf{C}$.

- Norm is nondegenerate: $|v| > 0$ if $v \neq 0$ and $|0| = 0$.
- Norm satisfies the triangle inequality: $|v + w| \leq |v| + |w|$.
- Norm respects the scalar multiplication: $|c \cdot v| = |c| \cdot |v|$. ■

With the notion of a norm, we can define special types of linear maps.

Definition 5.6.9. Let V and W be complex inner product space with norms $|\ \ |_V$ and $|\ \ |_W$, respectively. A linear map $T\colon V \longrightarrow W$ is **bounded** if T does not stretch or shrink a vector too much. In detail, for T, there is a constant $r_T > 0 \in \mathbf{R}$ that depends on T, such that for all $v \in V$, we have

$$|T(v)|_W \leq r_T |v|_v. \tag{5.143}$$

We will not be very bothered with bounded linear maps because we will mostly deal with finite dimensional vector spaces and linear maps between them are always bounded.

With the notion of inner product and a norm, we can define special types of bases of a vector space:

Definition 5.6.10. A basis $\mathcal{B} = \{b_1, b_2, b_3, \ldots\}$ is defined as follows:

- **Orthogonal** if any two different vectors in the basis are orthogonal (i.e., for any $i \neq j$, we have $\langle b_i, b_j \rangle = 0$).
- **Normal** if the norm of every vector in the basis is 1 (i.e., for all i, $\langle b_i, b_i \rangle = 1$).
- **Orthonormal** if it is both orthogonal and normal (i.e., for all i and j, $\langle b_i, b_j \rangle = \delta_{i,j}$, where $\delta_{i,j} = 1$ if $i = j$ and 0 if $i \neq j$). The function $\delta_{i,j}$ is called the **Kronecker delta**.

With a norm, we can define a distance function. The intuition is that $d(v, w)$ is the length between the end of vector v and the end of vector w.:

Definition 5.6.11. Let $(V, \langle\ \ ,\ \ \rangle)$ be a complex inner product space. We define a **distance function** as

$$d(\ \ ,\ \ ) : V \times V \longrightarrow \mathbf{R}, \tag{5.144}$$

where

$$d(v_1, v_2) = |v_1 - v_2| = \sqrt{\langle v_1 - v_2, v_1 - v_2 \rangle}. \tag{5.145}$$

Exercise 5.6.12. Show that the distance function has the following properties for all $v, w, x \in V$:

- Distance is nondegenerate: $d(v, w) > 0$ if $v \neq w$ and $d(v, v) = 0$.
- Distance satisfies the triangle inequality: $d(v, x) \leq d(v, w) + d(w, x)$.
- Distance is symmetric: $d(v, w) = d(w, v)$. ■

Let us use the inner product to describe vectors in a vector space. Let V be a finite-dimensional orthonormal basis $\mathcal{B} = \{b_1, b_2, b_3, \ldots, b_n\}$. Consider an arbitrary element $v = k_1 b_1 + k_2 b_2 + \cdots + k_n b_n$. The inner product of v with an element of the basis is

$$\langle k_1 b_1 + k_2 b_2 + \cdots + k_n b_n, b_i \rangle, \tag{5.146}$$

which by linearity reduces to

$$\langle k_1 b_1, b_i \rangle + \langle k_2 b_2, b_i \rangle + \cdots + \langle k_n b_n, b_i \rangle. \tag{5.147}$$

By orthonormality, all but one of the terms are 0, and the entire expression is $\langle k_i b_i, b_i \rangle = k_i$. This means that when we compare v to b_i, we get the scalar multiple in the b_i direction. This fact can be used to express v as

$$v = \langle v, b_1 \rangle + \langle v, b_2 \rangle + \cdots + \langle v, b_n \rangle = \Sigma_i \langle v, b_i \rangle, \tag{5.148}$$

which will be very helpful.

We need the notion of a sequence of vectors getting closer and closer together.

Definition 5.6.13. Let $(V, \langle \ , \ \rangle)$ be an inner product space. With the norm and a distance function, we can go on to define special sequences. A **Cauchy sequence** is a sequence of vectors $v_0, v_1, v_2, \ldots$, such that for every $\epsilon > 0$, there is an $N_0 \in \mathbf{N}$ with the property that

$$\text{for all } m, n \geq N_0, \quad d(v_m, v_n) \leq \epsilon. \tag{5.149}$$

What happens when we take a limit of a Cauchy sequence?

Definition 5.6.14. A complex inner product space is called **complete** if for any Cauchy sequence of vectors $v_0, v_1, v_2, \ldots$, there is a vector $\hat{v} \in V$ such that

$$lim_{n \to \infty} |v_n - \hat{v}| = 0. \tag{5.150}$$

The intuition behind this is that a vector space with an inner product is complete if any sequence of vectors that gets closer and closer will eventually converge to a point.

Definition 5.6.15. A **Hilbert space** is a complex inner product space that is complete. The category of Hilbert spaces and linear bounded maps between them is denoted as $\mathbb{Hilb}$. The central focus will be the subcategory of finite-dimensional Hilbert spaces and bounded linear maps between them, denoted as $\mathbb{FDHilb}$.

Completeness might seem like an overly complicated, calculus type of notion. Fear not, dear reader. All the inner product spaces that we will see here will be complete and hence will be Hilbert spaces. In particular, every inner product on a finite-dimensional complex vector space is automatically complete; hence every finite-dimensional complex vector space with an inner product is automatically a Hilbert space.

There is an obvious inclusion functor from $\mathbb{FDHilb}$ to $\mathbb{Hilb}$, and there are also obvious forgetful functors from categories of Hilbert spaces to categories of complex vector spaces. This can be summarized by the following commutative diagram:

$$\begin{array}{ccc} \mathbb{CVect} & \overset{U}{\twoheadleftarrow} & \mathbb{Hilb} \\ \uparrow & & \uparrow \\ \mathbb{CFDVect} & \underset{U}{\twoheadleftarrow} & \mathbb{FDHilb}. \end{array} \tag{5.151}$$

The notion of direct product and tensor product of vector spaces extends to a direct product and tensor product of Hilbert spaces. Let $(V, \langle\ ,\ \rangle_V)$ and $(W, \langle\ ,\ \rangle_W)$ be two Hilbert spaces. The direct product of the Hilbert spaces is $(V \oplus W, \langle\ ,\ \rangle_{V\oplus W})$, where the inner product is defined as

$$\langle (v, w), (v', w') \rangle_{V\oplus W} = \langle v, v' \rangle_V + \langle w, w' \rangle_W. \tag{5.152}$$

The tensor product of the Hilbert spaces is a completion of the space generated by $(V \otimes W, \langle\ ,\ \rangle_{V\otimes W})$, where the inner product is defined as

$$\langle (v \otimes w), (v' \otimes w') \rangle_{V\otimes W} = \langle v, v' \rangle_V \cdot \langle w, w' \rangle_W. \tag{5.153}$$

(For the most part, we will work with finite-dimensional Hilbert spaces and will not require this completion.) It can be shown that these functions satisfy all the requirements of being an inner product space. Furthermore, the completeness of each of the inner product operators ensures that the combined inner products are also complete. Thus, we have shown that the category of Hilbert spaces have two monoidal category structures $(\mathbb{Hilb}, \oplus, 0)$ and $(\mathbb{Hilb}, \otimes, \mathbf{C})$, which we call the "direct product" and the "tensor product," respectively. Similarly, there are monoidal category structures $(\mathbb{FDHilb}, \oplus, 0)$ and $(\mathbb{FDHilb}, \otimes, \mathbf{C})$.

Operators on Hilbert Spaces

An operator on a Hilbert space is a linear map in $\mathbb{Hilb}$ ($\mathbb{FDHilb}$) from a Hilbert space to itself. We are interested in two special types of operators within the category: Hermitian operators and unitary operators.

First, we take a brief detour and deal with matrices.

Definition 5.6.16. A real n by n matrix A is as follows:

- **Symmetric** if $A^T = A$
- **Orthogonal** if $A^T = A^{-1}$ (i.e., $A^T \cdot A = Id_n = A \cdot A^T$).

A matrix is orthogonal if all its rows are orthogonal to each other (and each row is of norm 1) and all its columns are orthogonal to each other (and each column is of norm 1). By multiplying $A \cdot A^T$, we see all the rows multiplied with each other. By multiplying $A^T \cdot A$, we see all the columns multiplied by each other. If both results are the identity, then the matrix is orthogonal (in fact, it is orthonormal). A unitary matrix is the complex version of this.

A complex n by n matrix A is as follows:

- **Hermitian** if $A^\dagger = A$
- **unitary** if $A^\dagger = A^{-1}$ (i.e., $A^\dagger \cdot A = Id_n = A \cdot A^\dagger$)

How do the matrix operations respect these sets of matrices?

Theorem 5.6.17. The set of Hermitian matrices is closed under matrix addition, direct sum, and Kronecker product. In general, the set of Hermitian matrices is not closed under scalar multiplication and matrix multiplication. ★

Proof.

- Matrix addition: If A and B are both Hermitian, then $(A+B)^\dagger = A^\dagger + B^\dagger = A + B$ is also Hermitian.
- Scalar multiplication: If A is Hermitian and c is an arbitrary complex number, then $(c \cdot A)^\dagger = \overline{c} \cdot A^\dagger = \overline{c} \cdot A \neq c \cdot A$. So it is not closed.
- Matrix multiplication: If A and B are both Hermitian, then

$$(A \cdot B)^\dagger = B^\dagger \cdot A^\dagger = B \cdot A \neq A \cdot B. \tag{5.154}$$

  So it is not closed.
- Direct sum: If A and B are both Hermitian, then $(A \oplus B)^\dagger = A^\dagger \oplus B^\dagger = A \oplus B$.
- Kronecker product: If A and B are both Hermitian, then $(A \otimes B)^\dagger = A^\dagger \otimes B^\dagger = A \otimes B$. ♣

Theorem 5.6.18. The set of unitary matrices is closed under matrix multiplication, direct sum, and Kronecker product. In general, the set of unitary matrices is not closed under matrix addition and scalar multiplication. ★

Proof.

- Matrix addition. If A and B are both unitary, then

$$(A+B)^\dagger \cdot (A+B) = (A^\dagger + B^\dagger) \cdot (A+B) = A^\dagger A + A^\dagger B + B^\dagger A + B^\dagger B$$
$$= Id + A^\dagger B + B^\dagger A + Id.$$

  This, in general, is not equal to the identity. So it is not closed under this operation.
- Scalar multiplication. If A is unitary and c is an arbitrary complex number, then $(c \cdot A)^\dagger \cdot (c \cdot A) = \overline{c} \cdot A^\dagger \cdot (c \cdot A) = \overline{c} \cdot c \cdot A^\dagger \cdot A = \overline{c} \cdot c Id \neq Id$. So it is not closed.

- Matrix multiplication. If A and B are both unitary, then

$$(A \cdot B)^{\dagger} \cdot (A \cdot B) = B^{\dagger} \cdot A^{\dagger} \cdot A \cdot B = B^{\dagger} \cdot Id \cdot B = B^{\dagger} \cdot B = Id. \quad (5.155)$$

  So it is closed.
- Direct sum: If A and B are both unitary, then

$$(A \oplus B)^{\dagger} \cdot (A \oplus B) = (A^{\dagger} \oplus B^{\dagger}) \cdot (A \oplus B) = (A^{\dagger} A) \oplus (B^{\dagger} B) = Id \oplus Id = Id. \quad (5.156)$$

  So it is closed.
- Kronecker product: If A and B are both unitary, then

$$(A \otimes B)^{\dagger} \cdot (A \otimes B) = (A^{\dagger} \otimes B^{\dagger}) \cdot (A \otimes B) = (A^{\dagger} A) \otimes (B^{\dagger} B) = Id \otimes Id = Id. \quad (5.157)$$

  So it is closed. ♣

Definition 5.6.19. Since unitary (orthogonal) matrices are closed under matrix multiplication and the identity matrix is unitary (and orthogonal), the collection of unitary (and orthogonal) matrices form a subcategory of matrices, which we denote as $\mathbb{UMat}$ ($\mathbb{OMat}$).

In contrast, Hermitian and symmetric matrices are not closed under matrix multiplication and do not form a subcategory of matrices. We symbolize the collection of Hermitian and symmetric matrices as collections of arrows $\{Hermitian\}$ and $\{symmetric\}$. Putting all this together, we have the following inclusions:

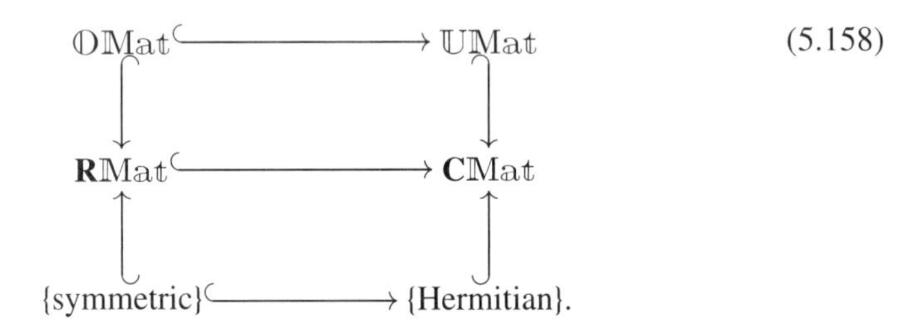

(5.158)

Since unitary matrices are closed under direct product and tensor product, they form strict monoidal category structures: $(\mathbb{UMat}, \oplus, 0)$ and $(\mathbb{UMat}, \otimes, 1)$.

Technical Point 5.6.20. There is a technical issue with the direct sum in $\mathbb{UMat}$. The category $\mathbf{K}\mathbb{Mat}$ has a lot of structure. In Example 5.1.9, we showed that has a strict monoidal structure, in Example 5.2.15, we showed that it has a Cartesian monoidal structure, and in Exercise 5.2.21, we showed that it has a co-Cartesian monoidal structure. However, in $\mathbb{UMat}$ where we have only unitary morphisms, there is no longer any Cartesian or co-Cartesian structure. The reason for this is that the projections given in Example 5.2.15 and the inclusions given in Exercise 5.2.21 are not unitary. In general, they are not even square. While the product and coproduct still exist in the strict monoidal category, they no longer satisfy universal properties. ♡

Let us move the discussion from matrices to operators.

Definition 5.6.21. Let V be a Hilbert space and $T\colon V \longrightarrow V$ be a bounded linear map. An **adjoint** or **dagger** of T is a unique function $T^\dagger\colon V \longrightarrow V$ that satisfies the following equation for all $v, w \in V$:

$$\langle T(v), w\rangle = \langle v, T^\dagger(w)\rangle. \tag{5.159}$$

(The reader should see a resemblance of this definition to the definition of adjoint functors. In fact, adjoint functors got their name because they are similar to an adjoint linear map.)

Exercise 5.6.22. Prove that the adjoint of a bounded linear map is also a bounded linear map. ■

Exercise 5.6.23. Prove that the adjoint linear map satisfies the following properties with respect to operations on linear transformations:

Operation	Dagger
Addition	$(T+T')^\dagger = T^\dagger + T'^\dagger$
Scalar multiplication	$(c\cdot T)^\dagger = \overline{c}\cdot T^\dagger$
Composition	$(T\circ T')^\dagger = T'^\dagger \circ T^\dagger$
Direct sum	$(T\oplus T')^\dagger = T^\dagger \oplus T'^\dagger$
Tensor product	$(T\otimes T')^\dagger = T^\dagger \otimes T'^\dagger$

■

The adjoint is used to define special types of operators.

Definition 5.6.24. A bounded linear operator $T\colon V \longrightarrow V$ is as follows:

- **Hermitain** or **self-adjoint** if it is its own adjoint, that is, $T^\dagger = T$.
- **Unitary** if its adjoint is its inverse, that is, $T^\dagger = T^{-1}$. That is, $T^\dagger \circ T = Id_V = T\circ T^\dagger$.

Notice that a unitary operator is, by definition, invertible.

Theorem 5.6.25. $T\colon V \longrightarrow V$ is Hermitian if and only if $\langle Tv, w\rangle = \langle v, Tw\rangle$. ★

Proof.
($\Longrightarrow$) Since $T^\dagger = T$.
($\Longleftarrow$) By the uniqueness of the adjoint. ♣

Theorem 5.6.26. If $T\colon V \longrightarrow V$ is a unitary operator, then it preserves norms. That is, $|v| = |Tv|$. ★

Proof.

$$\begin{aligned} |v| &= \sqrt{\langle v, v\rangle} && \text{by definition of norm.}\\ &= \sqrt{\langle v, I(v)\rangle} && I \text{ is the identity operator} \end{aligned}$$

$$= \sqrt{\langle v, T^\dagger T v\rangle} \qquad \text{because } T^\dagger T = I$$
$$= \sqrt{\langle Tv, Tv\rangle} \qquad \text{by definition of adjoint}$$
$$= |Tv| \qquad \text{by definition of norm.}$$

♣

Exercise 5.6.27. Show that unitary operators are closed under composition, direct product, and tensor product. ■

Definition 5.6.28. Since unitary operators are closed under composition, and identity morphisms are unitary, the collection of Hilbert spaces and unitary operators form a subcategory $\mathbb{UHilb}$ of $\mathbb{Hilb}$. Similarly, one can talk about the subcategory of finite-dimensional Hilbert spaces and unitary operators $\mathbb{UFDHilb}$.

Since all unitary operators are invertible, $\mathbb{UHilb}$ and $\mathbb{UFDHilb}$ are actually groupoids. There are inclusion functors from $\mathbb{UHilb}$ to $\mathbb{Hilb}$ and from $\mathbb{UFDHilb}$ to $\mathbb{FDHilb}$. Since unitary matrices are closed under direct sum and tensor product, there are monoidal category structures $(\mathbb{UHilb}, \oplus, 0)$, $(\mathbb{UFDHilb}, \oplus, 0)$, $(\mathbb{UHilb}, \otimes, \mathbf{C})$, and $(\mathbb{UFDHilb}, \otimes, \mathbf{C})$.

Technical Point 5.6.29. The strict monoidal categories $(\mathbb{UHilb}, \oplus, 0)$, $(\mathbb{UFDHilb}, \oplus, 0)$ have a similar problem to $(\mathbb{UMat}, \oplus, 0)$ that we saw in Technical Point 5.6.20. Namely, these categories have a strictly associative product and coproduct, but these operations do not satisfy the any universal properties because they lack projections and inclusions. ♡

Let us relate Hermitian and unitary matrices with Hermitian and unitary linear operators.

Theorem 5.6.30.

- If A is a Hermitain matrix, then T_A is Hermitian operator.
- Let V be a finite-dimensional Hilbert space with basis $\mathcal{B}$ and a Hermitian operator $T\colon V \longrightarrow V$. Then there is a Hermitian matrix A such that for all v in V, $T(v) = Av$.
- If A is a unitary matrix, then T_A is a unitary operator.
- Let V be a finite-dimensional Hilbert space with basis $\mathcal{B}$ and a unitary operator $T\colon V \longrightarrow V$. Then there is a unitary matrix A such that for all v in V, $T(v) = Av$. ★

Exercise 5.6.31. Prove Theorem 5.6.30. ■

With the last two statements of this theorem, we can prove that there is an equivalence of categories between $\mathbb{UMat}$ and $\mathbb{UFDHilb}$.

Let us summarize all the functors that we have been dealing with relating sets, vector spaces, matrices, and Hilbert spaces:

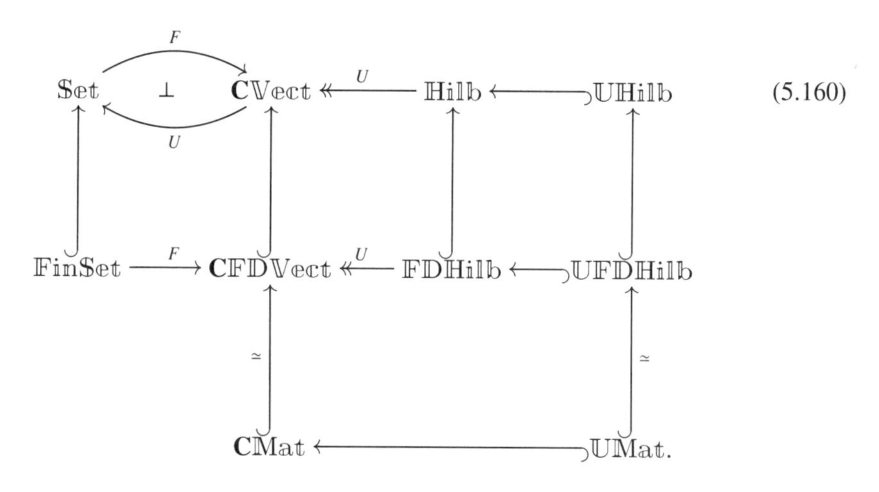

(5.160)

Eigenvalues and Eigenvectors

In this part, we deal with linear maps from a complex vector space to itself. Sometimes such operators only change the magnitude of a vector and leave the direction fixed. That is, the operator changes the vector only by a scalar multiple. Those vectors are almost a fixed point of the operator. We will see that such vectors and the amount that they are changed by the operators are very important for our study of such operators.

Definition 5.6.32. For a linear map $T\colon V \longrightarrow V$, if there is a $v \in V$ and a $\lambda \neq 0$ in $\mathbb{C}$ such that

$$T(v) = \lambda \cdot v, \tag{5.161}$$

then v is called an **eigenvector** of T and λ is called the **eigenvalue** of v. (The word *eigen* is derived from the German word meaning "own" or "self.")

For every $\lambda \in \mathbf{C}$, there is an operator $\lambda I\colon V \longrightarrow V$ that is defined for $v \in V$ as $\lambda I(v) = \lambda \cdot v$. It is not hard to see that λI is a linear map. Since the subtraction of two operators is still an operator, for any operator $T\colon V \longrightarrow V$, we have that $T - \lambda I$ is a linear map and is defined as $(T - \lambda I)(v) = T(v) - \lambda I(v) = T(v) - \lambda \cdot v$. For every eigenvalue λ, $T - \lambda I$ is a linear map and its kernel consists of those vectors that are eigenvectors for λ:

$$V_\lambda = \{v \in V : T(v) = \lambda \cdot v\} \subseteq V \tag{5.162}$$

and is called the **eigenspace** of T belonging to λ. If V is a vector space of functions, then an eigenvector will be called an **eigenfunction**. If a basis consists of eigenvectors for an operator, then the basis is called an **eigenbasis**.

From a categorical point of view, an eigenspace is simply the equalizer in the diagram

$$V_\lambda \xhookrightarrow{\ inc\ } V \underset{T}{\overset{\lambda I}{\rightrightarrows}} V. \tag{5.163}$$

What type of eigenvalues and eigenvectors do our two favorite operators have?

Theorem 5.6.33. The eigenvalues of a Hermitian operator are real. ★

Proof. Let $T: V \longrightarrow V$ be a Hermitian operator and v be a vector of norm 1 (i.e., $\langle v, v\rangle = 1$). Say that λ is the eigenvalue of T (i.e., $T(v) = \lambda v$):

$$\begin{aligned}
\lambda &= \lambda\langle v, v\rangle && \text{because} \langle v, v\rangle = 1 \\
&= \langle \lambda v, v\rangle && \text{by linearity of the inner product} \\
&= \langle T v, v\rangle && \text{by definition of an eigenvalue} \\
&= \langle v, T v\rangle && T \text{ is Hermitian} \\
&= \langle v, \lambda v\rangle && \text{by definition of an eigenvalue} \\
&= \bar{\lambda}\langle v, v\rangle && \text{by anti-linearity of the inner product} \\
&= \bar{\lambda} && \text{because } \langle v, v\rangle = 1.
\end{aligned}$$

Since $\lambda = \bar{\lambda}$, it is real. ♣

Theorem 5.6.34. The eigenvectors of distinct eigenvalues for a Hermitian operator are orthogonal. ★

Proof. Let $T: V \longrightarrow V$ be a Hermitian operator, and let v and w be vectors such that $T(v) = \lambda v$ and $T(w) = \mu w$ with $\lambda \neq \mu$.

$$\begin{aligned}
\lambda\langle v, w\rangle &= \langle \lambda v, w\rangle && \text{by linearity of the inner product} \\
&= \langle T v, w\rangle && \text{by definition of an eigenvalue} \\
&= \langle v, T w\rangle && T \text{ is Hermitian} \\
&= \langle v, \mu w\rangle && \text{by definition of an eigenvalue} \\
&= \bar{\mu}\langle v, w\rangle && \text{by anti-linearity of the inner product} \\
&= \mu\langle v, w\rangle && \mu \text{ is real.}
\end{aligned}$$

Since $\lambda\langle v, w\rangle = \mu\langle v, w\rangle$ and $\lambda \neq \mu$, it must be that $\langle v, w\rangle = 0$. ♣

Notice that the converse of this theorem is not necessarily true. That means that there could be a Hermitian operator with two eigenvectors that are not orthogonal but their respective eigenvalues are equal.

Theorem 5.6.35. The eigenvalues of a unitary operator have modulus 1. ★

Proof. Let $T: V \longrightarrow V$ be a unitary operator and v be an eigenvector with eigenvalue λ:

$$\begin{aligned}
|v|^2 &= |T v|^2 && \text{by Theorem 5.6.26} \\
&= \langle T v, T v\rangle && \text{by definition of norm} \\
&= \langle \lambda v, \lambda v\rangle && \text{by definition of eigenvalue} \\
&= \lambda\langle v, \lambda v\rangle && \text{linear of the first variable}
\end{aligned}$$

$$= \lambda\bar{\lambda}\langle v, v\rangle \qquad \text{anti-linear of the second variable}$$

$$= \lambda\bar{\lambda}|v|^2. \qquad \text{definition of norm.}$$

So $|v|^2 = \lambda\bar{\lambda}|v|^2$, and by dividing out, we get that $|\lambda| = 1$. ♣

Suggestions for Further Study

There are many fine linear algebra textbooks that emphasize different aspects of the field, for example, [197, 87, 154, 205]. This presentation gained much from [198] and [23].

6

Relationships between Monoidal Categories

The trend of mathematics and physics toward unification provides the physicist with a powerful new method of research into the foundations of his subject, a method which has not yet been applied successfully, but which I feel confident will prove its value in the future. The method is to begin by choosing that branch of mathematics which one thinks will form the basis of the new theory. One should be influenced very much in this choice by considerations of mathematical beauty. It would probably be a good thing also to give a preference to those branches of mathematics that have an interesting group of transformations underlying them, since transformations play an important role in modern physical theory, both relativity and quantum theory seeming to show that transformations are of more fundamental importance than equations. Having decided on the branch of mathematics, one should proceed to develop it along suitable lines, at the same time looking for that way in which it appears to lend itself naturally to physical interpretation.

Paul A. M. Dirac

The Relation Between Mathematics and Physics (1939)

In this chapter, we show how monoidal categories are related to each other. In context, Chapter 4 taught how categories are related with functors, while Chapter 5 showed that some of these categories have monoidal structures. Now we show that there are functors that describe the relationships between categories that also respect the monoidal structures. These functors highlight how the categories have shared properties. This is very important for our unification goal. We also prove important theorems relating monoidal categories to strict monoidal categories.

Section 6.1 introduces the many types of functors between monoidal categories and the natural transformations between such functors. In Section 6.2, we prove a fundamental result concerning the relationship between monoidal categories and strict monoidal categories. This is central to coherence theory. Section 6.3 is a short excursion dealing with weakening of coherence conditions. The chapter ends with Section 6.4, which presents a mini-course on duality theory.

6.1 Monoidal Functors and Natural Transformations

We have introduced monoidal categories. Now we will discuss how monoidal categories relate to each other (see Important Categorical Idea 1.4.9.) We saw that for a set function $f\colon M \longrightarrow M'$ to be a monoid homomorphism $f\colon (M,\star,e) \longrightarrow (M',\star',e')$, it must respect the operations:

$$f(x \star y) = f(x) \star' f(y) \text{ and } f(e) = e'. \tag{6.1}$$

In Definition 6.1.1, we describe a category theoretic version of these conditions.

Definition 6.1.1. Given monoidal categories $(\mathbb{A},\otimes,\alpha,I,\lambda,\rho)$ and $(\mathbb{B},\otimes',\alpha',I',\lambda',\rho')$, a **monoidal functor** $(F,\tau,\upsilon)\colon (\mathbb{A},\otimes,\alpha,I,\lambda,\rho) \longrightarrow (\mathbb{B},\otimes',\alpha',I',\lambda',\rho')$ is

- A functor $F\colon \mathbb{A} \longrightarrow \mathbb{B}$
- A natural transformation called a **mapping funnel**:

$$\tau\colon\ \otimes' \circ (F \times F) \Longrightarrow F \circ \otimes \tag{6.2}$$

  That is, for all a and a' in $\mathbb{A}$, there is a morphism:

$$\tau_{a,a'}\colon F(a) \otimes' F(a') \longrightarrow F(a \otimes a'), \tag{6.3}$$

  which funnels all the elements into the parentheses.
- Let $u\colon \mathbf{1} \longrightarrow \mathbb{A}$ and $u'\colon \mathbf{1} \longrightarrow \mathbb{B}$ be the functors that pick out the units I and I' of $\mathbb{A}$ and $\mathbb{B}$, respectively. Then there is a natural transformation called a **unital funnel**:

$$\upsilon\colon u' \Longrightarrow F \circ u. \tag{6.4}$$

  That is, a morphism in $\mathbb{B}$:

$$\upsilon_*\colon I' \longrightarrow F(I). \tag{6.5}$$

These natural transformations must satisfy the following coherence requirements:

- The mapping funnel τ must cohere with itself and with the reassociators α and α', as in this **hexagon coherence condition**:

$$\begin{array}{ccc} Fa \otimes' (Fb \otimes' Fc) & \xrightarrow{\alpha'_{Fa,Fb,Fc}} & (Fa \otimes' Fb) \otimes' Fc \\ \Big\downarrow{\scriptstyle id_{Fa} \otimes' \tau_{b,c}} & & \Big\downarrow{\scriptstyle \tau_{a,b} \otimes' id_{Fc}} \\ Fa \otimes' F(b \otimes c) & & F(a \otimes b) \otimes' Fc \\ \Big\downarrow{\scriptstyle \tau_{a,b\otimes c}} & & \Big\downarrow{\scriptstyle \tau_{a\otimes b,c}} \\ F(a \otimes (b \otimes c)) & \xrightarrow[F(\alpha_{a,b,c})]{} & F((a \otimes b) \otimes c). \end{array} \tag{6.6}$$

- The mapping funnel τ and the unital funnel υ must cohere with the right and left unitors ρ and λ:

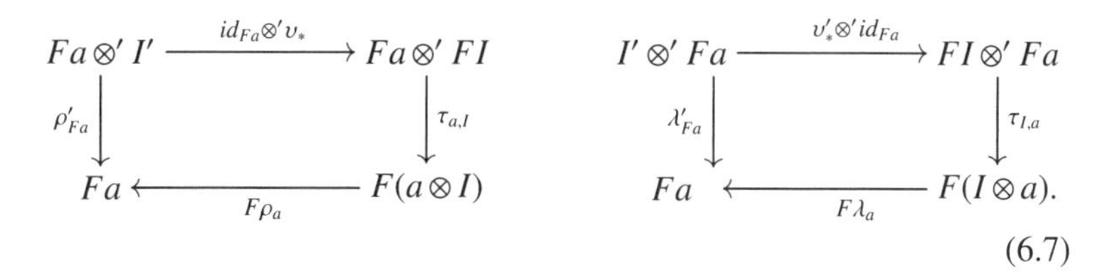

(6.7)

Remark 6.1.2. The τ and υ natural transformations can be seen as a weakening of the commuting square definition of a monoid homomorphism (see Diagram (2.11)). Rather than insisting that the diagrams commute, we insist that there are natural transformations from the composite of one side of the diagram to the other:

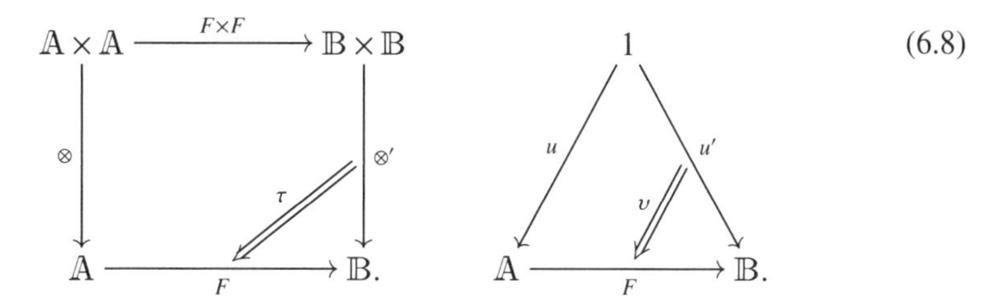

(6.8)

It also should be highlighted that the τ and the υ are, in general, not isomorphisms. We will see, that in many examples, these morphisms do not have inverses. ♠

Remark 6.1.3. In the event that $\mathbb{A}$ and $\mathbb{B}$ are strict monoidal categories, then Diagram (6.6) reduces to

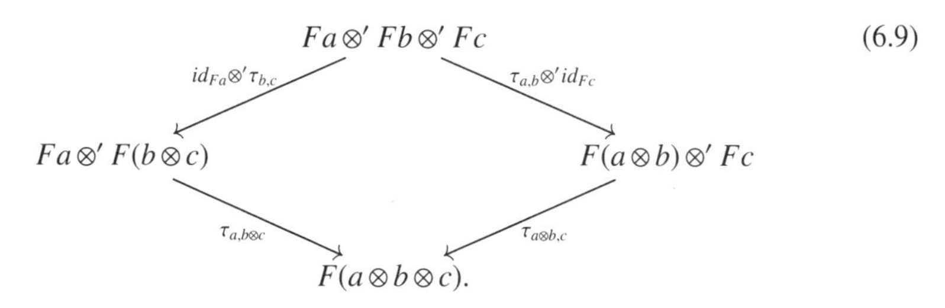

(6.9)

♠

Monoidal functors come in different flavors.

Definition 6.1.4.

- The above definition of a monoidal functor is also called a **weak monoidal functor** or a **lax monoidal functor**.

- If the τ and the υ natural transformations go the other way (i.e., we have

$$\tau_{a,a'}\colon F(a\otimes a')\longrightarrow F(a)\otimes' F(a') \tag{6.10}$$

$$\upsilon_*\colon F(I)\longrightarrow I'), \tag{6.11}$$

then F is called a **oplax monoidal functor** (also called a **comonoidal functor**, **opmonoidal functor** and **colax monoidal functor**).
- If τ and υ are isomorphisms, then F is called a **strong monoidal functor**.
- If τ and υ are identity morphisms, then we have $F(a)\otimes' F(a') = F(a\otimes a')$ and $I' = F(I)$. Such an F is called a **strict monoidal functor**.

Exercise 6.1.5. Give the appropriate definition for the identity monoidal functor. ■

Every elementary school child knows Example 6.1.6.

Example 6.1.6. In Example 5.1.12 we met the two strict monoidal categories $(\mathbf{R}, +, 0)$ and $(\mathbf{R}^+, \cdot, 1)$. For every real number $b > 1$, there is a strict monoidal functor

$$b^{(\ )}\colon (\mathbf{R}, +, 0)\longrightarrow(\mathbf{R}^+, \cdot, 1). \tag{6.12}$$

The fact that $b^{x+y} = b^x \cdot b^y$ and $b^0 = 1$ means that $b^{(\ )}$ strictly preserves the operations. There is also a strict monoidal functor:

$$Log_b(\ )\colon (\mathbf{R}^+, \cdot, 1)\longrightarrow(\mathbf{R}, +, 0). \tag{6.13}$$

The fact that $Log_b(x\cdot y) = Log_b(x) + Log_b(y)$ and $Log_b(1) = 0$ means that this functor strictly preserves the operations.

This brings to light a more general statement. If there are monoids M and M' that form strict monoidal categories, then a homomorphism from M to M' forms a strict monoidal functor. □

Example 6.1.7. In Example 4.4.28 we met the free functor $F\colon \mathbb{S}\mathrm{et}\longrightarrow \mathbf{K}\mathbb{V}\mathrm{ect}$, which takes a set S to $F(S)$, the vector space that has S as its basis. We also met the right adjoint to this functor, $U\colon \mathbf{K}\mathbb{V}\mathrm{ect}\longrightarrow\mathbb{S}\mathrm{et}$, that takes a vector space V to $U(V)$, its underlying set. Remember that the category of sets has two monoidal category structures: $(\mathbb{S}\mathrm{et}, +, \emptyset)$ and $(\mathbb{S}\mathrm{et}, \times, \{*\})$. There are also two monoidal structures on $\mathbf{K}\mathbb{V}\mathrm{ect}$. The adjunctions respect both monoidal structures as follows

$$(\mathbb{S}\mathrm{et}, +, \emptyset)\ \underset{(U,\tau_1,\upsilon_1)}{\overset{(F,\tau,\upsilon)}{\rightleftarrows}}\!\!\perp\ (\mathbf{K}\mathbb{V}\mathrm{ect}, \oplus, 0) \qquad (\mathbb{S}\mathrm{et}, \times, \{*\})\ \underset{(U',\tau'_1,\upsilon'_1)}{\overset{(F',\tau',\upsilon')}{\rightleftarrows}}\!\!\perp\ (\mathbf{K}\mathbb{V}\mathrm{ect}, \otimes, \mathbf{K})\ . \tag{6.14}$$

Let us examine each of these four functors and see how they respect the monoidal structures:

- $F(S+S')$ is isomorphic to $F(S)\oplus F(S')$, and $F(\emptyset)=0$. This is a strong monoidal functor.
- There is a map $U(V)+U(V')\longrightarrow U(V+V')$. This map is not an isomorphism because $U(V)+U(V')$ has two 0's and $U(V+V')$ has only one. As for the unit, we have $U(0)=0$.
- $F'(S\times S')\cong F'(S)\otimes F'(S')$ and $F'(\{*\})=\mathbf{K}$. This is a strong monoidal functor.
- There is a map $U'(V)\times U'(V')\longrightarrow U'(V\otimes V')$ because $V\otimes V'$ is defined by an equivalence relation on $V\times V'$. for the unit, we have $U'(\mathbf{K})=\mathbf{K}\neq\{*\}$. This means U' is a lax monoidal functor.

What about finite sets and finite-dimensional vector spaces? Notice that although there is a free functor from $\mathbb{FinSet}$ to $\mathbf{K}\mathbb{FDVect}$, the forgetful functor from finite-dimensional vector spaces does not output finite sets. In other words, in general, the underlying set of a finite dimensional vector space is not a finite set. □

Monoidal functors compose as follows. Given the monoidal functors

$$(F,\tau,\upsilon)\colon(\mathbb{A},\otimes,\alpha,I,\lambda,\rho)\longrightarrow(\mathbb{B},\otimes',\alpha',I',\lambda',\rho') \tag{6.15}$$

and

$$(F',\tau',\upsilon')\colon(\mathbb{B},\otimes',\alpha',I',\lambda',\rho')\longrightarrow(\mathbb{C},\otimes'',\alpha'',I'',\lambda'',\rho''), \tag{6.16}$$

we form

$$(F'\circ F,\tau'',\iota'')\colon(\mathbb{A},\otimes,\alpha,I,\lambda,\rho)\longrightarrow(\mathbb{C},\otimes'',\alpha'',I'',\lambda'',\rho''). \tag{6.17}$$

The component of the mapping funnel τ'' at elements a and a' is defined as the composition of the two maps:

$$\overset{\tau''_{a,a'}}{\overbrace{F'Fa\otimes'' F'Fa' \xrightarrow{\tau'_{Fa,Fa'}} F'(Fa\otimes' Fa') \xrightarrow{F'\tau_{a,a'}} F'F(a\otimes a')}}. \tag{6.18}$$

The natural transformation τ'' satisfies the hexagon condition (Diagram (6.6)), as can be seen in Figure 6.1. All the parts of this figure commute:

- The top hexagon commutes because τ' is a mapping funnel.
- The bottom hexagon commutes because τ is a mapping funnel and because of the functoriality of F'.
- The four triangles commute because of the definition of τ''.
- The left and right quadrilaterals commute because of the naturality of τ and τ'.

Since all the inner parts of the diagram commute, the outer hexagon of the diagram commutes. This ensures that the composed mapping funnel, τ'', satisfies the coherence condition.

Exercise 6.1.8. How do the unital funnels of monoidal functors compose? Show that the composition respects the appropriate coherence conditions. ■

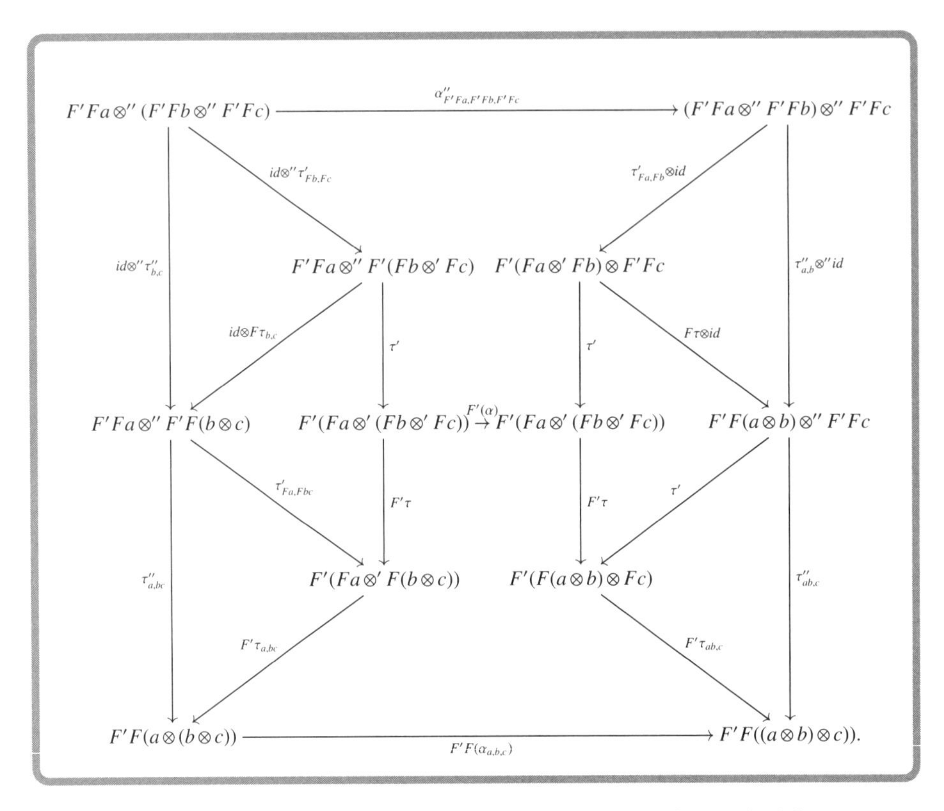

Figure 6.1. The coherence condition for the composition of monoidal functors.

With composition of monoidal functors, one can formulate the notion of an isomorphism made of monoidal functors. For example, the two strict monoidal functors in Example 6.1.6 are inverse to each other and form a monoidal isomorphism. Another example of isomorphic monoidal categories are the two strict monoidal categories $(\mathbf{N}, +, 0)$ and $(\{1\}^*, \bullet, \emptyset)$. There is clearly a functor $n \mapsto 11 \cdots 1(n$ times). This functor preserves the tensor product and is an isomorphism. It should be noted that just as isomorphism of categories are a rarity, isomorphism of monoidal categories are a rarity.

What about functors between symmetric monoidal categories?

Definition 6.1.9. Let $\mathbb{A}$ and $\mathbb{B}$ be symmetric monoidal categories with braidings γ and γ', respectively. A monoidal functor (F, τ, υ) from $\mathbb{A}$ to $\mathbb{B}$ is a **symmetric monoidal functor** if the mapping funnels and the braidings cohere with each other as follows:

$$\begin{array}{ccc} Fa \otimes' Fa' & \xrightarrow{\gamma'_{Fa,Fa'}} & Fa' \otimes' Fa \\ {\scriptstyle \tau_{a,a'}} \downarrow & & \downarrow {\scriptstyle \tau_{a',a}} \\ F(a \otimes a') & \xrightarrow[F\gamma_{a,a'}]{} & F(a' \otimes a). \end{array} \tag{6.19}$$

Definition 6.1.10. The collection of monoidal categories and strong monoidal functors form the category $\mathbb{MonCat}$. There is a subcategory of strict monoidal categories and strong monoidal functors denoted as $\mathbb{StrMonCat}$. The collection of symmetric monoidal categories and symmetric monoidal functors form a category called $\mathbb{SymMonCat}$. There is a subcategory of strictly associative symmetric monoidal categories and symmetric monoidal functors between them denoted as $\mathbb{StrSymMonCat}$. There are forgetful functors from the collections of symmetric monoidal categories to the collections of monoidal categories. These categories are related as follows:

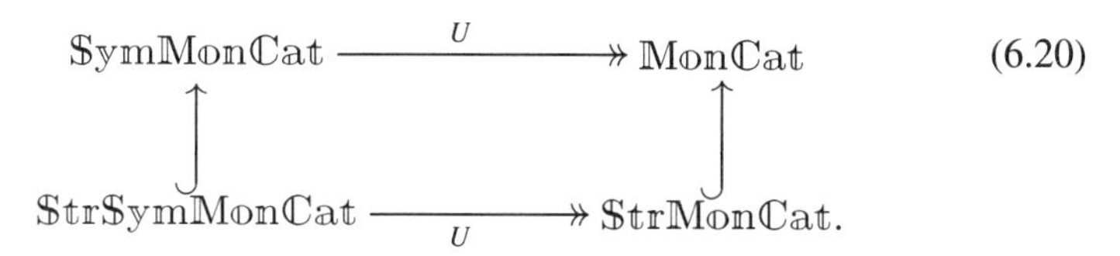

(6.20)

Let us go up one level and define a natural transformation between monoidal functors.

Definition 6.1.11. Let $\mathbb{A}$ and $\mathbb{B}$ be monoidal categories and let (F, τ, υ) and (F', τ', υ') be monoidal functors from $\mathbb{A}$ to $\mathbb{B}$. A **monoidal natural transformation** from (F, τ, υ) to (F', τ', υ') is a natural transformation $\mu\colon F \Longrightarrow F'$; that is, for every a in $\mathbb{A}$, there is a morphism $\mu_a\colon F(a) \longrightarrow F'(a)$, which coheres with the mapping funnel and unital functors as follows:

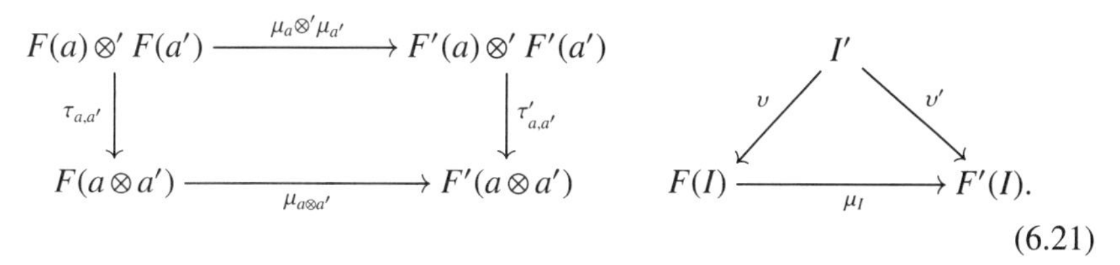

(6.21)

A **symmetric monoidal natural transformation** is a monoidal transformation between symmetric monoidal functors. No additional requirements are needed.

Definition 6.1.12. The collection of monoidal categories, strong monoidal functors, and monoidal natural transformations forms a 2-category called $\overline{\mathbb{MonCat}}$. There is a sub-2-category of strict monoidal categories, strong monoidal functors, and monoidal tranformations denoted as $\overline{\mathbb{StrMonCat}}$. There are similar statements about the 2-categories of a symmetric monoidal category $\overline{\mathbb{SymMonCat}}$ and a strictly associative symmetric monoidal category $\overline{\mathbb{StrSymMonCat}}$. These 2-categories fit together as shown in Diagram (6.20):

$$\begin{array}{ccc} \overline{\mathbb{SymMonCat}} & \xrightarrow{\;U\;} & \overline{\mathbb{MonCat}} \\ \uparrow & & \uparrow \\ \overline{\mathbb{StrSymMonCat}} & \xrightarrow[\;U\;]{} & \overline{\mathbb{StrMonCat}}. \end{array} \tag{6.22}$$

Exercise 6.1.13. Define vertical and horizontal composition of symmetric monoidal natural transformations. ■

Given the notion of a symmetric monoidal natural transformation, we can easily describe what it means for two symmetric monoidal categories to be equivalent.

Definition 6.1.14. A **monoidal equivalence** between $(\mathbb{A}, \otimes, \alpha, I, \lambda, \rho)$ and $(\mathbb{A}', \otimes', \alpha', I', \lambda', \rho')$ means that there is a monoidal functor (F, τ, υ) from $\mathbb{A}$ to $\mathbb{A}'$ and a monoidal functor (F', τ', υ') from $\mathbb{A}'$ to $\mathbb{A}$ with monoidal natural isomorphisms $\mu\colon Id_{\mathbb{A}} \longrightarrow F' \circ F$ and $\mu'\colon F \circ F' \longrightarrow Id_{\mathbb{A}'}$. An equivalence that uses strong monoidal functors has the property that the functor is full, faithful, and essentially surjective.

Diagram (6.3) shows how two objects funnel into one. Diagram (6.6) shows how three objects funnel into one. What about more objects? In Figure 6.2, there are four objects funneling into one. Some orientation around this diagram is needed. All vertical maps are instances of τ's:

- The top is Mac Lane's pentagon condition.
- The bottom is the image of Mac Lane's pentagon condition under F.
- The center diamond commutes because of the naturality of τ.
- The top quadrilaterals of both the front left and the front right commute out of the naturality of α.
- The bottom hexagons of both the front left and the front right commute because of the hexagon coherence condition.
- The back top left and top right are the hexagon coherence condition tensored with the identity on each side.
- The bottom of the back left and the back right commute by the naturality of τ.
- The top of the back commutes by naturality of α, and the bottom of the back commutes because it is an instance of the hexagon coherence condition.

The main point is that this diagram consists of naturality squares, pentagons, and hexagons. If one assumes the pentagons and the hexagons commute, then between any two objects in the diagram, there is at most one morphism between them. It can be shown that for any n, the mapping funnel that combines all the n elements into one pair of parentheses is also made of pentagons, hexagons, and naturality squares. This is similar to our other coherence conditions, where we assume that some set of smaller diagrams commute and show that all the larger ones commute.

Exercise 6.1.15. Fill in the missing names and subscripts to the morphisms in Figure 6.2. ■

The rest of this section consists of examples of monoidal functors.

Examples from Linear Algebra

In Diagram (5.160) on page 277, we summarized all the functors that we have been dealing with relating sets, vector spaces, matrices, unitary operators, unitary matrices, and Hilbert spaces. We described monoidal structures on these categories in

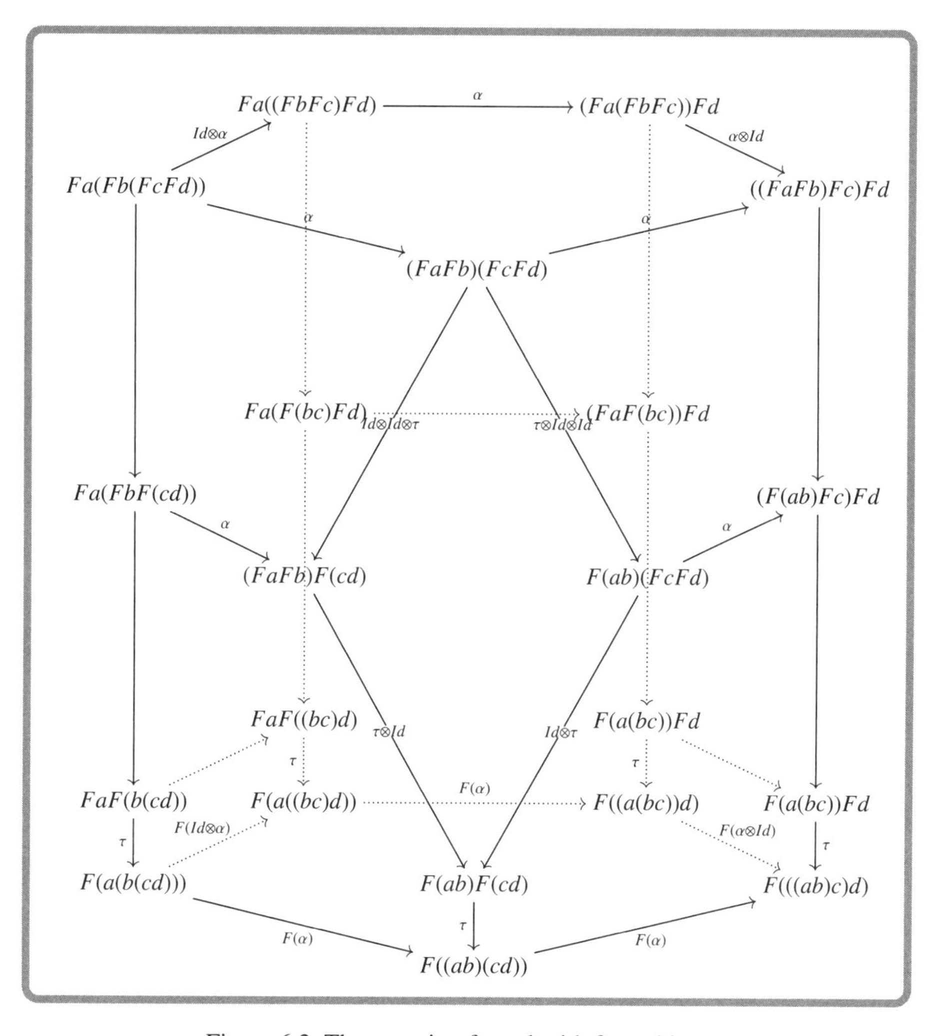

Figure 6.2. The mapping funnel with four objects.

Section 5.6. Figures 6.3 and 6.4 extends that diagram to include the monoidal structures and the monoidal functors. All the categories are strictly associative symmetric monoidal categories and all the functors are symmetric monoidal functors (not neccessarily strict). Figure 6.3 has the tensor product and Cartesian product monoidal structures, and Figure 6.4 has the coproducts and directs sums monoidal structures.

Examples with Circuits, Functions, and Matrices

The following three examples of monoidal functors are connected, as can be seen in Figure 6.5. Generalizations of these monoidal functors will be used in our mini-course on quantum computing in Chapter 10.

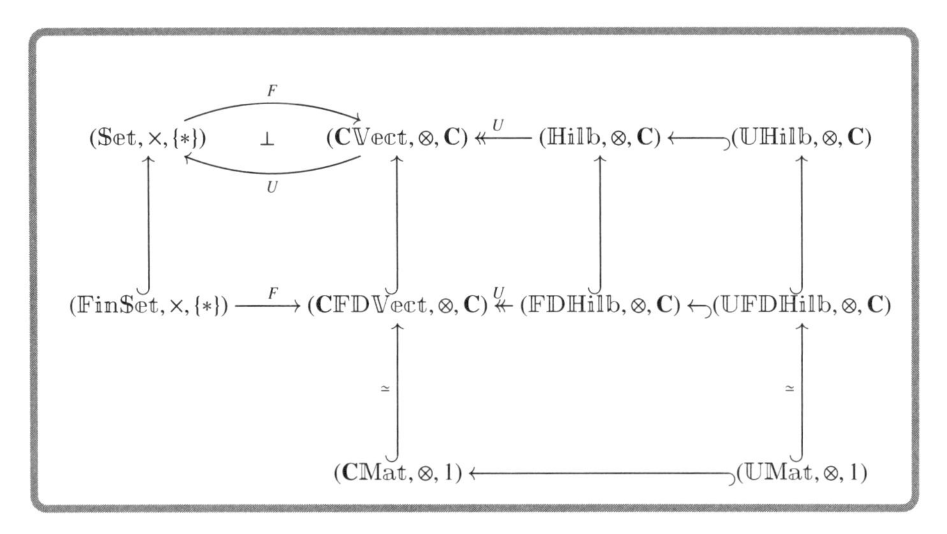

Figure 6.3. Monoidal categories with tensors and functors from linear algebra.

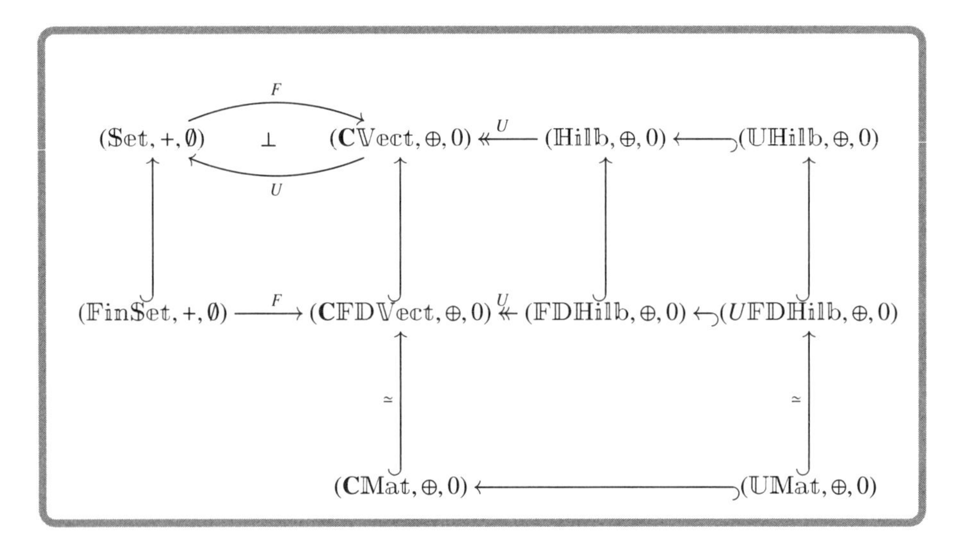

Figure 6.4. Monoidal categories with direct sums and functors from linear algebra.

Example 6.1.16. In Example 4.1.57 in Chapter 4, we introduced the functor *FuncDesc*: $\mathbb{Circuit} \longrightarrow Bool\mathbb{Func}$. We discussed the monoidal structure of $\mathbb{Circuit}$ in Example 5.3.18 and the monoidal structure of $Bool\mathbb{Func}$ in Example 5.2.19. The functor is a strict monoidal functor:

$$FuncDesc(C \oplus C') = FuncDesc(C) \oplus FuncDesc(C'). \tag{6.23}$$

The $\oplus$ on the left refers to the disjoint parallel processing of the two circuits, while the $\oplus$ on the right means the disjoint union of two functions. □

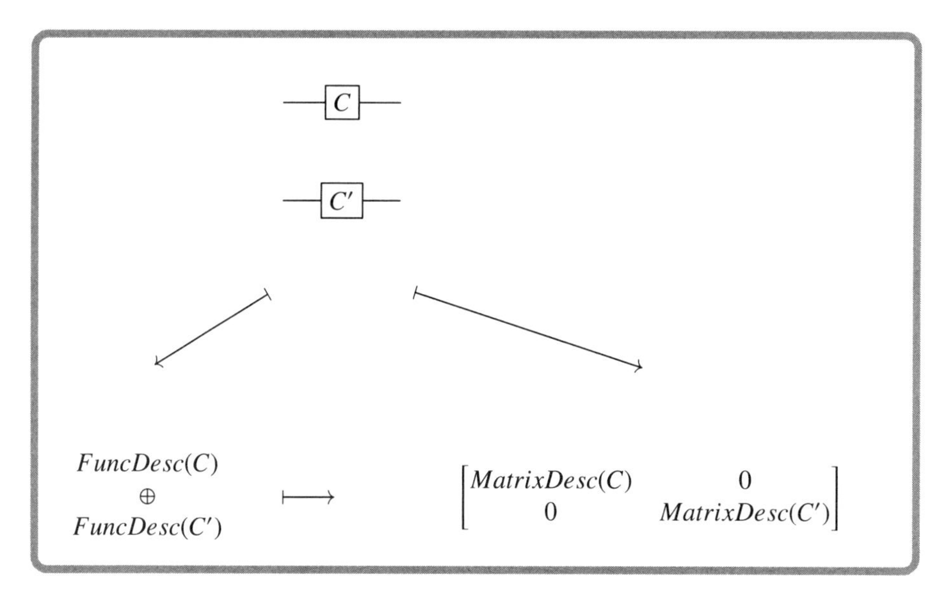

Figure 6.5. Monoidal functors with circuits, functions, and matrices.

Example 6.1.17. The functor $MatrixDesc\colon \mathbb{Circuit} \longrightarrow Bool\mathbb{Mat}$ was introduced in Example 4.1.58. We discussed the monoidal structure of $\mathbb{Circuit}$ in Example 5.3.18 and the monoidal structure of $Bool\mathbb{Mat}$ in Example 5.1.9. The functor is a strict monoidal functor because

$$MatrixDesc(C \oplus C') = MatrixDesc(C) \oplus MatrixDesc(C'). \qquad (6.24)$$

□

Example 6.1.18. The functor $FuncEval\colon Bool\mathbb{Func} \longrightarrow Bool\mathbb{Mat}$ was introduced in Example 4.1.60. The category $Bool\mathbb{Mat}$ has a monoidal structure, which is just addition on objects. On morphisms (Boolean matrices), this is disjoint union as described in Example 5.1.9. The functor *FuncEval* respects the monoidal structure. □

An Example with Logical Formulas

Logical circuits are intimately related to logical formulas. We describe a category called $\mathbb{Logic}$, whose morphisms are sequences of logical formulas. We will then describe a functor:

$$L\colon \mathbb{Circuit} \longrightarrow \mathbb{Logic}, \qquad (6.25)$$

which will take every circuit to the sequence of logical formulas that is associated with it. For example, the logical circuit given in Figure 2.9 on page 54 corresponds to the sequence of logical formulas:

$$((A \wedge B) \vee (\neg(C \wedge E))) \wedge (D \wedge E), \qquad A \vee ((D \wedge E) \wedge (\neg F)). \qquad (6.26)$$

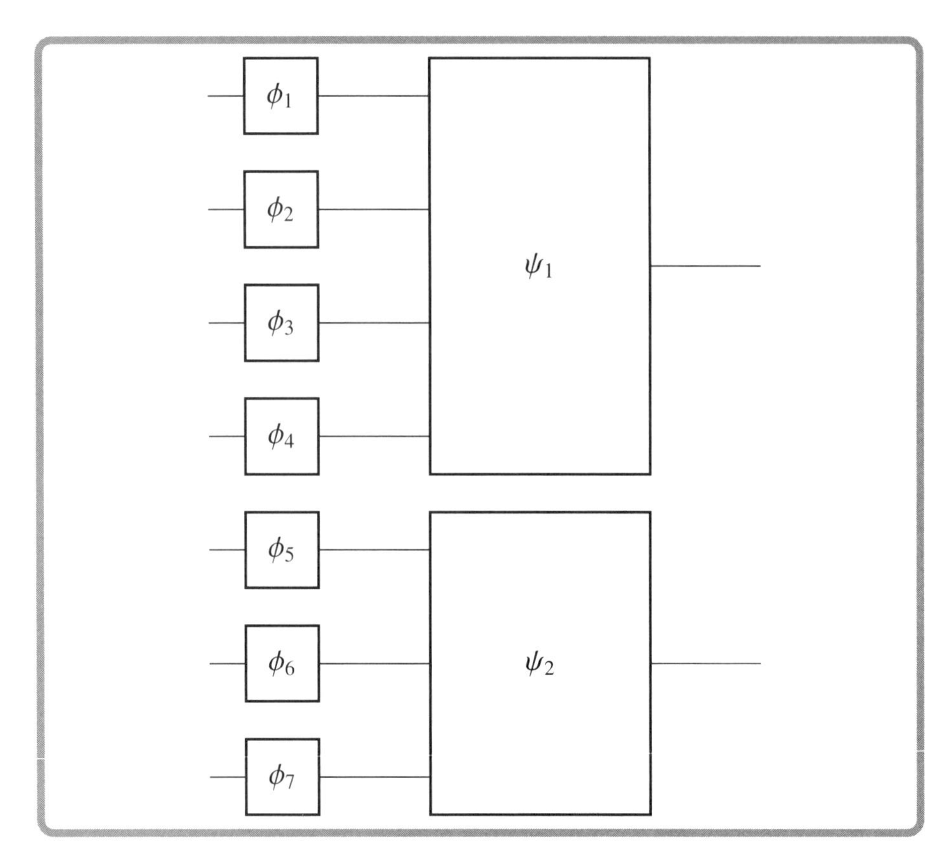

Figure 6.6. Composition of logical circuits.

Definition 6.1.19. The collection of logical formulas form a category called $\mathbb{Logic}$. The objects of $\mathbb{Logic}$ are the natural numbers. The morphisms from m to n are equivalence classes of n-tuples of logical formulas where each formula uses at most m variables. Two n-tuples are considered equivalent if they are the same formulas except for an exchange of variable names. For example, the sequence described in line (6.26) represents an element of $Hom_{\mathbb{Logic}}(6,2)$ because there are 6 variables and 2 formulas. A general morphism will be written as $\Phi\colon m \longrightarrow n$, which corresponds to logical formulas $(\phi_1, \phi_2, \ldots, \phi_n)$, where each ϕ_i has at most m variables. A map $\Phi\colon 0 \longrightarrow m$ corresponds to an m-tuple of true and false values. A map $\Phi\colon m \longrightarrow 0$ corresponds to the empty sequence of formulas.

Composition in $\mathbb{Logic}$ can be understood by looking at an example of composition of circuits, as shown in Figure 6.6. There are seven circuits on the left, which will correspond to a morphism in $\mathbb{Logic}$ written as $\Phi\colon m \longrightarrow 7$ or $\Phi = (\phi_1, \phi_2, \ldots, \phi_7)$. These compose into two circuits on the right. The four top wires enter the circuit that corresponds to ψ_1. The four variables in ψ_1 will be changed to the formulas ϕ_1, ϕ_2 , ϕ_3 and ϕ_4, which correspond to the four circuits on

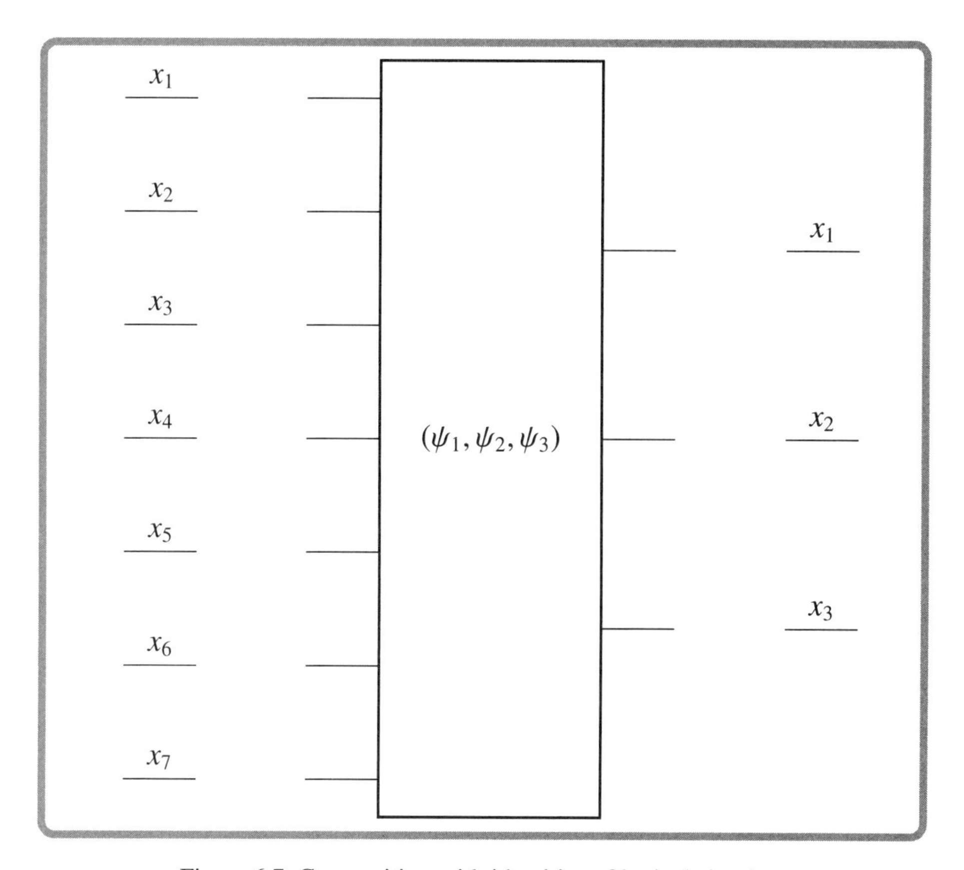

Figure 6.7. Composition with identities of logical circuits.

the left. It is similar with the three wires entering the bottom circuit. Let us be formal about composition. Let $\Phi\colon m \longrightarrow n$ correspond to the logical formulas $(\phi_1, \phi_2, \ldots, \phi_n)$ and $\Psi\colon n \longrightarrow p$ correspond to the logical formulas $(\psi_1, \psi_2, \ldots \psi_p)$. The composition $\Psi \circ \Phi\colon m \longrightarrow p$ corresponds to the logical formulas $(\xi_1, \xi_2, \ldots, \xi_p)$, where each ξ_i uses at most m variables. Here, ξ_i is defined by substituting the jth variable with ϕ_j.

The identity map $id_n\colon n \longrightarrow n$ in $\mathbb{Logic}$ is the sequence of n logical formulas $(x_1, x_2, \ldots, x_n)$. The fact that composition with such identity maps does not change the equivalence classes of logical formulas can be seen by looking at Figure 6.7.

Exercise 6.1.20. Show that the composition in this category is associative. ∎

Exercise 6.1.21. Show that the identity acts like a unit to the compostion. ∎

This category of logical formulas has a symmetric monoidal structure that corresponds to the disjoint union of logical circuits, $(\mathbb{Logic}, \oplus, 0)$. The monoidal structure on the objects is simply addition (i.e., $m \oplus n = m + n$). If $\Phi\colon m \longrightarrow n$ correspond to the logical formulas $(\phi_1, \phi_2, \ldots, \phi_n)$ and $\Psi\colon m' \longrightarrow n'$ corresponds to the logical formulas $(\psi_1, \psi_2, \ldots, \psi_{n'})$, then $\Phi \oplus \Psi\colon m + m' \longrightarrow n + n'$ corresponds to the logical formulas $(\phi_1, \phi_2, \ldots, \phi_n, \psi_1, \psi_2, \ldots, \psi_{n'})$, where each of the ϕ's and ψ's use at most $m + m'$ variables.

Since the category and monoidal structure of $\mathbb{Logic}$ was made to be similar to $\mathbb{Circuit}$, it is not surprising that there is a strict monoidal functor $L\colon \mathbb{Circuit} \longrightarrow \mathbb{Logic}$ that takes every logical circuit to the logical formulas that it describes. In detail, on objects, $L(m) = m$ and L takes a circuit with m inputs and n outputs to n logical formulas each using up to m variables. The generators of the category are the logical gates AND, OR, NOT, NAND, etc. The functor takes these gates to the logical formulas $A \wedge B$, $A \vee B$, $\neg A$, and $\neg(A \wedge B)$, etc. The functor L respects the composition in the categories because the composition in $\mathbb{Logic}$ was made to mimic the composition in $\mathbb{Circuit}$. Similarly, L is a strict monoidal functor. Interestingly, L is not a symmetric monoidal functor because—as we will see in Example 7.1.5—$\mathbb{Circuit}$ is not a symmetric monoidal category.

Exercise 6.1.22. Show that L is a strict monoidal functor. ■

6.2 Coherence Theorems

In this section, we prove properties of monoidal categories and theorems about the relationship of monoidal categories and strict monoidal categories:

- We start by proving a coherence theorem that says that $\mathbb{Assoc}$ is the paradigm of a monoidal category in the sense that it is the free monoidal category on one generator. This will lead to profound statements about every monoidal category.
- Then we will state and prove a strictification theorem that says that every monoidal category is monoidally equivalent to a strict monoidal category. Since this is a fundamental theorem about monoidal categories, we prove it in two ways. In both proofs, we associate a strict monoidal category with a monoidal category. We will then show that the original monoidal category is monoidally equivalent to the strict one. The tensor products of the strict monoidal categories are (i) string concatenation, and (ii) function composition. These are two paradigms of strictly associative operations.
- Then the strictification theorem is used to prove the coherence theorem.
- We conclude with similar theorems about symmetric monoidal categories.

Theorem 6.2.1 is the first coherence theorem in the book and future coherence theorems will follow the same form as this one. We strongly recommend that before tackling the first statement of the theorem, the reader goes back to Example 4.4.15 and meditate on its ideas. In particular, think about the free monoid on a one-object set.

Theorem 6.2.1. The Coherence Theorem for Monoidal Categories. We state this theorem in four equivalent ways:

- For every monoidal category $(\mathbb{C}, \otimes, I, \alpha, \rho, \lambda)$ and every object x in $\mathbb{C}$, there is a unique, strict monoidal functor $F_x\colon \mathbb{Assoc} \longrightarrow \mathbb{C}$, such that $F(\bullet) = x$.

- There is an isomorphism of categories:

$$Hom(\mathbb{Assoc}, \mathbb{C}) \cong Hom_{\mathbb{Cat}}(\mathbf{1}, \mathbb{C}), \tag{6.27}$$

where the *Hom* set on the left is strict monoidal functors, and the Hom set on the right is from $\mathbb{Cat}$.
- The monoidal category $\mathbb{Assoc}$ is the free monoidal category on one generator.
- Two morphisms in $\mathbb{C}$ in the image of F_x, generated by identities, α, ρ, λ and their inverses and built up by $\circ$ and $\otimes$, are equal. ★

Proof. Remember the association category $\mathbb{Assoc}$. The objects are bracketed words, and between any two objects, there is exactly one morphism. Let $(\mathbb{C}, \otimes, I, \alpha, \lambda, \rho)$ be a monoidal category. For any object x in $\mathbb{C}$, we will show that there is a strict monoidal functor $F_x\colon \mathbb{Assoc} \longrightarrow \mathbb{C}$. For the association of one letter $\bullet$, the functor $F_x(\bullet) = x$. It is obvious how to define F_x on objects of $\mathbb{Assoc}$. For an association $\bullet\bullet$, the functor will go to $x \otimes x = xx$. The association $\bullet(\bullet\bullet)$ will go to $x(xx)$. Within $\mathbb{Assoc}$, there is a unique map $\bullet(\bullet\bullet) \longrightarrow (\bullet\bullet)\bullet$ that F_x takes to α in $\mathbb{C}$. There are also maps λ and ρ in $\mathbb{Assoc}$, which F_x takes to λ and ρ in $\mathbb{C}$.

We start by considering graph $\mathbb{C}_{n,x}$ whose vertices are all words in w of length n that do not have the unit. A morphism $v \longrightarrow w$ is a certain type of map in $\mathbb{C}$ from $F_x(v)$ to $F_x(w)$. The maps are built out of α's, α^{-1} 's, and identities in $\mathbb{C}$. Also, if β is such a map, then so are $\beta \otimes id$ and $id \otimes \beta$. Two typical objects of this graph are (written without the tensors) $((ab)c)((de)f)$ and $c(f(a(b(de))))$. A typical morphism should look like $id \otimes ((\alpha \otimes id) \otimes \alpha^{-1}$. Notice that since we are restricting the morphisms like this, the only morphisms will be between objects that are the same permutation with different associations of the letters.

The core of this proof is that this graph is commutative. That is, any two paths that start and finish in the same vertices are equal.

Let $w^{(n)} = (((\cdots(ab)c)d)\cdots z)$ be a word with all the parentheses at the beginning. There is a way to measure how far any bracketing is from this bracketing. For a word w, let $rank(w)$ be a natural number that tells how far the parentheses are from all in the beginning. In detail, $rank(a) = 0$, and for words v and w,

$$rank(v \otimes w) = rank(v) + rank(w) + length(w) - 1. \tag{6.28}$$

Let us look at some examples:

- $rank(ab) = rank(a) + rank(b) + 1 - 1 = 0 + 0 + 1 - 1 = 0.$
- $rank((ab)c) = rank(ab) + rank(c) + 1 - 1 = 0 + 0 + 1 - 1 = 0.$
- $rank(a(bc)) = rank(a) + rank(bc) + 2 - 1 = 0 + 0 + 2 - 1 = 1.$
- $rank(((ab)c)d) = rank((ab)c) + rank(d) + 1 - 1 = 0 + 0 + 1 - 1 = 0.$
- $rank((ab)(cd)) = rank(ab) + rank(cd) + 2 - 1 = 0 + 0 + 2 - 1 = 1.$
- $rank(a(b(cd))) = rank(a) + rank(b(cd)) + 3 - 1 = 0 + 1 + 3 - 1 = 3.$

Of course, $rank(w^{(n)}) = 0$.

We focus on the subcategory of $\mathbb{C}_{n,x}$ that contains $(((\cdots(ab)c)d)\cdots z)$. For any bracketing v_i made of these letters in this order, there is a unique morphism $v_i \longrightarrow w^{(n)}$ made of α's. This map will bring the outermost parentheses into correct form first.

A map $v_1 \longrightarrow v_n$ consists of the composition of α's and α^{-1}'s as follows:

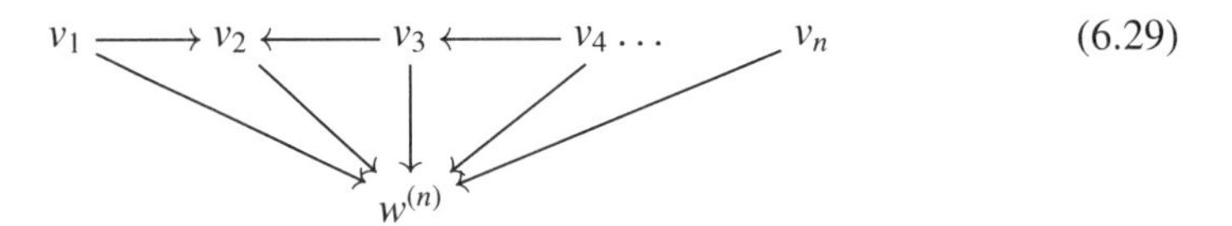

(6.29)

Since there are unique maps $v_i \longrightarrow w^{(n)}$, to show that there is only one map $v_1 \longrightarrow v_n$, it suffices to show that for any two maps lead to the same map to $w^{(n)}$. Assume that there are two different morphisms β and γ coming out of $v = v_i$. We must show that these two maps can be completed to get a commutative diagram as follows:

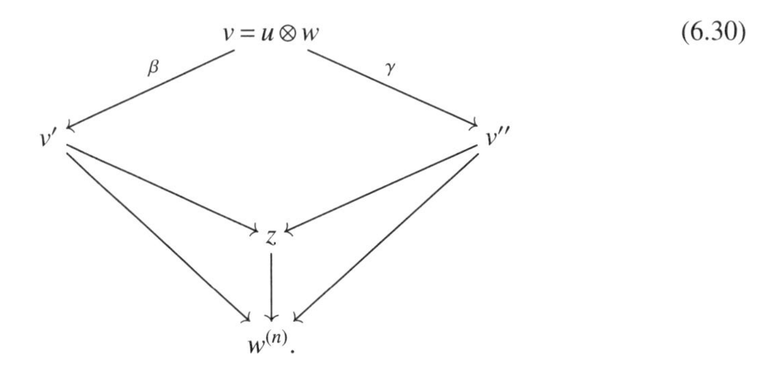

(6.30)

Notice that both the β's and the γ's consist of α's and take words of higher rank to lower rank.

There are three possible types of β:

- $v = s \otimes t$ and β can act on s (i.e., $\beta = \beta' \otimes id_t$).
- $v = s \otimes t$ and β can act on t (i.e., $\beta = id_s \otimes \beta'$).
- $v = s \otimes (t \otimes u)$ and β is $\alpha_{s,t,u}$.

Similarly, there are three possibilities for γ. All nine possibilities are shown in Figure 6.8. As you can see, each of the possibilities can be completed:

- (i) and (iii) follow by induction on the rank of v.
- (ii) and (iv) follow by the bifunctoriality of $\otimes$.
- (ii), (vi), (vii), and (viii) follow from the naturality of α.
- (ix) is completed by forming a pentagon coherence condition.

Thus, there is a unique isomorphism $v \longrightarrow w^{(n)}$.

That takes care of each $\mathbb{C}_{n,x}$. What about the relationship between $\mathbb{C}_{n,x}$ and $\mathbb{C}_{n-1,x}$? If we permit the unit I as a potential object in $\mathbb{C}_{n,x}$ then the λ's and the ρ's are isomorphisms from certain objects in $\mathbb{C}_{n,x}$ to objects in $\mathbb{C}_{n-1,x}$. These isomorphisms satisfy the triangle coherence condition. Hence, between any two objects of the form $F_x(v)$ and $F_x(w)$ in $\mathbb{C}$, there is a unique isomorphism of the structure maps of the monoidal category.

Let us deal with the other three ways to state the coherence theorem:

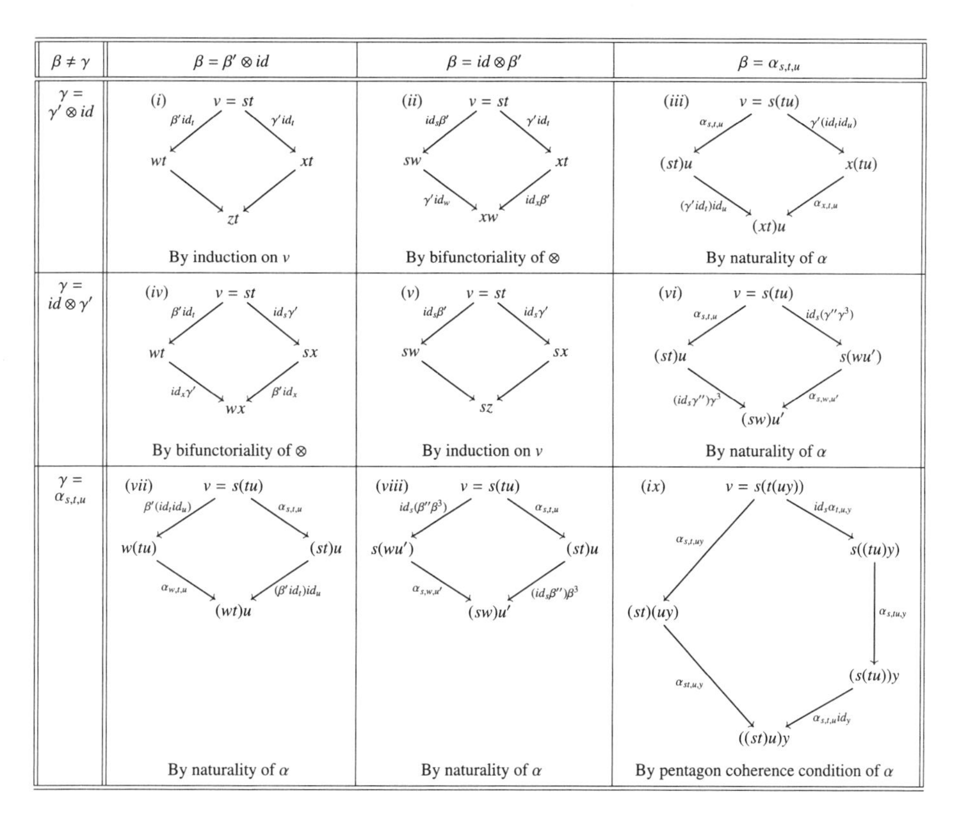

Figure 6.8. The nine cases of completing two maps coming out of a tensor product.

- The natural isomorphism given in Equation (6.27) follows from the fact that x in $\mathbb{C}$ is chosen by a $F'\colon \mathbf{1} \longrightarrow \mathbb{C}$. This functor determines and is determined by the unique functor $F_x\colon \mathbb{Assoc} \longrightarrow \mathbb{C}$.
- By "free monoidal category," we mean that there is a forgetful functor $U\colon \mathbb{MonCat} \longrightarrow \mathbb{Cat}$. This functor has a left adjoint $Free\colon \mathbb{Cat} \longrightarrow \mathbb{MonCat}$, which takes a category to its free monoidal category. Equation (6.27) says that $Free(\mathbf{1}) = \mathbb{Assoc}$.
- Since in $\mathbb{Assoc}$ there is only one morphism between each bracketing in $\mathbb{Assoc}$, the functor F_x only goes to one morphism. Notice that in an arbitrary monoidal category, there need not be any morphism between bracketing of words where the underlying order is not the same. This means that there might not be a morphism of the form $a \otimes b \longrightarrow b \otimes a$. ♣

Remark 6.2.2. Many books and papers state the coherence result as follows: any two arrows within a monoidal category that start and finish at the same objects and are generated by α, λ and ρ and their inverses are equal. But this is simply not true. Consider a monoidal category with objects a, b, b', and c, where $b \otimes c = b' \otimes c$ and $a \otimes b = a \otimes b'$. Such a category might have the following noncommutative diagram made of identities and α's:

$$\begin{array}{ccc} a\otimes(b\otimes c) & \xrightarrow{\alpha_{a,b,c}} & (a\otimes b)\otimes c \\ \| & & \| \\ a\otimes(b'\otimes c) & \xrightarrow[\alpha_{a,b',c}]{} & (a\otimes b')\otimes c. \end{array} \tag{6.31}$$

The problem is that when you have those objects equal to each other, the morphisms might not be equal to each other. In contrast, when we are dealing with a free category, such equalities cannot happen. We can make general statements about commutative diagrams if we dealing with any free structures such as $\mathbb{A}\mathbf{ssoc}$. We can also use the freeness of $\mathbb{A}\mathbf{ssoc}$ and universal functors from $\mathbb{A}\mathbf{ssoc}$ to make statements about any structure.

I am grateful to Ross Street for help in this area. ♠

Theorem 6.2.3. Strictification Theorem for Monoidal Categories. Every monoidal category is monoidally equivalent to a strict monoidal category. ★

Proof. Using string concatenation. Let $(\mathbb{C},\otimes,\alpha,I,\lambda,\rho)$ be a monoidal category. As we saw in Example 5.1.3, the category of strings on an alphabet with the concatenation operation is a strict monoidal category. We will form such a strict monoidal category $(\mathbb{C}^\bullet,\bullet,\emptyset)$ that will be monoidally equivalent to $(\mathbb{C},\otimes,\alpha,I,\lambda,\rho)$.

The objects of $\mathbb{C}^\bullet$ are strings of objects in $\mathbb{C}$. A typical object will look like this: $c_1c_2\ldots c_m$. The morphisms in $\mathbb{C}^\bullet$ are derived from $\mathbb{C}$. They are defined as the morphisms between the object in $\mathbb{C}$ with all the parentheses on the left; that is,

$$Hom_{\mathbb{C}^\bullet}(c_1c_2\cdots c_m, c_1'c_2'\cdots c_n') = \tag{6.32}$$

$$Hom_{\mathbb{C}}(((\cdots(c_1\otimes c_2)\otimes\cdot)\cdots)\otimes c_m, ((\cdots(c_1'\otimes c_2')\otimes\cdots)\cdots)\otimes c_n').$$

Composition in $\mathbb{C}^\bullet$ is inherited from the composition in $\mathbb{C}$.

Category $\mathbb{C}^\bullet$ has a strict monoidal category structure where the tensor $\bullet$ is the concatenation of strings. This means that

$$c_1c_2\ldots c_m \bullet c_1'c_2'\ldots c_n' = c_1c_2\ldots c_mc_1'c_2'\ldots c_n'. \tag{6.33}$$

For morphisms, it is almost the same. If $f\colon c_1c_2\ldots c_m \longrightarrow c_1'c_2'\ldots c_n'$ and $g\colon d_1d_2\ldots d_{m'} \longrightarrow d_1'd_2'\ldots d_{n'}'$, then

$$f\bullet g\colon (c_1c_2\ldots c_m \bullet d_1d_2\ldots d_{m'}) \longrightarrow (c_1'c_2'\ldots c_n' \bullet d_1'd_2'\ldots d_{n'}'), \tag{6.34}$$

which corresponds to the morphism in $\mathbb{C}$ with all the parentheses on the left side of each of the two terms. The $\bullet$ is easily seen to be strictly associative, and the unit is the empty string $\emptyset$.

There is a functor $F\colon \mathbb{C}^\bullet \longrightarrow \mathbb{C}$, which is defined on objects as

$$F(c_1c_2\ldots c_m) = ((\cdots((c_1\otimes c_2)\otimes\ldots)\otimes c_m. \tag{6.35}$$

This functor is full and faithful because

$$Hom_{\mathbb{C}^\bullet}(c_1c_2\dots c_m, c_1'c_2'\dots c_n') = Hom_{\mathbb{C}}(F(c_1,c_2,\dots c_m), F(c_1',c_2',\dots c_n')). \quad (6.36)$$

The functor is also essentially surjective because for every bracketing of $c_1c_2\dots c_m$, there is an isomorphism from $((\cdots(c_1,c_2),\dots)c_m$ to that bracketing. (The coherence theorem tells us that the isomorphism is actually unique.) This makes F an equivalence of categories.

Functor F is a strong monoidal functor because there is a unique isomorphism:

$$\begin{aligned} F(c_1c_2\dots c_m)\otimes F(c_1'c_2'\dots c_n') &= ((\cdots((c_1\otimes c_2)\otimes\cdots)\otimes c_m \otimes ((\cdots((c_1'\otimes c_2')\otimes\cdots)\otimes c_n' \\ &\longrightarrow \\ &((\cdots((c_1\otimes c_2)\otimes\cdots)\otimes c_m)\otimes c_1')\otimes c_2')\cdots c_m')) \\ &= F(c_1c_2\dots c_m \bullet c_1'c_2'\dots c_n'). \end{aligned}$$

Functor F takes $\emptyset$, the unit of $\mathbb{C}^\bullet$, to I, the unit of $\mathbb{C}$. This means that the unital funnel is the identity.

Although we already proved that F is a monoidal equivalence, it pays to examine the quasi-inverse of F. The functor $G\colon \mathbb{C} \longrightarrow \mathbb{C}^\bullet$ takes any element c of $\mathbb{C}$ to the list containing that single element. For clarity, the element $c_1\otimes((c_2\otimes c_3)\otimes c_4)$ will go to the list of the single element $c_1\otimes((c_2\otimes c_3)\otimes c_4)$. The functor is defined similarly for morphisms. This functor is monoidal because there is the identity morphism

$$id_{c_1\otimes c_2}\colon G(c_1)\bullet G(c_2) = c_1c_2 \longrightarrow c_1\otimes c_2 = G(c_1\otimes c_2), \quad (6.37)$$

which is in the Hom set

$$Hom_{\mathbb{C}^\bullet}(c_1c_2, c_1\otimes c_2) = Hom_{\mathbb{C}}(c_1\otimes c_2, c_1\otimes c_2). \quad (6.38)$$

There is also a morphism in $\mathbb{C}^\bullet$ that is also an identity morphism $\emptyset \longrightarrow G(I) = I$.

Notice that G takes any element c in $\mathbb{C}$ to the one-object list in $\mathbb{C}^\bullet$ and that F in turn takes that one-object list to itself. This means that $FG = Id_{\mathbb{C}}$. In contrast, if we start with a list of elements $c_1,c_2,\dots c_m$ in $\mathbb{C}^\bullet$ and apply F to it, we get the element $((\cdots(c_1\otimes c_2),\cdots)\otimes c_m$ in $\mathbb{C}$. Applying G to this gives us the one-element list with $((\cdots(c_1\otimes c_2)\cdots)\otimes c_m$. There is an isomorphism $id_{((\cdots(c_1,c_2),\dots)c_m}$. So we have $GF \longrightarrow Id_{\mathbb{C}^\bullet}$. ♣

Remark 6.2.4. Before we move to the next proof, it pays to understand some historical background. The notion of a group goes back to the work of Évariste Galois in the early 1830. For him, a group was a concrete set of transformations or rearrangements (of polynomial roots) and a way of composing such transformations. As time went on, researchers formulated the notion of an abstract group (Definition 1.4.56) as a set with a binary operation, a unit, and an inverse operation. In the 1850s, Arthur Cayley proved what became known as **Cayley's theorem**. This representation theorem says that any abstract group is isomorphic to a subset of a collection of concrete transformations of objects (as first described by Galois). Cayley's theorem shows that even though groups can get fairly abstract, they can all still be seen as a more concrete set of transformations.

While Cayley first described his theorem for groups, we will examine Cayley's theorem for monoids. This theorem says that every monoid $(M, \star, e)$ is isomorphic to a concrete monoid of transformations. If we are to find such a concrete set of transformations, the first question to ask is: What elements should this monoid transform? In other words, what are the objects that are transformed? The answer is the set of elements in M. We look at all functions from M to M and select some of them. When we multiply an element $m \in M$ by an element $m' \in M$, we get $m'' = m \star m'$. Thus, m corresponds to a transformation or a set function $\hat{m}\colon M \longrightarrow M$, which performs the transformation $m' \mapsto m''$. The function $\hat{m}$ is defined as $\hat{m}(m') = m \star m' = m''$. Not all functions from M to M are relevant. Only those functions $f\colon M \longrightarrow M$ that respect the monoid operation; that is,

$$f(m') \star m'' = f(m' \star m''). \tag{6.39}$$

Such functions form a monoid M° under composition. If f corresponds to multiplying by m and f' corresponds to multiplying by m', then $f' \circ f$ corresponds to multiplying by $m' \star m$. This collection of transformations and this composition is the monoid isomorphic to the original M.

In a sense, Theorem 6.2.3 states Cayley's theorem for monoidal categories. Every abstract monoidal category is equivalent to a concrete strict monoidal category where the tensor is composition.

In a similar vein, the Yoneda embedding theorem discussed in Section 4.7 of Chapter 4 is also a Cayley-type theorem. It showed that every abstract category is isomorphic to—or can be embedded in—a more concrete category of functors $\mathbb{Set}^{\mathbb{A}^{op}}$. In fact, Cayley's theorem on groups and monoids, as well as Theorem 6.2.3, can all be seen as special instances of a general Yoneda lemma. ♠

Proof. Using function composition. As we saw in Example 5.1.7, the collection of endomorphisms of a category with composition is a strict monoidal category. Let $(\mathbb{C}, \otimes, \alpha, I, \lambda, \rho)$ be a monoidal category. We will form such a strict monoidal category $(\mathbb{C}^\circ, \circ, Id_{\mathbb{C}})$. The monoidal category $\mathbb{C}$ will be monoidally equivalent to a subcategory of this strict monoidal category.

First, let us describe category $\mathbb{C}^\circ$. The objects are pairs (F, τ), where $F\colon \mathbb{C} \longrightarrow \mathbb{C}$ is a functor and τ is a natural isomorphism whose components are given as

$$\tau_{a,b}\colon F(a) \otimes b \longrightarrow F(a \otimes b). \tag{6.40}$$

This natural isomorphism is the categorical version of Equation (6.39). In this equation, the τ is there to ensure that F is not just any endofunctor, but one that respects $\otimes$. The morphisms from (F, τ) to (F', τ') are natural transformations $\theta\colon F \Longrightarrow F'$ such that

$$\begin{array}{ccc} F(a) \otimes b & \xrightarrow{\theta_a \otimes id_b} & F'(a) \otimes b \\ \downarrow{\scriptstyle \tau_{a,b}} & & \downarrow{\scriptstyle \tau'_{a,b}} \\ F(a \otimes b) & \xrightarrow[\theta_{a \otimes b}]{} & F'(a \otimes b). \end{array} \tag{6.41}$$

The composition of morphisms in $\mathbb{C}^\circ$ is as follows. Given $\theta\colon (F,\tau) \longrightarrow (F',\tau')$ and $\theta'\colon (F',\tau') \longrightarrow (F'',\tau'')$, they are composed as $\theta' \circ \theta$, and the coherence condition is satisfied because the two squares compose.

Category $\mathbb{C}^\circ$ has a strict monoidal category structure where the tensor is functor composition. In detail,

$$(F,\tau)\otimes(F',\tau') = (F\circ F', \tau'\cdot\tau), \tag{6.42}$$

where $\tau'\cdot\tau$ is defined as

$$F(F'(a))\otimes b \xrightarrow{\tau_{F'(a),b}} F(F'(a)\otimes b) \xrightarrow{F(\tau'_{a,b})} F(F'(a\otimes b)). \tag{6.43}$$

The main point is that we use this monoidal product because it is strictly associative. That is,

$$(F,\tau)\otimes((F',\tau')\otimes(F'',\tau'')) = (F\circ F'\circ F'', \tau_4) = ((F,\tau)\otimes((F',\tau'))\otimes(F'',\tau''), \tag{6.44}$$

where τ_4 is

$$\begin{aligned} F(F'(F''(a)))\otimes b &\xrightarrow{\tau_{F'(F''(a)),b}} F(F'(F''(a))\otimes b) \xrightarrow{F(\tau'_{F''(a),b})} F(F'(F''(a)\otimes b)) \\ &\qquad\qquad\qquad\qquad\qquad\quad \xrightarrow{F(F'(\tau''_{a,b}))} F(F'(F''(a\otimes b))). \end{aligned} \tag{6.45}$$

The tensor product on morphisms is done similarly. Let $\theta\colon (F,\tau) \Longrightarrow (F',\tau')$ and $\theta'\colon (F'',\tau'') \Longrightarrow (F^3,\tau^3)$, as in

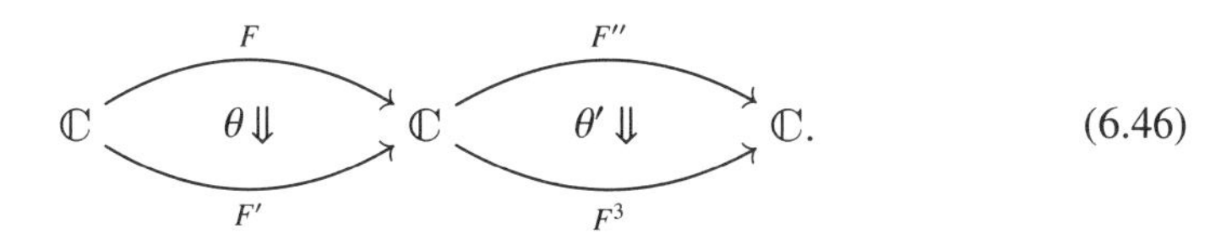

(6.46)

Then

$$\theta'\otimes\theta = \theta'\circ_H \theta\colon (F''\circ F) \Longrightarrow (F^3\circ F'). \tag{6.47}$$

To see that this composition satisfies Diagram (6.41), consider the square in Figure 6.9. The square commutes for the following reasons:

- The top-left square commutes because it is an instance of Diagram (6.41).
- The top-right square commutes because of the naturality of τ^3.
- The bottom-left square commutes because of the naturality of θ'.
- The bottom-right square commutes because of the functoriality of F^3 and Diagram (6.41).
- The outside triangles commute because of the various definitions of composition.

The unit element is $(Id_{\mathbb{C}}, id_{a\otimes b})$.

There is a functor $H\colon \mathbb{C} \longrightarrow \mathbb{C}^\circ$. The original monoidal category $\mathbb{C}$ will be monoidally equivalent to the subcategory of the image of the H functor. For an object

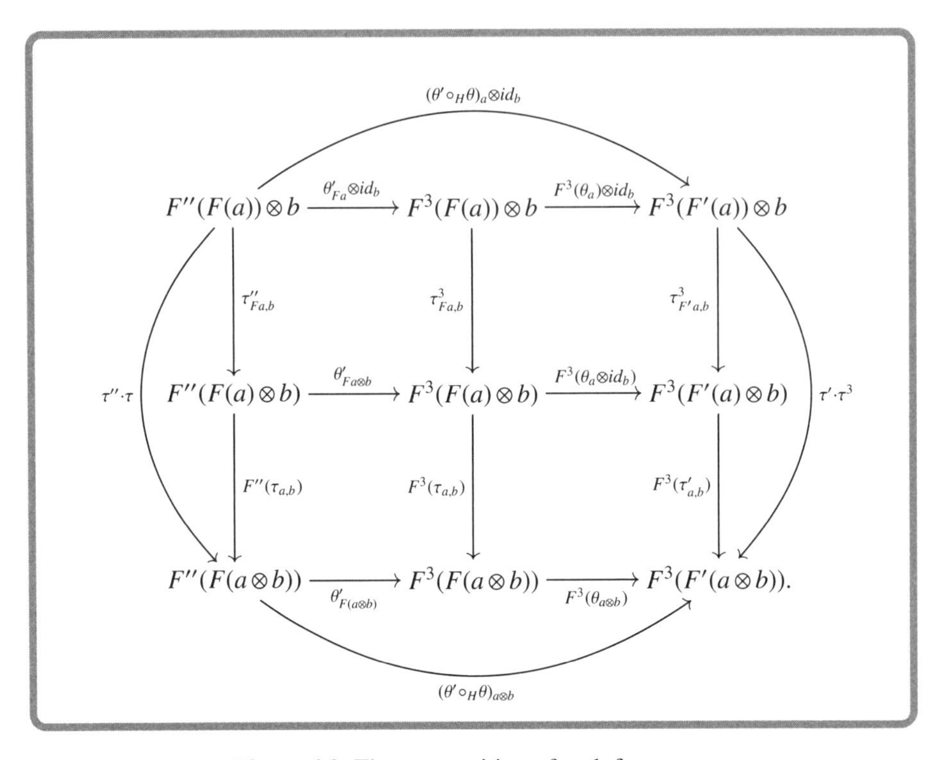

Figure 6.9. The composition of endofunctors.

a in $\mathbb{C}$, we define $H(a)$ as the pair $(a \otimes (\ ), \alpha_{a,(\ ),(\ )})$. The first part of the pair is the endofunctor $a \otimes (\ ) \colon \mathbb{C} \longrightarrow \mathbb{C}$, which tensors every element b with a to get $a \otimes b$. The second part of the pair is a natural transformation whose component on b and c is the reassociation of $\mathbb{C}$–namely, $\alpha_{a,b,c} \colon a \otimes (b \otimes c) \longrightarrow (a \otimes b) \otimes c$. Here, α is doing the job of Diagram (6.40). On a morphsim $f \colon a \longrightarrow a'$, functor H is defined as

$$H(f) = f \otimes (\ ) \colon a \otimes (\ ) \Longrightarrow a' \otimes (\ ). \tag{6.48}$$

Diagram (6.41) becomes

$$\begin{array}{ccc} a \otimes (b \otimes c) & \xrightarrow{f \otimes (id_b \otimes id_c)} & a' \otimes (b \otimes c) \\ {\scriptstyle \alpha_{a,b,c}} \downarrow & & \downarrow {\scriptstyle \alpha_{a',b,c}} \\ (a \otimes b) \otimes c & \xrightarrow[(f \otimes id_b) \otimes id_c]{} & (a' \otimes b) \otimes c. \end{array} \tag{6.49}$$

Functor H is faithful. Let f and g be morphisms in $\mathbb{C}$ and $H(f) = H(g)$. This means that $f \otimes (\ ) = g \otimes (\ )$. In particular, this is true for the I component of this natural transformation, which means that $f \otimes I = g \otimes I$. The naturality of ρ means that if the left two arrows are equal, then the right two arrows are equal:

$$\begin{array}{ccc} a \otimes I & \xrightarrow{\rho_a} & a \\ {\scriptstyle g\otimes I}\downarrow\ \downarrow{\scriptstyle f\otimes I} & & {\scriptstyle g}\downarrow\ \downarrow{\scriptstyle f} \\ b \otimes I & \xrightarrow[\rho_b]{} & b \end{array} \tag{6.50}$$

that is $f = g$.

Functor H is full. Let $f\colon H(a) \Longrightarrow H(b)$. That is, $f\colon a \otimes (\ ) \Longrightarrow b \otimes (\ )$. Consider $\hat{f}$ to be the composition in $\mathbb{C}$ of

$$a \xrightarrow{\rho_a^{-1}} a \otimes I \xrightarrow{f_I} b \otimes I \xrightarrow{\rho_b} b. \tag{6.51}$$

We claim that $f = H(\hat{f})$ thus making H full. To see this, consider the following commutative diagram:

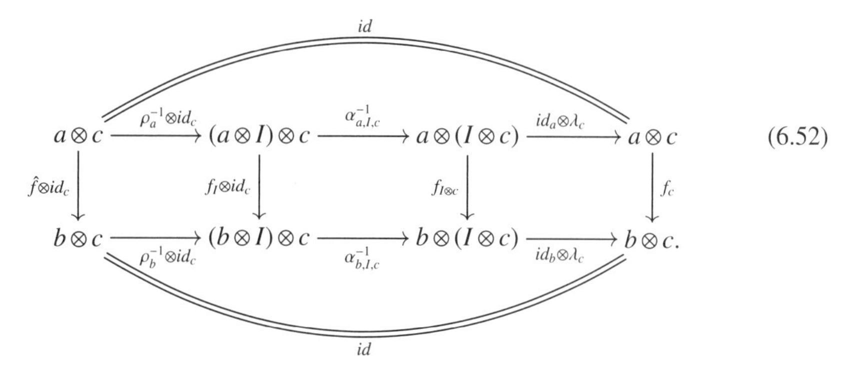

(6.52)

- The left square commutes because the square is the definition of $\hat{f}$ tensored with id_c.
- The middle square commutes because it is an instance of Diagram (6.49).
- The right square commutes by the naturality of f.
- The top and bottom semi-ellipses commute because of Diagram (5.35).

The fact that the entire diagram commutes means that $\hat{f} \otimes id_c = f_c$ for every object c. Hence, we have proved that $H(\hat{f}) = f$.

The functor H is a strong monoidal functor. The object $H(a) \otimes H(b)$ is the pair

$$((a \otimes (b \otimes (\ )), \tau_{(\ ),(\ )} = (id_a \otimes \alpha^{-1}_{b,(\ ),(\ )}) \circ (\alpha^{-1}_{a,b\otimes(\ ),(\ )})). \tag{6.53}$$

This means that for all objects c and d, we have

$$\tau_{c,d}\colon (a(bc))d \xrightarrow{\alpha^{-1}_{a,bc,d}} a((bc)d) \xrightarrow{id_a \otimes \alpha^{-1}_{b,c,d}} a(b(cd)). \tag{6.54}$$

The object $H(a \otimes b)$ is the pair

$$((a \otimes b) \otimes (\ ), \tau'_{(\ ),(\ )} = \alpha^{-1}_{a\otimes b,(\ ),(\ )}). \tag{6.55}$$

This means that for all objects c and d, we have

$$\tau'_{c,d} = \alpha^{-1}_{ab,c,d} \colon ((ab)c)d \longrightarrow (ab)(cd). \tag{6.56}$$

The mapping funnel from $H(a) \otimes H(b)$ to $H(a \otimes b)$ is a

$$\theta = \alpha_{a,b,(\ )} \colon a \otimes (b \otimes (\ )) \Longrightarrow (a \otimes b) \otimes (\ ). \tag{6.57}$$

The fact that $\theta = \alpha_{a,b,(\ )}$ is a map of pairs means that it must satisfy Diagram (6.41). This is true because its commutativity is exactly the commutativity of the pentagon condition, Diagram (5.34), as can be seen in the following diagram:

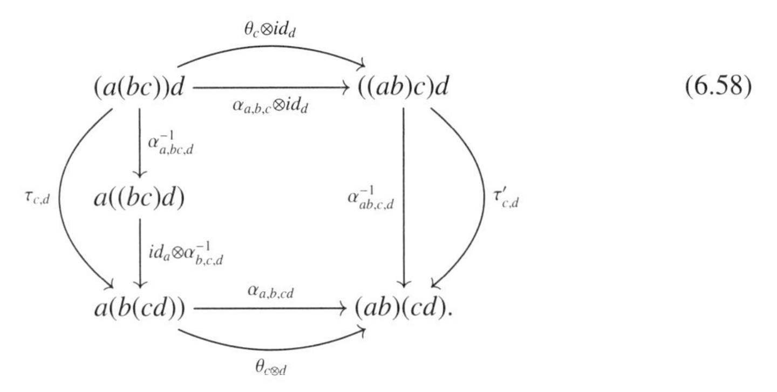

(6.58)

The unit funnel is simply the map $\rho^{-1} \colon I \Longrightarrow H(I) = I \otimes (\ )$.

The monoidal category $\mathbb{C}$ is not monoidally equivalent to the whole strict monoidal category $(\mathbb{C}^\circ, \circ, Id)$. Rather, there is a subcategory of $\mathbb{C}^\circ$ that consists of all functors in the image of H. Thus, H is onto—and certainly essentially surjective—to this subcategory. Thus, we have shown that H is a strong monoidal equivalence of categories. ♣

Remark 6.2.5. Notice that in contrast to the previous proof of the strictification theorem, this proof did not need to assume the coherence theorem. It proves the strictification theorem without the coherence theorem. ♠

This theorem shows that the inclusion 2-functor

$$\overline{\mathbb{S}\mathbb{t}\mathbb{r}\mathbb{M}\mathbb{o}\mathbb{n}\mathbb{C}\mathbb{a}\mathbb{t}} \hookrightarrow \overline{\mathbb{M}\mathbb{o}\mathbb{n}\mathbb{C}\mathbb{a}\mathbb{t}} \tag{6.59}$$

is not only full and faithful, but it is almost essentially surjective in the sense that every monoidal category is monoidally *equivalent* (not necessarily isomorphic) to a strict monoidal category.

How does this theorem help us? When we use our favorite category like $\mathbb{S}\mathbb{e}\mathbb{t}$, $\mathbb{G}\mathbb{r}\mathbb{o}\mathbb{u}\mathbb{p}$, $\mathbb{G}\mathbb{r}\mathbb{a}\mathbb{p}\mathbb{h}$, or $\mathbb{T}\mathbb{o}\mathbb{p}$, we know that the Cartesian product is not strict. This theorem tells us that although it is not strict, it is strongly monoidally equivalent to another category that does have a strict monoidal product. Since this equivalence is monoidal, most of the properties of the strict monoidal category are the same as the original category. So what is true about the strict monoidal category is true about the equivalent original monoidal category. Therefore, when dealing with a monoidal category, you might as well imagine that it is a strict monoidal category.

Remark 6.2.6. The above theorem is subtle. We have shown that every monoidal category is monoidally equivalent to a strict monoidal category. Most people use this fact so as not to worry that the product in their favorite categories like $\mathbb{Set}$ and $\mathbb{Top}$ are not strictly associative. The reasoning is that it is well known that all "mathematically relevant" properties are preserved and reflected by functors that are part of an equivalence. This means that if a property is true in a strict monoidal category, then it must also be true in the monoidal category to which it is equivalent. Hence, they might as well use the properties of the strict monoidal category.

Are we really sure this is true for *all* mathematically relevant properties? As we saw in Remark 4.6.8, equivalence of categories can take a diagram that is not an equalizer to an equalizer. Now we have to ask about monoidal equivalence. Does a monoidal equivalence show that the source and the target have the same mathematical relevant properties? I do not know.

Peter Freyd describes what properties are the same when there is an equivalence of categories (see [81] and Section 1.3 of [82]). This was done in the 2-category of $\mathbb{Cat}$. We are asking about the 2-category of interest here (namely, $\mathbb{MonCat}$). What is exactly true about a strict monoidal category and not true about its equivalent monoidal category? As far as I know, no one has worked out the details of this. It is worthy of further study. ♠

Now we use the strictification theorem to prove the coherence theorem.

Theorem 6.2.7. Let $\mathbb{C}$ be a monoidal category and $\mathbb{C}^x$ be a strict monoidal category that is equivalent to $\mathbb{C}$ with a monoidal functor $L\colon \mathbb{C} \longrightarrow \mathbb{C}^x$. ($\mathbb{C}^x$ can be $\mathbb{C}^\bullet$ or $\mathbb{C}^\circ$, as given in the two proofs of the strictification theorem.) Let u and v be two ways of bracketing any objects in $\mathbb{C}$ (i.e., they are two functors $\mathbb{C}^n \longrightarrow \mathbb{C}$, which only use the tensor product, identities, and composition). If ϕ and ψ are two natural transformations from u to v that are generated by identities, α, ρ, λ and their inverses, and are built up using $\circ$ and $\otimes$, then $\phi = \psi$. ★

Proof. Let $c_1, c_2, \ldots, c_n$ be a sequence of elements of $\mathbb{C}$. Then consider u to be a particular bracketing of elements and $u(c_1, c_2, \ldots, c_n)$ to be the bracketing of those elements. We are interested in natural transformations $u \Longrightarrow v$ between two bracketings and maps, which are instances of such natural transformations $u(c_1, c_2, \ldots, c_n) \longrightarrow v(c_1, c_2, \ldots, c_n)$. For every bracketing u, there is an isomorphism

$$L_u\colon u(L(c_1), L(c_2), \ldots, L(c_n)) \longrightarrow L(u(c_1, c_2, \ldots, c_n)), \tag{6.60}$$

which is built out of the mapping funnels of L. We can see instances of such funneling in Diagram (6.6) and Figure 6.2. Let $\phi, \psi\colon u \Longrightarrow v$ be two natural transformations of the type described in the theorem. We then have the following diagram of $\mathbb{C}$, $\mathbb{C}^x$, and L:

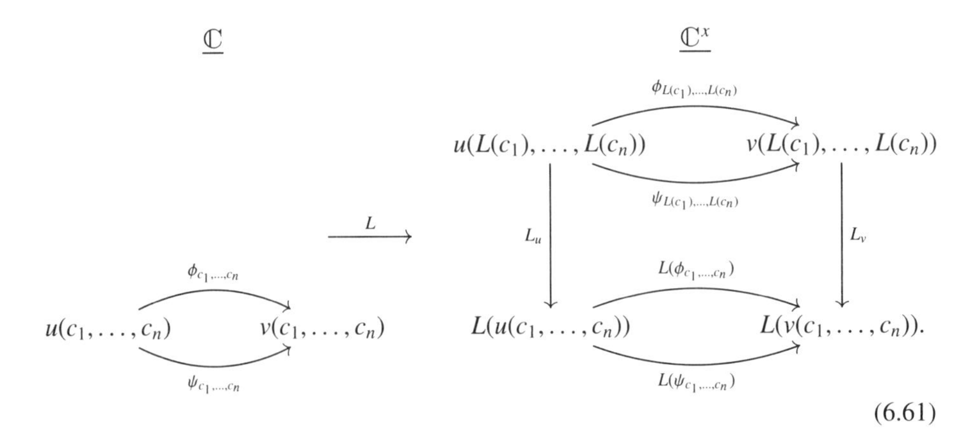

(6.61)

In the right part of the diagram, the top two maps are equal because $\mathbb{C}^x$ is a strict monoidal category. From the naturality of ϕ and ψ, the bottom two maps are also equal. Hence, we have that $L(\phi_{c_1,\ldots,c_n}) = L(\psi_{c_1,\ldots,c_n})$. Since L is faithful, this means that $\phi_{c_1,\ldots,c_n} = \psi_{c_1,\ldots,c_n}$. This is true for all $c_1,\ldots c_n$, so $\phi = \psi$. ♣

Now that we have dealt with coherence for monoidal categories, we can move on to coherence theory for symmetric monoidal categories. Here, we work with the category $\mathbb{Sym}$ rather than the category $\mathbb{Assoc}$.

Theorem 6.2.8. The Coherence Theorem for Symmetric Monoidal Categories. We state it in four equivalent ways:

- For every symmetric monoidal category $(\mathbb{C}, \otimes, I, \alpha, \rho, \lambda, \gamma)$ and every object x in $\mathbb{C}$, there is a unique, strict, symmetric monoidal functor $F_x\colon \mathbb{Sym} \longrightarrow \mathbb{C}$ such that $F_x(1) = x$.
- There is an isomorphism of categories:

$$Hom(\mathbb{Sym}, \mathbb{C}) \cong Hom_{\mathbb{Cat}}(\mathbf{1}, \mathbb{C}), \tag{6.62}$$

  where the *Hom* set on the left is a strict symmetric monoidal functor and the Hom set on the right is from $\mathbb{Cat}$.
- The symmetric monoidal category $\mathbb{Sym}$ is the free symmetric monoidal category on one generator.
- Two morphisms in $\mathbb{C}$ generated by images of F_x of identities $\alpha, \rho, \lambda, \gamma$, and their inverses, built up by $\circ$ and $\otimes$, are equal if the two underlying permutations are the same. ★

Proof. This proof is the same format as the last proof. We will only comment when the ingredients are different. Remember that the objects in $\mathbb{Sym}$ are the natural numbers. For every object x in $\mathbb{C}$, there is a strictly associative symmetric monoidal functor F_x such that $F_x(1) = x$. In $\mathbb{Sym}$, $2 = 1 + 1$, and so $F_x(2) = x \otimes x$. In general, $F_x(n) = x^{\otimes n}$.

The braiding $\gamma_{1,1}\colon 2 \longrightarrow 2$ in $\mathbb{S}\mathrm{ym}$ must go to $F_x(\gamma_{1,1}) = \gamma_{x,x}\colon x \otimes x \longrightarrow x \otimes x$. Every map $n \longrightarrow n$ in $\mathbb{S}\mathrm{ym}$ is an element of the nth symmetric group. In Example 5.3.10 we saw that every element in the nth symmetric group can be written as a composition of the small permutations $s_1, s_2, \ldots, s_{n-1}$. The small permutation can be written as $s_i = id_{i-1} \otimes \gamma_{1,1} \otimes id_{n-i-1}$. The functor F_x takes this morphism to $id_{x^{\otimes(i-1)}} \otimes \gamma_{x,x} \otimes id_{x^{\otimes(n-i-1)}}$. We must show that the three defining relations of the symmetric group are satisfied by the category $\mathbb{C}$ as follows:

- $s_i s_i = e$ corresponds to the symmetry coherence condition shown in Diagram (5.38).
- $s_i s_j = s_j s_i$ for $|i - j| > 1$ corresponds to the fact that $\otimes$ is a bifunctor:

$$\begin{array}{ccc} x^{\otimes n} & \xrightarrow{id_{x^{\otimes(i-1)}}\otimes\gamma_{x,x}\otimes id_{x^{\otimes(n-i-1)}}} & x^{\otimes n} \\ {\scriptstyle id_{x^{\otimes(j-1)}}\otimes\gamma_{x,x}\otimes id_{x^{\otimes(n-j-1)}}}\downarrow & & \downarrow{\scriptstyle id_{x^{\otimes(j-1)}}\otimes\gamma_{x,x}\otimes id_{x^{\otimes(n-j-1)}}} \\ x^{\otimes n} & \xrightarrow[id_{x^{\otimes(i-1)}}\otimes\gamma_{x,x}\otimes id_{x^{\otimes(n-i-1)}}]{} & x^{\otimes n}. \end{array} \tag{6.63}$$

- $s_i s_{i+1} s_i = s_{i+1} s_i s_{i+1}$ corresponds to the fact that Diagram (5.108) commutes.

Functor F_x is a monoidal category that takes side-by-side braids to tensor products in $\mathbb{C}$. It is also not hard to show that this evaluation at 1 for every F_x gives the equivalence of categories in Diagram (6.62). ♣

Notice that there is a major difference between the coherence of monoidal categories and the coherence of symmetric monoidal categories. With monoidal categories, every diagram of a certain class of morphisms generated by the structure maps commutes. In contrast, with symmetric monoidal categories diagrams commute only if the underlying permutations commute. For example, the following diagram commutes in a symmetric monoidal category:

$$\begin{array}{ccc} a \otimes b & \xrightarrow{\gamma} & b \otimes a \\ {\scriptstyle\gamma}\downarrow & & \downarrow{\scriptstyle\gamma} \\ b \otimes a & \xleftarrow[\gamma]{} & a \otimes b \end{array} \tag{6.64}$$

because

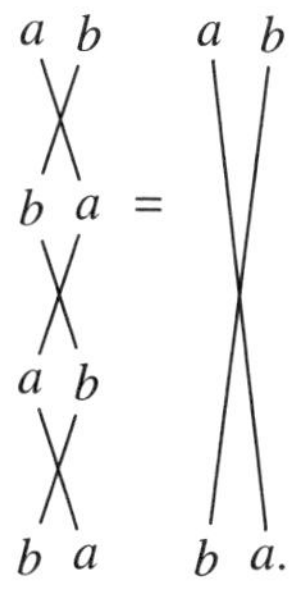

$$\begin{array}{ccc} a\ b & & a\ b \\ \times & & \\ b\ a & = & \times \\ \times & & \\ a\ b & & \\ \times & & \\ b\ a & & b\ a. \end{array} \tag{6.65}$$

In contrast, the following diagram does not commute in a symmetric monoidal category:

$$a \otimes a \underset{\gamma_{a,a}}{\overset{id_{a\otimes a}}{\rightrightarrows}} a \otimes a. \tag{6.66}$$

The point of the coherence theorem is that if two morphisms in the category describe the same permutation in the symmetric group, then the two morphisms are the same.

The strictification theorem for monoidal categories extends to symmetric monoidal categories.

Theorem 6.2.9. Strictification Theorem for Symmetric Monoidal Categories. Every symmetric monoidal category is monoidally equivalent to a strictly associative, symmetric monoidal category. This equivalence is via symmetric monoidal functors and symmetric monoidal natural transformations. This is done by making the underlying monoidal category strict and then transporting the symmetric structure to the strict monoidal structure. ★

This theorem shows that the inclusion 2-functor

$$\overline{\mathbb{S}\mathrm{tr}\mathbb{S}\mathrm{ym}\mathbb{M}\mathrm{on}\mathbb{C}\mathrm{at}} \hookrightarrow \overline{\mathbb{S}\mathrm{ym}\mathbb{M}\mathrm{on}\mathbb{C}\mathrm{at}} \tag{6.67}$$

is not only full and faithful, but it is almost essentially surjective in the sense that every symmetric monoidal category is monoidally equivalent (not necessarily isomorphic) via symmetric monoidal functors and symmetric monoidal natural transformations to a strictly associative, symmetric monoidal category.

6.3 When Coherence Fails

While the coherence theorem is one of the central ideas in the world of monoidal categories, it is only the beginning of the story. We will see that there are many other structures, and for each structure, there are many different coherence conditions.

Let us examine the pentagon coherence condition. Consider the category of vector spaces and let V, W, and X be three vector spaces. Using the tensor product, we form the vector spaces $V \otimes (W \otimes X)$ and $(V \otimes W) \otimes X$. There is a linear isomorphism $\alpha\colon V \otimes (W \otimes X) \longrightarrow (V \otimes W) \otimes X$, defined by

$$\langle v, \langle w, x\rangle\rangle \longmapsto \langle\langle v, w\rangle, x\rangle, \tag{6.68}$$

which satisfies the pentagon coherence condition. Formally, the fact that the pentagon commutes means that

$$(\alpha_{V,W,X} \otimes id_Y) \circ (\alpha_{V,WX,Y}) \circ (id_V \otimes \alpha_{W,X,Y}) = (\alpha_{VW,X,Y}) \circ (\alpha_{V,W,XY}). \tag{6.69}$$

Since the α's are isomorphisms and we can work with the inverses of the morphisms, we can rewrite this as

$$(\alpha_{VW,X,Y})^{-1} \circ (\alpha_{V,W,XY})^{-1} \circ (\alpha_{V,W,X} \otimes id_Y) \circ (\alpha_{V,WX,Y}) \circ (id_V \otimes \alpha_{W,X,Y}) = id_{V(W(XY))}. \tag{6.70}$$

That is, rather than saying that the composite of the three maps in the pentagon coherence condition is the same map as the composite of the two other maps, we can say that going around the pentagon completely brings one right back to the element where one started.

Now let us look at an example where the pentagon coherence condition fails. Consider the linear isomorphism $\alpha' \colon V \otimes (W \otimes X) \longrightarrow (V \otimes W) \otimes X$, defined by

$$\langle v, \langle w, x \rangle\rangle \longmapsto (-1)\langle\langle v, w\rangle, x\rangle. \tag{6.71}$$

While this is a legitimate isomorphism, the pentagon coherence condition is *emphatically* not satisfied. Using the three composable maps means taking the elements to the composition of three -1's, meaning that

$$\langle v, \langle w, \langle x, y\rangle\rangle \longmapsto (-1)\langle\langle\langle v, w\rangle, x\rangle, y\rangle. \tag{6.72}$$

In contrast, using the other two maps has two (an even number) (-1)'s, which gives the map

$$\langle v, \langle w, \langle x, y\rangle\rangle \longmapsto \langle\langle\langle v, w\rangle, x\rangle, y\rangle. \tag{6.73}$$

These maps are different. Another way of saying this is that going all the way around the pentagon condition gives us

$$(\alpha'_{VW,X,Y})^{-1} \circ (\alpha'_{V,W,XY})^{-1} \circ (\alpha'_{V,W,X} \times id_Y) \circ (\alpha'_{V,WX,Y}) \circ (id_V \times \alpha'_{W,X,Y}) = (-1)Id_{V(W(XY))}. \tag{6.74}$$

This is not the identity. Rather than saying that α' is coherent, we say that it has an "obstruction to coherence." Notice that going around the pentagon with α' does not give the identity, but going around the pentagon with α' *twice* is a composition of ten (an even number) linear maps $(-1^{10} = 1)$, and hence it does give the identity. This example can be generalized: for every n, there are reassociators that commute when going around the pentagon n times and no less.

In my thesis [266, 267], I studied different types of pentagon coherence conditions and worked with general reassociators that have no requirement of making the pentagon commute at all. That is, they will not commute going around the pentagon any amount of times. This reassociator is simply an isomorphism and has no commuting axiom. This is like an associahedron with open pentagons instead of filled-in pentagons. These shapes are interesting. Such an associahedron for four letters will be equivalent to a circle and not a (commutative) disk. Such an associahedron for five letters is equivalent to a sphere with six pentagons removed and only four naturality squares in place. This is different than Figure 5.7 where we saw that there is a unique morphism between any two vertices as a filled-in sphere.

I was able to describe the shapes (actually the fundamental homotopy groups) of all such degenerate associahedra. One can learn about how uncoherent a reassociator is by looking at these shapes.

While working on these issues, I came upon an interesting shape. Consider the associahedron on seven letters, A_7. There are 132 ways of bracketing 7 letters, and hence, this shape has 132 vertices. Figure 6.10 shows a part of A_7 that looks like 36 objects. Notice that each row and column has three arrows one way and two arrows the other way. That is, each row and column is actually an instance of Mac Lane's pentagon. For readability, Figure 6.11 shows a "zoomed-in" view of the upper-left corner of Figure 6.10. The horizontal α maps move the round parentheses and the vertical β maps move the square parentheses. By examining the subscripts of the α and β maps, one sees that each square commutes because of naturality. Notice also that the top and the bottom rows of Figure 6.10 are exactly the same. Similarly, the left column is exactly the same as the right column. (This is similar to the Pac-Man game screen, where the players can go out the top and come in at the bottom and go out the left and come in on the right, or vice versa) Since the top and the bottom rows are the same, we can "bend the paper" and glue the edges together, as in Figure 6.12. Since the left and right columns are the same, we can "bend the tube" and paste them together as depicted. This means that within A_7, there is a hollow doughnut or torus. Because the top row is the same as the bottom row and the left column is the same as the right collumn, there are really 25 distinct objects in Figure 6.10. In the case where we assume that the Mac Lane pentagon commutes, then from any vertex in the torus to any other vertex, there is a unique map. This is similar to when the doughnut is full. However, if we do not assume that the pentagon commutes, there is at least one hollow doughnut built into A_7. This is true for A_7 and also true for any higher associahedra. There are many other interesting shapes within the associahedra, permutohedra, and permuto-associahedra.

There are many other coherence conditions discussed in the literature. For example, we saw that we can require the reassociator to be the identity (strict monoidal categories) or an isomorphism (monoidal categories). What if we require that the reassociator just be a morphism that satisfies the pentagon condition? A category with a monoidal product featuring a reassociator that need not be an isomorphism is called a **pre-monoidal category** or a **skew monoidal category**. There is a lot of applications for such structures in diverse areas—including an application to quark confinement in the physics literature [120].

We will see more coherence conditions in Section 7.1 where we will discuss symmetric monoidal categories that fail the symmetry condition in Diagram (5.38). Such a categorical structure is called a **braided monoidal category** and has applications in many upcoming sections.

Not only are there many different types of coherence conditions for categories, there are also many different coherence conditions for functors. We outlined some of the varieties in Definition 6.1.4. But that is not the end of the story. In my thesis [266], I studied functors F with a natural transformation mapping funnel $\tau\colon \otimes' \circ (F \times F) \Longrightarrow F \circ \otimes$ that do not necessarily satisfy the hexagon coherence condition, Diagram (6.6). With this diagram not commuting, there is no reason why Figure 6.2 should commute. I studied various types of mapping funnels for n letters.

The point that we are making in this section is that coherence conditions are not toggle switches that are either on or off. Rather, there are a whole spectrum of coherence conditions, and every coherence condition implies different properties. The

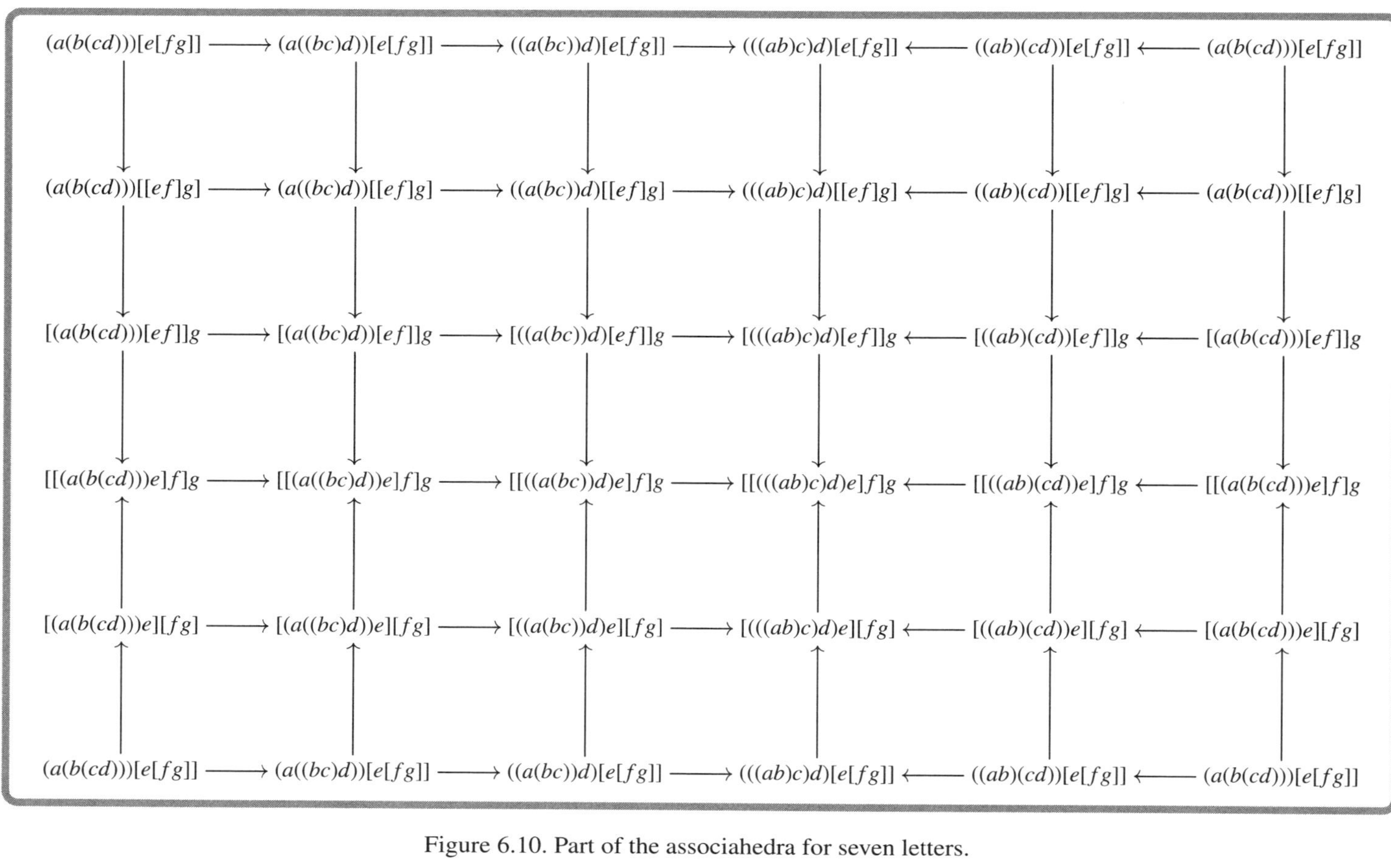

Figure 6.10. Part of the associahedra for seven letters.

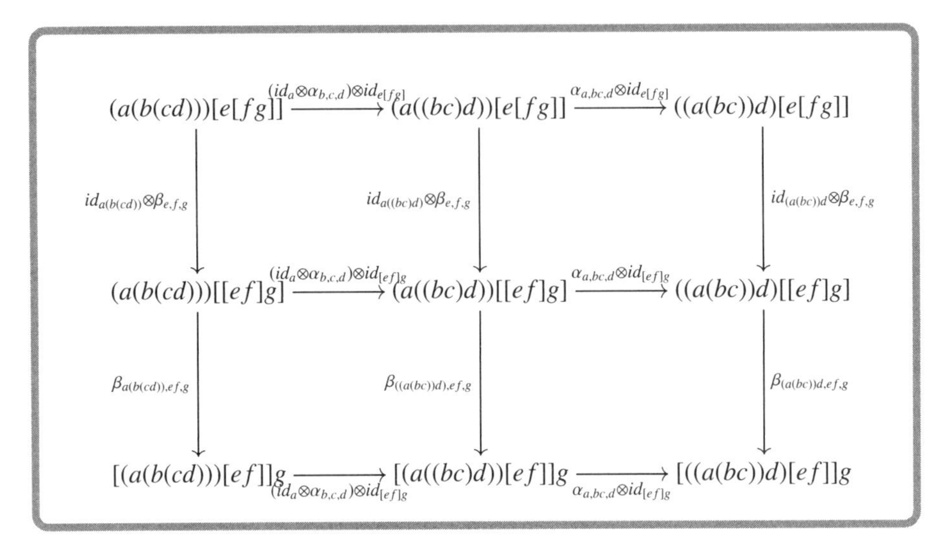

Figure 6.11. The top-left part of the torus for seven letters.

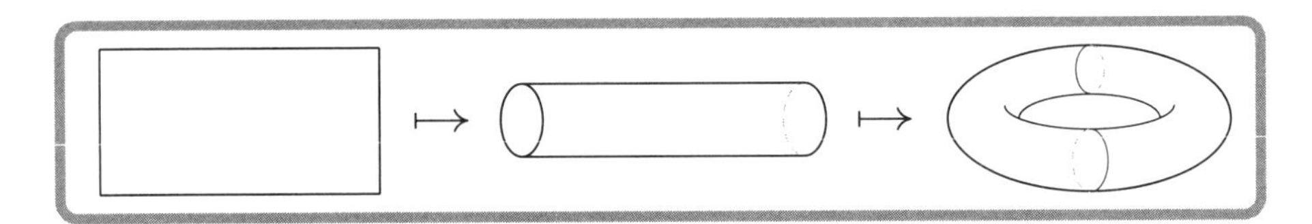

Figure 6.12. Folding part of the associahedra for seven letters into a torus.

different structures are used to describe various mathematical and real-world phenomena. Just as algebra is used in every aspect of mathematics and science, coherence theory—also called **higher-dimensional algebra**—arises in many areas of mathematics and science. It is certain that we will see much more coherence theory in the coming decades.

Suggestions for Further Study

The notion of a monoidal functor was first formulated by D. B. A. Epstein in [70].

The two proofs of the strictification theorem for monoidal categories can be found in various texts:

- The proof using string concatenation can be found in Section XI.3 of [180] and Section XI.5 of [134].
- The proof using function composition can be found in a version of [131], Section 1.3.3 of [110], Section 2.8 of [71], and Exercises 2 and 3 of Section XI.3 of [180].

The coherence theorem for monoidal categories can be found in Section XI.3 of [180], Section 1.3.4 of [110], and Section 2.9 of [71].

6.4 Mini-Course: Duality Theory

Category theory unifies many areas. One of the best ways to unify two subjects is to show that two seemingly different categories essentially have the same structure. A **duality theorem** is a statement showing that there exists an equivalence of one category with another, for example, $\mathbb{A} \simeq \mathbb{B}$, or more typically, an equivalence of one category with the opposite of a category, for example, $\mathbb{A}^{op} \simeq \mathbb{B}$. This shows that the two categories are two sides of the same coin or two ways of describing the same structure. Categories that are dual to each other are mirror images of each other, and we can study one category by looking at the properties of the other. For example, a monomorphism in one category corresponds to an epimorphism in the other category. An initial object in one category corresponds to a terminal object in the other category. Many times, duality theorems are stated as adjunctions between $\mathbb{A}^{op}$ and $\mathbb{B}$, and we use Theorem 4.4.23 to find the two subcategories that are equivalent.

We have been fortunate to have seen many examples of duality in these pages. In Examples 4.3.6 through 4.3.10, we met several instances of equivalences of categories. In Examples 4.1.28 and 4.2.12, we saw the following equivalences: $\mathbb{Rel}^{op} \cong \mathbb{Rel}$, $\mathbf{K}\mathbb{Mat}^{op} \cong \mathbf{K}\mathbb{Mat}$, and $\mathbf{K}\mathbb{FDVect}^{op} \cong \mathbf{K}\mathbb{FDVect}$. Notice that these four examples are actually stronger than the usual duality theorems because they are isomorphisms rather than equivalences, and furthermore, they are instances of **self duality**, that is, they show a category equivalent to the opposite of itself.

Many other examples of duality theorems will be shown in this mini-course. We focus on a group of duality theorems collectively known as **Stone duality**. These theorems show that certain types of topological structures are related to certain types of algebraic structures. The algebraic structures are centered around Boolean algebras.

Boolean Algebras

One of the main purposes of this text is to show how different fields are related and unified. Boolean algebra is a subject that has been unifying different fields long before category theory came on the scene. A Boolean algebra is a structure that is used in mathematical logic, partial orders, logical circuits, and algebra.

There are two equivalent definitions of Boolean algebras. We first define a Boolean algebra as a special type of partial order.

Definition 6.4.1. There are various types of partial orders. See Figure A.4 in Appendix A for a clear Venn diagram, and Diagram (6.77) for the relationship in terms of categories and functors:

- In Example 5.1.6, we saw that a partial order P with a strict Cartesian category structure $(P, \wedge, 1)$ is a **bounded meet semilattice**. Between such partial orders, we are interested in maps that preserve the $\wedge$ (i.e., $f(x \wedge y) = f(x) \wedge f(y)$), and preserve the 1 (i.e., $f(1) = 1$). The category of such partial orders and maps will be denoted as $\mathbb{BMslattice}$.

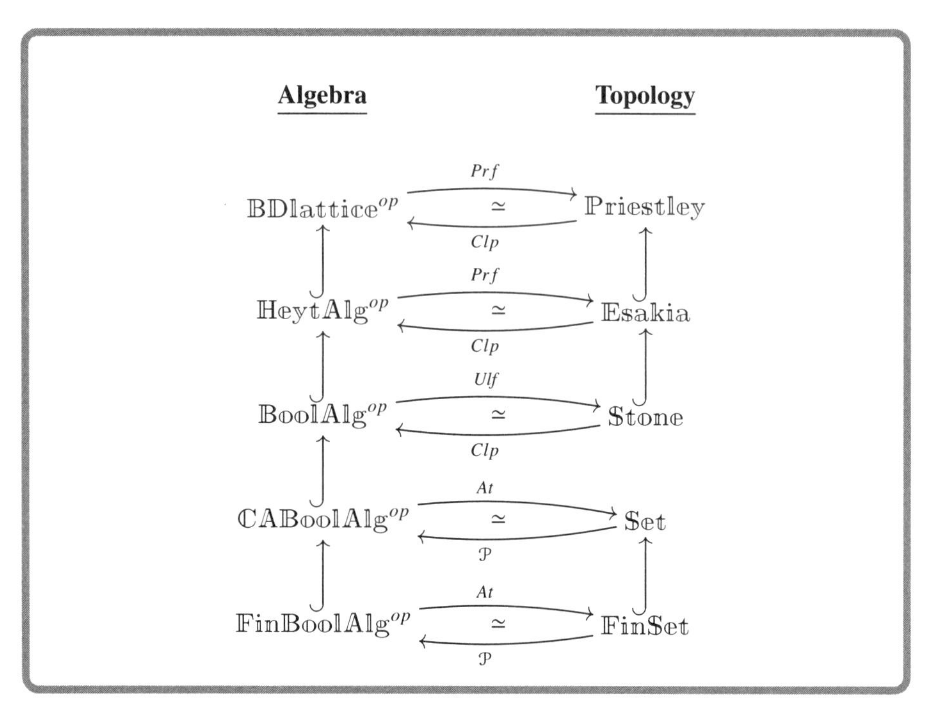

Figure 6.13. A hierarchy of Stone duality. (There is an inclusion from $\mathbb{Set}$ into $\mathbb{Stone}$, but that square does not commute.)

- In Example 5.1.6 we also saw that a partial order P with a strict co-Cartesian category structure $(P, \vee, 0)$ is a **bounded join semilattice**. Between such partial orders, we are interested in maps that preserve the $\vee$ (i.e., $f(x \vee y) = f(x) \vee f(y)$), and preserve the 0 (i.e., $f(0) = 0$). The category of such partial orders and maps will be denoted as $\mathbb{BJslattice}$. There is an isomorphism of categories $\mathbb{BMslattice} \longrightarrow \mathbb{BJslattice}$ that takes every bounded meet lattice $(P, \wedge, 1)$ to the same underlying set of the partial order with the opposite order $(P, \leq^R, 1)$ where $x \leq^R y$ iff $y \leq x$. This isomorphism swaps $\wedge$ for $\vee$, and 0 for 1.
- A partial order P with a Cartesian and a co-Cartesian structure $(P, \wedge, 1, \vee, 0)$ is called a **bounded lattice**. The category of all such partial orders with maps that preserves the two operations and the 0 and 1 is denoted as $\mathbb{Blattice}$. There are obvious forgetful functors: $\mathbb{Blattice} \longrightarrow \mathbb{BMslattice}$ and $\mathbb{Blattice} \longrightarrow \mathbb{BJslattice}$.
- A bounded lattice that satisfies

$$x \wedge (y \vee z) = (x \wedge y) \vee (x \wedge z) \qquad \text{and} \qquad x \vee (y \wedge z) = (x \vee y) \wedge (x \vee z) \tag{6.75}$$

is called a **bounded distributive lattice**. Each of these equations implies the other, so only one needs to be stated. The category of bounded distributive lattices has the same maps as bounded lattices and is denoted as $\mathbb{BDlattice}$. There is an inclusion functor: $\mathbb{BDlattice} \hookrightarrow \mathbb{Blattice}$.

- A **Heyting algebra** is a bounded distributive lattice with a binary operation $\Rightarrow$ that satisfies the following for all x, y, and z in P:

$$(x \wedge y) \leq z \qquad \text{if and only if} \qquad y \leq (x \Rightarrow z). \tag{6.76}$$

Categorically, this says that for every object x in the partial order, the map $x \wedge (\ )$ has a right adjoint $x \Rightarrow (\ )$. (In Section 7.2, we call a category where the product has such a right adjoint a "Cartesian closed category.") The category of all Heyting algebras and maps that preserves $\wedge$, $\vee$, $\Rightarrow$, 0, and 1 is denoted as $\mathbb{HeytAlg}$. There is a forgetful functor: $\mathbb{HeytAlg} \longrightarrow \mathbb{BDlattice}$.

- Within a bounded lattice, the **complement** of an element x is an element y such that $x \wedge y = 0$ and $x \vee y = 1$. If a complement of an element exists, then it is unique, and we denote the complement of x as x'. A bounded lattice where every element has a complement is called a **complemented lattice**. The category of all complemented lattices and maps that preserve the 0 and 1 and the operations is denoted as $\mathbb{BClattice}$. There is a forgetful functor: $\mathbb{BClattice} \longrightarrow \mathbb{Blattice}$.
- A **Boolean algebra** is a complemented, distributive lattice. A map of Boolean algebras is a set function that preserves $\wedge$, $\vee$, 0, and 1. (Stating some of these properties is redundant since some can be proven from others.) The category of all Boolean algebras and maps between them is denoted as $\mathbb{BoolAlg}$. There are the inclusions $\mathbb{BoolAlg} \hookrightarrow \mathbb{BClattice}$ and $\mathbb{BoolAlg} \hookrightarrow \mathbb{BDlattice}$. There is a full subcategory $\mathbb{FinBoolAlg}$ of Boolean algebras with only a finite number of elements. Every Boolean algebra can be seen as a Heyting algebra by defining $x \Rightarrow y$ as $x' \vee y$. This entails an embedding: $\mathbb{BoolAlg} \hookrightarrow \mathbb{HeytAlg}$.

The relationship between these categories can be summarized as

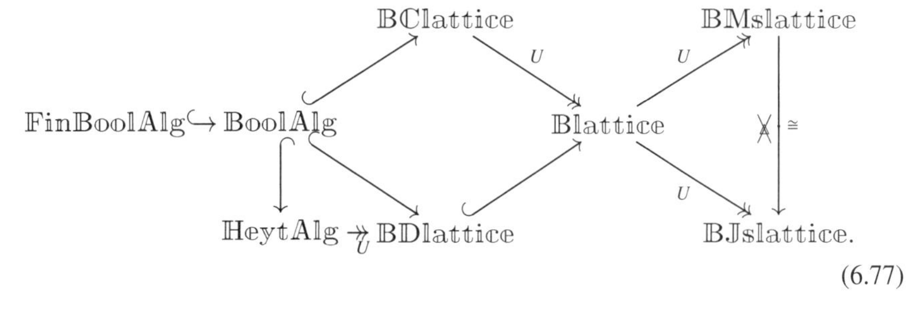

(6.77)

Notice that the right triangle need not commute.

We started with notions of a partial order and found algebraic operations. We can also go the other way and start by defining algebraic operations and show that these entail a partial order structure.

Definition 6.4.2. A **Boolean algebra** $(B, \wedge, \vee, (\ )', 0, 1)$ is a set B with the following operations $\wedge\colon B \times B \longrightarrow B$ and $\vee\colon B \times B \longrightarrow B$, $(\ )'\colon B \longrightarrow B$, and two constants, 0 and 1. Here, $\wedge$ and $\vee$ are commutative and associative. All three operations are idempotent (i.e., $x \wedge x = x$, $x \vee x = x$, and $x'' = x$).

Theorem 6.4.3. Definitions 6.4.1 and 6.4.2 of Boolean algebras are equivalent. ★

Proof. We saw that given the partial order, we can form the meet and the join using products and coproducts, respectively. One can also go the other way: given the meet and the join, we determine the partial order by defining $x \leq y$ when $x \wedge y = x$. Equivalently, we can define $x \leq y$ when $x \vee y = y$. ♣

Theorem 6.4.4. The following are some equations that a Boolean algebra satisfies:

(i) $(x \vee y)' = x' \wedge y'$	(ii) $(x \wedge y)' = x' \vee y'$	(iii) $x \wedge x = x$	(iv) $x \vee x = x$
(v) $x \wedge (x \vee y) = x$	(vi) $x \vee 1 = 1$	(vii) $x \wedge 1 = x$	(viii) $x \vee 0 = x$
(ix) $x \wedge (y \wedge z) = (x \wedge y) \wedge z$	(x) $x \wedge x' = 0$	(xi) $x \wedge y = y \wedge x$	(xii) $x \wedge 0 = 0$
(xiii) $x \vee (y \vee z) = (x \vee y) \vee z$	(xiv) $x \vee x' = 1$	(xv) $x'' = x$	(xvi) $0' = 1$

Example 6.4.5. There are many examples of Boolean algebras:

- The world's smallest example of a Boolean algebra is the partial order $\{*\}$ with one element. In this Boolean algebra, $0 = 1 = *$. This Boolean algebra is the terminal object in $\mathbb{BoolAlg}$.
- The next smallest Boolean algebra is $\{0, 1\}$ with $0 < 1$ (i.e., category $\mathbf{2}$ with two objects and a nontrivial morphism from 0 to 1). We denote this Boolean algebra as $\mathbf{2}_{BA}$ for reasons that will become apparent. The $\wedge$ operation is defined as $0 \wedge 0 = 1 \wedge 0 = 0 \wedge 1 = 0$, and $1 \wedge 1 = 1$. The $\vee$ operation is defined as $1 \vee 0 = 0 \vee 1 = 1 \vee 1 = 1$, and $0 \vee 0 = 0$. We also have $0' = 1$ and $1' = 0$. This Boolean algebra is the initial object in $\mathbb{BoolAlg}$.
- For any set S, the powerset of S, $\mathcal{P}(S)$, is a partial order and a Boolean algebra. The operations are the intersection, $\cup$, the union, $\cap$, and the complement, $(\ )^c$, defined for $T \subseteq S$ as $T^c = S - T$.
- Another way to view the powerset of a set as a Boolean algebra is by thinking of characteristic functions. Let $\mathbf{2}_{Set}$ be the set with two elements $\{0, 1\}$. As we saw in Section 1.4, for any set S, the $\mathcal{P}(S) = Hom_{\mathbb{Set}}(S, \mathbf{2}_{Set})$. This Hom set inherits the structure of a Boolean algebra from $\mathbf{2}_{Set}$. Let us be explicit with the operations. If $f\colon S \longrightarrow \mathbf{2}_{Set}$ and $g\colon S \longrightarrow \mathbf{2}_{Set}$, then $(f \wedge g)\colon S \longrightarrow \mathbf{2}_{Set}$ is defined as $(f \wedge g)(s) = f(s) \wedge g(s)$ for all $s \in S$. The $\wedge$ operation is defined on $\mathbf{2}_{Set}$ as $\mathbf{2}_{BA}$. There is a similar definition for $\vee$. Function $f'\colon S \longrightarrow \mathbf{2}_{Set}$ is defined as $f'(s) = 1 - f(s)$.

- For any natural number n, the set $(\mathbf{2}_{BA})^n$ (the product of n copies of $\mathbf{2}_{BA}$) is also a Boolean algebra. The operations are done pointwise. For example, $(0,1,1,0,1) \vee (1,1,0,0,0) = (1,1,1,0,1)$ and $(0,1,1,0,1)' = (1,0,0,1,0)$. Notice that these Boolean algebras are instances of the previous examples; that is, $(\mathbf{2}_{BA})^n$ is the same as $Hom(\{1,2,\ldots,n\}, \mathbf{2}_{Set})$.
- Let N be a square-free number (i.e., a number whose prime decomposition does not have a prime number that is squared or any higher power). For example, $105 = 3 \cdot 5 \cdot 7$ or $715 = 5 \cdot 11 \cdot 13$. The set of all the factors of N forms a Boolean algebra. Here, $\wedge$ is the highest common factor. For example, if $N = 105$, then $15 \wedge 35 = (3 \cdot 5) \wedge (5 \cdot 7) = 5$. The $\vee$ is the least common multiple. For example, if $N = 105$, then $15 \vee 35 = (3 \cdot 5) \vee (5 \cdot 7) = 105$. If x is a factor, then x' is the product of all the prime factors not in x. For example, if $N = 105$ then $3' = 5 \cdot 7 = 35$. If $N = 715$, then $11' = 5 \cdot 13 = 65$. The 0 of this Boolean algebra is the factor 1. The 1 of this Boolean algebra is N. In essence, this example can be seen as the Boolean algebra $\mathcal{P}(S)$, where S is the set of prime factors of N.
- In a topological space, a **closed set** is a set whose complement is an open set. A set is called **closed-open set** or a **clopen set** if it is both open and closed (i.e., both the set and its complement are open). Consider the set of clopen sets of any topological space. Using DeMorgan's law, one can see that the union and intersection of clopen sets are also clopen. Similarly, the complements of clopen sets are clopen. The empty set is clopen and is 0. The whole topological space is clopen and is 1. We conclude that the clopen sets of a topological space form a Boolean algebra.
- If you know about **finite automata** (sometimes called **finite state machines**), then you know that **regular languages** are the languages that finite automata recognize. The collection of all regular languages forms a Boolean algebra. This is because the intersection, union, and complement of regular languages are regular languages. The 0 of this Boolean algebra is the empty language, and the 1 is Σ^*, the language consisting of all the words in the alphabet. □

Before we start exploring Stone duality, there is another duality associated with Boolean algebras. The **duality principle** says that any true statement about Boolean algebras is also true when all the joins are swapped for meets, and the 0s are swapped for 1s. For example, in Theorem 6.4.4, properties (i) and (ii) are dual to each other. Similarly, (iii) and (iv), (vi) and (xii), (vii) and (viii), (ix) and (xiii), and (x) and (xiv) are dual to each other. This duality comes about when we swap a partial order $\leq$ with its reverse partial order $\leq^R$, where $x \leq^R y$ iff $y \leq x$. Such a change in the partial order swaps $\wedge$ for $\vee$ and swaps 0 for 1. Categorically, this means that there is a functor $D\colon \mathbb{BoolAlg} \longrightarrow \mathbb{BoolAlg}$ that takes every partial order $(P, \leq)$ to $(P, \leq^R)$. On morphisms, the functor D takes an order-preserving map to the same order-preserving map. This gives us a nontrivial isomorphism $\mathbb{BoolAlg} \cong \mathbb{BoolAlg}$. The dual of a statement concerning partial order $(P, \leq)$ is a statement about $D(P, \leq) = (P, \leq^R)$.

Baby Stone Duality. Let us begin our journey toward Stone duality theory with easy cases. We shall spend time on these cases because all the core ideas are here. We will then use these ideas to build up to more and more general cases. It is helpful to keep an eye on Figure 6.13, which is the map of our journey and a summary of this mini-course.

Theorem 6.4.6. The opposite of the category of finite Boolean algebras is equivalent to the category of finite sets. That is, $\mathbb{FinBoolAlg}^{op} \simeq \mathbb{FinSet}$. ★

Proof. There is a functor $\mathcal{P}\colon \mathbb{FinSet} \longrightarrow \mathbb{FinBoolAlg}^{op}$ that takes a finite set S to the powerset of S, $\mathcal{P}(S)$, which is a finite Boolean algebra. This is the contravariant powerset functor (preimage functor) that we saw in Example 4.1.34 which takes the set function $f\colon S \longrightarrow T$, to $\mathcal{P}(f)\colon \mathcal{P}(T) \longrightarrow \mathcal{P}(S)$. Such functions respect all the operations of a Boolean algebra. For example, if X and Y are subsets of T, then

$$\begin{aligned}\mathcal{P}(f)(X \cap Y) &= \{x \in S : f(x) \in X \cap Y\} \\ &= \{x \in S : f(x) \in X\} \cap \{x \in S : f(x) \in Y\} \\ &= \mathcal{P}(f)(X) \cap \mathcal{P}(f)(Y).\end{aligned}$$

The functor also respects the complement operation because

$$\begin{aligned}\mathcal{P}(f)(Y^c) &= \{x \in S : f(x) \in Y^c\} \\ &= \{x \in S : f(x) \notin Y\} \\ &= \{x \in S : f(x) \in Y\}^c = (\mathcal{P}(f)(Y))^c.\end{aligned}$$

Hence, the target of this functor is $\mathbb{FinBoolAlg}^{op}$.

There is an **atom functor** $At\colon \mathbb{FinBoolAlg}^{op} \longrightarrow \mathbb{FinSet}$ that is the quasi-inverse of $\mathcal{P}$. First, a definition: within a Boolean algebra, an element a is called an **atom** if there is nothing smaller than it other than 0. Formally, a is an atom if $0 \lneq a$ and for all x, if $x \leq a$, then $x = a$ or $x = 0$. If you think of a Boolean algebra as a type of lattice with 1 on the top and 0 on the bottom, then the set of atoms are those elements right above 0. For a Boolean algebra made of the subsets of a set, the atoms are the singleton sets and any subset is made of the union of those singleton sets. In general, for any finite Boolean algebra, every element b is made of the finite join of all the atoms that are less than b. Expressed in symbols,

$$b = \bigvee \{x : x \text{ is an atom, and } x \leq b\}. \tag{6.78}$$

The functor At takes a finite Boolean algebra B to $At(B)$, the finite set of atoms of B. A map of Boolean algebras $g\colon B_1 \longrightarrow B_2$ induces a set function $At(g)\colon At(B_2) \longrightarrow At(B_1)$.

These two functors form the equivalence stated in the theorem. For a finite Boolean algebra B, there is an isomorphism $\phi\colon B \longrightarrow \mathcal{P}(At(B))$. In detail, $\mathcal{P}(At(B))$ is the powerset of atoms of B. The isomorphism ϕ is defined as

$$b \quad \mapsto \quad \{x : x \text{ is an atom, and } x \leq b\}. \tag{6.79}$$

The map ϕ has the following properties:

- ϕ is a surjection because for every subset $S \subseteq At(B)$, the join of all the elements of S map to S. Expressed in symbols,

$$\bigvee \{\text{atom } \{s\} : s \in S\} \quad \mapsto \quad S. \tag{6.80}$$

- ϕ is an injection because if $b_1 \neq b_2$, then b_1 and b_2 differ by some atom. That means that there is some atom that is in one and not in the other. Hence, $\phi(b_1) \neq \phi(b_2)$.
- ϕ preserves $\wedge$ and $\vee$ because

$$\phi(b_1 \wedge b_2) = \{x : x \leq (b_1 \wedge b_2)\} = \{x : x \leq b_1\} \cap \{x : x \leq b_2\} = \phi(b_1) \cap \phi(b_2). \tag{6.81}$$

  It is similar for $\vee$.
- ϕ preserves 0 and 1 because $\phi(0) = \emptyset$ and $\phi(1) = B$.

For every finite set S, there is an isomorphism $\psi \colon S \longrightarrow At(\mathcal{P}(S)) = \{\{s\} : s \in S\}$. This isomorphism is the function defined by

$$s \in S \quad \mapsto \quad \{s\} \in At(\mathcal{P}(S)). \tag{6.82}$$

It is easy to see that this function is an injection and a surjection of sets. ♣

This duality theorem shows that every finite Boolean algebra has 2^n elements for some n. Furthermore, any two finite Boolean algebras of the same size are isomorphic. The way to see this is that if there are two Boolean algebras, B_1 and B_2, both of size 2^n, then there are two sets of atoms, S_1 and S_2, each of size n. Two finite sets of the same size entail a set isomorphism $f \colon S_1 \longrightarrow S_2$. The functor $\mathcal{P}$ takes isomorphisms to isomorphisms. By composing the isomorphsims as follows:

$$B_2 \longrightarrow \mathcal{P}(S_2) \longrightarrow P(S_1) \longrightarrow B_1, \tag{6.83}$$

we have shown that the Boolean algebras are isomorphic.

Before we move on to the next step, it will be useful to look at the functors from a more categorical prospective. The functor $\mathcal{P}$ is defined as $\mathcal{P}(S) = Hom_{\mathbb{Set}}(S, \mathbf{2}_{Set})$, where $\mathbf{2}_{Set}$ is the set with two elements.

The functor At can also be seen from a more categorical perspective. Consider an atom a in Boolean algebra B. We can view this atom as a Boolean algebra map $f_a \colon B \longrightarrow \mathbf{2}_{BA}$, defined as

$$f_a(b) = \begin{cases} 1 & : a \leq b \\ 0 & : a \nleq b. \end{cases} \tag{6.84}$$

Stated in words, f_a takes all those elements above a to 1 and the rest to 0. Thus, the set of atoms $At(B)$ for a finite Boolean algebra B can be seen as a subset of $Hom_{\mathbb{FinBoolAlg}}(B, \mathbf{2}_{BA})$. We will generalize this in the coming pages.

Juvenile Stone Duality. While finite sets are fine, let us generalize to all sets. To do that, we will need to explore beyond finite Boolean algebras. The powerset of an infinite set is an infinite Boolean algebra. What other properties will the powerset of an infinite set have?

Definition 6.4.7. A Boolean algebra must contain meets and joins of two elements. Taking the meets and joins a finite number of times gives us all finite meets and joins. A Boolean algebra that has arbitrary (not just finite) meets and joins is called **complete**. Categorically, this says that the partial order category has all products and coproducts. If X is an arbitrary subset of elements of a Boolean algebra, then we write $\bigvee_{x \in X} x$ for their join, and $\bigwedge_{x \in X} x$ for their meet.

A Boolean algebra is **atomic** if every element is the join of a set of its atoms (0 is the join of the empty set of atoms).

The collection of all complete, atomic Boolean algebras and Boolean algebra homomorphisms is denoted as $\mathbb{CABoolAlg}$.

All finite Boolean algebras are complete and atomic. Thus, we have the following embeddings:

$$\mathbb{FinBoolAlg} \hookrightarrow \mathbb{CABoolAlg} \hookrightarrow \mathbb{BoolAlg}. \tag{6.85}$$

Theorem 6.4.8. The opposite of the category of complete atomic Boolean algebras is equivalent to the category of sets, that is $\mathbb{CABoolAlg}^{op} \simeq \mathbb{Set}$. ★

Proof. The proof follows almost exactly like the previous proof.

There is a functor $\mathcal{P}\colon \mathbb{Set} \longrightarrow \mathbb{CABoolAlg}^{op}$, which is defined for set S as $\mathcal{P}(S) = Hom_{\mathbb{Set}}(S, \mathbf{2}_{Set})$. For any set S—finite or infinite—the powerset $\mathcal{P}(S)$ is complete because it has arbitrary large unions and intersections. It is also atomic because the singletons are atoms, and every set is the arbitrary union of its single elements. Thus, we have that the powerset of any set is a complete atomic Boolean algebra.

There is a functor $At\colon \mathbb{CABoolAlg}^{op} \longrightarrow \mathbb{Set}$ that takes any complete atomic Boolean algebra to its set of atoms.

For a complete atomic Boolean algebra B, there is an isomorphism $\phi\colon B \longrightarrow \mathcal{P}(At(B))$. For an arbitrary set S, there is an isomorphism $\psi\colon S \longrightarrow At(\mathcal{P}(S)) = \{\{s\} : s \in S\}$. Both of these isomorphisms work almost exactly the same as the finite case. ♣

Remark 6.4.9. A consequence of this theorem is that we now have a way of thinking about the category $\mathbb{Set}^{op}$. This category is not a typical category (technically called a **concrete category**), where every morphism corresponds to some type of function. In $\mathbb{Set}$, there is a unique function $\emptyset \longrightarrow \{*\}$, which corresponds to a unique map $\{*\} \longrightarrow \emptyset$ in $\mathbb{Set}^{op}$. What can this morphism mean? It takes the element $*$ to where? By meditating on our theorem, which says $\mathbb{CABoolAlg} \simeq \mathbb{Set}.^{op}$, we can think of $\mathbb{Set}^{op}$ as follows: the objects are sets and a morphism $S \longrightarrow T$ corresponds to the Boolean morphisms $\mathcal{P}(S) \longrightarrow \mathcal{P}(T)$. In particular, the morphism $\{*\} \longrightarrow \emptyset$ in $\mathbb{Set}^{op}$ corresponds to the Boolean algebra homomorphism $\mathcal{P}(\emptyset) = \{*\} \longrightarrow \mathcal{P}(\{*\}) = \{0, 1\}$, with $* \mapsto 1$.

This way of understanding $\mathbb{Set}^{op}$ provides an important idea that is useful for understanding the rest of this mini-course. A morphism $f_x\colon \{*\} \longrightarrow S$ in $\mathbb{Set}$ (and $\mathbb{Top}$) picks out the element x in S. Corresponding to f_x in $\mathbb{Set}$ is a morphism

$\bar{f}_x\colon S \longrightarrow \{*\}$ in $\mathbb{Set}^{op}$ and a Boolean algebra homomorphism $\hat{f}_x\colon \mathcal{P}(S) \longrightarrow \mathcal{P}(\{*\}) = \{0,1\}$. The homomorphism $\hat{f}_x$ is a characteristic function that picks out those subsets of S (elements of $\mathcal{P}(S)$) that contain x. This idea—that maps picking out an element are equivalent to maps that determine if that element is in a subset—is absolutely central to Stone duality and all its generalizations. ♠

Before we move on, it pays to think of sets from a more general point of view. Let S and T be sets. We can think of these sets as topological spaces with the discrete topologies (S,σ_d) and (T,τ_d). From this perspective, it is important to notice that $Hom_{\mathbb{Set}}(S,T) = Hom_{\mathbb{Top}}((S,\sigma_d),(T,\tau_d))$. In other words, one can think of sets as special types of topological spaces where every map between the topological spces is considered continuous. This entails an inclusion functor $\mathbb{Set} \hookrightarrow \mathbb{Top}$. Another way to say this is that the category of sets is equivalent to the subcategory of topological spaces where all the topologies are discrete. Notice also that when using the discrete topology, all open sets are also closed sets (i.e., every open set is clopen). We will continue our journey using the language of topology rather than sets.

Stone Duality. Let us generalize from complete atomic Boolean algebras to all Boolean algebras. To do that, we need to go beyond sets. We define special types of topological spaces next.

Definition 6.4.10.

- A topological space is **compact** if it has a type of finiteness condition. Formally, a topological space T is compact if every collection $C = \{V_i\}$ of open sets of T that has the property that the union of those open sets is equal to the entire T has a finite subset $\bar{C} \subset C$ such that the union of all the open sets in $\bar{C}$ is also equal to the entire space T. Intuitively, this says that the topological space does not need many open sets to cover it. Notice that an infinite topological space with the discrete topology is not compact because the cover consisting of the infinite set of elements does not have a finite subcover.
- A topological space is **Hausdorff** if any two points in the space can be separated by two disjoint open sets. Formally, a topological space T is Hausdorff if for any two points x and y, there are open sets A and B such that $x \in A$ and $y \in B$ with $A \cap B = \emptyset$. Intuitively, a Hausdorff space has a lot of open sets to separate points.
- A topological space is **totally disconnected** if the connected components are one-point sets.
- A topological space is a **Stone space** if it is compact, Hausdorff, and totally disconnected.

The category of Stone spaces and continuous maps between them is denoted as $\mathbb{Stone}$.

Theorem 6.4.11. Stone Duality. The opposite of the category of Boolean algebras is equivalent to the category of Stone spaces, that is $\mathbb{BoolAlg}^{op} \simeq \mathbb{Stone}$. ★

Proof. There is a functor $Clp\colon \mathbb{Stone} \longrightarrow \mathbb{BoolAlg}^{op}$, which is defined for Stone space T as $Clp(T) = Hom_{\mathbb{Stone}}(T, \mathbf{2}_{SS})$, where $\mathbf{2}_{SS}$ is the two-object Stone space $\{0, 1\}$. Within $\mathbf{2}_{SS}$, both $\{0\}$ and $\{1\}$ are open (and hence also closed) sets. Any continuous map $f\colon T \longrightarrow \mathbf{2}_{SS}$ will act like a characteristic function and choose a subset $f^{-1}(1)$ of X. This subset will be open and (because $f^{-1}(0)$ is open) closed. So $Hom_{\mathbb{Stone}}(T, \mathbf{2}_{SS})$ describes all the clopen subsets. In other words, functor Clp takes any Stone space to the Boolean algebra of its clopen sets. Given a continuous map of Stone space $g\colon T_1 \longrightarrow T_2$, the map $Hom_{\mathbb{Stone}}(g, \mathbf{2}_{SS})$ will be a homomorphism of Boolean algebras.

There is a functor $Ulf\colon \mathbb{BoolAlg}^{op} \longrightarrow \mathbb{Stone}$, which is defined for a Boolean algebra B as $Ulf(B) = Hom_{\mathbb{BoolAlg}}(B, \mathbf{2}_{BA})$. This Hom set is a Stone space. To ensure that the reader can better communicate with poor souls who do not already know category theory, we will describe these maps in a noncategorical language.

Definition 6.4.12. For a Boolean algebra map $f\colon B \longrightarrow \mathbf{2}_{BA}$, the set of elements that go to 1 (i.e., $F = f^{-1}(1)$), form a structure called an **ultrafilter**. Following Remark 6.4.9, we might think of these ultrafilters as types of elements of B. Set F satisfies the following list of properties:

- F is a **filter** because
  - F is nonempty: for example, $1 \in F$ (because $f(1) = 1$).
  - F is upward closed: if $a \in F$ and $a \leq b$, then $b \in F$ (because f is order preserving).
  - F is closed under meet: if $a \in F$ and $b \in F$, then $a \wedge b \in F$ (because f preserves the meet).
- F is a **proper filter**: F does not contain everything; that is, $F \neq B$ or, equivalently, $0 \notin F$ (because $f(0) = 0$).
- F is a **prime filter**: If $a \vee b \in F$, then $a \in F$ or $b \in F$ (because $f(a \vee b) = f(a) \vee f(b)$).
- F is a **maximal filter**. For all $a \in B$, either $a \in F$ or $a' \in F$ (because $f(a') = f(a)'$, so either $f(a) = 1$ or $f(a') = 1$).

For each $a \in B$, there is a special filter called a **principle filter generated by** a which is $\{b \in B : a \leq b\}$. This corresponds to the map $f_a\colon B \longrightarrow \mathbf{2}_{BA}$ defined in Equation (6.84). For a Boolean algebra, the principle filter generated by a is an ultrafilter if and only if a is an atom.

The set $Ulf(B) = Hom_{\mathbb{BoolAlg}}(B, \mathbf{2}_{BA})$ is the set of ultrafilters on B. For each $a \in B$, there is a collection $U_a \subseteq Ulf(B)$ of ultrafilters defined as $\{F \text{ is an ultrafilter} : a \in F\}$ or, in terms of maps, $\{f\colon B \longrightarrow \mathbf{2}_{BA} : f(a) = 1\}$.

This set $Ulf(B)$ has the structure of a Stone space:

- The points of the space are ultrafilters, or equivalently, Boolean algebra homomorphisms of the form $B \longrightarrow \mathbf{2}_{BA}$.
- The open sets are unions of sets of the form U_a. The empty set is $U_0 = \emptyset$ and $U_1 = Ulf(B)$.

- The space is compact. Say that $Ulf(B) = \bigcup_{i \in I} U_{a_i}$ for some infinite set I. If $Ulf(B)$ was not compact, then for every finite subset $J \subset I$, we have that $Ulf(B) \neq \bigcup_{i \in J} U_{a_i}$. This means $1 \neq \bigvee_{i \in J} a_i$. By DeMorgan's law, we infer that $0 \neq \bigwedge_{i \in J} a'_i$. This set $\{a'_i\}_J$ has what is called the **finite intersection property**. There is an infinite process by which we can add elements to this finite set and get an ultrafilter F. (This bridge from the finite to the infinite uses Zorn's lemma.) This $\{a'_i\}_J$ is a subset of F. We assumed that $F = U_a$ for some $a \in B$. Thus, F contains a and a', which contradicts F being proper and hence is not an ultrafilter.
- The space is Hausdorff. If there are two ultrafilters $F_1 \neq F_2$, then there is some $a \in B$ such that $a \in F_1 \backslash F_2$ or $a \in F_2 \backslash F_1$. Say that the first case is true. Then $F_1 \in U_a$ and $F_2 \in U_{a'}$ and $U_a \cap U_{a'} = \emptyset$.
- The space is totally disconnected. Assume that we are given a nontrivial set of points (ultrafilters) X and an ultrafilter $F \in X$ with an element a. This means $a \in F \in X \subseteq Ulf(B)$. Then U_a contains F, and $U_{a'} = \overline{U_a}$ is disjoint from U_a. Furthermore, X is within $U_a \cup U_{a'}$.

One part of the equivalence is given by the following isomorphisms. For a Boolean algebra B, there is an isomorphism of Boolean algebras:

$$\phi \colon B \longrightarrow Clp(Ulf(B)), \tag{6.86}$$

which is

$$\phi \colon B \longrightarrow Hom_{\mathbb{S}\mathrm{tone}}(Hom_{\mathbb{B}\mathrm{ool}\mathbb{A}\mathrm{lg}}(B, \mathbf{2}_{BA}), \mathbf{2}_{SS}). \tag{6.87}$$

This is defined as

$$b \quad \mapsto \quad ev[b] \colon Hom_{\mathbb{B}\mathrm{ool}\mathbb{A}\mathrm{lg}}(B, \mathbf{2}_{BA}) \longrightarrow \mathbf{2}_{SS}. \tag{6.88}$$

The continuous map $ev[b]$ evaluates a Boolean algebra homomorphism $f \colon B \longrightarrow \mathbf{2}_{BA}$ as follows:

$$ev[b](f) = \begin{cases} 1 & : f(b) = 1 \\ 0 & : f(b) = 0. \end{cases} \tag{6.89}$$

This shortens to

$$ev[b](f) = f(b). \tag{6.90}$$

In other words, $ev[b]$ acts like a characteristic function that picks out those f's such that $f(b) = 1$. In terms of prime ultrafilters, $ev[b]$ picks out those prime ultrafilters F with $b \in F$.

The other isomorphism is as follows. For Stone space T, there is an isomorphism of Stone spaces:

$$\psi \colon T \longrightarrow Ulf(Clp(T)), \tag{6.91}$$

which is

$$\psi \colon T \longrightarrow Hom_{\mathbb{B}\mathrm{ool}\mathbb{A}\mathrm{lg}}(Hom_{\mathbb{S}\mathrm{tone}}(T, \mathbf{2}_{SS}), \mathbf{2}_{BA}). \tag{6.92}$$

This is defined as

$$x \quad \mapsto \quad ev[x]\colon Hom_{\mathbb{Stone}}(T, \mathbf{2}_{SS}) \longrightarrow \mathbf{2}_{BA}. \tag{6.93}$$

The Boolean algebra map $ev[b]$ evaluates a continious map $g\colon T \longrightarrow \mathbf{2}_{SS}$ as follows:

$$ev[x](g) = \begin{cases} 1 & : g(x) = 1 \\ 0 & : g(x) = 0. \end{cases} \tag{6.94}$$

This shortens to

$$ev[x](g) = g(x). \tag{6.95}$$

In other words, $ev[x]$ acts like a characteristic function that picks out those g's such that $g(x) = 1$. In terms of open sets, $ev[x]$ picks out those open sets O with $x \in O$.

The fact that these two maps are isomorphisms is easy to see. The functoriality of the two isomorphisms follows from the fact that they are defined with Hom sets. ♣

We have seen the object **2** in many different contexts. It is the collection with two objects that form a set, a partial order, a Boolean algebra, and a Stone space. (In the coming pages, we will see even more incarnations of **2**.) In other words, many different categories have an object **2**. In Important Categorical Idea 2.1.55 we saw that we cannot just talk about an object. Rather, we must specify what context it is in. We distinguish the various **2**'s with subscripts, such as $\mathbf{2}_{Set}$, $\mathbf{2}_{BA}$ and $\mathbf{2}_{SS}$. Such an structure that has various incarnations in different categories is called a **dualizing object** or a **schizophrenic object**. One can see in this proof that the duality depends on such dualizing objects. It is again the power of category theory to see many different duality theorems as coming from one idea. The point of these dualizing objects is that category theory does not neatly separate different structures in different areas. Rather, these objects connect and unify different areas.

Remark 6.4.13. These proofs of the isomorphisms of the duality theorem are very similar to the proof in Example 4.2.12 where we showed a finite-dimensional vector space is naturally isomorphic to its double dual. This means that for a finite-dimensional vector space V, there is an isomorphism

$$V \longrightarrow Hom_{\mathbf{K}\mathbb{FDVect}}(Hom_{\mathbf{K}\mathbb{FDVect}}(V, \mathbf{K}), \mathbf{K}). \tag{6.96}$$

Equations (6.90) and (6.95) are exactly the same as Equation (4.58). Once again, category theory shows connections between disparate areas. The ideas of Stone duality take us back to where Eilenberg and Mac Lane started category theory. ♠

This fact that every Boolean algebra is isomorphic to some collection of sets of elements (subsets or clopens) is called **Stone's representation theorem** and was proved by Marshall H. Stone in 1936 (before category theory existed). It is the main theorem of this field. One of the consequences of the theorem is that one can easily check some fact about Boolean algebras (like the statements in Theorem 6.4.4) by simply testing them on collections of a set. If the statement is true for the collection of sets, then it is true for all Boolean algebras.

Esakia Duality. Let us generalize from Boolean algebras to Heyting algebras. This was done by Leo Esakia in 1974. We will now deal with partial orders that are not necessarily complemented but have an $\Rightarrow$ operation. To deal with Heyting algebras, we have to be concerned with topological spaces with more structure. The extra idea needed is an ordering. First, here are some definitions:

Definition 6.4.14. Given a partial order $(P, \leq)$, an **upward-closed set** X is a set such that for all $x \in X$, $x \leq y$ implies $y \in X$.

An **ordered topological space** $(T, \tau, \leq)$ is a set T, a topology τ on the set T, and a partial order $\leq$ on the set T. One can think of any topological space as a type of ordered topological space where the order is the identity order (i.e., only $x \leq x$ for all $x \in T$). In particular, a Stone space is an ordered topological space with such an ordering.

An **Esakia space** is a Stone space such that both the following are true:

- Upward-closed sets are closed: for all $x \in T$, $\uparrow x = \{y \in T : x \leq y\}$ is closed.
- Downward-clopen sets are clopen: for all clopen $C \subseteq X$, the set

$$\downarrow C = \{y \in T : y \leq x \text{ for some } x \in C\} \tag{6.97}$$

  is clopen.

A morphism of Esakia spaces is a map $f\colon (T_1, \tau_1, \leq_1) \longrightarrow (T_2, \tau_2, \leq_2)$ such that f is continuous and for all $x \in T_1$, we have $f(\uparrow x) = \uparrow f(x)$. The category of Esakia spaces and their morphisms is denoted as $\mathbb{Esakia}$.

There is an inclusion $\mathbb{Stone} \hookrightarrow \mathbb{Esakia}$.

Theorem 6.4.15. The opposite of the category of Heyting algebras is equivalent to the category of Esakia spaces (i.e., $\mathbb{HeytAlg}^{op} \simeq \mathbb{Esakia}$). ★

Proof. This proof is almost the same as the proof of Theorem 6.4.14; only the changes are pointed out here. There is a Heyting algebra $\mathbf{2}_{HA}$ with objects 0 and 1. The arrow is defined as $(0 \Rightarrow 0) = (0 \Rightarrow 1) = (1 \Rightarrow 1) = 1$ and $(1 \Rightarrow 0) = 0$. We begin with the functor $Prf\colon \mathbb{HeytAlg}^{op} \longrightarrow \mathbb{Esakia}$, which is defined for a Heyting algebra H as $Prf(H) = Hom_{\mathbb{HeytAlg}}(H, \mathbf{2}_{HA})$. Each Heyting algebra map $f\colon H \longrightarrow \mathbf{2}_{HA}$ corresponds to $F = f^{-1}(1)$, which is a prime filter (and not necessarily an ultrafilter) because the $\vee$ is preserved (and it need not be maximal because there is no complement to preserve).

The set $Prf(H)$ is an Esakia space:

- The elements are prime filters.
- Let U_a be the set of prime filters that contain a. The open sets of $Prf(H)$ are generated by U_a and by $Prf(H) - U_a$ for all $a \in H$. (Notice that we cannot talk of $U_{a'}$.)

- There is an order on prime filters. We have $F_1 \leq F_2$ if $F_1 \subseteq F_2$. $\uparrow h$ and *All filters*$- \uparrow h$. The partial order $\leq$ is an ordering of prime filters by inclusion (i.e., $F_1 \leq F_2$ if and only if $F_1 \subseteq F_2$).

There is a functor $Clp\colon \mathbb{E}\mathbb{s}\mathbb{a}\mathbb{k}\mathbb{i}\mathbb{a} \longrightarrow \mathbb{H}\mathbb{e}\mathbb{y}\mathbb{t}\mathbb{A}\mathbb{l}\mathbb{g}^{op}$ that takes every Esakia space X to its clopen upward-closed sets. Formally, $Clp(X) = Hom_{\mathbb{Esakia}}(X, \mathbf{2}_{ES})$ where $\mathbf{2}_{ES}$ is the Esakia space with two elements. Realize that since $0 \in \mathbf{2}_{ES}$ is open, $1 \in \mathbf{2}_{ES}$ is closed, and for any $f\colon X \longrightarrow \mathbf{2}_{ES}$, the set $f^{-1}(1)$ is clopen and is an upward-closed set. Given two clopen up sets U and V, we define

$$U \Rightarrow V \quad = \quad X - (\downarrow (U - V)). \tag{6.98}$$

The isomorphisms showing the equivalence are essentially the same as before:

$$\phi\colon H \longrightarrow Hom_{\mathbb{Esakia}}(Hom_{\mathbb{HeytAlg}}(H, \mathbf{2}_{HA}), \mathbf{2}_{ES}) \tag{6.99}$$

$$\psi\colon X \longrightarrow Hom_{\mathbb{HeytAlg}}(Hom_{\mathbb{Esakia}}(X, \mathbf{2}_{ES}), \mathbf{2}_{HA}). \tag{6.100}$$

♣

Priestley Duality. Let us generalize from Heyting algebras to bounded distributive lattices. This was done by Hilary Priestley in the early 1970s. In doing this, we deal with partial orders that do not even have a $\Rightarrow$ operation.

Definition 6.4.16. An ordered topological space is a **Priestley space** if it satisfies the **Priestley separation axiom**:

If $x \nleq y$, then there is a clopen up-set U such that $x \in U$ and $y \notin U$.

This can be understood as follows:

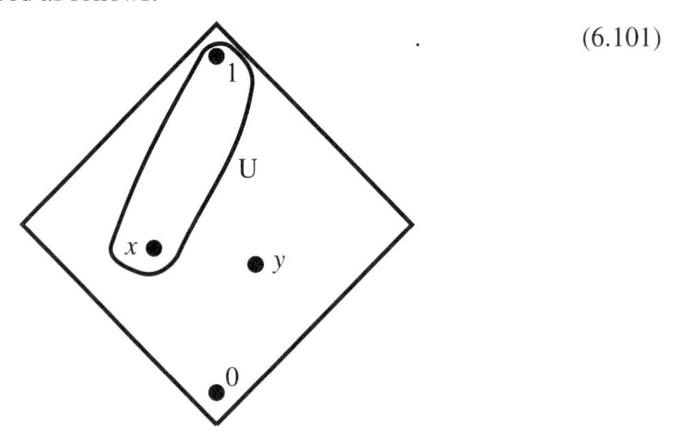

$$. \tag{6.101}$$

A morphism between two Priestley spaces is continuous and order preserving.

The collection of all Priestley spaces and their morphisms form the category $\mathbb{Priestley}$. There is an inclusion $\mathbb{Esakia} \hookrightarrow \mathbb{Priestley}$.

Theorem 6.4.17. The opposite of the category of bounded distributive lattice is equivalent to the category of Priestley spaces, that is $\mathbb{BDlattice}^{op} \simeq \mathbb{Priestley}$. ★

Proof. The proof is very similar to the last one. We begin with the two-element distributive lattice $\mathbf{2}_{DL}$, which is easy to define. Notice that there is no complement in $\mathbf{2}_{DL}$. That is, we do not have $0' = 1$ and $1' = 0$. There is also no $0 \Rightarrow 1$. There is a functor $Prf\colon \mathbb{BDlattice}^{op} \longrightarrow \mathbb{Priestley}$, which is defined for a bounded distributive lattice D as $Prf(D) = Hom_{\mathbb{BDlattice}}(D, \mathbf{2}_{DL})$. The set $Prf(D)$ is the ordered topological space of prime filters on D. The open sets are generated by principle filters and complements of principle filters. The partial order $\leq$ is an ordering of prime filters by inclusion (i.e., $F_1 \leq F_2$ if and only if $F_1 \subseteq F_2$).

There is a functor $Clp\colon \mathbb{Priestley} \longrightarrow \mathbb{BDlattice}^{op}$ that takes every Priestley space X to its clopen up-sets. Formally, $Clp(X) = Hom_{\mathbb{Priestley}}(X, \mathbf{2}_{PS})$, where $\mathbf{2}_{PS}$ is the Priestley space with two elements. Since $1 \nleq 0$ in $\mathbf{2}_{PS}$, the set $U = \{1\}$ satisfies the Priestley separation axiom.

The isomorphisms showing the equivalence are essentially the same as before:

$$\phi\colon D \longrightarrow Hom_{\mathbb{Priestley}}(Hom_{\mathbb{BDlattice}}(D, \mathbf{2}_{DL}), \mathbf{2}_{PS}) \tag{6.102}$$

$$\psi\colon X \longrightarrow Hom_{\mathbb{BDlattice}}(Hom_{\mathbb{Priestley}}(X, \mathbf{2}_{PS}), \mathbf{2}_{DL}). \tag{6.103}$$

♣

Figure 6.13 shows a summary of duality theorems that we have dealt with.

Pointless Topology. Stone duality is not the only duality that shows a relationship between algebraic and topological ideas. Pointless topology is related to Stone duality and shows another relationship.

We saw in Definition 2.1.18 in Chapter 2 that a topological space is a set of points with open sets that satisfy certain requirements. The central question is: To what extent are the properties of a topological space determined by the structure of the open sets (while ignoring the points)? This gives us the humorous title of "pointless topology."

In Example 4.1.49 in Chapter 4, we met a functor O that takes every topological space (T, τ) to the partial order $O(T)$. This partial order reflects the structure of the subsets of the topological space with the meet, $\wedge$, being the intersection of open sets and the join, $\vee$, being the union of open sets. As such, the partial order has finite meets and all joins. The partial order also satisfies a distributivity requirement. Let us make a formal definition of such a partial order.

Definition 6.4.18. A **locale** L is a partial order that has all finite meets and arbitrary joins. Furthermore, there is a requirement that an infinite distributive law be satisfied: for any element $U \in L$ and any set of elements $\{V_i\}$ in L, we have

$$U \wedge \bigvee_i V_i = \bigvee_i (U \wedge V_i). \tag{6.104}$$

Morphisms between locales are interesting. We want such morphisms to mimic continuous maps between topological spaces. Remember that $f\colon T_1 \longrightarrow T_2$ is a continuous map of topological spaces if f^{-1} takes open sets to open sets. The important point is that f^{-1} goes the other way. With this in mind, we define a morphism of locales $f\colon L_1 \longrightarrow L_2$ to be a map such that f^{-1} preserves finite meets and all joins. The collection of locales and locale maps forms a category $\mathbb{Locale}$.

A better way to deal with the awkwardness of the directions of the maps is to discuss the opposite of the category of locales. Define a **frame** exactly as a locale was defined. A morphism of frames $f\colon F_1 \longrightarrow F_2$ is a map of partially ordered sets that preserves finite meets and all joins. The collection of all frames and frame morphisms is denoted as $\mathbb{Frame}$, and by definition, $\mathbb{Locale} = \mathbb{Frame}^{op}$.

Now let's look at the duality of the topological structures and the partial orders.

Theorem 6.4.19. There is an adjunction between the category of locales—which is the opposite of the category of frames—and the category of topological spaces:

$$\mathbb{Frame}^{op} \underset{O}{\overset{Pt}{\rightleftarrows}} \mathbb{Top}. \quad \top \tag{6.105}$$

★

Proof. The functor $O\colon \mathbb{Top} \longrightarrow \mathbb{Frame}^{op}$ takes every topological space (T,τ) to $O(T)$, the frame of its open sets. Every map of topological spaces $f\colon (T_1,\tau_1) \longrightarrow (T_2,\tau_2)$ goes to the frame map $f^{-1}\colon O(T_2) \longrightarrow O(T_1)$.

The functor that goes the other way picks out the set of points of a frame $Pt\colon \mathbb{Frame}^{op} \longrightarrow \mathbb{Top}$. This functor takes every frame F to the topological space $Pt(F)$. The points of the topological space are defined as the frame maps from F to $\mathbf{2}_{Fr}$ which is the two-element partial order $0 < 1$; that is,

$$Pt(F) = Hom_{\mathbb{Frame}}(F, \mathbf{2}_{Fr}) = Hom_{\mathbb{Locale}}(\mathbf{2}_{Fr}, F). \tag{6.106}$$

The main idea is that the frame of the one-object topological space $*$ is the partial order $\emptyset < \{*\}$, i.e., $\mathbf{2}_{Fr}$. See Remark 6.4.9. The open sets of $Pt(F)$ are given as follows. Let a be an element of F, then an open set of $Pt(F)$ is all the points that contain a. That is, the set of all maps is

$$\{f\colon F \longrightarrow \mathbf{2}_{Fr} : f(a) = 1\} \subseteq Pt(F). \tag{6.107}$$

The following shows that the open sets are closed under finite union: for two points a and a' in F, we have

$$\{f\colon F \longrightarrow \mathbf{2}_{Fr} : f(a) = 1\} \cap \{f\colon F \longrightarrow \mathbf{2}_{Fr} : f(a') = 1\} = \{f\colon F \longrightarrow \mathbf{2}_{Fr} : f(a \wedge a') = 1\}. \tag{6.108}$$

There is a similar argument showing that the open sets are closed under arbitrary unions. The open set that corresponds to $1 \in F$ is the entire $Pt(F)$. A frame map $g\colon F_1 \longrightarrow F_2$ will induce a map of topological spaces:

$$Pt(g) = Hom_{\mathbb{Frame}}(g, \mathbf{2}_{Fr})\colon Hom_{\mathbb{Frame}}(F_2, \mathbf{2}_{Fr}) \longrightarrow Hom_{\mathbb{Frame}}(F_1, \mathbf{2}_{Fr}). \tag{6.109}$$

There is an adjunction as follows:

$$Hom_{\mathbb{Top}}(T, Pt(F)) \cong Hom_{\mathbb{Locale}}(\mathcal{O}(T), F) \cong Hom_{\mathbb{Frame}}(F, \mathcal{O}(T)). \tag{6.110}$$

In other words, a continuous function $f\colon T \longrightarrow Pt(F)$ of topological space corresponds to a frame map $\hat{f}\colon F \longrightarrow \mathcal{O}(T)$. See page 477 of [181].

The unit and the counit of this adjunction are important. The unit of the adjunction for a topological space T is

$$T \longrightarrow Hom_{\mathbb{Frame}}(\mathcal{O}(T), \mathbf{2}_{Fr}), \tag{6.111}$$

which is defined for $x \in T$ as

$$x \quad \mapsto \quad f_x\colon \mathcal{O}(T) \longrightarrow \mathbf{2}_{Fr}. \tag{6.112}$$

For an open set $O \in \mathcal{O}(T)$, $f_x(O) = 1$ if and only if $x \in O$. In other words, f_x picks out those open sets that contain x.

The counit of the adjunction for a frame F is

$$F \longrightarrow \mathcal{O}(Hom_{\mathbb{Frame}}(F, \mathbf{2}_{Fr})), \tag{6.113}$$

which is defined for $a \in F$ as

$$a \quad \mapsto \quad \{f\colon F \longrightarrow \mathbf{2}_{Fr} : f(a) = 1\}. \tag{6.114}$$

This means that a goes to all the frame maps that pick out a.

♣

The subcategories where the unit and counit of this adjunction are isomorphisms form an equivalence of categories. First, here are some definitions.

Definition 6.4.20. A **sober space** is a topological space T such that every nonempty closed subset of T that cannot be separated is the closure of exactly one point. (A space that is not sober will have some closed set that cannot be separated by two or more points. Perhaps we should call a space that is not sober a "tequila space.") Such a space has enough closed spaces.

The category of all sober spaces and continuous maps between them is denoted as $\mathbb{Sober}$. A space is sober exactly when the unit is an isomorphism (homeomorphism) of topological spaces.

A frame F is a **spatial frame** if for any a and b in F with $a \nleq b$, there is a $f\colon F \longrightarrow \mathbf{2}_{Fr}$ such that $f(a) = 1$ and $f(b) = 0$. In other words, there is a point of F that separates a and b. A frame is spatial exactly when the counit of the adjunction is an isomorphism of frames. One can think of such frames as having enough objects. The collection of all spatial frames and frame morphisms is denoted as $\mathbb{SFrame}$.

Thus, we have proved the following.

Theorem 6.4.21. The adjunction discussed here induces an equivalence of categories between the category of spatial frames and the category of sober spaces. In summary:

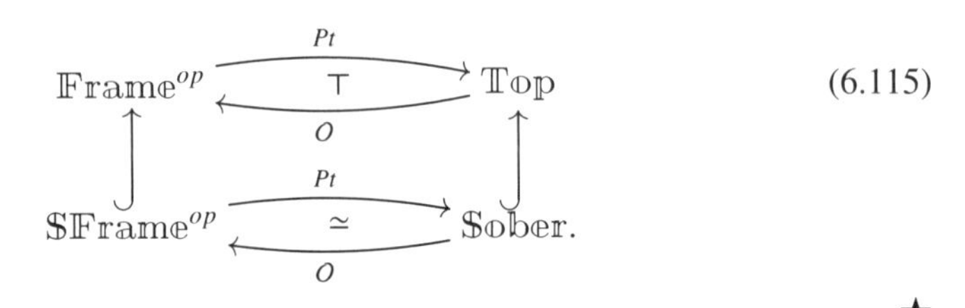

(6.115)

★

Another way to say this is that the topological spaces with enough open sets are the same as those frames with enough objects.

Suggestions for Further Study

This presentation gained much from [85], a paper by H.-E. Porst and Walter Tholen [214], and the slides from a first of a three-part talk [202] titled "The Mirror of Mathematics" by Prakash Panangaden.

Here are some sources for the topics discussed in this mini-course:

- Boolean algebras: Chapter 11 of [173], [89]
- Stone duality: [125, 157, 99]
- Dualizing objects: [30, 31, 214, 63]
- Esakia and Priestley dualities: [56, 85]
- Pointless topology: Chapter IX of [181], [124], [256], and Chapter II of [125].

The duality that was discussed here is just the beginning of the story. There are more advanced algebraic structures and topological structures that are related through duality. Other types of duality can be found in the following sources:

- Gelfand duality: [200] Chapter IV of [125].
- Pontryagin duality: [201]. (Both Gelfand and Pontragin duality have ramifications in quantum theory, see, e.g., Section C.15 of [152].)
- Tannaka–Krein duality: [130].

For a fascinating history of duality theory within category theory, see Jean-Pierre Marquis [189].

In Section 8.1 we will see a duality concerning algebraic structures and descriptions of algebraic structures.

For more on how computer scientists use duality, see the works of Samson Abramsky, such as [4, 3, 10]. See also [85].

Physicists work with many different types of duality. For example, electromagnetism is a duality that shows that an electrical field can be looked at as a magnetic field. There are various types of duality in string theory, which indicates that different theories are really duals of each other. A hot topic in current physics literature is something called ADS/CFT, which shows that there are seemingly two radically different ways of looking at the universe that are really the same. See Sir Michael F. Atiyah [19] for a broad view on duality in physics.

7

Variations of Monoidal Categories

In a metamathematical sense our theory provides general concepts applicable to all branches of mathematics, and so contributes to the current trend toward uniform treatment of different mathematical disciplines.
Samuel Eilenberg and Saunders Mac Lane
[68], page 236.

Monoidal categories and symmetric monoidal categories are only the beginning of the story. In this chapter, we describe many types of monoidal categories. For each type, we will describe the functors, natural transformations, and coherence conditions associated with that structure. Many examples will be provided, and there will be more examples in the upcoming chapters and mini-courses.

Use Venn diagram A.6 in Appendix A as an aid for this chapter. We began building this Venn diagram in Chapter 5, where we saw that every Cartesian category is a symmetric monoidal category, and in turn, every symmetric monoidal category is a monoidal category with extra structure. We also saw that every strict monoidal category is a monoidal category. This chapter will further develop that Venn diagram.

First, a disclaimer: there are many variations of monoidal categories. Peter Selinger describes at least twenty-five variations in his 2009 survey paper [233]. Many more structures have accrued in the literature since then. The situation is confusing because of the many types of structures and because various authors use different names for the same structure. Here, we only describe the major variations. However, after reading this chapter, one should be able to understand most of the definitions in the literature.

Section 7.1 deals with the commutativity of the monoidal product and is about a variation of a symmetric monoidal category called a "braided monoidal category." Section 7.2 deals with variations that concern duality. Section 7.3 deals with ribbon categories. The chapter ends with Section 7.4, which presents a mini-course about a modern branch of algebra called "quantum groups." This field deals with algebraic structures that are related to various types of monoidal categories.

7.1 Braided Monoidal Categories

We begin with a type of commutativity of the monoidal structure. We already have encountered symmetric monoidal categories, which are monoidal categories where one can uniformly switch $a \otimes b$ with $b \otimes a$. We drew this as

a b

b a. (7.1)

Now we want to discuss *the way* that the elements are switched. We can draw two possibilities:

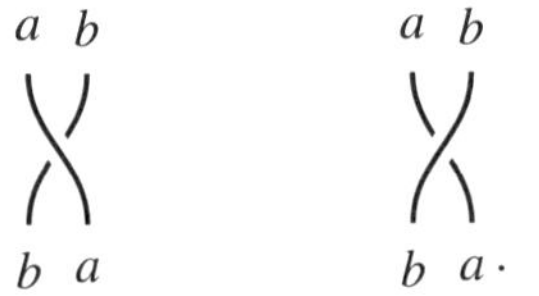

a b a b

b a b a. (7.2)

On the left, the a strand goes over the b strand and we denote the transition as the morphism

$$\gamma_{a,b} \colon a \otimes b \longrightarrow b \otimes a, \tag{7.3}$$

while on the right, the b strand goes over the a strand and we denote the transition as the morphism

$$\gamma^{-1}_{b,a} \colon a \otimes b \longrightarrow b \otimes a. \tag{7.4}$$

Where there is a crossing of strands, the lower strand is drawn with cutouts for the upper strand. These two diagrams correspond to two different morphisms from $a \otimes$b to $b \otimes a$. These two diagrams are the building blocks of the various ways of switching or braiding the elements that are tensored. The structures that we will be dealing with are called "braided monoidal categories."

One of the axioms of a symmetric monoidal category is the symmetry coherence condition $\gamma_{b,a} \circ \gamma_{a,b} = id_{a \otimes b}$. This was shown in terms of permutations in Diagram (5.67). In a braided monoidal category, we take into account how the elements are switched. Hence, this identity is not necessarily true, as in the following inequality:

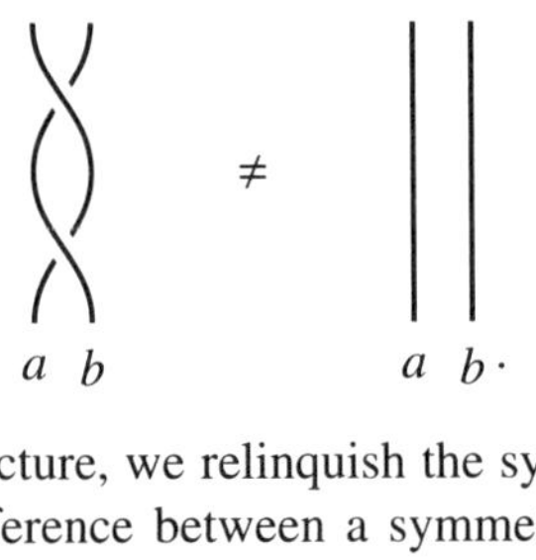

a b a b

$\neq$

a b a b. (7.5)

To define this type of structure, we relinquish the symmetry coherence condition. This idea is the essential difference between a symmetric monoidal category and a braided monoidal category.

The philosophical point is that idealized mathematical structures are symmetric. Physical objects are not symmetric and can twist around each other like braids. It is their physicality that keeps them from being symmetric. We will see more of these physical aspects of strings in the coming pages.

Definition 7.1.1. A **braided monoidal category** $(\mathbb{A}, \otimes, \alpha, I, \lambda, \rho, \gamma)$ is a monoidal category $(\mathbb{A}, \otimes, \alpha, I, \lambda, \rho)$ with a way of changing the order of the elements; that is, a **braiding** natural isomorphism $\gamma\colon \otimes \longrightarrow \otimes \circ br$. For all objects a, b in $\mathbb{A}$, there is an isomorphism $\gamma_{a,b}\colon a \otimes b \longrightarrow b \otimes a$. This braiding must satisfy the two hexagon coherence conditions in Diagrams (5.39) and (5.40). But—in contrast to a symmetric monoidal category—it need not satisfy the symmetry condition: $\gamma_{b,a} \circ \gamma_{a,b} = id_{a\otimes b}$.

Technical Point 7.1.2. As stated in Technical Point 5.3.3, when the symmetry condition is satisfied, only one of the hexagon coherence conditions needs to be stated because the other is automatically implied. However, since braided monoidal categories need not satisfy the symmetry condition, we must require both hexagon conditions. ♡

Even though the braiding fails to satisfy the symmetry condition, the natural transformation γ is still an isomorphism. In a symmetric monoidal category, the inverse is given as $\gamma_{a,b}^{-1} = \gamma_{b,a}$, while in a braided monoidal category, the inverse of is given as $\gamma_{a,b}^{-1} = \gamma_{b,a}^{-1}$. We can see this more clearly with the string diagrams:

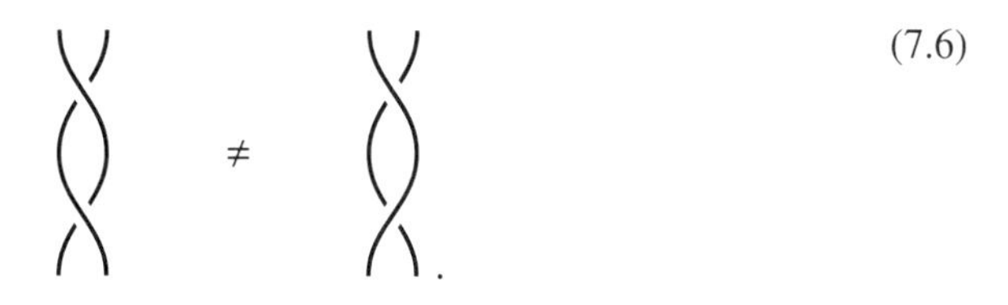
(7.6)

On the left is $\gamma_{b,a} \circ \gamma_{a,b}$, which is not equal the identity, and on the right is $\gamma_{b,a}^{-1} \circ \gamma_{a,b}$, which is equal to the identity.

Example 7.1.3. Every symmetric monoidal category is automatically a braided monoidal category. □

Example 7.1.4. The monoidal category of vector spaces $(\mathbf{K}\mathbb{V}\mathrm{ect}, \otimes, \mathbf{K})$ has the usual braiding $\gamma_{V,W}\colon V \otimes W \longrightarrow W \otimes V$, which is defined as $v \otimes w \mapsto w \otimes v$. This satisfies the symmetry condition. In Section 7.4, we meet nonstandard braidings for this category, which are defined as $v \otimes w \mapsto Rw \otimes v$, where R is a matrix that changes the elements in interesting ways. Such braidings need not satisfy the symmetry condition. This makes such a category of **K** vector spaces into a braided monoidal category. □

Example 7.1.5. The category $\mathbb{C}\mathrm{ircuit}$ is a braided monodial category, not a symmetric monoidal category. One can see this by realizing that the physical wires must go around each other and cannot cross each other. □

Example 7.1.6. In our discussion of symmetric monoidal categories, we introduced the symmetry category $\mathbb{S}\mathrm{ym}$. For braided monoidal categories, we introduce the **braid category** $\mathbb{B}\mathrm{raid}$. The presentation mostly parallels the presentation of $\mathbb{S}\mathrm{ym}$ given in Section 5.3.

There is a sequence of groups $B_0, B_1, B_2, B_3, \ldots, B_n, \ldots$, where B_n is called the n**th braid group** and corresponds to the various ways that n strands can wrap around each other in three dimensions. For example, a typical element of B_5 looks like this:

(7.7)

Such a two-dimensional diagram of a three-dimensional braid is called a **braid projection**. "Don't confuse the map for the territory!" We are interested in braids, but we describe them using braid projections. A single braid can have many braid projections. In other words, two different braid projections can represent the same braid. For example, notice that although the following two braid projections look different from the braid projection in Diagram (7.7), they are really the same because the strings can be pushed around:

(7.8)

Here is another element of B_5:

(7.9)

Here are depictions of braids from B_7 and B_9:

(7.10)

The composition in each of the braid groups is given by attaching the strands. So the composition of the two braids described in Diagrams (7.7) and (7.9) is

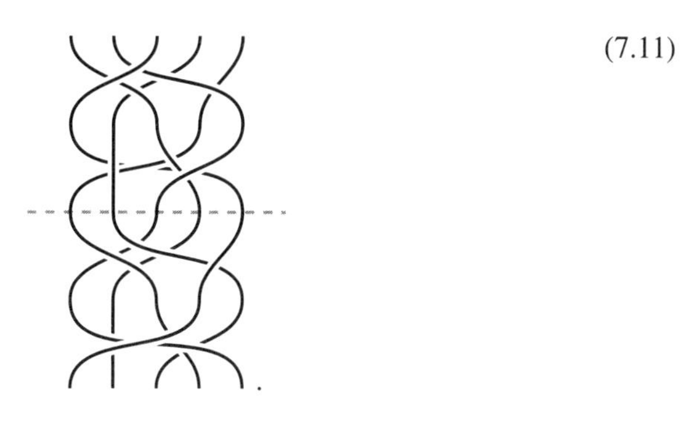
(7.11)

The identity of B_5 is

(7.12)

which corresponds to not swapping any objects. Here is a nontrivial braid projection of the identity element of B_5:

(7.13)

The inverse of a braid is given by taking the mirror image of the braid from the bottom. This undoes all the crossings. The inverse of Diagram (7.7) is

(7.14)

One can compose Diagrams (7.7) and (7.14) to see that it gives yet another braid projection of the identity braid:

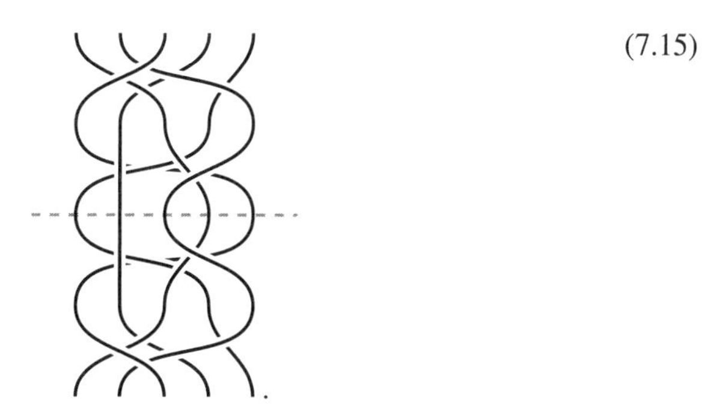
(7.15)

The groups B_0 and B_1 are each the trivial group since there aren't two strands to wrap around. In contrast, group B_2 corresponds to the integers because the two strings cannot wrap around each other or they wrap around each other a positive or negative integer amount.

Just as we saw that every permutation in $\mathbb{S}\mathrm{ym}$ can be built up from small permutations, every braid can be built up from small braids. For group B_n, we denote the generators as $\sigma_1, \sigma_2, \ldots, \sigma_{n-1}$. For B_5, the generator σ_3 looks like this:

(7.16)

The inverse of this generator is

(7.17)

These generators satisfy the following two rules:

- $\sigma_i \sigma_j = \sigma_j \sigma_i$ for $|i - j| > 1$.
- $\sigma_i \sigma_{i+1} \sigma_i = \sigma_{i+1} \sigma_i \sigma_{i+1}$.

These two equations can be understood graphically in the same way that the rules for s_n can be understood (see Equations (5.68) and (5.69)).

(7.18)

(7.19)

Notice that we do not have the rule $\sigma_i \sigma_i = e$ as we had for group S_n. This is because we do not assume the symmetry condition.

The fact that all braids can be generated by these simple braids with these two simple braid rules means that it is easy to tell when two braid projections represent the same braid. Two braid projections represent the same braid if there is a sequence of braid projections between them where each differ by such simple moves.

The same way as the symmetry groups come together to form the strictly associative symmetric monoidal category $\mathbb{S}\mathrm{ym}$, these braid groups come together to form the strictly associative braided monoidal category $\mathbb{B}\mathrm{raid}$. The objects are the natural numbers; and the morphisms are $Hom_{\mathbb{B}\mathrm{raid}}(m,n) = B_m$ if $m = n$, and the empty set otherwise. The strict monoidal structure on objects is addition, that is, $m \otimes n = m + n$. On the morphisms, the strict monoidal structure comes from placing the braids side by side. For example, the tensor of the two braids described in Diagram (7.10) is

(7.20)

It is easy to see that this monoidal structure is strictly associative. The unit of the monoidal structure is the zero of the natural numbers. This will correspond to putting a braid side by side with the empty braid. The braiding is a natural transformation whose components are $\gamma_{m,n} : m + n \longrightarrow n + m$, which will have the strings cross each other. For example, if $m = 4$ and $n = 3$, then $\gamma_{4,3}$ looks like

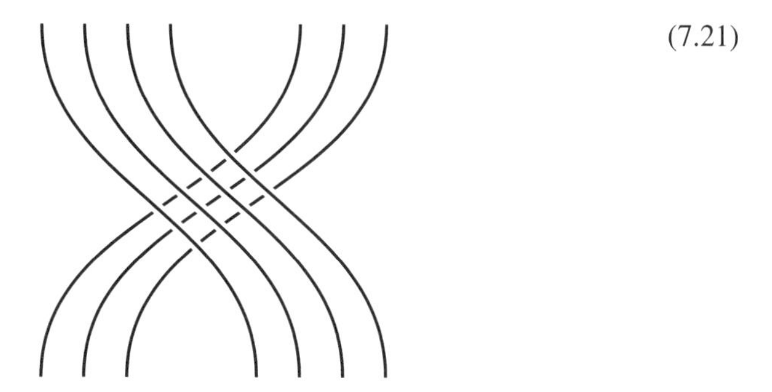

(7.21)

This braiding does not satisfy the symmetry conditions because the strands go over each other. The braidings satisfy the hexagon coherence conditions for the strictly associative symmetric monoidal category structure (see Diagram 5.42). This is seen by thinking of the many strands as a thick cable. The two triangles then become

$a \quad b \; c \qquad\qquad a \quad b \quad c$

$=$ (7.22)

$b \; c \quad a \qquad\qquad b \quad c \quad a$

which is $\gamma_{a,b\otimes c} = (id_b \otimes \gamma_{a,c}) \circ (\gamma_{a,b} \otimes id_c)$, and

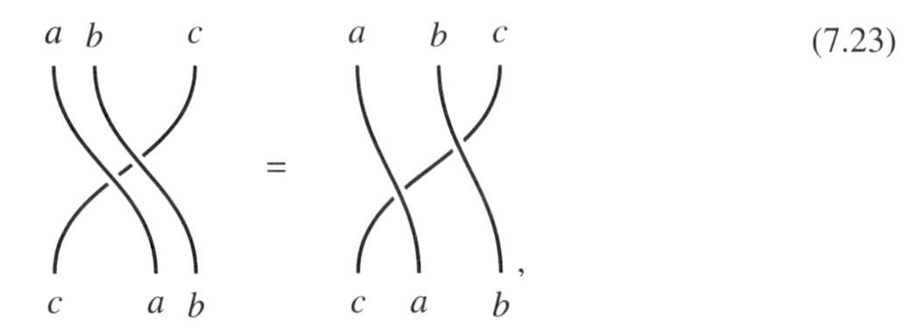

(7.23)

which is $\gamma_{a\otimes b,c} = (\gamma_{a,c} \otimes id_b) \circ (id_a \otimes \gamma_{b,c})$.

We will see that $\mathbb{Braid}$ is the paradigm of braided monoidal categories. □

Let us examine the relationship between the braid category and other categories. For this, we have to look at the braid groups. Whereas each S_n is finite, the braid groups (except for B_0 and B_1) are infinite. Just think of B_2, where you can twist the two strings around each other any number of times. For each n, there is a surjective group homomorphism $B_n \longrightarrow S_n$. This homomorphism takes a braid to the permutation that it represents. In other words, it looks at where all the strands start and end. For example,

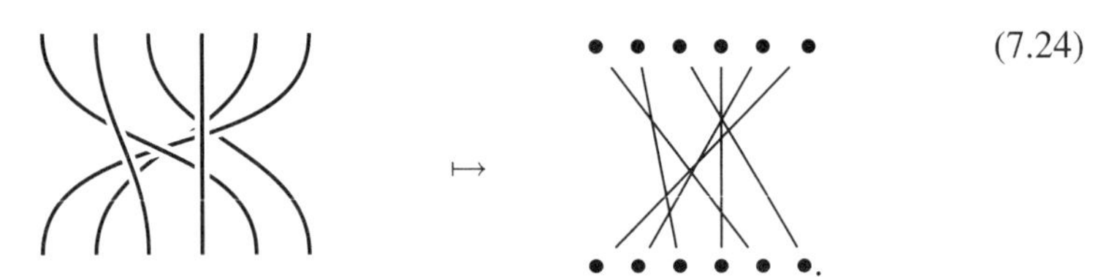

(7.24)

For example, the braid in Diagram (7.20) will go to the permutation

$$(1, 2, 3, \ldots, 15, 16) \mapsto (5, 3, 2, 6, 4, 7, 1, 9, 13, 8, 12, 15, 14, 16, 11, 10).^1 \qquad (7.25)$$

These homomorphisms take the generators of the braid groups to the generators of the symmetry groups (i.e., $\sigma_i \mapsto s_i$). The map is a group homomorphism because (i) the composition of two braids goes to the composition of permutation, (ii) the inverse of a braid goes to the inverse of the permutation, and (iii) the identity braid goes to the identity permutation.

These homomorphisms from the braid groups to the symmetry groups extend to a functor $\mathbb{Braid} \longrightarrow \mathbb{Sym}$. The functor is the identity on objects, and the surjectivity of the homomorphisms ensure that the functor is full (every permutation can be written as at least one braid) but not faithful (many braids give the same permutation). This functor is a strict braided monoidal functor.

Definition 7.1.7. For every B_n there is a subset (kernel) of elements that the homomorphism $B_n \longrightarrow S_n$ takes to the identity permutation. We call such braids **pure braids**. They have strands that twist and turn but in the end come back to where they started. Here are three examples of pure braids:

[1] I am indebted to Rivka, Boruch, and Miriam Yanofsky for providing me with this mapping.

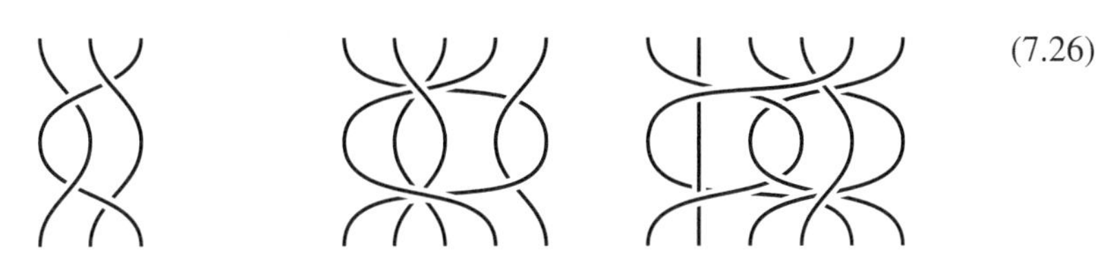

(7.26)

For each n, there is a subgroup of pure braids. They combine in the same way as braids to form the category of pure braids, denoted $\mathbb{PBraid}$.

Notice that $\mathbb{PBraid}$ is not a braided monoidal category because a braid like Diagram (7.21) is not a pure braid. It is only a strictly monoidal category. The category of pure braids is a submonoidal category of braids; that is, $\mathbb{PBraid} \hookrightarrow \mathbb{Braid}$ (see Venn diagram A.5.) This monoidal functor is the identity on objects, faithful but not full. The composition of the two functors

$$\mathbb{PBraid} \hookrightarrow \mathbb{Braid} \twoheadrightarrow \mathbb{Sym} \tag{7.27}$$

is the identity on objects and takes every pure braid on n strings to id_n in $\mathbb{Sym}$. This has a feel of being a short exact sequence, which we saw in Definition 2.4.24 (although, of course, it is not a short exact sequence because we are not dealing with vector spaces).

Functors between braided monoidal categories satisfy the same requirement as symmetric monoidal functors (i.e., Diagram (6.19)). Natural transformations between such functors satisfy the same requirements as natural transformations between monoidal functors (i.e., Diagram (6.21)).

Definition 7.1.8. The collection of all braided monoidal categories, functors between them, and natural transformations between them, form a 2-category $\overline{\mathbb{BraMonCat}}$.

There is a forgetful 2-functor to $\overline{\mathbb{MonCat}}$. There is also an inclusion 2-functor from the 2-category of symmetric monoidal categories.

For each of these 2-categories, there are full sub-2-categories of strictly associative versions.

Definition 7.1.9. There are the sub-2-categories $\overline{\mathbb{StrSymMonCat}}$, $\overline{\mathbb{StrBraMonCat}}$, and $\overline{\mathbb{StrMonCat}}$, which are versions of $\overline{\mathbb{SymMonCat}}$, $\overline{\mathbb{BraMonCat}}$, and $\overline{\mathbb{MonCat}}$, where the monoidal structure is strict.

Thus, we can extend Diagram (6.22) as follows:

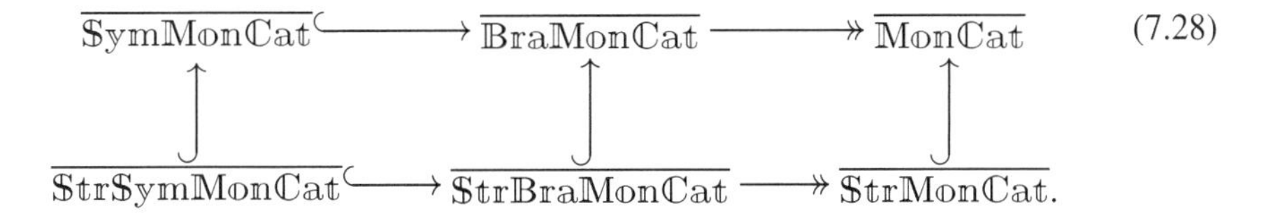

(7.28)

See Venn diagram A.6.

The coherence theorem for braided monoidal categories is very similar to Theorems 6.2.1 and 6.2.8, which are the coherence theorem for monoidal categories and symmetric monoidal categories, respectively. Here, rather than working with the categories $\mathbb{Assoc}$ or $\mathbb{Sym}$, we work with the category $\mathbb{Braid}$.

Theorem 7.1.10. The Coherence Theorem for Braided Monoidal Categories. We state it in four equivalent ways:

- For every braided monoidal category $(\mathbb{C}, \otimes, I, \alpha, \rho, \lambda, \gamma)$ and every object x in $\mathbb{C}$, there is a unique braided monoidal functor $F_x \colon \mathbb{Braid} \longrightarrow \mathbb{C}$ such that $F_x(1) = x$.
- There is an isomorphism of categories:

$$Hom(\mathbb{Braid}, \mathbb{C}) \cong Hom_{\mathbb{Cat}}(\mathbf{1}, \mathbb{C}), \tag{7.29}$$

  where the Hom set on the left is for strict, braided monoidal functors and the Hom set on the right is for $\mathbb{Cat}$.
- The braided category, $\mathbb{Braid}$, is the free, braided monoidal category on one generator.
- Two morphisms in $\mathbb{C}$ that are in the image of F_x of identities, $\alpha, \rho, \lambda, \gamma$ and their inverses, and built up with $\circ$ and $\otimes$, are equal if the two underlying braids are the same. ★

Proof. The proof is almost the same as the proof of Theorem 6.2.8. We will comment here only when the ingredients are different. Remember that the objects in $\mathbb{Braid}$ are the natural numbers. For every object x in $\mathbb{C}$, there is a strictly associative, braided, monoidal functor F_x such that $F_x(1) = x$. In $\mathbb{Braid}$, $2 = 1 + 1$ and so $F_x(2) = x \otimes x$. In general, $F_x(n) = x^{\otimes n}$. The braiding $\gamma_{1,1} \colon 2 \longrightarrow 2$ in $\mathbb{Braid}$ must go to $F_x(\gamma_{1,1}) = \gamma_{x,x} \colon x \otimes x \longrightarrow x \otimes x$. Every map $n \longrightarrow n$ in $\mathbb{Braid}$ is an element of the nth braid group. In Example 7.1.6, we saw that every element in the nth braid group can be written as a composition of the small braids $\sigma_1, \sigma_2, \ldots, \sigma_{n-1}$. These small braids can be written as $\sigma_i = id_{i-1} \otimes \gamma_{1,1} \otimes id_{n-i-1}$. The functor F_x takes this morphism to $id_{x^{\otimes(i-1)}} \otimes \gamma_{x,x} \otimes id_{x^{\otimes(n-i-1)}}$. The functor F_x is a braided monoidal category that takes side-by-side braids to tensor products in $\mathbb{C}$. It is also not hard to show that this evaluation at 1 for every F_x gives the equivalence of categories. ♣

We must look at the meaning of this theorem in the context of Theorems 6.2.1 and 6.2.8. With monoidal categories, we saw that all diagrams made of α, ρ, and λ must commute. With symmetric monoidal categories, we saw that diagrams commute only if their underlying permutations are equal. Here, with braided monoidal categories, diagrams commute only if their underlying braids are equal. In other words, since there is

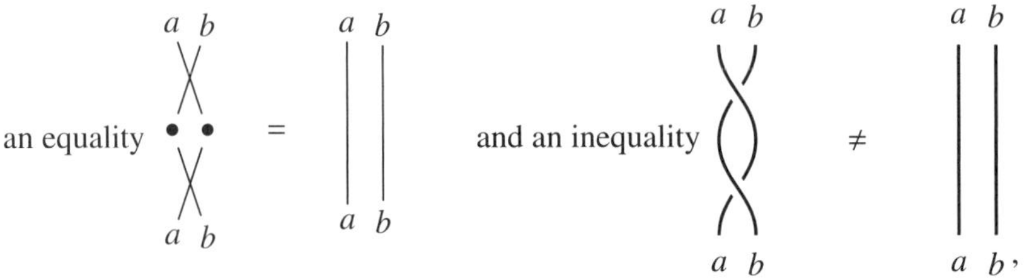

an equality (braid with strands a b) = (two straight strands a b) and an inequality (braid with strands a b) $\neq$ (two straight strands a b), (7.30)

the diagram

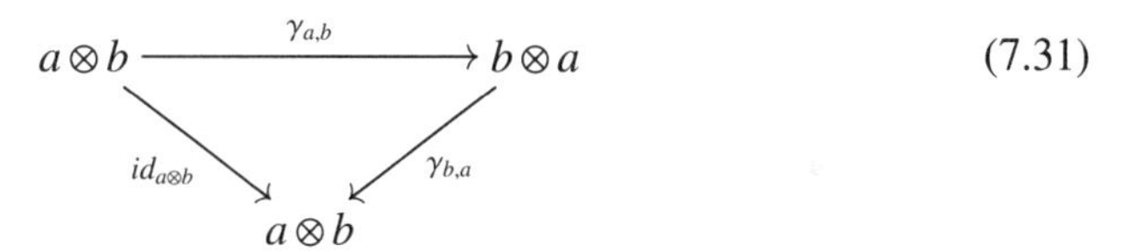

(7.31)

commutes in a symmetric monoidal category and does not commute in a braided monoidal category. In a similar way, two strings wrapped around each other two and a half times are not equal to a single twist. This corresponds to the fact that

$$\begin{array}{ccccc} a\otimes b & \xrightarrow{\gamma_{a,b}} & b\otimes a & \xrightarrow{\gamma_{b,a}} & a\otimes b \\ \downarrow{\scriptstyle \gamma_{a,b}} & & & & \downarrow{\scriptstyle \gamma_{a,b}} \\ b\otimes a & \xleftarrow[\gamma_{a,b}]{} & a\otimes b & \xleftarrow[\gamma_{b,a}]{} & b\otimes a \end{array} \tag{7.32}$$

commutes in a symmetric monoidal category and does not commute in a braided monoidal category. The main point of the coherence theorem is that if two morphisms in the category describe the same braid in the braid group, then the two morphisms are the same.

7.2 Closed Categories

We are interested in when the collection of morphisms between two objects is itself an object in the category. This means that the Hom sets are also objects of the category.

Definition 7.2.1. A category $\mathbb{A}$ is a **closed category** if for all objects a and b in $\mathbb{A}$, the Hom set $Hom_{\mathbb{A}}(a,b)$ is not only a set, but also an object in the category. These Hom sets have to satisfy many properties showing that they cohere with composition and the unit morphisms. We will not elaborate here because these sets are not our central focus. When we have a closed category, the function $Hom\colon \mathbb{A}^{op} \times \mathbb{A} \longrightarrow \mathbb{A}$ is called an **internal Hom**.

While closed categories are interesting in and of themselves, and we will revisit them in Section 9.1, our focus here is on monoidal categories and how the internal Hom relates to the monoidal structures.

We begin with closed categories that have a simple type of monoidal structure that is, Cartesian categories.

Definition 7.2.2. Category $\mathbb{A}$ is a **Cartesian closed category** if it is a Cartesian category (i.e., it has all finite products including a terminal object) and has an internal Hom functor such that for any object b, the functor $-\times b$ is left adjoint to $Hom_{\mathbb{A}}(b,-)$. In detail, for any three objects a, b, and c of $\mathbb{A}$, there is a natural isomorphism:

$$Hom_{\mathbb{A}}(a \times b, c) \cong Hom_{\mathbb{A}}(a, Hom_{\mathbb{A}}(b,c)). \tag{7.33}$$

For completeness, let us spell out the unit and counit of these adjunctions. For every b, the counit ev^b is the familiar **evaluation map**, whose a component is

$$ev_a^b\colon Hom(b,a)\times b \longrightarrow a. \tag{7.34}$$

The unit $coev^b$ is the **coevaluation map**, whose a component is

$$coev_a^b\colon a \longrightarrow Hom(b, a\times b). \tag{7.35}$$

The evaluation and coevaluation maps must satisfy the triangle identities in Diagram (4.76) because of the adjunction.

Exercise 7.2.3. Write the triangle identities for the unit and the counit. ■

The following are examples of Cartesian closed categories.

Example 7.2.4.

- $\mathbb{S}\mathrm{et}$ and $\mathbb{F}\mathrm{in}\mathbb{S}\mathrm{et}$. The product is the Cartesian product. Way back in Section 1.4, we saw that internal Homs are sets of functions from one set to another. Equation (1.22) showed us that the internal Hom is right adjoint to the product.
- $\mathbb{C}\mathrm{at}$. The product is the Cartesian product of categories. In Remark 4.2.7, we saw that the internal Hom is the category of functors and natural transformations. The fact that the categorical product is left adjoint to the internal Hom is similar to $\mathbb{S}\mathrm{et}$. In detail,

$$Hom_{\mathbb{C}\mathrm{at}}(\mathbb{A}\times\mathbb{B},\mathbb{C}) \cong Hom_{\mathbb{C}\mathrm{at}}(\mathbb{A}, Hom_{\mathbb{C}\mathrm{at}}(\mathbb{B},\mathbb{C})). \tag{7.36}$$

  Functor $F\colon \mathbb{A}\times\mathbb{B}\longrightarrow\mathbb{C}$ corresponds to functor $F'\colon \mathbb{A}\longrightarrow Hom_{\mathbb{C}\mathrm{at}}(\mathbb{B},\mathbb{C})$, where $F'(a)(b)=F(a,b)$ for a in $\mathbb{A}$ and b in $\mathbb{B}$. For $f\colon a\longrightarrow a'$ in $\mathbb{A}$, $F'(f)$ is a natural transformation from functor $F'(a)$ to functor $F'(a')$. At b in $\mathbb{B}$, the component $F'(f)_b$ is $F(f, id_b)$, which is a morphism in $\mathbb{C}$. The inverse mapping is left to the reader.
- $\mathbb{P}\mathrm{rop}$ and $\mathbb{P}\mathrm{red}(n)$. The product is the meet $\wedge$. We saw that the internal Hom of propositions P and Q is $P\Rightarrow Q$, and in Theorem 4.8.2, we saw that the internal Hom is right adjoint to the product $\wedge$. Similarly, a Heyting algebra, thought of as a partial order category, is actually a Cartesian closed category.
- $\mathcal{P}(S)$ for set S. The product is intersection $\cap$. Given subsets X, Y, and Z of S, we have that

$$(X\cap Y)\leq Z \text{ if and only if } X\leq (Y^c\cup Z). \tag{7.37}$$

  This is similar to the situation in logic if we remember that $P\Rightarrow Q \equiv \neg P\vee Q$.
- The unit interval [0, 1] with multiplication. The total order of real numbers from 0 to 1 is a Cartesian category where the tensor is multiplication $\cdot$. The unit is 1. The internal Hom $[b/c]$ is an operation that we call **bounded division**, which is defined as

$$[b/c] = \begin{cases} b/c & : b\leq c \\ 1 & : b>c. \end{cases} \tag{7.38}$$

The adjunction can be seen from

$$a \cdot b \leq c \qquad \text{if and only if} \qquad a \leq [b/c]. \tag{7.39}$$

- $\mathbb{S}\mathrm{et}^{\mathbb{A}^{op}}$ for any small category $\mathbb{A}$. The Cartesian product is the pointwise product, that is, for $F, G\colon \mathbb{A} \longrightarrow \mathbb{S}\mathrm{et}$, $(F \times G)(a) = F(a) \times G(a)$. The internal Hom is a little more complicated. First, let us show what does not work. With Important Categorical Idea 2.4.4 in mind, one might think that since $\mathbb{S}\mathrm{et}$ is Cartesian closed, so $\mathbb{S}\mathrm{et}^{\mathbb{A}^{op}}$ inherits the property of being Cartesian closed from $\mathbb{S}\mathrm{et}$. In that case, the obvious definition of an internal Hom is that $F^G\colon \mathbb{A} \longrightarrow \mathbb{S}\mathrm{et}$ is defined for $a \in \mathbb{A}$ as $(F^G)(a) = Hom(F(a), G(a)) = F(a)^{G(a)}$. This fails because such a mapping is not functorial. Given $f\colon a \longrightarrow a'$, the mapping $(F^G)(f)$ does not make sense because it is contravariant in F and covariant in G.

So what does work? Take F^G on input a to be the set of natural transformations from the functor $Hom(\ \ , a) \times G$ to the functor F. Expressed in symbols,

$$(F^G)(a) = Hom_{\mathbb{S}\mathrm{et}^{\mathbb{A}^{op}}}(Hom(-, a) \times G(-), F(-)). \tag{7.40}$$

Notice that all three functors, $Hom(-, a)$, $G(-)$, and $F(-)$, are contravariant on $\mathbb{A}$. For $f\colon a \longrightarrow a'$ in $\mathbb{A}$, we have

$$(F^G)(f)\colon (F^G)(a') \Longrightarrow (F^G)(a). \tag{7.41}$$

This is defined on $\theta_{a'}\colon Hom(-, a') \times G(-) \Longrightarrow F(-)$ as

$$\begin{aligned}(F^G)(f)(\theta_{a'}) = \theta_{a'} \circ (Hom-, f) \times id_G)\colon\ & Hom(-, a) \times G(-) \\ & \Longrightarrow Hom(-, a') \times G(-) \Longrightarrow F(-).\end{aligned} \tag{7.42}$$

Let $\theta_a\colon Hom(-, a) \times G(-) \Longrightarrow F(-)$ be such a natural transformation. At object b in $\mathbb{A}$, we have the following map:

$$(\theta_a)_b\colon Hom(b, a) \times G(b) \longrightarrow F(b). \tag{7.43}$$

To show that the category is Cartesian closed, we need to prove that there is a natural isomorphism:

$$Hom(H \times G, F) \cong Hom(H, F^G). \tag{7.44}$$

Let $\alpha\colon H \times G \Longrightarrow F$ be a natural transformation on the left. This means that for all objects a and a' in $\mathbb{A}$, there are the morphisms $\alpha_a\colon H(a) \times G(a) \longrightarrow F(a)$ and $\alpha_{a'}\colon H(a') \times G(a') \longrightarrow F(a')$. Furthermore, for any map $g\colon a' \longrightarrow a$ in $\mathbb{A}$, there is a commutative naturality square:

$$\begin{array}{ccc} H(a) \times G(a) & \xrightarrow{\alpha_a} & F(a) \\ {\scriptstyle H(g)\times G(g)}\downarrow & & \downarrow{\scriptstyle F(g)} \\ H(a') \times G(a') & \xrightarrow[\alpha_{a'}]{} & F(a'). \end{array} \tag{7.45}$$

There is a corresponding $\bar{\alpha}: H \Longrightarrow F^G$. In detail, for object a in $\mathbb{A}$, we have

$$\bar{\alpha}_a: H(a) \longrightarrow Hom_{\mathbb{S}\mathrm{et}^{\mathbb{A}^{op}}}(Hom(-,a) \times G(-), F(-)). \tag{7.46}$$

For $x \in H(a)$, there is a natural transformation:

$$(\bar{\alpha}_a)(x): Hom(-,a) \times G(-) \Longrightarrow F(-). \tag{7.47}$$

For a' in $\mathbb{A}$, there is a component:

$$(\bar{\alpha}_a)(x)_{a'}: Hom(a',a) \times G(a') \longrightarrow F(a'). \tag{7.48}$$

This is defined for map $g: a' \longrightarrow a$ and $y \in G(a')$ as

$$\alpha_{a'}(H(g)(x), y) \in F(a'). \tag{7.49}$$

We leave the rest of the details for the reader's leisure time.

Special cases of this is when $\mathbb{A}$ is a one-object category (group or monoid). This means that $\mathbb{S}\mathrm{et}^G$ (i.e., the category of G-sets), and $\mathbb{S}\mathrm{et}^M$ (i.e., the category of M-sets), are Cartesian closed categories.

- **K**$\mathbb{V}\mathrm{ect}$. The product (not the tensor product) is the Cartesian product. The internal Hom is just the set of functions (not linear maps) from V to W. But this is not that interesting.
- The category of topological spaces, $\mathbb{T}\mathrm{op}$, does not form a Cartesian closed category. However, there is a full subcategory of "nice" topological spaces called **compactly generated Hausdorff spaces** that do form a Cartesian closed category. The subcategory of such spaces is denoted as $\mathbb{CGHS}$. □

Examples of Cartesian closed categories that we will *not* discuss are categories related to **typed λ-calculus**. These are formal representations of computation that are the basis of functional programming languages. We chose not to discuss them because they are not well known even to most computer scientists. However, they play a prominent role in other books on category theory and its relation to computers.

Cartesian closed categories will play a major role in our introduction to topos theory (discussed in Section 9.5).

Remark 7.2.5. The internal Hom is right adjoint to $- \times b$. What about a right adjoint to $b \times -$? In a Cartesian category, there is an isomorphism between $a \times b$ and $b \times a$, so their right adjoints are isomorphic. This follows from a slight generalization of Theorem 4.4.25. ♠

Let us generalize to when the internal Hom coheres with a monoidal product on a category.

Definition 7.2.6. A **left closed monoidal category** is a monoidal category $(\mathbb{A}, \otimes, I)$ such that for every object b, the functor $-\otimes b$ has a right adjoint $[b,-]$. This means that there is a natural isomorphism:

$$Hom_{\mathbb{A}}(a\otimes b, c) \cong Hom_{\mathbb{A}}(a, [b,c]). \tag{7.50}$$

In detail, there are evaluations ev^b and coevaluations $coev^b$:

$$ev^b_a\colon [b,a]\otimes b \longrightarrow a \qquad \text{and} \qquad coev^b_a\colon a \longrightarrow [b, a\otimes b], \tag{7.51}$$

which satisfy the triangle identities.

A **right closed monoidal category** is a monoidal category $(\mathbb{A}, \otimes, I)$ such that for every object b, the functor $b\otimes -$ has a right adjoint $[b,-]'$. This means that there is a natural isomorphism:

$$Hom_{\mathbb{A}}(b\otimes a, c) \cong Hom_{\mathbb{A}}(a, [b,c]'). \tag{7.52}$$

In detail, there are evaluations ev'^b and coevaluations $coev'^b$:

$$ev'^b_a\colon b\otimes [b,a]' \longrightarrow a \qquad \text{and} \qquad coev'^b_a\colon a \longrightarrow [b, b\otimes a]', \tag{7.53}$$

which satisfy the triangle identities.

A generalization of Remark 7.2.5 implies that if there is a braiding (or a symmetry), then $-\otimes b$ is isomorphic to $b\otimes -$ and

$$Hom_{\mathbb{A}}(a, [b,c]) \cong Hom_{\mathbb{A}}(a\otimes b, c) \cong Hom_{\mathbb{A}}(b\otimes a, c) \cong Hom_{\mathbb{A}}(a, [b,c]'). \tag{7.54}$$

It follows that the two internal Homs are isomorphic. When both $b\otimes -$ and $-\otimes b$ have isomorphic right adjoints, we call the structure a **closed monoidal category**.

Example 7.2.7. Every Cartesian closed category is a closed monoidal category. □

Example 7.2.8. Consider the natural numbers as a partially ordered category with addition $(\mathbf{N}, +, 0)$. It is a closed monoidal category with the internal Hom as the **bounded subtraction** operation, defined as

$$[b,c] = \begin{cases} c-b & : b\leq c \\ 0 & : \text{otherwise.} \end{cases} \tag{7.55}$$

The adjunction can be seen by noticing that

$$a+b\leq c \qquad \text{if and only if} \qquad a\leq c-b. \tag{7.56}$$

□

Example 7.2.9. The category of relations, $\mathbb{Rel}$, is a closed monoidal category. The monoidal product and the internal Hom are both the tensor product of sets. The adjunction can be seen for sets S, T, and U by considering the following correspondence:

$$S\times T \not\longrightarrow U \qquad \text{corresponds to} \qquad S \not\longrightarrow T\times U. \tag{7.57}$$

This means that there is a relation between the sets $S \times T$ and U if and only if there is a relation between the sets S and $T \times U$. In terms of elements of sets, these two relations are correlated as follows:

$$(s,t) \sim u \qquad \text{if and only if} \qquad s \sim (t,u). \tag{7.58}$$

□

Example 7.2.10. The category of finite dimensional vector spaces, $\mathbf{K}\mathbb{FDV}\mathbb{ect}$, with the usual tensor product forms a closed monoidal category. For vector spaces V and W, the internal Hom is defined as $Hom(V,W) = V^* \otimes W$, where $V^* = Hom(V, \mathbf{K})$. □

Related to closed monoidal categories is the notion of a dual object in a category.

Definition 7.2.11. Let $(\mathbb{A}, \otimes, I, \alpha)$ be a monoidal category (assume that ρ and λ are the identity) and a be an object of $\mathbb{A}$. We say a^* is a **right dual of** a and that a is a **left dual of** a^* if there are the morphisms

$$\epsilon_a \colon a \otimes a^* \longrightarrow I \qquad \text{and} \qquad \eta_a \colon I \longrightarrow a^* \otimes a \tag{7.59}$$

such that the following diagrams commute:

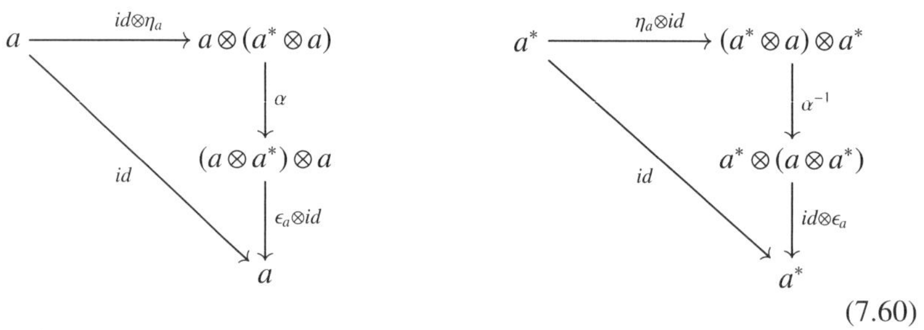

(7.60)

When $\mathbb{A}$ is a strict monoidal category, then these two diagrams reduce to

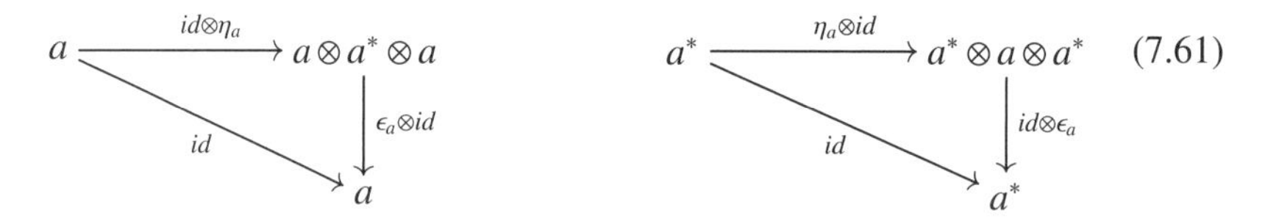

(7.61)

These are like the triangle identities of adjoint functors. Just as adjoint functors, when duals exist, they are unique up to a unique isomorphism.

As string diagrams (from left to right), we can draw ϵ and η as follows:

a^*, a and a, a^*. (7.62)

The triangle identities for an adjunction was given in Diagram (5.127).

Let a and b be objects with right duals a^* and b^* in a strict monoidal category $\mathbb{A}$. Consider $f\colon a \longrightarrow b$ be a morphism in $\mathbb{A}$. We construct a right dual of f denoted as $f^*\colon b^* \longrightarrow a^*$, as the composition

$$b^* \xrightarrow{id_{b^*}\otimes\eta_{a^*}} b^*\otimes a\otimes a^* \xrightarrow{id_{b^*}\otimes f\otimes id_{a^*}} b^*\otimes b\otimes a^* \xrightarrow{\epsilon_{b^*}\otimes id_{a^*}} a^*. \quad (7.63)$$

There are similar constructions of a left dual. Thus, there are a right dual functor $(\ )^*\colon \mathbb{A} \longrightarrow \mathbb{A}^{op}$ and a left dual functor $^*(\ )\colon \mathbb{A} \longrightarrow \mathbb{A}^{op}$.

Definition 7.2.12. A monoidal category is **right rigid** if every object a has a right dual that we denote as a^*. A monoidal category is **left rigid** if every object a has a left dual that we denote as $^*a$. A monoidal category where every object has both a right dual and a left dual is called a **rigid category**, a **category with duals**, or an **autonomous category**. (To keep this text uncluttered, we shall focus on rigid categories.)

Example 7.2.13. Any right closed monoidal category has right duals and any left closed monoidal category has left duals. We can use internal Homs to form dual objects in a category. For an object a, the object $a^* = [a, I]$ is a dual object. □

Definition 7.2.14. Let $\mathbb{A}$ and $\mathbb{B}$ be rigid monoidal categories. A functor $F\colon \mathbb{A} \longrightarrow \mathbb{B}$ is a **rigid monoidal functor** if F preserves the left and right duals up to isomorphism. This means that for all a in $\mathbb{A}$, we have $F(a^*) \cong F(a)^*$ and $F(^*a) \cong {}^*F(a)$. There also exists a definition of a rigid monoidal transformation. The collection of all rigid categories, rigid functors, and rigid natural transformations forms a 2-category called $\mathbb{R}\mathbb{igid}\mathbb{M}\mathbb{on}\mathbb{C}\mathbb{at}$.

Example 7.2.15. Along the lines of the categories $\mathbb{S}\mathbb{ym}$ and $\mathbb{B}\mathbb{raid}$, there is a category of tangles, written as $\mathbb{T}\mathbb{ang}$, which will be the prototypical rigid monoidal category. While in $\mathbb{S}\mathbb{ym}$ and $\mathbb{B}\mathbb{raid}$, the morphisms always go from top to bottom, in the category $\mathbb{T}\mathbb{ang}$, the morphisms can also go from bottom to top. A **tangle** is a collection of strings that can go down or up and twist around each other. A typical example of a tangle is given in Figure 7.1.

Notice that there are two strands that go from + to + and one strand that goes upward from − to −. There is a strand that starts at the bottom and ends at the bottom. There is also a strand that does not touch the top or the bottom. It is important to realize that this diagram is actually a **tangle projection** that represents a tangle. Two projections represent the same tangle if one can be smoothly transformed into the other while keeping the end points fixed.

Let us look at the details of the category $\mathbb{T}\mathbb{ang}$. The objects are not natural numbers, but rather sequences of +'s and −'s. A typical object is $-++-+--$. The monoidal structure on the objects is simply concatenation. For example,

$$++---+-\otimes-++--+++++ \quad = \quad ++---+--++--+++++. \quad (7.64)$$

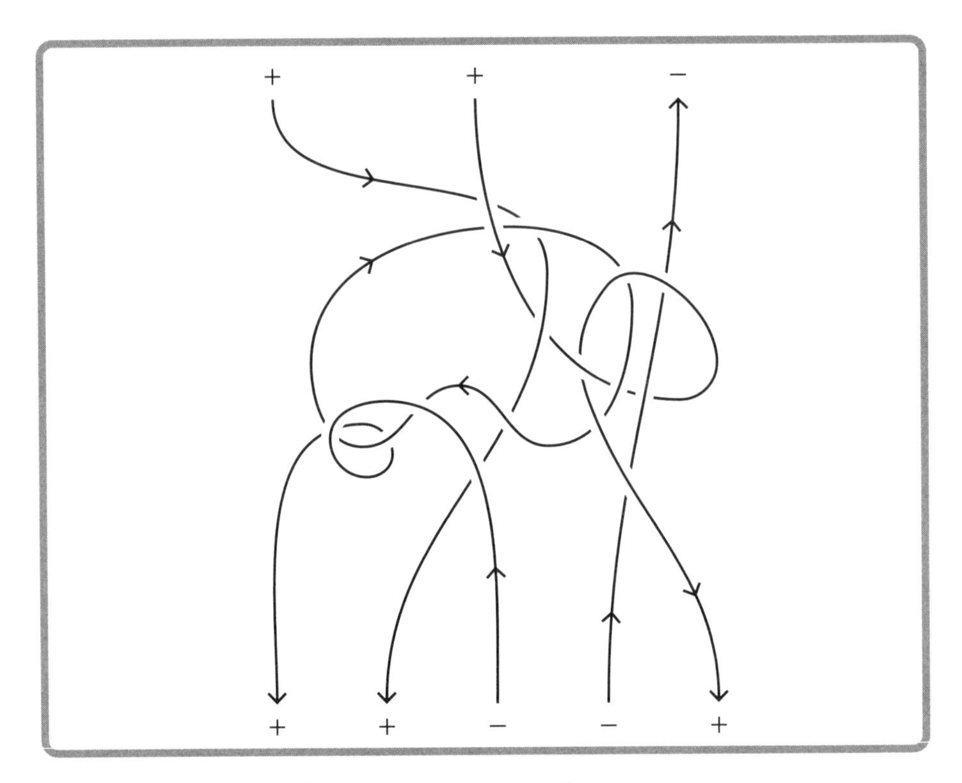

Figure 7.1. An example of a tangle.

The unit of the monoidal structure is the empty sequence written as $\emptyset$. The tangle in Figure 7.1 is from $++-$ to $-++-+$. Figure 7.2 shows a tangle from $-++-+$ to $++--+$. The composition of the tangle in Figure 7.1 with the tangle in Figure 7.2 goes from $++-$ to $++--+$ and is given in Figure 7.3.

This composition combined two strands that were connected to ends, and now their composition is not connected to any end. Notice that each of the Hom sets need not be a group because tangles need not have inverses. The composition is clearly associative. The identity tangle for $+--+-$ is simply this:

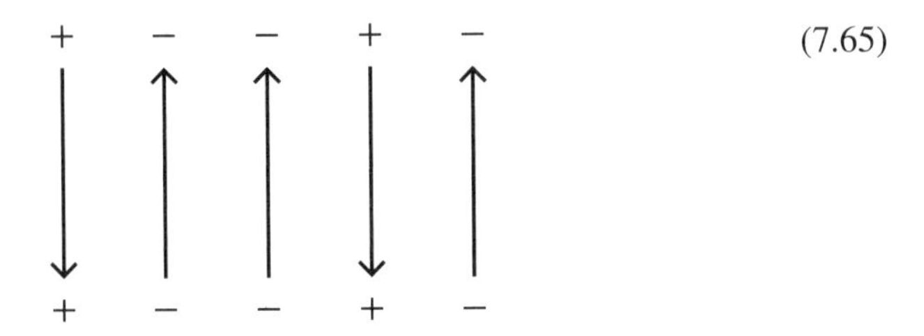

(7.65)

The monoidal structure on this category is simply placing the tangles side by side, as we did with permutations and braids (see Diagrams (5.75) and (7.20)). This category has a braiding just like the category $\mathbb{Braid}$, but some morphisms can go up. The reason

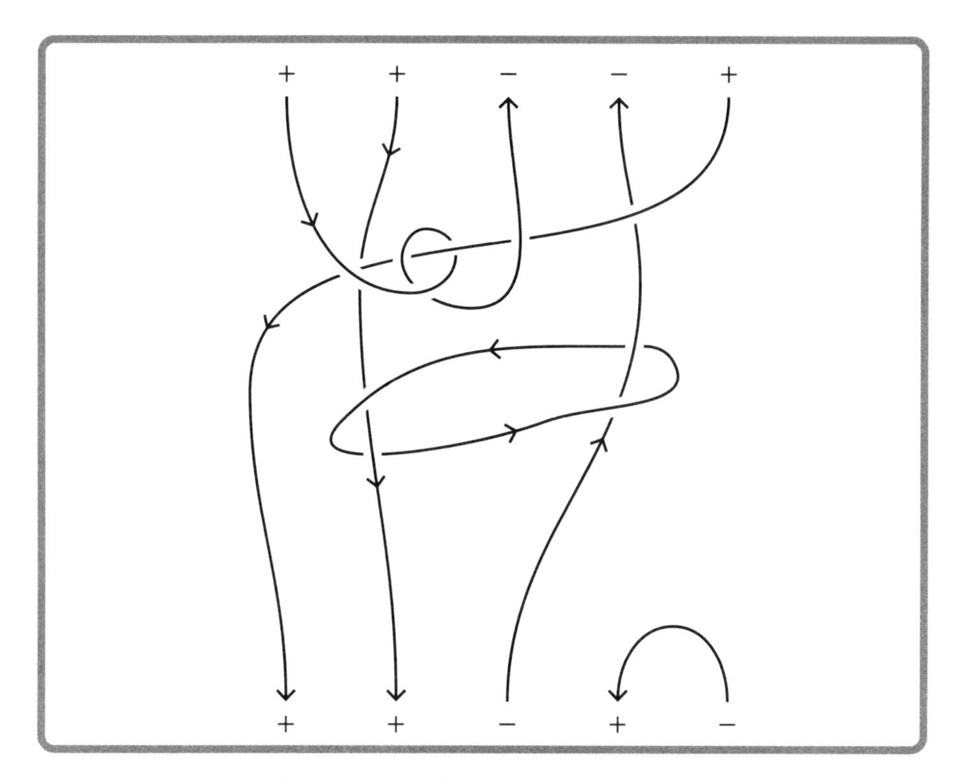

Figure 7.2. Another example of a tangle.

why this category is a rigid monoidal category is that every tangle can be turned upside down.

We saw that the category $\mathbb{Sym}$ is generated by the morphism s_i, and the category $\mathbb{Braid}$ is generated by the morphism σ_i. What about the category $\mathbb{Tang}$? This category is generated by σ_i and a few others. In total, there are six types of generating morphisms:

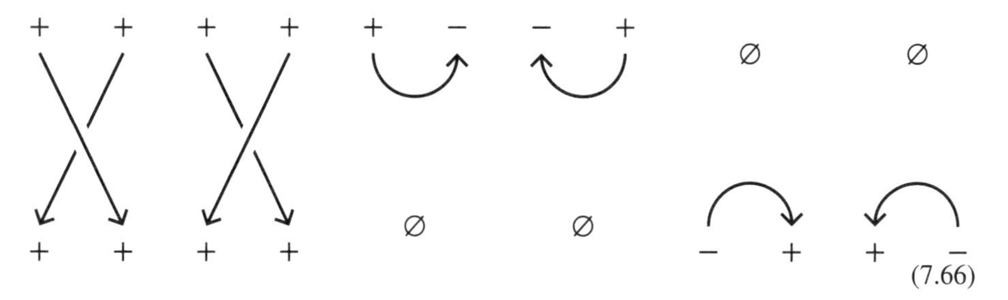

(7.66)

We denote these morphisms as σ_+, σ_-, $\cup^{\rightarrow}$, $\cup^{\leftarrow}$, $\cap^{\rightarrow}$, and $\cap^{\leftarrow}$. There are relations that tell when two tangle projections are the same. Several such relations are given in Figure 7.4. For a complete listing of all the relations, see page 301 of [134]. □

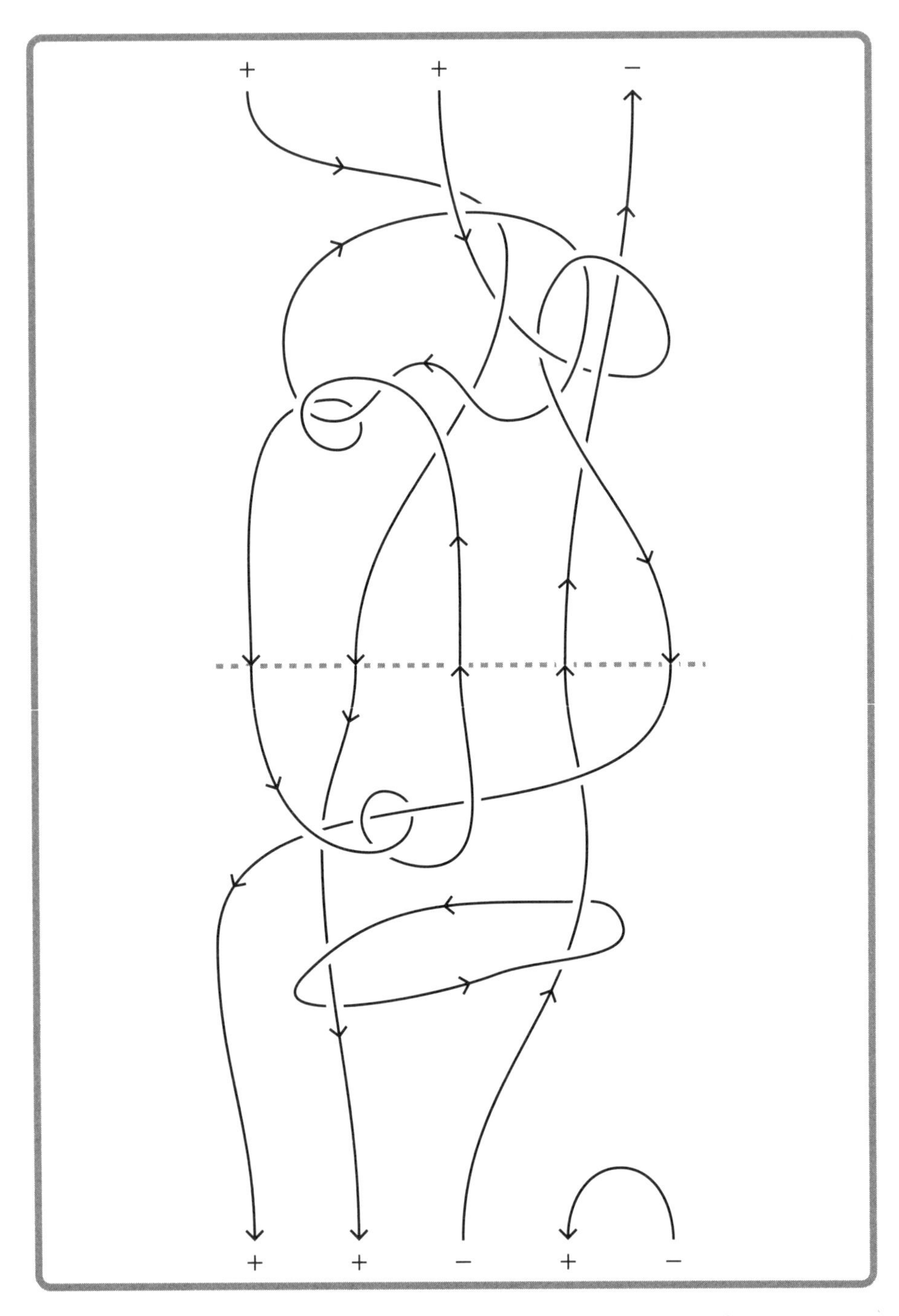

Figure 7.3. The composition of Figure 7.1 with Figure 7.2.

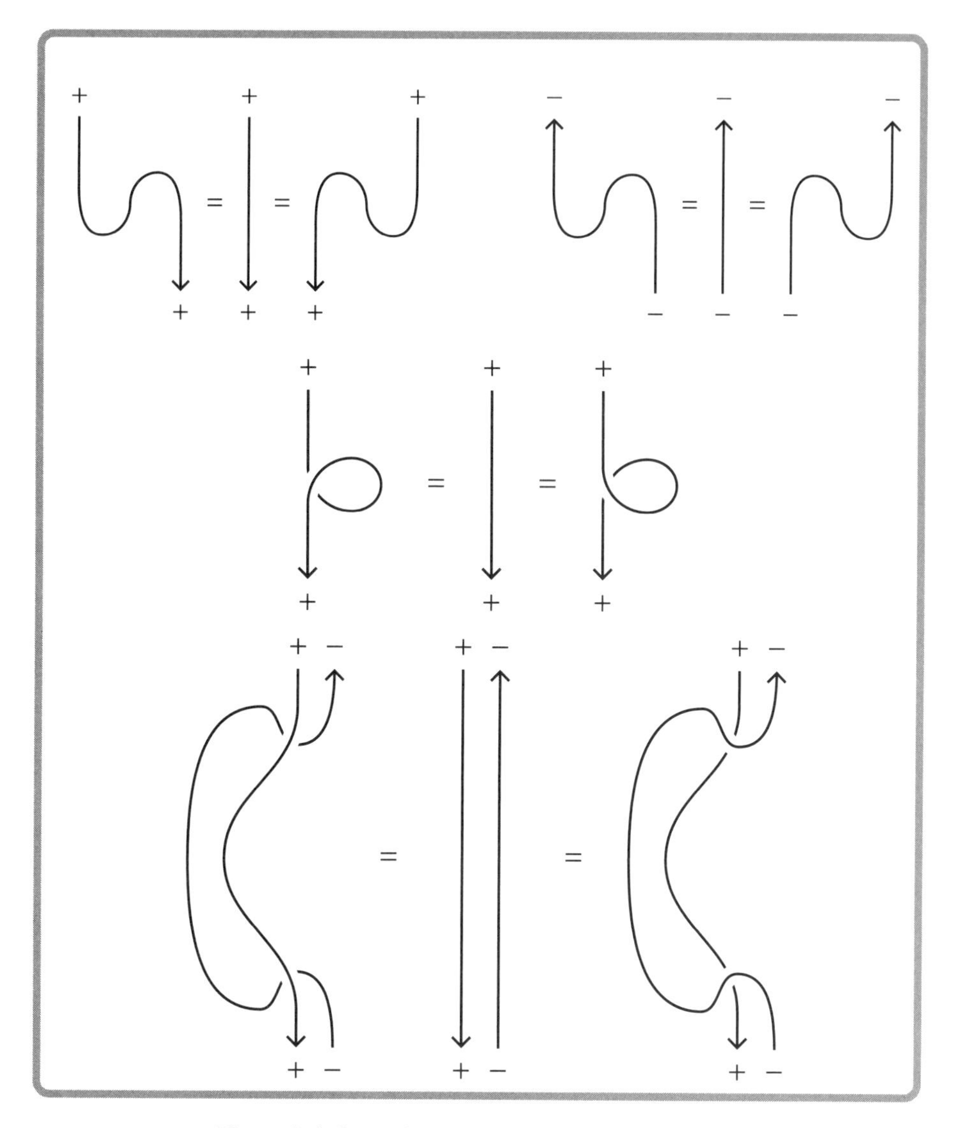

Figure 7.4. Several relations of tangle projections.

Let us examine the relationship of braids to tangles. Notice that a braid is a type of tangle. Braids are tangles that only go down. A typical element of the fifth braid group will be a tangle from $+++++$ to $+++++$. This means that there is a braided monoidal inclusion functor, $\mathbb{B}\text{raid} \hookrightarrow \mathbb{T}\text{ang}$. On objects, this functor takes n to $+++\cdots+$ (with n +'s) and the braids go to the tangles that are like the braids. We can include pure braids and get the following inclusions:

$$\mathbb{PB}\text{raid} \hookrightarrow \mathbb{B}\text{raid} \hookrightarrow \mathbb{T}\text{ang}. \tag{7.67}$$

Definition 7.2.16. A tangle that consists of only strands that go around each other and are not connected to the top or bottom is called a **link**. The morphisms from $\emptyset$ to $\emptyset$ are links. In general, they are made of more than one strand wrapping around each other. There is a submonoidal category called $\mathbb{Link}$, which has only one object, $\emptyset$. It has a monoidal structure because $\emptyset \otimes \emptyset = \emptyset$. This implies that there is a monoidal functor, $\mathbb{Link} \hookrightarrow \mathbb{Tang}$. This submonoidal category will be of extreme importance in our mini-course on knot theory (Section 10.1.) If there is only one strand in a link, it is called a **knot**. The collections of knots form a set, but not a category. They go from $\emptyset$ to $\emptyset$, but the compositions of two knots is not a knot. We denote the set of knots as **Knot**, and there is an inclusion of the set of knots into $Hom_{\mathbb{Tang}}(\emptyset, \emptyset)$, which we write as **Knot** $\hookrightarrow \mathbb{Link}$.

We can extend Diagram 7.67 to include **Knot** and $\mathbb{Link}$:

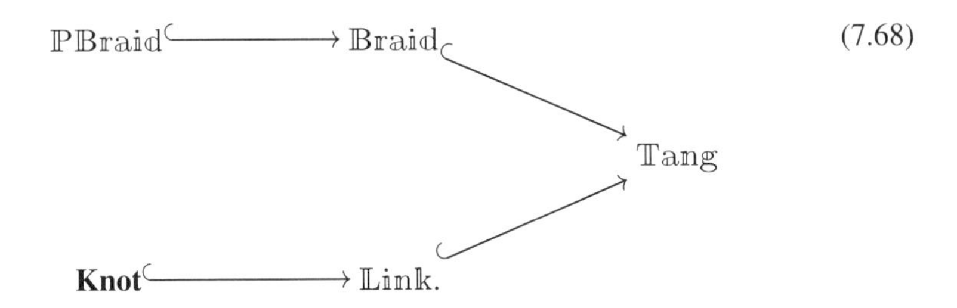

(7.68)

See Venn diagram A.5.

Given a rigid monoidal category, one can forget its duals. This entails a forgetful 2-functor from rigid monoidal categories to monoidal categories. So we have the following:

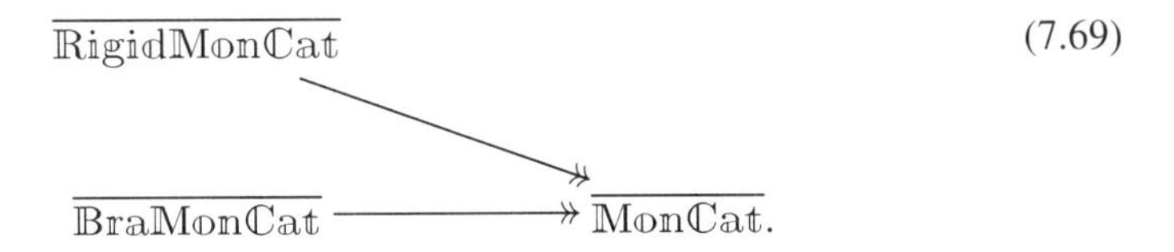

(7.69)

Rather than give the coherence theorem for rigid monoidal categories, we look at rigid monoidal categories with more structure called "ribbon categories" and give the universal properties in Theorem 7.3.9.

Related to closed and rigid categories are dagger categories. We met an example of such a category in Section 5.6 when we spoke of matrices with the adjoint operation (see page 268). Remember that the adjoint operation (like the transpose operation) takes an m by n matrix, and outputs take an n by m matrix. Expressed in symbols:

$$(\ )^{\dagger} : \mathbf{C}\mathbb{Mat} \longrightarrow \mathbf{C}\mathbb{Mat}^{op}. \tag{7.70}$$

Notice also that for a matrix A, we have $((A)^{\dagger})^{\dagger} = A$.

Let us make a formal definition now.

Definition 7.2.17. A **dagger category** (also called an **involutive category**) $(\mathbb{A}, (\ )^\dagger)$ is a category $\mathbb{A}$ and a contravariant functor

$$(\ )^\dagger\colon \mathbb{A} \longrightarrow \mathbb{A}^{op}. \tag{7.71}$$

(Remember that the contravarience means that for composable morphisms f and g, we have $(f \circ g)^\dagger = g^\dagger \circ f^\dagger$) such that

- $(\ )^\dagger$ is the identity on objects: for all a, we have $a^\dagger = a$.
- $(\ )^\dagger$ is involutive, this means that $((\ )^\dagger)^\dagger = Id_{\mathbb{A}}$.

Notice that $(\ )^\dagger$ being involutive means that the functor is an isomorphism of categories where it is its own inverse.

A **dagger monoidal category** is a strict monoidal category such that the dagger functor coheres with the tensor product: $(f \otimes g)^\dagger = f^\dagger \otimes g^\dagger$.

A **dagger symmetric category** or **dagger braided category** is a symmetric or braided monoidal category such that the dagger functor coheres with the braiding: for objects a and b, the dagger of the braiding isomorphism $\gamma_{a,b}\colon a \otimes b \longrightarrow b \otimes a$ is its inverse (i.e., $\gamma^\dagger_{a,b} = \gamma^{-1}_{a,b}\colon b \otimes a \longrightarrow a \otimes b$).

One must be careful not to confuse the dual of a morphism and the dagger of a morphism. For a morphism $f\colon a \longrightarrow b$, its dual is $f^*\colon b^* \longrightarrow a^*$ while its dagger is $f^\dagger\colon b \longrightarrow a$. In terms of string diagrams, these three morphisms are as follows:

$$a \longrightarrow \boxed{f} \longrightarrow b \qquad a^* \longleftarrow \boxed{f^*} \longleftarrow b^* \qquad a \longleftarrow \boxed{f^\dagger} \longleftarrow b \tag{7.72}$$

Example 7.2.18. Some examples of dagger categories are in order:

- If $\mathbb{A}$ is any discrete category, then $(\mathbb{A}, Id_{\mathbb{A}})$ is a dagger category simply because the only morphisms in the category are identities.
- Any groupoid (and hence any group, thought of as a one-object groupoid), can be given the structure of a dagger category by setting $f^\dagger = f^{-1}$ for any morphism in the groupoid.
- As we have already seen, the category of $\mathbf{K}\mathbb{M}\mathbb{a}\mathbb{t}$ is a dagger category. In Exercise 5.6.3, we saw that the dagger respects the Kroenecker product, and hence $\mathbf{K}\mathbb{M}\mathbb{a}\mathbb{t}$ is a dagger monoidal category.
- The category of finite-dimensional $\mathbf{K}$ vector spaces, $\mathbf{K}\mathbb{FDV}\mathbb{ect}$ is equivalent to $\mathbf{K}\mathbb{M}\mathbb{a}\mathbb{t}$ and also has a dagger structure. The linear operator $T\colon V \longrightarrow V'$ goes to the linear operator $T^\dagger\colon V' \longrightarrow V$ (a slight generalization of Definition 5.6.21). This is similar, for $\mathbb{FDH}\mathbb{ilb}$. In fact, they are all dagger symmetric monoidal categories.
- The category of relations $\mathbb{R}\mathbb{el}$ is a dagger symmetric category. The functor that is the identity on objects (sets) and takes the relation $R\colon S \not\longrightarrow T$ to $R^{-1}\colon T \not\longrightarrow S$ makes $\mathbb{R}\mathbb{el}$ into a dagger monidal category. We leave the fact that it is a dagger symmetric monoidal category to the reader. □

If $\mathbb{A}$ has a dagger category structure, we can define special morphisms in the category.

Definition 7.2.19. A morphism $f: a \longrightarrow a$ is

- **Hermitian** or **self-adjoint** if it is its own dagger (i.e., $f = f^{\dagger}$).
- **unitary** if its dagger is its own inverse (i.e., $f^{\dagger} = f^{-1}$).

These definitions were inspired by Definition 5.6.24.

7.3 Ribbon Categories

In our discussion of different types of monoidal categories, we have progressively gone more and more physical. We started talking about idealized strings that can pass through each other. Then we had strings that pass above or below each other. After that, we discussed strings that also go up. Here, we deal with strings that are not infinitesimally thin but have some thickness to them. This thickness will permit the strands to twist, and this twisting operation is important in geometry and physics.

Definition 7.3.1. A **balanced monoidal category** is a braided monoidal category $(\mathbb{A}, \otimes, I, \alpha, \rho, \lambda\gamma)$ with a natural transformation:

$$\theta\colon Id_{\mathbb{A}} \Longrightarrow Id_{\mathbb{A}}, \tag{7.73}$$

which is called a **twist**. One should think of this twist at component a as a wide string that twists several times:

(7.74)

We will take to calling such wide strings **framed strings**. The twist must respect the braided monoidal structure:

- Twist must cohere with the braiding; that is,

$$\begin{array}{ccc} a \otimes b & \xrightarrow{\gamma_{a,b}} & b \otimes a \\ {\scriptstyle \theta_{a\otimes b}}\downarrow & & \downarrow{\scriptstyle \theta_b \otimes \theta_a} \\ a \otimes b & \xleftarrow[\gamma_{b,a}]{} & b \otimes a, \end{array} \tag{7.75}$$

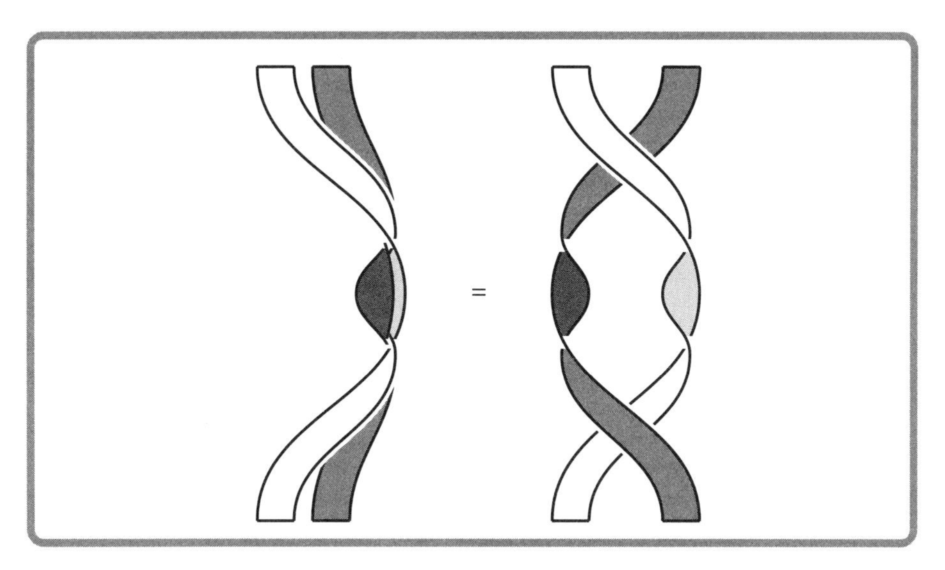

Figure 7.5. The main axiom for balanced monoidal categories. Twisting two ribbons around each other is the same as swapping the two, twisting each one separately, and then swapping again.

which can be visualized as Figure 7.5.

- Twist must respect the unit I:

$$\theta_I = id_I. \tag{7.76}$$

This second requirement actually follows the first one (see page 350 of [134]).

Example 7.3.2 presents the paradigm example of a strictly associative balanced monoidal category.

Example 7.3.2. One can generalize the notion of braids to the notion of **framed braids** or **wide braids**. These are braids that not only turn around each other, but also can twist. The category of framed braids, $\mathbb{fBraid}$, has natural numbers as objects and the morphisms are framed braids that can turn around and twist. A typical framed braid with six framed strings is the following:

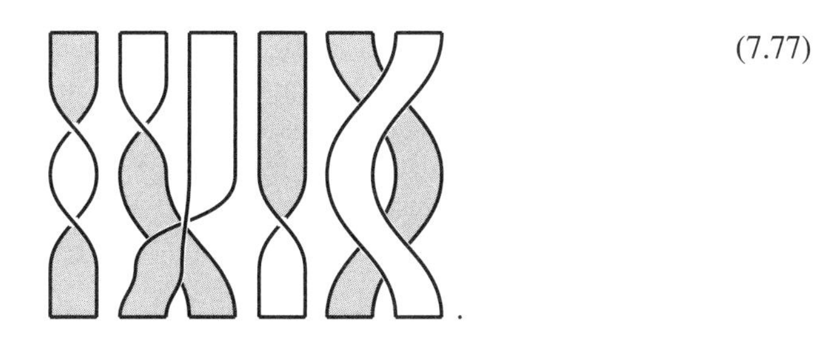
(7.77)

Similar to the generators of the symmetric category and braid category, there are generators in the category of framed braids. They are of the following form:

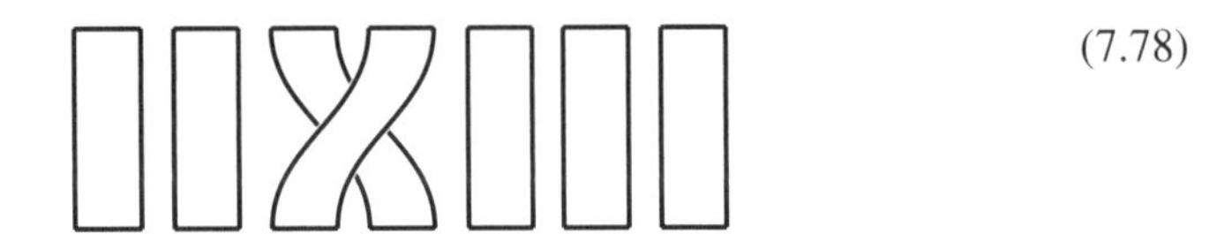

(7.78)

with one more generator:

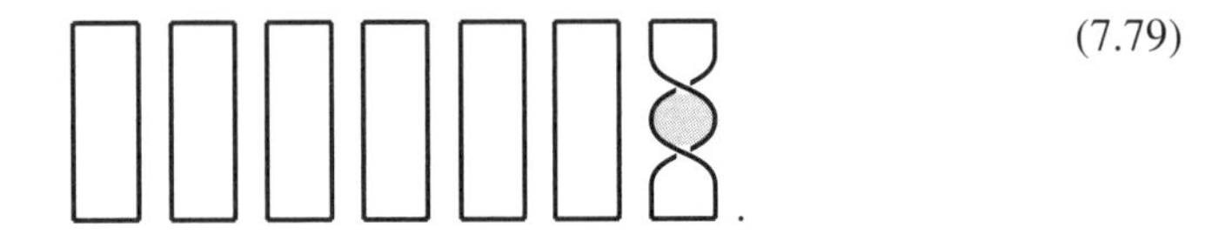

(7.79)

There are equations that show relations between these generators (see page 67 of [131]).

There is also a generalization of the notion of pure braids to the notion of framed pure braids. We denote this category as $\mathbb{fPBraid}$. □

There are obvious forgetful functors $\mathbb{fBraid} \longrightarrow \mathbb{Braid}$ and $\mathbb{fPBraid} \longrightarrow \mathbb{PBraid}$ that takes a braid and forgets all its twists. We can extend Diagram (7.27) to include these framed strings as follows:

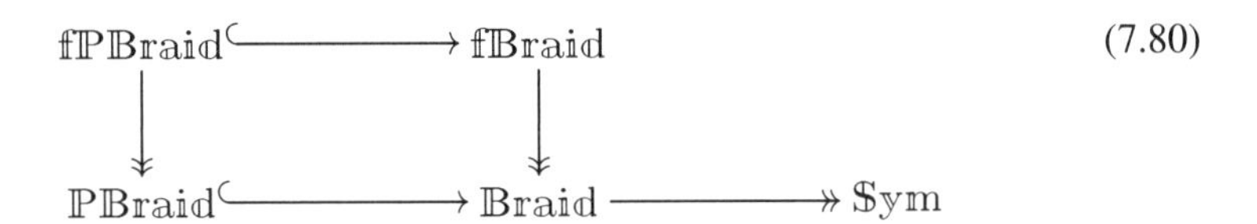

(7.80)

Definition 7.3.3. A functor from one strictly associative balanced monoidal category to another is one that preserves the twisting (i.e., $F(\theta_a) = \theta_{F(a)}$). The collection of balanced monoidal categories, their functors, and natural transformations form the 2-category $\overline{\mathbb{BalMonCat}}$. There is a sub-2-category of strictly associative balanced monoidal categories called $\overline{\mathbb{StrBalMonCat}}$ with an obvious inclusion from that category into $\overline{\mathbb{BalMonCat}}$.

Example 7.3.4. Every symmetric monoidal category is automatically a balanced monoidal category with $\theta = \iota_{Id_{\mathbb{A}}}$, the identity natural transformation from $Id_{\mathbb{A}}$ to $Id_{\mathbb{A}}$. One can see that Diagram (7.75) is satisfied because both $\theta_{a \otimes b}$ and $\theta_a \otimes \theta_b$ are the identity natural transformation, and the composition of the γs in a symmetric monoidal category is the identity. This means there is an inclusion 2-functor $\overline{\mathbb{SymMonCat}} \hookrightarrow \overline{\mathbb{BalMonCat}}$ of the 2-category of symmetric monoidal categories into the 2-category of balanced monoidal categories. In particular, the category of vector spaces with its typical braiding is a symmetric monoidal category, and hence is a balanced monoidal category. □

Given any balanced monoidal category, one can forget the twist and get a braided monoidal category. This entails a forgetful 2-functor $\overline{\mathbb{BalMonCat}} \longrightarrow \overline{\mathbb{BraMonCat}}$ from the 2-category of balanced monoidal categories to the 2-category of braided monoidal categories. Balanced monoidal categories should be seen as a level between braided monoidal categories and symmetric monoidal categories. There are similar 2-categories and 2-functors for strictly associative versions of these structures. This all summarizes to the extension of Diagram (7.28) as follows:

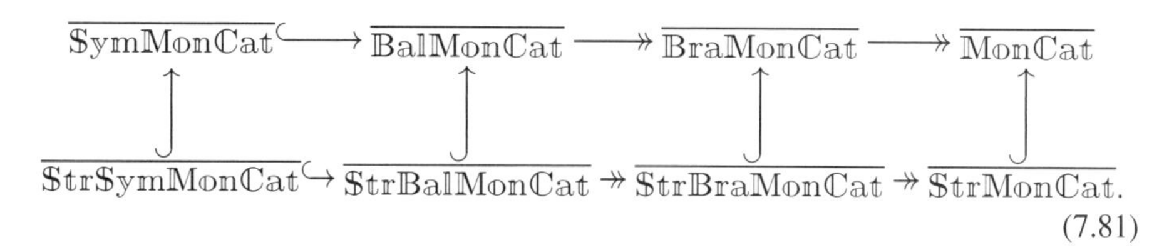

(7.81)

See Venn diagram A.6.

Example 7.3.5. The category of braids has a balanced monoidal structure. Here, θ is comprised of the strands themselves. The twists θ_0 and θ_1 are the identity. The morphism θ_n for $n > 1$ takes the n strands and twists them 360 degrees around each other. For example, θ_5 looks like this:

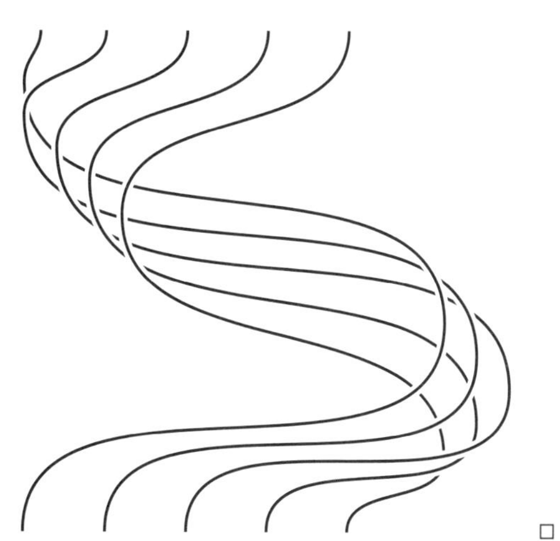

□

There is a coherence theorem for balanced monoidal categories that says that framed braids are the free balanced monoidal category on one generator.

We are also interested in the framed strings going up.

Definition 7.3.6. A **ribbon category**, also called a **tortile category** ("tortile" means twisted or coiled), is a rigid monidal category with a twist such that

$$(\theta_a)^* = \theta_{a^*} : a^* \longrightarrow a^*. \tag{7.82}$$

This requirement can be graphically represented as follows:

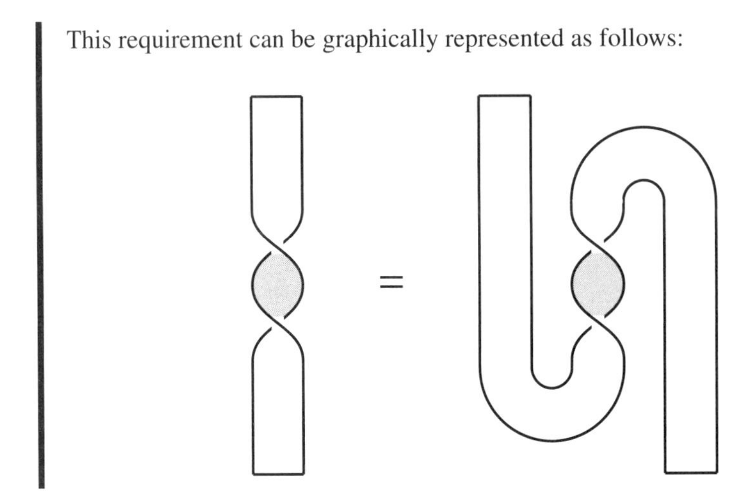

The paradigm example of a ribbon category is the collection of framed tangles.

Example 7.3.7. In the last section, we met the category of tangles called $\mathbb{T}\mathbb{ang}$. We now widen the strings to get a category of **framed tangles** or **ribbons** denoted as $\mathbb{f}\mathbb{T}\mathbb{ang}$. The objects of the category are sequences of +'s and -'s, while the morphisms are wide tangles where the strands can twist. A typical morphism in the category of framed tangles is shown in Figure 7.6. When composing such framed tangles, we must make sure that the framed strands are pasted together so their orientation is the same (i.e., going up or going down). □

There is an inclusion of framed braids into framed tangles. There is also an inclusion of framed links into framed tangles. One can forget the twisting of any framed tangle and just get a tangle. This entails a forgetful functor from the category of framed tangles to the category of tangles. We extend Diagrams (7.80) and (7.68)to include these new collections:

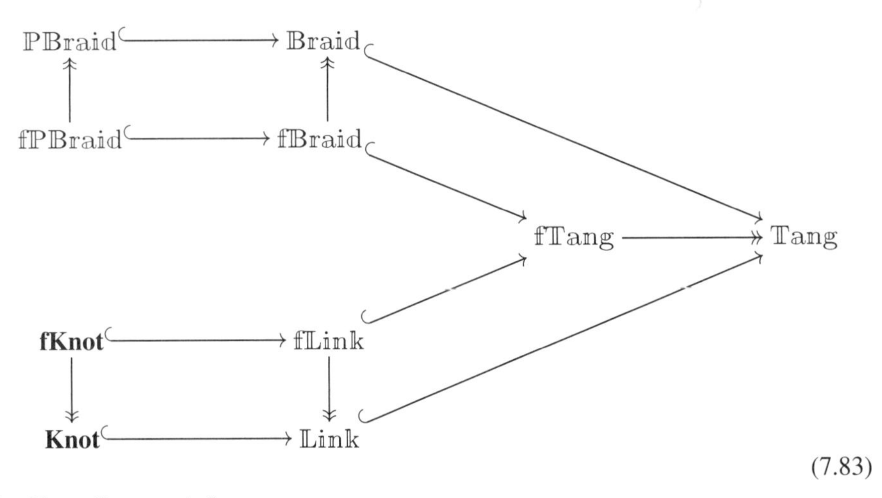

(7.83)

See Venn diagram A.5.

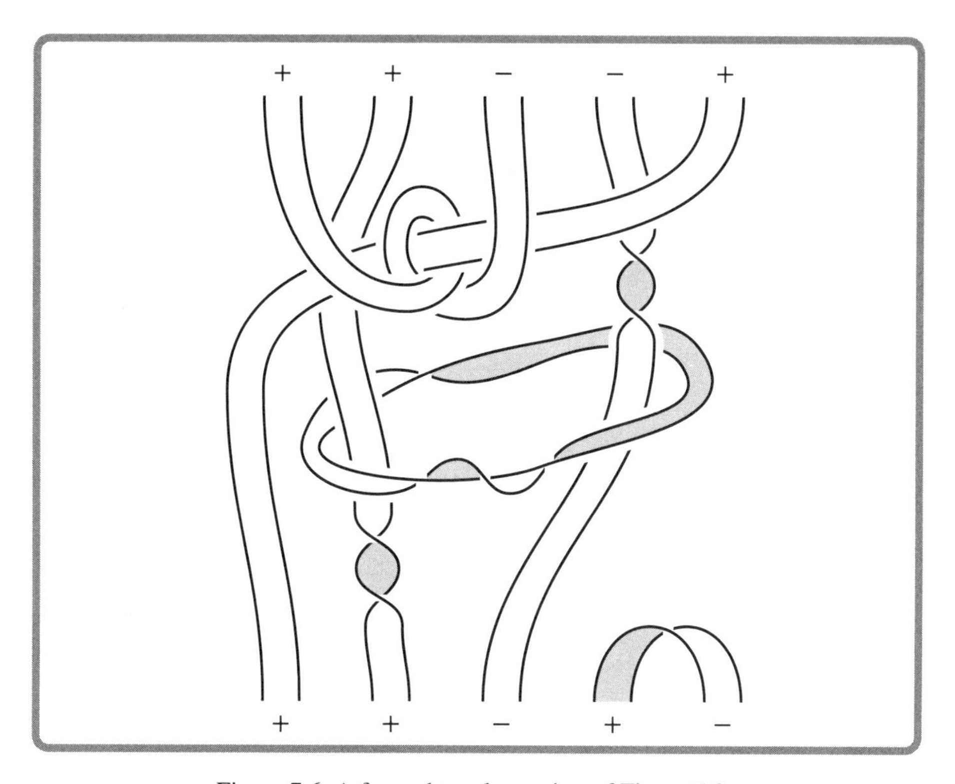

Figure 7.6. A framed tangle version of Figure 7.2.

Definition 7.3.8. One can go on and form the 2-category of ribbon categories, their functors, and their natural transformations, $\overline{\mathbb{RibMonCat}}$.

One forgets the duality to get a forgetful 2-functor, $\overline{\mathbb{RibMonCat}} \longrightarrow \overline{\mathbb{BalMonCat}}$. One can also forget "the requirement stated in Line (7.82)" to get a forgetful 2-functor, $\overline{\mathbb{RibMonCat}} \longrightarrow \overline{\mathbb{RigidMonCat}}$. We can see how ribbon monoidal categories sit in the context of all the other structures by combining Diagrams (7.69) and (7.81) to get

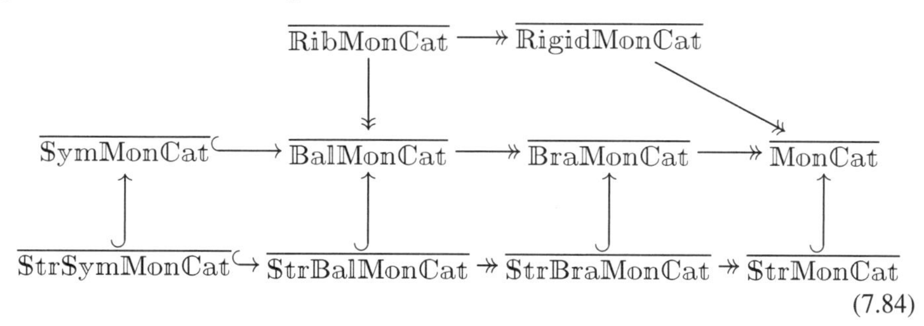

(7.84)

The coherence theorem for ribbon categories is similar to earlier coherence theorems. Rather than stating it in four equivalent ways, we simply state it as in Theorem 7.3.9.

Theorem 7.3.9. The Coherence Theorem for Ribbon Categories also called **Shum's Theorem.** The category $\mathbb{fTang}$ is monoidally equivalent to the free ribbon category on one generator. ★

Proof. The proof can be found in [239]. See also Chapter 9 of [278] and Theorem 2.5 on page 39 of [253]. ♣

Suggestions for Further Study

This chapter has covered many topics. More can be learned about these topics from the following sources:

- Braided monoidal categories: Section 2 of [131], Chapter XIII of [134], Section 8.1 of [71], Chapter 11 of [247], and Section 1.2 of [110].
- Cartesian closed categories: Section I.6 of [181] and Section 6.2 of [21].
- Rigid categories: Section 7 of [131], Section XIV.2 of [134], Chapter 12 of [247], and Section 3.1 of [110].
- Dagger categories: Section 7 of [233] and Section 2.3 of [110].
- Balanced monoidal categories: Section 6 of [131] and Chapter 11 of [247].
- Ribbon categories: Section 7 of [131], Section XIV.3 of [134], and Section 8.10 of [71].

An excellent survey of many of these structures—and much more—can be found in Ross Street [248].

7.4 Mini-Course: Quantum Groups

The field of quantum groups is a modern part of algebra that deals with algebraic structures that are not exactly commutative or associative. Such structures play a major role in modern mathematics.

We highlight a small part of this large field. For certain types of algebraic structures, there is an associated category of representations or modules of that algebraic structure. We look at various types of algebraic structures and examine the types of categorical structure of the associated category of representations. Many of the variations of monoidal categories that we learned about earlier in this chapter are found in this section. The end of the mini-course points to some of the many applications these structures have in physics and number theory.

Algebras

Let us begin with an analogy. Whereas the algebraic structures we saw in Chapter 2 were sets with extra structure, here we will be dealing with complex vector spaces with extra structure. The first few structures are similar to the structure of monoids and are called "algebras." For each monoid M, we examined the category $\mathbb{Set}^M$. By analogy, for each algebra A, we examine the category of A-modules, which we denote as $\mathbb{Mod}(A)$. Keep the following analogy in mind for the next few pages:

	Sets	Vector Spaces	
$\mathbb{Monoid}$	monoid Monoid homomorphism	algebra Algebra homomorphism	$\mathbb{Algebra}$
$\mathbb{Set}^M$	M-set M-set homomorphism	A-module A-homomorphism	$\mathbb{Mod}(A)$

Remark 7.4.1. Some foreshadowing is needed first. Our presentation is somewhat novel. We examine the collection of algebras. Rather than just looking at the category of algebras and algebra homomorphisms, we introduce 2-cells between algebra homomorphisms and formulate the 2-category $\overline{\mathbb{Algebra}}$. This 2-category has a monoidal 2-category structure. (Although monoidal 2-categories were not formally defined, we employ them because their definition is obvious. This is especially true because we are dealing with a strict version, so the coherence issues are easy.) We also examine the 2-category $\overline{\mathbb{Cat}/\mathbb{CVect}}$, where categories of modules and their forgetful functors to vector spaces exist. Connecting these two 2-categories is a monoidal 2-functor:

$$MOD\colon \overline{\mathbb{Algebra}}^{op} \longrightarrow \overline{\mathbb{Cat}}/\mathbb{CVect}. \tag{7.85}$$

This 2-functor will be helpful in relating the structure of the algebras to the structures of the categories of modules. (Again, monoidal 2-functors have not been formally defined. But one can easily do it on one's own.) As an aid, keep your eyes on the variations of algebraic structures in Venn diagram A.7 and the variations of categorical structures in Venn diagram A.6. ♠

The objects of $\overline{\mathbb{Algebra}}$: the algebras. We begin at the ground level with algebras, which are like monoids, but in the monoidal category of vector spaces rather than in the Cartesian category of sets.

Definition 7.4.2. An **algebra** is a triple (A, μ, υ) where

- A is a complex vector space.
- μ is a multiplication (i.e., a linear map $\mu\colon A \otimes A \longrightarrow A$), and
- υ chooses a unit for the multiplication (i.e., a linear map $\upsilon\colon \mathbb{C} \longrightarrow A$).

These ingredients must satisfy the following requirements: μ must be associative, and υ must choose an element that behaves like a unit for the multiplication. This means that the following diagrams must commute:

$$\begin{array}{ccc} A\otimes A\otimes A & \xrightarrow{id_A\otimes\mu} & A\otimes A \\ {\scriptstyle \mu\otimes id_A}\downarrow & & \downarrow{\scriptstyle \mu} \\ A\otimes A & \xrightarrow[\mu]{} & A \end{array} \qquad \begin{array}{ccccc} \mathbf{C}\otimes A & \xrightarrow{\upsilon\otimes id_A} & A\otimes A & \xleftarrow{id_A\otimes\upsilon} & A\otimes\mathbf{C} \\ & {\scriptstyle\cong}\searrow & \downarrow{\scriptstyle\mu} & \swarrow{\scriptstyle\cong} & \\ & & A. & & \end{array} \tag{7.86}$$

These two diagrams are very similar to Diagram (2.10), where we saw the definition of a monoid.

Algebras are vector spaces with more structure. Many examples of algebras will be taken from Example 2.4.2, where we listed types of vector spaces.

Here are some examples of algebras.

Example 7.4.3.

- The vector space of complex numbers is an algebra. The multiplication is regular complex number multiplication, and the unit is $1 \in \mathbf{C}$. We write this algebra as $(\mathbf{C}, \cdot, 1)$.
- Square matrices of any particular size form an algebra. That is, the complex vector space $\mathbf{C}^{n\times n}$ for any n is an algebra. The multiplication is the matrix multiplication, and the unit is the identity matrix. We write this algebra as $(\mathbf{C}^{n\times n}, \cdot, I_n)$.
- The vector space of polynomials forms an algebra where the multiplication is polynomial multiplication and the unit is the polynomial 1, which always evaluates to 1. We denote this algebra as $(Poly_{\mathbf{C}}, \cdot, 1)$.
- We saw in Example 2.1.12 that the set of endomorphisms of an object in a category with the composition operation forms a monoid. In particular, for any vector space V, the set of linear maps from V to V with the composition form an algebra. We write this algebra as $(End(V), \circ, id_V)$.
- We saw in Exercise 2.4.3 that for any set S, the set $Hom_{\mathbb{Set}}(S, \mathbf{C})$ of all functions from S to the complex numbers $\mathbf{C}$ forms a vector space. The multiplication is inherited from the $\cdot$ multiplication in $\mathbf{C}$. In detail, given $f\colon S \longrightarrow \mathbf{C}$ and $g\colon S \longrightarrow \mathbf{C}$, we can form $(f \cdot g)\colon S \longrightarrow \mathbf{C}$ defined by $(f \cdot g)(s) = f(s) \cdot g(s)$. The unit is the constant function c_1 that always outputs $1 \in \mathbf{C}$. We write this algebra as $(Hom_{\mathbb{Set}}(S, \mathbf{C}), \cdot, c_1)$. □

The morphisms of $\overline{\mathbb{Algebra}}$: the algebra homomorphisms. Analogous to a monoid homomorphism is an algebra homomorphism.

Definition 7.4.4. Let (A, μ, υ) and (A', μ', υ') be algebras. An **algebra homomorphism** is a linear map $f\colon A \longrightarrow A'$ that respects the multiplications and the units, which means that the following diagrams commute:

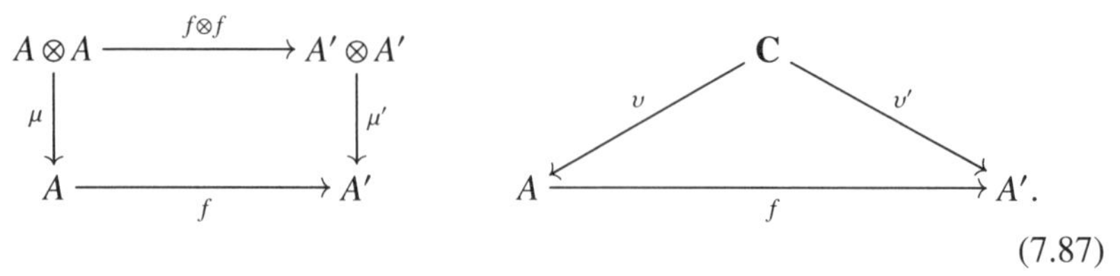

(7.87)

These two diagrams are very similar to Diagrams (2.11), where we saw the definition of a homomorphism of monoids.

The 2-cells of $\overline{\mathbb{Algebra}}$: the algebra conjugates. Before going further, go back and meditate on Example 4.2.14. Realize that the following is a generalization based on that example.

Definition 7.4.5. Let $f, g\colon (A, \mu, \upsilon) \longrightarrow (A', \mu', \upsilon')$ be two algebra homomorphisms. An **algebra conjugate** (or just **conjugate**) from f to g is an element Φ of A' such that for all a in A, we have

$$\Phi f(a) = g(a)\Phi, \tag{7.88}$$

where both multiplications are in algebra A'. The reason for the name is that if Φ is an invertible element, then the inverse can be multiplied on the right of both sides of the equation to get

$$\Phi f(a)\Phi^{-1} = g(a). \tag{7.89}$$

If Φ has an inverse, Φ^{-1}, then Φ^{-1} goes from g to f. A conjugate can be viewed as a 2-cell:

$$A \underset{g}{\overset{f}{\Downarrow \Phi}} A'. \tag{7.90}$$

These conjugates are 2-cells in a 2-category $\overline{\mathbb{Algebra}}$ and as such have both a vertical and horizontal composition. Vertical composition is defined as follows: Let $f, g, h\colon A \longrightarrow A'$ be algebra homomorphisms and let $\Phi\colon f \Longrightarrow g$ and $\Psi\colon g \Longrightarrow h$ be conjugates. This means that for all a in A, we have $\Phi f(a) = g(a)\Phi$ and $\Psi g(a) = h(a)\Psi$. We can form $\Psi \circ_v \Phi\colon f \Longrightarrow h$, which satisfies

$$\Psi\Phi f(a) = \Psi g(a)\Phi = h(a)\Psi\Phi. \tag{7.91}$$

It is not hard to see that this composition is associative and the unit is the Id_f.

Conjugates also have a horizontal composition. Let $f, g\colon A \longrightarrow B$ and $f', g'\colon B \longrightarrow C$ be algebra homomorphisms, and let $\Phi\colon f \Longrightarrow g$ and $\Psi\colon f' \Longrightarrow g'$. This means that for all a in A, we have $\Phi f(a) = g(a)\Phi$, and for all b in B, we have $\Psi f'(b) = g'(b)\Psi$. Then we can form $\Psi \circ_h \Phi\colon (f' \circ f) \Longrightarrow (g' \circ g)$, where the conjugate is $\Psi f'(\Phi)$. In detail, for all a in A:

$$\begin{aligned}
\Psi f'(\Phi) f'(f(a)) &= \Psi f'(\Phi f(a)) && \text{because } f' \text{ is a homomorphism}\\
&= g'(\Phi f(a))\Psi && \text{because } \Psi \text{ is a conjugate from } f' \text{ to } g'\\
&= g'(g(a)\Phi)\Psi && \text{because } \Psi \text{ is a conjugate from } f \text{ to } g\\
&= g'(g(a))g'(\Phi)\Psi && \text{because } g' \text{ is a homomorphism}\\
&= g'(g(a))\Psi f'(\Phi) && \text{because } \Psi \text{ is a conjugate from } f' \text{ to } g'.
\end{aligned}$$

Exercise 7.4.6. Show that we could have also set the conjugate of the composition to be $g'(\Phi)\Psi$. ■

The 2-cells must satisfy an interchange law. This means that given the following diagram of conjugates,

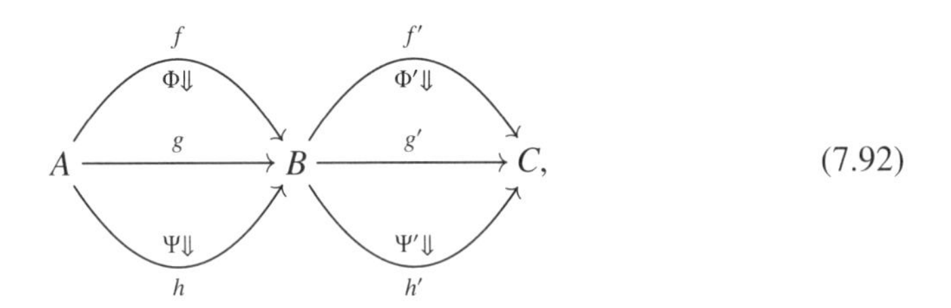

(7.92)

we have

$$
\begin{aligned}
(\Psi' \circ_V \Phi') \circ_H (\Psi \circ_V \Phi) &= (\Psi'\Phi') \circ_H (\Psi\Phi) && \text{by the def. of vertical composition}\\
&= \Psi'\Phi' f'(\Psi\Phi) && \text{by the def. of horizontal composition}\\
&= \Psi'\Phi' f'(\Psi) f'(\Phi) && \text{because } f' \text{ is a homomorphism}\\
&= \Psi' g'(\Psi)\Phi' f'(\Phi) && \text{because } \Phi' \text{ is a conjugate from } f' \text{ to } g'\\
&= (\Psi' \circ_H \Psi)(\Phi' \circ_H \Phi) && \text{by the def. of horizontal composition}\\
&= (\Psi' \circ_H \Psi) \circ_V (\Phi' \circ_H \Phi) && \text{by the def. of vertical composition.}
\end{aligned}
$$

The monoidal structure of $\overline{\mathbb{Algebra}}$. The 2-category of algebras has a monoidal structure. Given two algebras (A_1, μ_1, υ_1) and (A_2, μ_2, υ_2), we can form $A_1 \otimes A_2$. The multiplication is given as

$$A_1 \otimes A_2 \otimes A_1 \otimes A_2 \xrightarrow{id \otimes br \otimes id} A_1 \otimes A_1 \otimes A_2 \otimes A_2 \xrightarrow{\mu_1 \otimes \mu_2'} A_1 \otimes A_2. \tag{7.93}$$

The unit of $A_1 \otimes A_2$ is given by the map $\upsilon_1 \otimes \upsilon_2 \colon \mathbf{C} \cong \mathbf{C} \otimes \mathbf{C} \longrightarrow A_1 \otimes A_2$.

The unit of the monoidal structure is the algebra $(\mathbf{C}, \cdot, 1)$. This monoidal structure will be strictly associative since we can assume that the monoidal structure in $\mathbf{C}\mathbb{Vect}$ is strictly associative. There is no reason to think that this monoidal category is symmetric or even braided.

Given two algebra homomorphisms $f \colon A_1 \longrightarrow A_2$ and $g \colon A_3 \longrightarrow A_4$, their monoidal product is $f \otimes g \colon (A_1 \otimes A_3) \longrightarrow (A_2 \otimes A_4)$, which is an algebra homomorphism because the following two diagrams commute:

$$
\begin{array}{ccc}
A_1 \otimes A_3 \otimes A_1 \otimes A_3 & \xrightarrow{f\otimes g\otimes f\otimes g} & A_2 \otimes A_4 \otimes A_2 \otimes A_4\\
\downarrow{\scriptstyle id\otimes br\otimes id} & & \downarrow{\scriptstyle id\otimes br\otimes id}\\
A_1 \otimes A_1 \otimes A_3 \otimes A_3 & \xrightarrow{f\otimes f\otimes g\otimes g} & A_2 \otimes A_2 \otimes A_4 \otimes A_4\\
\downarrow{\scriptstyle \mu_1\otimes\mu_3} & & \downarrow{\scriptstyle \mu_2\otimes\mu_4}\\
A_1 \otimes A_3 & \xrightarrow{f\otimes g} & A_2 \otimes A_4
\end{array}
\qquad
\begin{array}{ccc}
& \mathbf{C} \cong \mathbf{C} \otimes \mathbf{C} & \\
{\scriptstyle \upsilon_1\otimes\upsilon_3}\swarrow & & \searrow{\scriptstyle \upsilon_2\otimes\upsilon_4}\\
A_1 \otimes A_3 & \xrightarrow[f\otimes g]{} & A_2 \otimes A_4.
\end{array}
\tag{7.94}
$$

Given two algebra conjugates, their monoidal product is given as follows:

$$A_1 \underset{f_2}{\overset{f_1}{\Downarrow \Phi}} A_2 \quad \otimes \quad A_3 \underset{f_4}{\overset{f_3}{\Downarrow \Psi}} A_4 \quad = \quad A_1 \otimes A_3 \underset{f_2 \otimes f_4}{\overset{f_1 \otimes f_3}{\Downarrow \Phi \otimes \Psi}} A_2 \otimes A_4, \tag{7.95}$$

where $\Phi \otimes \Psi$ is an element of $A_2 \otimes A_4$. The $\Phi \otimes \Psi$ must satisfy the following requirement:

$$\text{for all } a \in A_1 \otimes A_3, \text{ we have } (\Phi \otimes \Psi)((f_1 \otimes f_3)(a)) = ((f_2 \otimes f_4)(a))(\Phi \otimes \Psi). \tag{7.96}$$

The 2-category $\overline{\mathbb{Cat}}/\mathbf{C}\mathbb{Vect}$. Remember that $\mathbb{Cat}$ is the category of locally small categories and functors. The collection $\overline{\mathbb{Cat}}$ is 2-category of locally small categories, functors, and natural transformations. The slice category $\overline{\mathbb{Cat}}/\mathbf{C}\mathbb{Vect}$ has objects that are functors, such as $\mathbb{A} \longrightarrow \mathbf{C}\mathbb{Vect}$; morphisms are commutative triangles; and 2-cells are natural transformation such that the following diagram commutes:

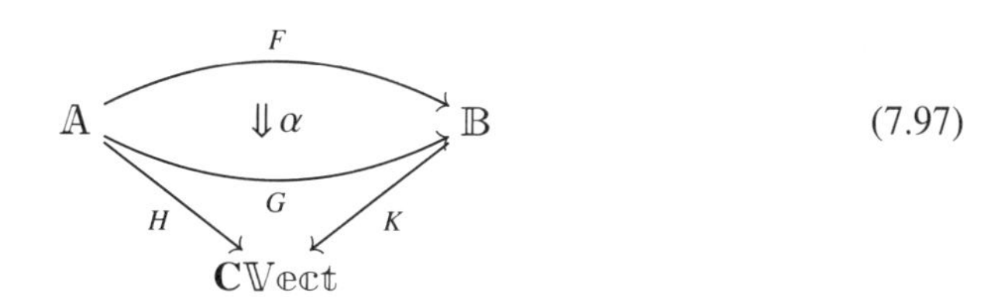

(7.97)

The 2-category $\overline{\mathbb{Cat}}/\mathbb{C}\mathbb{Vect}$ has a monoidal structure. Given objects $H\colon \mathbb{A} \longrightarrow \mathbf{C}\mathbb{Vect}$ and $K\colon \mathbb{B} \longrightarrow \mathbf{C}\mathbb{Vect}$, their tensor product is

$$H \boxtimes K\colon \mathbb{A} \times \mathbb{B} \longrightarrow \mathbf{C}\mathbb{Vect}, \tag{7.98}$$

which is defined on the objects a of $\mathbb{A}$ and b of $\mathbb{B}$ as

$$(H \boxtimes K)(a,b) = H(a) \otimes K(b), \tag{7.99}$$

where the $\otimes$ on the right is the tensor product of complex vector spaces. The unit is the functor $0\colon \mathbf{1} \longrightarrow \mathbf{C}\mathbb{Vect}$, which takes the single-element category to $\mathbf{C}$ the trivial complex vector space. The monoidal structure on morphisms and 2-cells are defined in the obvious way.

The Category of Modules for an Algebra

In Example 4.1.47 we saw that monoids act on sets. Since algebras are like monoids in the category of vector spaces, we should examine how algebras act on vector spaces.

The 2-Functor $MOD\colon \overline{\mathbb{Algebra}}^{op} \longrightarrow \overline{\mathbb{Cat}}/\mathbf{C}\mathbb{Vect}$ on objects.

Definition 7.4.7. Let (A, μ, υ) be an algebra. A **module** (or a **representation**) for A is $(M, \cdot)$, where

- M is a complex vector space.
- $\cdot \colon A \otimes M \longrightarrow M$ is a linear map that describes the action of A on M, that is, how the elements of A change the elements of M.

The action must satisfy the following two requirements:

- The action must respect the algebra multiplication. This means, for all a, a' in A, and for all $m \in M$, we have that $(a * a') \cdot m = a \cdot (a' \cdot m)$, where $a * a' = \mu(a, a')$ is the algebra multiplication.
- The action must respect the identity of the algebra. That is, for e, the identity of the algebra, and for all $m \in M$, we have $e \cdot m = m$.

These two requirements are the same as requiring the following diagrams to be commutative:

(7.100)

These commuting diagrams are very similar to Diagram (4.24), where we saw how to define an M-set with commutative diagrams.

How do modules relate to each other?

Definition 7.4.8. Let $(M, \cdot)$ and $(M', \cdot')$ be two A-modules. An A-**module homomorphism** $f \colon (M, \cdot) \longrightarrow (M', \cdot')$ is a linear map $f \colon M \longrightarrow M'$ such that f respects the actions. That is, the following diagram commutes:

$$\begin{array}{ccc} A \otimes M & \xrightarrow{id_A \otimes f} & A \otimes M' \\ \Big\downarrow{\scriptstyle \cdot} & & \Big\downarrow{\scriptstyle \cdot'} \\ M & \xrightarrow[f]{} & M'. \end{array} \qquad (7.101)$$

This diagram is similar to the diagram in Example 4.2.4, where we saw the requirements for a function to be an M-set homomorphism.

For every algebra (A, μ, υ), the collection of A-modules and A-module homomorphisms forms a category, denoted as $\mathbb{M}\mathrm{od}(A, \mu, \upsilon)$ or simply $\mathbb{M}\mathrm{od}(A)$. What type of structure does such a category have? A large part of this mini-course is dedicated to determining how the extra structure of algebra A relates with the extra categorical structure of category $\mathbb{M}\mathrm{od}(A)$.

For every algebra A, there is a forgetful functor $U_A \colon \mathbb{M}\mathrm{od}(A) \longrightarrow \mathbf{C}\mathbb{V}\mathrm{ect}$ that takes a module $(M, \cdot)$ to its vector space M, and a module homomorphism to its underlying linear map. This forgetful functor is an object in $\overline{\mathbb{C}\mathrm{at}}/\mathbf{C}\mathbb{V}\mathrm{ect}$. The 2-functor

MOD is defined on objects as follows:

$$A \quad \longmapsto \quad U_A \colon \mathbb{M}\mathrm{od}(A) \longrightarrow \mathbf{C}\mathbb{V}\mathrm{ect}. \tag{7.102}$$

The 2-Functor $MOD\colon \overline{\mathbb{A}\mathrm{lgebra}}^{op} \longrightarrow \overline{\mathbb{C}\mathrm{at}}/\mathbf{C}\mathbb{V}\mathrm{ect}$ **on morphisms.** What relationship does algebra homomorphisms induce on the category of modules? First, let us look at a module from a different point of view: rather then regarding an action as a linear map $\cdot\colon A \otimes M \longrightarrow M$, consider M as a vector space and look at $End(M)$, which is an algebra. Then an A-module is simply the algebra homomorphism $A \longrightarrow End(M)$.

Exercise 7.4.9. Prove that the two notions of an A-module are equivalent. Hint: Use the fact that the category of complex vector spaces is a closed monoidal category and

$$Hom(A \otimes M, M) \cong Hom(A, Hom(M, M)). \tag{7.103}$$

■

The 2-functor MOD is defined for morphisms as follows:

$$f\colon A \longrightarrow A' \quad \longmapsto \quad \mathbb{M}\mathrm{od}(A') \xrightarrow{MOD(f)} \mathbb{M}\mathrm{od}(A), \quad MOD(A')=U_{A'}\colon \mathbb{M}\mathrm{od}(A') \searrow \mathbf{C}\mathbb{V}\mathrm{ect}., \quad U_A=MOD(A)\colon \mathbb{M}\mathrm{od}(A) \swarrow \mathbf{C}\mathbb{V}\mathrm{ect}. \tag{7.104}$$

The functor $MOD(f)$ takes module M with the action

$$A' \otimes M \xrightarrow{\ \cdot'\ } M \tag{7.105}$$

to module M with the action

$$A \otimes M \xrightarrow{f \otimes id_M} A' \otimes M \xrightarrow{\ \cdot'\ } M. \tag{7.106}$$

This is even simpler when one looks at a module as $A \longrightarrow End(M)$. In that case, $MOD(f)$ takes $A' \longrightarrow End(M)$ to $A \xrightarrow{f} A' \longrightarrow End(M)$. The forgetful functors are also respected by $MOD(f)$ because both take the module M to its vector space.

The 2-Functor $MOD\colon \overline{\mathbb{A}\mathrm{lgebra}}^{op} \longrightarrow \overline{\mathbb{C}\mathrm{at}}/\mathbf{C}\mathbb{V}\mathrm{ect}$ **on 2-cells.** The 2-functor MOD takes conjugates to natural transformations as follows:

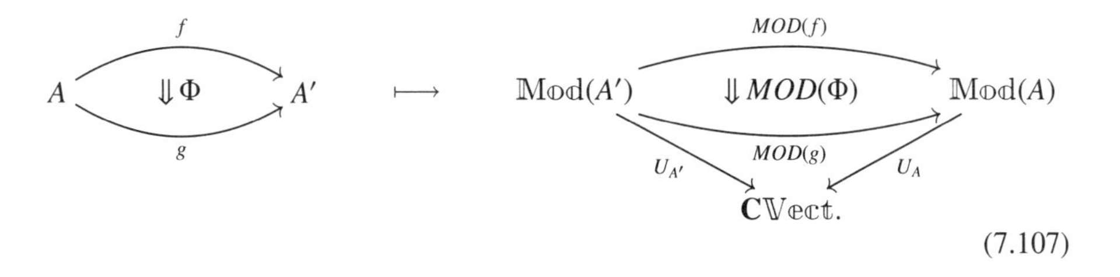

(7.107)

The natural transformation $MOD(\Phi)$ is defined at the component A'-module $(M,\cdot)$ as the A-module homomorphism

$$MOD(\Phi)_{(M,\cdot)}\colon MOD(f)(M,\cdot) \longrightarrow MOD(g)(M,\cdot), \tag{7.108}$$

which sends m to $\Phi\cdot m$. In detail, Φ is an element of A', and m is an element of M. For this to be an A-module homomorphism, the following instance of Diagram (7.101) must commute:

$$\begin{array}{ccc} A\otimes M & \xrightarrow{id_A\otimes\Phi\cdot(\ )} & A\otimes M \\ {\scriptstyle f\otimes id_m}\downarrow & & \downarrow{\scriptstyle g\otimes id_m} \\ A'\otimes M & & A'\otimes M \\ {\scriptstyle \cdot}\downarrow & & \downarrow{\scriptstyle \cdot} \\ M & \xrightarrow[\Phi\cdot(\ )]{} & M. \end{array} \tag{7.109}$$

For this to commute, we need that for all a in A, we have $\Phi\cdot(f(a)\cdot m) = g(a)\cdot(\Phi\cdot m)$. This is true because

$$\begin{aligned} \Phi\cdot(f(a)\cdot m) &= (\Phi * f(a))\cdot m && \text{because of the left side of Diagram (7.100)} \\ &= (g(a) * \Phi)\cdot m && \text{because } \Phi \text{ is a conjugate from } f \text{ to } g \\ &= g(a)\cdot(\Phi\cdot m) && \text{because of the left side of Diagram (7.100).} \end{aligned}$$

Notice that the underlying vector space M does not change, which means that the triangle in Diagram (7.107) commutes.

The 2-Functor $MOD\colon \overline{\mathbb{Algebra}}^{op} \longrightarrow \overline{\mathbb{Cat}}/\mathbf{C}\mathbb{Vect}$ **is a monoidal 2-functor.** To be a monoidal 2-functor, we need that for algebras A and A', there is a functor in $\overline{\mathbb{Cat}}/\mathbf{C}\mathbb{Vect}$:

$$\begin{array}{ccc} \mathbb{Mod}(A)\times\mathbb{Mod}(A') & \xrightarrow{T_{A,A'}} & \mathbb{Mod}(A\otimes A') \\ & {\scriptstyle U_A\boxtimes U_{A'}}\searrow \quad \swarrow{\scriptstyle U_{A\otimes A'}} & \\ & \mathbf{C}\mathbb{Vect}. & \end{array} \tag{7.110}$$

In detail, $T_{A,A'}$ is defined for a given A-module $(M,\cdot)$ and an A'-module $(M',\cdot')$ as the $A\otimes A'$ module $M\otimes M'$, with the action

$$A\otimes A'\otimes M\otimes M' \xrightarrow{id_A\otimes br\otimes id_{M'}} A\otimes M\otimes A'\otimes M' \xrightarrow{(\cdot)\otimes(\cdot')} M\otimes M'. \tag{7.111}$$

In general, there is no functor the other way. The monoidal structures for morphisms and 2-cells are similar.

Variations of Algebras

We now have the tools in our toolbox to examine interesting types of algebras and their categories of modules. What type of extra structure does the category $\mathbb{Mod}(A)$ have for a given algebra A?

Does $\mathbb{Mod}(A)$ have a tensor product? We are asking if the tensor product of two A-modules is an A-module. Formally, given two A-modules, $(M, \cdot)$ and $(M', \cdot')$ tell if there is an action:

$$A \otimes M \otimes M' \longrightarrow M \otimes M'. \tag{7.112}$$

It is easy to see that $M \otimes M'$ is an $A \otimes A$-module with the following action:

$$A \otimes A \otimes M \otimes M' \xrightarrow{id \otimes br \otimes id} A \times M \otimes A \otimes M' \xrightarrow{(\cdot) \otimes (\cdot')} M \otimes M'. \tag{7.113}$$

If there is a morphism in $\overline{\mathbb{Algebra}}$ of the form $\Delta\colon A \longrightarrow A \otimes A$, then it induces a functor

$$MOD(\Delta)\colon \mathbb{Mod}(A \otimes A) \longrightarrow \mathbb{Mod}(A), \tag{7.114}$$

which takes the $A \otimes A$ action on $M \otimes M'$ in Diagram (7.113) to the A action on $M \otimes M'$, given by

$$A \otimes M \otimes M' \xrightarrow{\Delta \otimes id \otimes id} A \otimes A \otimes M \otimes M' \xrightarrow{id \otimes br \otimes id} A \times M \otimes A \otimes M' \xrightarrow{(\cdot) \otimes (\cdot')} M \otimes M'. \tag{7.115}$$

This gives us the following bifunctor:

$$\mathbb{Mod}(A) \times \mathbb{Mod}(A) \longrightarrow \mathbb{Mod}(A \otimes A) \xrightarrow{Mod(\Delta)} Mod(A)\,. \tag{7.116}$$

What about the unit of $\mathbb{Mod}(A)$? A morphism in $\overline{\mathbb{Algebra}}$ of the form $\varepsilon\colon A \longrightarrow \mathbf{C}$ induces a functor

$$MOD(\varepsilon)\colon \mathbb{Mod}(\mathbf{C}) \longrightarrow \mathbb{Mod}(A). \tag{7.117}$$

The category $\mathbb{Mod}(\mathbf{C})$ has only one object: complex numbers. Hence, $\mathbb{Mod}(\mathbf{C}) \cong \mathbf{1}$ is the trivial category and the functor $MOD(\varepsilon)$ chooses an object in $\mathbb{Mod}(A)$, which is the unit. Just having these morphisms alone does not make category $\mathbb{Mod}(A)$ into a strict monoidal category. We need the bifunctor and the unit to satisfy certain axioms.

Definition 7.4.10. The collection $(A, \mu, \upsilon, \Delta, \varepsilon)$ is a **bialgebra**, where

- (A, μ, υ) is an algebra
- Δ is a way of factoring element or a **comultiplication** (i.e., a linear map $\Delta\colon A \longrightarrow A \otimes A$).
- ε is a way of evaluating the elements of A or a **counit** (i.e., a linear map $\varepsilon\colon A \longrightarrow \mathbf{C}$).

These extra ingredients must have the following requirements: Δ must be coassociative, and ε must behave like a counit. This means that the following two diagrams must commute:

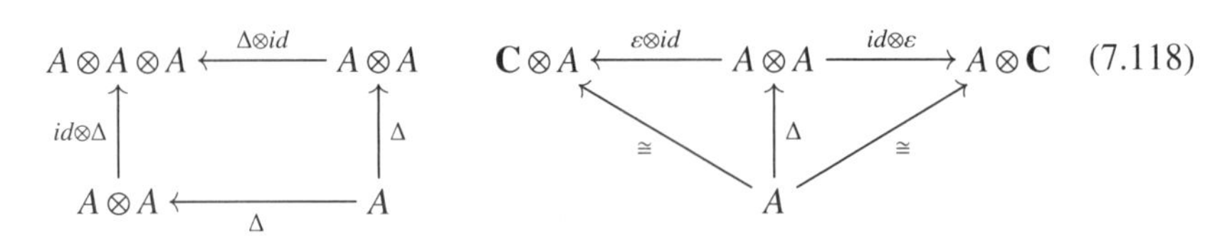

(7.118)

These diagrams are very similar to Diagram (2.10) which gave us the definition of a monoid. Here, the arrows are reversed, giving us the definition of a comonoid.

In terms of equations, these commutative diagrams are

$$(id\otimes\Delta)\Delta=(\Delta\otimes id)\Delta \quad \text{and} \quad (id\otimes\varepsilon)\Delta=id_A=(\varepsilon\otimes id)\Delta. \tag{7.119}$$

With this coassociativity and the axioms for the counit, we can relate algebras with extra structure and categories with extra structure.

Theorem 7.4.11. Let (A,μ,υ) be an algebra with algebra homomorphisms $\Delta\colon A\longrightarrow A\otimes A$ and $\varepsilon\colon A\longrightarrow\mathbf{C}$. The morphisms Δ and ε satisfy Diagram (7.118) if and only if $MOD(\Delta)$ and $MOD(\varepsilon)$ induce a strict monoidal category structure on $\mathbb{M}\text{od}(A)$, the category of modules of A. ★

Proof. ($\Longrightarrow$) Simply apply the $MOD(\ )$ 2-functor to Diagrams (7.118). Let us work out the details of the associativity because we will see variations of this in the coming pages. Applying the 2-functor to the left diagram gives us the following diagram:

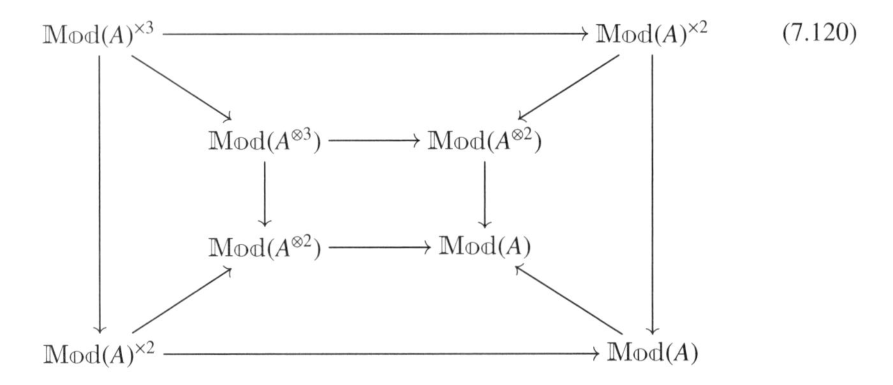

(7.120)

The four outer quadrilaterals commute because of the naturality of the mapping funnels. The inner quadrilateral commutes because it was assumed to commute in the definition of a bialgebra. Hence, the entire diagram commutes, which means that the tensor product on category $\mathbb{M}\text{od}(A)$ is strictly associative.

($\Longleftarrow$) Consider algebra A and let $(\mathbb{M}\text{od}(A),\otimes,I)$ be a strict monoidal category induced by $MOD(\Delta)$ and $MOD(\varepsilon)$. We will show that the special object A, which is in category $\mathbb{M}\text{od}(A)$, has the required comultiplication and counit. Algebra A is an object of $MOD(A)$. That is, algebra A acts on the module A with multiplication. The category $\mathbb{M}\text{od}(A)$ also contains an object $A\otimes A\otimes A$. From the strict monoidal structure, there is an equality

$$A \otimes (A \otimes A) = (A \otimes A) \otimes A. \tag{7.121}$$

That is, for all u, v, w in A, there is the equality

$$u \otimes (v \otimes w) = (u \otimes v) \otimes w. \tag{7.122}$$

This equality is linear in the sense that for all a in A, the following action is true:

$$au \otimes (v \otimes w) = a(u \otimes v) \otimes w. \tag{7.123}$$

Setting $u = v = w = 1$, the unit of A, gives us

$$(id \otimes \Delta)(\Delta(a)) = (\Delta \otimes id)(\Delta(a) \tag{7.124}$$

for all a. This says that Δ is coassociative. Similarly, we can show that for all a in A, we have

$$(\varepsilon \otimes id)(\Delta(a)) = a \qquad \text{and} \qquad (id \otimes \varepsilon)(\Delta(a)) = a, \tag{7.125}$$

which shows that the counit acts appropriately. ♣

Let us move on to another algebraic structure. We weaken the definition of a bialgebra so that the category of modules is not a strict monoidal category, but rather a general monoidal category.

Definition 7.4.12. The collection $(A, \mu, \upsilon, \Delta, \varepsilon, \Phi, \mathrm{P}, \Lambda)$ is a **quasi-bialgebra** or a **Drinfeld algebra**, where the following are true:

- (A, μ, υ) is an algebra.
- Δ is a comultiplication (i.e., $\Delta\colon A \longrightarrow A \otimes A$ is a linear map).
- ε is a counit (i.e., $\varepsilon\colon A \longrightarrow \mathbf{C}$ is a linear map).
- Φ, P, and Λ are invertible conjugates, which are used to weaken the coassociativity, the right counit rule, and the left counit rule. (The conjugate Φ is called a **Drinfeld associator**.) The conjugates go between the following morphisms in the (not necessarily commutative) diagrams

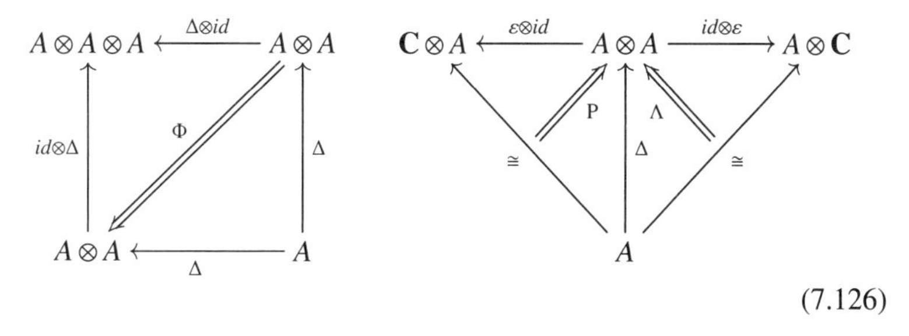

(7.126)

These linear maps and conjugates must satisfy the following requirements:

- **Drinfeld's pentagon equation**:

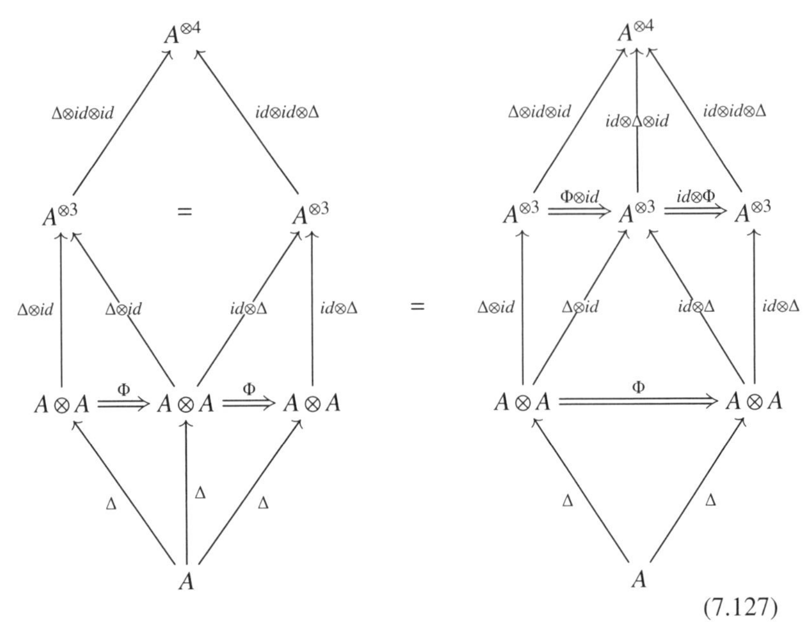

(7.127)

Notice that this is related to Mac Lane's pentagon coherence condition.

- **Drinfeld's triangle equation**:

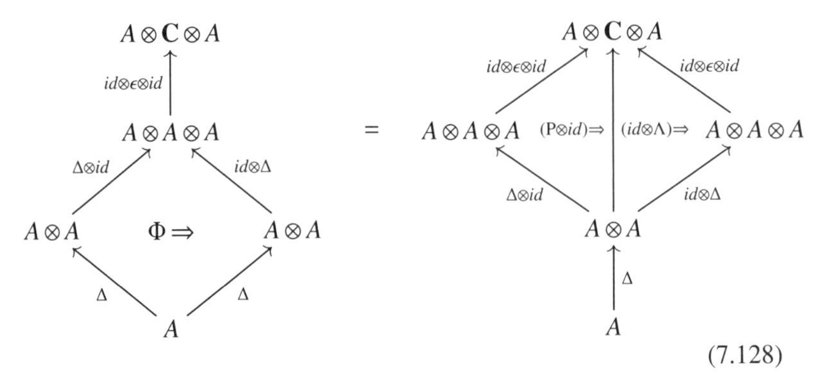

(7.128)

We will see in the upcoming proof that this is related to Mac Lane's triangle condition.

To help the reader communicate with the poor souls that have not yet read this book, we state the three conjugates of Diagram (7.126) with what is found in the literature: for all $a \in A$, we have

$$(id \otimes \Delta)(\Delta(a)) = \Phi(\Delta \otimes id)(\Delta(a)\Phi^{-1} \tag{7.129}$$

$$(id \otimes \varepsilon)(\Delta(a) = \mathrm{P}^{-1}a\mathrm{P} \tag{7.130}$$

$$(\varepsilon \otimes id)(\Delta(a) = \Lambda^{-1}a\Lambda. \tag{7.131}$$

The requirements on the conjugates are stated in the literature as

$$(id \otimes id \otimes \Delta)(\Phi)(\Delta \otimes id \otimes id)(\Phi) = (id \otimes \Phi)(id \otimes \Delta \otimes id)(\Phi)(\Phi \otimes id) \tag{7.132}$$

and

$$(id \otimes \varepsilon \otimes id)(\Phi) = \mathrm{P} \otimes \Lambda^{-1} \tag{7.133}$$

It is important to realize that Vladimir Drinfeld did not just come up with these requirements out of thin air. He had in mind the definition of a monoidal category when he created this definition of a quasi-bialgebra. We will see in the proof of Theorem 7.4.13 that the conjugates of the algebra will take the place of the natural transformations of the monoidal category.

Theorem 7.4.13. Let (A, μ, υ) be an algebra with morphisms Δ, and ε and with conjugates Φ, P, and Λ as discussed previously. The conjugates satisfy Diagrams (7.127) and (7.128) if and only if $MOD(\Delta)$, $MOD(\varepsilon)$, $MOD(\Phi)$, $MOD(\mathrm{P})$, and $MOD(\Lambda)$ induce a monoidal category structure on $\mathbb{M}\mathrm{od}(A)$, the category of modules of A. ★

Proof. ($\Longrightarrow$) The 2-functor MOD takes A to $\mathbb{M}\mathrm{od}(A)$ with its forgetful functor to $\mathbf{C}\mathbb{V}\mathrm{ect}$. Here, Δ and ε become the tensor product and the unit of the monoidal category, and the Φ, P, and Λ become the α, ρ, and λ of the monoidal category. The Drinfeld pentagon condition becomes the Mac Lane pentagon condition. The Drinfeld triangle condition becomes the Mac Lane triangle condition. This last point can be seen as follows: The usual way to look at the Mac Lane's triangle coherence condition is to write it as

$$(\rho_a \otimes id_b) \circ \alpha_{a,I,b} = (id_a \otimes \lambda_b). \tag{7.134}$$

This can be rewritten as

$$\alpha_{a,I,b} = (\rho_a \otimes id_b)^{-1} \circ (id_a \otimes \lambda_b), \tag{7.135}$$

which is equivalent to

$$\alpha_{a,I,b} = \rho_a^{-1} \otimes \lambda_b. \tag{7.136}$$

($\Longleftarrow$) Given a monoidal category $(\mathbb{M}\mathrm{od}(A), \otimes, I, \alpha, \lambda, \rho)$, we can form the conjugates as follows:

$$\Phi = \alpha_{A,A,A}(1 \otimes 1 \otimes 1) \in A \otimes A \otimes A \tag{7.137}$$

$$\Lambda = \lambda_A(1 \otimes 1) \in A \tag{7.138}$$

$$\mathrm{P} = \rho_A(1 \otimes 1) \in A. \tag{7.139}$$

These conjugates satisfy the algebra requirements because α, λ, and ρ satisfy the categorical coherence requirements. ♣

Let us move on from associativity to commutativity.

Definition 7.4.14. A bialgebra $(A,\mu,\upsilon,\Delta,\varepsilon)$ is **cocommutative bialgebra** if

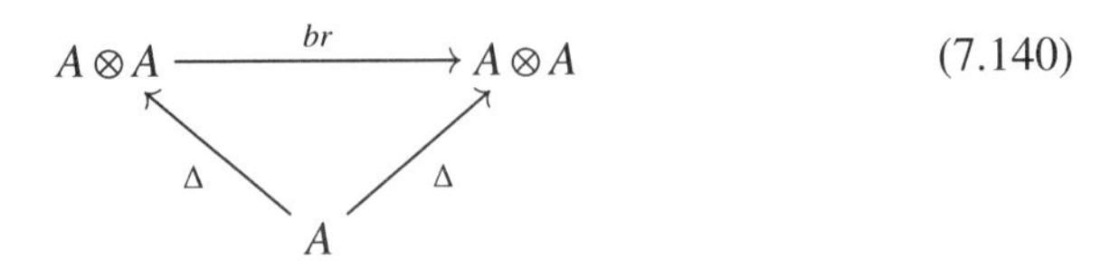

(7.140)

that is, $\Delta = br \circ \Delta$, where $br\colon A\otimes A \longrightarrow A\otimes A$ swaps the elements of the tensor.

Theorem 7.4.15. Let $(A,\mu,\upsilon,\Delta,\varepsilon)$ be a bialgebra. The bialgebra A is cocommutative if and only if $MOD(\Delta)$ and $MOD(\varepsilon)$ induce a strictly symmetric monoidal category structure on $\mathbb{M}\mathrm{od}(A)$, the category of modules of A. ★

Proof. Take MOD of Diagram (7.140) and follow along the same way as in the previous proofs. ♣

The problem is that the strictly symmetric strict monoidal categories are a rarity. We need to weaken the cocommutative requirement just as we weakened the definition of a strictly symmetric monoidal category into a symmetric monoidal category.

Definition 7.4.16. Let $(A,\mu,\upsilon,\Delta,\varepsilon)$ be a bialgebra. Consider the invertible conjugate R:

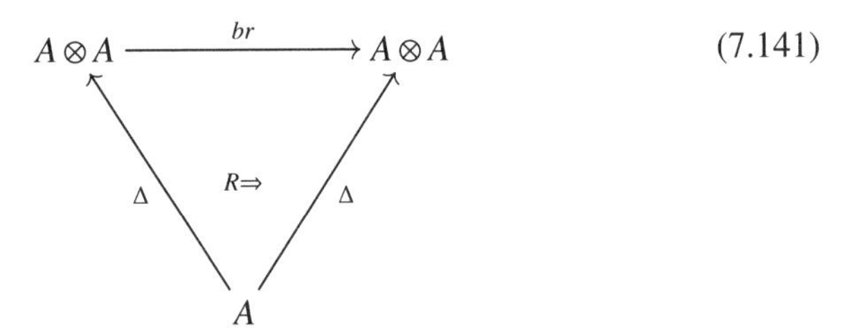

(7.141)

We say that R is a **universal R-matrix** if the following two diagrams of conjugates are equal:

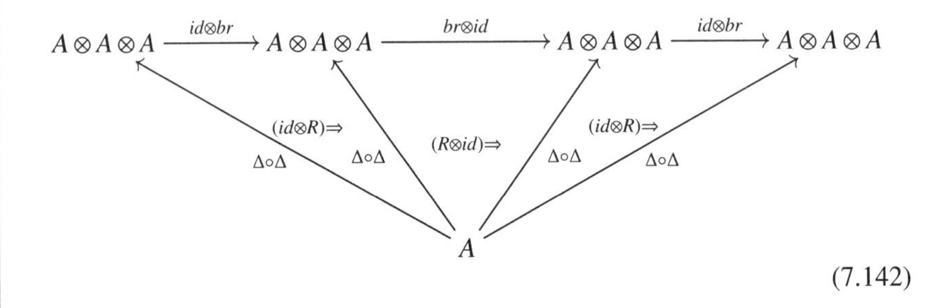

(7.142)

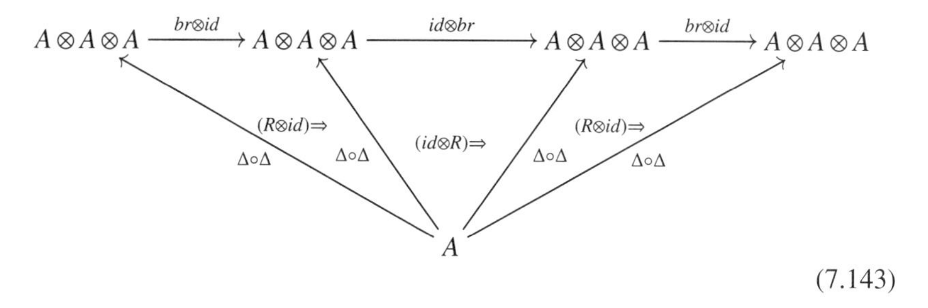

(7.143)

This is called the **Yang-Baxter equation**.

A **triangular bialgebra** $(A, \mu, \upsilon, \Delta, \varepsilon, R)$ is a bialgebra $(A, \mu, \upsilon, \Delta, \varepsilon)$ with a universal R-matrix that satisfies the following symmetry axiom:

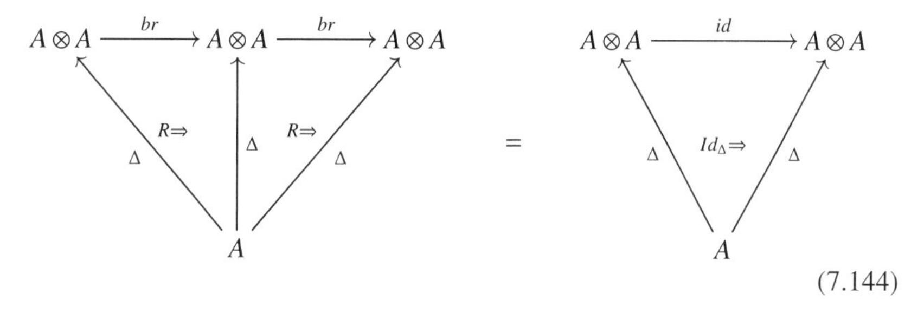

(7.144)

If we do not assume that R satisfies the symmetry axiom, then A is called a **quasi-triangular bialgebra**.

The conjugate in Diagram (7.141) is an invertible element of $A \otimes A$ and is written for all a in A as $R(\Delta(a))R^{-1} = (br \circ \Delta)(a)$.

To express the Yang-Baxter equation as elements of algebras, we need to change some notation. As R is an element of $A \otimes A$, it can be written as $\Sigma_i a_i \otimes a'_i$. Element R can also be embedded into $A \otimes A \otimes A$ in various ways. For example, it can be embedded as $\Sigma_i a_i \otimes a'_i \otimes 1$, and we denote such an element of $A^{\otimes 3}$ as R_{12}. Another embedding is $\Sigma_i a_i \otimes 1 \otimes a'_i$, which we denote as R_{13}. We can even switch the order of the elements so $\Sigma_i a'_i \otimes a_i \otimes 1$, which is an element of $A \otimes A \otimes A$ and is denoted as R_{21}. Using this notation, we write the Yang-Baxter equation as

$$R_{12}R_{13}R_{23} = R_{23}R_{13}R_{12}. \quad (7.145)$$

The Yang-Baxter equation is equivalent to

$$(\Delta \otimes Id)(R) = R_{13}\, R_{23} \qquad \text{and} \qquad (Id \otimes \Delta)(R) = R_{13}\, R_{12}. \quad (7.146)$$

This is similar to the equivalence of the commuting of Diagrams (5.108) and (5.42) in Chapter 5.

The symmetry condition is written as

$$R^{-1} = brR \qquad \text{or} \qquad R^{-1} = R_{21}. \quad (7.147)$$

Theorem 7.4.17. Let $(A, \mu, \upsilon, \Delta, \varepsilon)$ be a bialgebra, and let R be a conjugate. This R satisfies the Yang-Baxter equation and the symmetry condition—A is a triangular bialgebra—if and only if $MOD(\Delta)$ and $MOD(R)$ induce a symmetric monoidal category structure on $\mathbb{M}\mathrm{od}(A)$, the category of modules of A.

Let $(A, \mu, \upsilon, \Delta, \varepsilon)$ be a bialgebra, and let R be a conjugate. This R satisfies the Yang-Baxter equation—A is a quasi-triangular bialgebra—if and only if $MOD(\Delta)$ and $MOD(R)$ induce a braided monoidal category structure on $\mathbb{M}\mathrm{od}(A)$, the category of modules of A. ★

Proof. The Yang-Baxter equation is clearly a conjugate version of Diagrams (5.42). The symmetry condition is clearly a conjugate version of the symmetry coherence condition given in Diagram (5.38). ♣

Of course one can weaken the coassociativity *and* the cocommutativity at the same time.

Definition 7.4.18. Let $(A, \mu, \upsilon, \Delta, \varepsilon, \Phi, \mathrm{P}, \Lambda)$ be a quasi-bialgebra and R be a conjugate. If R satisfies

$$(id \otimes \Delta)(R) = (\Phi_{231})^{-1} R_{13} \Phi_{213} R_{12} (\Phi_{123})^{-1} \tag{7.148}$$

and

$$(\Delta \otimes id)(R) = \Phi_{312} R_{13} (\Phi_{132})^{-1} R_{23} \Phi_{123}, \tag{7.149}$$

then A is a **quasi-triangular quasi-bialgebra**. These equations are very similar to Diagrams (5.39) and (5.40). One can also see these in the string diagrams in Equation (5.79).

Exercise 7.4.19. Draw the diagram that corresponds to Equations (7.148) and (7.149). ■

Theorem 7.4.20. In a quasi-triangular quasi-bialgebra, the following equation is satisfied:

$$R_{12} \Phi_{312} R_{13} (\Phi_{132})^{-1} R_{23} \Phi_{123} = \Phi_{321} R_{23} (\Phi_{231})^{-1} R_{13} \Phi_{213} R_{12}. \tag{7.150}$$

★

Proof. This is the same proof that the dodecagon in Figure 5.8 commutes. Namely, it follows from either the commutativity of Diagram (5.39) or (5.39), as well as naturality conditions. ♣

Definition 7.4.21. A universal R matrix acts on the elements of $A \otimes A$. Let us generalize this definition. Let a be an object in a monoidal category $\mathbb{A}$. An isomorphism $\gamma \colon a \otimes a \longrightarrow a \otimes a$ is a **Yang-Baxter operator on** a if it satisfies the following diagram:

$$\begin{array}{ccccccccccc} a\otimes(a\otimes a) & \overset{\alpha}{\to} & (a\otimes a)\otimes a & \overset{\gamma\otimes id}{\to} & (a\otimes a)\otimes a & \overset{\alpha^{-1}}{\to} & a\otimes(a\otimes a) & \overset{id\otimes\gamma}{\to} & a\otimes(a\otimes a) & \overset{\alpha}{\to} & (a\otimes a)\otimes a \\ \downarrow{\scriptstyle id\otimes\gamma} & & & & & & & & & & \downarrow{\scriptstyle \gamma\otimes id} \\ a\otimes(a\otimes a) & \underset{\alpha}{\to} & (a\otimes a)\otimes a & \underset{\gamma\otimes id}{\to} & (a\otimes a)\otimes a & \underset{\alpha^{-1}}{\to} & a\otimes(a\otimes a) & \underset{id\otimes\gamma}{\to} & a\otimes(a\otimes a) & \underset{\alpha}{\to} & (a\otimes a)\otimes a \end{array} \tag{7.151}$$

This is based on Equation (7.150). If $\mathbb{A}$ is a strict monoidal category, this requirement reduces to

$$\begin{array}{ccccc} a\otimes a\otimes a & \xrightarrow{\gamma\otimes id} & a\otimes a\otimes a & \xrightarrow{id\otimes\gamma} & a\otimes a\otimes a \\ \downarrow{\scriptstyle id\otimes\gamma} & & & & \downarrow{\scriptstyle \gamma\otimes id} \\ a\otimes a\otimes a & \xrightarrow[\gamma\otimes id]{} & a\otimes a\otimes a & \xrightarrow[id\otimes\gamma]{} & a\otimes a\otimes a \end{array} \tag{7.152}$$

This is based on Equation (7.145).

Theorem 7.4.22 shows that Yang-Baxter operators are preserved by strong monoidal functors.

Theorem 7.4.22. Let $(\mathbb{A}, \otimes, I, \alpha, \lambda, \rho)$ and $(\mathbb{B}, \otimes', I', \alpha', \lambda', \rho')$ be monoidal categories, and let

$$(F, \tau, \upsilon)\colon (\mathbb{A}, \otimes, I, \alpha, \lambda, \rho) \longrightarrow (\mathbb{B}, \otimes', I', \alpha', \lambda', \rho') \tag{7.153}$$

be a strong monoidal functor. If a in $\mathbb{A}$ and $\gamma\colon a\otimes a \longrightarrow a\otimes a$ is a Yang-Baxter operator, then $\gamma'\colon F(a)\otimes F(a) \longrightarrow F(a)\otimes F(a)$ is defined as follows:

$$\begin{array}{ccc} F(a)\otimes F(a) & \xrightarrow{\gamma'} & F(a)\otimes F(a) \\ \downarrow{\scriptstyle \tau_{a,a}} & & \uparrow{\scriptstyle \tau^{-1}_{a,a}} \\ F(a\otimes a) & \xrightarrow[F(\gamma')]{} & F(a\otimes a) \end{array} \tag{7.154}$$

which is a Yang-Baxter operator on $F(a)$. ★

Proof. We have to show that γ' satisfies the requirement of Yang-Baxter operator. The proof is essentially Figure 7.7. Some orientation is needed. The inner commuting dodecagon is the image under F of the dodecagon, showing that γ is a Yang-Baxter operator. Some of the commuting quadrilaterals are instances of the naturality of τ, and some are commuting quadrilaterals from Diagram (7.154). The commuting hexagons are instances of Diagram (6.6). Since the inner dodecagon and all the other diagrams commute, the outer dodecagon commutes. This shows that γ' is a Yang-Baxter operator. ♣

What type of algebra has a category of representations with duals?

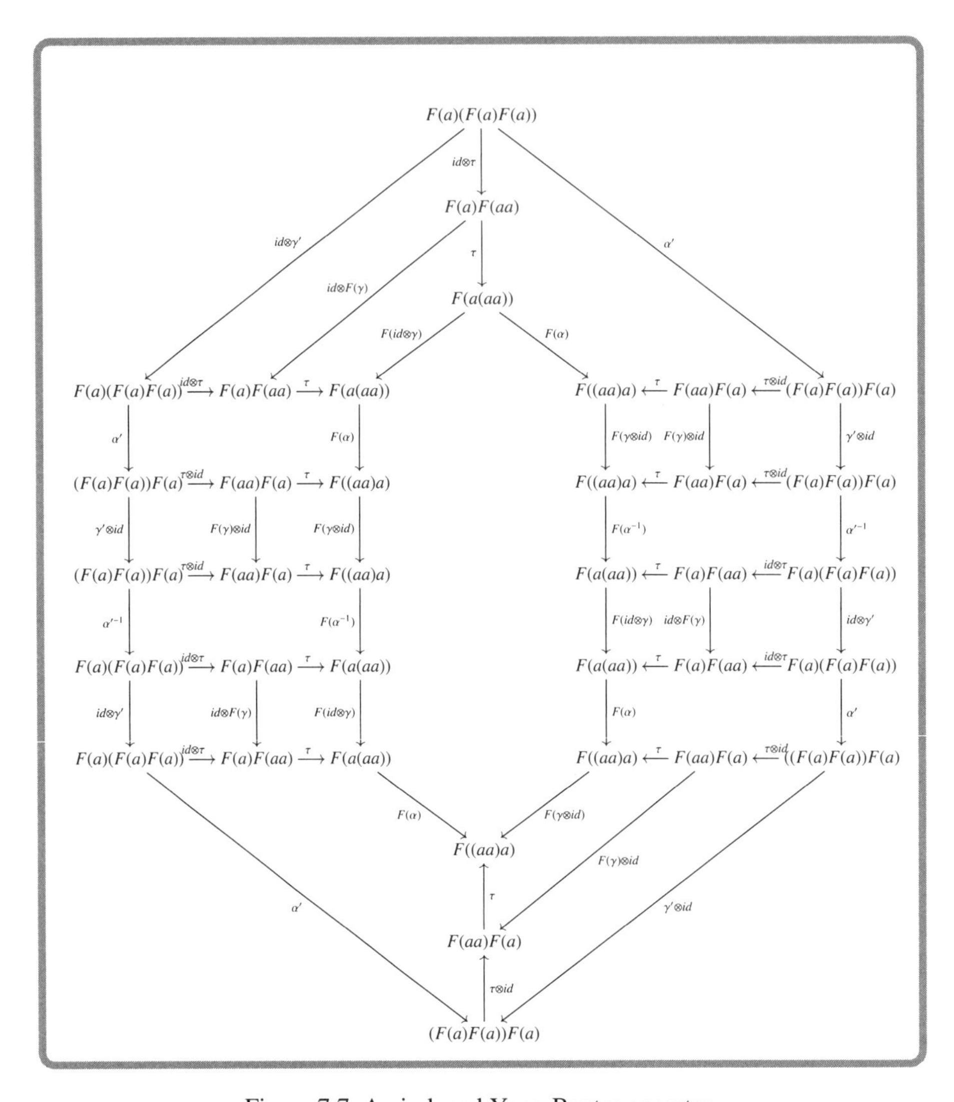

Figure 7.7. An induced Yang-Baxter operator.

Definition 7.4.23. A **Hopf algebra** $(H, \mu, \upsilon, \Delta, \varepsilon, S)$ is a bialgebra with a linear map $S : A \longrightarrow A$ such that the following commutes:

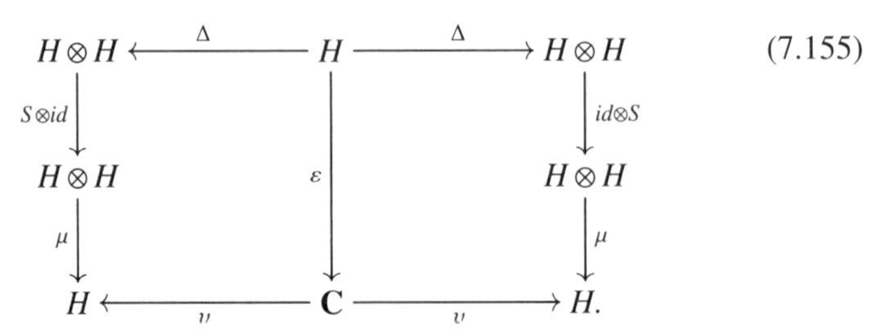

(7.155)

If you think of S as the map that takes an element to its inverse (i.e., $x \mapsto x^{-1}$), then the left part of the this diagram is similar to the bottom part of Figure 1.4 where we saw the categorical way of describing a group. This means that a Hopf algebra is like a group, but in the category of vector spaces rather than the category of sets. Homage must be paid to Important Categorical Idea 1.4.60.

The dual of a map of modules is found using the $S : A \longrightarrow A$ of the Hopf algebra. We will be sketchy with the details. We simply state the main theorem of Hopf algebras for our purposes as follows:

Theorem 7.4.24. The category of finite-dimensional modules for a Hopf algebra is a rigid monoidal category. ★

One can easily go on and weaken the coassociativity of a Hopf algebra to form a **quasi-Hopf algebra**. One can also weaken the coassociativity and cocommuttativity of a Hopf algebra to define a **quasi-triangular quasi-Hopf algebra**. For such an algebra H, the category of finite-dimensional modules of H, $\mathbb{M}\mathrm{od}(H)$, forms a rigid, braided, monoidal category.

There are other algebras that are related to balanced categories and ribbon categories.

Definition 7.4.25. A quasi-triangular bialgebra A is a **balanced algebra** if there is an element θ in A that commutes with all the elements in A (such an element is called **central** and is related to naturality) such that

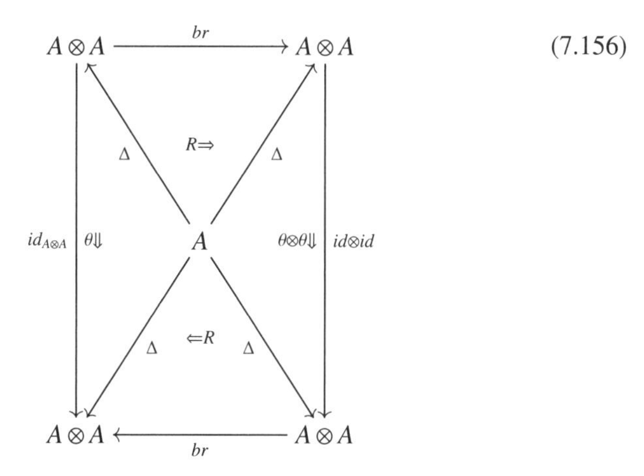

(7.156)

This diagram can be written as

$$\Delta(\theta) = R(\theta \otimes \theta)R. \tag{7.157}$$

In the literature, this equation is sometimes written as the equivalent:

$$\Delta(\theta) = (R_{21}R)^{-1}(\theta \otimes \theta). \tag{7.158}$$

One goes on to define a **balanced Hopf algebra**, which is also called a **ribbon algebra** (see the sources at the end of this section for more details).

As expected, we have the following theorems about the categories of modules for such algebra.

Theorem 7.4.26. The category of modules for a balanced algebra is a balanced monoidal category. The category of finite-dimensional modules for a ribbon algebra is a ribbon monoidal category. ★

A few moments of thought must be given to the 2-functor $MOD\colon \overline{\mathbb{Algebra}}^{op} \longrightarrow \overline{\mathbb{Cat}}/\mathbf{C}\mathbb{Vect}$. This 2-functor makes the connection between the algebraic structure and the categorical structure. The 2-functor is not surjective on objects because there are many categories $\mathbb{C}$ and functors $\mathbb{C} \longrightarrow \mathbf{C}\mathbb{Vect}$ that are not module categories. The main point is that this 2-functor respect and reflects the structure in each of the 2-categories it connects. One must meditate for long periods of time on the Venn diagrams A.6 and A.7 and see the symmetry between the categorical structures and the algebraic structures.

Applications

The field of quantum groups has many applications in mathematics and science. We just mention a few here.

Quantum groups were formulated to deal with quantum theory. They are related to structures called **deformed Lie algebras**, **quantum doubles**, and **quantum integrable systems**, which we will not get into in this text. For more on these, see the textbooks listed at the end of this mini-course.

They also arise in statistical mechanics. One way they are used is in a simple way of looking at collections called an **Ising model**. The Yang-Baxter equation in this context becomes the **star-triangle relation**. See more about it in Section 7.4 of [13] and Chapter 7 of [237].

The Yang-Baxter equations also arise in electrical engineering. There is a star-triangle relation with electrical resistors. See Section 7.9 of [237].

There is also a very deep and mysterious relationship between the ideas of quantum groups and number theory. The researcher who started quantum groups, Vladimir Drinfeld, formulated a structure called the **Grothendieck–Teichmüller group**, which consists of automorphisms that respect the associativity and commutativity [65]. This group is related to a group called the **Galois group of Q-bar over Q** and is denoted as $Gal(\bar{\mathbf{Q}}/\mathbf{Q})$. This group has profound implications about number theory and the rational numbers. One can get a glimpse into this world and its relationship with coherence theory in [29].

Suggestions for Further Study

There are several textbooks on the subject [134, 184, 57, 237]. The most categorical presentation is Ross Street's wonderful book [247].

8

Describing Structures

The point of these observations is not the reduction of the familiar to the unfamiliar . . . but the extension of the familiar to cover many more cases.
Saunders Mac Lane [180], page 226.

Category theory is the study of structures and processes. Monoidal category theory is the study of combining structures and processes. Over the decades, researchers have come up with various ways of describing structures and processes. In this chapter, we present some of the methods used to describe various categories of structures, and we show how the same methods of descriptions are used in mathematics, physics, and computing.

The literature has many theorems discussing the strength of each method, that is, what type of structure can and cannot be described by each method. We will omit most of that. We are going to just present the major definitions and examples of these methods. Our aim is to entice the reader to learn more about the methods.

This chapter is organized as follows:

- Section 8.1: Algebraic Theories
- Section 8.2: Operads
- Section 8.3: Monads
- Section 8.4: Algebraic 2-Theories

The chapter ends (Section 8.5) with a mini-course on Databases and Scheduling.

8.1 Algebraic Theories

Algebraic theories were introduced in F. William Lawvere's thesis [158] (and in his further papers, such as [163]) as a method of describing various algebraic structures. Algebraic theories are the centerpiece of an entire field called **functorial semantics**. An algebraic theory is a category that describes the operations of an algebraic structure and what axioms it satisfies. (This is related to what logicians call the **signature** of an algebraic structure.) Thus, in category theory, not only does a collection of objects that have the same type of structure form a category, but *the description of the structure itself forms a category.*

Before we get into technical details and definitions, let us focus on a motivating example.

Example 8.1.1. There is a category $\mathbf{T}_{Group}$, which is the **algebraic theory of groups**. The objects of the category are the natural numbers $0, 1, 2, \ldots$, and the category has the following important morphisms:

- $+ \colon 2 \longrightarrow 1$, corresponding to the main operation in a group that takes two inputs and gives one output
- $-(\ ) \colon 1 \longrightarrow 1$, corresponding to the inverse operation in a group that takes one input and gives one output
- $e \colon 0 \longrightarrow 1$, corresponding to the unit in a group which has zero inputs and just gives the constant unit u as the output

The category $\mathbf{T}_{Group}$ is a Cartesian category. Using the product $\times$, the composition $\circ$, and the identities, these morphisms generate many other morphisms. For example,

- The map $(+ \times id_1) \colon 2 + 1 \longrightarrow 1 + 1$ is a morphism from 3 to 2 that corresponds to performing $+$ on the first two elements and leaving the third input alone. That is, $(a, b, c) \mapsto (a + b, c)$.
- The map $(e \times +) \colon 0 + 2 \longrightarrow 1 + 1$, which corresponds to the operation $(a, b) \mapsto (u, a + b)$.
- The map

$$\begin{aligned}(-(\ ) \times id \times -(\ ) \times id) \circ (+ \times + \times + \times +) : 2 + 2 + 2 + 2 &\longrightarrow 1 + 1 + 1 + 1 \\ &\longrightarrow 1 + 1 + 1 + 1,\end{aligned} \tag{8.1}$$

  which corresponds to the operation

$$(a, b, c, d, e, f, g, h) \mapsto (a + b, c + d, e + f, g + h) \mapsto (-(a+b), c+d, -(e+f), g + h). \tag{8.2}$$

- The map $+ \circ (+ \times id) : 2 + 1 \longrightarrow 1 + 1 \longrightarrow 1$, which corresponds to the operation $(a, b, c) \mapsto (a + b, c) \mapsto (a + b) + c$.
- The map $+ \circ (id \times +) : 1 + 2 \longrightarrow 1 + 1 \longrightarrow 1$, which corresponds to the operation $(a, b, c) \mapsto (a, b + c) \mapsto a + (b + c)$.

Notice that in a group, the last two operations are equal because the $+$ operation is associative.

These operations must satisfy the axioms of being a group. We met those axioms as diagrams when we first introduced the notion of a group in Figure 1.4. In terms of maps within an algebraic theory, these axioms are stated as the commutative diagrams in Figure 8.1.

The category that has the natural numbers as objects and the operations that are generated by $+$, $-(\ )$, and e and that satisfies the axioms of Figure 8.1 is the theory of groups, $\mathbf{T}_{Group}$. A way of thinking of this category is that it is the "shape" or "template"

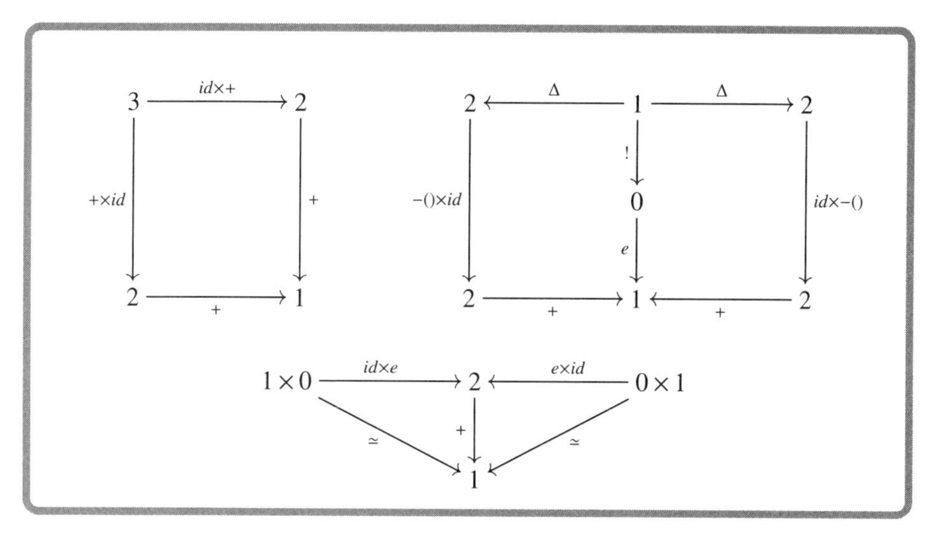

Figure 8.1. The axioms of the theory of groups.

of groups. One can visualize a small part of this shape as

$$\cdots \xrightarrow{+} 4 \xrightarrow{+} 3 \;\underset{+(id\times+)}{\overset{+(+\times id)}{\rightrightarrows}}\; \longrightarrow 2 \xrightarrow{+} 1 \circlearrowleft^{-} \xleftarrow{e} 0, \tag{8.3}$$

where the two morphisms from 3 to 1 are the two equal ways of adding three objects. □

Let us get formal about defining an algebraic theory. Consider the category whose objects are $0, 1, 2, 3, \ldots, n, \ldots$ and where the morphisms are set up so that every n is isomorphic to $1 \times 1 \times 1 \times \cdots \times 1$ (n times). This category is called the **theory of sets** and is denoted as $\mathbf{T}_{Set}$. The category $(\mathbf{T}_{Set}, \times, 0)$, is a Cartesian category with $\times$ as the tensor and the empty set as the unit. This category is called the theory of sets because it describes all the projection and twisting maps needed for sets and has no more operations. All algebraic theories will be built on this category.

Definition 8.1.2. An **algebraic theory** $(\mathbf{T}, \times, 0)$ is a Cartesian category where the objects are natural numbers and where n is isomorphic to the n-ary product of 1; that is, $n = 1 \times 1 \times 1 \times \cdots \times 1$ (n times). One should think of an arbitrary algebraic theory as having more morphisms than $\mathbf{T}_{Set}$.

Example 8.1.3. The theory of abelian (commutative) groups $\mathbf{T}_{AbGroup}$ is like $\mathbf{T}_{Group}$, but we insist that the two operations:

- $+ \colon 2 \longrightarrow 1$
- $+ \circ br \colon 2 \longrightarrow 2 \longrightarrow 1$, where br is the braiding operation that swaps the two entries in a product

are the same (i.e., the operation is commutative). □

Example 8.1.4. Another example is the theory of rings $\mathbf{T}_{Ring}$, which is similar to $\mathbf{T}_{AbGroup}$ but has another 2-ary operation $\cdot \colon 2 \longrightarrow 1$ and a nullary operation $e' \colon 0 \longrightarrow 1$ satisfying the ring axioms. □

Example 8.1.5. The theory of complex vector spaces $\mathbf{T}_{CVect}$ has every operation of the theory of abelian groups $\mathbf{T}_{AbGroup}$, but much more. Remember that for a vector space V, there is a scalar multiplication $\cdot \colon \mathbf{C} \times V \longrightarrow V$ that has to satisfy certain requirements. This can also be viewed as a function $V \longrightarrow V$ for every $c \in \mathbf{C}$, which describes the action that c does to each vector in V. In the language of algebraic theories, we write the scalar multiplication as a function $c \colon 1 \longrightarrow 1$ for each $c \in \mathbf{C}$. In other words, for every c, there is an endomorphism of 1 in the theory, i.e., $Hom_{\mathbf{T}_{CVect}}(1, 1) \cong \mathbf{C}$, the complex numbers. □

One can go on to give the definitions of other theories of algebraic structures such as monoids, commutative monoids, Boolean algebras, and lattices.

We are not only interested in the algebraic theory, but we are also interested in how the algebraic theory gives the actual structures it describes. How do we go from the theory of groups to the category of groups and group homomorphisms? In general, how do we go from a theory of a certain structure to the category of such structures?

Definition 8.1.6. Given an algebraic theory, $\mathbf{T}$, a **$\mathbf{T}$-algebra**, or **algebra** of $\mathbf{T}$, or **model** of $\mathbf{T}$, is a functor in $\mathbb{Set}^{\mathbf{T}}$ that is product preserving. In detail, a $\mathbf{T}$-algebra is a strict monoidal functor from the Cartesian category $(\mathbf{T}, \times, 0)$ to the Cartesian category $(\mathbb{Set}, \times, \{*\})$.

Example 8.1.7. Let us examine a product-preserving functor $F \colon \mathbf{T}_{Group} \longrightarrow \mathbb{Set}$. The functor on input 1 is $F(1) = S$, where S is some set. Since F is product preserving and $2 = 1 \times 1$, $F(2) = S \times S = S^2$. In general, $F(n) = S^n$, and $F(0) = S^0 = \{*\}$, the one object set. Morphisms go where they are supposed to go, e.g.,

$$F(+ \colon 2 \longrightarrow 1) \quad = \quad + \colon S^2 \longrightarrow S. \tag{8.4}$$

These operations on S must satisfy the same relations that $\mathbf{T}_{Group}$ satisfies. Thus, we have within $\mathbb{Set}$ a similar diagram to Diagram (8.3):

$$\cdots S^4 \xrightarrow{+} S^3 \underset{+(+\times id)}{\overset{+(id\times+)}{\rightrightarrows}} \xrightarrow{+} S^2 \xrightarrow{+} S^1 \circlearrowleft^{-} \xleftarrow{e} \{*\}. \tag{8.5}$$

In a sense, $\mathbf{T}_{Group}$ is the "cookie cutter," and the product-preserving functors from $\mathbf{T}_{Group}$ to $\mathbb{S}\mathrm{et}$ are the cookies. In other words, $\mathbf{T}_{Group}$ is the "shape" of groups, and the $\mathbf{T}_{Group}$-algebras are the representations of that shape. □

Every product-preserving functor from $\mathbf{T}$ to $\mathbb{S}\mathrm{et}$ determines a $\mathbf{T}$-algebra. Given two such $\mathbf{T}$-algebras, $F, G\colon \mathbf{T} \longrightarrow \mathbb{S}\mathrm{et}$, a natural transformation $\alpha\colon F \Longrightarrow G$ looks like this:

$$\mathbf{T} \underset{G}{\overset{F}{\rightrightarrows}} \mathbb{S}\mathrm{et}. \quad \alpha \Downarrow \tag{8.6}$$

The component at 1 assigns to 1 in $\mathbf{T}$ a set function $\alpha_1\colon F(1) \longrightarrow G(1)$. This is just a function of the underlying sets of the $\mathbf{T}$-algebras. From the fact that $n \cong 1^n$ and from the product-preserving nature of F and G, we have that $\alpha_n = (\alpha)^n\colon F(n) \longrightarrow G(n)$. The naturality of α implies that for any operation $\oplus\colon m \longrightarrow n$ in $\mathbf{T}$, the square

$$\begin{array}{ccc} F(m) & \xrightarrow{\alpha_m} & G(m) \\ {\scriptstyle F(\oplus)}\downarrow & & \downarrow{\scriptstyle G(\oplus)} \\ F(n) & \xrightarrow[\alpha_n]{} & G(n) \end{array} \tag{8.7}$$

commutes. This translates into saying that the set function α_1 respects the $\oplus$ operation (i.e., α_1 is a homomorphism of the structures). We conclude that for a given theory $\mathbf{T}$, product-preserving functors from $\mathbf{T}$ to $\mathbb{S}\mathrm{et}$ and natural transformations between such functors form the category of $\mathbf{T}$-algebras, $Alg(\mathbf{T}, \mathbb{S}\mathrm{et})$. This category is a full subcategory of $\mathbb{S}\mathrm{et}^{\mathbf{T}}$.

Example 8.1.8. The collection of product-preserving functors from $\mathbf{T}_{Ring}$ to $\mathbb{S}\mathrm{et}$ and natural transformations between such functors forms the category $Alg(\mathbf{T}_{Ring}, \mathbb{S}\mathrm{et})$. This category is isomorphic to the category $\mathbb{R}\mathrm{ing}$. The same construction is true for most of the algebraic structures that we saw in Definition 2.1.8. The categories of magmas, semigroups, monoids, groups, and abelian groups can all be constructed in this way. This is not true for fields. Why? Because the inverse operation (i.e., the function that sends x to $\frac{1}{x}$) is not always defined since there is no inverse of 0. One needs a more sophisticated method than an algebraic theory to describe the structure of fields. □

Exercise 8.1.9. Show that $Alg(\mathbf{T}_{Set}, \mathbb{S}\mathrm{et})$ is the category $\mathbb{S}\mathrm{et}$.
(Hint: $\mathbf{T}_{Set}$ does not have any other operations other than projection maps.) ■

Algebraic Functors

As you know by now, category theory is not only interested in "things," but in morphisms between "things." Algebraic theories do not stand alone. We are interested in functors between algebraic theories. These functors describe relationships between different types of structures.[1]

Definition 8.1.10. Let $\mathbf{T}_1$ and $\mathbf{T}_2$ be algebraic theories. An **algebraic theory morphism** or **theory morphism**, $G\colon \mathbf{T}_1 \longrightarrow \mathbf{T}_2$ is a functor that is the identity on objects, i.e., $G(n) = n$, and it respects the product (i.e., $G(n) = G(1)^n$). In other words, it is a strict monoidal functor from $(\mathbf{T}_1, \times, 0)$ to $(\mathbf{T}_2, \times, 0)$ that is the identity on objects. Any such G induces a map

$$Hom_{\mathbf{T}_1}(m,n) \longrightarrow Hom_{\mathbf{T}_2}(G(m), G(n)) = Hom_{\mathbf{T}_2}(m,n). \tag{8.8}$$

This means operations from m to n go to operations from m to n.

Since theory morphisms are identity on objects, we call full theory morphisms "surjections" and faithful theory morphisms "inclusions."

For a theory morphism $G\colon \mathbf{T}_1 \longrightarrow \mathbf{T}_2$, a set of operations in $\mathbf{T}_1$ might (i) include into a set of operations in $\mathbf{T}_2$, in which case $\mathbf{T}_2$ has more operations than $\mathbf{T}_1$. Or G might (ii) take two different operations in $\mathbf{T}_1$ to a single operation in $\mathbf{T}_2$, which means $\mathbf{T}_2$ will have to satisfy more axioms than $\mathbf{T}_1$. These two cases are related to the two cases in Remark 4.1.18 on page 146. These cases correspond to when we have (i) forgetful functors and (ii) inclusion functors. In general, a theory morphism will have both of these types of mappings of operations.

Example 8.1.11. Here are several examples of theory morphisms:

- There is an inclusion of the theory of abelian groups into the theory of rings. The theory of rings is the theory of abeilian groups with another binary operation and another unit. The binary operations have to properly distribute as in Equation (2.9).
- Similarly, there is an inclusion of the theory of monoids into the theory of groups.
- There is a surjection of the theory of groups onto the theory of abelian groups. The two operations $+\colon 2 \longrightarrow 1$ and $+ \circ br\colon 2 \longrightarrow 2 \longrightarrow 1$ (br is a twisting operation that swaps its two elements) in the theory of groups have to go to the same operation $+\colon 2 \longrightarrow 1$ in the theory of abelian groups.
- Similarly, there is a surjection of the theory of monoids onto the theory of commutative monoids. □

Exercise 8.1.12. Show that for every algebraic theory $\mathbf{T}$, there is a unique inclusion from the theory of sets $\mathbf{T}_{Set}$ into $\mathbf{T}$. In other words, every theory $\mathbf{T}$ is basically $\mathbf{T}_{Set}$, which can have more operations. ■

Theory morphisms can be composed, and there is a notion of a identity theory morphism.

[1]There are actually three fields of mathematics that study structures from a general point of view: (1) functorial semantics, which is part of category theory; (2) model theory, which is part of logic; and (3) universal algebra, which is part of algebra. Each field has their own specialty that makes it important. Functorial semantics specialty is relationships between different types of structures.

Definition 8.1.13. The collection of algebraic theories and theory morphisms form the category of theories, denoted as $\mathbb{Theory}$.

Exercise 8.1.12 shows that $\mathbf{T}_{Set}$ is an initial object in this category.

Definition 8.1.14. Given $G\colon \mathbf{T}_1 \longrightarrow \mathbf{T}_2$ and a $\mathbf{T}_2$-algebra $F : \mathbf{T}_2 \longrightarrow \mathbb{Set}$, we can compose and get a $\mathbf{T}_1$ algebra $F \circ G$. Hence, G induces the functor

$$G^* = Alg(G, \mathbb{Set})\colon Alg(\mathbf{T}_2, \mathbb{Set}) \longrightarrow Alg(\mathbf{T}_1, \mathbb{Set}). \tag{8.9}$$

Functors of the form G^* are called **algebraic functors**.

This should remind the reader of our discussion of induced functors involving exponentials from Section 4.5. After all, a $G\colon \mathbf{T}_1 \longrightarrow \mathbf{T}_2$ induces $G^*\colon \mathbb{Set}^{\mathbf{T}_2} \longrightarrow \mathbb{Set}^{\mathbf{T}_1}$.

Example 8.1.15. Let us examine what algebraic functors are induced by the theory morphisms described in Example 8.1.11:

- The inclusion theory morphism $G\colon \mathbf{T}_{AbGroup} \longrightarrow \mathbf{T}_{Ring}$ induces the forgetful functor $G^*\colon \mathbb{Ring} \longrightarrow \mathbb{AbGp}$.
- The inclusion theory morphism $G\colon \mathbf{T}_{Monoid} \longrightarrow \mathbf{T}_{Group}$ induces the forgetful functor $G^*\colon \mathbb{Group} \longrightarrow \mathbb{Monoid}$.
- The surjection theory morphism $G\colon \mathbf{T}_{Group} \longrightarrow \mathbf{T}_{AbGroup}$ induces the inclusion functor $G^*\colon \mathbb{AbGp} \longrightarrow \mathbb{Group}$.
- The surjection theory morphism $G\colon \mathbf{T}_{Monoid} \longrightarrow \mathbf{T}_{common}$ induces the inclusion functor $G^*\colon \mathbb{ComMonoid} \longrightarrow \mathbb{Monoid}$.

Notice that there is a duality going on here. What is an inclusion in $\mathbb{Theory}$ is mirrored by a surjection of categories of algebras. On the other hand, a surjection in $\mathbb{Theory}$ mirrors an injection of categories of algebras. □

One of the main theorems of functorial semantics is the following:

Theorem 8.1.16. Every theory morphism $G\colon \mathbf{T}_1 \longrightarrow \mathbf{T}_2$ induces an adjunction:

$$Alg(\mathbf{T}_2, \mathbb{Set}) \underset{G^*}{\overset{G_*}{\rightleftarrows}} \perp \; Alg(\mathbf{T}_1, \mathbb{Set}). \tag{8.10}$$

That is, every algebraic functor G^* has a left adjoint G_*. ★

Proof. We will not go through all the details, but suffice it to say that category theory provides us with a mechanism for constructing the left adjoint to every algebraic functor in a unified and clear way. Given $G\colon \mathbf{T}_1 \longrightarrow \mathbf{T}_2$ and a $\mathbf{T}_1$-algebra $F\colon \mathbf{T}_1 \longrightarrow \mathbb{Set}$, one completes the following diagram:

with a special type of limit called a "Kan extension." We will meet Kan extensions in Section 9.2. The functor $G_*(F)$ preserves products and is a $\mathbf{T}_2$-algebra. ♣

Example 8.1.17. Let us examine the left adjoints to the algebraic functors described in Example 8.1.15:

- The left adjoint of $G^*\colon \mathbb{R}\mathrm{ing} \longrightarrow \mathbb{A}\mathrm{b}\mathbb{G}\mathrm{p}$ is a functor that takes an abelian group to the free ring over it.
- The left adjoint of $G^*\colon \mathbb{G}\mathrm{roup} \longrightarrow \mathbb{M}\mathrm{onoid}$ is a functor that takes a monoid and freely adds an inverse to every element making it a group.
- The left adjoint of $G^*\colon \mathbb{A}\mathrm{b}\mathbb{G}\mathrm{p} \longrightarrow \mathbb{G}\mathrm{roup}$ takes every group H to the quotient group $H/\{ab - ba\}$, where a and b are any elements iin H. Such a functor is called an **abelianization functor**.
- The left adjoint of $G^*\colon \mathbb{C}\mathrm{om}\mathbb{M}\mathrm{onoid} \longrightarrow \mathbb{M}\mathrm{onoid}$ takes every monoid M to the quotient monoid $M/\{ab - ba\}$, where a and b are any elements in M (i.e., an abelianization functor for monoids). □

Since the theory of sets is the initial object in $\mathbb{T}\mathrm{heory}$, for all theories $\mathbf{T}$, there is a unique theory morphism $\mathbf{T}_{Set} \hookrightarrow \mathbf{T}$. This theory morphism induces a forgetful functor:

$$U\colon Alg(\mathbf{T}, \mathbb{S}\mathrm{et}) \longrightarrow Alg(\mathbf{T}_{Set}, \mathbb{S}\mathrm{et}) \cong \mathbb{S}\mathrm{et}. \tag{8.12}$$

This algebraic functor take an algebra to its underlying set (i.e., it forgets all the operations). The left adjoint of this functor takes every set to the free algebra of that type.

We have shown that $Alg(\ ,\mathbb{S}\mathrm{et})$ is a functor that takes an algebraic theory to a category of algebras, and this category of algebras has a forgetful functor to $\mathbb{S}\mathrm{et}$. The functor $Alg(\ ,\mathbb{S}\mathrm{et})$ is contravariant for the same reason that $Hom(\ ,a)$ is contravariant (Example 4.1.32). Thus, we have a functor:

$$Alg(\ ,\mathbb{S}\mathrm{et})\colon \mathbb{T}\mathrm{heory}^{op} \longrightarrow \mathbb{C}\mathrm{at}/\mathbb{S}\mathrm{et}. \tag{8.13}$$

Remarkably, this functor has a left adjoint. We will describe the significance of this adjoint presently.

Algebraic Theory Reconstruction

We have seen how to go from an algebraic theory $\mathbf{T}$ to its category of algebras $Alg(\mathbf{T}, \mathbb{S}\mathrm{et})$. For example, we can go from the theory of groups $\mathbf{T}_{Group}$ to the category of groups with its forgetful functor to $\mathbb{S}\mathrm{et}$, $U\colon \mathbb{G}\mathrm{roup} \longrightarrow \mathbb{S}\mathrm{et}$. Functorial semantics shows us how to go the other way. We can go from a category of algebras with its forgetful functor to $\mathbb{Set}$ and reconstruct the algebraic theory that gives that category. That is, we can go from $U: \mathbb{Group} \longrightarrow \mathbb{Set}$ and form the algebraic theory $\mathbf{T}_{Group}$.

Technical Point 8.1.18. There might be many algebraic theories that describe equivalent categories of algebras. That is, there might be $\mathbf{T}_1 \neq \mathbf{T}_2$ but $Alg(\mathbf{T}_1, \mathbb{Set}) \simeq Alg(\mathbf{T}_2, \mathbb{Set})$. Such theories are called **Morita equivalent**. The process that we are presenting here is a way of forming an algebraic theory for a category of algebras. We might say this is the "correct" description because it is the one that is formed in an organic way from the category of algebras with its forgetful functor to the category of sets. ♡

Let us get into the details now. Consider the slice category $\mathbb{Cat}/\mathbb{Set}$. The objects of this category are of the form $U: \mathbb{C} \longrightarrow \mathbb{Set}$. Call such a functor **tractable** if for all natural numbers m and n, the collection of natural transformations $Nat(U^m, U^n)$ form a set. Here, $U^m: \mathbb{C} \longrightarrow \mathbb{Set}$ is a functor that takes $c \in \mathbb{C}$ to the mth product of $U(c)$, written as $(U(c))^m$. There is a full subcategory of $\mathbb{Cat}/\mathbb{Set}$ that consists of such tractable functors denoted as $Tract(\mathbb{Cat}/\mathbb{Set})$. The functor $Alg(\ \ , \mathbb{Set})$ takes values in $Tract(\mathbb{Cat}/\mathbb{Set})$.

There is an a left adjoint to $Alg(\ \ , \mathbb{Set})$ called

$$Struct: Tract(\mathbb{Cat}/\mathbb{Set}) \longrightarrow \mathbb{Theory}^{op} \tag{8.14}$$

which takes a tractable functor $U: \mathbb{C} \longrightarrow \mathbb{Set}$ and forms the algebraic theory $Struct(U) = \mathbf{T}_U$. The objects of $\mathbf{T}_U$ are, of course, the natural numbers. The morphisms of $\mathbf{T}_U$ are defined as

$$Hom_{\mathbf{T}_U}(m, n) = Nat(U^m, U^n), \tag{8.15}$$

which by tractability form a set. Start with an algebraic theory $\mathbf{T}$ and then take $U: Alg(\mathbf{T}, \mathbb{Set}) \longrightarrow \mathbb{Set}$. The reconstructed theory $Struct(U)$ is isomorphic to $\mathbf{T}$. This means that $Struct \circ Alg(\ \ , \mathbb{Set}) \cong Id_{\mathbb{Theory}^{op}}$ but $Alg(\ \ , \mathbb{Set}) \circ Struct$ might be very different than Id. However, as with all adjunctions, there are subcategories that form an equivalence of categories (Theorem 4.4.23). In this case, it is as follows:

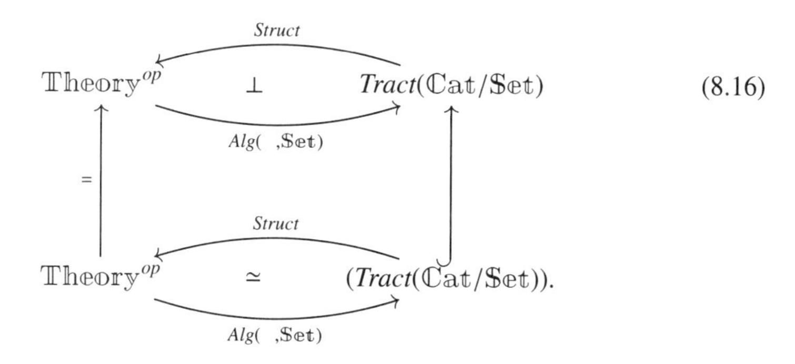

(8.16)

The category $(Tract(\mathbb{Cat}/\mathbb{Set}))$ is the full subcategory of $Tract(\mathbb{Cat}/\mathbb{Set})$ consisting of those objects that are in the essential image of $Alg(\ \ , \mathbb{Set})$.

Here is the intuition as to why this reconstruction works. Let an algebraic theory have an operation $\oplus\colon m \longrightarrow n$, and let the sets A and B have the structure described by this algebraic theory. This means that there are functions $\oplus\colon A^m \longrightarrow A^n$ and $\oplus'\colon B^m \longrightarrow B^n$. A homomorphism from algebra A to algebra B is a set function $\alpha\colon A \longrightarrow B$, which makes the following diagram commute:

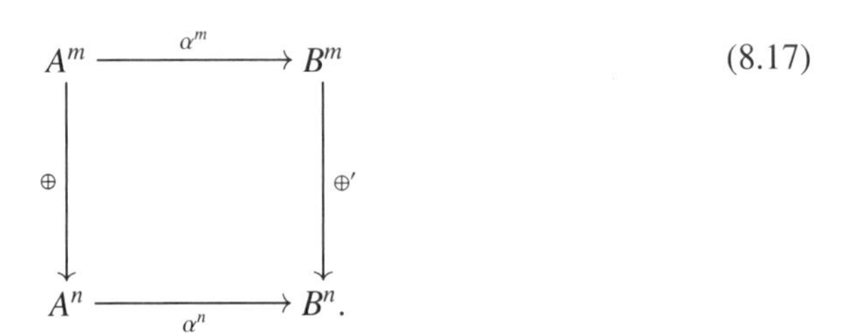

(8.17)

We can describe this commutative square with the following motto:

<u>A homomorphism respects the operation.</u>

There is, however, another way to look at this commutative square. We can just as easily say that the commutative square means the following:

<u>An operation respects the homomorphism.</u>

In other words, to find the operations, examine all the set functions that respect the homomorphisms. Such maps will be the reconstructed operations of the algebraic theory. A natural transformation $\alpha\colon U^m \Longrightarrow U^n$ at component A is a set function $\alpha_A : U(A)^m \longrightarrow U(A)^n$. The naturality condition means that for a homomorphism of algebras $f\colon A \longrightarrow B$, the usual square commutes.

This reconstruction gives a beautiful duality between theories (syntax, the descriptions of structure) and algebras (semantics, the actual structures).

(This idea of reconstructing the theory by looking at the forgetful functor is the beginning of an important part of quantum groups called **Tannaka-Krein duality**.)

Generalizations

Functorial semantics has been around for more than fifty years and has been generalized in many directions. With these generalizations, the ideas of functorial semantics have become applicable in many areas of mathematics, physics, and computer science.

The first generalization comes from using different categories instead of $(\mathbb{S}\mathrm{et}, \times, \{*\})$. The only property of $\mathbb{S}\mathrm{et}$ that we used is that it has finite products. Any other category with finite products can also be used. In other words, an algebraic theory can take algebras in an arbitrary Cartesian category $(\mathbb{C}, \times, t)$, where t is the terminal object. This gives us many more contexts in which we can describe algebraic structures. Any product-preserving Cartesian functor from $\mathbf{T}$ to $\mathbb{C}$ gives us an algebra in $\mathbb{C}$. The collection of all algebras and their morphisms gives us the category $Alg(\mathbf{T}, \mathbb{C})$.

For example, consider $\mathbf{T}_{Monoid}$ the theory of monoids, and $\mathbb{C}$ a category with products. A product-preserving functor $F\colon \mathbf{T}_{Monoid} \longrightarrow \mathbb{C}$ describes a **monoid object** in the category $\mathbb{C}$. In particular, $F\colon \mathbf{T}_{Monoid} \longrightarrow \mathbb{T}\mathrm{op}$ describes a topological monoid, which is a space X, a multiplication that is a continuous map $m\colon X \times X \longrightarrow X$, and a special object u in X picked out by the continuous map $\upsilon\colon \{*\} \longrightarrow X$. This extra structure of the topological space makes the following diagrams commute:

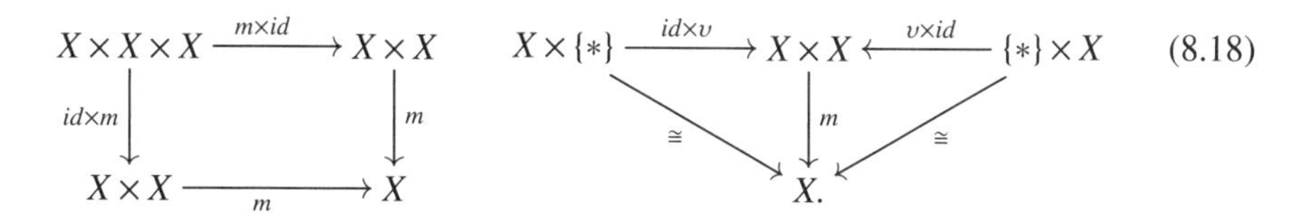

These diagrams look very much like the definition of a monoid.

Researchers go further by generalizing from a Cartesian category $(\mathbb{C}, \times, t)$ to a strictly associative monoidal category $(\mathbb{C}, \otimes, I)$. In this case, algebras are strict monoidal functors from $\mathbf{T}$ to $\mathbb{C}$ and the collection of all such algebras and morphisms between them gives us a category also denoted as $Alg(\mathbf{T}, \mathbb{C})$. A strrict monoidal functor from $\mathbf{T}_{Monoid}$ to a monoidal category $(\mathbb{C}, \otimes, I)$ describes a monoid object in the monoidal category $\mathbb{C}$. In the mini-course in quantum groups (Section 7.4,) we saw that a monoid object in complex vector spaces, $(\mathbf{C}\mathbb{V}\mathrm{ect}, \otimes, \mathbf{C})$, is called an **algebra** (a word that has way too many different meanings). Such algebras satisfy requirements that can be described as the following—now familiar—commutative diagrams:

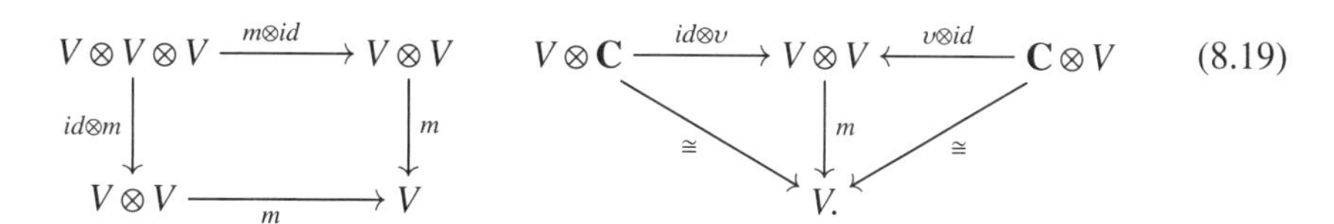

We can describe many algebraic structures within any monoidal category. We have already seen many examples of algebraic objects within a monoidal category, but we did not call them that. Let us call out a few examples:

Example 8.1.19.

- A monoid (commutative monoid, group, abelian group, ring) object in $(\mathbb{S}\mathrm{et}, \times, \{*\})$ is a monoid (commutative monoid, group, abelian group, ring).
- A (commutative) monoid object in $(\mathbb{A}\mathrm{b}\mathbb{G}\mathrm{p}, \otimes, \mathbf{Z})$ is a (commutative) ring.
- A monoid object in $(\mathbb{M}\mathrm{onoid}, \otimes, 1)$ is a commutative monoid. (This is the content of Theorem 5.1.16.)
- A monoid object in $(\mathbb{C}\mathrm{at}, \times, \mathbf{1})$ is a strict monoidal category.
- A commutative monoid object in $(\mathbb{C}\mathrm{at}, \times, \mathbf{1})$ is a strictly symmetric strict monoidal category.
- A bialgebra is a monoid object in the opposite category of algebras.
- A Hopf algebra is a group object in the opposite category of algebras. □

This terminology ("an x object in the category y") is used in the literature and will be employed in the rest of this text.

Another way of generalizing algebraic theories is to discuss **multisorted theories**, which are sometimes called **colored theories**. Such an algebraic theory is a category whose objects are the products of different elements. So there might be $a^0, a^1, a^2, \ldots,$ $b^0, b^1, b^2, \ldots, c^0, c^1, c^2, \ldots,$ etc. Algebras (in, e.g., $\mathbb{Set}$) for such theories will have a go to one set, b go to another set, and c go to yet another set. A typical operation might be represented by a function $f\colon b^4 \longrightarrow a^2 \times c^7$. This permits all types of structures.

There are also ways of combining descriptions of algebraic structures. That is, if $\mathbf{T}_1$ is a description of one type of algebraic structure and $\mathbf{T}_2$ is a description of another type of algebraic structure, then we can combine $\mathbf{T}_1$ and $\mathbf{T}_2$ to get a description of a type of algebraic structure that contains both structures. This entails a monoidal category structure on the category of theories. There are, in fact, two such monoidal category structures. We start here with the simpler one.

Definition 8.1.20. The **sum of theories** or **coproduct of theories** is the algebraic theory that contains the descriptions of both structures. Let $\mathbf{T}_1$ and $\mathbf{T}_2$ be two algebraic theories, then $\mathbf{T}_1 \oplus \mathbf{T}_2$ is the theory obtained by taking the pushout of the following diagram:

$$\begin{array}{ccc} \mathbf{T}_{Set} & \xrightarrow{inc} & \mathbf{T}_1 \\ {\scriptstyle inc}\downarrow & & \downarrow \\ \mathbf{T}_2 & \longrightarrow & \mathbf{T}_1 \oplus \mathbf{T}_2. \end{array} \qquad (8.20)$$

As always, $\mathbf{T}_{Set}$ includes into $\mathbf{T}_1 \oplus \mathbf{T}_2$ along with all the extra operations of $\mathbf{T}_1$ and the extra operations of $\mathbf{T}_2$. Applying the contravariant algebra functor $Alg(\ \ , \mathbb{Set})$ to Diagram (8.20) gives us the pullback

$$\begin{array}{ccc} Alg(\mathbf{T}_{Set}, \mathbb{Set}) \cong \mathbb{Set} & \xleftarrow{U} & Alg(\mathbf{T}_1, \mathbb{Set}) \\ {\scriptstyle U}\uparrow & & \uparrow \\ Alg(\mathbf{T}_2, \mathbb{Set}) & \longleftarrow & Alg(\mathbf{T}_1 \oplus \mathbf{T}_2, \mathbb{Set}). \end{array} \qquad (8.21)$$

This means that an algebra for $\mathbf{T}_1 \oplus \mathbf{T}_2$ in $\mathbb{Set}$ will be a single set with both structures.

The sum of theories construction makes the category of theories into a monoidal category $(\mathbb{Theory}, \oplus, \mathbf{T}_{Set})$.

There is another more important monoidal structure on the category of algebraic theories.

Definition 8.1.21. The **Kronecker product of algebraic theories** combines two theories but insists that each operation of one theory respects the operations of the other theory. Let $\mathbf{T}_1$ and $\mathbf{T}_2$ be two algebraic theories whose sum is formed as $\mathbf{T}_1 \oplus \mathbf{T}_2$. Now form the quotient category by making the following congruence relation: every operation in one algebraic theory must respect (i.e., be a homomorphism of) every operation in the other algebraic theory (i.e., for $f: m \longrightarrow n$, a morphism in $\mathbf{T}_1$, and $g: m' \longrightarrow n'$, and a morphism in $\mathbf{T}_2$), we insist that the following diagram commutes:

$$\begin{array}{ccc} m \cdot m' & \xrightarrow{g^m} & m \cdot n' \\ {\scriptstyle f^{m'}}\downarrow & & \downarrow{\scriptstyle f^{n'}} \\ n \cdot m' & \xrightarrow[g^n]{} & n \cdot n'. \end{array} \tag{8.22}$$

(Actually, the diagram is a little more complicated, but we do not have to go into detail here.) With this congruence relation, we form the algebraic theory $\mathbf{T}_1 \otimes \mathbf{T}_2 = (\mathbf{T}_1 \oplus \mathbf{T}_2)/\sim$. This is called the "Kronecker product of algebraic theories" because it is similar to the Kronecker product of matrices, which we saw in Figure 5.2 where every element of one matrix is multiplied by every element of the other matrix. Analogously, here we insist that every operation of one theory is a homomorphism of the other theory.

What do the algebras of $\mathbf{T}_1 \otimes \mathbf{T}_2$ in $\mathbb{Set}$ look like? The operations of $\mathbf{T}_1$ will induce homomorphisms in the category $Alg(\mathbf{T}_2, \mathbb{Set})$. Similarly, the operations of $\mathbf{T}_2$ will induce homomorphisms in $Alg(\mathbf{T}_1, \mathbb{Set})$. In other words,

$$Alg(\mathbf{T}_1 \otimes \mathbf{T}_2, \mathbb{Set}) \cong Alg(\mathbf{T}_1, Alg(\mathbf{T}_2, \mathbb{Set})) \cong Alg(\mathbf{T}_2, Alg(\mathbf{T}_1, \mathbb{Set})). \tag{8.23}$$

This is similar to the usual statement about exponentiation that we saw about sets in Theorem 1.4.26. In terms of categories and functors, this becomes

$$\mathbb{Set}^{\mathbf{T}_1 \otimes \mathbf{T}_2} \cong (\mathbb{Set}^{\mathbf{T}_1})^{\mathbf{T}_2} \cong (\mathbb{Set}^{\mathbf{T}_2})^{\mathbf{T}_1}. \tag{8.24}$$

However, here we are not looking at all functors. Rather, we are only looking at product-preserving functors.

In summary, there is a symmetric monoidal category $(\mathbb{Theory}, \otimes, \mathbf{T}_{Set})$.

Thus, we have shown that not only can structure be described with categories, but there are at least two ways of combining such descriptions to get other descriptions.

A **sketch** is similar to an algebraic theory, but it has more structure than just products. Algebras are functors from sketches to other categories that preserve the structure. For example, we can ask our sketch to have finite limits. There is a whole hierarchy of such sketches, one more powerful than the other. A nice survey is given by Charles Wells [261]. Many different types of sketches are described in Barr and Wells ([33] and [32]). See also [148].

Example 8.1.22. Let us describe the **sketch of categories**. What is a category? A category can be described by two collections. The collection of objects will be represented by C_0, while the collection of morphisms will be represented by C_1. (The fact that there is more than one collection means that the sketch of categories is **two-sorted**;

that is, there are two sorts of elements to deal with. This is in contrast to the single-sorted theories that we looked at earlier.) There are three morphisms between these representatives:

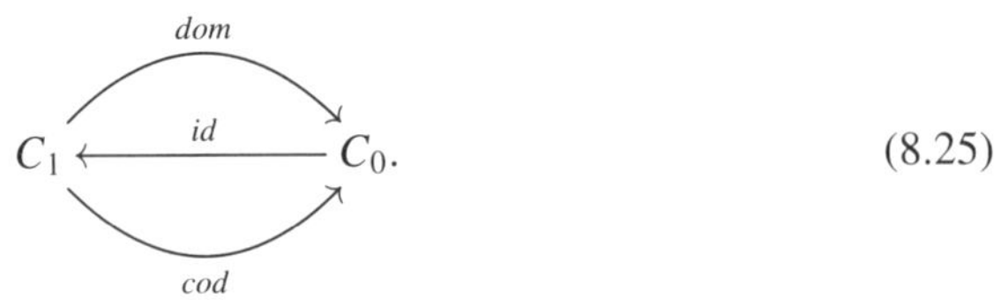

(8.25)

These morphisms correspond to the domain of an arrow, the codomain of an arrow (which is similar to what we saw in Example 4.5.4, where we saw how to represent graphs), and the morphism *id*, which corresponds to assigning an identity morphism to every object.

Given C_1, C_0, dom, cod, and *id*, how does one describe the composition of morphisms? The wrong answer to this question is that composition corresponds to a morphism $\circ\colon C_1 \times C_1 \longrightarrow C_1$; that is, composition takes two morphisms and composes them to get one morphism. This is wrong because such a function would take *every* pair of morphisms and compose them. However, for a category, we can compose only composable morphisms. The correct answer is to first talk about composable morphisms. This set of composable morphisms can be described by the following pullback:

$$\begin{array}{ccc} C_1 \times_{C_0} C_1 & \longrightarrow & C_1 \\ \downarrow & & \downarrow{\scriptstyle dom} \\ C_1 & \xrightarrow[cod]{} & C_0. \end{array} \tag{8.26}$$

The object $C_1 \times_{C_0} C_1$ corresponds to pairs of morphisms (g, f) such that $dom(g) = cod(f)$. With this at hand, we can then describe composition as $\circ\colon C_1 \times_{C_0} C_1 \longrightarrow C_1$. Associativity of composition entails first creating $C_1 \times_{C_0} C_1 \times_{C_0} C_1$, which corresponds to triples of composable morphisms, and then insisting that the following diagram commutes:

$$\begin{array}{ccc} C_1 \times_{C_0} C_1 \times_{C_0} C_1 & \xrightarrow{id \times \circ} & C_1 \times_{C_0} C_1 \\ {\scriptstyle \circ \times id}\downarrow & & \downarrow{\scriptstyle \circ} \\ C_1 \times_{C_0} C_1 & \xrightarrow[\circ]{} & C_1. \end{array} \tag{8.27}$$

We leave proving that the identity maps act like units to the reader.

The main point is that we needed pullbacks (not just products) to describe the structure of categories. This indicates that categories are a more sophisticated notion of structure than the usual algebraic structures (group, monoids, rings, etc.).

With the sketch of categories in our toolbox, we can talk about categories within various categories. A category will be a pullback-preserving functor from the sketch of categories (suitably defined) to a category $\mathbb{A}$ that has pullbacks. Categories in category $\mathbb{Set}$ give us categories where the collection of objects and the collection of morphisms are sets. We can examine categories in $\mathbb{T}\mathrm{op}$ called "topological categories" and categories in various other categories. We can even look at categories in the category of ... categories. This gives us the rich notion of a **double category**. □

Exercise 8.1.23. Give a commutative diagram to describe what it means for the identity morphisms to act like units to composition. ■

Exercise 8.1.24. Describe the single sorted sketch for categories as they are defined in Definition 2.1.61. ■

There are many more generalizations. We mention some of them here:

- While usual algebraic operations are finite, one can generalize and form algebraic theories with infinitary operations. This was first formulated by Fred Linton [172].
- In an algebraic theory, every Hom set $Hom_{\mathbf{T}}(m, n)$ is a set of operations. There are **enriched theories**, which are theories where every Hom set is an object in some other category. For example, the Hom sets might belong to $\mathbb{T}\mathrm{op}$, and then there will be a whole topological space of operations. In Section 8.4, we will discuss theories where the Hom sets are categories. Such theories are called **2-theories**. We will see more about enriched categories in Section 9.1 of Chapter 9.
- Computer scientists have long since coopted algebraic theories for many purposes. (Wagner [257] is a survey article about **ordered theories**, **iteration theories**, **rational theories**, **iterative theories**, and other topics.) Such generalizations have been used in context-free grammars, flowchart semantics, recursion schemata, algebraic databases, recursively defined domains, and other fields.)
- At the end of the next section, we will meet even more generalizations such as **PROPs** (for "PROduct and Permutation categories") and **colored PROPs**. These are like algebraic theories, but they are based on monoidal categories, not simply on Cartesian categories.

Suggestions for Further Study

There are many places to read about Lawvere's thesis and a follow-up paper[159][161], which have been reprinted by TAC and have a new foreword by the author. See [11] and Chapter 3 of volume 2 of Francis Borceux [41]. See also the many places mentioned in the generalizations.

8.2 Operads

Operads are another way of describing structure. They show how to combine operations. However, we will see that operads are applicable in many areas.

Basic Definitions

First, a motivating example. Consider the operation

$$(a \div b) \cdot (c - (6d)). \tag{8.28}$$

This operation takes four real numbers and outputs one real number. We say that this function has arity 4 and can be viewed as a tree with four leaves:

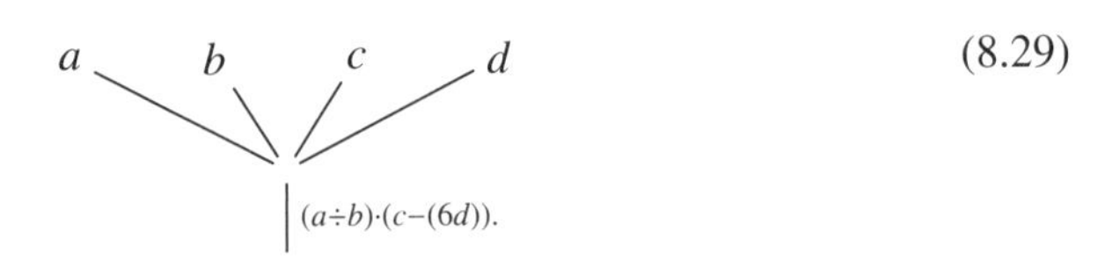

Now consider the function with 3 inputs $r + s \cdot t$, which we can write as

We can substitute this operation into the c of the original operation and get

$$(a \div b) \cdot ((r + s \cdot t) - (6d)), \tag{8.31}$$

which is an operation that takes $4 + 3 - 1 = 6$ inputs. This can be depicted as a combination of the two previous trees to get

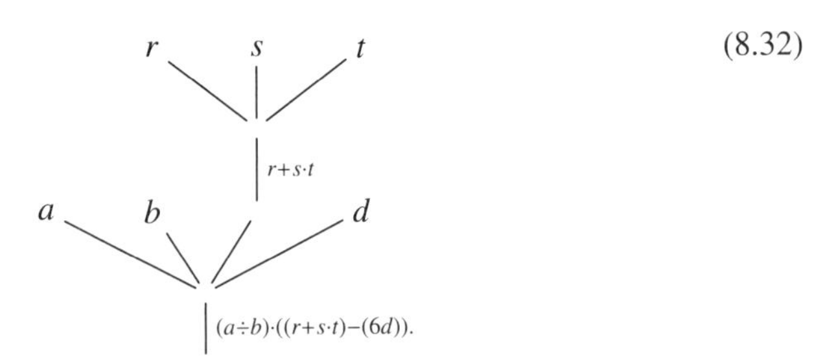

This is what operads are all about. Operads are formal ways of showing how operations are combined. In this example, we inserted one operation into another operation. We are going to talk about many operations that have different arities and combine them. An operad is going to consist of a sequence of collections $O(0), O(1), O(2), O(3), \ldots$. Each $O(i)$ consists of those operations of arity i. In this example, we took an element of $O(4)$ and combined it with an element of $O(3)$ to form an element of $O(6)$. We combined it by replacing the third variable (c) of of an operation in $O(4)$. We could replace other variables with other operations as well. For each $O(4)$ operation and each $O(3)$ operation, there is a $O(6)$ operation that tells what happens when these operations are combined. This entails a function of the following form:

$$\circ_3 : O(4) \times O(3) \longrightarrow O(6). \tag{8.33}$$

In general, there are functions that take an element of $O(n)$ and an element of $O(j)$ for $j \leq n$ and output an element of $O(n+j-1)$. These functions have to satisfy certain axioms.

For the formal definition of an operad, we will generalize our motivating example in two ways:

1. Rather than thinking of $O(n)$ as a set, think of it as an object in a category $\mathbb{A}$.

2. Furthermore, rather than taking the Cartesian product $\times$ of the objects in the category, consider some strict monoidal category $(\mathbb{A}, \otimes, I)$ and take the monoidal product.

We are thus led to a simple definition of an operad.

Definition 8.2.1. An **operad** O in a monoidal category $(\mathbb{A}, \otimes, I)$ consists of the following:

- A sequence of objects in category $\mathbb{A}$: $O(0), O(1), O(2), O(3), \ldots$.
- For all n, $1 \leq i \leq n$, and $1 \leq j \leq n$, there are morphisms

$$\circ_i : O(n) \times O(j) \longrightarrow O(n+j-1). \tag{8.34}$$

- There is a map in $\mathbb{A}$, $\upsilon \colon I \longrightarrow O(1)$ that picks out an element id of $O(1)$, which plays the role of the identity.

These objects and morphisms must satisfy the following requirements:

- An associativity axiom for doing both a $\circ_i$ and a $\circ_j$ operation. This means that for all A operations in $O(n)$, B operations in $O(m)$, and C operations in $O(p)$, we have

$$(A \circ_j B) \circ_i C = \begin{cases} (A \circ_i C) \circ_{j+p-1} B & \text{if } 1 \leq i \leq j-1. \\ A \circ_j (B \circ_{i-j+1} C) & \text{if } j \leq i \leq m+j-1. \\ (A \circ_{i-m+1} C) \circ_j B & \text{if } i \geq m+j. \end{cases} \tag{8.35}$$

  These three cases are explained in Figures 8.2, 8.3, and 8.4.
- Composition must respect the identity: for every A operation in $O(m)$, and for every $1 \leq i \leq m$, we have $A \circ_i id = A = id \circ_1 A$. These equations are explained in Figure 8.5.
- In the literature, some require that the function $\circ_i$ essentially remains the same if the entries are permuted. We do not make that requirement. Here, that is a special type of operad called a "symmetric operad."

Technical Point 8.2.2. While this definition is satisfactory, the literature has another, more sophisticated but equivalent, definition of an operad. Rather than just substituting

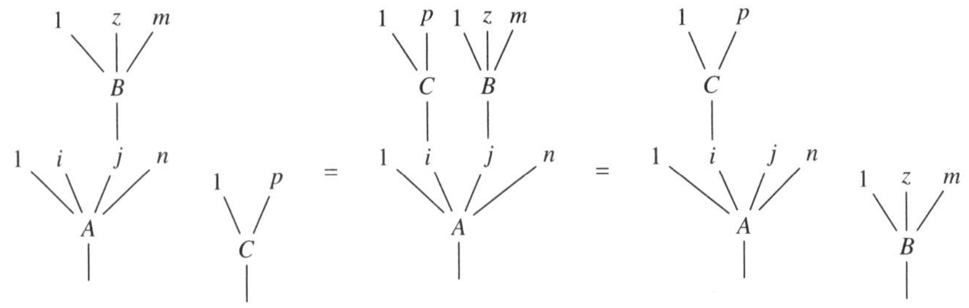

$$(A \circ_j B) \circ_i C = (A \circ_i C) \circ_{j+p-1} B$$

Figure 8.2. The first case of the associativity axiom for operads.

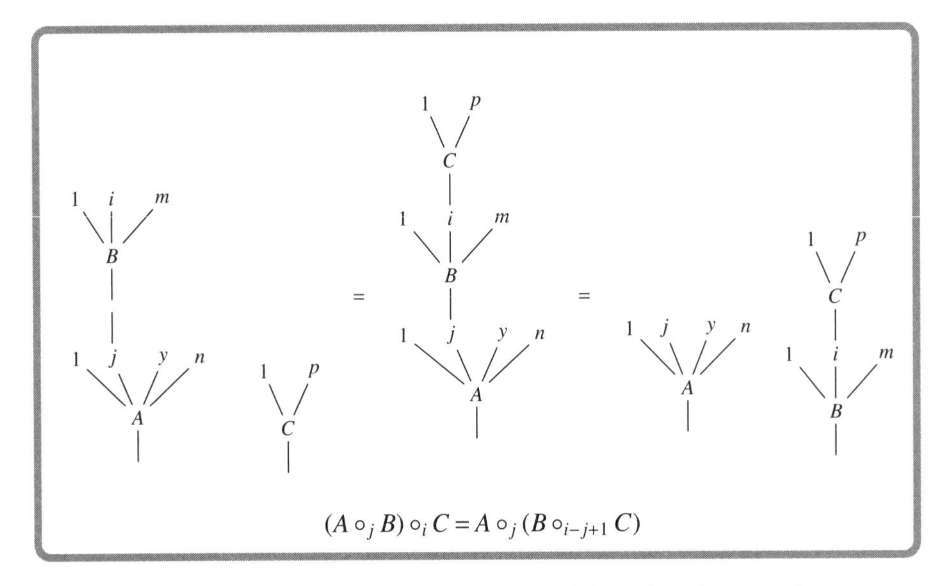

Figure 8.3. The second case of the associativity axiom for operads.

one of the inputs into $O(n)$, let us allow for substituting all the inputs. For example, consider again Equation (8.28). Rather than substituting a single operation for c, substitute all the operations in the following way:

- For a, substitute $x-(3y)$, an operation that accepts two inputs.
- For b, substitute $r \cdot s \cdot t$, an operation that accepts three inputs.
- For c, substitute 5, an operation that accepts no inputs.
- For d, substitute $(((g \div h)+i)-5j) \cdot k$, an operation that accepts five inputs.

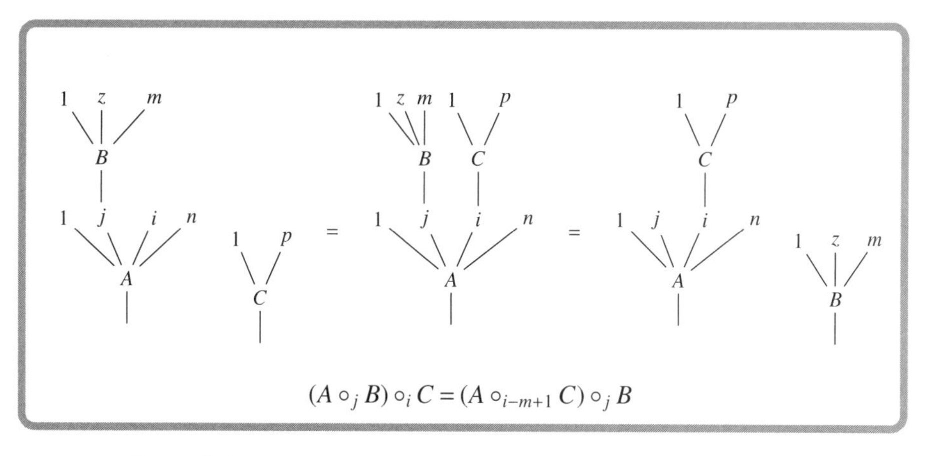

Figure 8.4. The third case of the associativity axiom for operads.

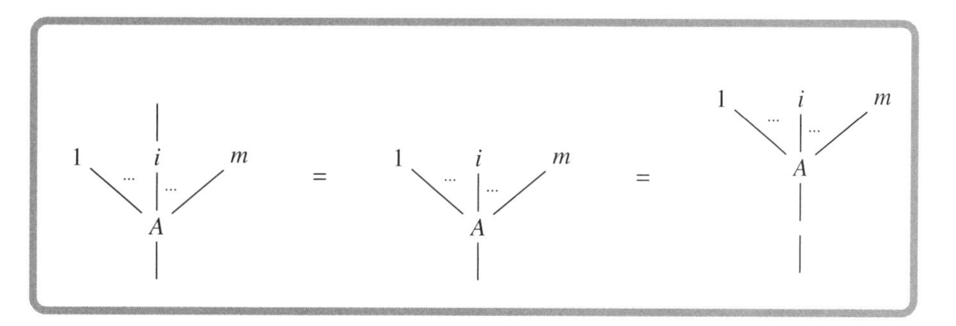

Figure 8.5. The identity axiom for an operad: $A \circ_i id = A = id \circ_1 A$.

After the substitution, we get the following operation:

$$(x - (3y)) \div (r \cdot s \cdot t)) \cdot (5 - (6((((g \div h) + i) - 5j) \cdot k))). \tag{8.36}$$

This operation takes $2 + 3 + 0 + 5 = 10$ inputs and outputs a number. What we did here was take an operation of $\mathcal{O}(4)$ and operations from $\mathcal{O}(2)$, $\mathcal{O}(3)$, $\mathcal{O}(0)$, and $\mathcal{O}(5)$ and produce an operation from $\mathcal{O}(2 + 3 + 0 + 5) = \mathcal{O}(10)$. This entails the function

$$\circ\colon \mathcal{O}(4) \times (\mathcal{O}(2) \times \mathcal{O}(3) \times \mathcal{O}(0) \times \mathcal{O}(5)) \longrightarrow \mathcal{O}(10). \tag{8.37}$$

Using this idea, we come to the other definition of an operad. An operad is again a sequence of objects in a category $\mathcal{O}(0), \mathcal{O}(1), \mathcal{O}(2), \mathcal{O}(3), \ldots$. Then, for every n and for every sequence of integers $j_1, j_2, j_3, \ldots, j_n$, there is the function

$$\circ\colon \mathcal{O}(n) \times \mathcal{O}(j_1) \times \mathcal{O}(j_2) \times \cdots \times \mathcal{O}(j_n) \longrightarrow \mathcal{O}(j_1 + j_2 + \ldots + j_n). \tag{8.38}$$

These composition functions must satisfy more sophisticated associativity and unity conditions, which we omit here. ♡

Just as algebraic theories have algebras, so do operads have algebras. First, let's look at a motivating example. We saw before how an operad is used to compose functions. Think of the previous motivating example as functions of real numbers. Then there is a map $\theta_n\colon O(n)\times \mathbf{R}^n \longrightarrow \mathbf{R}$ that takes the function of arity n and n real numbers and evaluates the function to output a real number.

Definition 8.2.3. Given an operad $(O, \{\circ_i\}_n, id)$ in a monoidal category $(\mathbb{A}, \otimes, I)$, an **algebra** $(a, \{\theta\}_n)$ is an object a in $\mathbb{A}$ and a family of maps:

$$\theta_n\colon O(n)\otimes a^{\otimes n} \longrightarrow a, \tag{8.39}$$

where $a^{\otimes n}$ is the tensor product of a with itself n times. These maps must satisfy the following requirements: (i) the θ cohere with the $\circ_i$, and (ii) the θ cohere with the $id \in O(1)$. These two requirements are depicted as the following commutative diagrams: for all i,

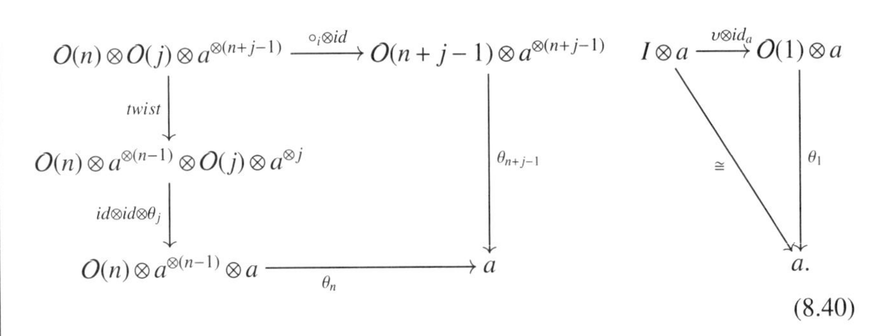

(8.40)

These diagrams are similar to Diagram (4.24), where we saw how to define an M-set.

Definition 8.2.4. Given two algebras for O, $(a, \{\theta\}_n)$ and $(a', \{\theta'\}_n)$, an **algebra homomorphism** is a map $f\colon a \longrightarrow a'$ such that

$$\begin{array}{ccc} O(n)\otimes a^{\otimes n} & \xrightarrow{id_{O(n)}\otimes f^{\otimes n}} & O(n)\otimes a'^{\otimes n} \\ \Big\downarrow{\scriptstyle \theta_n} & & \Big\downarrow{\scriptstyle \theta'_n} \\ a & \xrightarrow[f]{} & a'. \end{array} \tag{8.41}$$

This diagram is similar to the diagram in Example 4.2.4, where we saw the definition of a homomorphism of M-sets.

As expected, operads do not stand by themselves.

Definition 8.2.5. Let $(O, \{\circ_i\}_n, id)$ and $(O', \{\circ'_i\}_n, id')$ be operads in a monoidal category $(\mathbb{A}, \otimes, I)$. An **operad homomorphism** is a family of morphisms

$f_n: O(n) \longrightarrow O'(n)$ such that for all n, j, and i, we have

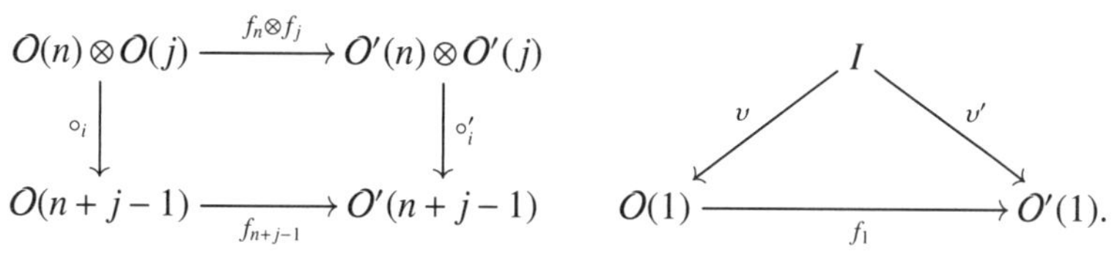

$$\begin{array}{ccc} O(n) \otimes O(j) & \xrightarrow{f_n \otimes f_j} & O'(n) \otimes O'(j) \\ \downarrow \circ_i & & \downarrow \circ'_i \\ O(n+j-1) & \xrightarrow[f_{n+j-1}]{} & O'(n+j-1) \end{array} \qquad \begin{array}{ccc} & I & \\ \swarrow_{\upsilon} & & \searrow^{\upsilon'} \\ O(1) & \xrightarrow[f_1]{} & O'(1). \end{array} \tag{8.42}$$

These two diagrams are very similar to Diagram (2.11), where we saw the definition of a homomorphism of monoids.

Definition 8.2.6. The collection of all operads and operad homomorphisms form a category called $\mathbb{Operad}$.

We will not get into it, but this category has a monoidal structure similar to monoidal structure on algebraic theories. This permits us to combine structures described by operads.

Examples

Example 8.2.7. We began this section with a motivating example consisting of functions of real numbers. Formally, this operad is in the monoidal category $(\mathbb{Set}, \times, \{*\})$. The collections are given as $O(n) = Hom_{\mathbb{Set}}(\mathbf{R}^n, \mathbf{R})$. The composition functor

$$\circ_i: Hom(\mathbf{R}^n, \mathbf{R}) \times Hom(\mathbf{R}^j, \mathbf{R}) \longrightarrow Hom(\mathbf{R}^{n+j-1}, \mathbf{R}) \tag{8.43}$$

is defined as $\circ_i(f, g) = \langle id, g, id \rangle \circ f$ as follows:

$$\mathbf{R}^{n+j-1} = \mathbf{R}^{i-1} \times \mathbf{R}^j \times \mathbf{R}^{n-i} \xrightarrow{id \times g \times id} \mathbf{R}^n \xrightarrow{f} \mathbf{R}. \tag{8.44}$$

The map $\upsilon: \{*\} \longrightarrow Hom_{\mathbb{Set}}(\mathbf{R}, \mathbf{R})$ picks out the $id_{\mathbf{R}}$. □

Example 8.2.8. Example 8.2.7 did not use any feature of real numbers. We can generalize it from $\mathbf{R}$ to any set S in $\mathbb{Set}$. The collections are

$$O(n) = Hom_{\mathbb{Set}}(S^n, S), \tag{8.45}$$

and the composition is the same as Example 8.2.7. □

Example 8.2.9. The above two examples can be generalized from the Cartesian category $\mathbb{Set}$ to any monoidal category $(\mathbb{A}, \otimes, I)$. For any object a in $\mathbb{A}$, there is an **endomorphism operad** at a, denoted as $End(a)$. The sequence of objects is given as

$$End(a)(n) = Hom_{\mathbb{A}}(a^{\otimes n}, a). \tag{8.46}$$

The composition morphism

$$\circ_i\colon Hom_{\mathbb{A}}(a^{\otimes j}, a) \otimes Hom_{\mathbb{A}}(a^{\otimes n}, a) \longrightarrow Hom_{\mathbb{A}}(a^{n+j-1}, a) \tag{8.47}$$

is defined as $\circ_i(g, f) = f \circ (id^{\otimes(i-1)} \otimes g \otimes id^{\otimes(n-i)})$. □

In Section 7.4 we saw that for an algebra A, a module can be defined as an algebra homomorphism $A \longrightarrow End(M)$. Similarly, assuming that $\mathbb{A}$ is a monoidally closed category, for an operad O, an algebra a for the operad is an operad homomorphism $f\colon O \longrightarrow End(a)$. In detail, such an f is a family of morphisms $f_n\colon O(n) \longrightarrow Hom_{\mathbb{A}}(a^{\otimes n}, a)$. By the monoidal closedness of $\mathbb{A}$, such maps are equivalent to $\hat{f}_n\colon O(n) \otimes a^{\otimes n} \longrightarrow a$, which is the usual definition of an algebra of an operad.

Example 8.2.10. As we saw in the motivating example at the beginning of this section, trees are helpful in describing operads. In fact, there is a **tree operad**. The objects $O(n)$ comprise the collection of all rooted trees with n leaves. The composition, as we saw, is given as follows:

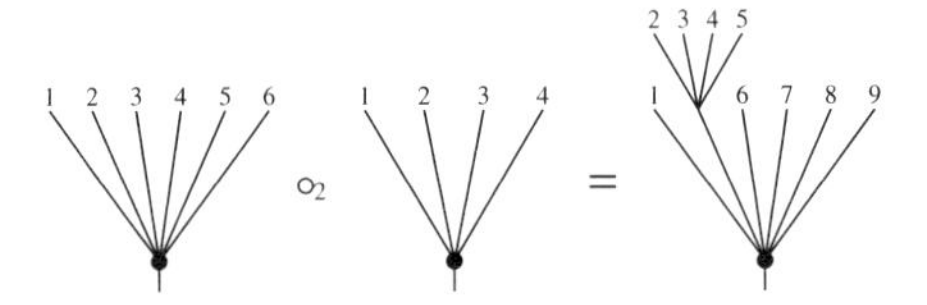

The unit is the obvious element of $O(1)$. □

Trees are one-dimensional topological spaces, and they play an important role in the theory of operads. Just as other categories of the low-dimensional topological spaces $\mathbb{Braid}$, $\mathbb{Tang}$, and $\mathbb{Ribbon}$ play important roles in describing monoidal categories with structure, the tree operad plays an important role in the category of operads.

Example 8.2.11. One can think of an operad as a collection of operations with a single output. In this context, we must examine the relationship between operads and algebraic theories. Given an algebraic theory $\mathbf{T}$, we can form an operad $O_{\mathbf{T}}$ with $O_{\mathbf{T}}(n) = Hom_{\mathbf{T}}(n, 1)$. The composition

$$\circ_i\colon Hom_{\mathbf{T}}(n, 1) \times Hom_{\mathbf{T}}(j, 1) \longrightarrow Hom_{\mathbf{T}}(n + j - 1, 1) \tag{8.48}$$

is given by the composition of $\mathbf{T}$. It must be stressed that this works only when we are dealing with an operad that uses the Cartesian product, and not a more general monoidal product.

We can also go the other way. Given a Cartesian operad O, we can form an algebraic theory $\mathbf{T}_O$, where

$$Hom_{\mathbf{T}_O}(n, m) = O(n)^{\times m}. \tag{8.49}$$

In other words, an operation $n \longrightarrow m$ is the same as m operations of the form $n \longrightarrow 1$.

This forms a correspondence between algebraic theories and Cartesian monads. We will discuss operads with arbitrary monoidal products at the end of this section. □

Example 8.2.12. In Section 5.4, we met the associahedra $A_1, A_2, A_3, \ldots$, the permutohedra $P_1, P_2, P_3, \ldots$, and the permuto-associahedra $PA_1, PA_2, PA_3, \ldots$, each of these sequences form an operad. For example, given an association of four letters, $[ab][cd]$, and an association of five letters, $(r(st))(uv)$, we can compose the five letters into the third letter c of the $[ab][cd]$ association to get an eight-letter association: $[ab][((r(st))(uv))d]$. There are similar compositions with reassociations. The operad structure of the permutohedra and the permuto-associahedra follow in the same way. □

Some of these operads are related by operad homomorphisms.

Example 8.2.13. There is an operad homomorphism from the tree operad to the associahedra operad. Because coherence and associahedra are at the core of monoidal categories, it pays to look at this homomorphism in depth. One can assign a tree to every association. For example, the binary trees

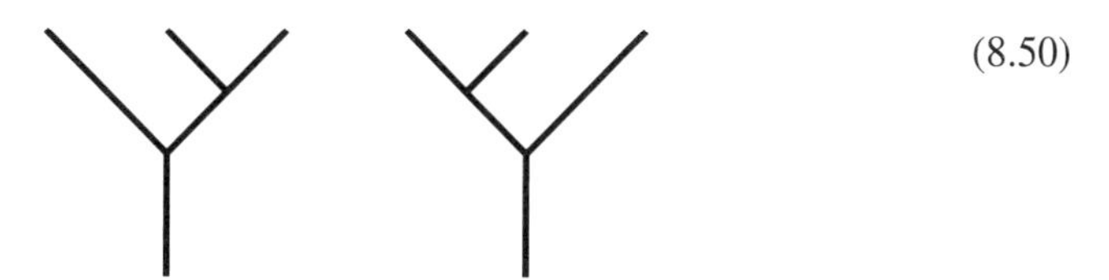

(8.50)

can be assigned to the associations $a(bc)$ and $(ab)c$, respectively. This correspondence can be extended to reassociations between associations. For example, the tree

(8.51)

which is halfway between the two previous trees can be assigned to the reassociation $a(bc) \longrightarrow (ab)c$. This idea can be seen in Figure 8.6. The tree with four leaves coming out of one stump corresponds to the entire associahedra of four letters. □

Example 8.2.14. There is an operad homomorphism from associahedra to permuto-associahedra. In detail, there is an inclusion $A_i \longrightarrow PA_i$ that takes the association to the permutassociation without switching the order of the elements. In other words, $(ab)(cd)$ goes to $(ab)(cd)$, not to $(bd)(ca)$. Furthermore, there is a map from permuto-associahedra to the permutohedra that takes a permuted association like $((fh)(e(bc)))(d(ag))$ to the permuted elements $fhebcdag$ without parentheses. This map is clearly surjective. We can combine these two maps to get a type of short exact sequence (see Definition 2.4.24) of operads:

$$0 \longrightarrow \{A_i\} \longrightarrow \{PA_i\} \longrightarrow \{P_i\} \longrightarrow 0. \tag{8.52}$$

(It is not really an exact sequence because these are not vector spaces and there is no clear definition of a kernel, but it is similar.) Continuing the analogy and keeping

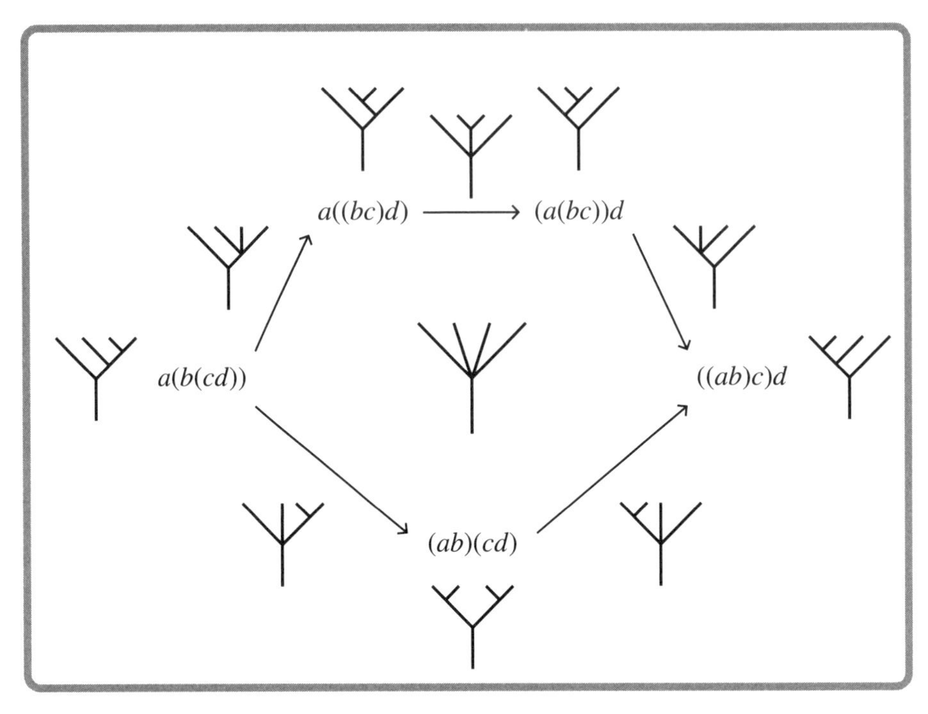

Figure 8.6. The pentagon coherence condition with related rooted trees.

Theorem 5.2.12 in mind, we can say that

$$\{PA_i\} = \{A_i\} \times \{P_i\}. \tag{8.53}$$

Or, in English, a permuted association is both an association and a permutation. □

Example 8.2.15. Andy Tonks [252] has shown a fascinating connection between associations and permutations that extends the short exact sequence of the previous example. Consider the permutation of five letters, *ecabd*. This corresponds to the association of the six-letter (five-tensor) word

$$((a \otimes b) \otimes (c \otimes d)) \otimes (e \otimes f). \tag{8.54}$$

The permutation *ecabd* tells us to first parenthesize the fifth tensor (*e* is the fifth letter of the alphabet), then parenthesize the third tensor, then parenthesize the first tensor, etc. With this correspondence, the permutation *edcba* corresponds to

$$a \otimes (b \otimes (c \otimes (d \otimes (e \otimes f)))) \tag{8.55}$$

and the permutation *abcde* corresponds to

$$((((a \otimes b) \otimes c) \otimes d) \otimes e) \otimes f. \tag{8.56}$$

Notice that this correspondence between permutations of five letters and associations of six letters is onto but not one-to-one. The permutation *acebd* also gives the association in Line (8.54). This correspondence can be extended to any permutation of n letters to an association of $n+1$ letters. Furthermore, a repermutation of n letters corresponds to a reassociation of $n+1$ letters. This means that there is a map $P_n \longrightarrow A_{n-1}$. Because of the change in subscripts, this is not a operad homomorphism, but we can extend the parts of Diagram (8.52) to

$$0 \longrightarrow A_1 \longrightarrow PA_1 \longrightarrow P_1 \longrightarrow A_2 \longrightarrow PA_2 \longrightarrow P_2 \longrightarrow A_3 \longrightarrow PA_3 \longrightarrow \cdots. \quad (8.57)$$

□

Example 8.2.16. The **small disk operad** $\mathbb{D}$ has many applications in mathematics and is used in string theory. An element of $\mathbb{D}(n)$ is the placement of n numbered, non-overlapping discs in a circle of size 1. The composition can be understood from the following diagram:

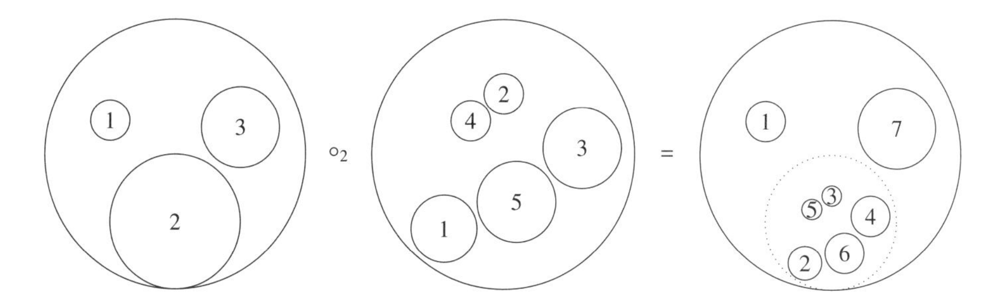

This shows an instance of $\circ_2 \colon \mathbb{D}(3) \times \mathbb{D}(5) \longrightarrow \mathbb{D}(7)$. The operation takes all discs in $\mathbb{D}(5)$, shrinks them to the size of the disk marked 2 in the left circle, and puts them in that circle. The disk that it is inserted into is then erased (shown with a dotted circle), and the numbers are reshuffled. The unit in $\mathbb{D}(1)$ is the single disk of size 1. □

Example 8.2.17. Similar to Example 8.2.16 is the **little square operad**. The elements in $O(n)$ are numbered, nonoverlapping squares that fit into a unit square. The composition is similar to that example. This example is the two-dimensional version of a whole family of operads. Its three-dimensional version is called the **little cube operad**. The collection $O(n)$ is the set of all n nonoverlapping smaller cubes that fit into a cube of size 1. The compositions $\circ_i$ shrinks the cube to the right size and inserts all the smaller cubes into the ith cube.

The one-dimensional version of this is called the **little interval operad**. The collection $O(n)$ consists of the unit interval with $n-1$ points separating n nonoverlapping intervals. The composition $\circ_i \colon O(n) \times O(j) \longrightarrow O(n+j-1)$ takes as inputs a particular splitting of the unit interval into n intervals and a particular splitting of the unit interval into j intervals. The second interval is shrunk to the size of the ith interval of the first input and inserted into the first interval. This returns the unit interval with $n+j-1$ intervals. The unit of this composition is the unit interval seen as one interval. This operad has use in string theory as it deals with strings that are open (as opposed to little closed rubber bands). □

Example 8.2.18. The small disks operad and the little interval operad combine to form something called the **Swiss cheese operad**. There are collections $O(m, n)$ consisting of configurations of m small labeled disks and n semidisks in a large semidisk as follows:

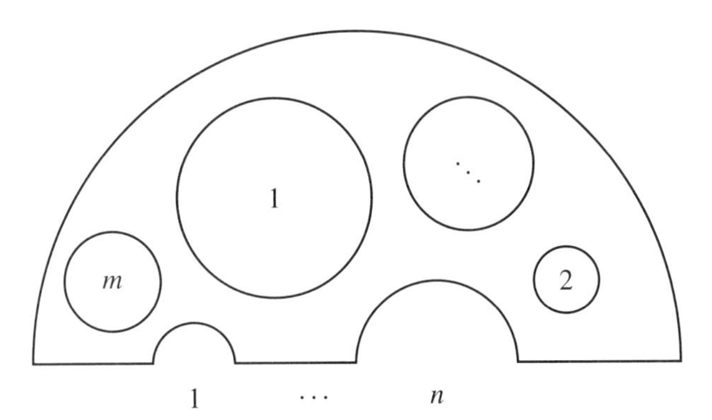

There are actually two types of composition here:

- Swiss cheese can be combined with Swiss cheese:

$$\circ_i\colon O(m,n) \times O(j,k) \longrightarrow O(m+j-1, n+k-1). \tag{8.58}$$

  This means that the ith interval in the first Swiss cheese gets the rescaled second Swiss cheese.
- Swiss cheese can be combined with little discs:

$$_i\colon O(m,n) \times \mathbb{D}(j) \longrightarrow O(m+j-1, n). \tag{8.59}$$

  This means that configuration in $\mathbb{D}(j)$ is scaled down and put into the jth circle in the semidisk.

The Swiss cheese operad is used in string theory, where it combines the theory of both closed and open strings.

□

The following is an example of a generalization of an operad for computer science.

Example 8.2.19. We are going to assume that the reader knows a little about recursion. If not, skip this example. Consider functions from powers of the natural numbers to the natural numbers $f\colon \mathbf{N}^n \longrightarrow \mathbf{N}$. There are special functions called **basic functions**:

- The zero function: $z\colon \mathbf{N}^0 = \{*\} \longrightarrow \mathbf{N}$, which is defined as $z(*) = 0$.
- The successor function: $s\colon \mathbf{N} \longrightarrow \mathbf{N}$, which is defined as $s(n) = n+1$.
- The projections functions: for each n and for all $1 \le i \le n$, the function $\pi_i^n\colon \mathbf{N}^n \longrightarrow \mathbf{N}$, defined as $\pi_i^n(x_1, x_2, \ldots, x_n) = x_i$.

We focus on two ways of generating new functions from old functions:

- Composition of functions: for $f\colon \mathbf{N}^n \longrightarrow \mathbf{N}$ and n functions $g_1\colon \mathbf{N}^t \longrightarrow \mathbf{N}$, $g_2\colon \mathbf{N}^t \longrightarrow \mathbf{N}, \ldots, g_n\colon \mathbf{N}^t \longrightarrow \mathbf{N}$, we can form function

$$h\colon \mathbf{N}^t \longrightarrow \mathbf{N}, \tag{8.60}$$

defined as

$$h = f \circ (g_1 \times g_2 \times \cdots \times g_n). \tag{8.61}$$

We can illustrate this composition operation with the following tree:

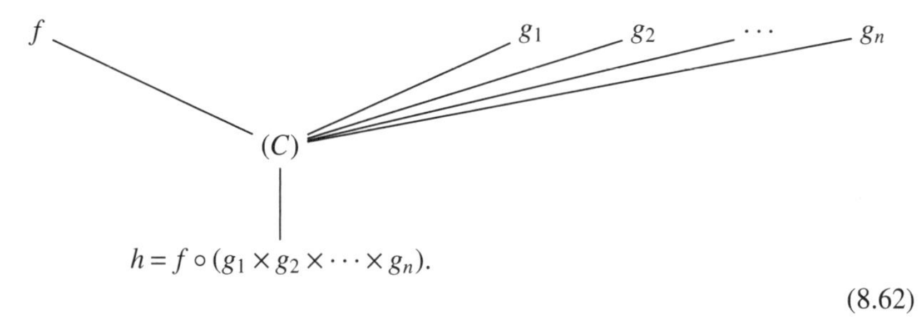

(8.62)

- Recursion of functions: for $f\colon \mathbf{N}^n \longrightarrow \mathbf{N}$ and $g\colon \mathbf{N}^{n+2} \longrightarrow \mathbf{N}$, we can form a function $h\colon \mathbf{N}^{n+1} \longrightarrow \mathbf{N}$, defined as

$$h(0, x_1, x_2, \cdots, x_n) = f(x_1, x_2, \cdots, x_n) \tag{8.63}$$

and

$$h(m+1, x_1, x_2, \cdots, x_n) = g(m, x_1, x_2, \ldots x_n, h(m, x_1, x_2, \cdots, x_n)). \tag{8.64}$$

Since h depends on f and g, we write it as $h = f\sharp g$ and describe this recursion operation with the following tree:

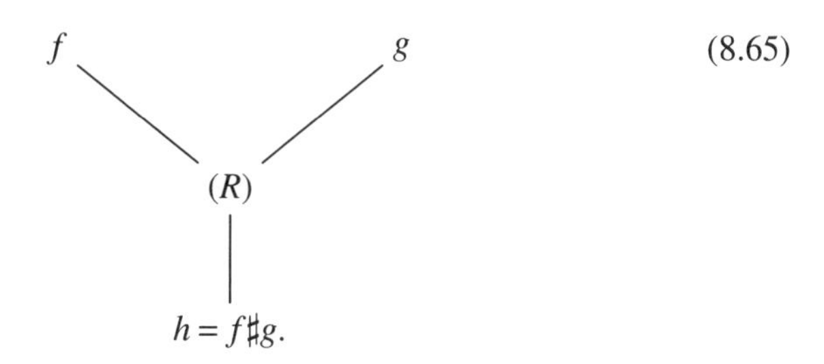

(8.65)

Starting with the basic functions, we generate labeled, rooted trees. The leaves of these trees are labeled with basic functions, and their internal nodes will be either (C) for composition or (R) for recursion. Each such tree is a description of a primitive recursive function. There might be many descriptions of one primitive recursive function.

We can think of such descriptions of primitive functions as a type of operad, and the $O(n)$ are descriptions with n leaves. The compositions then become as follows:

- Composition is $\circ_C : O(n) \times O(t) \longrightarrow O(t)$.
- Recursion is $\circ_R : O(n) \times O(n+2) \longrightarrow O(n+1)$.

The intended algebra for this operad is comprised of the powers of natural numbers. In other words, there is a $\theta_n : O(n) \times \mathbf{N}^n \longrightarrow \mathbf{N}$, which takes a description of a primitive recursive function and a sequence of natural numbers and gives the output of that function for those inputs.

It turns out that such descriptions are like a program. However, this programming language only has three types of constants (basic functions) and two operations (composition and recursion). It might be the world's smallest programming language. It can be shown that these descriptions can describe almost any function that any other programming language can describe.

Given two descriptions of primitive recursive functions of n variables (i.e., $h, h' \in O(n)$), say that h is equivalent to h' if they perform the same function. This forms an equivalence relation on each $O(n)$. This equivalence relation respects the operations, and hence there is an equivalence relation on the operad. The quotient operad is the operad of primitive recursive functions. This operad has a product and something called a natural number object (we will look at this concept more closely in Section 9.5.) This operad has certain universal properties that make it special.

Primitive recursive functions go back to David Hilbert and Kurt Gödel. They are a well-established part of theoretical computer science. Looking at descriptions of primitive recursive functions as an operad was first done in my article [271] and was taken up and advanced in Chapter IX of Yuri Manin's book [185]. □

We have just mentioned a few operads. There is a veritable zoo of operads to describe every type of structure on spaces, algebras, vector spaces, and other structures. Some of the famous types of operads are called A_∞, E_∞, **cacti**, **Lie**, and **Lie**$_\infty$. See the end of this section for suggestions on where to learn more about them.

Generalizations

We gave the simplest definition of an operad. There are many types of operads depending on what extra axioms the maps $\circ$ satisfy. They go by names like **symmetric operads**, **cyclic operads**, and **spherical operads**.

For an operad O, a typical element of $O(n)$ corresponds to an operation $a^{\otimes n} \longrightarrow a$. An obvious generalization of this is to consider descriptions where a typical element of $\mathcal{P}(m,n)$ corresponds to an operation $a^{\otimes m} \longrightarrow a^{\otimes n}$. Such a generalization is referred to as a **PROP** (for "PROduct and Permutation category")[2] or a **monoidal theory**. To be exact, a PROP is a monoidal category $\mathcal{P}$ where the the objects are the natural numbers, and the monoidal structure on objects is simply addition (i.e., $m \otimes n = m + n$). The morphisms $m \longrightarrow n$ correspond to operations $a^{\otimes m} \longrightarrow a^{\otimes n}$. One should also look at a PROP as a generalization of an algebraic theory that is in a Cartesian category. Within a Cartesian algebraic theory, a morphism $f : m \longrightarrow n$ can be written as n morphisms

[2]While a PROP is a generalization of an operad, historically PROPs were developed first.

$f_1: m \longrightarrow 1, f_2: m \longrightarrow 1, \ldots, f_n: m \longrightarrow 1$. In other words, in an algebraic theory $\mathbf{T}$,

$$Hom_{\mathbf{T}}(m,n) \cong (Hom_{\mathbf{T}}(m,1))^n. \tag{8.66}$$

Within a PROP, there can be morphisms $f: m \longrightarrow n$ that cannot be written as a monoidal product of n morphisms from m to 1. This makes PROPs much more interesting. Algebras for monoidal theories can occur in any $\mathbb{C}$ with a strict monoidal category structure. Algebras for a monoidal theory are no longer Cartesian product-preserving functors. Rather, they are strict monoidal-preserving functors (i.e., strict monoidal functors where $F(n) = F(1)^{\otimes n}$).

With the notion of a PROP, we can describe structures with **cooperations** and **coalgebraic structures**. For example, a comultiplication is described by an element of a PROP $\mathcal{P}(1,2)$. We say that an operation has the property of being "coassociative" if the two ways of using the comultiplication twice are set equal in collection $\mathcal{P}(1,3)$. Turning around the arrows in the definitions of a monoid and a group gives the notions of a comonoid and a cogroup, respectively. We met some of these operations, properties, and structures in Section 7.4. For example, we saw

- A bialgebra is a comonoid object in the category of algebras.
- A Hopf algebra is a cogroup in the category of algebras.

A generalization of PROPs are **multisorted PROPs** or **colored PROPs**. These arise in many areas of mathematical physics. Colored operads are applied to wiring diagrams as well [277].

The collection of certain space-time diagrams called **cobordisms** form PROPs. Algebras of these PROPs in certain categories of vector spaces are called "topological field theories." They are important in mathematical physics and geometry.

Suggestions for Further Study

Some nice introductory articles on operads are [244, 55]. See Chapter 12 of [219] as well. There is a wonderful two-part blog post by Tai-Danae Bradley [45]. The first two chapters of [169] is an excellent introduction to the whole field.

There are at least two books [174, 186], which are collections of papers showing how operads are used all over.

8.3 Monads

Monads and their dual cousins, comonads, are systematic ways of broadening the structure of the objects and morphisms of a category. The systematic broadening is done with an endofunctor of the category. In other words, let $\mathbb{A}$ be a category and $M: \mathbb{A} \longrightarrow \mathbb{A}$ be an endofunctor. For object a in $\mathbb{A}$, the broadened structure of a is $M(a)$, and for morphism $f: a \longrightarrow a'$ in $\mathbb{A}$, the broadened structure of f is $M(f)$. Related to this endofunctor are some natural transformations that are required to satisfy some simple rules. With a monad (or comonad) at hand, one then examines the relationship of the added structure to the original structure. This is done in two ways: Eilenberg-Moore algebras and Kleisli morphisms.

The goal of this discussion is to explore the many places where monads and comonads arise. This section is high on definitions and examples and low on theorems and proofs. More theorems and proofs can be learned from the sources given at the end of the section.

Monads

First, let's look at a motivating example.

Example 8.3.1. List monad. Take a set of letters and consider all the words made up of the elements of the set. This entails a functor $List\colon \mathbb{S}\mathrm{et} \longrightarrow \mathbb{S}\mathrm{et}$, which takes the set S to the set, $List(S)$, of all words with the letters of S. We already met this endofunctor in Example 4.2.3. There we saw that there is a natural transformation, $Flatten\colon List \circ List \Longrightarrow List$, which takes lists of lists of letters and makes them into a list. This shows that when we endow the set twice with this new structure, there is a way of seeing it as endowing it only once. We also saw that there is a natural transformation $Unit\colon Id_{\mathbb{S}\mathrm{et}} \Longrightarrow List$, which, for a set S, amounts to the function $Unit_S : S \longrightarrow List(S)$. This function takes a single letter and treats it as a list of length 1 (i.e., $Unit_S(s) = s$). From this, we see that the original structure can be nicely embedded in the enhanced structure.

Let us consider some properties of this functor and these natural transformations. For a given set, say $S = \{a, b, c\}$, some typical elements of $List(S)$ are $aaaa$, abc, or $ccaaaa$. A typical element of $List(List(S))$ is a list of such words (e.g., (abc, a, ccc, ba)). A typical element of $List(List(List(S))) = List^3(S)$, which is a list of lists of lists of letters, is

$$(abc, a, ccc, ba), (cba, cca, cc), (a, , baaaa, babab). \tag{8.67}$$

This can be "flattened," or the "elements can be combined" in two different ways:

- On the one hand, we can first flatten in terms of the outermost grouping and get

$$(abc, a, ccc, ba, cba, cca, cc, a, baaaa, babab). \tag{8.68}$$

  We can then still flatten out this list of lists to get a single list:

$$abcacccbacbaccaccabaaaababab. \tag{8.69}$$

- On the other hand, we can first flatten or combine the innermost level by removing the inner commas. This gives us

$$(abcacccba), (cbaccacc), (abaaaababab). \tag{8.70}$$

  We can then still flatten out this list by removing the parentheses separating the words, which gives us the same final result.

The *Unit* natural transformation satisfies two requirements. First, if you start with a list of letters and think of it as a single element, and then you flatten it, it becomes the original list of elements:

$$a \quad \mapsto \quad (a) \quad \mapsto \quad a. \tag{8.71}$$

Second, consider the map $Unit_S : S \longrightarrow List(S)$ and apply the *List* functor to this function to get $List(Unit_S)\colon List(S) \longrightarrow List(List(S))$. If you compose this map by flattening, you get

$$List(S) \xrightarrow{List(Unit_S)} List(List(S)) \xrightarrow{Flatten} List(S), \tag{8.72}$$

which is defined as the following example:

$$abc \quad \mapsto \quad (abc) \quad \mapsto \quad abc. \tag{8.73}$$

This map takes a list of elements and thinks of it as a list of lists, and then it flattens it to get the original list. These two rules say that the *Unit* map acts like a unit to the *Flatten* multiplication. □

With this motivating example understood, we can formally define a monad.

Definition 8.3.2. Let $\mathbb{A}$ be a category. A **monad** or a **triple** (M, μ, η) on $\mathbb{A}$ is all of the following:

- An endofunctor $M\colon \mathbb{A} \longrightarrow \mathbb{A}$ that endows objects and morphisms with extra structure.
- A multiplication $\mu\colon M \circ M = M^2 \Longrightarrow M$ natural transformation, which shows that there is a systematic way of showing how one can view an added structure to an added structure as just an added structure.
- A unit $\eta\colon Id_{\mathbb{A}} \Longrightarrow M$ natural transformation, which shows how the original object can be, in a sense, included in the extra structure. (Notice that we do not actually require η to be a monomorphism.)

These natural transformations must satisfy the following associativity and unit laws:

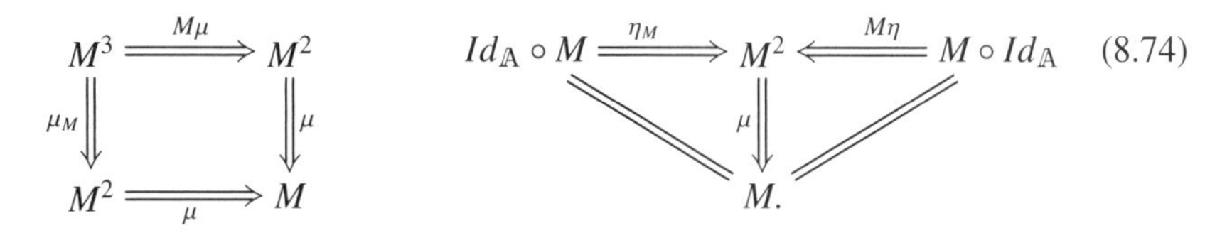

(8.74)

Notice that M^3 is $M \circ M \circ M$ and $M\mu$ means $Id_M \circ_H \mu$. Along the same lines, μ_M means that $\mu \circ_H M$. These requirements are very similar to the requirements given in Diagram (2.10), which are part of the definition of a monoid.

Exercise 8.3.3. Prove that a monad on a category $\mathbb{A}$ is a monoid object in the strict monoidal category $(\mathbb{A}^{\mathbb{A}}, \circ, Id_{\mathbb{A}})$.

(Hint: The functor M is an object in $\mathbb{A}^{\mathbb{A}}$. The μ is the monoid operation, and η is the unit of the monoid.) ■

We begin with perhaps (maybe) one of the simplest monads around. This example is given with all the details, but as we progress through the examples, we will describe fewer and fewer details.

Example 8.3.4. Maybe monad. The simplest way to broaden a set is to add one element to the set. We define the "maybe" monad (the name will be explained in a few pages) as (M, μ, η) on the category $\mathbb{S}\mathrm{et}$. The endofunctor $M\colon \mathbb{S}\mathrm{et} \longrightarrow \mathbb{S}\mathrm{et}$ adds an element to a set (i.e., it is defined on a set S as $M(S) = S + \{*\}$, where $\{*\}$ is just a one-element set and + is the disjoint union of sets). On a set function $f\colon S \longrightarrow S'$, the functor acts by taking the function to the same function but where the extra element goes to the extra element (i.e., $M(f)\colon S + \{*\} \longrightarrow S' + \{*\}$, where $M(f)$ takes the elements of S to S' and $f(*) = *$). The multiplication of the monad is a function that takes the set with two extra elements to the set with a single extra element. To make this more understandable, we shall put subscripts on $*$ to distinguish them. The natural transformation $\mu\colon M \circ M \Longrightarrow M$, whose component on set S is $\mu_S : S + \{*_x\} + \{*_y\} \longrightarrow S + \{*_x\}$ and is defined as $\mu_S(s) = s$ for $s \in S$ and $\mu_S(*_x) = \mu_S(*_y) = *_x$. The unit of the monad is the inclusion of the set into the extended set. That is, $\eta\colon Id_{\mathbb{S}\mathrm{et}} \Longrightarrow M$ is a natural transformation whose component on set S is the inclusion $\eta_S : S \hookrightarrow S + \{*_x\}$.

Let us show that the associativity axiom is satisfied. The natural transformation $M\mu$ at set S is the function $M\mu_S : S + \{*_x\} + \{*_y\} + \{*_z\} \longrightarrow S + \{*_a\} + \{*_b\}$. This function takes the elements of S to the same elements of S, while $*_x \mapsto *_a$, $*_y \mapsto *_a$, and $*_z \mapsto *_b$. In contrast, function μ_M works as follows: $*_x \mapsto *_a$, $*_y \mapsto *_b$, and $*_z \mapsto *_b$. Regardless, the composition of each map with μ gives the function $*_x \mapsto *_a$, $*_y \mapsto *_a$, and $*_z \mapsto *_a$.

The two unit laws are also satisfied. The natural transformation η_M takes the set $S + \{*_x\}$ to the set $S + \{*_a\} + \{*_b\}$, where $*_x \mapsto *_a$. The natural transformation $M\eta$ takes the set $S + \{*_x\}$ to the set $S + \{*_a\} + \{*_b\}$, where $*_x \mapsto *_b$. After composing both maps with μ, we get the identity function that takes $S + \{*_x\}$ to itself.

When we look at the algebras for this simple monad, we will see that it has many applications in mathematics and also will be important for a programming paradigm called "functional programming." □

Example 8.3.5. Powerset monad. Consider the covariant power set function $\mathcal{P}\colon \mathbb{S}\mathrm{et} \longrightarrow \mathbb{S}\mathrm{et}$, which we saw in Example (4.1.5). This endofunctor is the direct image functor, which takes every set S to $\mathcal{P}(S)$, the set of subsets of S. It also takes every function $f\colon S \longrightarrow T$ to $\mathcal{P}(f)\colon \mathcal{P}(S) \longrightarrow \mathcal{P}(T)$. For a set S, a typical element of $\mathcal{P}(\mathcal{P}(S))$ is a subset of subsets of S. For example, for the set $S = \{a, b, c\}$, an element of $\mathcal{P}(\mathcal{P}(S))$ is $\{\{a\}, \{b, a\}, \{b\}\}$. The multiplication $\mu\colon \mathcal{P} \circ \mathcal{P} \longrightarrow \mathcal{P}$ does the same thing that the natural transformation *flatten* did in the list monad (it also deletes repetitions). For a set S, the unit $\eta_S : S \longrightarrow \mathcal{P}(S)$ takes an element s to the set $\{s\}$. The fact that these natural transformations satisfy the monad axioms follows from the same reasoning as we saw with the list monad in Example 8.3.1. □

Example 8.3.6. Writer monad or the **action monad**. Let $(M, \star, e)$ be any monoid. The functor $M \times (\ ): \mathbb{S}\mathrm{et} \longrightarrow \mathbb{S}\mathrm{et}$ is defined on set S as $M \times S$. On function $f: S \longrightarrow S'$, it is defined as $(id_M \times f): (M \times S) \longrightarrow (M \times S')$. The multiplication on set S is a function $\mu_S: M \times M \times S \longrightarrow M \times S$, which takes (m, m', s) to $(m \star m', s)$. The unit on S is the function $\eta_S: S \longrightarrow M \times S$, which takes s to (e, s), where e is the unit of the monoid. The associativity law is true because $\star$ is associative. The two unit laws are true because e is a unit of the monoid. We will see that such monads are used in many areas, including computer science. □

Example 8.3.7. Equivalence class monad. Consider the category of equivalence relations $\mathbb{E}\mathrm{quiv}$. The objects are pairs $(S, \sim)$, where S is a set and $\sim$ is an equivalence relation on S. The morphisms are functions that respect the equivalence relation, which means that $f: (S, \sim) \longrightarrow (S', \sim')$ is a morphism if $f: S \longrightarrow S'$ such that $s_1 \sim s_2$ implies $f(s_1) \sim' f(s_2)$. The endofunctor $M: \mathbb{E}\mathrm{quiv} \longrightarrow \mathbb{E}\mathrm{quiv}$ is defined as $M(S, \sim) = (S/\sim, id)$. This means that M takes an equivalence relation to its quotient set with the identity relation. The endofunctor M is defined on map $f: (S, \sim) \longrightarrow (S', \sim')$ as $M(f): (S/\sim, id) \longrightarrow (S'/\sim', id)$ and is well defined as $M(f)([s]) = [f(s)]$. The multiplication $\mu: M^2 \longrightarrow M$ is an isomorphism because $M(S/\sim, id) = (S/\sim, id)$. Another way to say this is that M is idempotent. The unit for $(S, \sim)$ is the quotient map $(S, \sim) \longrightarrow (S/\sim, id)$, which takes every element to its equivalence class. The axioms are easily seen to be satisfied. The associativity follows from the idempotency of M. The unit laws follow from the fact that μ is an isomorphism. □

Example 8.3.8. Distribution monad. In this example, we deal with some simple ideas of probability theory. Consider the functor $Prob: \mathbb{S}\mathrm{et} \longrightarrow \mathbb{S}\mathrm{et}$, which is defined on S as the set of all nice ways of assigning probabilities to the elements of S (i.e., the finite probability distributions on S). This means that

$$Prob(S) = \{g: S \longrightarrow [0,1] : g(s) = 0 \text{ for all but a finite subset of } S \text{ and } \sum_{s \in S} g(s) = 1\}. \tag{8.75}$$

For a function $f: S \longrightarrow T$, there is a function of probability distributions $Prob(f): Prob(S) \longrightarrow Prob(T)$. For example, consider the following function of sets, probability distribution on S, and induced probability distribution on T:

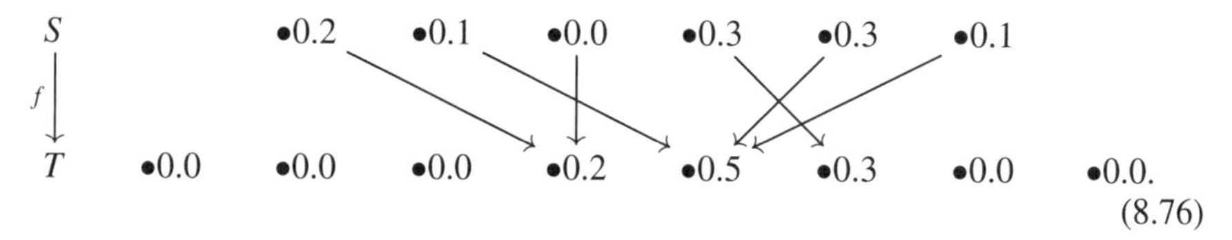

(8.76)

Notice that the probability given to any element of T is the sum of the probabilities of the elements of S that point to that particular element. Formally, the function $Prob(f)$

is defined for probability distribution $g\colon S \longrightarrow [0,1]$ as $Prob(f)(g) = \bar{g}\colon T \longrightarrow [0,1]$ where

$$\bar{g}(t) = \sum_{s \in f^{-1}(t)} g(s). \tag{8.77}$$

The multiplication μ is all about weighted averages. It takes into account a probability distribution on a set of probability distributions. As a simple example, consider the following game. Assume that there are six urns containing various numbers of blue and red marbles as follows:

Urn	1	2	3	4	5	6
Blue	4	5	1	0	4	9
Red	6	4	10	1	1	10

This describes six probability distributions. Now we work with a probability distribution on this set of probability distributions. Consider an unfair die where even numbers are twice as likely to turn up as odd numbers. This means that the chance of getting each even face is $\frac{2}{9}$, and the chance of getting each odd face is $\frac{1}{9}$. One plays the game by throwing the die and then selecting a marble from the urn described by the die. The question is what color marble gets picked. We can graphically show this game as follows:

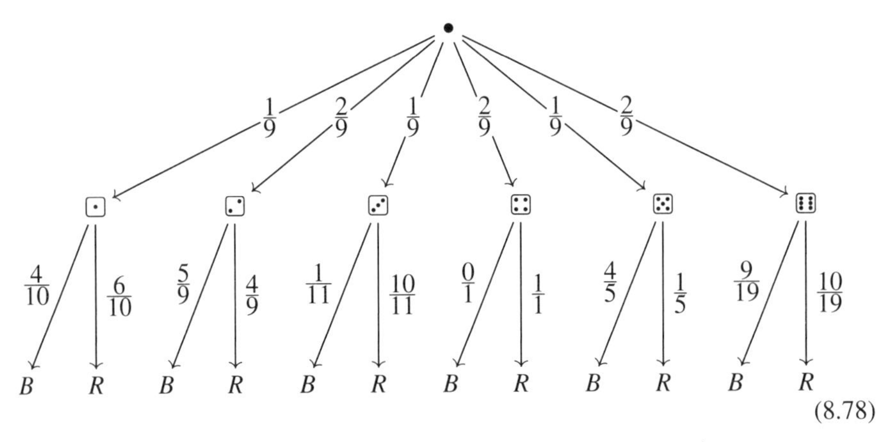

(8.78)

To get a blue marble in this game, one needs to get to the first urn ($\frac{1}{9}$) and then get a blue ($\frac{4}{10}$), OR get to the second urn ($\frac{2}{9}$) and then get a blue ($\frac{5}{9}$), OR ..., OR get to the sixth urn ($\frac{2}{9}$) and then get a blue ($\frac{9}{19}$). That is, the chance of getting a blue is

$$\frac{1}{9}\frac{4}{10} + \frac{2}{9}\frac{5}{9} + \frac{1}{9}\frac{1}{11} + \frac{2}{9}\frac{0}{1} + \frac{1}{9}\frac{4}{5} + \frac{2}{9}\frac{9}{19} = 0.3722. \tag{8.79}$$

We can do a similar analysis for getting a red. The final result is the probability distribution given by the weighted average:

(8.80)

With this example in mind, we can formulate the notion of the multiplication of the monad. Think of $Prob(S)$ as consisting of distributions $g\colon S \longrightarrow [0,1]$ and $Prob(Prob(S))$ as consisting of distributions $g'\colon Prob(S) \longrightarrow [0,1]$. For set S, the function $\mu_S : Prob(Prob(S)) \longrightarrow Prob(S)$ takes a function $g'\colon Prob(S) \longrightarrow [0,1]$ and forms the probability distribution $\mu_S(g')\colon S \longrightarrow [0,1]$, which is defined for $s \in S$ as

$$\mu_S(g')(s) = \sum_{g \in Prob(S)} (g'(g) \cdot g(s)). \tag{8.81}$$

We leave it for the reader to prove that μ is natural.

The unit of this monad is easy. For set S, the unit $\eta_S : S \longrightarrow Prob(S)$ is defined as

$$s \quad \mapsto \quad \delta_s\colon S \longrightarrow [0,1], \tag{8.82}$$

where δ_s takes s to 1 and everything else to 0. This function is called the **Dirac delta**. We leave it to the reader to show that η is natural. The reader must also show that the axioms of being a monad are satisfied.

There are more sophisticated versions of the probability monad that take into account fancier notions of probability (measure theory). One of those is called the **Giry monad**. □

Exercise 8.3.9. Show the naturality of the multiplication and the unit of the previous example. Also, show that μ and η satisfy the requirements of being a monad.

(Hint: The fact that μ satisfies the requirement essentially says that taking weighted averages is associative. The fact that η satisfies the requirements basically shows that the Dirac delta acts like a unit for taking weighted averages.) ■

Example 8.3.10. Monad on a preorder category. Consider a preorder category $\mathbb{A}$. A functor $M\colon \mathbb{A} \longrightarrow \mathbb{A}$ is monotonic (i.e., if $x \leq y$ then $M(x) \leq M(y)$). The multiplication natural transformation shows that for all objects x in $\mathbb{A}$, we have that $M(M(x)) \leq M(x)$. The unit natural transformation shows that $x \leq M(x)$. If we apply M to both sides of this inequality, we get $M(x) \leq M(M(x))$. Hence, $M(x) = M(M(x))$. A function of preorders that satisfies these properties is called a **closure operator**. The endofunctor takes an element x to a larger element $M(x)$. □

Example 8.3.11. Monad induced by an adjunction. A left adjoint followed by its right adjoint creates an endofunctor. Such an endofunctor has a monad structure. Let $L\colon \mathbb{A} \longrightarrow \mathbb{B}$ be left adjoint to $R\colon \mathbb{B} \longrightarrow \mathbb{A}$ with unit η and counit ε. Then there is the following monad: $(M = R \circ L, R\varepsilon_L, \eta)$. The natural transformation $R\varepsilon_L\colon RLRL \Longrightarrow RL$, which can be rewritten as $R\varepsilon_L\colon MM \Longrightarrow M$ is the monad multiplication. The natural transformation $\eta\colon Id_{\mathbb{A}} \Longrightarrow R \circ L = M$ is the monad unit. These natural transformations satisfy the monad requirements (Diagram (8.74)) because of the commutativity of the following diagrams:

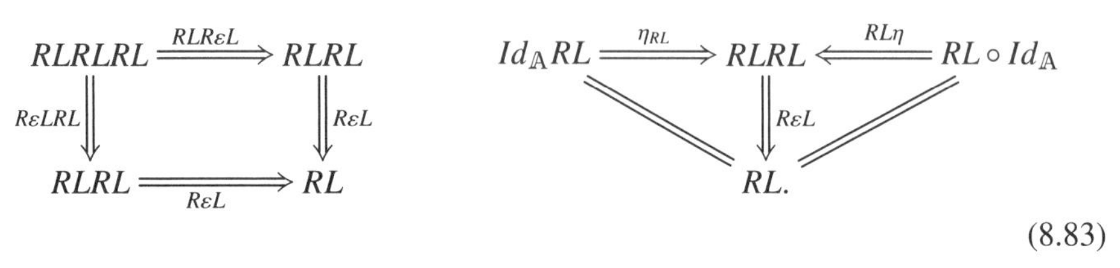

(8.83)

The left diagram commutes from the horizontal composition of natural transformations, and the right diagram commutes from the triangle identities of an adjunction. □

A special case of this example is the following.

Example 8.3.12. Monad induced by the free-forgetful adjunction of monoids. Consider the functor $F\colon \mathbb{Set} \longrightarrow \mathbb{Monoid}$ and its right adjoint $U\colon \mathbb{Monoid} \longrightarrow \mathbb{Set}$. The composition $U \circ F$ takes set S and returns the set of all the words with the elements of S. The natural transformation $U\varepsilon F$ is exactly the *flatten* natural transformation, and the unit of the adjuntction is exactly the *Unit* natural transformation. All this means that the monad induced by this adjunction is exactly the list monad. □

There are many other examples of monads that we will not mention. In particular, for every algebraic theory, there is a monad that describes the same structure as the algebraic theory. Similarly, for every operad, there is a monad that describes the same structure.

How do monads relate to each other?

Definition 8.3.13. Given two monads (M, μ, η) and (M', μ', η') on a category $\mathbb{A}$, a natural transformation $\varphi\colon M \Longrightarrow M'$ is a **morphism of monads** if the following two diagrams commute:

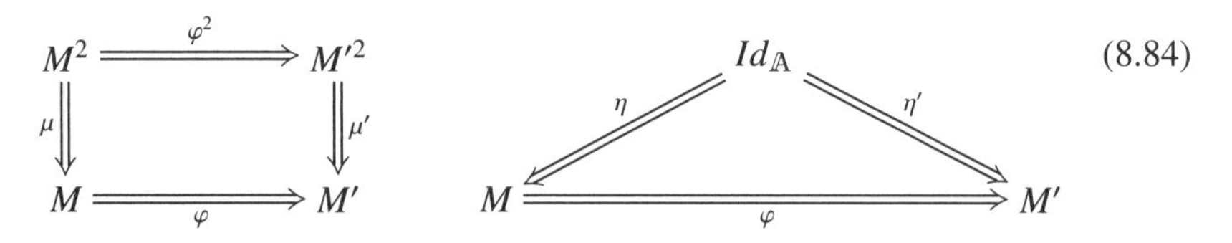

(8.84)

These are very similar to Diagram (2.11) which shows a monoid homomorphism.

Definition 8.3.14. The collection of monads on $\mathbb{A}$ and morphisms of monads forms a category called $\mathbb{Monad}(\mathbb{A})$.

Algebras of a Monad

We are interested in how the extra structure that a monad gives an object interacts with the original structure. That is, we examine how the extra structure "acts" on the structure. Similar to how a monad is a generalization of the notion of a monoid, the definition of an algebra for a monad will be a generalization of an M-set.

Definition 8.3.15. Let (M, μ, ν) be a monad on $\mathbb{A}$. An **Eilenberg-Moore algebra** or an M**-algebra** is a pair (a, h) where a is an object of $\mathbb{A}$ and $h\colon M(a) \longrightarrow a$ is a morphism in $\mathbb{A}$. These pairs must satisfy the following commutative diagrams:

$$\begin{array}{ccc} M^2(a) & \xrightarrow{Mh} & M(a) \\ {\scriptstyle \mu_a}\downarrow & & \downarrow{\scriptstyle h} \\ M(a) & \xrightarrow[h]{} & a \end{array} \qquad \begin{array}{ccc} a & \xrightarrow{\eta_a} & M(a) \\ & {\scriptstyle id}\searrow & \downarrow{\scriptstyle h} \\ & & a \end{array} \tag{8.85}$$

These two diagrams are similar to Diagram (4.24) which gave conditions for an M-set.

Given two M-algebras (a, h) and (a', h'), a **homomorphism of Eilenberg-Moore algebras** $f\colon (a, h) \longrightarrow (a', h')$ is a morphism $f\colon a \longrightarrow a'$ in $\mathbb{A}$ such that the following diagram commutes:

$$\begin{array}{ccc} M(a) & \xrightarrow{M(f)} & M(a') \\ {\scriptstyle h}\downarrow & & \downarrow{\scriptstyle h'} \\ a & \xrightarrow[f]{} & a'. \end{array} \tag{8.86}$$

This condition is very similar to the condition given in Diagram (4.2.4), which was the definition of a homomorphism of M-sets.

The collection of Eilenberg-Moore algebras and Eilenberg-Moore algebra homomorphisms forms the **Eilenberg-Moore category** of M and is denoted as $\mathbb{A}^M$. There is a forgetful functor $U^M\colon \mathbb{A}^M \longrightarrow \mathbb{A}$ that takes (a, h) to a. There is also a free functor $F^M\colon \mathbb{A} \longrightarrow \mathbb{A}^M$ that takes a in $\mathbb{A}$ to $(M(a), \mu_a)$.

Let us determine the Eilenberg-Moore algebras for some of the monads that we already have seen.

Example 8.3.16. List monad. An Eilenberg-Moore algebra for the list monad is a monoid. An algebra is a pair $(S, h\colon \mathit{List}(S) \longrightarrow S)$, where S is a set and h assigns to every sequence of elements of S, an element of S. One should think of multiplying all the elements in the list to get the output of h. The left requirement in Diagram (8.85) means that h respects taking lists of lists of elements of S. The right requirement says that for an element $s \in S$, we have $h(s) = s$. Hence, (S, h) has the structure of the monoid over S. It is for this reason that the list monad is also called the **Monoid monad**. □

Example 8.3.17. Maybe monad. An Eilenberg-Moore algebra of the maybe monad is a pointed set (which we met in Example 3.3.5) and a homomorphism of such algebras is a set function that preserves the designated point. Let us get into the details. An algebra is a pair $(S, h\colon S + \{*\} \longrightarrow S)$. The function h chooses a distinguished element $h(*)$. The left requirement of Diagram (8.85) says that there is only one distinguished element, while the right requirement says that h is the identity on the elements of the set. Given two such pointed sets $(S, h(*))$ and $(S', h'(*))$, a homomorphism is a set function $f\colon S \longrightarrow S'$ such that Diagram (8.86) commutes. This amounts to saying $f(h(*)) = h'(*)$ (i.e., the distinguished point is preserved). In other words, the category of Eilenberg-Moore algebras for the maybe monad is the category of pointed sets. □

Example 8.3.18. Powerset monad An Eilenberg-Moore algebra of the power set monad is a complete semilattice. This is very similar to Example 8.3.16, where $h\colon \mathcal{P}(S) \longrightarrow S$ assigns the value of the collection. However, here we do not have a list of elements; rather, we have a subset of elements. Notice that in the subset, there are no repetitions of elements. This corresponds to the fact that $x \wedge x = x$ and $x \vee x = x$. Furthermore, in a subset, there is no order of elements. This corresponds to the fact that $x \wedge y = y \wedge x$ and $x \vee y = y \vee x$. Another way to say this is that a semilattice is an idempotent and commutative monoid. The morphisms of the Eilenberg-Moore algebra are functions that preserve one of $\wedge$ or $\vee$ operations. This means that the category of Eilenberg-Moore algebra of the power set monad is the category of semilattices. □

Example 8.3.19. Writer monad. An algebra for the writer monad is a pair $(S, h\colon M \times S \longrightarrow S)$, where S is a set and h is a set function. The h function is equivalent to an action $\bar{h}\colon M \longrightarrow Hom(S, S)$. The fact that h must satisfy the requirements of being an Eilenberg-More algebra is exactly the requirements of S being an M-set (i.e., M acting on S; hence the name "action monad"). Being the homomorphism of algebras is exactly the same requirement for being an M-set homomorphism. We conclude that the Eilenberg-Moore category for the writer monad of monoid M is isomorphic to the category of M-sets. □

Example 8.3.20. Distribution monad. An Eilenberg-Moore algebra for the distribution monad assigns to every probability distribution on set S, an element of the set. This assignment would like to be the expected value of the probability distribution, but there is a problem with that. Consider a fair coin. The expected value of flipping a fair coin is $\frac{1}{2}$ heads $+\frac{1}{2}$ tails. This is not a member of the set { heads, tails }. In other words, the expected value is not a member of the set. Similarly, consider a fair die. The expected value of a fair die is $\frac{1+2+3+4+5+6}{6} = 3.5$, which again is not an element of the set of faces. Hence, the only types of probability distributions that give the expected value are continuous sets with all their intermediate values. Such sets are called **convex spaces**. The primary example of such a set is the unit interval $[0, 1]$.

In contrast to the distribution monad, the Eilenberg-Moore algebra for the more sophisticated Giry monad is the expected value. □

Example 8.3.21. Monads on a preorder category. An Eilenberg-Moore algebra of the closure operator is the set of closed sets. Let P be a preorder and M be a closure operator. Then, an M-algebra is an object a in P such that $M(a) \leq a$. Since we already know that $a \leq M(a)$, we have that $M(a) = a$. In other words, the M-algebras are fixed points of M. □

Exercise 8.3.22. Let $\varphi\colon M \Longrightarrow M'$ be a morphism of monads. Show that φ induces a functor $\varphi^*\colon \mathbb{A}^{M'} \longrightarrow \mathbb{A}^M$. This is very similar to what we saw with functorial semantics, where an algebraic theory morphism induces a morphism of algebras. ■

We have talked about adding structure to an object a of $\mathbb{A}$ by forming $M(a)$, and adding structure to a morphism $f\colon a \longrightarrow a'$ by forming $M(f)\colon M(a) \longrightarrow M(a')$. We can also add structure to the possible outputs of morphisms by considering morphisms of the form $f\colon a \longrightarrow M(a')$. This gives us a rich class of morphisms.

Definition 8.3.23. For a given category $\mathbb{A}$ and a monad (M, μ, η) on $\mathbb{A}$, a **Kleisli morphism** is a morphism in $\mathbb{A}$ of the form $f\colon a \longrightarrow M(a')$. Morphism $f\colon a \longrightarrow M(a')$ can be composed with a morphism $g\colon a' \longrightarrow M(a'')$ as follows:

$$a \xrightarrow{f} M(a') \xrightarrow{M(g)} M(M(a'')) \xrightarrow{\mu_{a''}} M(a'')\,. \tag{8.87}$$

This corresponds to using f to give a its extra structure, then using $M(g)$ to give its extra structure through $M(g)$, and then "crunching down" the extra structure with μ. For every object a of $\mathbb{A}$, there is an identity Kleisli morphism, $\eta\colon a \longrightarrow M(a)$.

With the notion of a Kleisli morphism, we can form the **Kleisli category** of M, which is denoted as $\mathbb{A}_M$. The objects of $\mathbb{A}_M$ are exactly the same as the objects of $\mathbb{A}$, and the morphisms are Kleisli morphisms with their composition. In other words, $f\colon a \longrightarrow a'$ in $\mathbb{A}_M$ will be the morphism $\bar{f}\colon a \longrightarrow M(a')$. Composition of the morphisms in $\mathbb{A}_M$ is as explained previously. There is a forgetful functor, $U_M\colon \mathbb{A}_M \longrightarrow \mathbb{A}$, and a left adjoint free functor, $F_M\colon \mathbb{A} \longrightarrow \mathbb{A}_M$.

Exercise 8.3.24. Prove that the composition of Kleisli morphisms is associative. ■

Example 8.3.25. Maybe monad. A Kleisli morphism for the maybe monad corresponds to a partial function. In detail, the morphism is a set function $f\colon S \longrightarrow S' + \{*\}$. One thinks of this as a partial function, $\hat{f}\colon S \longrightarrow S'$. The elements that f takes to $*$ will be exactly those elements for which $\hat{f}$ is not defined. The Kleisli category of the maybe monad is the category $\mathbb{P}\mathrm{ar}$ of sets and partial functions.

The maybe monad is used in functional programming. Programs in a functional programming language consist of the composition of functions where each function goes from a type to a type. However, sometimes there are problems, and the function will not give any output. For example, there could be a function $g\colon Type_1 \longrightarrow Type_2$. We look at the Kleisli morphism of the function $\hat{g}\colon Type_1 \longrightarrow Type_2 + *$. If there is a problem, function g will output $*$. This is called **exception handling**.

It is interesting to note that the two categories of algebras for the maybe monad—namely, the category of pointed sets $\{*\}/\mathbb{S}\mathrm{et}$ of Example 8.3.17, and the category of partial functions $\mathbb{P}\mathrm{ar}$—were shown to be equivalent in Example 4.3.9. □

Example 8.3.26. Powerset monad. Consider the category $\mathbb{S}\mathrm{et}$ and the covariant power set monad $(\mathcal{P}, \mu, \eta)$. A Kleisli morphism for this monad is a set function $f\colon S \longrightarrow \mathcal{P}(S')$ for sets S and S'. Such a function corresponds to a relation $\hat{f}\colon S \not\longrightarrow S'$. The relation $\hat{f}$ can be understood by thinking of f as taking an element $s \in S$ to the subset of S' containing all the elements that are related to s. Expressed in symbols, $f(s) = \{t : (s,t) \in \hat{f}\}$. In other words, f are $\hat{f}$ are related by

$$(s, r) \in \hat{f} \text{ if and only if } r \in f(s). \tag{8.88}$$

Remember that the image of s can be the empty set. The composition of two Kleisli morphisms is the composition of the two relations that correspond to the Kleisli morphisms. In detail, given $f\colon S \longrightarrow \mathcal{P}(S')$ and $g\colon S' \longrightarrow \mathcal{P}(S'')$, then $f(s)$ is the set of all elements that are related to s, and $g(f(s))$ is the set of all elements that are related to any element of $f(s)$. This will be a set of sets. The μ in the Kleisli morphism composition crunches down or flattens the set of sets to one set. We summarize by stating that the Kleisli category of the covariant power set monad is the category $\mathbb{R}\mathrm{el}$ of sets and relations. □

Example 8.3.27. Distribution monad. A Kleisli morphism for the distribution monad is a function $f\colon S \longrightarrow \mathit{Prob}(S')$ for the sets S and S'. This means that f assigns to an $s \in S$ a probability distribution on S'. One can think of it as the probability of s going to some element of S'. Such maps are called **stochastic maps**. □

Comonads

At this point in the text, you should be used to the fact that if you turn around all the morphisms in a structure, you get a dual structure that is equally interesting. The dual of a monad is a "comonad." As with all dual constructions, the details and proofs of a comanad are the same as the details and proofs of monads if you stand on your head. We will be a little swifter with the details and the proofs than we have up to now.

As with a monad, a comonad is a method of systematically adding extra structure to the elements of a category. However, as a comonad, the nature of the extra structure is different.

Definition 8.3.28. Let $\mathbb{A}$ be a category. A **comonad** or a **cotriple** on $\mathbb{A}$ is (C, δ, η), which consists of the following:

- An endofunctor $C\colon \mathbb{A} \longrightarrow \mathbb{A}$, which endows objects and morphisms with the extra structure.
- A comultiplication $\delta\colon C \Longrightarrow C \circ C = C^2$ natural transformation which shows that there is a systematic way of viewing the added structure as structure that was added to added structure.
- A counit $\eta\colon C \Longrightarrow Id_{\mathbb{A}}$ natural transformation which systematically shows how one can forget the extra structure. (Notice that we do not actually require η to be an epimorphism.)

These natural transformations must satisfy the following coassociativity and counit laws:

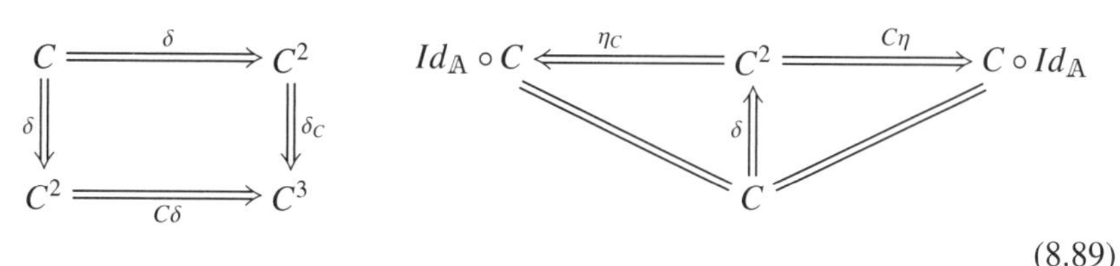

(8.89)

These commutative diagrams are simply the requirements for the definition of a comonoid (i.e., the requirements for the definition of a monoid with the arrows inverted).

Along the lines of Exercise 8.3.3, we have the following exercises.

Exercise 8.3.29. Prove that a comonad on a category $\mathbb{A}$ is a comonoid object in the strict monoidal category $(\mathbb{A}^{\mathbb{A}}, \circ, Id_{\mathbb{A}})$, or equivalently, it is a monoid object in the strict monoidal category $(\mathbb{A}^{\mathbb{A}}, \circ^{op}, Id_{\mathbb{A}})$. Here, $\circ^{op}$ is defined as $a \circ^{op} b = b \circ a$. ■

Example 8.3.30. The reader comonad. In the dual sprit of the writer monad, there is the reader comonad. Let E be a set, then $C\colon \mathbb{S}\mathrm{et} \longrightarrow \mathbb{S}\mathrm{et}$, defined as $C(S) = S \times E$ is an endofunctor. On function $f\colon S \longrightarrow S'$, the endofunctor is defined as $C(f) = f \times id_E$. The comultiplication at set S is $\delta_S : S \times E \longrightarrow S \times E \times E$, which is defined as $\delta_S(s, e) = (s, e, e)$. In other words, δ doubles the extra structure. The counit on set S is $\eta_S : S \times E \longrightarrow S$, which is defined as $\eta_S(s, e) = s$, the simple projection function. That is, η forgets the extra structure. The coassociativity axiom is satisfied for the following reason: there are two maps of the form

$$(s, e) \quad \mapsto \quad (s, e, e) \quad \mapsto \quad (s, e, e, e). \tag{8.90}$$

One such map doubles the first e, and the other such map doubles the second e. Coassociativity means that both these maps are the same. The counit axioms basically say that after performing the map $(s, e) \mapsto (s, e, e)$, you can forget either the first or the second e and get back to the original (s, e). □

Dual to Example 8.3.10 is the following.

Example 8.3.31. Comonads on preorder categories. Consider a preorder category $\mathbb{A}$. A functor $C\colon \mathbb{A} \longrightarrow \mathbb{A}$ is monotonic. A comultiplication natural transformation shows that for all objects x in $\mathbb{A}$ we have that $C(x) \leq C(C(x))$. A counit natural transformation shows that $C(x) \leq x$. If we apply C to both sides of this inequality, we get $C(C(x)) \leq C(x)$. Hence, $C(x) = C(C(x))$. A function of preorders that satisfies these properties is called an **interior operator**. Here, C takes an element to itself or a smaller element. Such a smaller element might be called an "open element." □

Example 8.3.32. The stream comonad. Let $\mathbf{N}$ be the monoid of natural numbers. We will define the stream comonad (C, δ, η) on $\mathbb{S}\mathrm{et}$. For set S, the endofunctor is defined as $C(S) = Hom_{\mathbb{S}\mathrm{et}}(\mathbf{N}, S)$. We can think of $C(S)$ as the collection of histories of elements.

In other words, each history is a sequence of elements of S, such as $x_0, x_1, x_2, \ldots$. We can think of each such sequence as telling what the state of a system is at each time click. The comultiplication at set S is a function

$$\delta_S : Hom(\mathbf{N}, S) \longrightarrow Hom(\mathbf{N}, Hom(\mathbf{N}, S)). \tag{8.91}$$

This takes a history $x_0, x_1, x_2, \ldots$ and outputs a sequence of histories:

$$(x_0, x_1, x_2, \ldots), (x_1, x_2, x_3, \ldots), (x_2, x_3, x_4, \ldots), \ldots, \tag{8.92}$$

where each history is shifted by one time click. The counit η takes a history and outputs the first element of the history; that is,

$$x_0, x_1, x_2, \ldots \qquad \mapsto \qquad x_0. \tag{8.93}$$

The coassociativity axiom is satisfied because the two ways of building a three-dimensional history are the same. The two counit axioms state that the first element of a history of a history, is the original history.

There is nothing special about the natural numbers. We could have worked with **Z**, and then a typical element would be

$$\ldots, x_{-3}, x_{-2}, x_{-1}, x_0, x_1, x_2, x_3, \ldots, \tag{8.94}$$

and we could also use the set of nonnegative real numbers, $\mathbf{R}^*$. In that case, a stream would be a continuous history of an object. □

In Example 8.3.11, we saw that the composition of adjunctions in one order gives a monad. The composition of the adjoint functors in the opposite order gives a comonad.

Example 8.3.33. Comonad induced by an adjunction. Let $L\colon \mathbb{A} \longrightarrow \mathbb{B}$ be left adjoint to $R\colon \mathbb{B} \longrightarrow \mathbb{A}$ with unit η and counit ε. Then there is the following comonad: $(C = L \circ R, L\eta_R, \varepsilon)$. In detail, $C = L \circ R\colon \mathbb{B} \longrightarrow \mathbb{A} \longrightarrow \mathbb{B}$ is an endofunctor. The natural transformation $L\eta_R\colon C = LR \Longrightarrow LRLR = CC$ is the comonad comultiplication. The natural transformation $\varepsilon\colon C = LR \Longrightarrow Id_{\mathbb{A}}$ is the comonad counit. These natural transformations satisfy the comonad requirements (Diagram (8.89)). We leave the details to the reader. □

Within logic, there is a comonad.

Example 8.3.34. Modality comonad. Consider a comonad on the category $\mathbb{P}\mathrm{rop}$ of propositions and entailment. We write the comonad as $\Box\colon \mathbb{P}\mathrm{rop} \longrightarrow \mathbb{P}\mathrm{rop}$. For a proposition P, we interpret $\Box P$ as meaning "S is necessary." From the functoriality of $\Box$ we have

$$\Box(P \longrightarrow Q) \text{ implies } \Box P \longrightarrow \Box Q. \tag{8.95}$$

This implication is called the **distribution axiom**. The comultiplication at proposition P, δ_P, gives us the rule

$$\Box P \text{ implies } \Box\Box P. \tag{8.96}$$

This says that if P is necessarily true, then it is necessary that its necessarily true. For historical reasons, this axiom is called **4**. The counit of the comonad gives us

$$\Box P \text{ implies } P. \tag{8.97}$$

This says that if P is necessarily true, then P is true. This axiom is called the **reflexivity axiom**. A logical system with these three axioms is called **S4**. This is the beginning of modal logic. □

By turning around the arrows in Definition 8.3.13, we can formulate the notion of a morphism of comonads. The collection of comonads on $\mathbb{A}$ and morphisms of comonads on $\mathbb{A}$ forms a category called $Comonad(\mathbb{A})$.

Coalgebras of a Comonad

Just as we had monads acting on Eilenberg-Moore algebras, we have comonads acting on the dual versions of Eilenberg-Moore algebras.

Definition 8.3.35. Let $\mathbb{A}$ be a category and (C, δ, ε) be a comonad on $\mathbb{A}$. An **Eilenberg-Moore coalgebra** or an M-**coalgebra** is a pair (a, h), where a is an object of $\mathbb{A}$ and $h\colon a \longrightarrow C(a)$ is a morphism in $\mathbb{A}$. These pairs must satisfy the following commutative diagrams:

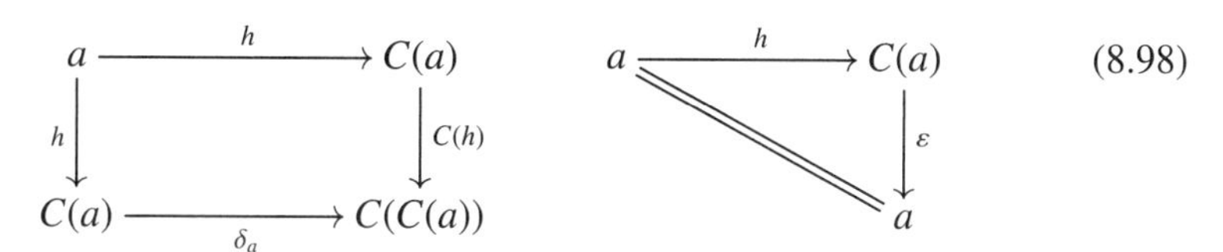

(8.98)

The definition of homomorphism between Eilenberg-Moore coalgebras is an obvious dual of the definition of a homomorphism between Eilenberg-Moore algebras of a monad. The collection of all Eilenberg-Moore coalgebras, and their homomorphisms forms the **Eilenberg-Moore category** of the comonad, which is denoted as $\mathbb{A}^C$.

There is a forgetful functor, $U^C\colon \mathbb{A}^C \longrightarrow \mathbb{A}$, which takes (a, h) to a. This functor has a left adjoint, $F^C\colon \mathbb{A} \longrightarrow \mathbb{A}^C$.

Let us look at Eilenberg-Moore coalgebras for some of the comonads that we have seen.

Example 8.3.36. Comonad for a partial order. For the comonad of an interior operator on a partial order, the Eilenberg-Moore coalgebras are the fixed points, which are open sets. □

Example 8.3.37. Stream comonad. An Eilenberg-Moore coalgebra for the stream comonad is a set S and set function $h\colon S \longrightarrow C(S)$. We will write h as

$$x \quad \mapsto \quad h_0(x), h_1(x), h_2(x), h_3(x), \ldots. \tag{8.99}$$

The fact that $\varepsilon_S \circ h = id_S$ means that h_0 is the identity. This means that h is defined as

$$x \quad \mapsto \quad x, h_1(x), h_2(x), h_3(x), \ldots. \tag{8.100}$$

The right side of the $C(h) \circ h = \delta_S \circ h$ axiom is the map

$$\begin{array}{lllll} x & \mapsto & x, h_1(x), h_2(x), h_3(x), \ldots & \mapsto & (x, h_1(x), h_2(x), h_3(x), \ldots) \\ & & & & (h_1(x), h_2(x), h_3(x), h_4(x), \ldots) \\ & & & & (h_2(x), h_3(x), h_4(x), h_5(x), \ldots) \\ & & & & \vdots \end{array}$$

The left side is the map

$$\begin{array}{lllll} x & \mapsto & x, h_1(x), h_2(x), h_3(x), \ldots & \mapsto & (x, h_1(x), h_2(x), h_3(x), \ldots) \\ & & & & (h_1(x), h_1(h_1(x)), h_1(h_2(x)), h_1(h_3(x)), \ldots) \\ & & & & (h_2(x), h_2(h_1(x)), h_2(h_2(x)), h_2(h_3(x)), \ldots) \\ & & & & (h_3(x), h_3(h_1(x)), h_3(h_2(x)), h_3(h_3(x)), \ldots) \\ & & & & \vdots \end{array}$$

Saying that these two maps are equal means that all the entries in the sequences are equal. In other words, $h_1(h_1(x)) = h_2(x)$ and $h_2(h_3(x)) = h_5(x)$. In general,

$$h_i(h_j(x)) = h_{i+j}(x) = (h_1 \circ h_1 \circ \cdots \circ h_1)(x). \tag{8.101}$$

This means that the h map is

$$x \quad \mapsto \quad x, h_1(x), h_1(h_1(x)), h_1(h_1(h_1(x))), \ldots. \tag{8.102}$$

Such a system describes a **dynamical system**. They tell how an object changes in a set over time. We have shown that an Eilenberg-Moore coalgebra for the stream comonad describes a dynamical system. It works the other way as well: any dynamical system gives an Eilenberg-Moore coalgebra. There is a definition of a morphism of dynamical systems that aligns perfectly with the notion of a homomorphism of a coalgebra for the stream comonad. Hence, the category of Eilenberg-Moore coalgebras for the stream comonad is the category of dynamical systems.

One can go further with this. Consider the monoid inclusion of the natural numbers into the nonnegative real numbers $\mathbf{N} \hookrightarrow \mathbf{R}^*$. This inclusion induces a morphism of the stream comonad based on $\mathbf{R}^*$ to the stream comonad based on $\mathbf{N}$; that is, there is a natural transformation

$$Hom(\mathbf{R}^*, S) \longrightarrow Hom(\mathbf{N}, S). \tag{8.103}$$

This morphism of comonads further induces a forgetful functor from the category of Eilenberg-Moore coalgebras over the $\mathbf{R}^*$ stream comonad to the category of Eilenberg-Moore coalgebras over the $\mathbf{N}$ stream comonad. This forgetful functor takes a continuous time dynamical system and looks at what happens at the discrete time steps $t_0, t_1, t_2, \ldots$. In fact, scientists do this all the time. They take a continuous dynamical system and measure it at discrete time steps. The strength of this functor and the relationship between these two categories of dynamical systems tell us a lot about how complicated the continuous dynamical systems are. There is much study in this direction. □

There is also a comonad version of Kleisli morphisms and the categories that they form.

Definition 8.3.38. Let (C, δ, ε) be a comonad on $\mathbb{A}$. A **Kleisli morphism** is a map in $\mathbb{A}$ of the form $f\colon C(a) \longrightarrow b$. Think of these as morphisms that have extra structure in the source rather than the target. Given two Kleisli morphisms $f\colon C(a) \longrightarrow b$ and $g\colon C(b) \longrightarrow c$, we can compose them as

$$C(a) \xrightarrow{\delta_a} C(C(a)) \xrightarrow{C(f)} C(b) \xrightarrow{g} c. \tag{8.104}$$

For a comonad C on $\mathbb{A}$, we can form the **Kleisli category**, which is denoted as $\mathbb{A}_C$. The objects are the same objects in $\mathbb{A}$, and the morphisms $Hom(a, b)$ are the Kleisli morphisms $C(a) \longrightarrow b$. Composition of morphisms is as previously described.

There is a forgetful functor, $U_C\colon \mathbb{A}_C \longrightarrow \mathbb{A}$, and a left adjoint, $F_C\colon \mathbb{A} \longrightarrow \mathbb{A}_C$. Let us examine Kleisli morphisms for some of the comonads that we have seen.

Example 8.3.39. Reader comonad. A Kleisli morphism for the reader comonad is the set function $C(S) = S \times E \longrightarrow T$. Such a morphism "reads" the information in E to decide a value for S. □

Example 8.3.40. Stream comonad. A Kleisli morphism for the stream comonad is the set function $C(S) \longrightarrow T$. Such a morphism looks like this:

$$x_0, x_1, x_2, \ldots \qquad \mapsto \qquad t. \tag{8.105}$$

In other words, the Kleisli morphism is a function that takes into account the entire history—or trajectory—of an element x_0 rather then just the value of x_0 alone. This has many applications. □

How do all these descriptions of structures relate to each other? Figure 8.7 summarizes some of the structures that we have seen in this section. Each of the forgetful functors have adjoints. There are also relationships between these categories of algebras and coalgebras. Furthermore, one can compose the adjoint functors to get a derived monad and a derived comonand. The original monad and comonad are related to these derived monads and comonads. Much work has been done in this direction, and it remains a fruitful area of study.

Combining Monads

Often, we would like to combine descriptions of structures. We saw that with algebraic theories, there are methods to combine descriptions of structures. Similarly, there are methods to combine monads.

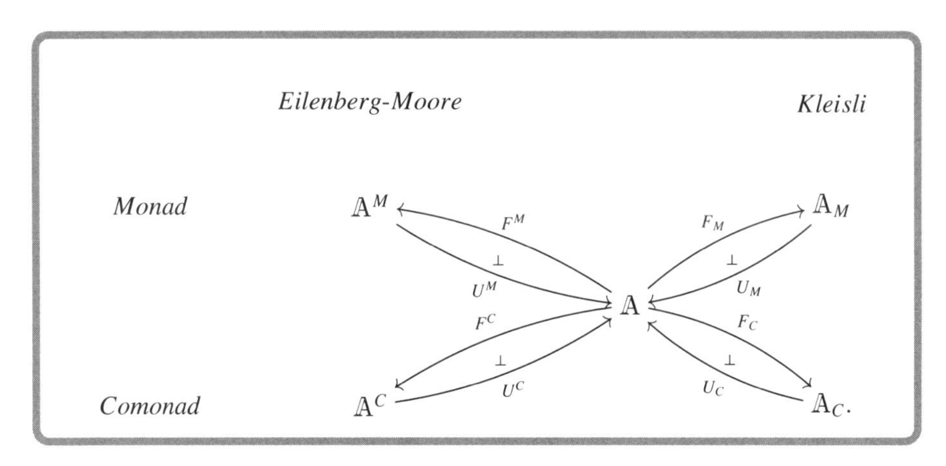

Figure 8.7. The categories of extra structures described by monads and comonads. The forgetful functor for each category forgets the extra structure.

Let us consider the following motivating example.

Example 8.3.41. Consider the algebraic structure of a ring. Remember that a ring $(M, \star, e, -, \odot, u)$ has the structure of an abelian group $(M, \star, e, -)$ and a monoid $(M, \odot, u)$ on the same set. The $\odot$ operation must distribute over the $\star$ operation, as in

$$x \odot (y \star z) = (x \odot y) \star (x \odot z) \qquad \text{and'} \qquad (y \star z) \odot x = (y \odot x) \star (z \odot x). \tag{8.106}$$

Given a monad on $\mathbb{Set}$ that describes abeliean groups, and a monad on $\mathbb{Set}$ that describes monoids, a distributive law is a rule as to how the operations of one structure relate to the other structure. □

Definition 8.3.42. Let (M, μ, η) and (M', μ', η') be monads on $\mathbb{A}$. A **distributive law** is a natural transformation:

$$\delta \colon MM' \Longrightarrow M'M, \tag{8.107}$$

such that δ coheres with μ's and the η's as follows:

$$\begin{array}{ccccc} MM'M' & \overset{\delta M'}{\Rightarrow} & M'MM' & \overset{M'\delta}{\Rightarrow} & M'M'M \\ {\scriptstyle M\mu'}\Downarrow & & & & \Downarrow{\scriptstyle \mu' M} \\ MM' & & \underset{\delta}{\Longrightarrow} & & M'M \end{array} \qquad \begin{array}{ccccc} & & M & & \\ & {\scriptstyle M\eta'}\swarrow & & \searrow{\scriptstyle \eta' M} & \\ MM' & & \underset{\delta}{\Longrightarrow} & & M'M \end{array} \tag{8.108}$$

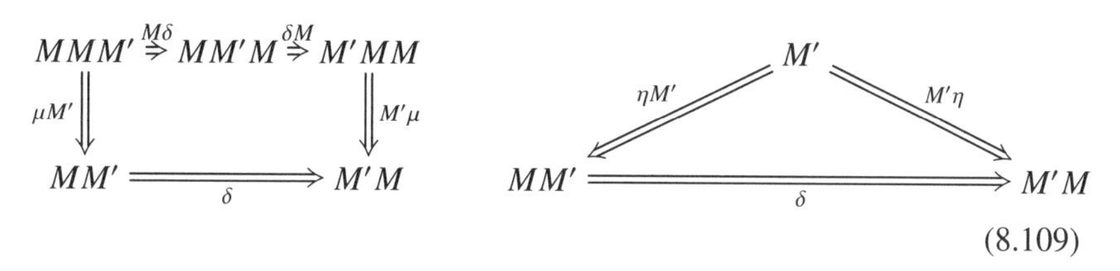

(8.109)

With such a distributive law, one can go on to describe a combined monad (M'', μ'', η'') defined as

- $M'' = M'M\colon \mathbb{A} \longrightarrow \mathbb{A}$.
- $\mu'' = (\mu'\mu) \circ (M'\delta M)\colon M'MM'M \Longrightarrow M'M'MM \Longrightarrow M'M$.
- $\eta'' = \eta' \circ \eta\colon Id_{\mathbb{A}} \Longrightarrow M'M$.

There also are distributive laws between two comonads, and even distributive laws between monads and comonads. This is just the beginning of the story.

There are generalizations of monads to 2-categories called "2-monads." For example, there are 2-monads on the 2-category $\mathbb{C}\mathrm{at}$ that endow a category with the extra structure so that its algebras are monoidal categories or symmetric monoidal categroies. (See, e.g., [37].)

There are large schools of mathematics and computer science that deal with algebras and coalgebras of endofunctors (without the natural transformations). They get interesting results without even assuming the axioms of monads and comonads. (See, e.g., [12].)

Suggestions for Further Study

There is a beautiful presentation of most of these ideas in Chapter 5 of [207]. Nice presentations of advanced ideas about monads and comonads can be found in Chapter VI of [180], Chapter 10 of [21], Chapter 14 of [33], and Chapter 5 of [222].

Paolo Perrone gave a beautiful introduction to probability monads at the New York City Category Theory Seminar [206].

8.4 Algebraic 2-Theories

Before Chapter 5, the central theme of this book was sets with extra structure. From that chapter on, we have been dealing with categories with extra structure. In this section, we will explore a way of describing categories with extra structure.

Certain categories with extra structure can be described using algebraic theories. Simply examine models of such algebraic theories in the Cartesian category $(\mathbb{C}\mathrm{at}, \times, \mathbf{1})$. We have already mentioned such structures, such as a monoid object in $\mathbb{C}\mathrm{at}$ that is a strict monoidal category, and a commutative monoid object in $\mathbb{C}\mathrm{at}$ that is a strictly symmetric strict monoidal category. While such structures are important, we have seen that there are many other types of categories with extra structure. We would like to describe categories with extra structure that include natural transformations (like reassociations and braidings).

To describe categories with extra structure that include natural transformations, we need algebraic theories with another dimension. The new dimension is formulated in the language of 2-categories (or equivalently, in the language of algebraic theories enriched over $\mathbb{C}\mathrm{at}$). We have seen 2-categories several times in these pages and will look at them formally in Section 9.4.

Definition 8.4.1. A **2-theory** or an **algebraic 2-theory** $\overline{\mathbf{T}}$ is a 2-category whose objects are the natural numbers with the property that $n \cong 1^n$.

Within a 2-theory, there are the following elements:

- *0-cells*: A natural number n corresponds to the category $\mathbb{C}^n$ for some category $\mathbb{C}$.
- *1-cells*: A morphism $f\colon m \longrightarrow n$ corresponds to a functor $F\colon \mathbb{C}^m \longrightarrow \mathbb{C}^n$ for some category $\mathbb{C}$.
- *2-cells*: A 2-cell $\alpha\colon f \Longrightarrow g$ corresponds to a natural transformation between functors. For example, α in

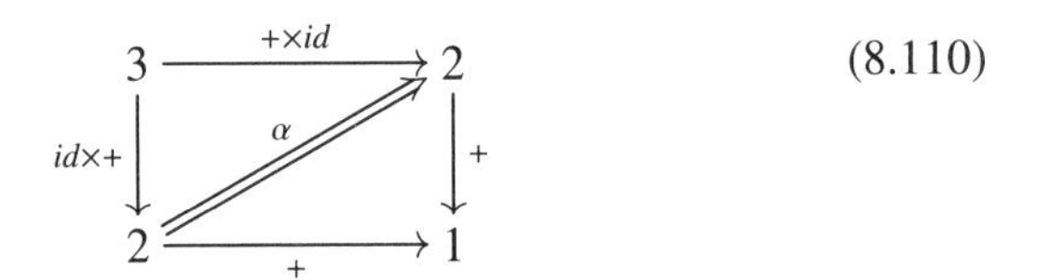

(8.110)

corresponds to the reassociativity morphism between the two ways of associating three objects in $\mathbb{C}$.

When two morphisms in a 2-theory are set to be equal to each other, this corresponds to two functors that are set to be equal to each other. However, in 2-theories, there is a new dimension. Rather than having an equality of morphisms, we can have a 2-cell between them. We then go on and ask when two 2-cells are equal to each other. This corresponds to a coherence rule. For example, the 2-cell in Diagram (8.110) corresponds to a reassociation natural transformation. One glues together these α as in Figure 8.8 to form the now-familiar pentagon condition for monoidal categories, as shown in Diagram (5.34). Let us examine Figure 8.8 more carefully. There are two shapes. Both are about morphisms from 4 to 1 (i.e., functors that accept four inputs and output one output. The shape on the left has two α's, while the shape on the right has three α's. The central equal sign says that the two sides of the pentagon are equal to each other.

Example 8.4.2. Here are some examples of 2-theories:

- The 2-theory of strict monoidal categories $\overline{\mathbf{T}_{stmc}}$. (This is really the theory of monoids $\mathbf{T}_{Monoid}$ with only identity 2-cells.)
- The 2-theory of strictly symmetric strict monoidal categories $\overline{\mathbf{T}_{sssmc}}$. (This is really the theory of commutative monoids with only identity 2-cells.)
- The 2-theory of monoidal categories $\overline{\mathbf{T}_{mc}}$.
- The 2-theory of symmetric monoidal categories $\overline{\mathbf{T}_{smc}}$.

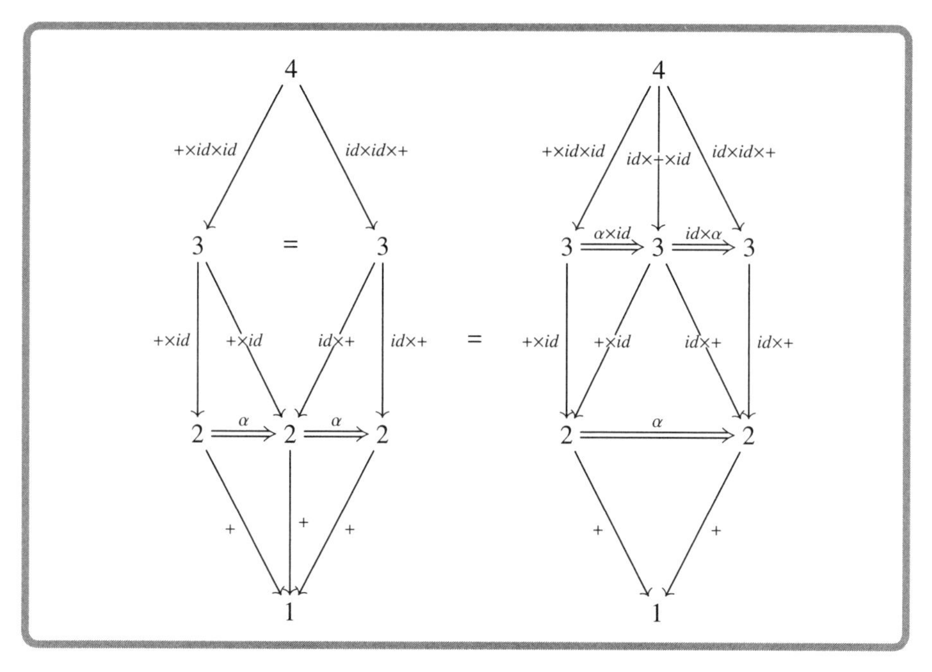

Figure 8.8. Mac Lane's pentagon as an equality in the 2-theory of monoidal categories.

- The 2-theory of braided monoidal categories $\overline{\mathbf{T}_{bmc}}$.
- The 2-theory of balanced monoidal categories $\overline{\mathbf{T}_{balmc}}$. □

Most of the rest of this section will just follow the same yoga as Section 8.1, but in a higher dimension. We will move a little swifter because the ideas have already been discussed.

Definition 8.4.3. For every algebraic 2-theory, there is a 2-category of **algebras**.

- Categories with extra structure correspond to algebras for a 2-theory $\overline{\mathbf{T}}$. They are product-preserving 2-functors from the 2-theory $\overline{\mathbf{T}}$ to the 2-category $\overline{\mathbb{Cat}}$.
- Functors between categories with extra structure correspond to natural transformations between such 2-functors.
- Natural transformations between such functors correspond to morphisms between natural transformations (sometimes called **modifications**).

In summary, for a 2-theory $\overline{\mathbf{T}}$, the collection of all three levels forms a 2-category of algebras: $\overline{Alg(\overline{\mathbf{T}}, \overline{\mathbb{Cat}})}$.

Given two 2-theories, $\overline{\mathbf{T}}$ and $\overline{\mathbf{T}'}$, it is easy to define a 2-theory morphism $G\colon \overline{\mathbf{T}} \longrightarrow \overline{\mathbf{T}'}$. It is a 2-functor that is the identity on 0-cells and preserves the product structure. There is also the notion of a 2-theory natural transformation.

Definition 8.4.4. We gather 2-theories, 2-theory morphisms, and 2-theory natural transformations to form the 2-category of 2-theories, $\overline{2\mathbb{Theory}}$.

There are four adjoint functors between the one-dimensional version of algebraic theories, $\mathbb{Theory}$ and the two-dimensional version, $\overline{2\mathbb{Theory}}$. This is similar to the relationship between $\mathbb{Cat}$ and $\mathbb{Set}$ (and $\mathbb{Top}$ and $\mathbb{Set}$) that we saw in Section 4.4. The adjunctions are

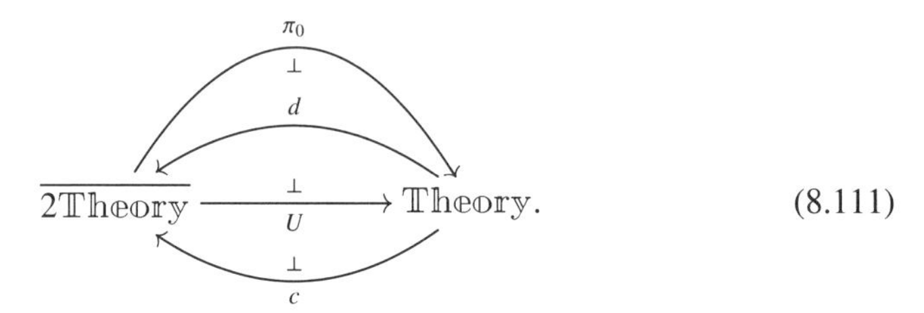

(8.111)

These four functors work as follows:

- Functor U takes a 2-theory to the 1-theory which has the same 1-cells and no 2-cells (forget the 2-cells).
- Functor d takes a 1-theory to the 2-theory, which has the same 1-cells and only has identity 2-cells.
- Functor c takes a 1-theory to the 2-theory, which has the same 1-cells and a unique 2-cell between any two parallel 1-cells in the 1-theory.
- Functor π_0 takes a 2-theory to the 1-theory, which considers any two 1-cells to be the same if there is a 2-cell between them.

A 2-theory morphism $G\colon \overline{\mathbf{T}} \longrightarrow \overline{\mathbf{T}'}$ induces a 2-functor:

$$G^*: \overline{Alg(\overline{\mathbf{T}'}, \overline{\mathbb{Cat}})} \longrightarrow \overline{Alg(\overline{\mathbf{T}}, \overline{\mathbb{Cat}})}. \tag{8.112}$$

Example 8.4.5. Here are some examples of 2-theory morphisms G and the G_* that they induce:

- There is an injection of the 2-theory of monoidal categories to the 2-theory of symmetric monoidal categories. (All 2-theory morphisms are the identity on objects. So an injection means on 1-cells.) This induces the forgetful 2-functor from the 2-category of symmetric monoidal categories onto the 2-category of monoidal categories.
- There is an injection of the 2-theory of monoidal categories to the 2-theory of braided monoidal categories. This induces the forgetful 2-functor from the 2-category of braided monoidal categories onto the 2-category of monoidal categories.
- There is a surjection of the 2-theory of monoidal categories to the 2-theory of strict monoidal categories. This induces the inclusion of the 2-category of strict monoidal categories into the 2-category of monoidal categories.
- There is a surjection of the 2-theory of braided monoidal categories to the 2-theory of symmetric monoidal categories (the two ways of braiding are set

equal). This induces the inclusion of the 2-category of symmetric monoidal categories into the 2-category of braided monoidal categories.

Notice that just as in Example 8.1.15, there is a duality here between 2-theories and their 2-categories of algebras. □

A 2-theory morphism $G\colon \overline{\mathbf{T}} \longrightarrow \overline{\mathbf{T'}}$ also induces a 2-functor:

$$G_*\colon \overline{Alg(\overline{\mathbf{T}}, \overline{\mathbb{C}\mathrm{at}})} \longrightarrow \overline{Alg(\overline{\mathbf{T'}}, \overline{\mathbb{C}\mathrm{at}})}. \tag{8.113}$$

The 2-functors G^* and G_* are a type of adjunction for 2-categories. See [92] and [268] for details.

Ever since Mac Lane's classic paper [180], coherence questions have played a major role when studying categories with extra structure. Let us revisit coherence theory from the perspective of 2-theories and 2-theory morphisms. Every 2-theory-morphism G between two 2-theories induces an algebraic 2-functor G^* and its left 2-adjoint (a 2-categorical version of adjunction) G_*. The left 2-adjoint may be of differing strengths (we will not go into details here). The types of the unit, $\eta\colon id \longrightarrow (G^* \circ G_*)$, and the counit, $\varepsilon\colon (G_* \circ G^*) \longrightarrow id$, tell us to what extent one structure can replace another. Whereas in one-dimensional functorial semantics, the left adjoint is an equivalence of categories if and only if the theory-morphism is an isomorphism, in two-dimensional functorial semantics, there are many intermediate levels of the adjunction. A 2-theory morphism can induce a weak-, quasi-, strict-, equivalence-, or bi-equivalence-adjunction (see [268] and [269] for details). Different levels of adjunction tell us how much one structure can replace another structure. For example, there is an obvious 2-theory morphism from the 2-theory of monoidal categories to the 2-theory of strict monoidal categories. This 2-theory-morphism induces an inclusion from the 2-category of strict monoidal categories into the 2-category of monoidal categories. The left 2-adjoint assigns to every monoidal category a strict monoidal category. The unit of this adjunction is an equivalence at each monoidal category. This is a restatement of Mac Lane's main coherence theorem (namely, that every monoidal category is equivalent to a strict monoidal category). Another example is the 2-theory morphism from the 2-theory of braided monoidal categories to the 2-theory of symmetric monoidal categories. This 2-theory morphism induces an inclusion of symmetric monoidal categories into braided monoidal categories. The left 2-adjoint assigns to every braided monoidal category a symmetric monoidal category. However, in this case, the unit of the adjunction is a *quasi*-equivalence, and hence not every braided monoidal category is equivalent to a symmetrical monoidal category.

This process gives us a recipe for solving coherence questions: one must examine the two 2-theories and at the 2-theory-morphisms between them; determine the strength of the adjunction induced by this 2-theory morphism; and "replace" one structure by the other structure to the extent that the 2-adjunction allows. This recipe gives us a universal and organic manner of handling coherence questions.

Let us get a little spacey and broadly discuss what was done in Sections 7.4, 8.1, and this section. From a bird's-eye perspective, these three sections are related in an interesting way.

In Section 8.1, we met algebraic theories. There are generalizations of such theories, which can be used to describe the types of algebras needed in the study of quantum groups (e.g., bialgebras, quasi-bialgebras, quasi-triangular bialgebras, Hopf algebras, quasi-triangular Hopf algebras, etc.). Without going into detail about such generalized theories, call the category of such generalized algebraic theories and their appropriate theory morphisms $\mathbb{Theory}'$.

In Section 7.4, we saw the 2-category of algebras $\overline{\mathbb{Algebra}}^{op}$, which actually has the algebras needed for quantum groups. There is some type of functor or 2-functor $Alg(\ ,\mathbb{Vect})\colon \mathbb{Theory}' \longrightarrow \overline{\mathbb{Algebra}}$ that takes a generalized theory to the algebras that it describes.

In Section 7.4, we saw the 2-category $\overline{\mathbb{Cat}}/\mathbf{C}\mathbb{Vect}$, which contains the categories of modules for those algebras. We also looked at the monoidal 2-functor MOD that connects them.

In this section, we will examine the collection of 2-theories, $2\mathbb{Theory}$, that describe categorical structure. There is a variation of the functor $2Alg(\ ,\mathbb{Cat})$ that takes such a 2-theory to the categorical structure it describes.

We can see all these connections as follows:

Description of algebras	Algebras	Categories of modules	Descriptions of categoricial structure

$$\mathbb{Theory}' \xrightarrow{Alg(\ ,\mathbb{Vect})} \overline{\mathbb{Algebra}}^{op} \xrightarrow{MOD(\ )} \overline{\mathbb{Cat}}/\mathbf{C}\mathbb{Vect} \xleftarrow{2Alg(\ ,\mathbb{Cat})} \overline{2\mathbb{Theory}'} \tag{8.114}$$

These functors have various types of adjoints. This shows a profound relationship between algebraic structures and categorical structures.

Suggestions for Further Study

John W. Gray introduced 2-theories in [93]. Two-dimensional functorial semantics is described in my articles [268] and [269]. In these papers, many examples of 2-theories and their algebras are given. The relationship between theories and 2-theories is exploited to give many examples of each. The papers also show how one can combine two 2-theories to get a new 2-theory, called the "Kronecker product," such that the algebras of the new 2-theory have both structures of the two 2-theories.

Notice that the 2-theories that we described are not very powerful. We did not talk about describing categorical structure with contravariance or with a closed structure. One must generalize and expand the notion of 2-theory to describe more sophisticated types of categorical structure.

There are many generalizations that need to be mentioned. Again, one need not take algebras only in $\mathbb{Cat}$. One can also take algebras in any 2-category that has finite products. For example, let $\overline{\mathbb{Top}}$ be the 2-category of topological spaces, continuous maps, and homotopies between continuous maps. The 2-theories can then be used to describe many structures on topological spaces. For other work on various notions of

2-theories and their relation to physics, see Thomas M. Fiore's book [74] and the papers [115, 114, 116, 75].

It should be mentioned that there are other ways of describing extra structure on categories. We already mentioned 2-monads [37]. Max Kelly formulated the notion of a **club** to describe categories with extra structure [139, 141, 142].

8.5 Mini-Course: Databases and Schedules

Applied category theory is a new and exciting area of category theory whose goal is to use the tools and ideas of category theory for industry and various other applications. Some of the leaders in this area are John C. Baez, Bob Coecke, and David Spivak. They have each established schools that are focused on showing the broad utility of categorical language and methods.

This mini-course will highlight one such area. We examine databases and how to solve scheduling and optimization problems with the tools of category theory. The categorical ideas needed have already been learned in Section 4.5 where we covered exponentiation, and Section 8.1, where we looked at algebraic theories.

Databases

Databases are at the core of any large business or institution. They are ways of storing structured information. The database does not just have the information; it is structured in a desirable way. This chapter taught us how to describe structures. Let us use some of those methods to describe the structures of databases.

Some foreshadowing is in order. We will see that a category is a way of describing a database. We call such a category a **database schema**. In the same vein as functorial semantics, just as a functor from an algebraic theory to $\mathbb{S}\mathrm{et}$ is an algebraic structure, a functor from a database schema to $\mathbb{S}\mathrm{et}$ is an instance of the database. A functor between two database schema is a way of describing how information flows. Such a functor will induce functors between databases. One of the hard parts of database theory is integrating various databases. These functors will help in database integration. In this mini-course, we go through several toy examples showing these methods. We end with an example that uses these tools to describe and solve a scheduling problem.

Let us start with a simple database.

Example 8.5.1. Consider the following database of a class with two types of students: regular students and honor students. (The students are characters from *Crime and Punishment*.)

Regular Student	First Name	Last Name	Grade
001	Rodion	Romanovitch	A
002	Sofya	Marmeladova	A
003	Dmitry	Prokofyich	C
004	Avdotya	Raskolnikova	B
005	Pyotr	Petrovich	A

Honor Student	First Name	Last Name	Grade
a	Arkady	Ivanovich	F
b	Pulkheria	Raskolnikova	A
c	Semyon	Marmeladov	B
d	Andrey	Lebezyatnikov	A

Every vertical column is an attribute or property and every horizontal row is an entry in the database.

How can we describe this database with a category? First, meditate on Diagrams (8.3) and (8.5). Just think about the objects and the morphisms and forget the fact that the objects are products. Within Diagram (8.3), the objects correspond to types of information and the morphisms correspond to functions that take as input one type of information and output the other type of information. Within Diagram (8.5), the objects correspond to the actual information (of a type) and the morphisms correspond to functions that take the information (of a type) and output other information (of a type). Let us employ this insight into our student database. The attributes (or the names of the columns) of the database correspond to types of information. One type of information is "Regular Student" and another is "First Name." The fact that every row that has a regular student also has a first name means that there is a function

$$\begin{matrix}\text{Regular}\\ \text{Student}\end{matrix} \longrightarrow \begin{matrix}\text{First}\\ \text{Name.}\end{matrix} \tag{8.115}$$

Similarly, the fact that every "Honor Student" has a "Grade" shows that there is a function

$$\begin{matrix}\text{Honor}\\ \text{Student}\end{matrix} \longrightarrow \text{Grade.} \tag{8.116}$$

We can put all these functions together to form a directed graph. By adding identities and the obvious compositions, we can form a category:

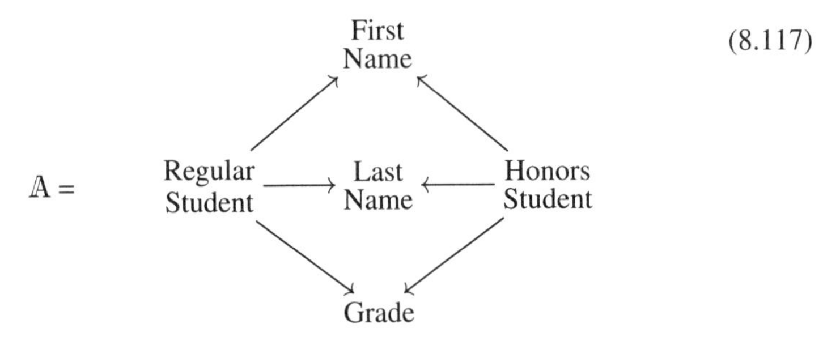

This category, $\mathbb{A}$, describes the structure of the database and is called a "database schema." In a sense, $\mathbb{A}$ is the shape of the database.

In the same way that an algebraic structure corresponds to a functor $P\colon \mathbf{T} \longrightarrow \mathbb{Set}$ from an algebraic theory to $\mathbb{Set}$, an instance of this database will correspond to a functor $P\colon \mathbb{A} \longrightarrow \mathbb{Set}$. When we were dealing with the algebraic theory $\mathbf{T}$, we insisted that P be a product-preserving functor. For the database schema, where there need not be any products, no such requirement is made. Any functor—not just a product-preserving functor—is an instance of the database schema. A typical functor $P\colon \mathbb{A} \longrightarrow \mathbb{S}\mathrm{et}$ takes "Regular Student" to a set. There are many sets, but in this case, we have

$$P(\text{Regular Student}) = \{001, 002, 003, 004, 005\} \tag{8.118}$$

and

$$P(\text{Grade}) = \{A, B, C, D, F\}. \tag{8.119}$$

Notice that although none of the wonderful students of our database received a "D" in the class, it is possible for a student to get that grade. The set function

$$P(\text{Regular Student}) \longrightarrow P(\text{Grade}) \tag{8.120}$$

assigns to every regular student a grade as described by our table.

The collection of all functors from $\mathbb{A}$ to $\mathbb{S}\mathrm{et}$, and natural transformation between such functors, form the category $\mathbb{S}\mathrm{et}^{\mathbb{A}}$, which we call **$\mathbb{A}$-Instances**.

Let us move on and formulate an even simpler database that has only one table of "Students" and does not make a distinction between regular students and honor students. The top of this single-table database looks like this:

Student	First Name	Last Name	Grade

This database can be described with a similar category $\mathbb{B}$, which is on the right side of Diagram (8.121). There is an obvious surjective functor $G\colon \mathbb{A} \longrightarrow \mathbb{B}$, which is the identity on "First Name," "Last Name," and "Grade." The functor G takes both "Regular Student" and "Honor Student" to "Student."

(8.121)

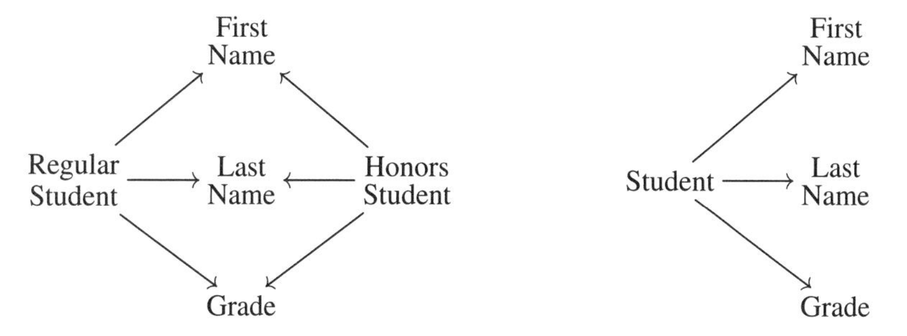

Just as in functorial semantics, this functor G induces a **pullback functor** $\Delta_G\colon \mathbb{S}\mathrm{et}^{\mathbb{B}} \longrightarrow \mathbb{S}\mathrm{et}^{\mathbb{A}}$ from $\mathbb{B}$-instances to $\mathbb{A}$-instances. The functor Δ_G is defined by precomposition; that is,

$$P\colon \mathbb{B} \longrightarrow \mathbb{S}\mathrm{et} \qquad \mapsto \qquad P \circ G\colon \mathbb{A} \longrightarrow \mathbb{B} \longrightarrow \mathbb{S}\mathrm{et}. \tag{8.122}$$

In other words, given an instance of the database,

Student	First Name	Last Name	Grade
1-	Jack	Baxter	A
2-	Phyllis	Wahl	A

The output will simply copy the information to both "Regular Student" and "Honor Student" as follows:

Regular Student	First Name	Last Name	Grade
1-	Jack	Baxter	A
2-	Phyllis	Wahl	A

Honor Student	First Name	Last Name	Grade
1-	Jack	Baxter	A
2-	Phyllis	Wahl	A

This is not a very interesting operation. After all, a student is regular or honors, not both. So in this case, Δ_G is not useful.

The main idea of this mini-course is that for every functor G between two database schema, the induced functor Δ_G has two adjoints as follows:

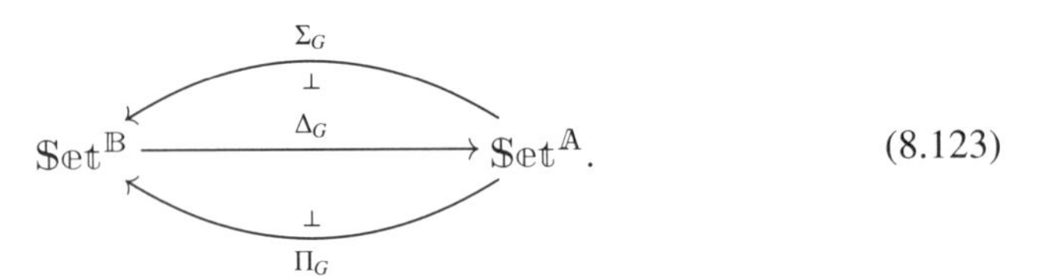

(8.123)

The functor Σ_G is called the **left pushforward functor**, while Π_G is called the **right pushforward functor**. Presently, we will show how these functors are defined by example. The formal definition of these functors will be seen in Section 9.2 when we explore Kan extensions. Whenever there is a functor $G\colon \mathbb{A} \longrightarrow \mathbb{B}$ between two schema (categories), we will explore how the data migrates with the three induced functors $\Pi_G \vdash \Delta_G \vdash \Sigma_G$.

The functor Σ_G is going to take instances of regular and honors students and map them all to students. That is, it takes records that are in "Regular Student" or "Honor Student" and puts them into "Student." This functor is used to merge two databases. The functor Π_G will take those students that are in both "Regular Student" and "Honor Student" to "Student." In other words, Π_G is like a limit and picks out the duplicates. This example shows how categorical methods can be used to unite two databases that have the same attributes.

The astute reader should notice that the functors Σ_G and Π_G are similar to the quantifiers (functors) $\exists$ and $\forall$ that we saw in Section 4.8. These, in turn, are related to the logical operations (functors) $\vee$ and $\wedge$. □

Let us explore other functors between database schemas.

Example 8.5.2. One of the simplest database schemas is the category **1**, with just one object and one identity morphism. The set of **1**-instances is $\mathbb{Set}^{\mathbf{1}}$, which is simply $\mathbb{Set}$. We might write a set as a silly database as follows:

1
Jack
Jill
x
blue

For any database schema $\mathbb{A}$, there is a unique functor $!\colon \mathbb{A} \longrightarrow \mathbf{1}$. The induced functor $\Delta_!$ takes any set and makes a silly database out of it, where the entries for $\mathbb{A}$ become the elements of the set. The functor $\Sigma_!$ takes any $\mathbb{A}$-instance and returns the colimit (like the union) of all the elements in the instance. (The functorr $\Pi_!$ is the limit, but here it is not interesting.) □

Example 8.5.3. Let us look at merging two databases with different attributes. Consider the following two tables of spies:

Spy ID	First Name	Last Name	Birth Year	Death Year
001	Klaus	Fuchs	1911	1988
002	Whittaker	Chambers	1901	1961
003	Moe	Berg	1902	1972
004	Benedict	Arnold	1741	1801
005	Nathan	Hale	1755	1776
006	Mata	Hari	1876	1917

Spy Letter	First Name	Last Name	Country For	Country Against
a	Klaus	Fuchs	USSR	US
b	Whittaker	Chambers	USSR	US
c	Nathan	Hale	US	Britain
d	Kim	Philby	USSR	Britain
e	Benedict	Arnold	Britain	US
f	Mata	Hari	Germany	France
g	Sidney	Reilly	Britain	Russia

The first table has information about the lifespan of the spies, while the second table has information about which country the spies worked for and which one they worked against. These two tables can be described by the database schema on the left side of Figure 8.9.

A simpler database schema is on the right of Figure 8.9, which will correspond to a single table with all the information. There is a functor G from the left to the right database schema. Here, G is the identity on "Birth Year," "Death Year," etc., and $G(\text{Spy ID}) = G(\text{Spy Letter}) = \text{Spy}$. The functor Δ_G simply takes all the information and duplicates it in two tables. The more interesting functor is Σ_G. It takes the two tables and makes one table out of them with all the information.

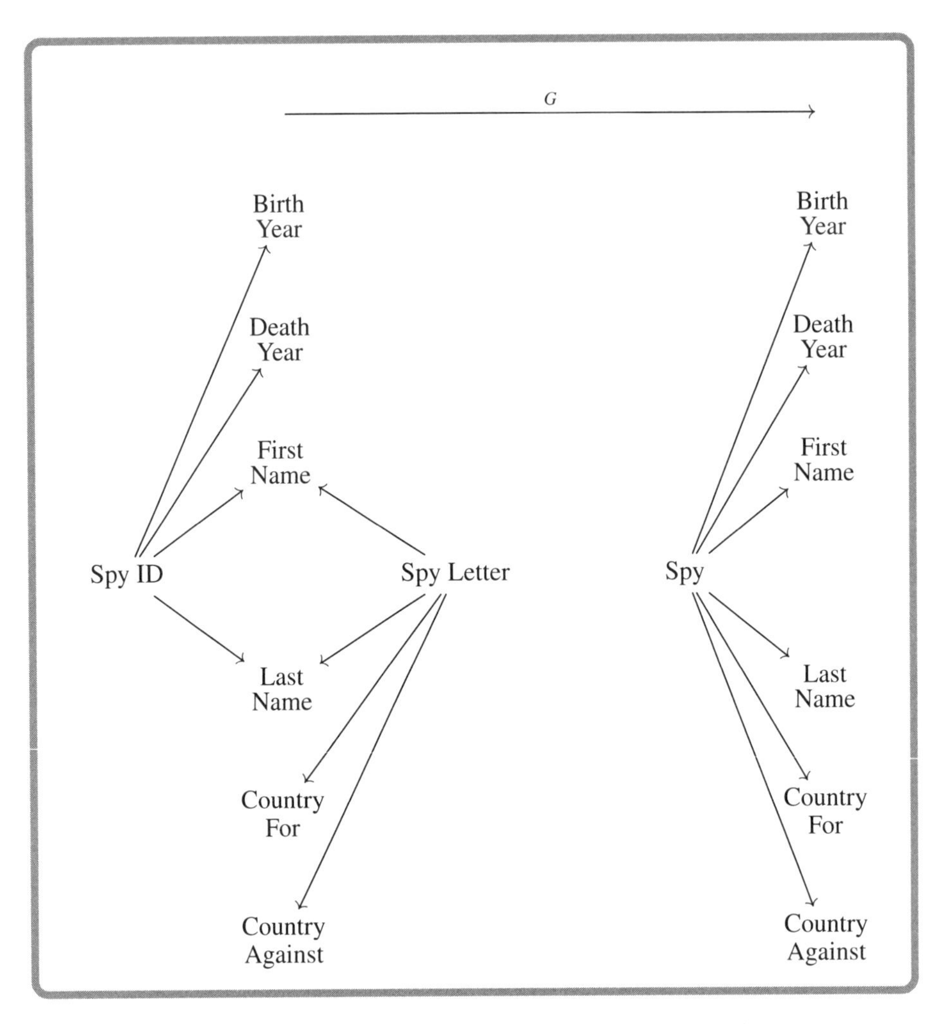

Figure 8.9. Two database schema and a functor relating them.

Spy	First Name	Last Name	Birth Year	Death Year	Country For	Country Against
1	Klaus	Fuchs	1911	1988	USSR	US
2	Whittaker	Chambers	1901	1961	USSR	US
3	Moe	Berg	1902	1972	3.For	3.Against
4	Benedict	Arnold	1741	1801	Britain	US
5	Nathan	Hale	1755	1776	US	Britain
6	Mata	Hari	1876	1917	Germany	France
7	Kim	Philby	7.Birth	7.Death	USSR	Britain
8	Sidney	Reilly	8.Birth	8.Death	Britain	Russia

What is to be done with information that we do not have? As you can see, the database does not have information as to what country Moe Berg worked for or against. The database also does not have the birth or death year of Kim Philby and Sidney Reilly. The Σ_G functor makes up the information and fills in the missing pieces. There is nothing universal about how the information is filled in. In other words, Σ_G is not unique, but it is unique up to isomorphism. □

We conclude our discussion of databases with a variation of Example 8.5.3, which deals with isomorphisms.

Example 8.5.4. Consider the previous database schemas and the functor G between them. Now extend the one on the left by adding an isomorphism between "Spy ID" and "Spy Letter." The changed part looks like this:

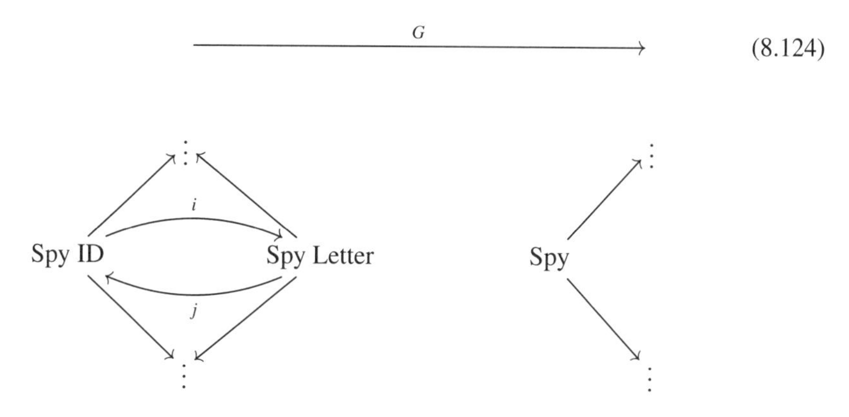

with $i \circ j = id$ and $j \circ i = id$. We furthermore assume that i and j commute with the name morphisms in the schema.

The first thing to notice is that the isomorphism ensures that the two tables have the same number of rows. The fact that the isomorphism commutes with the other morphisms means that the two tables have the same entries. Another point is that the isomorphisms are part of the data. This means that the two tables will have information about each other. A typical list can then be as follows:

Spy ID	**First Name**	**Last Name**	**Birth Year**	**Death Year**	i
001	Klaus	Fuchs	1911	1988	c
002	Whittaker	Chambers	1901	1961	b
003	Benedict	Arnold	1741	1801	a

Spy Letter	**First Name**	**Last Name**	**Country For**	**Country Against**	j
a	Benedict	Arnold	Britain	US	003
b	Whittaker	Chambers	USSR	US	002
c	Klaus	Fuchs	USSR	US	001

If we applied Δ_G to an instance of the right schema, we would get such an output with the information of the isomorphisms in the tables. □

Scheduling

Researchers have taken these categorical ideas much further. They use some of these methods to describe and actually solve scheduling and optimization problems. To solve such problems, one must go beyond just looking at the data as simply information and look at it as actual values that can be manipulated and computed.

Our first step is to increase the expressiveness of our database schema. To fix our ideas and ensure that we can express the usual types of data, we introduce a category $\mathbb{Type}$ whose objects are all the usual types like *Strings*, *Nats*, *Ints*, *Char*, and *Reals*. The morphisms will be all computable functions. For example, there will be the following morphisms:

- $+\colon Reals \times Reals \longrightarrow Reals$, which corresponds to addition
- $|\ |\colon Strings \longrightarrow Nat$, which corresponds to taking a string and returning its length
- $(3y+2x)/7z\colon Nats \times Ints \times Reals \longrightarrow Reals$, which corresponds to the function that takes these three types of numbers and returns the stated operation
- $Sqrt\colon Reals^* \longrightarrow Reals$, which describes the positive square root operation
- $-\colon Ints \times Ints \longrightarrow Ints$, which corresponds to subtraction

One should think of $\mathbb{Type}$ as the basic schema. The usual interpretation of $\mathbb{Type}$ is the functor $P_{\mathbb{Type}}\colon \mathbb{Type} \longrightarrow \mathbb{Set}$, which takes *Strings* to the set of strings (for some alphabet) and takes *Ints* to $\mathbf{Z}$, *Nats* to $\mathbf{N}$, etc. The functor $P_{\mathbb{Type}}$ also takes functions to their standard interpretation in $\mathbb{Set}$. For example, the subtraction morphism $-\colon Ints \times Ints \longrightarrow Ints$ goes to the actual subtraction function $-\colon \mathbf{Z} \times \mathbf{Z} \longrightarrow \mathbf{Z}$ in $\mathbb{Set}$.

While a database schema is simply a category, an **enhanced database schema** is a category $\mathbb{A}$ with an inclusion functor $Inc\colon \mathbb{Type} \hookrightarrow \mathbb{A}$. This means that all the usual types are part of $\mathbb{A}$ and can be used. While an instance for a database schema is a functor from the category to $\mathbb{Set}$, an instance of an enhanced database schema $\mathbb{A}$ is a functor $P\colon \mathbb{A} \longrightarrow \mathbb{Set}$ such that the following diagram commutes:

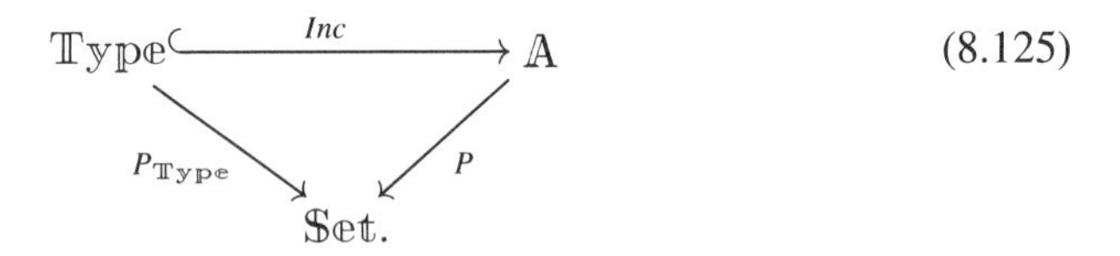

(8.125)

In other words, P takes the shape of the database to various sets while taking the types to their usual interpretations. A functor from one enhanced database schema to another must commute with the inclusions; that is,

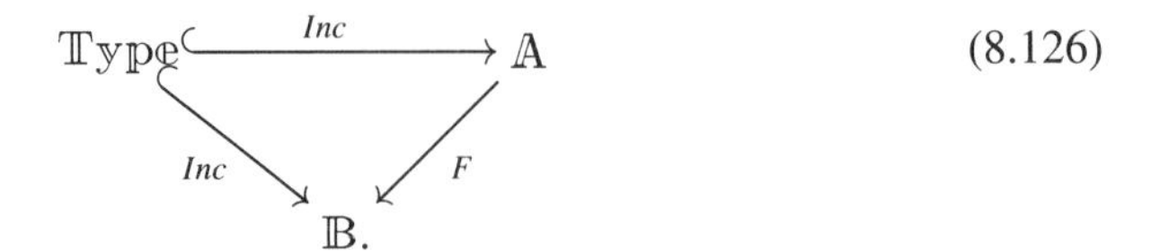

(8.126)

The fact that every enhanced database schema has $\mathbb{T}\mathrm{ype}$ embedded in it gives us another way of writing the arrows of the schema. Rather than writing Diagrams (8.115) and (8.116), we write them as follows:

$$\text{Regular Student} \xrightarrow{\text{First Name}} Strings \qquad \text{Honor Student} \xrightarrow{\text{Grade}} Char. \tag{8.127}$$

Part of the increased expressiveness of the enhanced schema is that there is a special relationship with logic. Let $Type_1$ be an object in an enhanced database schema, and let ϕ be a logical formula whose domain of discourse is $Type_1$. In other words, ϕ is an expression that is true for certain values of $Type_1$. Then there is an object $[\![\phi]\!]$ and an inclusion $[\![\phi]\!] \hookrightarrow Type_1$. This new object is called a **subobject** of $Type_1$. (We will formally meet subobjects in Definition 9.5.2.) Within $\mathbb{S}\mathrm{et}$, if $Type_1$ corresponds to set $P(Type_1)$, then $[\![\phi]\!]$ corresponds to $\{x : \phi(x) \text{ is true}\} \subseteq P(Type_1)$. If logical formula ϕ implies ψ, then $\{x : \phi(x) \text{ is true}\} \subseteq \{x : \psi(x) \text{ is true}\} \subseteq P(Type_1)$. (The connection between logic, categories, and sets will be further elaborated along these lines in Section 9.5, when we meet topos theory.)

Another way to increase the expressiveness of our enhanced database schema requires the instances of the schema to not only be functors from the schema to $\mathbb{S}\mathrm{et}$, but to also preserve some of the structure of the schema. For example, if there is a product $a \times b$ in $\mathbb{A}$, then an instance of $\mathbb{A}$ is a product-preserving functor $P\colon \mathbb{A} \longrightarrow \mathbb{S}\mathrm{et}$ such that $P(a \times b) = P(a) \times P(b)$. This is not new because we saw such requirements when dealing with algebraic theories. In the examples that follow, we might also insist that an inclusion $a \hookrightarrow b$ goes to an inclusion $P(a) \hookrightarrow P(b)$.

Example 8.5.5. Let us look at a first example with a computation in it. Consider our spies from Example 8.5.3. A simplified version of this only looks at each name, birth year, and death year. The schema is as follows:

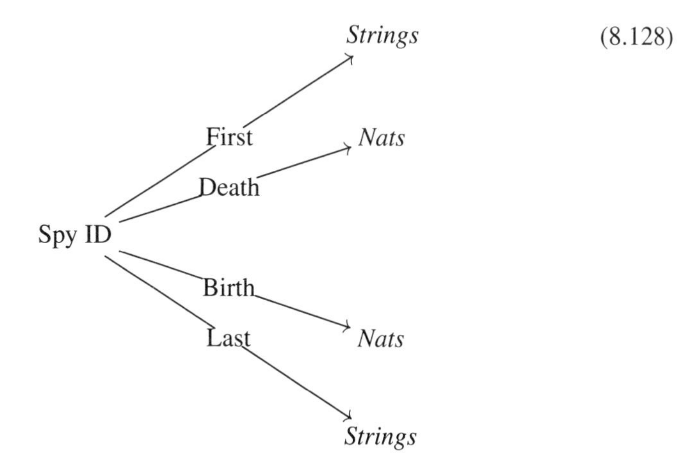

(8.128)

A typical instance of this schema is the first table in Example 8.5.3.

Let us extend this schema to figure out the age when the spy died. Obviously, the spy's final age is the subtraction of the birth year from the death year. This can be described by the following schema:

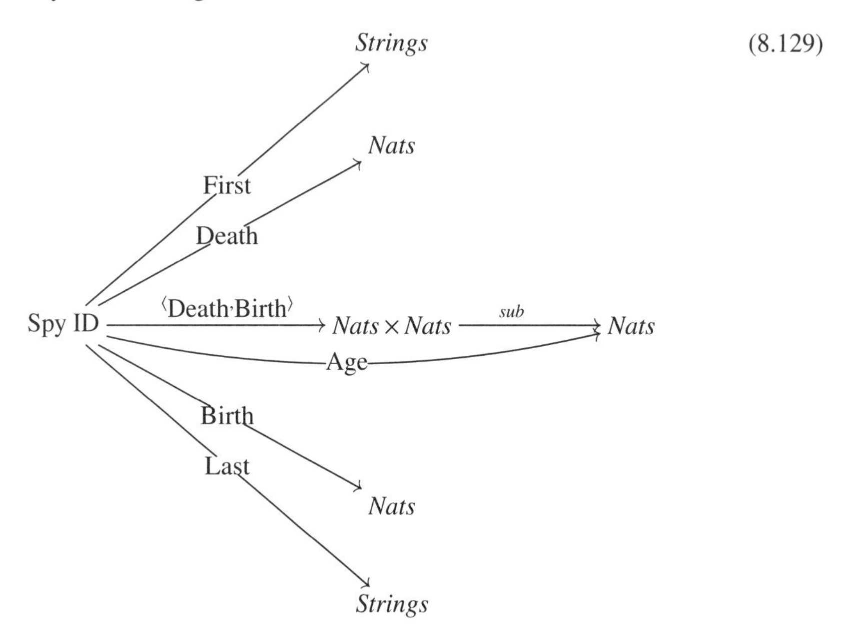

(8.129)

where *sub* is subtraction. (Projection maps are left out of the diagram for readability.) Instances of this schema are required to preserve products. There is an obvious inclusion functor *Inc* from the first schema to this schema. The Δ_{Inc} functor takes an instance with the age information and outputs an instance without that information. In contrast, the functor Σ_{Inc} takes an instance like the first table in Example 8.5.3 to the following table:

Spy ID	First Name	Last Name	Birth Year	Death Year	Death Age
001	Klaus	Fuchs	1911	1988	77
002	Whittaker	Chambers	1901	1961	60
003	Moe	Berg	1902	1972	70
004	Benedict	Arnold	1741	1801	60
005	Nathan	Hale	1755	1776	21
006	Mata	Hari	1876	1917	41

The main point is that the Σ_{Inc} functor adds the extra information in the correct way. □

Example 8.5.6 was taken from [242].

Example 8.5.6. Consider the following simple schema:

$$ID \xrightarrow{Hrs} Reals^*, \tag{8.130}$$

where the type *Real** is nonnegative real numbers. Such numbers are a subobject of the real numbers given by the logical formula $r \geq 0$. A typical instance of this schema is as follows:

ID	Hrs
001	3.14
002	15.9
003	2.653
004	58.9

Think of the ID as the name of an employee and Hrs as the amount of hours worked. If every employer gets paid fifty dollars an hour, then we can extend the schema as follows to get their pay:

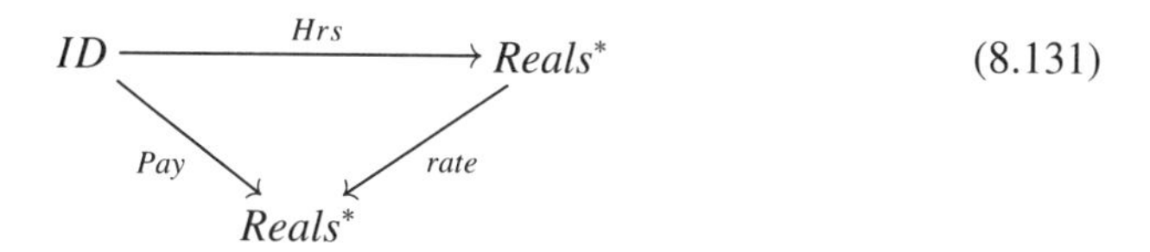

(8.131)

where the morphism $rate\colon Reals^* \longrightarrow Reals^*$ corresponds to the set function that multiplies every number by 50 (i.e., $rate(x) = 50 \cdot x$). There is an inclusion *Inc* of the first schema into the second. The Δ_{Inc} functor forgets the extra pay information, while Σ_{Inc} adds in the extra information. For the $\mathbb{A}$ instance given here, Σ_{Inc} will output the following table:

ID	Hrs	Pay
001	3.14	157
002	15.9	795
003	2.653	132.65
004	58.9	2,945

Again, the Σ_{Inc} performed a computation. □

We close with Example 8.5.7, which is shamelessly stolen from [48].

Example 8.5.7. Here, we use categories to solve a simple version of the **open-scheduling problem**. The problem involves several jobs that have to be completed on several machines. Each job needs to occupy some machine for a specified amount of time. Each job can occupy only one machine at a time, and each machine can be occupied by one job at a time. A solution for such a problem is specific start times and end times for each job on each machine.

In detail, an instance of this problem might look like this: There is a set of jobs that have to be done, $J = \{j_1, j_2, j_3, j_4\}$. These jobs have to be done on machines $M = \{$saw, drill, lathe, mill$\}$. For each job j and each machine m, there is a duration $\tau_{j,m} \geq 0$ of hours that the job must occupy the machine. In this case, the sixteen τ's are given as follows:

		Jobs			
	τ	j_1	j_2	j_3	j_4
Machine	saw	2hr	2hr	2hr	2hr
	drill	2hr	3hr	0hr	3hr
	lathe	2hr	3hr	3hr	0hr
	mill	2hr	2hr	2hr	3hr

A solution for such a scheduling problem is a way of assigning to each job j and each machine m a starting time $s_{j,m}$ and a terminal time $t_{j,m}$. In other words, a solution is an assignment of a time when each job occupies the machine. We denote the interval for a job on a machine as $[s_{j,m}, t_{j,m}]$. The solution times cannot overlap. That means no two jobs are occupying the same machine at one time. This means that for any two jobs, j and j', and for any machine m, we either have $[s_{j,m}, t_{j,m}] \cdots [s_{j',m}, t_{j',m}]$ or $[s_{j',m}, t_{j',m}] \cdots [s_{j,m}, t_{j,m}]$. Similarly, no job is occupying two machines at one time. These requirements can be written in logical notation as

$$\forall j \neq j' \quad \forall m \quad [(t_{j,m} \leq s_{j',m}) \quad \vee \quad (t_{j',m} \leq s_{j,m})] \tag{8.132}$$

and

$$\forall j \quad \forall m \neq m' \quad [(t_{j,m} \leq s_{j,m'}) \quad \vee \quad (t_{j,m'} \leq s_{j,m})]. \tag{8.133}$$

One solution for Example 8.5.7 is given in the following table:

<table>
<tr><th></th><th></th><th colspan="10">Time</th></tr>
<tr><th></th><th></th><th>1</th><th>2</th><th>3</th><th>4</th><th>5</th><th>6</th><th>7</th><th>8</th><th>9</th><th>10</th></tr>
<tr><td rowspan="4">Machine</td><td>saw</td><td colspan="2">j_1</td><td></td><td colspan="2">j_3</td><td></td><td colspan="2">j_2</td><td colspan="2">j_4</td></tr>
<tr><td>drill</td><td colspan="3">j_2</td><td colspan="3">j_4</td><td colspan="2">j_1</td><td></td><td></td></tr>
<tr><td>lathe</td><td colspan="3">j_3</td><td colspan="3">j_2</td><td></td><td></td><td colspan="2">j_1</td></tr>
<tr><td>mill</td><td colspan="3">j_4</td><td colspan="2">j_1</td><td colspan="2">j_3</td><td></td><td colspan="2">j_2</td></tr>
</table>

Now that we understand the open-scheduling problem, let us put it into categorical language. The enhanced database schema to describe the problem is as follows:

$$\mathbb{A} = \qquad J \times M \xrightarrow{\quad \tau \quad} Real^*. \tag{8.134}$$

An instance of this schema must preserve the product. This means that τ assigns a nonnegative real number to any ordered pair consisting of a job and a machine.

The enhanced database schema for the solution is as follows:

$$\mathbb{B} = \qquad J \times M \overset{s}{\underset{t}{\rightrightarrows}} Real^* \tag{8.135}$$

The category also contains the information that $s \leq t$ and the logical formulas that express that the solutions cannot overlap in Equations (8.132) and (8.133). An instance of this schema in sets must preserve the product.

How is the problem related to the solution? There is a functor $F\colon \mathbb{A} \longrightarrow \mathbb{B}$, which is the identity on the objects. The only nontrivial morphism in $\mathbb{A}$ is τ, which goes to the morphism related to

$$(- \circ \langle s, t\rangle)\colon J \times M \longrightarrow Reals^* \times Reals^* \longrightarrow Real^*. \tag{8.136}$$

Stated in words, τ goes to the subtraction of the ending time minus the starting time. The functor F induces Δ_F, which takes a solution of a scheduling problem back to its problem.

The most interesting part of this project is that functor Σ_F takes instances of the problem to its solution. This is actually fairly complicated mathematics. In our example, there are sixteen input variables and thirty-two output variables. While we are here given an intuition of how Σ_F works, in Section 9.2 it will be formally defined as a Kan extension.

This is just the beginning. For more information, see [48] for many ways that these ideas have been extended and modified so they can be used in industry and real-life problems. □

Suggestions for Further Study

This is a very new and exciting topic. A good introduction can be found in Chapter 3 of [77]. This is also a central theme in David Spivak's book [243]. More can be read in recent research papers such as [48, 242, 231].

9

Advanced Topics

Rigidity leads to death, flexibility results in survival . . .
Lao Tzu
Tao Te Ching, Chapter 76

So far, this text has only revealed the minute tip of the top of the iceberg that is category theory. Here, we examine some subfields of category theory in broad strokes. We describe areas of category theory that are important now and will probably be important in the unfathomable future.

It must be stressed that there is no way that these topics can be taught in a few pithy pages. Each subject is an entire field, with many books and hundreds of papers written about it. Rather than teaching each topic, think of these introductions as showing what the field is about. The sincere hope is to whet the reader's appetite and that the presentation inspires one to learn more about the topic in question.

This chapter is organized as follows:

- Section 9.1: Enriched Category Theory
- Section 9.2: Kan Extensions
- Section 9.3: Homotopy Theory
- Section 9.4: Higher Category Theory
- Section 9.5: Topos Theory
- Section 9.6: A mini-course on Homotopy Type Theory (HOTT)

9.1 Enriched Category Theory

Categories were defined to have Hom sets that are sets or classes. In a sense, the definition of categories was based on sets and classes. Many times, however, we saw that some categories have Hom sets with more structure. Here are some examples we saw:

Example 9.1.1.

- The Hom sets in the category of matrices $\mathbf{K}\mathbb{M}\mathtt{at}$ are $\mathbf{K}$ vector spaces (Example 2.4.2).

- The Hom sets in the category of vector spaces $\mathbf{K}\mathbb{Vect}$ are $\mathbf{K}$ vector spaces (Theorem 2.4.38).
- The Hom sets in the 2-category of categories $\overline{\mathbb{Cat}}$ are categories (Remark 4.2.7).
- The Hom sets in a closed categories (which we met in Section 7.2) are, by definition, objects in the same category. □

We will see many more examples of categories with interesting Hom sets in the coming pages.

When a category has Hom sets as objects of a particular category, they are enhanced or "richer" than just having Hom sets in $\mathbb{Set}$. We say that the first category is "enriched" over the other category. Such enrichment is powerful.

In a sense, enriched category theory is all about putting category theory into the language of category theory. An explanation is needed. When categories, functors, natural transformations, adjunctions, limits, colimits, and other concepts were all introduced, we used English to describe them. However, throughout the text, we stressed the importance of defining structures in terms of arrows so we can use these structures in many different contexts (see Important Categorical Idea 1.4.60). In this section, we show how to describe the first few fundamental structures of category theory using the arrow notation of category theory. The very foundations of category theory are set out in the language of category theory.

Definitions

For a category to be enriched over $\mathbb{V}$, we will need $\mathbb{V}$ itself to have a certain amount of structure. For now, let $\mathbb{V}$ be a monoidal category.

Let us formally define what it means for a category to be enriched over another category.

Definition 9.1.2. Let $(\mathbb{V}, \otimes, I, \alpha, \lambda, \rho)$ be a monoidal category. Category $\mathbb{A}$ is **enriched over** $\mathbb{V}$ or is a $\mathbb{V}$-**category** if

- The Hom sets are objects of $\mathbb{V}$: for every a and b in $\mathbb{A}$, there is an object, $\mathbb{A}(a,b)$, in $\mathbb{V}$.
- Composition is a morphism in $\mathbb{V}$: for every $a, b, c \in \mathbb{A}$, there is a morphism in $\mathbb{V}$:
$$\circ\colon \mathbb{A}(b,c) \otimes \mathbb{A}(a,b) \longrightarrow \mathbb{A}(a,c). \tag{9.1}$$
- Identity maps are picked out by the unit of $\mathbb{V}$: for every a in $\mathbb{A}$, there is a map $\upsilon_a\colon I \longrightarrow \mathbb{A}(a,a)$, which corresponds to the identity morphism.

These maps must satisfy the following requirements:

- Composition is associative: for every a, b, c, and d in $\mathbb{A}$, the following diagram commutes:

$$\begin{array}{ccccc} \mathbb{A}(c,d)\otimes(\mathbb{A}(b,c)\otimes\mathbb{A}(a,b)) & & \xrightarrow{\alpha} & & (\mathbb{A}(c,d)\otimes\mathbb{A}(b,c))\otimes\mathbb{A}(a,b)) \\ \downarrow{\scriptstyle id\otimes\circ} & & & & \downarrow{\scriptstyle \circ\otimes id} \\ \mathbb{A}(c,d)\otimes\mathbb{A}(a,c) & \xrightarrow[\circ]{} & \mathbb{A}(a,d) & \xleftarrow[\circ]{} & \mathbb{A}(b,d)\otimes\mathbb{A}(a,b). \end{array} \tag{9.2}$$

- The identity maps chosen by υ's act as a unit: for all a and b in $\mathbb{A}$, the following two triangles commute:

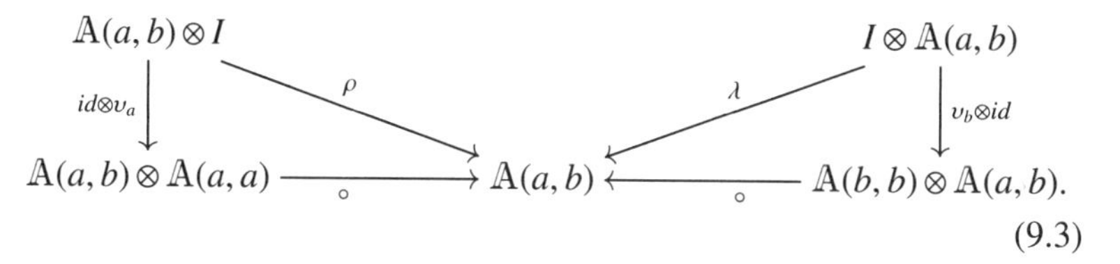

(9.3)

Here are some examples:

Example 9.1.3.

- Any locally small category is enriched over $(\mathbb{S}\mathrm{et}, \times, \{*\})$.
- Categories that are not necessarily locally small are enriched over $(\mathbb{SET}, \times, \{*\})$, where $\mathbb{SET}$ is the collection of all sets and classes. (Shhh. Most mathematicians do not consider this collection very kosher.)
- A 2-category is a category enriched over $(\mathbb{C}\mathrm{at}, \times, \mathbf{1})$.
- $\mathbf{K}\mathbb{M}\mathrm{at}$ and $\mathbf{K}\mathbb{V}\mathrm{ect}$ are enriched over $\mathbf{K}\mathbb{V}\mathrm{ect}$. □

We will see many more examples in the coming pages.

We go on and define a morphism between two $\mathbb{V}$-categories in Definition 9.1.4.

Definition 9.1.4. Let $\mathbb{A}$ and $\mathbb{B}$ be $\mathbb{V}$-categories. A map $F\colon \mathbb{A} \longrightarrow \mathbb{B}$ is a $\mathbb{V}$-**functor** if there is a function $F_0\colon Obj(\mathbb{A}) \longrightarrow Obj(\mathbb{B})$, and for every a, b in $\mathbb{A}$, a morphism in $\mathbb{V}$:

$$F_{a,b}\colon \mathbb{A}(a,b) \longrightarrow \mathbb{B}(F_0(a), F_0(b)). \tag{9.4}$$

This function and family of morphisms in $\mathbb{V}$ must satisfy the following requirements:

- The functor respects composition: for every triple of objects a, b, and c in $\mathbb{A}$, the following square commutes:

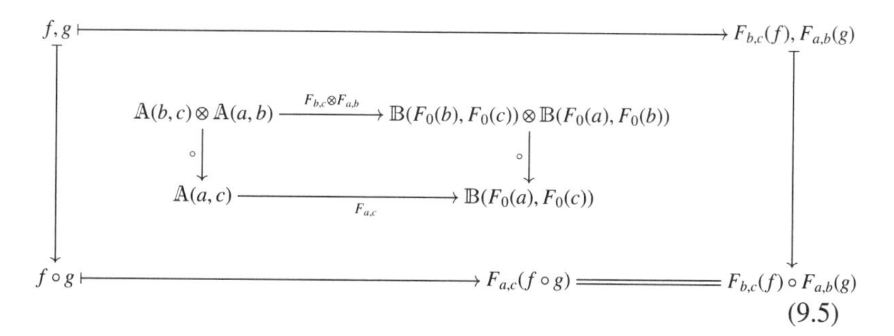

(9.5)

- The functor takes identities to identities: for every object a in $\mathbb{A}$, the following triangle commutes:

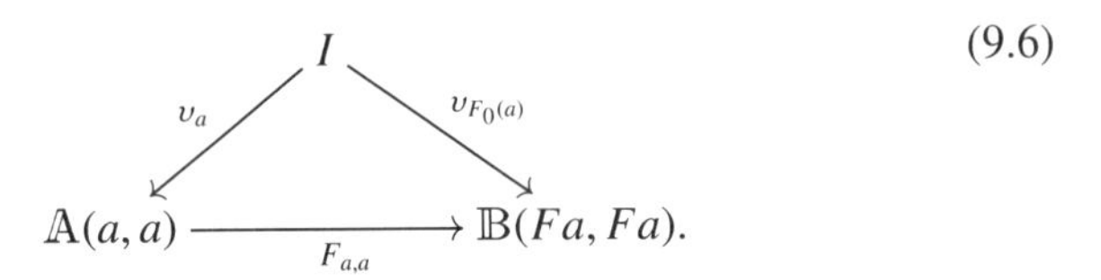

(9.6)

Exercise 9.1.5. Give the definition of a contravariant $\mathbb{V}$-functor. ■

Exercise 9.1.6. Show how to define the composition of two $\mathbb{V}$-functors. ■

Of course, there are morphisms between $\mathbb{V}$-functors.

Definition 9.1.7. Let $\mathbb{A}$ and $\mathbb{B}$ be $\mathbb{V}$-categories and $F\colon \mathbb{A} \longrightarrow \mathbb{B}$ and $G\colon \mathbb{A} \longrightarrow \mathbb{B}$ be two $\mathbb{V}$-functors. Then $\tau\colon F \Longrightarrow G$ is a $\mathbb{V}$-**natural transformation** if there is a family of morphisms in $\mathbb{V}$ indexed by the objects in $\mathbb{A}$:

$$\tau_a\colon I \longrightarrow \mathbb{B}(F_0(a), G_0(a)), \tag{9.7}$$

which satisfy the following commutative diagram (naturality condition):

$$\begin{array}{ccccc} \mathbb{A}(a,b) & \xrightarrow{\lambda^{-1}} & I \otimes \mathbb{A}(a,b) & \xrightarrow{\tau_b \otimes F_{a,b}} & \mathbb{B}(F_0(b), G_0(b)) \times \mathbb{B}(F_0(a), F_0(b)) \\ \downarrow{\scriptstyle \rho^{-1}} & & & & \downarrow{\scriptstyle \circ} \\ \mathbb{A}(a,b) \otimes I & \xrightarrow[G_{a,b}\otimes\tau_a]{} & \mathbb{B}(G_a(a), G_0(b)) \otimes \mathbb{B}(F_0(a), G_0(a)) & \xrightarrow[\circ]{} & \mathbb{B}(F_0(a), G_0(b)). \end{array} \tag{9.8}$$

Exercise 9.1.8. Formulate the definition of vertical and horizontal composition of $\mathbb{V}$-natural transformations. ■

Definition 9.1.9. Putting together $\mathbb{V}$-categories, $\mathbb{V}$-functors, and $\mathbb{V}$-natural transformations, we form the 2-category $\overline{\mathbb{V}\mathbb{C}\mathrm{at}}$. This is a whole universe based on the monoidal category $\mathbb{V}$.

Examples

Example 9.1.10. Consider the strict monoidal category $(\mathbf{2}, \wedge, \top)$, where the category $\mathbf{2}$ is a partial order with two elements, $0 \leq 1$. Let $\mathbb{A}$ be a category enriched over $\mathbf{2}$. For all elements a and b, the Hom set $Hom_{\mathbb{A}}(a,b)$ is either 0 or 1. Think of $Hom_{\mathbb{A}}(a,b) = 0$ as there being no morphism between a and b, and think of $Hom_{\mathbb{A}}(a,b) = 1$ as there being exactly one morphism between a and b. Given three objects a, b, and c of $\mathbb{A}$, there will be a composition of $a \longrightarrow b$ and $b \longrightarrow c$ if $Hom_{\mathbb{A}}(a,b) = 1$ and $(\wedge)$ $Hom_{\mathbb{A}}(b,c) = 1$. The main point is that the enrichment ensures that there is no more than one element in the Hom sets of $\mathbb{A}$. In other words, a category enriched over **2** is a preorder category. □

Example 9.1.11. Consider the symmetric strict monoidal category $(\mathbf{R}^+, +, 0)$. This category also has a preorder structure because $\mathbf{R}^+$ is totally ordered. Add one element, ∞, to this category and form the symmetric strict monoidal category $([0,\infty], +, 0)$. The new element is dealt with as follows: for all $r \in \mathbf{R}^+$, we have $r + \infty = \infty$ and $r \leq \infty$. Categories enriched over $[0,\infty]$ describe the minimum "cost" or "price" of getting from one object in the category to another object. In detail, let $\mathbb{A}$ be a $[0,\infty]$-category. For all a and b in $\mathbb{A}$, the value $Hom_{\mathbb{A}}(a,b) \geq 0$ tells the minimum cost of going from a to b. If $Hom_{\mathbb{A}}(a,b) = \infty$, then the cost of going from a to b is unattainable. If the cost of going from a to b is at least r_1 and the cost of going from b to c is at least r_2, then the cost of going from a to c is at least $r_1 + r_2$. This is written as follows:

$$Hom_{\mathbb{A}}(a,b) + Hom_{\mathbb{A}}(b,c) \geq Hom_{\mathbb{A}}(a,c). \tag{9.9}$$

This rule follows from the composition in the category and the preorder of the $[0,\infty]$. The rule is called the **triangle inequality** because it can be expressed as

$$\begin{array}{ccc} a & \xrightarrow{r_1} & b \\ & {\scriptstyle r_1+r_2} \searrow \quad \swarrow {\scriptstyle r_2} & \\ & c. & \end{array} \tag{9.10}$$

A $[0,\infty]$-category is called a **Lawvere metric space** because it satisfies some of the rules of being a metric space. (See Section 2.3.3 of [77].) □

The following definitions of special types of enriched categories are central to all advanced work in algebra and topology. Remember that $(\mathbb{A}\mathbb{b}\mathbb{G}\mathbb{p}, \otimes, \mathbf{Z})$ is the strict monoidal category of abelian groups.

Here are some definitions:

Definition 9.1.12.

- Category $\mathbb{C}$ is a **preadditive category** if it is enriched over abelian groups (i.e., $\mathbb{C}$ is an $\mathbb{A}\mathbb{b}\mathbb{G}\mathbb{p}$-category). Within a preadditive category, there is a zero morphism (for a, b in the category, $Hom(a,b)$ is an abelian group with an identity element), which permits one to talk about a kernel and a cokernel of a morphism. If $f_0 \colon a \longrightarrow b$ is the identity of $Hom(a,b)$ then for any $g \colon a \longrightarrow b$, the equalizer of f_0 and g is the kernel of g. The coequalizer of f_0 and g is the cokernel of g.
- Category $\mathbb{C}$ is an **additive category** if $\mathbb{C}$ is an $\mathbb{A}\mathbb{b}\mathbb{G}\mathbb{p}$-category with all finite products. In such a category, the finite product is also a finite coproduct, and hence a finite biproduct (we saw this for the direct sum of vector spaces in Exercises 5.2.7 and 5.2.22). Such categories also have zero objects (biproducts of the empty diagram).

- Category $\mathbb{C}$ is an **abelian category** if it is (i) an additive category; (ii) every morphism $f\colon a \longrightarrow b$ has a kernel included into its domain and a cokernel that is a quotient of its codomain:

$$Ker(f) \hookrightarrow a \xrightarrow{f} b \twoheadrightarrow Cok(f); \tag{9.11}$$

  and (iii) every monomorphism is the kernel of some map and every epimorphism is the cokernel of some map. With these kernels and cokernels, one can work with exact sequences.
- Category $\mathbb{C}$ is a **K-linear category** if it is enriched over $\mathbf{K}\mathbb{V}\mathbb{e}\mathbb{c}\mathbb{t}$. These are used in many aspects of physics.

Some examples are needed.

Example 9.1.13.

- The categories $\mathbb{A}\mathbb{b}\mathbb{G}\mathbb{p}$, $\mathbf{K}\mathbb{V}\mathbb{e}\mathbb{c}\mathbb{t}$, and $\mathbf{K}\mathbb{M}\mathbb{a}\mathbb{t}$ are preadditive, additive, and abelian.
- In Section 7.4, we defined a category of modules for an algebra. We can also define a category of modules for ring R. The collection of such modules form a category called $R\mathbb{M}\mathbb{o}\mathbb{d}$, which is an preadditive category.
- A ring is a one-object preadditive category. In detail, if $\mathbb{A}$ is a one-object preadditive category, then $Hom_{\mathbb{A}}(*,*)$ is an abelian group. There is also another binary operation given by composition. The distributivity follows from the fact that the second operation is the morphism

$$Hom(*,*) \otimes Hom(*,*) \longrightarrow Hom(*,*) \tag{9.12}$$

  in $\mathbb{A}\mathbb{b}\mathbb{G}\mathbb{p}$.
- The category $\mathbf{K}\mathbb{V}\mathbb{e}\mathbb{c}\mathbb{t}$ is K-linear. For all small $\mathbb{C}$, the category $\mathbf{K}\mathbb{V}\mathbb{e}\mathbb{c}\mathbb{t}^{\mathbb{C}}$ is $\mathbf{K}$-linear.
- The category of free abelian groups is an example of an additive category that is not an abelian category.
- An algebra (of Section 7.4) is a one-object $\mathbf{K}$-linear category. The multiplication is given by the composition.
- For a topological space X, the category $\mathbb{A}\mathbb{b}\mathbb{G}\mathbb{p}^{O(X)}$ is the category of presheaves on X (we will see this more in Section 9.5) is an abelian category and is the starting point of a lot of algebraic geometry.
- Is $\mathbb{A}$ is an abeilian category, then for any category $\mathbb{C}$, the functor category $\mathbb{A}^{\mathbb{C}}$ is an abelian category.
- $\mathbf{K}$-linear categories that have rigid monoidal structures are called **tensor categories**. (There is some confusion here because sometimes in the literature, a monoidal category is called a tensor category as well.) Special types of tensor categories are **fusion categories**. See pages 65 ff. of [71] for more on this topic. Such categories are fundamental to some advanced research in algebra and mathematical physics. □

Abelian categories may seem abstract. They are not. Several times in these pages, we have seen that abstract concepts have concrete representations. To recap:

- The Yoneda lemma (Section 4.7) taught us that every object of a small category $\mathbb{A}$ can be thought of as a concrete type of functor in $\mathbb{S}\mathrm{et}^{\mathbb{A}^{op}}$.
- Cayley's representation theorem (mentioned in Remark 6.2.4) taught us that every element of a group G can be thought of as a concrete transformation of elements.
- The two proofs of the coherence theorem (Section 6.2) showed us that every object in a monoidal category can be thought of as a concrete object in a strict monoidal category.
- The Stone duality theorem of Section 6.4 shows us that every object of a Boolean algebra can be thought of as a concrete subset of sets.

What about abelian categories? What type of concrete structures represent abelian categories?

Theorem 9.1.14. Freyd-Mitchell embedding theorem. Any small abelian category can be embedded into a category of R-modules for ring R. In detail, for every small abelian category $\mathbb{A}$, there is a ring R and a category of modules $R\mathbb{M}\mathrm{od}$ such that there is a faithful embedding $\mathbb{A} \hookrightarrow R\mathbb{M}\mathrm{od}$. The objects of $\mathbb{A}$ coincide with R-modules. This embedding respects short exact sequences. ★

This theorem helps because it tells us that when working with diagrams in abelian categories, you might as well think of diagrams in categories of modules. The proof of this theorem can be found in Section 1.6 of [259], in Section 1.14 of [41], and in [80].

The 2-category $\mathbb{V}\mathbb{C}\mathrm{at}$ does not stand by itself.

Definition 9.1.15. When dealing with enriched categories over $\mathbb{V}$ and $\mathbb{V}'$, a monoidal functor $F\colon \mathbb{V} \longrightarrow \mathbb{V}'$ is called a **change of basis**. Such a functor induces a 2-functor $F_*\colon \mathbb{V}\mathbb{C}\mathrm{at} \longrightarrow \mathbb{V}'\mathbb{C}\mathrm{at}$. For $\mathbb{A}$ as a $\mathbb{V}$-category, the objects of the $\mathbb{V}'$-category $F_*(\mathbb{A})$ are the same objects as $\mathbb{A}$; however, the morphisms of $F_*(\mathbb{A})$ are defined as

$$Hom_{F_*(\mathbb{A})}(a,b) = F(Hom_{\mathbb{A}}(a,b)). \tag{9.13}$$

That is, F_* takes the category to the same category with the same objects, but the Hom sets are the image of functor F.

Some examples are needed here.

Example 9.1.16.

- Let I be the unit of $\mathbb{V}$; then the functor $F = Hom_{\mathbb{V}}(I, \ )\colon \mathbb{V} \longrightarrow \mathbb{S}\mathrm{et}$ induces a forgetful 2-functor $F_*\colon \mathbb{V}\mathbb{C}\mathrm{at} \longrightarrow \mathbb{C}\mathrm{at}$ that takes every $\mathbb{V}$-category to its underlying category.
- The monoidal functor $F\colon [0,\infty] \longrightarrow \mathbf{2}$ is defined as $F(r) = 0$ if $r = 0$ and $F(r) = 1$ if $r > 0$. This functor takes a Lawvere metric space to the preorder for which there is an arrow from a to b if the cost of going from a to b is 0.
- The monoidal functor $G\colon [0,\infty] \longrightarrow \mathbf{2}$ is defined as $G(r) = 0$ if $r < \infty$ and $G(r) = 1$ if $r = \infty$. This functor takes a Lawvere metric space to the preorder for which there is an arrow from a to b if and only if the cost of going from a to b less than ∞.

- The forgetful functor $U\colon \mathbf{K}\mathbb{Vect} \longrightarrow \mathbb{AbGp}$ induces a forgetful functor from the 2-category of **K**-linear categories to the 2-category of abelian categories. □

See Section 6.4 of [41] for more on the change of base functors.

This is just the beginning of the story. Researchers have gone on to formulate every major categorical structure in an enriched context. So there are enriched adjunctions, enriched limits and colimits, the enriched Yoneda lemma, etc. For many of these constructions, one needs $\mathbb{V}$ to be more than just a monoidal category, such as a symmetric monoidal closed category.

While all these enriched concepts are important, enriched limits and colimits are very important and arise in many areas. They are usually called **weighted limits** and **weighted colimits**.

Researchers even look at enriched categories over an enriched category. The properties of the enriched category depend on the properties of the category over which it is enriched. This gives us universes within universes. Such thoughts have been known to cause headaches.

Suggestions for Further Study

For a nice introduction to the field, see the two-part blog post of Tai-Danae Bradley [43, 44]. For a more detailed introduction, see Chapter 2 of [77], Chapter 6 of [41], and Chapter 9 of [219]. Preadditive, additive, and abeilian categories can all be found in Chapter VIII of [180], Chapter 1 of [41], and any book of homological algebra, such as [259]. More about **K**-linear categories and tensor categories can be learned about in [71]. The bible of enriched category theory is [147], which has been retyped and put online as [143].

9.2 Kan Extensions

One of the most powerful tools in category theory was first formulated by Daniel Kan and is called a "Kan extension." This construction is a vast generalization of a limit or a colimit, and it can be used to describe so much more than that. Kan extensions are so powerful that Saunders Mac Lane famously titled Section X.7 of [180] "All Concepts Are Kan Extensions." That means that every construction with a universal property can be written as a type of Kan extension (see Important Categorical Idea 4.6.9).

The idea of a Kan extension is not simple, and a little categorical sophistication is required to understand what it is all about. One must struggle over the concept several times before it sinks in. Press on!

Let us begin with a simple example.

Example 9.2.1. Consider the totally ordered categories of the integer, **Z**, and the real number, **R**. There is an obvious inclusion $inc\colon \mathbf{Z} \hookrightarrow \mathbf{R}$. Use this arrow to create the following diagram:

(9.14)

Now pose the following question: What is the best functor $F\colon \mathbf{R} \longrightarrow \mathbf{Z}$ that completes this triangle. For an integer, say 3, the output, $F(3)$, should be 3. In that case, the triangle will commute. But what about the real number 3.14? Where should $F(3.14)$ be? While there are no order-preserving functors that make the triangle commute, there are many order-preserving functors where the triangle does not commute. There are, however, two "best-fitting" functors called the left and right Kan extensions. The left Kan extension takes a real number r to the floor of r, while the right Kan extension takes a real number r to the ceiling of r. A way to think about these functors is that they do not make the triangles commute, but they are the functors that are the closest to commuting. We will prove this fact about the ceiling and floor functor in a few pages. There are many such examples in this section.

The astute reader will remember that in Exercise 4.4.9 we saw that the ceiling functor is an adjunct. Also, in Exercise 4.4.10 we saw that the floor functor is an adjoint. We will see that all adjoints can be written as Kan extensions. □

Definitions

There are many ways to think about a Kan extension—and we have to use *all* possible intuitions to truly grasp it.

First intuition. Consider categories $\mathbb{A}$, $\mathbb{B}$, and $\mathbb{C}$, where $\mathbb{B}$ is a subcategory of $\mathbb{C}$ (see Figure 9.1). Let $Inc\colon \mathbb{B} \hookrightarrow \mathbb{C}$ be an inclusion functor and $G\colon \mathbb{B} \longrightarrow \mathbb{A}$ be any functor. Now we ask the simple question: What is the best way to extend (hence the

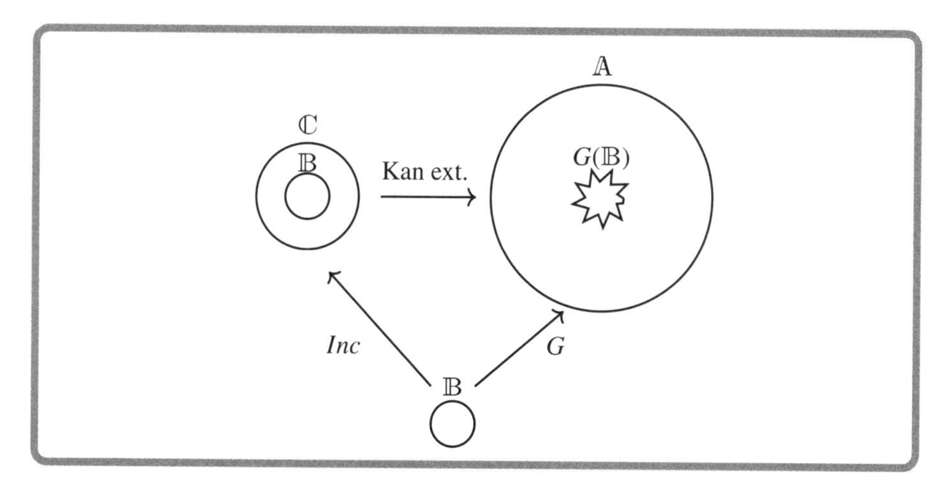

Figure 9.1. Extending the functor G from category $\mathbb{B}$ to the larger category $\mathbb{C}$.

word "extension") functor G to the whole category $\mathbb{C}$? This is the same idea we saw on page 18, which was about extending functions from one set to a larger set. Here, we are extending functors from one category to a larger category. Amazingly, Kan extensions work even for the functor $\mathbb{B} \longrightarrow \mathbb{C}$, which is an arbitrary functor and not an inclusion of categories.

Second intuition. A helpful way of thinking about Kan extensions is as a generalization of limits and colimits. As we saw in Chapter 3 and Section 4.6, a limit of a diagram $\mathbb{B}$ in a category $\mathbb{A}$ is an object of $\mathbb{A}$ that completes the diagram. Imagine that for every object c in $\mathbb{C}$, there is a subcategory of $\mathbb{B}$ written as $\mathbb{B}_c$, and a functor $\mathbb{B}_c \hookrightarrow \mathbb{B} \longrightarrow \mathbb{A}$ that describes a diagram in $\mathbb{A}$. In such a setup, there will be a functor that takes every object c of $\mathbb{C}$ to the limit of $\mathbb{B}_c$. Such a functor is called a "left Kan extension." We can also talk about colimits and cocompleting diagrams. In that case, we get a "right Kan extension."

While intuitions are fine and dandy, we need a clear definition. For pedagogical reasons, we are going to progressively unwrap the definition of a Kan extensions in three steps. The definitions get more and more detailed as we go on. We will then concentrate on the last, most detailed definition.

Definition 9.2.2. Let $\mathbb{A}$, $\mathbb{B}$, and $\mathbb{C}$ be categories. A functor $K : \mathbb{B} \longrightarrow \mathbb{C}$ induces

$$- \circ K = Hom_{\mathbb{C}\mathrm{at}}(K, \mathbb{A}) = \mathbb{A}^K : \mathbb{A}^{\mathbb{C}} \longrightarrow \mathbb{A}^{\mathbb{B}}. \tag{9.15}$$

Left and right adjoints of this functor, if they exist, are called **left** and **right Kan extensions along** K. A left Kan extension along K is denoted as Lan_K, while a right Kan extension along K is denoted as Ran_K. This is summarized by the following diagram:

$$\mathbb{A}^{\mathbb{C}} \underset{Ran_K}{\overset{Lan_K}{\underset{\perp}{\overset{\perp}{\rightleftarrows}}}} \mathbb{A}^{\mathbb{B}}. \quad (- \circ K : \mathbb{A}^{\mathbb{C}} \longrightarrow \mathbb{A}^{\mathbb{B}}) \tag{9.16}$$

We can see the first intuition here. When K is an inclusion map, the map $- \circ K$ takes any map $\mathbb{C} \longrightarrow \mathbb{A}$ to its restriction $\mathbb{B} \longrightarrow \mathbb{A}$. The other two maps take any map $\mathbb{B} \longrightarrow \mathbb{A}$ to the extensions of those maps $\mathbb{C} \longrightarrow \mathbb{A}$.

We can also see the second intuition here. The diagrams of $\mathbb{B}$ in $\mathbb{A}$ are found in $\mathbb{A}^{\mathbb{B}}$. The functor $K : \mathbb{B} \longrightarrow \mathbb{C}$ controls how the subcategories of $\mathbb{B}$ are related to the objects and morphisms of $\mathbb{C}$. The Kan extensions accept a functor in $\mathbb{A}^{\mathbb{B}}$ and return a functor that maps an object c of $\mathbb{C}$ to the limit or colimit of $\mathbb{B}_c$. This is a functor in $\mathbb{A}^{\mathbb{C}}$.

Example 9.2.3. Diagram (9.16) is very similar to other diagrams that we already saw on these pages: See Diagrams (4.107), (4.115), (4.178), (4.208), (4.227), (4.228), and (8.123). To appreciate the power and ubiquity of Kan extensions, it pays to see how these other diagrams are special instances of Diagram (9.16).

- Diagram (4.107) is a special instance of Diagram (9.16) in which the following are all true:

- Category $\mathbb{A}$ is any category.
- Category $\mathbb{B}$ is $\mathbf{2}_\circ$ the discrete category with two objects.
- Category $\mathbb{C}$ is $\mathbf{1}$.
- Functor $K\colon \mathbb{B} \longrightarrow \mathbb{C}$ is the unique functor : $\mathbf{2}_\circ \longrightarrow \mathbf{1}$.
- The induced functor is $-\circ K\colon \mathbb{A}^{\mathbf{1}} = \mathbb{A} \longrightarrow \mathbb{A}\times\mathbb{A} = \mathbb{A}^{\mathbf{2}_\circ}$.

• Diagram (4.115) is a special instance of Diagram (9.16). First, remember that

$$\mathbb{S}\mathrm{et}^{\,E \rightrightarrows V} \cong \mathbb{G}\mathrm{raph}.$$

and $\mathbb{S}\mathrm{et}^{\mathbf{1}} \cong \mathbb{S}\mathrm{et}$. There is a unique functor $!\colon \{\, E \rightrightarrows V \,\} \longrightarrow \mathbf{1}$. This functor has a left adjoint $L\colon \mathbf{1} \longrightarrow \{\, E \rightrightarrows V \,\}$, which takes $*$ to V. (It is a left adjoint because V is the terminal object in the category.) These two functors induce the following four functors:

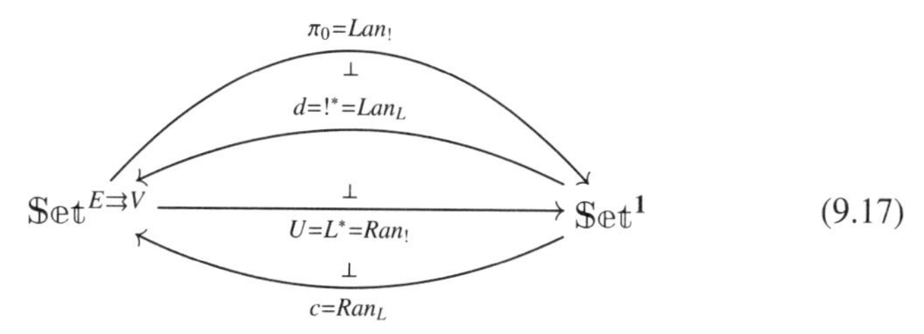

(9.17)

which is exactly Diagram (4.115). We will see a massive generalization of these ideas in Diagram (9.87).

• The fact that Diagram (4.178) is a special instance of Diagram (9.16) is elaborated on in Example 9.2.9.

• Diagram (4.208) is a special instance of Diagram (9.16) in which the following are all true:

- Category $\mathbb{A}$ is $\mathbb{P}\mathrm{rop}$.
- Category $\mathbb{B}$ is $\mathbf{2}_\circ$, the discrete category with two objects.
- Category $\mathbb{C}$ is $\mathbf{1}$.
- Functor $K\colon \mathbb{B} \longrightarrow \mathbb{C}$ is the unique functor : $\mathbf{2}_\circ \longrightarrow \mathbf{1}$.
- The induced functor is $-\circ K\colon \mathbb{P}\mathrm{rop}^{\mathbf{1}} = \mathbb{P}\mathrm{rop} \longrightarrow \mathbb{P}\mathrm{rop}\times\mathbb{P}\mathrm{rop} = \mathbb{P}\mathrm{rop}^{\mathbf{2}_\circ}$.

There is a similar construction for Diagram (4.219).

• Diagram (4.227) is a special instance of Diagram (9.16) in which the following are all true:

- Category $\mathbb{A}$ is $\mathbf{2}_\circ$, the discrete category with two objects.
- Category $\mathbb{B}$ is the discrete category whose objects are set S.
- Category $\mathbb{C}$ is the discrete category whose objects are set T.
- Functor $K\colon \mathbb{B} \longrightarrow \mathbb{C}$ is the functor between two discrete categories based on the function $f\colon S \longrightarrow T$.
- The induced functor is $-\circ K\colon \mathbf{2}_\circ^{\mathbb{C}} = \mathcal{P}(T) \longrightarrow \mathcal{P}(S) = \mathbf{2}_\circ^{\mathbb{B}}$.

• Diagram (4.228) is a Kan extension in which the following are all true:

- Category $\mathbb{A}$ is $\mathbf{2}_\circ$, the discrete category with two objects.
- Category $\mathbb{B}$ is the discrete category whose objects are the set $S \times T$.
- Category $\mathbb{C}$ is the discrete category whose objects are set T.
- Functor $K\colon \mathbb{B} \longrightarrow \mathbb{C}$ is the functor between two discrete categories based on the projection function $\pi\colon S \times T \longrightarrow T$.
- The induced functor is $- \circ K = \pi^{-1}\colon \mathbf{2}_\circ^{\mathbb{C}} = \mathcal{P}(T) \longrightarrow \mathcal{P}(S \times T) = \mathbf{2}_\circ^{\mathbb{B}}$.

- Diagram (4.233) is also a special instance of Diagram (9.16). However, the details are complicated, so we are not discussing them here.
- The fact that Diagram (8.123) is a special instance of Diagram (9.16) is elaborated on in Example 9.2.12. □

Spelling out the details of the adjunction in Definition 9.2.2 gives us the second definition.

Definition 9.2.4. Let $\mathbb{A}$, $\mathbb{B}$, and $\mathbb{C}$ be categories and $K\colon \mathbb{B} \longrightarrow \mathbb{C}$ be a functor. For every functor G in $\mathbb{A}^{\mathbb{B}}$ and P in $\mathbb{A}^{\mathbb{C}}$, a **left Kan extension along** K **of** G, written as $Lan_K(G)$, in $\mathbb{A}^{\mathbb{C}}$ satisfies the natural isomorphism

$$Hom_{\mathbb{A}^{\mathbb{C}}}(Lan_K(G), P) \cong Hom_{\mathbb{A}^{\mathbb{B}}}(G, P \circ K). \tag{9.18}$$

A **right Kan extension along** K **of** G, written as $Ran_K(G)$, in $\mathbb{A}^{\mathbb{C}}$ satisfies the natural isomorphism

$$Hom_{\mathbb{A}^{\mathbb{B}}}(P \circ K, G) \cong Hom_{\mathbb{A}^{\mathbb{C}}}(P, Ran_K(G)). \tag{9.19}$$

Spelling out the unit and counit description of these adjunctions with natural transformations gives us the following definition.

Definition 9.2.5. Let $\mathbb{A}$, $\mathbb{B}$, and $\mathbb{C}$ be categories with functors $K\colon \mathbb{B} \longrightarrow \mathbb{C}$ and $G\colon \mathbb{A} \longrightarrow \mathbb{B}$, as in

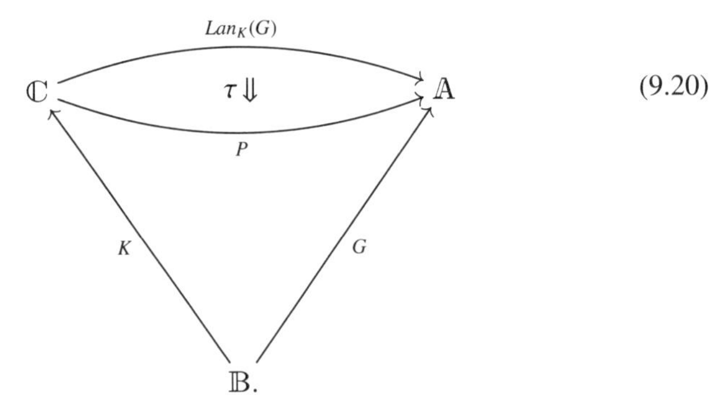

(9.20)

A **left Kan extension along** K **of** G is a pair $(Lan_K(G), \eta)$, where $Lan_K(G)\colon \mathbb{C} \longrightarrow \mathbb{A}$ is a functor, and $\eta\colon G \Longrightarrow Lan_K(G) \circ K$ is a natural transformation that is the unit of the adjunction. The natural transformation η describes how well $Lan_K(G)$ completes the triangle. Note that if η is the identity, then $Lan_K(G)$ makes the triangle commute. As the unit of the adjunction, η must satisfy the following universal property: for every pair (P, η') where P is a functor in $\mathbb{A}^{\mathbb{C}}$ (to be thought of as a "bad" completion of the triangle), and $\eta': G \Longrightarrow P \circ K$ a natural transformation (to be thought of as a measure how "bad" P is as a completion of the triangle), there is a unique natural transformation $\tau: Lan_K(G) \Longrightarrow P$ such that

$$\eta' = (\tau \circ_H K) \circ_V \eta. \tag{9.21}$$

We can clearly state the universal property of the unit of the adjunction as follows:

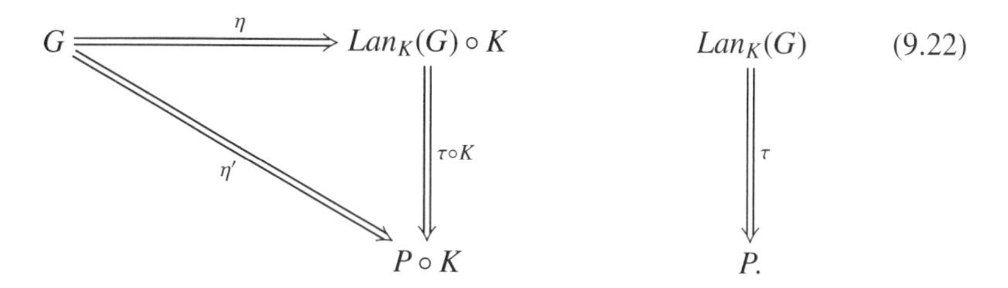

(9.22)

It also pays to go through the details for a right Kan extension. Let $\mathbb{A}$, $\mathbb{B}$, and $\mathbb{C}$ be categories with functors $K: \mathbb{B} \longrightarrow \mathbb{C}$ and $G: \mathbb{A} \longrightarrow \mathbb{B}$, as in

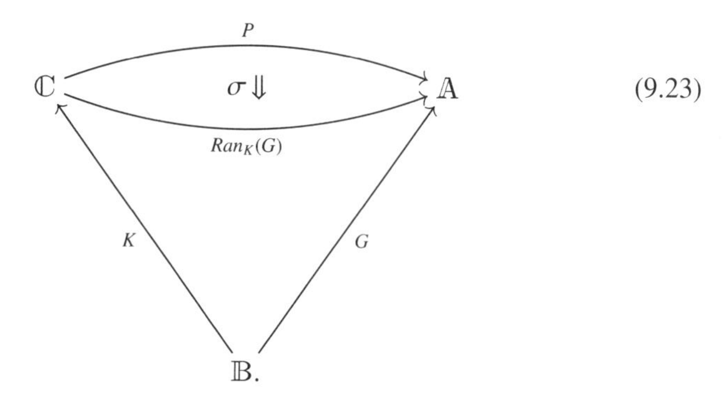

(9.23)

A **right Kan extension along** K **of** G is a pair $(Ran_K(G), \varepsilon)$ where $Ran_K(G)$: $\mathbb{C} \longrightarrow \mathbb{A}$ is a functor, and $\varepsilon: Ran_K(G) \circ K \Longrightarrow G$ is a natural transformation that is a counit of the adjunction. It satisfies the following universal property: for any pair (P, ε') where P is a functor in $\mathbb{A}^{\mathbb{C}}$ and $\varepsilon': P \circ K \Longrightarrow G$ is a natural transformation, there is a unique natural transformation $\sigma: P \Longrightarrow Ran_K(G)$ such that

$$\varepsilon' = \varepsilon \circ_V (\sigma \circ_H K). \tag{9.24}$$

We can clearly state the universal property of the counit of the adjunction as follows:

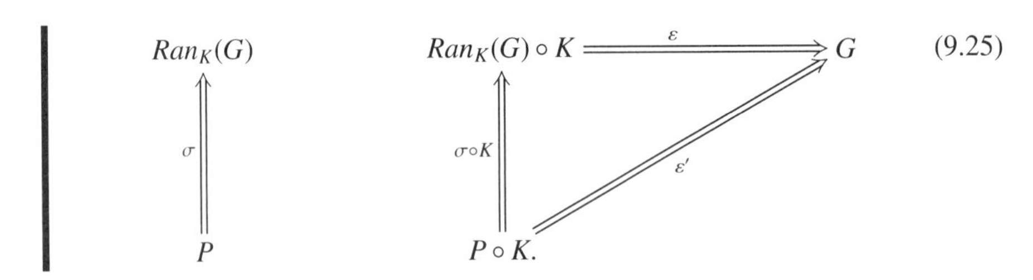

(9.25)

All these definitions are perfect. But how does one actually construct such a Kan extension? Some Kan extensions can be calculated with a **pointwise computation**.

Theorem 9.2.6. Let $\mathbb{A}$, $\mathbb{B}$, and $\mathbb{C}$ be categories with functors $K\colon \mathbb{B} \longrightarrow \mathbb{C}$ and $G\colon \mathbb{B} \longrightarrow \mathbb{A}$. Furthermore, let $\mathbb{A}$ have all colimits. A left Kan extension $Lan_K(G)$ at object c in $\mathbb{C}$ can be calculated as

$$Lan_K(G)(c) = Colim(G \circ U\colon (K \downarrow c) \longrightarrow \mathbb{B} \longrightarrow \mathbb{A}), \tag{9.26}$$

where $(K \downarrow c)$ is the comma category (Definition 4.5.5) whose objects are the pair $(b, f\colon K(b) \longrightarrow c)$ and $U\colon (K \downarrow c) \longrightarrow \mathbb{B}$ is the functor that forgets the f. For completeness, we mention how to calculate the right Kan extension. Assume that $\mathbb{A}$ has all limits; then the right Kan extension can be calculated as

$$Ran_K(G)(c) = Lim(G \circ U\colon (c \downarrow K) \longrightarrow \mathbb{B} \longrightarrow \mathbb{A}). \tag{9.27}$$

The proof that such a formula actually gives the Kan extension can be found in Section X.3 of [180]. ★

With this theorem, we can better understand the second intuition of this section. For every object c in $\mathbb{C}$, there is a subcategory of $\mathbb{B}_c$ of $\mathbb{B}$ for which a limit is taken. When $\mathbb{C}$ is a one-object category, the limit is taken of the diagram that is the entire category, i.e., $\mathbb{B}_* = \mathbb{B}$. This means that when $\mathbb{C} = \mathbf{1}$, then $K\colon B \longrightarrow \mathbf{1}$ is the unique functor to $\mathbf{1}$ and $Lan_K(*) = Lim(G)$. When $\mathbb{C}$ is not $\mathbf{1}$, a limit is taken of the diagram of part of $\mathbb{B}$.

This pointwise computation is particularly easy when the categories in question are partial orders. This is shown in the following simple example.

Example 9.2.7. Consider the Kan extension for the following simple situation:

$$\begin{array}{ccc} \mathbf{R} & \xrightarrow{Lan_{inc}(inc)} & \mathbf{Z} \\ & {\scriptstyle inc}\nwarrow \qquad \nearrow{\scriptstyle Id} & \\ & \mathbf{Z} & \end{array} \tag{9.28}$$

where inc is the obvious inclusion. Let $r \in \mathbf{R}$. Then, using Equation (9.26), we get

$$Lan_{inc}(inc)(r) = Colim(Id \circ U\colon (inc \downarrow r) \longrightarrow \mathbf{Z} \longrightarrow \mathbf{Z}). \tag{9.29}$$

The objects in category $(inc \downarrow r)$ consists of all integers less then r. The functor U takes all those integers to the category $\mathbf{Z}$ and the functor Id brings all those integers into $\mathbf{Z}$.

The limit of all those integers less then r is the largest integer less than r, which is the floor of r. A similar analysis shows that $Ran_{inc}(inc)(r)$ is the ceiling of r. □

Theorem 9.2.8. A Kan extension is unique up to a unique natural transformation. ★

Proof. Let us work with right Kan extensions here. Assume that there are two right Kan extensions, $Ran_K(G)$ and $Ran'_K(G)$. By the universal properties of both, there are natural isomorphisms going from one to the other. The rest of the proof is exactly the same as for Theorem 2.2.23, where we saw that two initial objects have a unique isomorphism between them. ♣

"All Concepts Are Kan Extensions"

Let us show that many of the constructions that we looked at in this text can be realized as Kan extensions.

Example 9.2.9. A Kan extension is a way of describing limits and colimits. Diagram (9.16), which gives the definition of a Kan extension, is the same as Diagram (4.178) if $\mathbb{C} = \mathbf{1}$. In that case, functor K can only be the unique functor to $\mathbf{1}$, and $- \circ K = -\circ !$, which we denote as Δ. Since adjoints of Δ are unique up to an isomorphism, we have that $Colim = Lan$ and $Lim = Ran$. This means that setting $\mathbb{B}$ to be the appropriate diagrams that we met in Chapter 3, Kan extensions provide us with products, coproducts, pullbacks, pushouts, equalizers, coequalizers, and other elements. □

Example 9.2.10. Kan extensions can be used to find adjoints of functors. Consider any functor $K\colon \mathbb{B} \longrightarrow \mathbb{C}$. Functor K has a left adjoint if and only if $Ran_K(Id_{\mathbb{B}})$ exists and is preserved by K. There is a similar construction for the right adjoint of K. For more, see Theorem X.7.2 of [180], Proposition 6.5.2 of [222], and Proposition 3.7.6 of [40]. □

One of the main theorems of functorial semantics, Theorem 8.1.16, was an instance of Kan exensions.

Example 9.2.11. Let $\mathbf{T}$ and $\mathbf{T}'$ be algebraic theories, and let $F\colon \mathbf{T} \longrightarrow \mathbf{T}'$ be an algebraic functor. As we saw in Section 8.1, this algebraic functor induces $F^*\colon Alg(\mathbf{T}', \mathbb{C}) \longrightarrow Alg(\mathbf{T}, \mathbb{C})$. When $\mathbb{C}$ is suitably complete, F^* has a left adjoint $F_*\colon Alg(\mathbf{T}, \mathbb{C}) \longrightarrow Alg(\mathbf{T}', \mathbb{C})$. The left adjoint is the Kan extension, as shown in Diagram (8.11). □

Applications

Let us look at several applications of Kan extensions that show the power of this concept.

Example 9.2.12. In Section 8.5, we saw that basic databases can be described with categories. Consider two categories (database schemas), $\mathbb{B}$ and $\mathbb{C}$. We then have the functor category ($\mathbb{B}$-instances) $\mathbb{Set}^{\mathbb{B}}$ and the functor category ($\mathbb{C}$-instances) $\mathbb{Set}.^{\mathbb{C}}$ A functor between two database schemas $K\colon \mathbb{B} \longrightarrow \mathbb{C}$ induces three data migration functors: (i) the pullback functor $\Delta_K = - \circ K$, (ii) the left pushforward migration functor $\Sigma_K = Lan_K$, and (iii) the right pushforward migration functor $\Pi_K = Ran_K$. The functors can be calculated using Equations (9.26) and (9.27). □

Example 9.2.13. Dan Shiebler [235] uses Kan extensions to do some important parts of machine learning, which is an exciting new area of computer science that gets machines to learn various skills. In his paper, Shiebler describes a worked-out experiment that gets a computer to determine if certain pictures of shirts are pictures of T-shirts or not. This task is called **classification**, which means the computer learns how to classify if a certain object is or is not a part of a desired subset.

The method used is called **supervised learning**. Basically, a computer is given a training data set that is a subset of all possible data. This training data comes with the answers. The computer uses this training data to learn how to classify all the data.

Shiebler shows that such supervised learning can be done with Kan extensions. Following the setup of a Kan extension, there will be three categories and functors between them. The category $\mathbb{A} = \mathbf{2}$ is the preorder category with two elements $\{0, 1\}$, where $0 < 1$. The category $\mathbb{C}$ is a preorder category that is going to correspond to all data to classify. The category $\mathbb{B}$ is a discrete category corresponding to the training data set. The functor $inc\colon \mathbb{B} \longrightarrow \mathbb{C}$ is the inclusion of the discrete category that has already been labeled into the larger preorder of all data. The functor $G\colon \mathbb{B} \longrightarrow \mathbf{2}$ tells how the training data set is classified. One can view this as a Kan extension diagram:

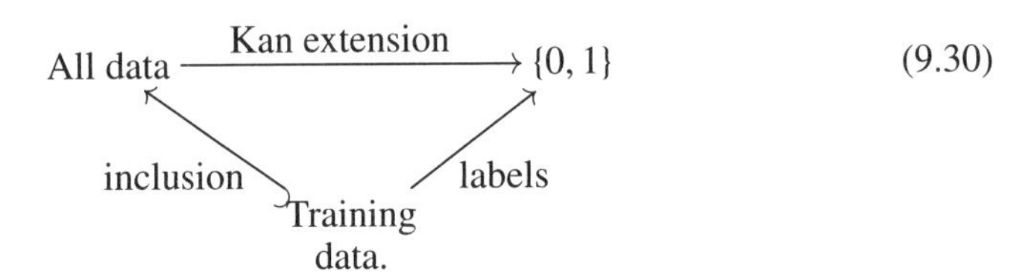

(9.30)

The Kan extensions can be calculated by Equations (9.26) and (9.27) to get

$$Lan_{inc}(G)(b) = \begin{cases} 1 & : \exists b' \in \mathbb{B}, b' \leq b, G(b') = 1 \\ 0 & : \text{else}. \end{cases} \tag{9.31}$$

$$Ran_{inc}(G)(b) = \begin{cases} 0 & : \exists b' \in \mathbb{B}, b \leq b', G(b') = 0 \\ 1 & : \text{else}. \end{cases} \tag{9.32}$$

The functor $Lan_{inc}(G)$ gives a true result whenever there is some test data that can make it true. In other words, this functor is trying to minimize false negatives. In contrast, $Ran_{inc}(G)$ gives a false result whenever there is some test data that can make it false. In other words, this functor is trying to minimize false positives.

The two Kan extensions partition all the data into three sections:

- Those c such that both Kan extensions agree that c is true (i.e., $Lan_{inc}(G)(c) = 1$ and $Ran_{inc}(G)(c) = 1$).

- Those c such that both Kan extensions agree that c is false (i.e., $Lan_{inc}(G)(c) = 0$ and $Ran_{inc}(G)(c) = 0$).
- Those c such that the Kan extensions disagree about c (i.e., $Lan_{inc}(G)(c) = 1$ and $Ran_{inc}(G)(c) = 0$).

Sheibler goes on to describe a method of minimizing the third region, where the Kan extensions disagree.

In the same paper, Shiebler uses Kan extensions to perform other common tasks in machine learning: (i) Clustering with supervision, (ii) meta-supervised learning, and (3) function approximation. For a general view of machine learning from a categorical perspective, see [52]. For a survey of category theory and machine learning, see [236]. □

Example 9.2.14. Let us talk a little about the world of particle physics. Researchers describe particles as certain types of **representations of groups**. Groups describe the symmetries that a particle must satisfy. If G is a group thought of as a one-object category, then a functor from G to $\mathbb{CV}\mathbb{ect}$ chooses a vector space, and for every element of G, an invertable linear map from that vector space to itself. (To be more exact, we are interested in G being a Lie group, vector spaces are Hilbert spaces, and invertable maps being unitary maps. But let us keep it simple.) Such a functor is a representation of G. A natural transformations between such representations is called an **intertwiner**, or an **intertwining operator**. As particles are described by representations, interactions of particles are described by intertwining operators. The collection of all such representations and intertwining operators (functors and natural transformations) is denoted as $\mathbb{CV}\mathbb{ect}^G$.

When H is a subgroup of G with inclusion $inc\colon H \longrightarrow G$, there is an induced functor called the **restriction representation functor** $Res^G_H\colon \mathbb{CV}\mathbb{ect}^G \longrightarrow \mathbb{CV}\mathbb{ect}^H$. This functor takes a G representation and forgets some of its symmetries to get an H representation. Depending on G and H, there might be a left adjoint called the **induced representation functor** $Ind^G_H\colon \mathbb{CV}\mathbb{ect}^H \longrightarrow \mathbb{CV}\mathbb{ect}^G$. This functor describes how a particle that satisfies a small group of symmetries will behave when we take into account the larger group of symmetries. This adjunction is called **Frobenius reciprocity**.

There also might be a right adjoint to Res^G_H, which is called the **coinduced representation functor** $Coi^G_H\colon \mathbb{CV}\mathbb{ect}^H \longrightarrow \mathbb{CV}\mathbb{ect}^G$. The functor describes another way that a particle that satisfies a small group of symmetries will behave when we take into account the larger group of symmetries.

The indcution and coinduction functors are Kan extensions along $inc\colon H \longrightarrow G$, as in the following diagram.

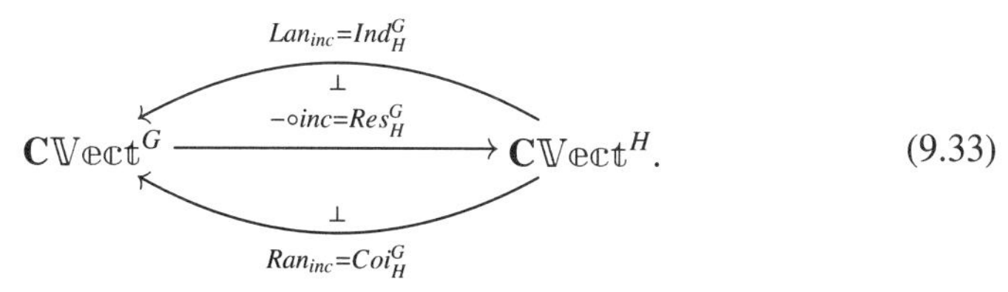

(9.33)

This is just the beginning of particle physics and the standard model of elementary particles. □

Only the language of category theory can show that major constructions in database theory, machine learning, and particle physics are all related!

Generalizations

Kan extensions have been around since the late 1950s and have been generalized in a number of different ways.

A generalization is to examine Kan extensions in 2-categories other than those of categories, functors, and natural transformations. Kan extensions can be defined in any 2-category. In particular, one can talk of Kan extensions in $\overline{\mathbb{MonCat}}$, the 2-category of monoidal categories, monoidal functors, and monoidal natural transformations. We can then ask questions as to when Kan extensions exist. Another question: Given the monoidal functors $K: (\mathbb{B}, \otimes_{\mathbb{B}}) \longrightarrow (\mathbb{C}, \otimes_C)$ and $G: (\mathbb{B}, \otimes_{\mathbb{B}}) \longrightarrow (\mathbb{A}, \otimes_{\mathbb{A}})$, what is the relationship between $Lan_K(G)(c \otimes_{\mathbb{C}} c')$ and $Lan_K(G)(c) \otimes_{\mathbb{A}} Lan_K(G)(c')$?

Another generalization has to do with inverting the functors. There is an old folk theorem (an idea that researchers say, but is not really written) in topology, which I believe is due to Witold Hurewicz. He said that many problems in topology are either extension problems or lifting problems, as depicted here:

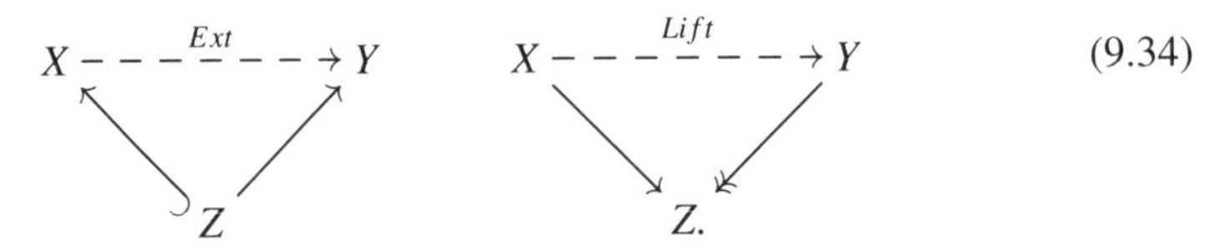

By an extension problem, we mean that given continuous inclusion function $Z \hookrightarrow X$ and continuous $Z \longrightarrow Y$, find a continuous extension function $X \longrightarrow Y$ that completes the triangle. By a lifting problem, we mean that given a continuous surjection $Y \twoheadrightarrow Z$ and a continuous $X \longrightarrow Z$, find a continuous lifting function $X \longrightarrow Y$ that completes the triangle. In this section, we have converted the extension problem into a categorical construction. What about a categorical version of the lifting problem?

We introduce Kan liftings in the same way that we introduced Kan extensions, just more quickly.

Definition 9.2.15. Consider the categories $\mathbb{A}$, $\mathbb{B}$, and $\mathbb{C}$, and the functor $K: \mathbb{C} \longrightarrow \mathbb{B}$. There is an induced functor $K \circ -: \mathbb{C}^{\mathbb{A}} \longrightarrow \mathbb{B}^{\mathbb{A}}$. Left and right adjoints of $K \circ -$, if they exist, are called **left** and **right Kan liftings along** K and are denoted as Lif_K and Rif_k, respectively. The adjoints are summarized as

$$\mathbb{C}^{\mathbb{A}} \underset{\underset{Rif_K}{\longleftarrow}}{\overset{\overset{Lif_K}{\longleftarrow}}{\xrightarrow{K\circ -}}} \mathbb{B}^{\mathbb{A}}. \qquad (9.35)$$

Working out their adjunctions in terms of Hom sets means that for all $G: \mathbb{A} \longrightarrow \mathbb{B}$ and all $P: \mathbb{A} \longrightarrow \mathbb{C}$, we have

$$Hom_{\mathbb{C}^{\mathbb{A}}}(Lif_K(G), P) \cong Hom_{\mathbb{B}^{\mathbb{A}}}(G, K \circ P), \tag{9.36}$$

$$Hom_{\mathbb{B}^{\mathbb{A}}}(K \circ P, G) \cong Hom_{\mathbb{C}^{\mathbb{A}}}(P, Rif_K(G)). \tag{9.37}$$

In terms of the units and counits, the following two diagrams are helpful:

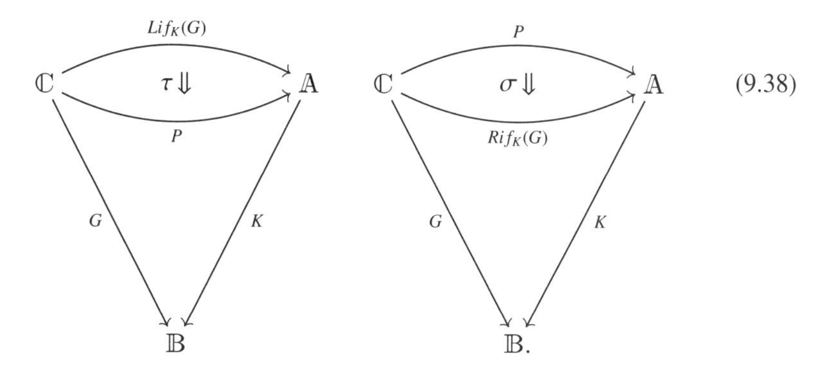

(9.38)

A left Kan lifting is a pair $(Lif_K(G), \eta)$ where $Lif_K(G)\colon \mathbb{C} \longrightarrow \mathbb{A}$ is a functor, and $\eta\colon G \Longrightarrow K \circ Lif_K(G)$ is a natural transformation that is the unit of the adjunction. As the unit of the adjunction, η must satisfy the following universal property: For every functor P in $\mathbb{A}^{\mathbb{C}}$ and every natural transformation $\eta'\colon G \Longrightarrow K \circ P$, there is a unique natural transformation $\tau\colon Lif_K(G) \Longrightarrow P$ such that

$$\eta' = (K \circ_H \tau) \circ_V \eta. \tag{9.39}$$

A right Kan lifting is a pair $(Rif_K(G), \varepsilon)$ where $Rif_K(G)\colon \mathbb{C} \longrightarrow \mathbb{A}$ is a functor, and $\varepsilon\colon K \circ Rif_K(G) \Longrightarrow G$ is a natural transformation that is a counit of the adjunction. It satisfies the following universal property: for any P in $\mathbb{A}^{\mathbb{C}}$ and any natural transformation $\varepsilon'\colon K \circ P \Longrightarrow G$, there is a unique natural transformation $\sigma\colon P \Longrightarrow Rif_K(G)$ such that

$$\varepsilon' = \varepsilon \circ_V (K \circ_H \sigma). \tag{9.40}$$

Kan liftings are not as popular in the literature as their extension cousins. Nevertheless, there are some simple examples where Kan liftings occur.

Example 9.2.16. Let $\mathbf{N}$ be the natural numbers thought of as a partial order category. Consider the functor $Succ\colon \mathbf{N} \longrightarrow \mathbf{N}$, which takes x to its successor, $Succ(x) = x + 1$. Then there is a predecessor $Pred\colon \mathbf{N} \longrightarrow \mathbf{N}$ functor, which takes x to $Pred(x) = x - 1$. (Note that this functor has $Pred(0) = 0$.) The predecessor is a Kan lifting as follows:

$$\begin{array}{ccc} \mathbf{1} & \xrightarrow{Lif_{Succ}(x)} & \mathbf{N} \\ & {}_{x}\searrow \quad \swarrow_{Succ} & \\ & \mathbf{N}. & \end{array} \tag{9.41}$$

□

Along with Example 9.2.11, where we found the left adjoint to an algebraic functor, we have Example 9.2.17.

Example 9.2.17. Let $\mathbf{T}$ be an algebraic theory and $\mathbb{C}$ be a Cartesian category that is complete. Then $Alg(\mathbf{T},\mathbb{C})$ is the category of algebras in $\mathbb{C}$. Let $\mathbb{C}'$ also be a complete Cartesian category, and $K\colon \mathbb{C} \longrightarrow \mathbb{C}'$ be a product-preserving functor. The functor K induces $K_* = K \circ -\colon Alg(\mathbf{T},\mathbb{C}) \longrightarrow Alg(\mathbf{T},\mathbb{C}')$. In detail, K_* takes an algebra $G\colon \mathbf{T} \longrightarrow \mathbb{C}$ to $K \circ G\colon \mathbf{T} \longrightarrow \mathbb{C} \longrightarrow \mathbb{C}'$. The functor K_* has a left adjoint $Lif_K = K^*$, which is constructed using a left Kan lifting as follows:

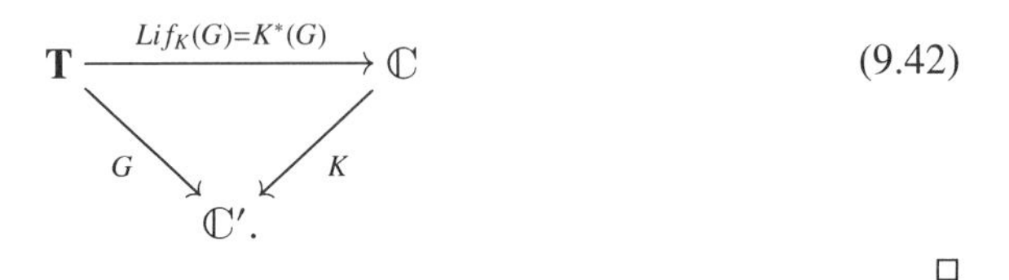

(9.42)

□

Before we take leave of the world of Kan extensions, it pays to go back to the first definition of a Kan extension and meditate on what was done in this section. All the constructions that we showed were examples of Kan extensions are the consequences of the adjunctions shown in Diagram (9.16). This means that much of category theory (and a lot of mathematics and more) follows from the simple notion of composition (from K to $- \circ K$) and the notion of an adjoint functor. That's it! Everything follows from those two simple notions. That's amazing!

Suggestions for Further Study

Kan extensions can be found in Chapter X of [180], and Chapter 6 of [222]. They are discussed in terms of Cartesian closed categories in Section 9.7 of [21]. Kan liftings were first mentioned (but not named as such) by Ross Street in [245] and in his work with Robert Walters [249].

9.3 Homotopy Theory

Homotopy theory is a field within topology that deals with the issue of when two entities are the "same" up to specified invariants. We begin by asking when two paths on a topological space can be nicely changed into one another. More generally, homotopy theory studies when two topological spaces can be changed into each other. Researchers have advanced and abstracted the field to include when two structures or processes are essentially the same. In its present incarnation, homotopy theory describes some very deep and profound ideas about how structures and processes change.

This section begins with some simple notions of homotopy of topological spaces and continuous maps. Although there are many topics that we can discuss, we cherry-pick important topics and what will be needed for the rest of the text. We then discuss simplicial sets, which are an easy way of describing topological spaces and so much more. We also discuss tools to deal with the homotopy of simplicial sets. The section ends with a discussion of **categorical homotopy theory** or **abstract homotopy**

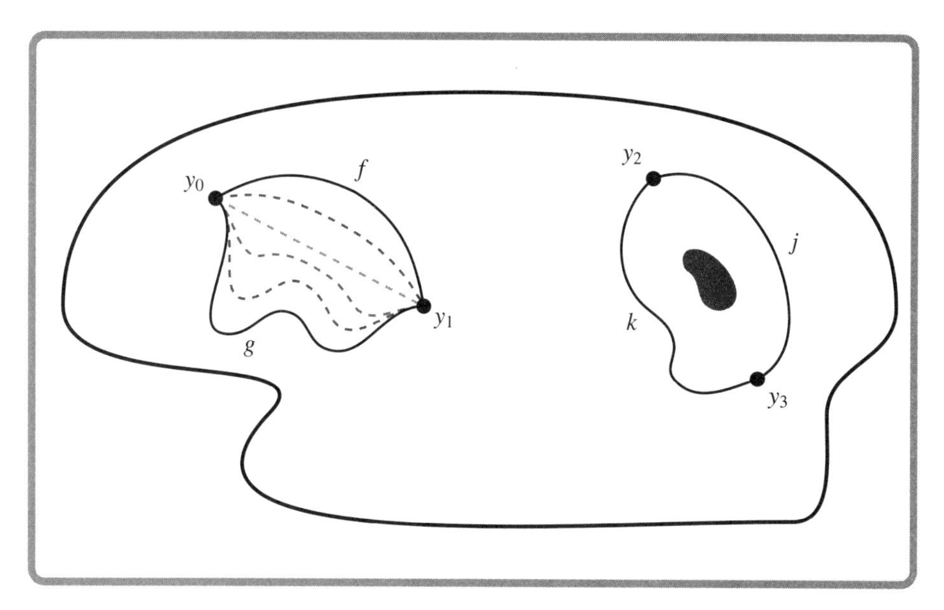

Figure 9.2. A topological space where it is possible to deform f to g, but not possible to deform j to k.

theory, which is a generalization of homotopy theory that deals with categories of various structures. We define **Quillen model categories**, which are categories with enough extra structure that one can formulate the tools of homotopy theory for that category.

Classical Homotopy Theory

We start in the category $\mathbb{Top}$, which consists of topological spaces and continuous maps between them. Consider the unit interval $[0,1]$ of all real numbers between 0 and 1. This is a topological space that we denote as I. A continuous map from I to any topological space Y (i.e., $f\colon I \longrightarrow Y$), describes a path in Y. The left side of Figure 9.2 depicts paths that start at a point $y_0 = f(0)$ and end at point $y_1 = f(1)$. Consider two such paths, f and g, and ask whether these two paths are essentially the same, or whether they can be changed into one another. In Figure 9.2, the two paths, f and g, which start at y_0 and end at y_1, can clearly be deformed into each other. In contrast, the two paths j and k, which start at y_2 and end at y_3, cannot be continuously deformed into each other because there is a hole in the topological space between j and k.

Deforming a path can be described with a **path homotopy between maps**. A path homotopy from f to g is a continuous map:

$$H\colon I \times I \longrightarrow X, \tag{9.43}$$

such that the following are true:

- The homotopy starts at f: $H(0,x) = f(x)$ for all x.
- The homotopy ends at g: $H(1,x) = g(x)$ for all x.

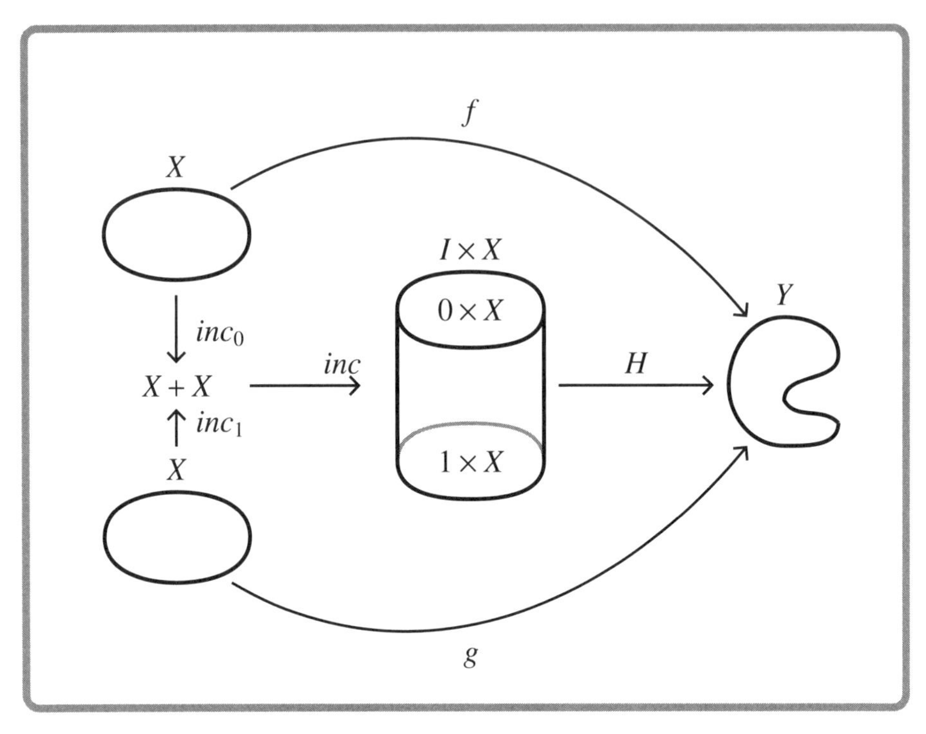

Figure 9.3. An illustration showing that f and g are left homotopic using a cylinder space. The space $X + X$ includes into the top and bottom of the cylinder space.

- All the paths start at y_0: $H(y, 0) = y_0$ for all y.
- All the paths end at y_1: $H(y, 1) = y_1$ for all y.

Such a path homotopy can be thought of as a map from the square $(I \times I)$ to Y. One side of the square goes to point y_0, while the opposite side goes to point y_1. The other two sides of the square go to f and g, respectively. There is no such continuous map between j and k because of the hole in the space between these paths. Two paths that have a path homotopy between them are called **path homotopic**.

We can generalize the notion of a homotopy to be a continuous deformation between any two continuous maps. Let $f\colon X \longrightarrow Y$ and $g\colon X \longrightarrow Y$. Then $H\colon I \times X \longrightarrow Y$, which satisfies $H(0, x) = f(x)$ and $H(1, x) = g(x)$, is a homotopy from f to g. The space $I \times X$ is called the **cylinder space** of X. We can see this clearly with the commutative diagrams in Figure 9.3. The map *inc* takes the first copy of X to $0 \times X$ and the second copy of X to $1 \times X$. Saying that the top half of the diagram commutes means that $H(0, x) = f(x)$. It is similar for the bottom half of the diagram. We call such an H a **left homotopy** for reasons that will become apparent later.

There is another way of showing two continuous maps $f, g\colon X \longrightarrow Y$ are homotopic. Similar to what we saw about sets, there are the functors $L_I, R_I\colon \mathbb{Top} \longrightarrow \mathbb{Top}$, which are defined as $L_I(X) = I \times X$ and $R_I(X) = X^I$. While the output of the L_I are called "cylinder spaces," the output of the R_I that are of the form Y^I (i.e., spaces of continuous

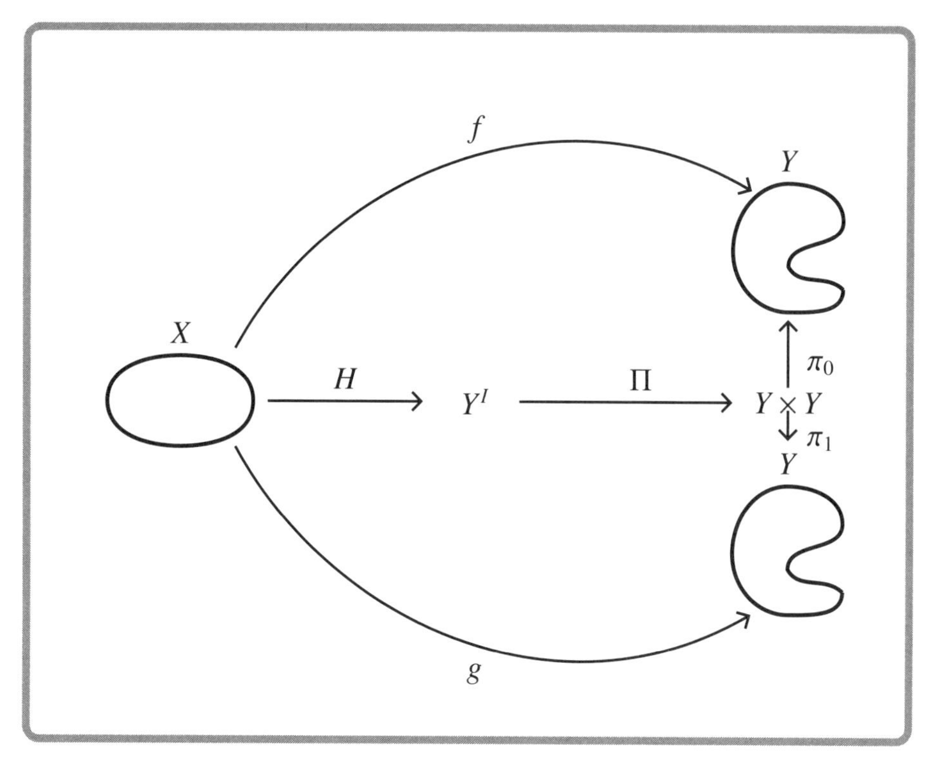

Figure 9.4. An illustration showing that f and g are right homotopic using a path space. The projection onto the space $Y \times Y$ takes a path to its two ends.

maps from I to a space Y) are called **path spaces**. When we are dealing with a "nice" category of spaces and maps,[1] the functors L_I and R_I are adjoint to each other. In other words, the maps $I \times X \longrightarrow Y$ will be in correspondence to the maps $X \longrightarrow Y^I$.

We can use path spaces to give another definition of a homotopy from f to g. A **right homotopy** from f to g is a continuous map $H\colon X \longrightarrow Y^I$, where Y^I is the path space on Y. This H must connect f and g as follows:

$$H(x) = P_x\colon I \longrightarrow Y \text{ where } P_x(0) = f(x) \text{ and } P_x(1) = g(x). \tag{9.44}$$

In other words, P_x is a path from f to g. We can see this definition in action by looking at Figure 9.4. The $\Pi\colon Y^I \longrightarrow Y \times Y$ map takes a path $P\colon I \longrightarrow Y$ to its value at 0 and value at 1. In other words, the path $P\colon I \longrightarrow Y$ goes to the ordered pair $(P(0), P(1))$. The commutativity of the diagram says that $H(x)$ is a path that starts at $f(x)$ and ends at $g(x)$.

[1]We avoid the discussion of what is a "nice" category of spaces and maps simply because it would distract from the discussion of homotopy theory. Any textbook on topology deals with such issues. Suffice it to say, that "nice" spaces are the typical spaces that one deals with have no pathologies like $(1/x)sin(1/x)$ or Alexander's "horned sphere." We also want the cylinder spaces and path spaces to be adjoint to each other. For those who know the lingo, CW complexes, or the category of compactly generated weak Hausdorff spaces, satisfy the requirements of being "nice."

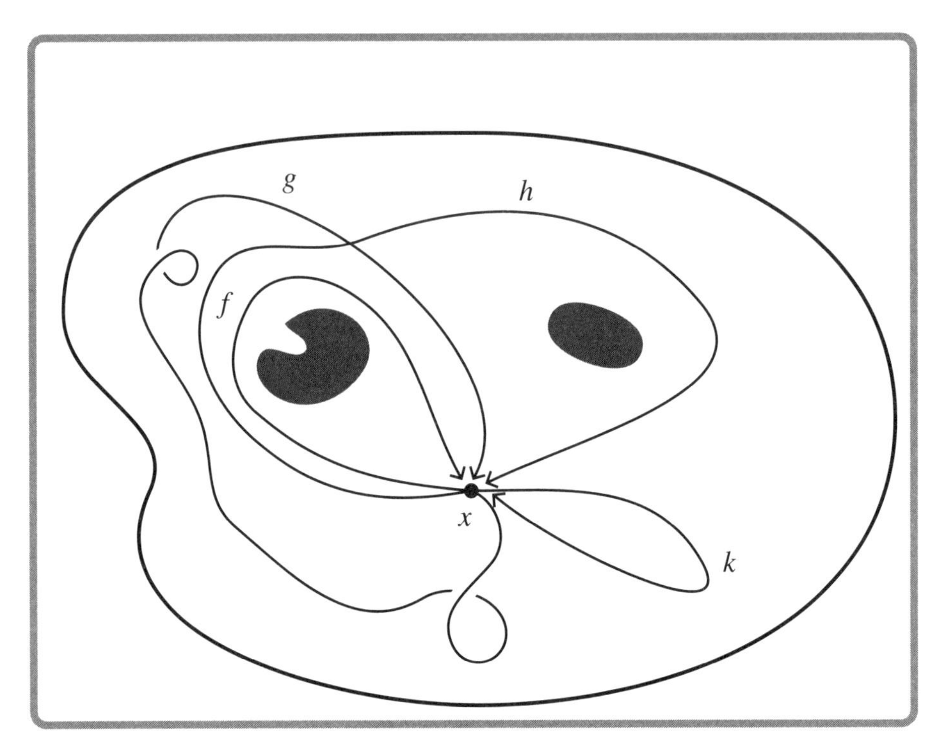

Figure 9.5. A space with several paths that start and finish at the same point.

It turns out that when X and Y are CW-complexes—which we will usually work with—there is a left homotopy between two continuous maps if and only if there is a right homotopy between them. We will not make a distinction between left and right homotopies.

We use the notion of homotopy to describe one of the most important ways of understanding topological spaces: the **fundamental group** of a space. Here, we are not interested in when two arbitrary paths are homotopic. Rather, we are interested in paths that start and end at the same point. Such paths are called **loops**. We are interested in when two loops are homotopically the same. Let X be a topological space and x be a point in that space. Consider maps $f: I \longrightarrow X$, such that $f(0) = x$ and $f(1) = x$. Two such loops are homotopic if there is a path homotopy between them. Look at Figure 9.5 for various examples. Loops f and g look very different, but one can make a homotopy between them, as they both go around the left hole of the space. In contrast, neither f nor g is homotopic to h because h goes around both holes. Loops f, g, and h are not homotopic to loop k, which does not go around any hole and can be shrunk to the map that is constant at x (i.e., the map $Const_x: I \longrightarrow X$, defined as $Const_x(r) = x$ for all $r \in [0, 1]$).

A better way of looking at such loops is as continuous maps from the pointed circle to the pointed space X. In detail, the unit circle is the space

$$S^1 = \{(x_1, x_2) \in \mathbf{R}^2 : x_1^2 + x_2^2 = 1\}. \tag{9.45}$$

Space S^1 with the point $(1,0)$, is a pointed space. A map from S^1 to X that takes $(1,0)$ to $x \in X$ is a loop based at x. We assume that the circle is oriented clockwise and the path in X is oriented clockwise.

Consider the set of all such maps. There is an equivalence relation on this set where two loops are considered equivalent if they are path homotopic. That is, we are looking at

$$Hom_{\mathbb{Top}_*}((S^1,(1,0)),(X,x))/\sim. \tag{9.46}$$

The set of equivalence classes is denoted as $\pi_1(X,x)$. These loops tell you all about the holes in the topological space X. The important feature of these equivalence classes is that they can be combined. Given two such loops, $j\colon S^1 \longrightarrow X$ and $k\colon S^1 \longrightarrow X$, one can combine them to form $(k \circ j)\colon S^1 \longrightarrow X$, which is defined as the equivalence class of going around the j loop at twice the speed and then going around the k loop at twice the speed. This creates one loop. The important point is that these loops do not only have a shape in X, they also have a speed at which they go around.

The set $\pi_1(X,x)$ with this composition operation forms a group that we call the **fundamental group of X with basepoint** x. The unit element is the constant function at x. Composition of any loop with this trivial loop is homotopic to the original loop. Given three loops represented by $f\colon S^1 \longrightarrow X$, $g\colon S^1 \longrightarrow X$, and $h\colon S^1 \longrightarrow X$, there are two compositions: $h \circ (g \circ f)\colon S^1 \longrightarrow X$ and $(h \circ g) \circ f\colon S^1 \longrightarrow X$. While these two loops have the same shape, they do not have the same speed. The first loop does f and g at four times the regular speed and h at twice the speed, while the second loop does g and h at four times the regular speed and f at twice the regular speed. Regardless of this, the loops are homotopic. The fundamental group is an extremely important algebraic structure associated to the topological space X. (A theorem of homotopy theory says that for any two basepoints x and x' in the same connected component of X, there is an isomorphism between $\pi_1(X,x)$ and $\pi_1(X,x')$.)

Given a continuous $f\colon (X,x) \longrightarrow (Y,y)$, the loops in X map to the loops in Y. Formally, a loop in $j\colon S^1 \longrightarrow X$ based at point x becomes a loop $f \circ j\colon S^1 \longrightarrow X \longrightarrow Y$ based at $f(x)$. If there is a homotopy between two loops $j, j'\colon S^1 \longrightarrow X$, then we can compose that homotopy with f to get a homotopy between $f \circ j$ and $f \circ j'$. This means that equivalence classes of loops go to equivalence classes of loops. From this, we see that the continuous map f induces a function of groups $\pi_1(f)\colon \pi_1(X,x) \longrightarrow \pi_1(Y,f(x))$. It is not hard to see that we have defined a functor $\pi_1 : \mathbb{Top}_* \longrightarrow \mathbb{Group}$. In general, π_1 of a topological space is not abelian.

The fundamental group is just the beginning of the story. It describes the one-dimensional holes in a topological space. One can also look at maps $j\colon S^2 \longrightarrow X$, where

$$S^2 = \{(x_1,x_2,x_3) \in \mathbf{R}^3 : x_1^2 + x_2^2 + x_3^2 = 1\} \tag{9.47}$$

is the three-dimensional sphere. Such maps will describe the two-dimensional holes of a topological space. Such maps up to homotopy also form a group that we denote as $\pi_2(X,x)$ and call the **second homotopy group of X based at** x. One can go on and talk about maps $j\colon S^n \longrightarrow X$, where S^n is the n-sphere

$$S^n = \{(x_1,x_2,x_3,\ldots,x_{n+1}) \in \mathbf{R}^{n+1} : x_1^2 + x_2^2 + x_3^2 + \cdots + x_{n+1}^2 = 1\}. \tag{9.48}$$

These maps can, in some cases, describe the n-dimensional holes of space X. However, the reader should exercise some caution. The third homotopy group of the 2-sphere, $\pi_3(S^2, x)$, is nontrivial. So we either need to expand our intuition so a space can have higher-dimensional holes or put some limitations on the intuitive notion of the homotopy groups measuring holes. For each $n > 0$, the set of equivalence classes of such maps form a group denoted as $\pi_n(X, x)$ and called the **n-th homotopy group of X based at x**. One can define a composition of such equivalence classes. Continuous maps of topological spaces induce homomorphisms of groups. For each $n > 1$, these groups are abelian. (The proof of this is a variation of the Eckmann-Hilton argument given in Theorem 5.1.16.) Thus, there are the functors $\pi_i : \mathbb{Top}_* \longrightarrow \mathbb{AbGp}$ for all $i > 1$.

We can also look at maps $j\colon S^0 \longrightarrow X$ from the 0th sphere, defined as

$$S^0 = \{(x_1 \in \mathbf{R}^1 : x_1^2 = 1\} = \{-1, 1\}. \tag{9.49}$$

Homotopy classes of such maps consist of one object for every part of X that is connected by a path. For CW-complexes, this is just the set of connected components of X. Such equivalence classes form a set and in general do not compose. For reasons similar to π_1, the assignment $X \mapsto \pi_0(X)$ extends to a functor $\pi_0\colon \mathbb{Top} \longrightarrow \mathbb{Set}$.

The functors π_n for $n > 0$ all demand a basepoint. We can also study topological spaces without basepoints. For a space X, we can talk about the **fundamental groupoid** of X, written as $\Pi_1(X)$. The objects of this groupoid are the points of the space and the invertible morphisms are equivalences classes of paths between the points.

Notice that when we are looking at $\Pi_1(X)$, we are losing information by looking at *equivalence classes* of paths. We simply declare two paths to be equal if there is some homotopy between them. We can remedy this situation by talking about the **infinity groupoid** $\Pi_\infty(X)$. We will formally define this type of structure at the end of Section 9.4. Suffice it to say that the objects of this structure are the points of the space, the morphisms are the paths between the points, the 2-cells are the homotopies between the paths, the 3-cells are the homotopies between the homotopies, the 4-cells are the homotopies between the 3-cells, etc.

Let us move beyond paths, loops, and higher-dimensional loops. We extend the notion of homotopy to not only determine when two continuous maps can be changed into one another, but to determine when two topological spaces can be changed into one another. Let us put this in context. There are three levels of "sameness" of topological spaces:

- A continuous map $f\colon X \longrightarrow Y$ is a **homeomorphism** if it is an isomorphism in $\mathbb{Top}$. This means that there is a continuous map $f^{-1}\colon Y \longrightarrow X$ and the compositions of f and f^{-1} with each other are *equal* to the identities. If there is a homeomorphism between two topological spaces, we say that the spaces are **homeomorphic**.
- Just as we weakened the notion of isomorphism of categories to an equivalence of categories, we similarly weaken the notion of isomorphism of topological spaces to the notion of homotopy equivalence of topological spaces. Formally, a continuous map $f\colon X \longrightarrow Y$ is a **homotopy equivalence** if there is a continuous map $g\colon Y \longrightarrow X$ such that there is a homotopy from $g \circ f$ to id_X and a homotopy

from $f \circ g$ to id_Y. See Figure 4.66 where we saw the concept in terms of equivalence of categories. If there is a homotopy equivalence between two topological spaces, we say that the spaces are **homotopy equivalent**.

- The π_n groups are used to describe properties of topological spaces. A continuous map $f\colon (X,x) \longrightarrow (Y,y)$ is a **weak homotopy equivalence** if f induces the isomorphisms

$$\pi_n(f)\colon \pi_n(X,x) \longrightarrow \pi_n(Y,f(x)) \qquad (9.50)$$

for all n. If there is a weak homotopy equivalence between topological spaces, we say that the topological spaces are **weakly homotopically equivalent**.

If two spaces are homeomorphic, then they are homotopy equivalent (the homotopies are the identity), and if they are homotopy equivalent, then they are weakly homotopy equivalent (because the maps going in both ways induce isomorphisms of groups). In general, the theorem does not go the other way: Two spaces can be weakly homotopically equivalent but not homotopically equivalent. However, the **Whitehead theorem** states that for two CW-complexes, the two spaces are homotopy equivalent if and only if they are weakly homotopy equivalent. This means that nice spaces are totally characterized by the various dimensional holes in the space.

Continuous maps that are homeomorphisms are one-to-one and onto. This means that the two homeomorphic spaces are really the same up to a change of shape, but without shrinking or crunching. The classical example of a homotopy equivalence is that a coffee mug is the same as a doughnut because the cup part can be pushed into the shape of the doughnut. Two spaces are homotopically equivalent if one can be shrunk or crunched into the other without closing holes or ripping parts of the space apart. We will not deal with any spaces that are not CW-complexes, so we will not see any examples of spaces that are weakly homotopy equivalent but not homotopically equivalent.

Example 9.3.1. We can understand these notions of sameness by looking at the twenty-six capital letters of the Latin alphabet. To make our task easier, consider the letters *sans serif.* (There are various types of *sans serif* type. In particular, there are variations of the letters R and K.)

Each of the following letters are homeomorphic to the other letters in the class:

- [A,R] (they are circles with two lines coming out)
- [D,O] (they are circles)
- [P] (it is a circle with a line coming out)
- [Q] (it is a circles with two lines coming out)
- [B] (it is two circles)
- [C,G,I,L,M,N,S,U,V,W,Z] (they are straight lines)
- [X] (it is four connected lines)
- [E,F,T,Y] they are a straight line with one line coming out)
- [H,K]

Each of the following letters are homotopy equivalent to the other letters in the class, and since all these letters are CW-complexes, they are weakly homotopy equivalent to the other letters in the class:

- [A,D,O,P,Q,R] (they are all the same as a circle)
- [B] (it is two circles)
- [C,E,F,G,H,I,J,K,L,M,N,S,T,U,V,W,X,Y,Z] (they can all be shrunk to a point) □

Exercise 9.3.2. Do a similar analysis for the lowercase Latin alphabet. ■

Let us use the ideas of homotopy to better classify topological spaces.

Definition 9.3.3.

- The category $\mathbb{Top}$ has topological spaces as objects and continuous maps as morphisms.
- There is a quotient category $\mathbb{hoTop}$, whose objects are topological spaces and whose morphisms are equivalence classes of continuous maps, where two maps are equivalent if there is a homotopy between them. There is an identity on objects, full functor $\mathbb{Top} \longrightarrow \mathbb{hoTop}$, which takes every topological space to the same space and every continuous map to its equivalence class.
- Similarly, we can consider pointed topological spaces to form $\mathbb{hoTop}_*$. The homotopy groups are then simply given as

$$\pi_n(X,x) = Hom_{\mathbb{hoTop}_*}((S^n,(1,0,0,\ldots,0)),(X,x)). \tag{9.51}$$

- Rather than simply declaring two maps equivalent if there is a homotopy, we can include the homotopies in the collection. Let us form the 2-category $\overline{\mathbb{Top}}$. The objects are topological spaces, the morphisms are continuous maps, and the 2-cells are equivalence classes of homotopies between continuous maps. In detail, for two continuous maps $f,g\colon X \longrightarrow Y$, we consider a homotopy H as a 2-cell from f to g or, more clearly,

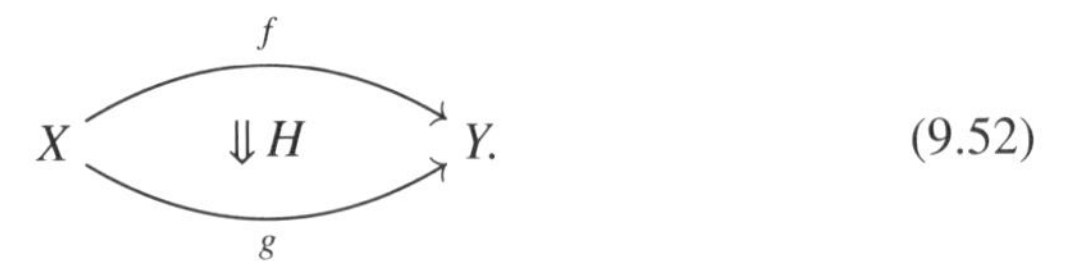

(9.52)

Let us elaborate on the 2-category structure of $\overline{\mathbb{Top}}$. A 2-cell is invertible because there is another homotopy from g to f, which is defined as

$$H'\colon I \times X \longrightarrow Y, \tag{9.53}$$

where $H'(t,x) = H(1-t,x)$, which means $H'(0,x) = g(x)$ and $H'(1,x) = f(x)$. To see how these 2-cells compose, keep the following picture in mind:

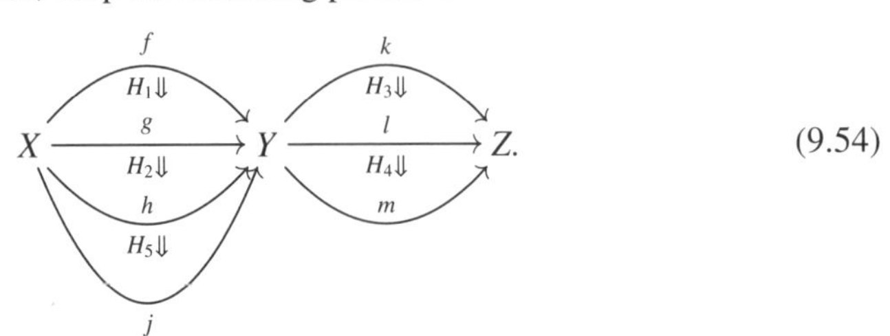

(9.54)

Homotopy H_1 goes from f to g. This means that $H_1\colon I\times X \longrightarrow Y$, such that $H_1(0,x)=f(x)$ and $H_1(1,x)=g(x)$. There is also a homotopy H_2 which goes from g to h. This means that $H_2\colon I\times X \longrightarrow Y$, such that $H_2(0,x)=g(x)$ and $H_2(1,x)=h(x)$. There is a vertical composition $H_2 \circ_V H_1$ that goes from f to h. It is defined as

$$(H_2 \circ_V H_1)(t,x) = \begin{cases} H_1(2t,x) & : t \leq \frac{1}{2} \\ H_2(2t-1,x) & : t \geq \frac{1}{2}. \end{cases} \tag{9.55}$$

This means that the composed homotopy does each homotopy at twice the speed. The problem is that this composition is not associative. The homotopy $H_5 \circ_V (H_2 \circ_V H_1)$ does H_1 and H_2 at four times the regular speed apiece, while doing H_5 at twice the regular speed. In contrast, the homotopy $((H_5 \circ_V H_2) \circ_V H_1)$ does H_2 and H_5 at four times the regular speed while doing H_1 at twice the regular speed.[2]

There is a similar problem with respect to horizontal composition of homotopies. Homotopy H_3 goes from k to l. This means that $H_3\colon I\times X \longrightarrow Y$, such that $H_3(0,x)=k(x)$ and $H_3(1,x)=l(x)$. There is a horizontal composition $H_3 \circ_H H_1$ that goes from $k \circ f$ to $l \circ g$ and is defined as

$$(H_3 \circ_H H_1)(t,x) = \begin{cases} k(H_1(2t,x)) & : t \leq \frac{1}{2} \\ H_3(2t-1,g(x)) & : t \geq \frac{1}{2}. \end{cases} \tag{9.56}$$

Horizontal composition of homotopies can also be defined as

$$(H_3 \circ_H H_1)(t,x) = \begin{cases} H_3(2t,f(x)) & : t \leq \frac{1}{2} \\ l(H_1(2t-1,x)) & : t \geq \frac{1}{2}. \end{cases} \tag{9.57}$$

These two definitions of horizontal composition of homotopies are very similar to the two equivalent definitions of horizontal composition of natural transformations that we saw in Section (4.2). This means that the composed homotopy does each homotopy at twice the speed. The same problem that we had for the associativity of vertical composition of homotopies occurs for the associativity of the horizontal composition of homotopies.

A further problem that we have is that the homotopies must satisfy the interchange law:

$$(H_4 \circ_V H_3) \circ_H (H_2 \circ_V H_1) = (H_4 \circ_H H_2) \circ_V (H_3 \circ_H H_1). \tag{9.58}$$

This law is not satisfied. Both sides are homotopic, but not identical. There are also problems with the composition with the identities. Because of all these problems, for

[2]There is an interesting solution to this problem that is related to coherence theory. Let us restate the problem in terms of the unit interval $[0,1]$: composition in the path space $X^{[0,1]}$ is not associative. A solution to this is sometimes called "Moore's trick," after the topologist John Moore. Rather than having path spaces exactly of the form $X^{[0,1]}$, permit the length of the paths to be of any length whatsoever. A path is now an element in $X^{[0,r]}$ for a real number $r>0$. Using this, we can compose a path in $X^{[0,r]}$ and a path in $X^{[0,r']}$ to be a path in $X^{[0,r+r']}$. This composition is associative. This is very much reminiscent of what we saw in Section 6.2 where the set of all strings—regardless of their length—composes to be associative. While this trick works, it has other drawbacks and is not conventionally done.

$\overline{\mathbb{Top}}$ to be a genuine 2-category, one must make the 2-cells be *equivalence classes* of homotopies. That is, two homotopies are set equivalent if there is a higher homotopy connecting them. With this definition, one forms the 2-category $\overline{\mathbb{Top}}$. Think of $\mathbb{hoTop}$ as a trivial 2-category with only identity 2-cells. There is a 2-functor $\overline{\mathbb{Top}} \longrightarrow \mathbb{hoTop}$, which is the identity on topological spaces, takes continuous maps to their equivalence class, and takes homotopies to identity 2-cells of $\mathbb{hoTop}$.

Let us foreshadow for a moment what we will see in the next section when we talk about higher category theory. Rather than simply declaring two homotopies equivalent if there is a higher homotopy between them, we can include the higher homotopies between them. There will be 3-cells that correspond to higher homotopies. These 3-cells go between homotopies (not equivalence classes of homotopies). Of course, you guessed it: these higher homotopies do not compose properly, so we need to have equivalence classes of higher homotopies to form the 3-category $\overline{\overline{\mathbb{Top}}}$. Continuing along the same vein, we get an infinite sequence

$$\cdots \longrightarrow \overline{\overline{\mathbb{Top}}} \longrightarrow \overline{\mathbb{Top}} \longrightarrow \mathbb{hoTop}. \tag{9.59}$$

Abandoning vagueness and being exact have their complexities.

There are special types of continuous maps and spaces that relate to homotopy theory. First, let's look at a definition.

Definition 9.3.4. For each n, the n**-disk** or the n**-ball** is the set of points around the origin in Euclidean space such that

$$D^n = \{(x_1, x_2, \ldots, x_n) \in \mathbf{R}^n : (\Sigma_{i=1}^n x_i^2) \leq 1\}. \tag{9.60}$$

We define a "nice" type of surjective map. A continuous map $p\colon X \longrightarrow Y$ is a **Serre fibration** if a homotopy in Y lifts to a homotopy in X. To be exact, it satisfies the following property: for all n, and for any f and H, if the following rectangle commutes (i.e., $p \circ f = H \circ inc_0$):

$$\begin{array}{ccc} D^n & \xrightarrow{f} & X \\ {\scriptstyle inc_0}\downarrow & \overset{\hat{H}}{\nearrow} & \downarrow{\scriptstyle p} \\ I \times D^n & \xrightarrow[H]{} & Y, \end{array} \tag{9.61}$$

then there is an $\hat{H}$ making the two triangles commute. There is another type of map called a **Hurewicz fibration** if there is a lifting over *all* spaces and not just the D^n. Since one can lift all spaces over a Hurewicz fibration, one can definitely lift the D^n over a Hurewicz fibration. This entails that every Hurewicz fibration is a Serre fibration. In general, not every Serre fibration is a Hurewicz fibration. However, if $f\colon X \longrightarrow Y$ is a map of CW-complexes, then it is true that f is a Serre fibration if and only if it is also a Hurewicz fibration. A space X is called **fibrant** if the unique map $X \longrightarrow *$ is a Serre fibration. It turns out that all spaces are fibrant.

There are also two types of very nice inclusion maps, which are called **Serre cofibrations** and **Hurewicz cofibrations** (the definitions are not needed here). A space X is called **cofibrant** if the unique map $* \longrightarrow X$ is a Serre cofibration. Intuitively, this also means that X is a nice space.

A space is called **bifibrant** if it is both fibrant and cofibrant. Such spaces are *very* nice.

Before we close our discussion of classical homotopy theory, we would like to examine two more ways of thinking of the fundamental group of a topological space. Both methods use the central features of category theory to arrive at the core of the idea.

The first way uses the 2-category, $\overline{\mathbb{Top}}$, of topological spaces, continuous morphisms, and equivalence classes of homotopies between continuous morphisms. Let X be a topological space and x be a point in that space. There is a unique continuous map, $pt_x\colon * \longrightarrow X$, which chooses x. The group $\pi_1(X,x)$ is nothing more than the automorphisms of pt_x. In other words, look at the homotopies $\alpha\colon pt_x \Longrightarrow pt_x$. We might envision this as the collection of invertible 2-cells of the form

$$* \xrightarrow[pt_x]{\ \circlearrowleft \alpha\ } X \qquad \text{or} \qquad * \underset{pt_x}{\overset{pt_x}{\underset{\Downarrow\alpha}{\rightrightarrows}}} X. \tag{9.62}$$

Each automorphism starts at x and makes a path back to x. Two paths are considered the same if there is a homotopy between them. These 2-cells/automorphisms compose to form a group. Automorphisms are ways of talking about symmetries. From the point of view afforded here, the fundamental group is the symmetries of a point on a space.

The second way is very clever and gets into another central idea of category theory. Let us go back and remember Important Categorical Idea 2.4.4 which says that $Hom(a,b)$ will inherit some of the structure of b. We saw this idea again in Exercise 2.4.3, where we showed that for any set S, the set $Hom(S,\mathbf{C})$ will inherit the structure of $\mathbf{C}$, which is a complex vector space. As an illustration, the vector space addition

$$\mathbf{C}\times\mathbf{C} \xrightarrow{\ +\ } \mathbf{C} \tag{9.63}$$

will be taken by the functor $Hom(S,\ \ )$ to the vector space addition:

$$Hom(S,\mathbf{C})\times Hom(S,\mathbf{C}) \xrightarrow{\ \cong\ } Hom(S,\mathbf{C}\times\mathbf{C}) \xrightarrow{Hom(S,+)} Hom(S,\mathbf{C}). \tag{9.64}$$

There is an equally important categorical idea that is dual to the idea that $Hom(a,b)$ sometimes inherits properties of b. It says that $Hom(a,b)$ will sometimes inherit the dual structure of a. For example, if a has a comultiplication (which we saw in Section 7.4)

$$a \xrightarrow{\ \Delta\ } a+a\ , \tag{9.65}$$

then the contravariant functor $Hom(\ \ ,b)$ will take that comultiplication to the multiplication

$$Hom(a,b) \xleftarrow{Hom(\Delta,b)} Hom(a+a,b) \xleftarrow{\cong} Hom(a,b)\times Hom(a,b). \tag{9.66}$$

From this perspective, it is easy to understand the fundamental group. Notice that the pointed circle S^1 has a comultiplication. One can take the circle and pinch it into two connected circles. This operation looks like this $O \mapsto \infty$ and is formally written as $S^1 \mapsto S^1 + S^1$. Within the homotopy category of pointed topological spaces, this operation makes S^1 into a **cogroup object**. The main idea emphasized here is that

$$\pi_1(X,x) = Hom_{\mathbb{hoTop}}((S^1,(1,0)),(X,x)) \tag{9.67}$$

inherits the dual structure of the circle and becomes a group. This is also true for the higher spheres: each of the pointed S^n is a cocommutative cogroup object and each of the homotopy groups $\pi_n(X,x)$ inherits the dual structure and becomes an abelian group. One can read more about this perspective of homotopy groups in Chapter 11 of [227].

Simplicial Sets

While topological spaces are intuitively clear, they are hard to describe. What are the sets of elements? What are the open sets? What is the shape being shown? Simplicial sets were invented as an easy way of describing types of topological spaces.

A simplicial set describes a topological space of any dimension with an infinite sequence of sets, $S_0, S_1, S_2, \ldots$ and maps between them. Set S_0 corresponds to the points of the space. Set S_1 corresponds to the one-dimensional lines of the space, set S_2 corresponds to the two-dimensional parts of the space, etc. In general, set S_n will describe the n-dimensional parts of the space. The maps between these sets tell how all these different parts are put together.

To get to the formal definition of a simplicial set, we have to do a little work. We begin by defining a monoidal category that has all the information needed to deal with how the various dimensional parts interact.

Definition 9.3.5. The **simplicial category** or the **simplex category** is denoted as $\mathbf{\Delta}$. The objects are totally ordered sets $[0]=\{0\}$, $[1]=\{0<1\}$, $[2]=\{0<1<2\}, \ldots,$ $[n]=\{0<1<2<\cdots<n\}, \ldots$. The morphisms are order-preserving functions; that is, functions of the form $f\colon [m] \longrightarrow [n]$ such that for all $i \leq j$, we have $f(i) \leq f(j)$. One can view the objects as partial-order categories and morphisms as order-preserving functors, and then think of $\mathbf{\Delta}$ as a subcategory of $\mathbb{Cat}$. There is an inclusion functor $Inc\colon \mathbf{\Delta} \hookrightarrow \mathbb{Cat}$.

A typical morphism from [6] to [9] looks like this:

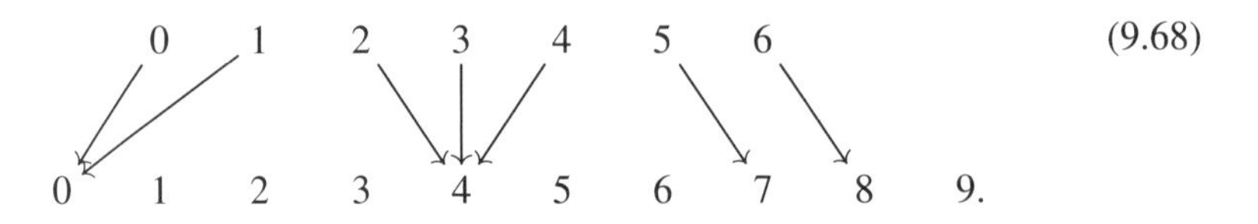

(9.68)

The fact that functions are order-preserving means that the arrows do not cross each other.

The category has a strict monoidal structure $(\mathbf{\Delta}, +, [0])$ in which the tensor product on objects is defined as $[n] + [m] = [n+m]$, and given $f\colon [n] \longrightarrow [n']$ and $g\colon [m] \longrightarrow [m']$, we have $f+g\colon [n+m] \longrightarrow [n'+m']$, defined as

$$(f+g)(i) = \begin{cases} f(i) & i = 0, 1, 2, \ldots, n \\ (n) + g(i) & i = n+1, \ldots, n+n' \end{cases}. \tag{9.69}$$

The monoidal structure on morphisms corresponds to placing diagrams like Diagram (9.68) side by side.

We saw on page 247 that the symmetry category is generated by a simple set of morphisms. Similarly, we saw on page 336 that the braid category is generated by a simple set of morphisms. In the same way, the morphisms in $\mathbf{\Delta}$ are generated by a simple set of morphisms.

Definition 9.3.6. For each n, and for each of the $n+2$ values of $i = 0, 1, 2, \ldots, n$, $n+1$, there is an injective map $\phi_i^n\colon [n] \longrightarrow [n+1]$ that skips the ith output. For example, $\phi_3^6\colon [6] \longrightarrow [7]$ looks like this:

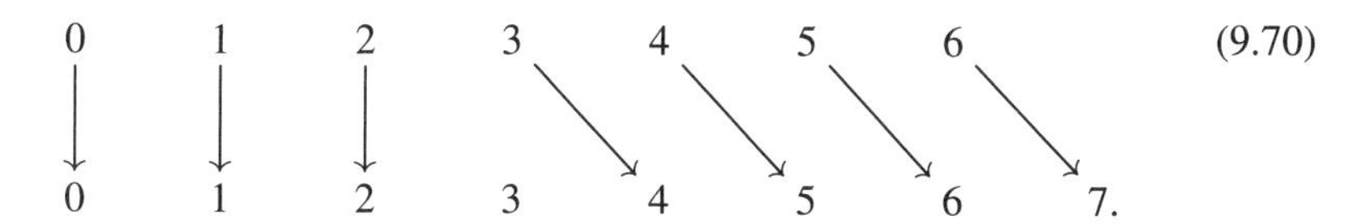

(9.70)

These morphisms are called **coface maps** for reasons that will become apparent after reading Definition 9.3.10.

For each n, and for each of the n values of $i = 0, 1, 2, \ldots, n-1$, there is a surjective map $\delta_i^n\colon [n] \longrightarrow [n-1]$ in which the ith output is the only output that has two inputs going to it. For example $\delta_2^6\colon [6] \longrightarrow [5]$ looks like this:

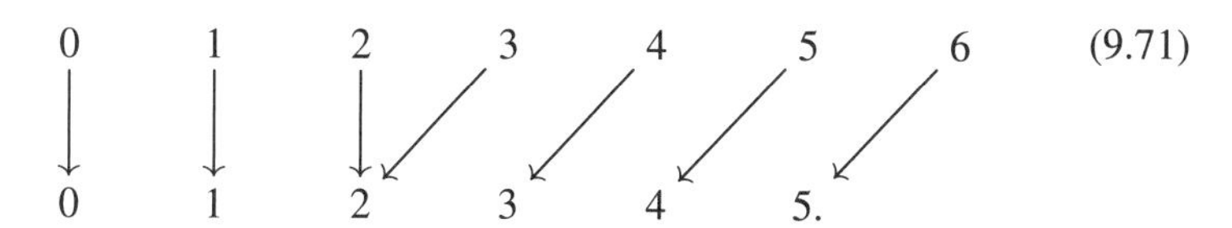

(9.71)

These morphisms are called **codegeneracy maps** for reason that will become apparent after reading Definition 9.3.10.

We can envision all these maps as shown in Figure 9.6.

Every map in $\mathbf{\Delta}$ factors as a sequence of codegeneracies followed by a sequence of coface maps. The best way to see this is with an example. The map in Diagram (9.68) factors as the maps shown in Figure 9.7.

Just as we saw that the generators of the symmetry category and the braid category satisfy certain rules, the compositions of coface and codegeneraciey maps do as well. These rules tell how (i) the coface maps cohere with each other, (ii) the codegeneracy

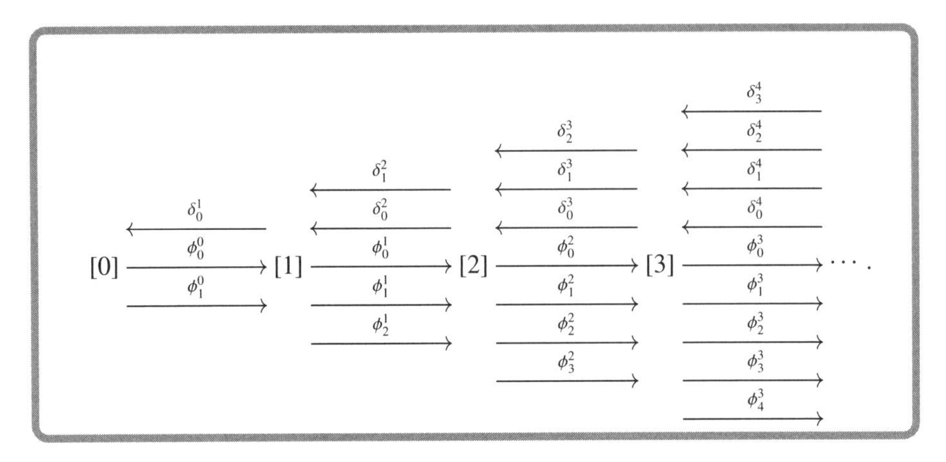

Figure 9.6. The first few coface and codegenerecy maps.

maps cohere with each other, and (iii) coface and codegeneracies cohere with each other:

(i) $\phi_j^{n+1} \circ \phi_i^n = \phi_i^{n+1} \circ \phi_{j-1}^n \qquad i \leq j$

(ii) $\delta_j^{n-1} \circ \delta_i^n = \delta_i^{n-1} \circ \delta_{j+1}^n \qquad i \leq j$

(iii) $\delta_j^{n+1} \circ \phi_i^n = \begin{cases} \phi_{j-1}^n \circ \delta_i & i < j \\ id_{[n]} & i = j \text{ or } i = j+1 \\ \phi_j^{n-1} \circ \delta_{i-1}^n & i > j+1. \end{cases}$

An example of rule (i) is $\phi_4^5 \circ \phi_2^4 = \phi_2^5 \circ \phi_3^4$, which can be seen as follows:

$$\begin{array}{ccc} 0\ 1\ 2\ 3\ 4 & & 0\ 1\ 2\ 3\ 4 \\ 0\ 1\ 2\ 3\ 4\ 5 & = & 0\ 1\ 2\ 3\ 4\ 5 \\ 0\ 1\ 2\ 3\ 4\ 5\ 6 & & 0\ 1\ 2\ 3\ 4\ 5\ 6. \end{array} \tag{9.72}$$

An example of rule (ii) is $\delta_3^4 \circ \delta_1^5 = \delta_1^4 \circ \delta_4^5$, which can be seen as follows:

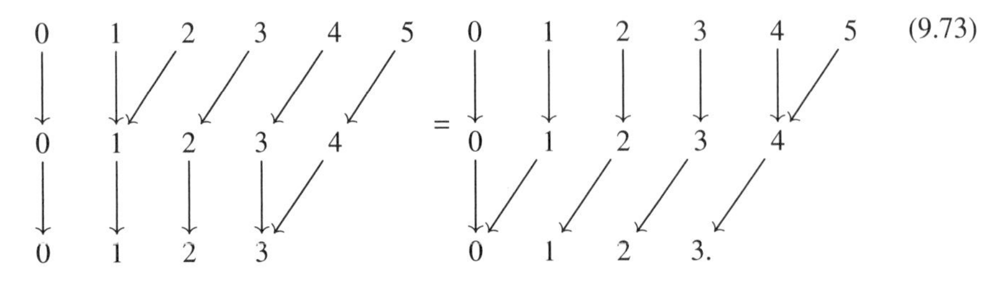

(9.73)

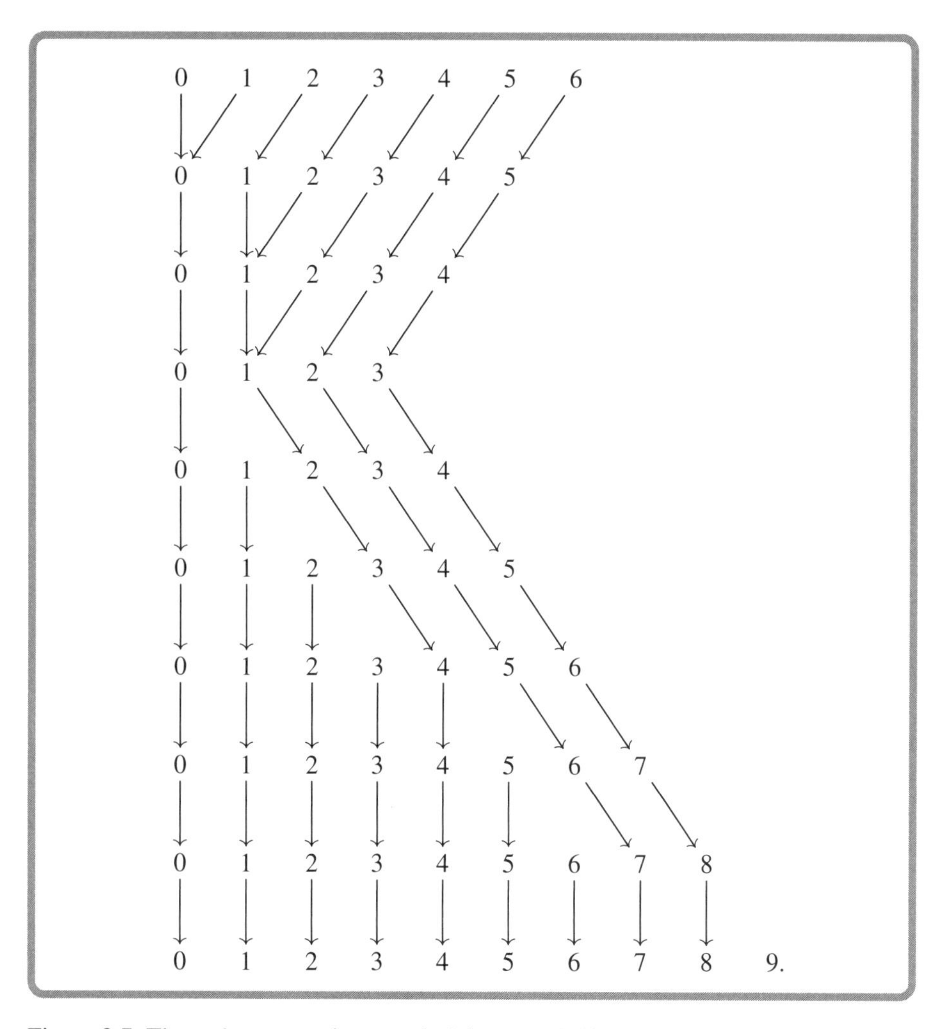

Figure 9.7. The order-preserving map in Diagram (9.68) as a composition of codegeneracy maps followed by coface maps.

Example 9.3.7. Here are examples of all three cases of rule (iii).

An example of the first case of rule (iii) is $\delta_3^5 \circ \phi_1^4 = \phi_1^3 \circ \delta_2^4$, which can be seen as follows:

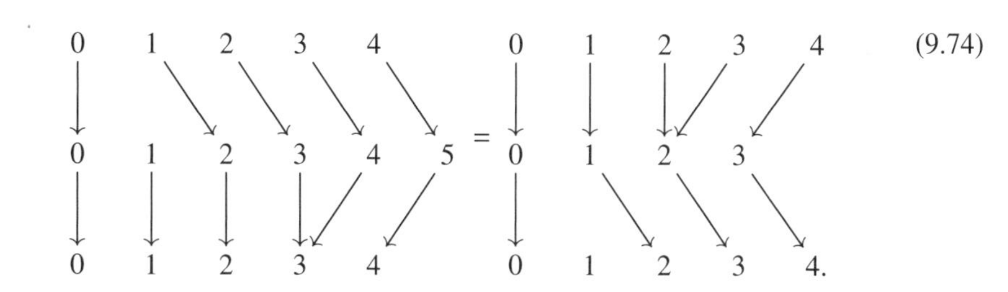

(9.74)

An example of the second case of rule (iii) is $\delta_3^5 \circ \phi_3^4 = id_{[n]}$, which can be seen as follows:

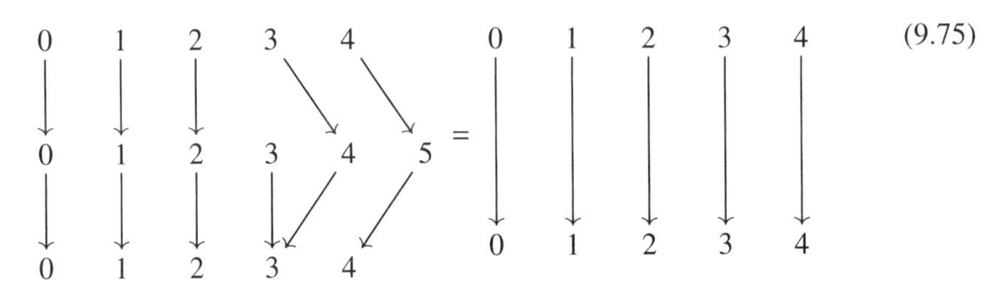

(9.75)

An example of the third case of rule (iii) is $\delta_1^5 \circ \phi_4^4 = \phi_3^3 \circ \delta_1^4$, which can be seen as follows:

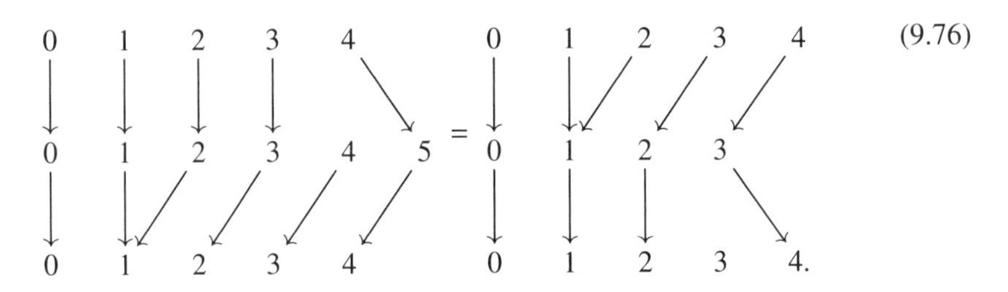

(9.76)

□

There is a functor $r\colon \mathbf{\Delta} \longrightarrow \mathbb{T}\mathrm{op}$ that realizes the objects as special topological spaces. That is, r takes each n to an n-dimensional **topological standard simplex**. Stated formally,

$$r([n]) = \Delta_n = \{(x_0, x_1, \ldots, x_n) \in \mathbf{R}^{n+1} | x_0 + x_1 + \cdots + x_n = 1, x_i \geq 0\}. \tag{9.77}$$

(The symbol Δ is used far too often in this book for too many different concepts. In the language of computer science, we say that Δ is overloaded. Unfortunately, there is little we can do about this since the literature on the topic uses this symbol in so many different ways.) The first few topological simplices look like this:

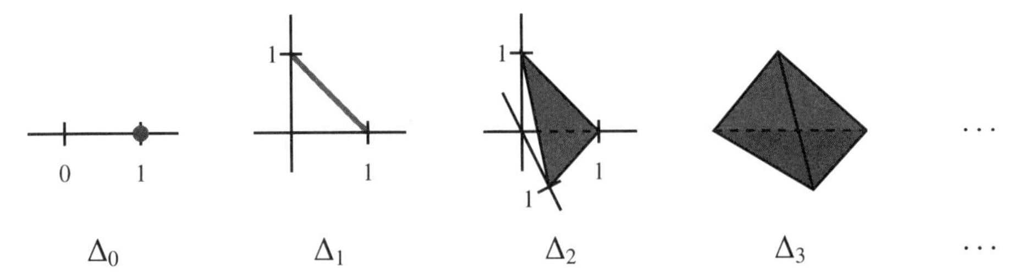

Δ_0 Δ_1 Δ_2 Δ_3 ...

The space Δ_3 is a subset of $\mathbf{R}^4$ that is hard to draw, and the higher simplices are even harder to draw.

The $r(\phi_i^n)\colon \Delta_n \longrightarrow \Delta_{n+1}$ takes an $n+1$-tuple and inserts a 0 in the ith position to get an $n+2$-tuple. This can be seen clearly in the following diagram, where equations can be be seen so that the edges of the empty triangle is included into the filled-in triangle. For example, a point on the edge is described as $(x,0,z)$ and has the property that $x+z=1$. One can think of it as a point $(x,0,z)$ within the entire triangle:

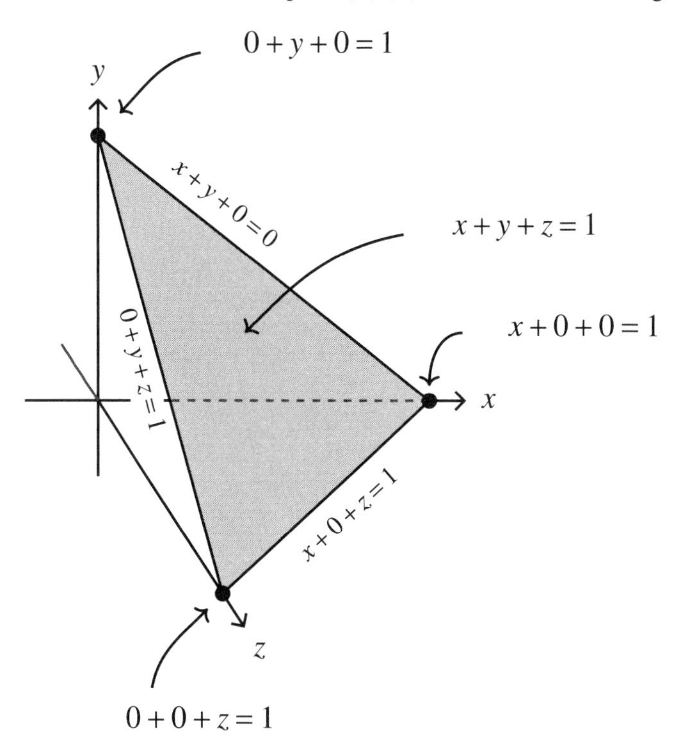

The $r(\delta_i^n)\colon \Delta_n \longrightarrow \Delta_{n-1}$ takes an $n+1$-tuple and adds the ith and the $i+1$th element to get an n-tuple. This too can be seen in this diagram. For example, a point (x,y,z) on the interior of the triangle has the property that $x+y+z=1$. We can add x and y and get the point $(x+y,0,Z)$. This point also has the property that $(x+y)+0+z=1$ and is on the edge.

Remark 9.3.8. The collection of topological standard simplices form an operad. This operad is important and is used in probability theory and information theory. The operad structure is also related to the distribution monad. First, here is an easy example. A typical point in Δ_2 is $\left(\frac{1}{3},\frac{1}{6},\frac{1}{2}\right)$ and a typical point in Δ_3 is $\left(\frac{3}{8},\frac{1}{8},\frac{0}{2},\frac{1}{2}\right)$. We can form an element of Δ_5 by "inserting" all the elements of Δ_3 into the 1th place the element in Δ_2. But before we insert the elements, we must first scalar-multiply by $\frac{1}{6}$. This entails an operad structure map:

$$\Delta_2 \circ_1 \Delta_3 \longrightarrow \Delta_{2+3}. \tag{9.78}$$

(Notice that there is a slight change in the usual operad notation because Δ_n has $n+1$ entries.) For our examples, we have

$$\left(\frac{1}{3},\frac{1}{6},\frac{1}{2}\right)\circ_1\left(\frac{3}{8},\frac{1}{8},\frac{0}{2},\frac{1}{2}\right)=\left(\frac{1}{3},\frac{1}{6}\cdot\frac{3}{8},\frac{1}{6}\cdot\frac{1}{8},\frac{1}{6}\cdot\frac{0}{2},\frac{1}{6}\cdot\frac{1}{2},\frac{1}{2}\right). \tag{9.79}$$

The final element of Δ_4 can be simplified as

$$\left(\frac{1}{3}, \frac{1}{16}, \frac{1}{48}, 0, \frac{1}{12}, \frac{1}{2}\right). \tag{9.80}$$

In general, the composition map

$$\Delta_m \circ_i \Delta_n \longrightarrow \Delta_{m+n} \tag{9.81}$$

is defined for $(x_0, x_1, \ldots, x_i, \ldots, x_m)$ and $(y_0, y_1, \ldots, y_n)$ as

$$(x_0, x_1, \ldots, x_{i-1}, x_i \cdot y_0, x_i \cdot y_1, \ldots, x_i \cdot y_n, x_{i+1}, \ldots, x_m). \tag{9.82}$$

Proving that these maps satisfy the operad properties is left for the reader. ♠

While $\mathbf{\Delta}$ is an interesting category in itself, its real usefulness comes from the fact that we use it to define simplicial sets.

Definition 9.3.9. A **simplicial set** is a functor $S: \mathbf{\Delta}^{\mathbf{op}} \longrightarrow \mathbb{Set}$. Let us unpack this definition. For every natural number n, there is a set $S([n])$ that we write as S_n, and for every order-preserving morphism $g: [n] \longrightarrow [m]$ in $\mathbf{\Delta}$, there is a set function $S(g): S_m \longrightarrow S_n$.

Morphisms between simplicial sets are natural transformations called **simplicial maps**. This means that a simplicial map $F: S \Longrightarrow S'$ is a sequence of set functions $F_n: S_n \longrightarrow S'_n$, which commute with the set maps $S(g)$ and $S'(g)$ for every map $g: [n] \longrightarrow [m]$ in $\mathbf{\Delta}$.

The collection of all simplicial sets and maps forms a category denoted as $\mathbb{Set}^{\Delta^{op}}$ or $s\mathbb{Set}$.

From the fact that every map in $\mathbf{\Delta}$ is generated by cofaces and codegeneracies, we can also define a simplicial set with them.

Definition 9.3.10. A **simplicial set** is all of the following:

- A sequence of sets $S_0, S_1, S_2, \ldots$.
- For each n, and for each of the $n+2$ values $i = 0, 1, 2, \ldots, n+1$, there is a map $S(\phi_i^n) = f_i^n: S_{n+1} \longrightarrow S_n$ called a **face map**.
- For each n, and for each of the n values $i = 0, 1, 2, \ldots n-1$, there is a map $S(\delta_i^n) = d_i^n: S_{n-1} \longrightarrow S_n$ called a **degeneracy map**.

We can envision the face and degeneracy maps as in Figure 9.8, which is dual to Figure 9.6.

Now, the definition of a simplicial map is slightly easier to state. We only need to make sure that the family of maps $F_n: S_n \longrightarrow S'_n$ satisfies the naturality condition with the faces and degeneracies. If the naturailty condition is satisfied for those maps, then it will be satisfied for all maps. The commuting of maps with face maps is a vast generalization of the requirement that we saw earlier–namely, that a graph homomorphism must commute with source and target maps.

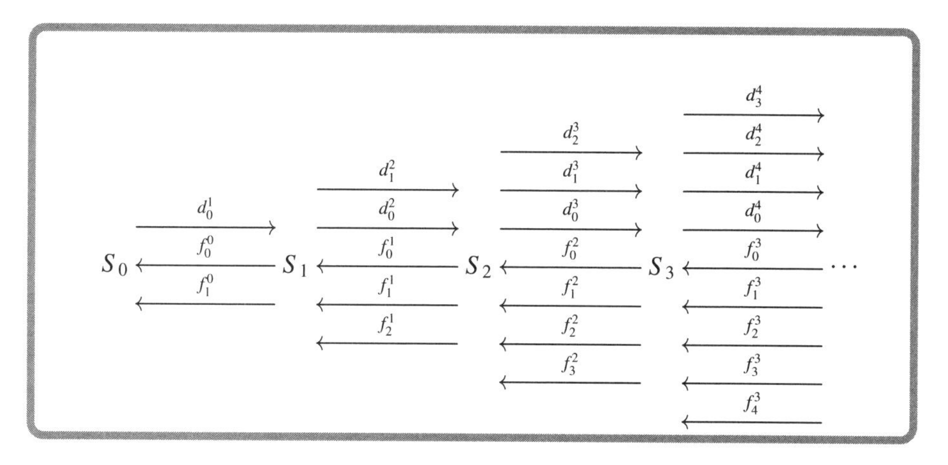

Figure 9.8. The first few face and degenerecy maps.

The face and degeneracies satisfy rules that are dual to the rules about cofaces and codegeneracies:

(i) $f_i^n \circ f_j^{n+1} = f_i^{n+1} \circ f_{j-1}^n \qquad i < j$

(ii) $d_i^n \circ d_j^n = d_{j+1}^n \circ d_i^n \qquad i \leq j$

(iii) $$f_i^n \circ d_j^{n+1} = \begin{cases} d_{j-1} \circ f_i & i < j \\ id_{S_n} & i = j \text{ and } i = j+1 \\ d_j^n \circ f_{i-1}^n & i > j+1 \end{cases} .$$

There are some very important classes of simplicial sets. Keep the following diagrams in mind while we define them:

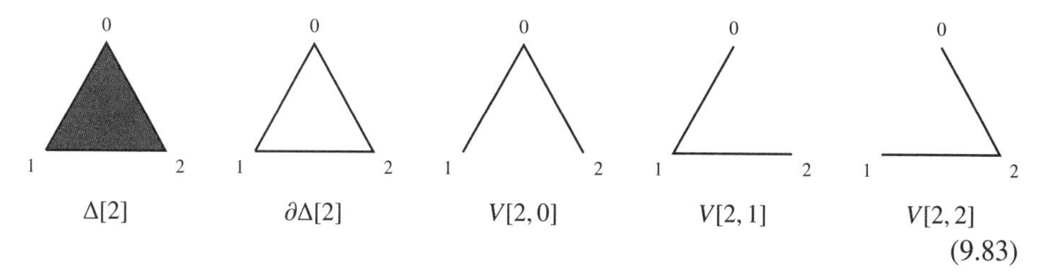

(9.83)

- The **standard simplex**, denoted as $\Delta[n]$, is a simplicial set that is the representable functor. This means that $\Delta[n]$ is defined as $Hom_{\mathbf{\Delta}}(\ ,[n])\colon \mathbf{\Delta}^{\mathbf{op}} \longrightarrow \mathbb{S}\mathrm{et}$. It corresponds to a complete n-dimensional object. The standard simplex will play a major role in the coming pages. Formally, it is also written as

$$\Delta[n]_m = \{g\colon [m] \longrightarrow [n]\} = Hom_{\mathbf{\Delta}}([m],[n]). \qquad (9.84)$$

These standard simplices are the image of the Yoneda embedding $y\colon \mathbf{\Delta} \longrightarrow \mathbb{S}\mathrm{et}^{\mathbf{\Delta}^{\mathbf{op}}}$.

- There is the **boundary** or **outer shell** of the n simplex, which we denote as $\partial\Delta[n]$:

$$\partial\Delta[n] = \bigcup_{i=0}^{n}(Image(d_i\colon \Delta[n-1] \longrightarrow \Delta[n])). \tag{9.85}$$

- There is the **horn** $V[n,k]$ (because in the two-dimensional case, it looks like the letter "V"), also denoted as Λ^n_k (that is also missing a side), which is the outer shell but without the k-th face. We can write this formally as

$$V[n,k] = \bigcup_{i=0, i\neq k}^{n}(Image(d_i\colon \Delta[n-1] \longrightarrow \Delta[n])). \tag{9.86}$$

Now that we have an understanding of what simplicial sets are, let us look at the various relationships between the category of simplicial sets and other categories. The following diagram of categories and functors literally shows the centrality of $\mathbb{S}\mathrm{et}^{\Delta^{op}}$:

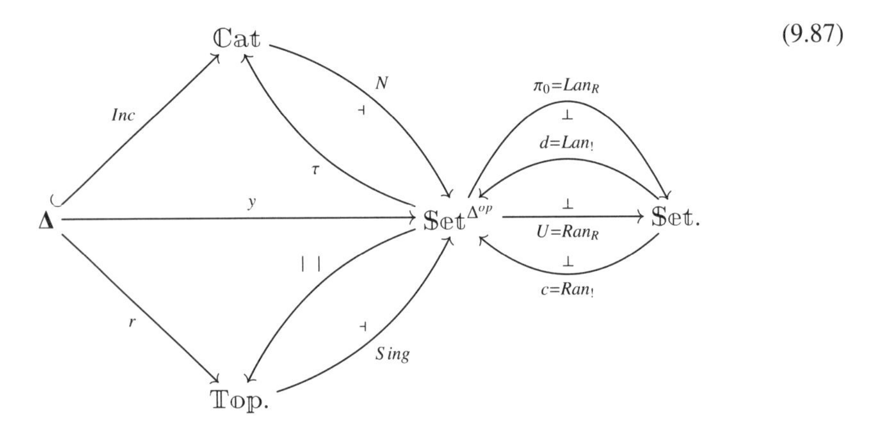

(9.87)

Let us start with the relationship between $\mathbb{S}\mathrm{et}^{\Delta^{op}}$ and $\mathbb{S}\mathrm{et}$. It might be helpful to review the functors in Diagrams (4.115), (4.116), and (9.17) because the functors described here are a vast generalization of the functors in those diagrams. There is an adjunction

$$\mathbf{\Delta^{op}} \underset{R}{\overset{!}{\rightleftarrows}} \mathbf{1}. \quad (\perp) \tag{9.88}$$

Here, R takes $*$ of $\mathbf{1}$ to $[0]$, the initial object of $\mathbf{\Delta}$. The functors R and $!$ induce the four Kan extensions given in the diagram. Let us look at each of these functors individually:

- The functor π_0 takes a simplicial set S to its set of connected components.
- The functor d takes a set V to the discrete simplicial set S with $S_0 = V$ and with only degeneracies in S_i for $i > 0$.

- The functor U takes a simplicial set S to the set S_0 and forgets all the other sets. (This functor is sometimes called ev_0; that is, evaluation at 0.)
- The functor c takes a set X to the simplicial set that has all dimensions $S_n = X$, with the faces and degeneracies being identity maps.

There is a **nerve functor** $N\colon \mathbb{C}\mathrm{at} \longrightarrow \mathbb{S}\mathrm{et}^{\Delta^{op}}$ that takes a small category and outputs a simplicial set which—in a sense—is the shape of the category. For a small category $\mathbb{A}$, the simplicial set $N(\mathbb{A})_n$ is the collection of $n-composible$ maps in $\mathbb{A}$. A typical element of $N(\mathbb{A})_n$ looks like this:

$$\bullet \xrightarrow{g_1} \bullet \xrightarrow{g_2} \bullet \xrightarrow{g_3} \cdots \xrightarrow{g_i} \bullet \xrightarrow{g_{i+1}} \bullet \longrightarrow \cdots \bullet \xrightarrow{g_n} \bullet. \tag{9.89}$$

In particular, for $n = 0$, the collection $N(\mathbb{A})_0$ consists of the identity maps of $\mathbb{A}$. The collection $N(\mathbb{A})_1$ is the set of morphisms in $\mathbb{A}$. The face map $f_i\colon N(\mathbb{A})_n \longrightarrow N(\mathbb{A})_{n-1}$ takes a string of n composible maps and actually composes the i-th and the $i+1$th map to get

$$\bullet \xrightarrow{g_1} \bullet \xrightarrow{g_2} \bullet \xrightarrow{g_3} \cdots \xrightarrow{g_{i+1} \circ g_i} \bullet \longrightarrow \cdots \bullet \xrightarrow{g_n} \bullet\,. \tag{9.90}$$

The degeneracy map $d_i\colon N(\mathbb{A})_n \longrightarrow N(\mathbb{A})_{n+1}$ takes a string of n composible maps and actually adds an identity map at the ith position to get a sequence of $n+1$ composible maps:

$$\bullet \xrightarrow{g_1} \bullet \xrightarrow{g_2} \bullet \xrightarrow{g_3} \cdots \xrightarrow{id} \bullet \xrightarrow{g_i} \bullet \longrightarrow \cdots \bullet \xrightarrow{g_n} \bullet. \tag{9.91}$$

The left adjoint of the nerve is the **first truncation functor**, which takes a simplicial set S and returns the category described by S_0, S_1, and S_2. The objects of the category are elements of set S_0. The arrows of the category are the elements of set S_1. The composition of two maps in S_1 is equal to the third if there is an element in S_2 that has these three maps as faces (in the correct order).

Let us move on to the relationship of simplicial sets and topological spaces. First, there is a functor $Sing\colon \mathbb{T}\mathrm{op} \longrightarrow \mathbb{s}\mathbb{S}\mathrm{et}$ that takes every topological space X to its **singular complex** or **singular simplicial set** . Formally,

$$Sing(X)_n = Hom_{\mathbb{T}\mathrm{op}}(\Delta_n, X); \tag{9.92}$$

that is, the set of all functions from the standard topological simplices to X. This idea is central from the category theory perspective: if you want to know everything about object X, look at all the sets of maps into X from a complete set of topological spaces. The face and degeneracies of $Sing$ are induced by the face and degeneracies of simplices. In other words,

$$f_i^n(g\colon \Delta_n \longrightarrow X) = g \circ \Delta(\phi_i^n)\colon \Delta_{n+1} \longrightarrow \Delta_n \longrightarrow X. \tag{9.93}$$

There is a similar definition for degeneracies.

The left adjoint of this is the **geometric realization functor** $|\ \ |\colon \mathbb{s}\mathbb{S}\mathrm{et} \longrightarrow \mathbb{T}\mathrm{op}$. It takes a simplicial set and produces the topological space that it describes. The construction of this function is very interesting. For a simplicial set S, set S_n is the set that corresponds to the number of n-dimensional parts of the simplicial set. Consider set $S_n \times \Delta_n$, which has one standard simplex for every element of S_n. Now consider

$$\amalg_{n=0}^{\infty} S_n \times \Delta_n. \tag{9.94}$$

This has copies of all the appropriate-size standard simplices for the simplicial set. These different simplices are put together with an equivalence relation. The faces and degenerereraciy maps tell us how they stick together.

It is important to notice that for any topological space X, the simplicial set $Sing(X)$ essentially has all the important information about X. The reason for this is that for any X, the space $|Sing(X)|$ is weakly homotopy equivalent to X.

Since simplicial sets are generalizations of the notion of topological spaces, there are ways of talking about notions of homotopy for simplicial sets. We use the geometric realization functor to discuss homotopy equivalence. A map $f\colon S_1 \longrightarrow S_2$ in $\mathbb{sSet}$ is a **homotopy equivalence** if $|f|\colon |S_1| \longrightarrow |S_2|$ is a weak homotopy equivalence in $\mathbb{Top}$.

The homotopy groups of a simplicial set can be calculated with simplicial sets. For a simplicial set S, a point will correspond to a simplicial map $pt\colon \Delta^0 \longrightarrow S$. The nth homotopy group will be about maps of the form $\Delta^n \longrightarrow S$. However, we want such maps to satisfy a requirement about the boundary. In detail, for a point s in S, we have

$$\pi_n(S,s) = \{f\colon \Delta^n \longrightarrow S \,|\, pt \circ ! = f \circ inc\}/\sim. \tag{9.95}$$

This condition means that

$$\begin{array}{ccc} \partial\Delta^n & \xrightarrow{\;\;!\;\;} & \Delta^0 \\ {\scriptstyle inc}\downarrow & & \downarrow{\scriptstyle pt} \\ \Delta^n & \xrightarrow[f]{} & S. \end{array} \tag{9.96}$$

The homotopy groups are equivalence classes of such maps. Two such maps $f, g\colon \Delta^n \longrightarrow S$ are deemed homotopically equivalent if there is a simplicial map $H\colon \Delta^{n+1} \longrightarrow S$ such that the two faces of this map are f and g.

For simplicial sets, a special type of surjection is called a **Kan fibration**. A simplicial map $f\colon S \longrightarrow S'$ is a Kan fibration if for all n, i, g, and g' in the commutative square

$$\begin{array}{ccc} V[n,i] & \xrightarrow{\;\;g\;\;} & S \\ {\scriptstyle inc}\downarrow & \overset{H}{\nearrow} & \downarrow{\scriptstyle f} \\ \Delta^n & \xrightarrow[g']{} & S', \end{array} \tag{9.97}$$

there is an H such that the two triangles commute. Let us examine this definition carefully. It says that if there is a standard simplex in S', then for every horn in S that goes down to that simplex, there is a way to lift the standard simplex (in other words, complete the horn) in S. Another way to say this is that there are no nontrivial holes in S that are not already in S'.

The terminal simplicial set $*$ has only one element in the bottom level, and all the other levels have degeneracies of only that single element. A fibrant simplicial set is called a **Kan complex**. It is a simplicial set S where the unique map $S \longrightarrow *$ is a Kan fibration. This means that the bottom triangle in Diagram (9.97) always commutes. This, in turn, says that every horn in S can be filled in.

A map $f\colon S \longrightarrow S'$ of simplicial sets is a **cofibration** if each f_n is an inclusion.

Before we close our discussion of simplicial sets, let us note that the definition has been generalized by researchers in many different directions as follows:

- As we said in Important Categorical Idea 4.5.3, the shape $\mathbf{\Delta^{op}}$ can take models in many contexts besides $\mathbb{Set}$. For any category $\mathbb{A}$, we can talk about the category $\mathbb{A}^{\Delta^{op}}$ which we call **simplicial objects in $\mathbb{A}$**. Several categories have shown to be very important in various areas: $\mathbb{Top}^{\mathbf{\Delta^{op}}}$ is called the category of **simplicial spaces**, $\mathbb{Cat}^{\mathbf{\Delta^{op}}}$ is called the category of **simplicial categories**, and $\mathbb{Group}^{\Delta^{op}}$ is called the category of **simplicial groups**.
- One can even take functors from Δ^{op} to the category of ... simplicial sets. We write this as
$$(\mathbb{Set}^{\mathbf{\Delta^{op}}})^{\mathbf{\Delta^{op}}} = \mathbb{Set}^{\mathbf{\Delta^{op}}\times\mathbf{\Delta^{op}}}. \tag{9.98}$$
The objects of this category are called **bisimplicial sets** and appear in many areas of homotopy theory and higher category theory. In detail, a bisimplicial set is a functor $S : \mathbf{\Delta^{op}} \times \mathbf{\Delta^{op}} \longrightarrow \mathbb{Set}$, and for every pair of natural numbers m and n, there is a set $S(m,n)$ with lots of faces and degeneracies going around.
- There is good reason to examine functors from category $\mathbf{\Delta}$ rather than from category $\mathbf{\Delta^{op}}$. For category $\mathbb{A}$, we have $\mathbb{A}^{\Delta}$, which is called the category of **cosimplicial objects in $\mathbb{A}$**.

Categorical Homotopy Theory

With the knowledge of classical homotopy theory and homotopy theory of simplicial sets in hand, Daniel Quillen wrote an important monograph in 1967 titled *Homotopical Algebra* [216], which looks at homotopy from a categorical, or abstract, point of view. Quillen described the structure that a category needs for there to exist the notions and tools of homotopy theory. Such a category is known as a "model category." The most amazing aspect of these ideas is that the examples of model categories extend far beyond topological examples. With a model structure on a category, one can form the homotopy category of a model category. The homotopy category has the same objects as the original category, but the objects that are deemed homotopicaly similar are made isomorphic in this category. One of the central goals of this field is to show when two different model categories have equivalent homotopy categories.

Definition 9.3.11. Category $\mathbb{A}$ is a **Quillen model category**, or simply a **model category**, if it has three classes of morphisms:

- **Weak equivalences**, denoted as WE, that go between objects that are homotopically the same and contain at least all the isomorphisms in the category (and hence the identities)

- **Fibrations**, denoted as Fib, which are to be thought of as nice surjective morphisms
- **Cofibrations**, denoted as Cof, which are to be thought of as nice injective morphisms

A map that is a weak equivalence and a fibration ($WE \cap Fib$) is called a **trivial fibration** and is to be thought of as a very nice surjection. Similarly, a map that is a weak equivalence and a cofibration ($WE \cap Cof$) is called a **trivial cofibration** and is to be thought of as a very nice injection.

The category and the three classes of morphisms must satisfy the following axioms:

Axiom I. Category $\mathbb{A}$ has all limits and colimits. In particular, it has an initial object $\emptyset$, a terminal object $*$, as well as products and coproducts.

Axiom II. This axiom has to do with when there are nice liftings. Consider the following diagram in $\mathbb{A}$:

$$\begin{array}{ccc} a & \xrightarrow{\quad f \quad} & a' \\ {\scriptstyle i}\downarrow & \overset{H}{\nearrow} & \downarrow{\scriptstyle p} \\ a'' & \xrightarrow[\quad g \quad]{} & a''' \end{array} \tag{9.99}$$

where $p \circ f = g \circ i$, i is a cofibration, and p is a fibration. If either i or p is also a weak equivalence, then there is an H such that the two triangles commute.

Axiom III. Every map $f\colon X \longrightarrow Y$ in $\mathbb{A}$ can be factored in two ways: (i) as a trivial cofibration followed by a fibration, and (ii) as a cofibration followed by a trivial fibration. We can write this as

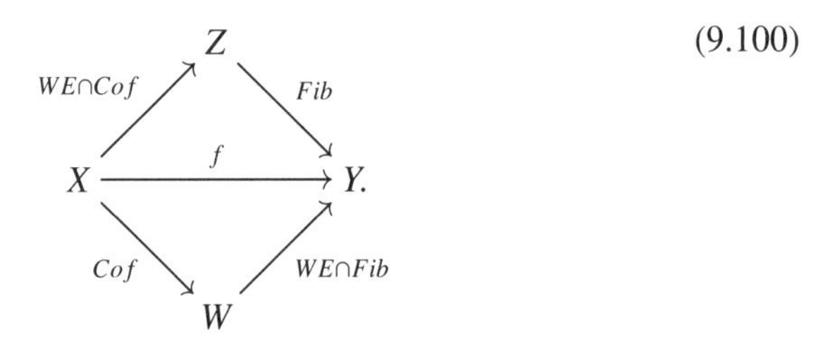

(9.100)

Axiom IV. Consider two composable maps in $\mathbb{A}$:

$$X \xrightarrow{\ f\ } Y \xrightarrow{\ g\ } Z, \qquad g \circ f\colon X \longrightarrow Z. \tag{9.101}$$

The maps f, g, and $g \circ f$ have the **two-out-of-three property**. This means that if any two of the three maps are weak equivalences, then the third is as well. The intuition is that if there are three objects and you have two maps showing that they are homotopic, then all three are homotopic.

Axiom V. An object X is a **retract** of an object Y if there is an inclusion $i\colon X \longrightarrow Y$ and a map $r\colon Y \longrightarrow X$ such that $r \circ i = id_X$. A map $f\colon X \longrightarrow X'$ is a **retract** of $g\colon Y \longrightarrow Y'$ if X is a retract of Y; X' is a retract of Y'; and the following diagram, which shows that f and g respect the retractions, commutes:

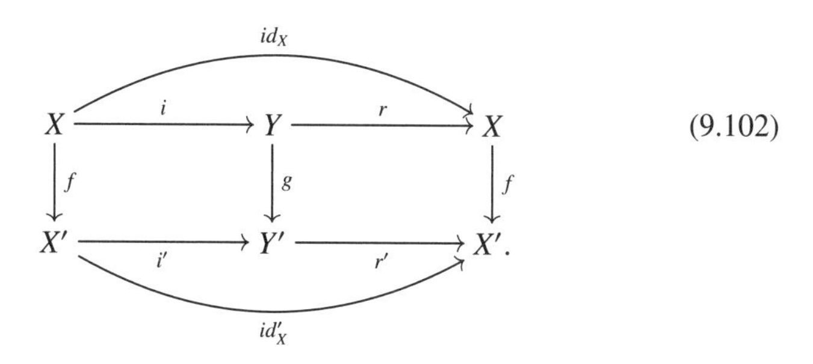

(9.102)

This axiom insists that the three classes of maps are closed under retracts. This means that if g is, say, a weak equivalence, then so is f. The same is true for fibrations and cofibrations.

Variations of these axioms were first stated in [216] and were given in final form in [215].

The main point of having a model category structure on $\mathbb{A}$ is that one can go on to form a **homotopy category**, denoted as $Ho(\mathbb{A})$, which has the same objects as $\mathbb{A}$ but is formed by making all the weak equivalences into isomorphisms. Another way of saying this is that the weak equivalences are inverted in $\mathbb{A}$ by including inverses for each weak equivalence. If $f\colon a \longrightarrow b$ is a weak equivalence, then a new morphism $f^*\colon b \longrightarrow a$ is added to the category. We furthermore insist that $f \circ f^*$ and $f^* \circ f$ are the appropriate identities. This process is also called **localizing the weak equivalences**. There is a **localization functor** $\gamma_{\mathbb{A}}\colon \mathbb{A} \longrightarrow Ho(\mathbb{A})$, which is the identity on objects and takes weak equivalences to isomorphisms. It is important to note that in general, $Ho(\mathbb{A})$ is not a quotient category of $\mathbb{A}$. It is more complicated.

The axioms are all very symmetric. This means that if $\mathbb{A}$ has a model category structure, then so does $\mathbb{A}^{op}$. In detail, if $\mathbb{A}$ has a model category structure, then $\mathbb{A}^{op}$ also has a model category structure in which the fibrations in $\mathbb{A}$ become the cofibrations in $\mathbb{A}^{op}$ and the cofibrations in $\mathbb{A}$ become the fibrations in $\mathbb{A}^{op}$. This means that many statements about homotopy are true if one swaps the fibrations and cofibrations. Another way to say this is that there is a contravariant functor from the category of model categories (suitably defined) to itself that takes $\mathbb{A}$ to $\mathbb{A}^{op}$, which preserves a lot of the structure. (All this is very similar to the duality that we saw in Section 6.4 concerning Boolean algebra. There, we saw that if we have a true statement and we swap the $\wedge$ and the $\vee$ as well as the 0s and the 1s, we once again get true statements.) This duality broadly goes by the name **Eckmann–Hilton duality**.

Before we get into the properties of Quillen model categories, let us look at some examples.

Example 9.3.12. The simplest example of such a category is any complete and cocomplete category where the special sets of morphisms are as follows:

- WE are isomorphisms. This corresponds to saying that isomorphic objects are already considered the same.
- Fib are all maps.
- Cof are all maps.

In such a model category, $Ho(\mathbb{A})$ is the same category as $\mathbb{A}$, as nothing new was inverted. □

Example 9.3.13. The motivating example of a Quillen model category is the category of topological spaces, $\mathbb{Top}$. There are actually two model category structures for $\mathbb{Top}$. The first model category was described by Quillen himself and is called the **classical model structure**:

- WE are weak homotopy equivalences.
- Fib are Serre fibrations.
- Cof are Serre cofibrations.

This is the model category structure that we will deal with for the rest of this section. The homotopy category is $Ho(\mathbb{Top})$. □

Example 9.3.14. The other model category structure of $\mathbb{Top}$ has described by Arne Strøm. The distinguished sets of maps for this structure are as follows:

- WE are homotopy equivalences.
- Fib are Hurewicz fibrations.
- Cof are Hurewicz cofibrations.

While this structure is interesting, we will not be dealing with it in this discussion. □

Example 9.3.15. All our work with $\mathbb{sSet}$ would be for naught if it did not have a model category structure. The special sets of morphisms are as follows:

- WE are maps whose geometric realizations are weak homotopy equivalences of topological spaces. That is, a map $f\colon S_1 \longrightarrow S_2$ is in WE if $|f|\colon |S_1| \longrightarrow |S_2|$ is a weak homotopy equivalence in $\mathbb{Top}$.
- Fib are Kan fibrations.
- Cof are maps $f\colon S \longrightarrow S'$ such that for each n, the set map $f_n\colon S_n \longrightarrow S'_n$ is an injective map.

This model category structure has the feeling of cheating because we are basically using the model structure of $\mathbb{Top}$. This is not such an egregious sin. After all, the category of $\mathbb{sSet}$ was created to be a clearer description of $\mathbb{Top}$. The homotopy category of $\mathbb{sSet}$ is $Ho(\mathbb{sSet})$. One of the main ideas that we will come to in this section is that the category $Ho(\mathbb{Top})$ is equivalent to $Ho(\mathbb{sSet})$. In other words, from the homotopy perspective, what can be described by topological spaces can be described by simplicial sets. □

These past three examples are about variations of topological spaces. There are other ways of describing topological spaces like chain complexes, CW-complexes, simplicial complexes, and pointed spaces. All these have Quillen model category structures, which we will not get into here. Abstract homotopy theory gets more interesting when the examples have less to do with topological spaces. We come to such examples now.

Example 9.3.16. The category of small categories, $\mathbb{C}\mathrm{at}$, has a Quillen model category structure. The distinguished sets of maps are as follows:

- WE are functors that are equivalences of categories (i.e., fully faithful functors that are essentially surjective).
- Fib are functors where isomorphisms lift. This means that they are functors $F\colon \mathbb{A} \longrightarrow \mathbb{B}$ such that for all $a \in \mathbb{A}$ and each isomorphism $g\colon F(a) \longrightarrow b$ in $\mathbb{B}$, there is an isomorphism $f\colon a \longrightarrow a'$ in $\mathbb{A}$ such that $F(f) = h$. Another way of saying this is that for the inclusion $inc\colon \mathbf{1} \longrightarrow \mathbf{2}_I$ of the one-object category into the two-object category with an isomorphism between the two objects, and for all functors G and G', there is an H that makes both diagrams commute:

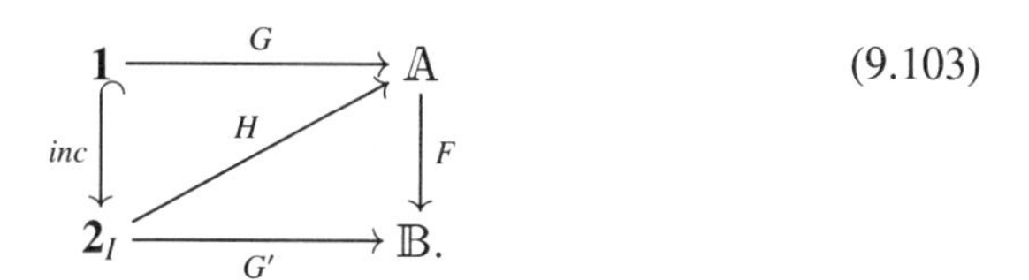

(9.103)

- Cof: are functors that are injective on objects.

The homotopy category is $Ho(\mathbb{C}\mathrm{at})$. For more details, see the paper by Charles Rezk [217]. □

Example 9.3.17. Equivalence relations are ways of describing notions of sameness. This is possibly the easiest way of talking about homotopy theory. We met the category of equivalence relations, $\mathbb{E}\mathrm{quiv}$. The objects are pairs $(S, \sim)$ where S is a set and $\sim$ is an equivalence relation on S. The morphisms $f\colon (S, \sim) \longrightarrow (S', \sim')$ are set functions that take equivalent elements to equivalent elements (i.e, if $s \sim s'$, then $f(s) \sim f(s')$). Finnur Larusson showed that $\mathbb{E}\mathrm{quiv}$ is a model category where the following are all true:

- WE are maps that induce a bijection of quotient sets.
- Fib: are maps $f\colon (S, \sim) \longrightarrow (S', \sim')$ that take each class of S onto a class of S'.
- Cof are maps where the set function is injective.

The homotopy category, $Ho(\mathbb{E}\mathrm{quiv})$, is equivalent to $\mathbb{S}\mathrm{et}$. More can be found about this in [155]. □

Now let us examine some properties of model categories.

We use the distinguished morphisms in a model category to talk about distinguished objects in the category.

Definition 9.3.18. An object a of a model category $\mathbb{A}$ is **fibrant** if the unique map $a \longrightarrow *$ is a fibration. Similarly, an object a is **cofibrant** if the unique map $\emptyset \longrightarrow a$ is a cofibration. We call an object **bifibrant** if it is both fibrant and cofibrant.

Consider the following diagram:

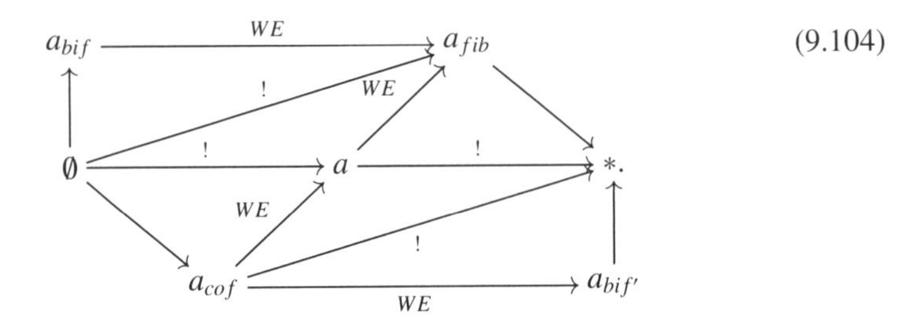

(9.104)

For every object a, one can use axiom III to factor the unique map $\emptyset \longrightarrow a$ and the unique map $a \longrightarrow *$. Notice that a_{fib} is fibrant and weakly equivlent to a. It is called a **fibrant replacement** of a. At the same time, a_{cof} is cofibrant and weakly equivalent to a. It is called a **cofibrant replacement** of a. Now take the unique map $\emptyset \longrightarrow a_{fib}$ and factor it. This gives us an object a_{bif}, which is bifibrant and weakly equivalent to a_{fib}. We call this a **bifibrant replacement** of a. We can also use the fibrant replacement of a_{cof} to get a bifibrant replacement of a. These two bifibrant replacements of a are weakly homotopic, but they need not be the same.

For a model category $\mathbb{A}$, we denote the full subcategory whose objects are fibrant as $\mathbb{A}_{Fib}$. Similarly, there is the full subcategory of cofibrant objects, $\mathbb{A}_{Cof}$, and the full subcategory of bifibrant objects, $\mathbb{A}_{Bif}$. They fit together with inclusions as follows:

$$\begin{array}{ccc} \mathbb{A}_{Bif} & \hookrightarrow & \mathbb{A}_{Fib} \\ \downarrow & & \downarrow \\ \mathbb{A}_{Fib} & \hookrightarrow & \mathbb{A}. \end{array} \qquad (9.105)$$

Since model categories have coproducts, there is a codiagonal map $\nabla\colon a+a \longrightarrow a$ for every object a. This map factors as a cofibration i followed by a trivial fibration. Denote the object that it factors through as $Cyl(a)$ and call it a **cylinder object** of a. (Think of it as $I \times a$.)

$$\begin{array}{ccc} a+a & \xrightarrow{\quad\nabla\quad} & a \\ & {}_{i,\,Cof}\searrow \quad \nearrow_{Fib\cap WE} & \\ & Cyl(a). & \end{array} \qquad (9.106)$$

We use the map $a+a \longrightarrow Cyl(a)$ to make the following definition: Let $f, g\colon a \longrightarrow b$ be two maps. We say that f and g are **left homotopic** if there is a map $H\colon Cyl(a) \longrightarrow b$ such that the following diagram, which is similar to Figure 9.3, commutes:

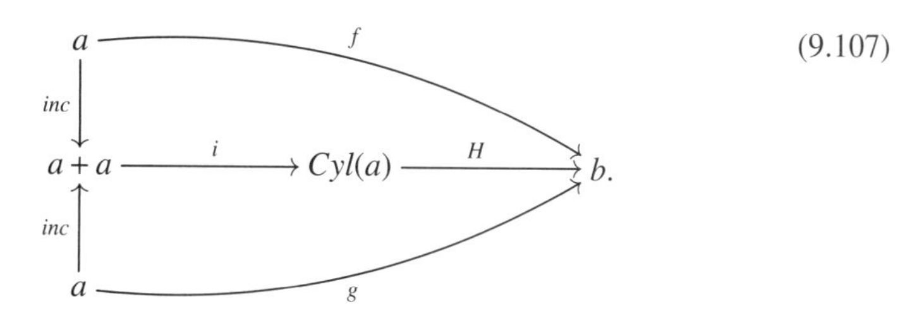

Another way to say this is that $f+g=i\circ H$. It can be shown that the relation of being left homotopic is an equivalence relation.

Since model categories have products, there is a diagonal map $\delta\colon b\longrightarrow b\times b$ for every object b. This map factors as a trivial cofibration followed by fibration p. Denote the object that it factors through as $Path(b)$ and call it a **path object** of b. (Think of it as a^I.)

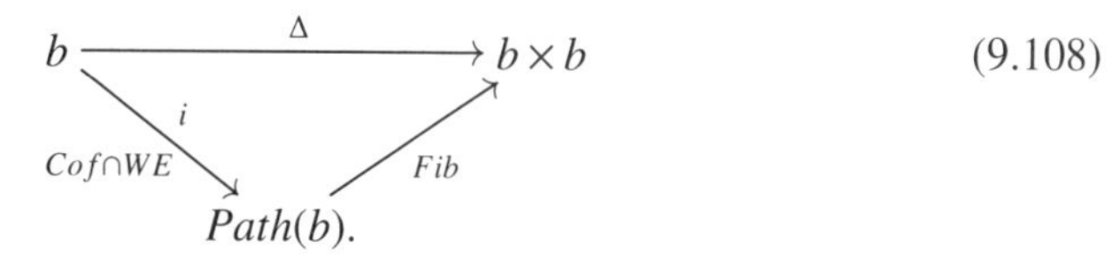

We use the map $p\colon Path(b)\longrightarrow b\times b$ to make the following definition. Let $f,g\colon a\longrightarrow b$ be two maps. We say f and g are **right homotopic** if there is a map $H\colon a\longrightarrow Path(b)$ such that the following diagram, which is similar to Figure 9.4, commutes:

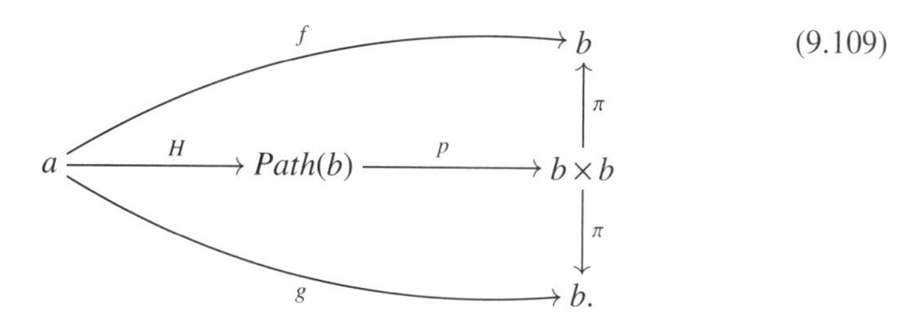

It can be shown that the relation of being right homotopic is an equivalence relation.

If a is cofibrant and b is fibrant, then the left homotopy equivalence relation and the right homotopy equivalence relation on the set $Hom(a,b)$ are the same. We use this fact to describe the morphisms in the homotopy category.

We are interested in the relationship between model categories and other categories.

Definition 9.3.19. Let $\mathbb{B}$ be a model category with $\gamma_{\mathbb{B}}:\mathbb{B}\longrightarrow Ho(\mathbb{B})$. If $F\colon\mathbb{B}\longrightarrow\mathbb{A}$, where $\mathbb{A}$ is any category, then when there is a left Kan extension as follows:

$$\begin{array}{ccc} Ho(\mathbb{B}) & \xrightarrow{Lan_{\gamma_{\mathbb{B}}}(F)} & \mathbb{A} \\ & {\scriptstyle\gamma_{\mathbb{B}}}\nwarrow \quad \nearrow{\scriptstyle F} & \\ & \mathbb{B} & \end{array} \tag{9.110}$$

the functor $Lan_{\gamma_{\mathbb{B}}}(F)$ is called a **left derived functor** of F. When a right Kan extension exists, it is called a **right derived functor** of F.

When the functor $F\colon \mathbb{B} \longrightarrow \mathbb{A}$ is "homotopy invariant" (i.e., it takes a weak equivalence to an isomorphism), then the left derived functor exists and the triangle commutes.

We are interested in the functors between model categories.

Definition 9.3.20. Let $\mathbb{B}$ and $\mathbb{A}$ be model categories with $\gamma_{\mathbb{B}}: \mathbb{B} \longrightarrow Ho(\mathbb{B})$ and $\gamma_{\mathbb{A}}: \mathbb{A} \longrightarrow Ho(\mathbb{A})$. If $F\colon \mathbb{B} \longrightarrow \mathbb{A}$, then when there is a left Kan extension as follows:

$$\begin{array}{ccc} Ho(\mathbb{B}) & \xrightarrow{Lan_{\gamma_{\mathbb{B}}}(\gamma_{\mathbb{A}} \circ F)} & Ho(\mathbb{A}) \\ {\scriptstyle \gamma_{\mathbb{B}}}\uparrow & & \uparrow{\scriptstyle \gamma_{\mathbb{A}}} \\ \mathbb{B} & \xrightarrow[F]{} & \mathbb{A}, \end{array} \tag{9.111}$$

the functor $Lan_{\gamma_{\mathbb{B}}}(\gamma_{\mathbb{A}} \circ F)$ is called a **total left derived functor** of F. When a right Kan extension exists, it is called a **total right derived functor** of F.

What about adjoint functors?

Definition 9.3.21. Let $\mathbb{A}$ and $\mathbb{B}$ be model categories, and let $L\colon \mathbb{A} \longrightarrow \mathbb{B}$ and $R\colon \mathbb{B} \longrightarrow \mathbb{A}$ with $L \dashv R$ being adjoint functors. These functors are called a **Quillen adjunction** if L preserves cofibrations and trivial cofibrations and R preserves fibrations and trivial fibrations. Such an adjunction induces an adjunction of homotopy categories as follows:

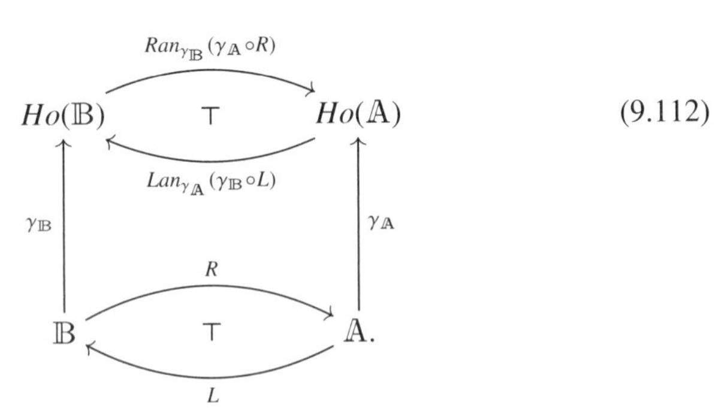

(9.112)

The Quillen adjunction is called a **Quillen equivalence** if we further have that every map $L(a) \longrightarrow b$ is a WE in $\mathbb{B}$ if and only if the corresponding map $a \longrightarrow R(b)$ under the adjunction is a WE in $\mathbb{A}$.

We now come to one of the main theorems of model categories:

Theorem 9.3.22. If there is a Quillen equivalence between $\mathbb{A}$ and $\mathbb{B}$, then $Ho(\mathbb{A})$ is equivalent to $Ho(\mathbb{B})$. ★

Example 9.3.23. The main example of Quillen equivalence is the relationship between topological spaces and a simplicial set. The singular complex/geometric realization is not only an adjunction, but a Quillen adjunction and a Quillen equivalence. Hence, $Ho(\mathbb{T}\mathrm{op}) \cong Ho(\mathbb{s}\$\mathrm{et})$. □

The focus of this text is monoidal categories, not just categories. Let $\mathbb{B}$ have a strict monoidal category structure and a Quillen model structure. Question: In what way do these structures have to respect each other so that the homotopy category, $Ho(\mathbb{B})$, inherits a monoidal structure? Notice first that it is not hard to describe the monoidal structure of $Ho(\mathbb{B})$ on objects since $\mathbb{B}$ and $Ho(\mathbb{B})$ have the same objects. The issue is only on morphisms. Consider two morphisms in $f\colon a \longrightarrow a'$ and $g\colon b \longrightarrow b'$ in $\mathbb{B}$. By bifunctoriality, the following commutes:

$$\begin{array}{ccc} a \otimes b & \xrightarrow{f \otimes id} & a' \otimes b \\ {\scriptstyle id \otimes g}\downarrow & & \downarrow{\scriptstyle id \otimes g} \\ a \otimes b' & \xrightarrow[f \otimes id]{} & a' \otimes b'. \end{array} \tag{9.113}$$

Consider the pushout of $f \otimes id$ and $id \otimes g$. This gives us the following diagram:

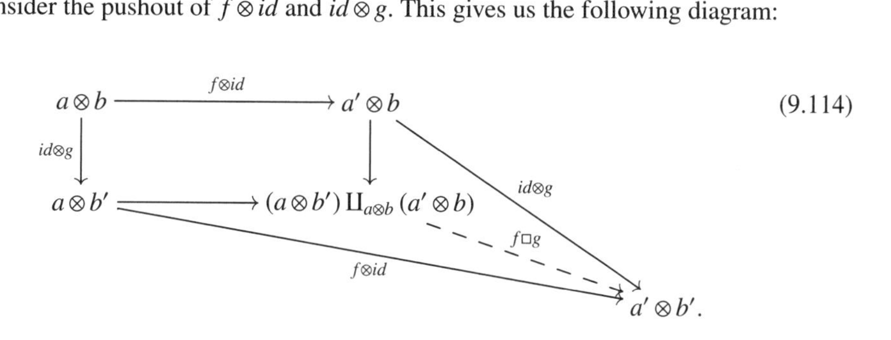

(9.114)

From the universal property of the pushout, there is a map $f \square g$. With these maps, we can state Theorem 9.3.24.

Theorem 9.3.24. Let $\mathbb{B}$ be a symmetric closed monoidal category that also has a Quillen model category structure. Consider the following two compatibility axioms:

- If f and g are cofibrations, then $f \square g$ is a cofibration. Furthermore, if f or g is also a weak equivalence, then $f \square g$ is a weak equivalence as well.
- Let I be the unit of the monoidal structure of $\mathbb{B}$, and let $q\colon I_{cof} \longrightarrow I$ be the cofibration replacement of I. For any object b in $\mathbb{B}$, we have the following two morphisms:

$$q \otimes id_b\colon I_{cof} \otimes b \longrightarrow I \otimes b \cong b, \qquad \text{and} \qquad id_b \otimes q\colon b \otimes I_{cof} \longrightarrow b \otimes I \cong b. \tag{9.115}$$

  Both of these morphisms need to be weak equivalences.

If both of these axioms are satisfied, then $Ho(\mathbb{B})$ has a symmetric monoidal structure induced by the symmetric monoidal structure in $\mathbb{B}$. We then say that $\mathbb{B}$ is a **monoidal model category**. ★

Many of the model categories that we have seen are monoidal model categories. For the proof and more on monoidal model categories, see Chapter 4 of [112].

Most of the constructions in this book are basically taking limits and colimits of diagrams in categories. We would like to discuss taking limits and colimits that respect the notions of homotopy. Such constructions are called **homotopy limits** and **homotopy colimits**. First, let us look at what does *not* work. The obvious idea is to take diagrams in the homotopy category $Ho(\mathbb{B})$. As we saw in Section 4.6, the usual definition of the limit and colimit are adjoints to the diagonal functor $\Delta\colon Ho(\mathbb{B}) \longrightarrow Ho(\mathbb{B})^{\mathbb{D}}$ for any diagram category $\mathbb{D}$:

$$Ho(\mathbb{B}) \underset{\underset{Lim}{\longleftarrow \bot}}{\overset{\overset{Colim}{\longleftarrow \bot}}{\xrightarrow{\Delta}}} Ho(\mathbb{B})^{\mathbb{D}}. \tag{9.116}$$

This does not work! Consider the following pushout:

$$\begin{array}{ccc} S^1 & \hookrightarrow & D^1 \\ \downarrow & & \downarrow \\ D^1 & \hookrightarrow & S^2. \end{array} \tag{9.117}$$

This says that if a circle is included in two disks (think of the equator of Earth S^1 included in the Northern Hemisphere D^1 and also included in the Southern Hemisphere D^1), then the pushout is S^2 (the surface of the Earth). The problem is that D^1 is a contractible space and is homotopic to the single-point topological space. If we take the same pushout with these homotopically equivalent spaces, we get a point:

$$\begin{array}{ccc} S^1 & \longrightarrow & * \\ \downarrow & & \downarrow \\ * & \longrightarrow & *. \end{array} \tag{9.118}$$

However, S^2 is not homotopically equivalent to $*$. In other words, given two different but homotopically equivalent diagrams, we get different homotopically nonequivalent pushouts. So regular pushouts do not respect homotopy.

Rather, what needs to be done is as follows: Instead of looking at $Ho(\mathbb{B})^{\mathbb{D}}$, we must look at the functor category $\mathbb{B}^{\mathbb{D}}$ and find a model category structure for it. Often, this can be done in several ways. With such a model category, we can look at $Ho(\mathbb{B}^{\mathbb{D}})$. We then look at the derived functor of Δ, *Lim*, and *Colim* to get the following functors, which are homotopy limits and homotopy colimits:

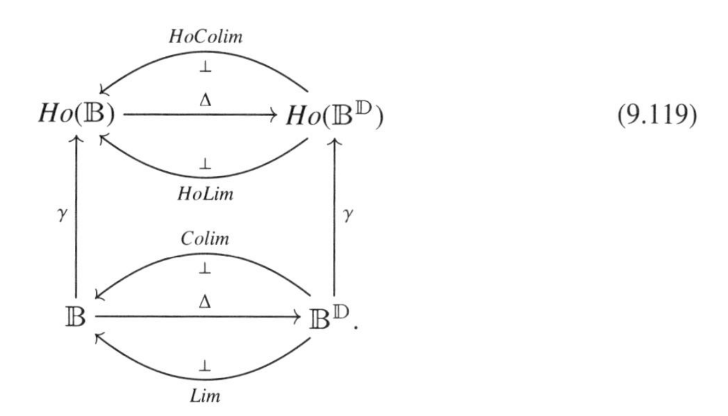

(9.119)

Homotopy limits and colimits arise in many places.

As we have seen, Kan extensions are vast generalizations of limits and colimits that encompass all categorical constructions. It is a short hop, skip, and jump from homotopy limits and colimits to left and right **homotopy Kan extensions**. Since homotopy theory deals with the essential nature of structures and all constructions are Kan extensions, it is no wonder that homotopy Kan extensions are so important.

These constructions are at the center of a deep project in higher mathematics dealing with a categorical framework called **derivators**. One of the greatest mathematicians of the twentieth century, Alexander Grothendieck, discussed derivators in his 1983 unpublished manuscript, *Pursuing Stacks*[96]. These ideas were further developed in an unpublished manuscript of almost 2,000 pages entitled *Les Dérivateurs* [97]. Independently, Alex Heller developed similar ideas in his memoir *Homotopy Theories* [103]. He advanced and applied these ideas to various areas in different works [105, 104, 106]. Others have progressed on their work. One can read more about derivators in Mike Shulman's exposition [238] and in Section A.3.3 of Jacob Lurie's book [176].

Before we close our discussion of homotopy theory, it is worth mentioning that there are very deep connections between monoidal categories, coherence theory, and homotopy theory. As we know, a strict monoidal category is a monoid object in $\mathbb{Cat}$. What about a nonstrict monoidal category? There are several ways to look at such a structure. As we outlined in Section 8.4, a monoidal category is an algebra for a 2-theory in the 2-category $\overline{\mathbb{Cat}}$. This 2-category is similar to $\overline{\mathbb{Top}}$. The natural transformations in the definition of a monoidal category have the feel of being homotopies between functors. In other words, the α reassociativity natural transformation is a homotopy from the functor $(Id \times \otimes)\otimes$ to the functor $(\otimes \times Id)\otimes$. The commutativity of the pentagon coherence condition might be considered a higher homotopy between homotopies. Following these lines of thinking, researchers have thought about monoidal categories as algebras in the homotopy category $Ho(\mathbb{Cat})$. Such **homotopical algebras** arise in algebra, algebraic topology, and physics.

Another connection between coherence theory and homotopy theory is evident from the relationship between the 2-category of strict monoidal categories, $\overline{\mathbb{StrMonCat}}$, and the larger 2-category of all monoidal categories, $\overline{\mathbb{MonCat}}$. There is a 2-functor inclusion $inc\colon \overline{\mathbb{StrMonCat}} \longrightarrow \overline{\mathbb{MonCat}}$. The content of the coherence theorem for monoidal categories is that there is a 2-functor $r\colon \overline{\mathbb{MonCat}} \longrightarrow \overline{\mathbb{StrMonCat}}$ such that $r \circ inc \cong id$ and $inc \circ r$ are naturally isomorphic to the identity. This brings to light a Quillen model category structure that was put on the category of 2-theories and 2-theory morphisms, $2\mathbb{T}\mathrm{heory}$. The weak equivalences in this model category structure induce an equivalence in the category of algebras. The paragon example of a weak equivalence is the 2-theory functor from the 2-theory of monoidal categories $\mathbf{T}_{Mon}$ to the 2-theory of strict monoidal categories $\mathbf{T}_{sMon}$. This 2-theory functor induces the essentially surjective $i\colon \overline{\mathbb{S}\mathrm{tr}\mathbb{M}\mathrm{on}\mathbb{C}\mathrm{at}} \longrightarrow \overline{\mathbb{M}\mathrm{on}\mathbb{C}\mathrm{at}}$. This work was carried out in [269]. It must be stressed that I do not know of any nontrivial Quillen model category structure on the category of algebraic theories. Algebraic theories are too rigid. Only the category of 2-theories, $2\mathbb{T}\mathrm{heory}$, has enough wiggle room and flexibility for there to be a notion of homotopy.

Suggestions for Further Study

Classical homotopy theory is a beautiful field with many geometric ideas. One can learn more about it in Chapter 4 of [111], Chapter 6 of [47], Chapters 1 and 4 of [102], Chapters 3 and 11 of [227], Chapters 2–4 of [251], and my favorite, Chapter 3 of [240].

Some nice presentations of simplicial sets can be found in Section VII.5 of [180], Section VIII.7 of [181], [220], [83], [62], and the books [90, 192].

Categorical homotopy theory can be found in many sources, including [90, 112, 221, 113, 132, 109]. Daniel Quillen's original book [216] is extremely readable. While Quillen's ideas dominate the field, his work is not the only game in town. There are other ways of talking about abstract homotopy theory, such as [16, 66].

Micah Miller gave an understandable introduction to homotopy limits at the New York City Category Theory Seminar [196].

Rick Jardine has some beautiful lecture notes on many aspects of homotopy theory available online [121].

9.4 Higher Category Theory

Categories have objects and morphisms that we sometimes called 0-cells and 1-cells. In our journey, we met 2-categories that have 0-cells, 1-cells, and 2-cells. There is no reason to stop there. As we saw in Important Categorical Idea 1.4.9 in Chapter 1, we can ask about morphisms between 2-cells, which we call 3-cells, morphisms between 3-cells, etc. The name 2-category comes from the fact that its highest nontrivial morphisms are 2-cells. Following this line of thought, we should call old-fashioned categories 1-categories. We should even call each set (thought of as a discrete category) a 0-category. In this section, we take a brief tour of higher categories. We will see various ways of dealing with 3-categories, 4-categories, and even infinity-categories.

Now, let's consider some motivation. When dealing with a set (0-category), the only notion that you can say about two elements is if they are equal or not. In a 1-category, one can also talk about two objects that are isomorphic. In a 2-category, one can discuss a weaker relationship between two objects, which is equivalence. As we saw in Important Categorical Idea 4.1.29, the weaker a notion is, the more phenomena can be modeled by that notion. So there is a general urge to weaken these relationships between objects in categories. When we go higher and higher in the categorical hierarchy, we are able to weaken the notion of relatedness between two objects.

And here's some more motivation. Old-fashioned mathematics is one-dimensional. The formulas are written in a one-dimensional line. In contrast, the diagrams in a category spread across a two-dimensional piece of paper. Such diagrams consist of objects and morphisms. However, as we have seen many times in this text, there are often 2-cells between morphisms. Such diagrams in 2-categories are three-dimensional. At each level, we go up in dimension. As we climb the categorical hierarchy, we get into higher geometrical notions.

Strict Higher Categories

How do all these different dimensional morphisms fit together? As always, we start at the simplest level and move up to the more complicated "ideas."

We started this text talking about 1-categories. The first time we met a 2-cell was in Section 4.2 when we introduced the concept of a natural transformation. We went from $\mathbb{Cat}$, the category of categories and functors, to $\overline{\mathbb{Cat}}$, the 2-category of categories, functors, and natural transformations. As we saw, there were two compositions for 2-cells: $\circ_H$ and $\circ_V$. Let us use those insights to give a formal definition of a 2-category along the lines of Definition 2.1.2. (See Section 4.2 for the clarifying diagrams.)

Definition 9.4.1. A **2-category** $\overline{\mathbb{A}}$ is a collection of objects (0-cells), $Ob(\overline{\mathbb{A}})$, a collection of morphisms (1-cells), $Mor(\overline{\mathbb{A}})$, and a collection of 2-cells $2Cell(\overline{\mathbb{A}})$, which has the following structure:

- Every morphism has a domain: $dom_1 \colon Mor(\overline{\mathbb{A}}) \longrightarrow Ob(\overline{\mathbb{A}})$.
- Every morphism has a codomain: $cod_1 \colon Mor(\overline{\mathbb{A}}) \longrightarrow Ob(\overline{\mathbb{A}})$.
- There is a composition of morphisms: $\circ_1 \colon Mor(\overline{\mathbb{A}}) \times Mor(\overline{\mathbb{A}}) \longrightarrow Mor(\overline{\mathbb{A}})$, which is only defined for adjoining morphisms; that is, f and g such that $cod_1(f) = dom_1(g)$. We have that $dom_1(g \circ_1 f) = dom_1(f)$ and $cod_1(g \circ_1 f) = cod_1(g)$.
- Every object has an identity morphism: $ident_0 \colon Ob(\overline{\mathbb{A}}) \longrightarrow Mor(\overline{\mathbb{A}})$.
- Every 2-cell has an domain: $dom_2 \colon 2Cell(\overline{\mathbb{A}}) \longrightarrow Mor(\overline{\mathbb{A}})$.
- Every 2-cell has a codomain: $cod_2 \colon 2Cell(\overline{\mathbb{A}}) \longrightarrow Mor(\overline{\mathbb{A}})$.
- The two domains and codomains cohere with each other: for every 2-cell α, we have
  - $dom_1(dom_2(\alpha)) = dom_1(cod_2(\alpha))$
  - $cod_1(dom_2(\alpha)) = cod_1(cod_2(\alpha))$
- There is a vertical composition of 2-cells: $\circ_V \colon 2Cell(\overline{\mathbb{A}}) \times 2Cell(\overline{\mathbb{A}}) \longrightarrow 2Cell(\overline{\mathbb{A}})$, which is only defined for 2-cells that are adjoined along a morphism; that is, α and β such that $cod_2(\alpha) = Dom_2(\beta)$. We have that $dom_2(\beta \circ_V \alpha) = dom_2(\alpha)$ and $cod_2(\beta \circ_V \alpha) = cod_2(\beta)$.
- There is a horizontal composition of 2-cells: $\circ_H \colon 2Cell(\overline{\mathbb{A}}) \times 2Cell(\overline{\mathbb{A}}) \longrightarrow 2Cell(\overline{\mathbb{A}})$, which is only defined for 2-cells that are adjoined along an object;

that is, α and β such that $cod_1(cod_2(\alpha)) = dom_1(dom_2(\beta))$. We have that $dom_2(\beta \circ_H \alpha) = dom_2(\beta) \circ_1 dom_2(\alpha)$ and $cod_2(\beta \circ_H \alpha) = cod_2(\beta) \circ_1 cod_2(\alpha)$.

- Every morphism has an identity 2-cell: there is a function $ident_1 : Mor(\overline{\mathbb{A}}) \longrightarrow 2Cell(\overline{\mathbb{A}})$.

This structure must satisfy the following axioms:

- All three compositions are associative:
  - $h \circ_1 (g \circ_1 f) = (h \circ_1 g) \circ_1 f$
  - $\alpha \circ_V (\beta \circ_V \gamma) = (\alpha \circ_V \beta) \circ_V \gamma$
  - $\alpha \circ_H (\beta \circ_H \gamma) = (\alpha \circ_H \beta) \circ_H \gamma$
- Composition with the identity does not change the original:
  - $f \circ_1 id = f = id \circ_1 f$
  - $\alpha \circ_V id = \alpha = id \circ_V \alpha$
  - $\alpha \circ_H id = \alpha = id \circ_H \alpha$
- The units must preserve compositions:

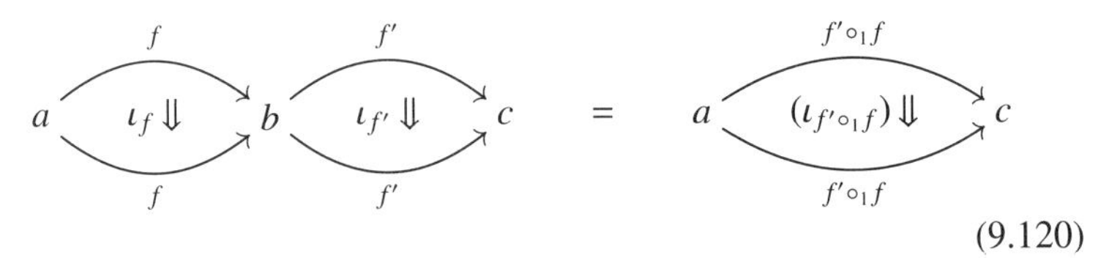

(9.120)

that is, $\iota_{f'} \circ_H \iota_f = \iota_{f' \circ_1 f}$.

- The interchange rule: when 2-cells are arranged as Diagram (4.52), then the following rule holds: $(\alpha \circ_V \beta) \circ_H (\gamma \circ_V \delta) = (\alpha \circ_H \gamma) \circ_V (\beta \circ_H \delta)$.

For reasons that will become apparent soon, we call such 2-categories **strict 2-categories**.

We met some examples of strict 2-categories.

Example 9.4.2.

- $\overline{\mathbb{Cat}}$ is the collection of small categories, functors, and natural transformations.
- $\overline{\mathbb{MonCat}}$ is the collection of monoidal categories, monoidal functors, and monoidal natural transformations.
- We met many other examples of such 2-categories in Chapters 5, 6, and 7, such as $\overline{\mathbb{RibMonCat}}$, $\overline{\mathbb{RigidMonCat}}$, $\overline{\mathbb{SymMonCat}}$, $\overline{\mathbb{BalMonCat}}$, $\overline{\mathbb{BraMonCat}}$, $\overline{\mathbb{StrSymMonCat}}$, etc.
- $\overline{\mathbb{Top}}$ is the collection of topological spaces, continuous maps, and equivalence classes of homotopies. □

Exercise 9.4.3. Formally define a **2-functor** between two 2-categories. ■

Definition 9.4.4. The collection of 2-categories and 2-functors form a category called $2\mathbb{Cat}$.

Just as there are adjoints between $\mathbb{C}\mathrm{at}$ and $\mathbb{S}\mathrm{et}$, as we saw in Example 4.4.16 in Chapter 4, there are the following adjoint functors:

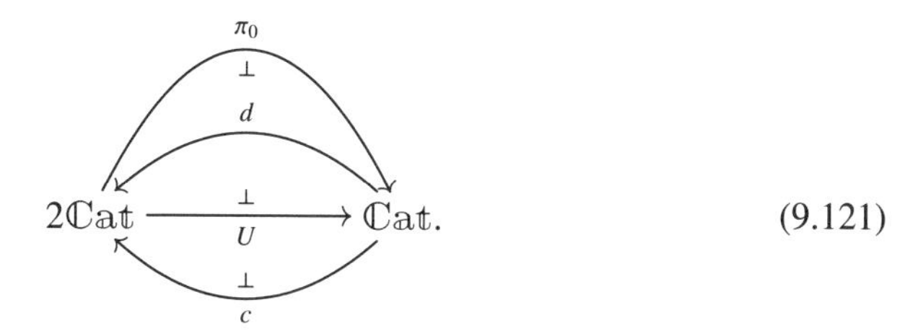

(9.121)

These functors are defined as follows:

- π_0 takes a 2-category and outputs a category that consists of only connected components of morphisms. This means that we consider any two morphisms to be the same if there is a 2-cell between them.
- d takes a category to the same 2-category, but with only identity 2-cells. This is the discrete 2-category of the input.
- U takes a 2-category to the category after it forgets the 2-cells.
- c takes a category to the 2-category with exactly one 2-cell between any two parallel morphisms.

One can move on and talk about **2-natural transformations** between 2-functors. One can also go on and talk about a 3-cell between 2-natural transformations, which is called a **modification**. Like a natural transformation, it has components, each of which is a 2-cell. The collection of 2-categories, 2-functors, 2-natural transformations, and modifications forms a **3-category**.

Remember that a one-object category is a monoid. The morphisms all come and go from the single object, so they can be composed. This composition is the monoidal multiplication. How does this generalize to higher categories?

Theorem 9.4.5. A one-object strict 2-category is a strict monoidal category. ★

Proof. Let the single object be $*$. The morphisms of the strict 2-category become the objects of the strict monoidal category. Since all the morphisms are of the form $f\colon * \longrightarrow *$, they form a monoid and the 2-cells go between any morphisms:

(9.122)

These 2-cells of the strict 2-category become the morphisms of the strict monoidal category. The tensor product of the objects are simply the composition of morphisms: $f \otimes g = f \circ g$. The tensor product of the morphisms is the horizontal composition: $\alpha \otimes \beta = \alpha \circ_H \beta$. Figure 9.9 will help.

Notice that the tensor product is strictly associative. ♣

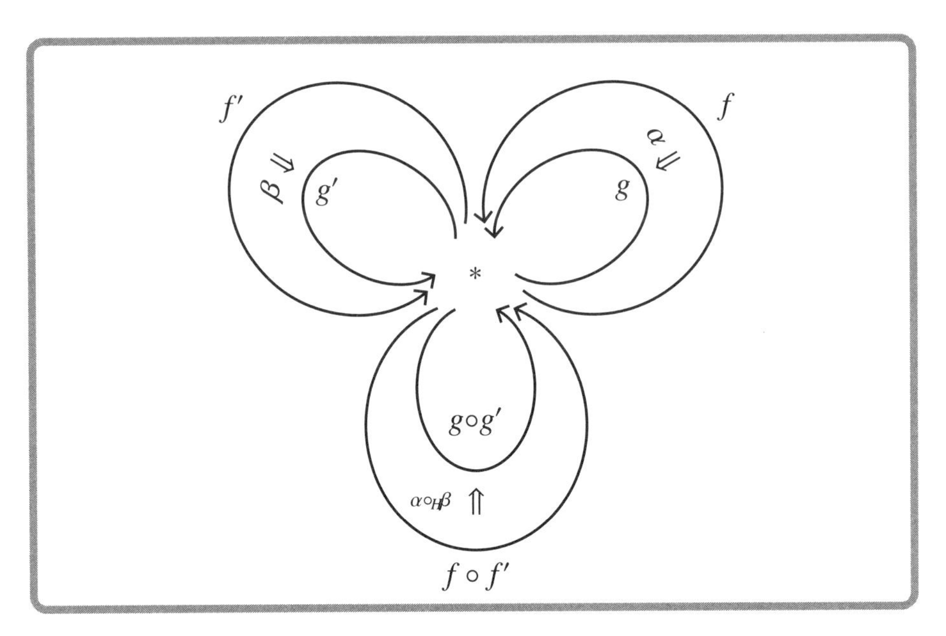

Figure 9.9. Composition in a one-object, strict 2-category.

Remark 9.4.6. What about a strict 2-category with only one object and one morphism? From Theorem 9.4.5, this will be a strict monoidal category with only one object. Furthermore, Theorem 5.1.16 used the interchange law to prove that this is a commutative monoid. ♠

We can go on and formally define a **strict 3-category** $\overline{\overline{\mathbb{A}}}$. These are collections of objects (0-cells), morphisms (1-cells), 2-cells, and 3-cells. There are several compositions associated with these morphisms and higher-dimensional cells. Rather then giving a formal definition, we give a formal definition of a strict n-category as a generalization of the definition of a category given in Definition 2.1.61. There, we defined a category as a set, two endofunctions on the set (*dom* and *cod*) and a composition. We can use the same type of definition for a strict n-category. Figure 9.10 is the higher-dimensional version of Diagram (2.22) and is helpful for understanding Definition 9.4.7.

Definition 9.4.7. For any positive integer n, a **strict n-category** $\widehat{\mathbb{A}}$ is a collection of morphisms, $Mor(\widehat{\mathbb{A}})$, with the following structure:

- For each $0 < i \leq n$, there are the following operations:
  - Every morphism has an ith domain morphism: $dom_i\colon Mor(\widehat{\mathbb{A}}) \longrightarrow Mor(\widehat{\mathbb{A}})$.
  - Every morphism has an ith codomain morphism: $cod_i\colon Mor(\widehat{\mathbb{A}}) \longrightarrow Mor(\widehat{\mathbb{A}})$.
  - There is an ith composition function $\circ_i$: if f and g satisfy $cod_i(f) = dom_i(g)$, then there is an associated morphism $g \circ_i f$.

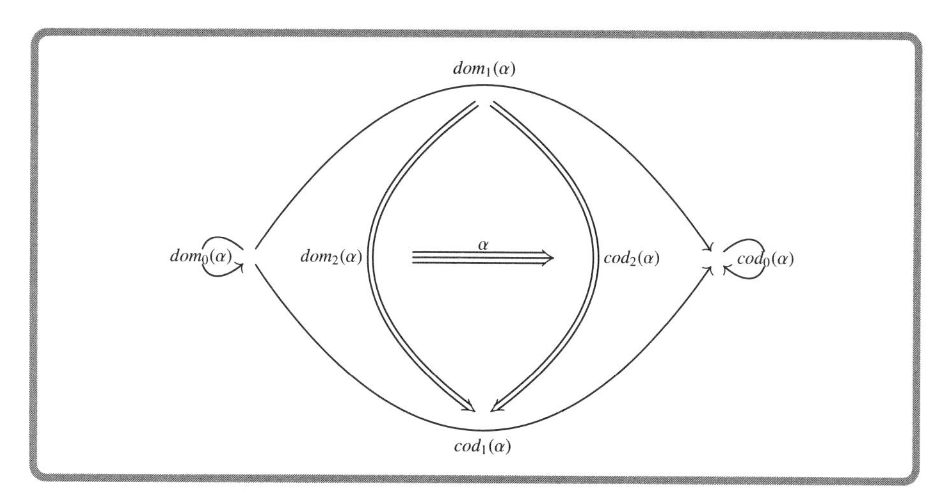

Figure 9.10. A morphism in a 3-category with all its domains and codomains.

- These operations must satisfy the following axioms for all $0 < i \leq n$:
  - The domain of a composite is the domain of the first: $dom_i(g \circ_i f) = dom_i(f)$.
  - The codomain of a composite is the codomain of the second: $cod_i(g \circ_i f) = cod_i(g)$.
  - The composition is associative: $h \circ_i (g \circ_i f) = (h \circ_i g) \circ_i f$.
  - The domain morphism acts like an identity: $f \circ_i dom_i(f) = f$.
  - The codomain morphism acts like an identity: $cod_i(f) \circ_i f = f$.
  - The domain morphisms come and go from themselves: $dom_i(dom_i(f)) = dom_i(f) = cod_i(dom_i(f))$.
  - The codomain morphisms come and go from themselves: $dom_i(cod_i(f)) = cod_i(f) = cod_i(cod_i(f))$.
- For each $0 < i < j \leq n$, the i- and j-category structures must cohere with each other.
  - For all f in $Mor(\widehat{\mathbb{A}})$, we have

$$dom_i(f) = dom_i(dom_j(f)) = dom_j(dom_i(f)) = dom_i(cod_j(f)). \tag{9.123}$$

  - For all f in $Mor(\widehat{\mathbb{A}})$, we have

$$cod_i(f) = cod_i(cod_j(f)) = cod_j(cod_i(f)) = cod_i(dom_j(f)). \tag{9.124}$$

  - For all a, b, c, and d in $Mor(\widehat{\mathbb{A}})$ with $dom_i(a) = cod_i(b)$, $dom_i(c) = cod_i(d)$, $dom_j(a) = cod_j(c)$, and $dom_j(b) = cod_j(d)$, we have,

$$(a \circ_i b) \circ_j (c \circ_i d) = (a \circ_j c) \circ_i (b \circ_j d), \tag{9.125}$$

    which is a generalization of the interchange law.

This definition is adopted from Ross Street's article [246].
One can easily extend this definition to a **strict infinity-category**.

Remark 9.4.8. Consider the hierarchy of strict n-categories. The bottom of the hierarchy is sets, then categories, then strict 2-categories, ..., then strict n-categories, ... There is a formal way to relate every level of the hierarchy to its neighbors:

- *Going down.* The Hom set of a category (strict 1-category) is a set (0-category), and the Hom set of a 2-category is a strict 1-category. In general, given any objects (0-cells) a and b in a strict n-category, $Hom(a, b)$ is a strict $n-1$-category.
- *Going up.* The collection of all sets (0-categories) and morphisms between them forms a strict 1-category. The collection of all strict 1-categories forms a strict 2-category. In general, the collection of all strict n-categories and all the morphisms between them forms a strict $n+1$-category. ♠

Weak Higher Categories

While strict higher categories are well defined and clearly understood, there are many more applications if we weaken the structure (see Important Categorical Idea 4.1.29). Let us weaken the definition of a category by weakening the requirements. Instead of requiring that the composition is associative, we only require that there is an invertible 2-cell between them. We make a similar change with respect to the identities as a unit. So instead of Diagram (2.23), we have the following:

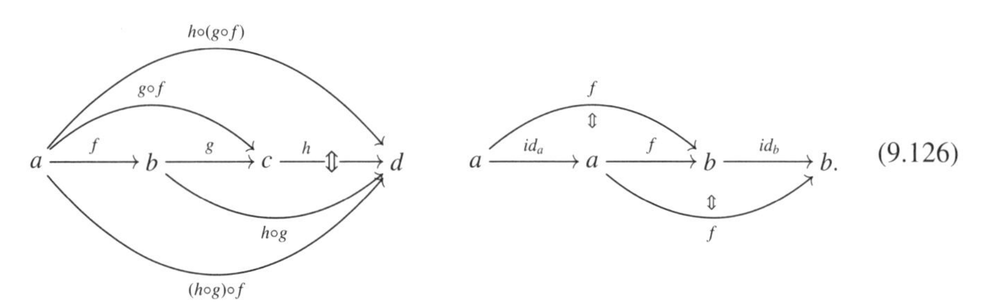

These invertible 2-cells must satisfy coherence conditions. Let us define it formally.

Definition 9.4.9. A **bicategory** $\widetilde{\mathbb{A}}$ has objects (0-cells), morphisms (1-cells), and 2-cells, with a composition $\circ$ such that for every three composable morphisms f, g, and h, there is an invertible 2-cell $\alpha_{h,g,f}\colon h \circ (g \circ f) \Longrightarrow (h \circ g) \circ f$. These associativity 2-cells must satisfy the following pentagon condition:

$$\begin{array}{ccccc} f \circ (g \circ (h \circ i)) & \overset{id_f \circ \alpha_{g,h,i}}{\Longrightarrow} & f \circ ((g \circ h) \circ i) & \overset{\alpha_{f,g\circ h,i}}{\Longrightarrow} & (f \circ (g \circ h)) \circ i \\ \Big\Downarrow{\scriptstyle \alpha_{f,g,h\circ i}} & & & & \Big\Downarrow{\scriptstyle \alpha_{a,b,c} \circ id_i} \\ (f \circ g) \circ (h \circ i) & & \underset{\alpha_{f\circ g,h,i}}{\Longrightarrow} & & ((f \circ g) \circ h) \circ i. \end{array} \tag{9.127}$$

There are also invertible 2-cells $\rho\colon f \circ id \Longrightarrow f$ and $\lambda\colon id \circ f \Longrightarrow f$, which satisfy

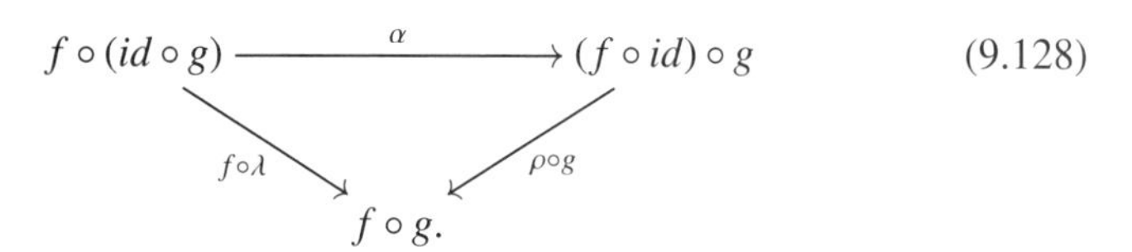

(9.128)

It is important to note that a bicategory is not a category. It is a weaker structure with higher cells.

Example 9.4.10. A category is a type of bicategory where all the 2-cells (including α, ρ and λ) form the identity. □

Following along the lines of Theorem 9.4.5 we have the following.

Theorem 9.4.11. A one-object bicategory is a monoidal category. ★

The proof is very similar to the proof of Theorem 9.4.5. However, here the composition of morphisms is not associative. This means that the monoidal product need not be associative.

Similar to the main coherence theorems of Chapters 5 and 6, there is a theorem that says that every bicategory is equivalent to a 2-category.

One can continue to go up the ladder of abstraction and define a weak notion of a strict 3-category called a **tricategory**. While we will not formally define such a structure, it pays to give a taste of the definition. A tricategory is a structure with objects (0-cells), morphisms (1-cells), 2-cells, and 3-cells. The pentagon condition, Diagram (9.127), does not necessarily commute, but there is a 3-cell from one side of the pentagon to the other. Now consider the tricategory analogy of Figure 5.7 in Chapter 5. The six pentagons are filled with 2-cells. Rather than saying that the three-dimensional sphere commutes, there is a requirement of a 3-cell from one side of the sphere to the other side. This 3-cell has to satisfy some larger coherence condition. Alas, things get very very complicated from here.

While every bicategory is equivalent to a strict 2-category, it is not true that every tricategory is equivalent to a strict 3-category. The higher we go, the messier things become.

What about higher weak categories? There were many researchers with different definitions about how to go higher. See [168] and the papers of John Baez and James Dolan, such as [26], for an introduction to this world.

Researchers have looked at adding a monoidal structure to a bicategory to get a **monoidal bicateogry**. One can also define a **symmetric monoidal bicategory**, a **braided monidal bicategory**, etc. Here, we need all the monoidal structures to cohere with the structure of a bicategory. This too gets a tad messy.

Example 9.4.12. There are several examples of monoidal bicategoires in my book [274]. Concrete models of computations such as circuits, Turing machines, and register machines all have such complicated structures. □

Along the lines of Theorems 9.4.5, 9.4.11, and Remark 9.4.6, we can ask about single-object, weak higher categories. Researchers have worked out a lot of it and put

	$n=0$	$n=1$	$n=2$	n=3
$k=0$	Sets	Categories	2-categories (bicategories)	3-categories (tricategories)
$k=1$	Monoids	Monoidal categories	Monoidal 2-categories	Monoidal bicategories
$k=2$	Commutative monoids	Braided monoidal categories	Braided monoidal 2-categories	Braided monoidal ???
$k=3$	"	Symmetric monoidal categories	Sylleptic monoidal 2-categories	
$k=4$	"	"	Symmetric monoidal 2-categories	
$k=5$	"	"	"	
$k=6$	"	"	"	"

Figure 9.11. The periodic table of all $(n+k)$–categories with only one j-morphism for $j<k$.

it into the "periodic table," seen in Figure 9.11. For every $n+k$ category, we can ask what happens when there is only one morphism for $j<k$. There are all types of exotic structures. See the papers of John Baez and James Dolan for more on this.

Remark 9.4.13. Before we close our discussion of weak higher categories, we must mention an interesting topic in current research called **categorification**. It is the process where you examine a structure and modify it so that the new structure is related to the old one, but at a higher level. For example, rather than having two objects that are equal to each other, we have them isomorphic to each other. An equation says that two objects are the same. In contrast, an isomorphism says there is a higher-dimensional isomorphism. We just saw this in our definition of a bicategory. Rather than having the two compositions of morphisms equal to each other, there is a higher-dimensional 2-cell that related the two. The higher cells usually demand a coherence condition.

Here are some examples of categorification that we have seen in these pages.

- In Section 1.4, within Remark 1.4.27, we saw that there is a set version of the simple fact about numbers: $m^{(n\cdot p)} = (m^n)^p$. This set version is a categorification of the equation.
- In Chapter 5, we saw that one can think of a monoidal category as a categorification of a monoid. The tensor product is weakened so it is no longer associative, but it is associative up to an isomorphism.
- In Section 6.2, we saw that the proof of the strictifaction theorem was a categorification of the proof of Cayley's representation theorem for monoids. See Remark 6.2.4.

- In Section 6.1, we saw in Diagram (6.8) that the definition of a monoidal functor is a categorifiaction of the usual definition of monoidal homomorphism, which is Diagram (2.11).
- In Section 6.1, we saw that by weakening Equation (6.39), we get Diagram (6.40).
- In Section 7.4, we saw many examples of weakening algebraic structures to get structures related to quantum groups. These weakened structures are optained through categorification. For example, a Drinfeld algebra is a weakening of a coaassociative algebra until the comultiplication is coassociative up to a 2-cell.
- In Section 8.4, we met algebraic 2-theories, which were the categorifications of algebraic theories.
- In Section 9.1, we saw that a tensor category as a categorification of a ring.
- In this section, we saw a weak category structure is a categorification of a strict categorical structure. Furthermore, a strict, weak $n+1$ category is the categorification of a strict, weak n-category.

The flip side of categorification is actually simpler and is called **decategorification**. This is when you modify a structure to be lower-dimensional and more concrete. The decategorificant of a finite set is the whole number of elements in the set. Relationships for sets become equations about numbers. The simplest example of this is that for any two sets S and T, there is an isomorphism $S \times T \cong T \times S$. Through decategorification, this become $m \cdot n = n \cdot m$.

These ideas have been very fruitful in dealing with different issues of mathematics and physics. See John Baez and James Dolan's book Chapter [27] for a nice introduction to this area. See also Aaron Lauda and Joshua Sussan [156]. ♠

Infinity Categories

In more recent times, researchers have gone on and formulated whole classes of interesting types of higher categories. An **(n, r)-category** is a collection with 0-cells, 1-cells, and 2-cells through n-cells, and where all the cells greater than r are invertible. Let us look at some common structures that we have seen or that arise in the literature:

- $(0,0)$-categories are sets or discrete categories, i.e., there are only 0-cells which are the elements of the sets.
- $(1,1)$-categories are categories.
- $(1,0)$-categories are groupoids, i.e., the 1-cells are all invertible.
- $(2,2)$-categories are 2-categories.
- $(2,0)$-categories are 2-groupoids, i.e., the 1-cells and 2-cells are all invertible.
- $(\infty,0)$-categories are ∞-groupoids, i.e., all higher morphisms are invertible.[3]
- $(\infty,1)$-categories are ∞-categories. This is the most popular form of higher category (at the moment). It has all higher cells, and only the 1-cells can be non-invertible.
- (∞,∞)-categories have all cells, and none need to be invertible. The literature is a bit confusing because these structures are also sometimes called ∞-categories.

[3] We already met an example of an $(\infty,0)$-category in Section 9.3: for a topological space X, the structure $\Pi_\infty(X)$ is an ∞-groupoid. In other words, Π_∞ is a functor from $\mathbb{Top}$ to the category of infinity-groupoids.

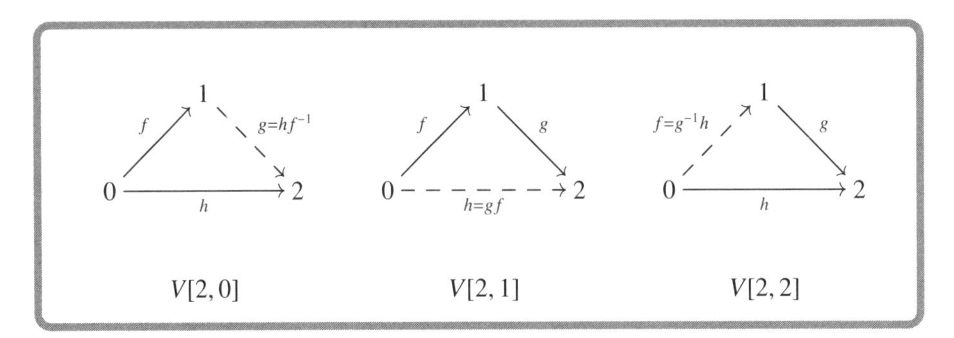

Figure 9.12. Inner and outer horns. The first and the third diagram form an outer horn, while the second is an inner horn.

Remark 9.4.14. In contrast to what was said in Remark 9.4.8, the Hom set of two objects in an (∞, ∞)-category is an (∞, ∞)-category. Also, the collection of all (∞, ∞)-categories and their morphisms is a (∞, ∞)-category. ♠

Much work goes into showing that there are infinity-category versions of 1-categorical notions like infinity-adjoints, infinity-limits, and infinity-colimits. See [223] for more on this and a general introduction to this world. See also [176].

Quasi-Categories

Recently, there has been a lot of work with a type of $(\infty, 1)$-categories called **quasi-categories**. They were originally defined by J. M. Boardman and R. M. Vogt in 1973 as "weak Kan complexes," but they were recently developed and rebranded by Andrei Joyal. Essentially, a quasi-category is a simplicial set that satisfies a special property. Let us slowly lean into this definition. Remember that the nerve of a category is given by the functor $N\colon \mathbb{C}\mathrm{at} \longrightarrow \mathbb{S}\mathrm{et}^{\Delta^{op}}$. It turns out that this functor is an equivalence of categories with its image. This means that every small category can be identified with a type of simplicial set. We characterize the image of the nerve functor, which will tell us what type of simplicial sets correspond to categories, and then generalize the notion to say what type of simplicial sets correspond to quasi-categories.

Recall the definition of a horn of a simplicial set $V[n,k]$. For each n, there is the standard simplex $\Delta[n]$ and its boundary $\partial\Delta[n]$. For each $0 \leq k \leq n$, there is the horn $V[n,k]$, which is the boundary of the standard n-simplex where the $n-1$–dimensional face across from the k vertex is missing. Thus, we have

$$V[n,k] \hookrightarrow \partial\Delta[n] \hookrightarrow \Delta[n]. \tag{9.129}$$

The set of horns is split into **outer horns** and **inner horns**. An outer horn is either $V[n,0]$ and $V[n,n]$. In contrast, each of $V[n,1], V[n,2], \ldots, V[n,n-1]$ is an inner horn. In Figure 9.12, we illustrate horns as directed commutative diagrams, where the dashed arrows are the missing part of the boundary.

Definition 9.4.15. For any simplical set S, we call a **horn in** S a map $f\colon V[n,k] \longrightarrow S$. For a horn f in S, we say that the horn is **filled in** if there exists a map $\hat{f}\colon \Delta[n] \longrightarrow S$ such that the following commutes:

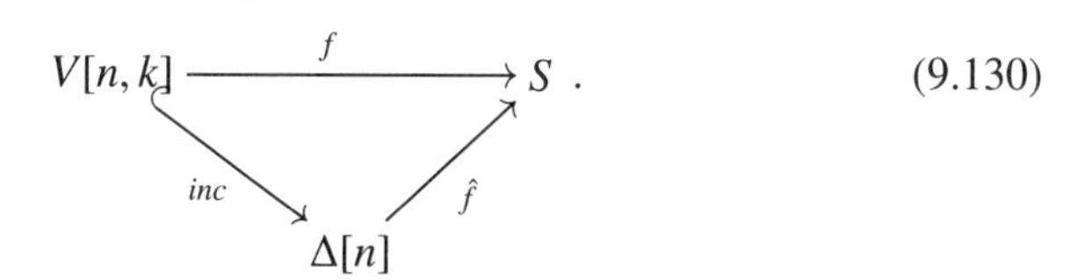

(9.130)

This is equivalent to saying that every map f extends to $\hat{f}$. When this happens, it means that the horn in the simplicial set S can be "completed."

If S is the nerve of a small category, then every inner horn $V[n,k] \longrightarrow S$ can uniquely be filled in. The maps in Figure 9.12 show how this works. Given f and g there is a unique $h = gf$.

If S is the nerve of a groupoid, then every horn (inner and outer) $V[n,k] \longrightarrow S$ can uniquely be filled in. The maps in Figure 9.12 show how this works. The unique extension provides not only the composition, but also the inverses.

Now we extend this definition to get more general notions. A simplical set S satisfies the **Kan condition** if every horn in S can be filled in. The main point is that there might be more than one way to fill in the horns. A **Kan complex** is a simplicial set that satisfies the Kan condition.

A **quasi-category** or a **weak Kan complex** is a simplicial set S such that all inner horns in S can be filled in (not necessarily uniquely). This means that there might be more than one way of completing the inner horns. The idea is that the many ways of completing the horns are all related by higher morphisms.

From this perspective, we can understand a quasi-category as a higher-dimensional generalization of a category. Whereas a category has a unique composition, a quasi–category can have many compositions. However, all the compositions have invertible higher cells connecting them. In the same way, a Kan complex is a higher-dimensional generalization of a groupoid.

For more on quasi-categories, see J. M. Boardman's and R. M. Vogt's amazingly prescient monograph [38] and Andrei Joyal's paper [128]. See also Jacob Lurie's readable introduction [175].

9.5 Topos Theory

Topos theory is an extremely large and rich branch of category theory. The ideas in this field spring from many parts of mathematics such as topology, logic, algebraic geometry, and set theory. This field has applications in many areas of mathematics, computer science, and physics. Consider the following as evidence of the breadth of topos theory: the main resource for this field is a two-volume text (with more than 1,500 pages) titled *Sketches of an Elephant: A Topos Theory Compendium* [126, 127] by Peter T. Johnstone. The title comes from the old Indian fable of several blind wise men who find an elephant. Each scholar touches a different part of the elephant and declares it a different object. So too, topos theory can be described in many ways (all of them expressed in the unifying language of category theory). Suffice it to say that there are both geometric and logical ideas at the core of topos theory.

While a topos is many different notions to many different people, the central idea of a topos is as a **universe of discourse** or a **domain of discourse**. It is a collection of objects, structures, and ideas assumed or implied for a topic of discussion. Different types of toposes (or topoi)[4] are about different universes. The two main types of toposes are "elementary toposes" and "Grothendieck toposes." Roughly speaking, an elementary topos is about a collection of sets and the concomitant logic that is associated with those sets. A Grothendieck topos is essentially a vast generalization of the notion of topological space. We will explore each of these in the coming pages. (It should be noted that a Grothendieck topos is a type of elementary topos.)

A central theme that we will meet over and over in this section is formulating mathematical concepts and constructions into the arrow language of category theory (see Important Categorical Idea 1.4.60). When this is successfully done, the concepts and constructions are generalized to many more contexts and is applicable in many more areas. This makes topos theory an area that promotes much cross-fertilization of ideas (see [50]).

Sets and Logic: Elementary Topoi

In the 1960s, one of the leaders of algebraic geometry, Alexander Grothendieck, generalized the notion of a topological space and came up with the notion of a Grothendieck topos. At the same time, one of the leaders of category theory, F. William Lawvere, was interested in the foundations of mathematics. Lawvere noticed that the notion of a Grothendieck topos had many features as the category of sets. As we saw in the our mini-course on Sets and Categorical Thinking (Section 1.4), many of the concepts and constructions of mathematics occur in the category of sets. A natural question is: What categories like Grothendiek toposes are sufficiently like the category of sets that one can perform typical constructions of mathematics? Lawvere answered with the notion of an elementary topos. (See [194] for the truthful historical motivations of topos theory.)

Definition 9.5.1. An **elementary topos** is a (i) Cartesian closed category with (ii) finite limits and a (iii) subobject classifier.

The main point is that many mathematical constructs can be done in categories with these structures.[5] We know what a Cartesian closed category is (see Definition 7.2.2)

[4]The word "topos" is Greek for "place." If the word is handled as in English, then its plural is "toposes." If the word is handled as in Greek, then its plural is "topoi." As "a foolish consistency is the hobgoblin of little minds," we refuse to be consistent in this discussion.

[5]It is important to realize that the notion of an elementary topos is not the only notion in town. Within the literature, there are many discussions of "weak toposes," "quasi-toposes," "pretoposes," etc. These are categories that do not have all the requirements of an elementary topos. On the other hand, there are also major discussions of elementary toposes with extra requirements such as "Boolean toposes," "toposes with natural number objects," "well-powered toposes," etc. The point is that within any category, certain constructions can be preformed. For every type of category, we examine which constructions can be performed. There is nothing particularly special about the notion of an elementary topos. It is one type on a whole spectrum of types of categories.

and we know what it means for a category to have finite limits (Definition 4.6.6). We are left to explain a subobject classifier.

One of the central ideas we learned about sets, is that subsets can be described by functions called characteristic functions (see Definition 1.4.16). Every subset of S can be characterized by a function $\chi\colon S \longrightarrow \{0, 1\}$. We can also go the other way: any function $\chi\colon S \longrightarrow \{0, 1\}$ describes a subset of S. For χ, the subset described is

$$T = \{s \in S : \chi(s) = 1\}. \tag{9.131}$$

Another way to say this is that $\chi\colon S \longrightarrow \{0, 1\}$ describes the set T that fits best into the following commutative diagram in $\mathbb{S}\mathrm{et}$:

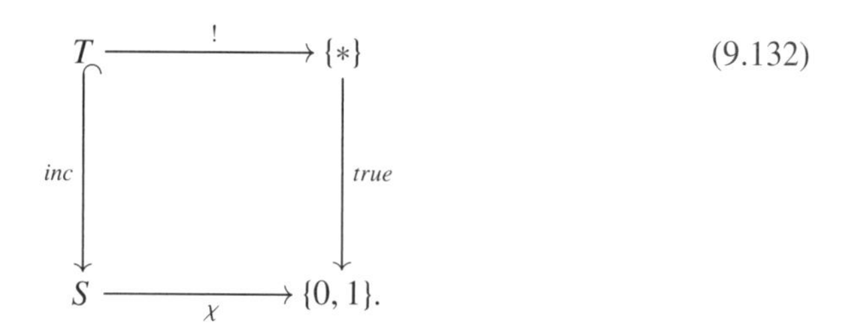

(9.132)

In detail, the top map ! takes every element in T to $*$ (the only function whose codomain is $\{*\}$). The left vertical map is the inclusion of subset T into set S. The right vertical map is the function that takes the single element $*$ to 1. The subset T consists of all those elements t such that $\chi(t) = 1$. Saying that T is the best-fitting set means that the square is a pullback.

Let us generalize this notion from $\mathbb{S}\mathrm{et}$ to other categories. In $\mathbb{S}\mathrm{et}$, we can talk about a "subset." In an arbitrary category, we must talk about a "subobject."

Definition 9.5.2. Let $\mathbb{A}$ be a category with an object a. Consider all monomorphisms into a. We put an equivalence relation on this collection. Say that a monomorphism $b \rightarrowtail a$ is equivalent to $c \rightarrowtail a$ if there is an isomorphism $b \longrightarrow c$ that makes the following diagram commute:

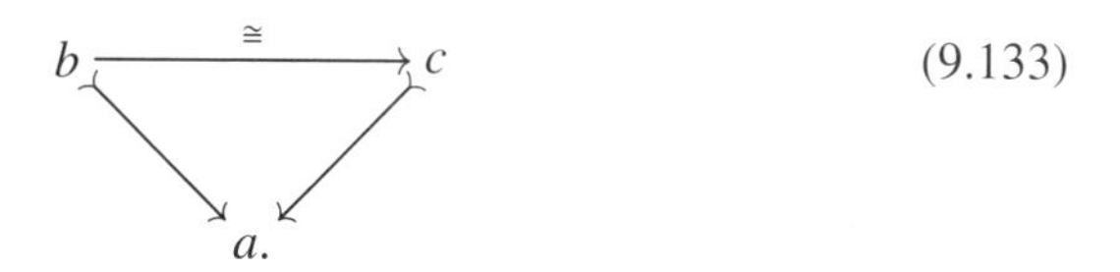

(9.133)

A **subobject** is an equivalence class of such monomorphisms. Within category $\mathbb{A}$, the collection of all subobjects of a is denoted as $Sub_{\mathbb{A}}(a)$.

Now that we have the notion of a subobject, we can talk about classifying them.

Definition 9.5.3. Within a category $\mathbb{A}$ with a terminal object 1 and pullbacks, a **suboject classifier** is an object Ω and a monomorphism $true\colon 1 \longrightarrow \Omega$ with the property that for every mono $f\colon b \rightarrowtail a$, there is a unique map $\chi_f\colon a \longrightarrow \Omega$ such that

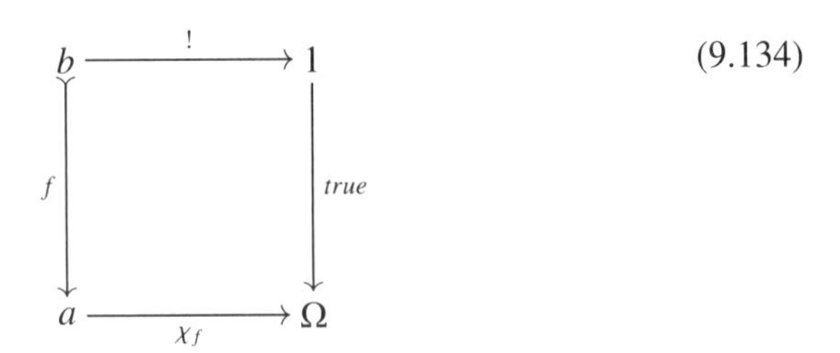

(9.134)

is a pullback.

It can be shown that for every equivalence class of monomorphisms (i.e., subobject) of a, there is a unique map $a \longrightarrow \Omega$. This correspondence becomes the following isomorphism:

$$Sub_{\mathbb{A}}(a) \cong Hom_{\mathbb{A}}(a, \Omega). \tag{9.135}$$

Now for some examples of elementary topoi. All the examples are of the form $\mathbb{Set}^{\mathbb{A}^{op}}$ for some small category $\mathbb{A}$. We will start off with simple and easy categories $\mathbb{A}$ and build up to the general case.

Example 9.5.4. The two most obvious examples of an elementary topos is $\mathbb{Set} = \mathbb{Set}^{\mathbf{1}}$ and $\mathbb{FinSet}$. After all, the definition of an elementary topos was made to mimic these categories. The fact that these categories are Cartesian closed was shown way back in this book, in Theorem 1.22. The fact that they have all limits was looked at in Chapter 3. The subobject classifier is $\Omega = \{0, 1\}$. And, as we saw, the fact that these categories have subobject classifiers is basically given by Definition 1.4.16.

For reasons that will become clear in a few minutes, let us look at the shape category $\mathbf{1} = \mathbf{1}^{op}$. It has only one object $*$. Now consider the contravariant representable functor $Hom_{\mathbf{1}}(\ ,*)\colon \mathbf{1} \longrightarrow \mathbb{Set}$. The only possible input to this functor is $*$ and the output is the set $\{id_*\}$. There are two subfunctors to this functor: one that outputs the empty set and one that outputs id_*. □

Example 9.5.5. Consider $\mathbb{Set}^{\mathbf{2}_0}$ where $\mathbf{2}_0$ is the category with two objects and only identity morphisms. Recall that $\mathbb{Set}^{\mathbf{2}_0}$ is isomorphic to $\mathbb{Set} \times \mathbb{Set}$. The fact that this category is Cartesian closed and has all limits is very similar to the previous example. The subobject classifier is

$$\Omega = \{0, 1\} \times \{0, 1\} = \{(0, 0), (0, 1), (1, 0), (1, 1)\}, \tag{9.136}$$

which has four elements and is a Boolean algebra.

Let us consider the shape category $\mathbf{2}_0 = \mathbf{2}_0^{op}$. This has two objects, a and b, and two identity morphisms. The representable functors are $Hom(\ ,a)$ and $Hom(\ ,b)$, and each of them have two possible subfunctors that output the empty set of the identity morphism. We conclude that there are four possible subfunctors. Notice that there are four elements in the subobject classifier. □

Example 9.5.6. Consider $\mathbb{Set}^{\longrightarrow}$ where $\longrightarrow$ is the category with two objects and a morphism from one to the other. Recall that the objects of $\mathbb{Set}^{\longrightarrow}$ are functions of sets and the morphisms are commutative squares. The fact that this category is Cartesian closed and has all limits is very similar to the previous example. The terminal object is the unique function $\{*\} \longrightarrow \{*\}$.

Let us look at the subobject classifier. Let $f\colon S_1 \longrightarrow S_2$ be an object in $\mathbb{S}\mathrm{et}^{\longrightarrow}$, and let $g\colon T_1 \longrightarrow T_2$ be a subject of f, as in

$$\begin{array}{ccc} T_1 & \xrightarrow{\quad g \quad} & T_2 \\ \big\downarrow & & \big\downarrow \\ S_1 & \xrightarrow[\quad f \quad]{} & S_2. \end{array} \tag{9.137}$$

This means that $T_1 \subseteq S_1$, and $T_2 \subseteq S_2$. The subobject classifier is the function $\Omega = h\colon \{0,1,2\} \longrightarrow \{0,1\}$, where $h(0)=0$, $h(1)=1$ and $h(2)=1$. The following diagram will help:

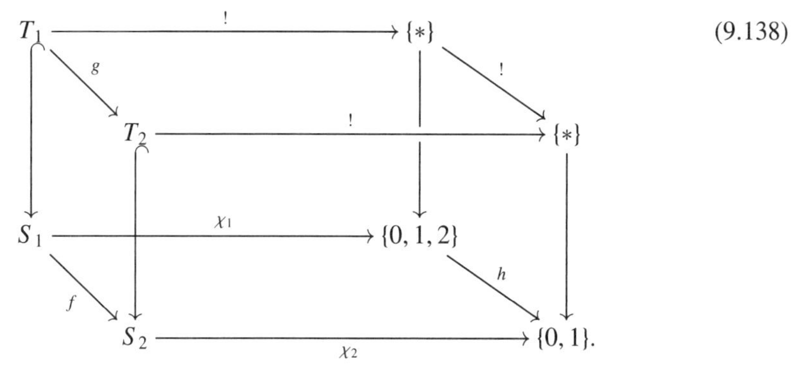

(9.138)

The back truth map takes $*$ to 2, and the front truth map takes $*$ to 1. Let us define the two characteristic maps. For $s \in S_1$ and $s' \in S_2$, we define

$$\chi_1(s) = \begin{cases} 0 & s \notin T_1 \\ 1 & s \notin T_1 \text{ and } f(s) \in T_2 \\ 2 & s \in T_1 \end{cases} \qquad \text{and} \qquad \chi_2(s') = \begin{cases} 0 & s' \notin T_2 \\ 1 & s' \in T_2 \end{cases}. \tag{9.139}$$

Again, let us look as subfunctors of representable functors. The shape category is $a \longrightarrow b = \{b \longrightarrow a\}^{op}$. The subfunctors of the representable functor $Hom(\ ,a)$ can output $\emptyset$ or id_a. This corresponds to the set $\{0,1\}$ of the subobject classifier. The subfunctors for $Hom(\ ,b)$ can be $\emptyset$, $t\colon a \longrightarrow b$, or id_b. This corresponds to the set $\{0,1,2\}$ of the subobject classifier. □

Example 9.5.7. We extend the previous example. Consider the shape category $* \longrightarrow * \longrightarrow *$. Or, more generally, consider category $\mathbf{N}$ as a total order. The category $\mathbb{S}\mathrm{et}^{\mathbf{N}}$ is called **variable sets through time**. The objects are infinite sequences of functions $S_0 \longrightarrow S_1 \longrightarrow S_2 \longrightarrow S_3 \longrightarrow \vdots$. The morphisms are of the form

$$\begin{array}{ccccccccc} S_0 & \longrightarrow & S_1 & \longrightarrow & S_2 & \longrightarrow & S_3 & \longrightarrow & \cdots \\ \downarrow & & \downarrow & & \downarrow & & \downarrow & & \\ S'_0 & \longrightarrow & S'_1 & \longrightarrow & S'_2 & \longrightarrow & S'_3 & \longrightarrow & \cdots \end{array} \tag{9.140}$$

The subobject classifier will be the sets

$$\Omega_0 \longrightarrow \Omega_1 \longrightarrow \Omega_2 \longrightarrow \cdots, \tag{9.141}$$

where each set is $\{0, 1, 2, \ldots, \omega\}$ and the morphisms subtract 1 from each entry except ω. We use the rule that $\omega - 1 = \omega$. The truth value takes $*$ to ω. The characteristic function tells when an object goes into the subsequence of variable sets through time. We leave exploring the details to the reader's leisure time. □

Exercise 9.5.8. Analyze the subfunctors of the representable functors of **N**. ■

Example 9.5.9. The category $\mathbb{G}\mathrm{raph} \cong \mathbb{S}\mathrm{et}^{A \overset{s}{\underset{t}{\rightrightarrows}} V}$ is a topos. We know that this category is Cartesian closed from Example 7.2.4. Finite limits in $\mathbb{G}\mathrm{raph}$ are inherited from $\mathbb{S}\mathrm{et}$ and are calculated pointwise. We are left to describe the subobject classifier. The terminal object is the graph with one object and one morphism coming and going to the single object. The Ω is the graph

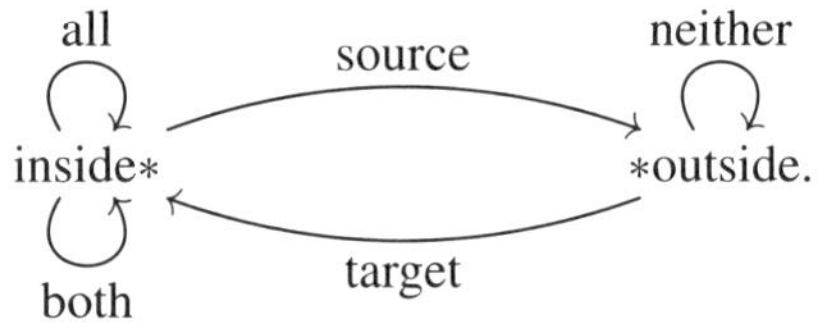

$$\begin{array}{ccc} \text{all} & & \text{neither} \\ \circlearrowright & \xrightarrow{\text{source}} & \circlearrowright \\ \text{inside}* & & *\text{outside.} \\ \circlearrowleft & \xleftarrow[\text{target}]{} & \\ \text{both} & & \end{array} \tag{9.142}$$

The map from the terminal object to Ω is the graph homomorphism that takes the single object to "inside" and the single morphism to "all." Let us see how this subobject classifier works. Consider the inclusion of one graph into another, $H \hookrightarrow G$. The characteristic graph homomorphism $\chi_H \colon G \longrightarrow \Omega$ takes all the objects in H to "inside" and all the other objects to "outside." For each edge of G, the map χ_H depends on the source, target and the edge. Formally, on edge e of G, we define

$$\chi_H(e) = \begin{cases} \text{all} & e \in H, src(e) \in H, trg(e) \in H. \\ \text{both} & e \notin H, src(e) \in H, trg(e) \in H. \\ \text{source} & e \notin H, src(e) \in H, trg(e) \notin H. \\ \text{target} & e \notin H, src(e) \notin H, trg(e) \in H. \\ \text{neither} & e \notin H, src(e) \notin H, trg(e) \notin H. \end{cases} \tag{9.143}$$

□

Exercise 9.5.10. Analyze the subfunctors of the representable functors of $A \overset{s}{\underset{t}{\rightrightarrows}} V$.

■

Now that we have seen some examples of toposes, let us look at a general case.

Example 9.5.11. For any category $\mathbb{A}$, category $\mathbb{S}\mathrm{et}^{\mathbb{A}^{op}}$ is an elementary topos. The terminal object is the functor $t\colon \mathbb{A}^{op} \longrightarrow \mathbb{S}\mathrm{et}$ such that $t(a) = \{*\}$ for all objects a in $\mathbb{A}$. The $\Omega\colon \mathbb{A}^{op} \longrightarrow \mathbb{S}\mathrm{et}$ is defined for an object $a \in \mathbb{A}$ as the set of all subfunctors of $Hom_{\mathbb{A}}(\ ,a)$. □

We said that the notion of a elementary topos is to be similar enough to the category of $\mathbb{S}\mathrm{et}$ that one can do many constructions. Every elementary topos $\mathbb{E}$ has the following properties, which are similar to $\mathbb{S}\mathrm{et}$:

- **Exponentiation.** In $\mathbb{S}\mathrm{et}$, there is exponentiation, which means that for sets S and T, there is a set S^T that corresponds to the collection of all functions from S to T. Similarly, in any topos $\mathbb{E}$, for any two objects a and b, there is an exponential object a^b that corresponds to the collection of functions from a to b. This follows from the Cartesian closedness of the topos.
- **Power object.** Just as every set S in $\mathbb{S}\mathrm{et}$ has a powerset $\mathcal{P}(S) \cong \{0,1\}^S$, every object a in $\mathbb{E}$ has a power object Ω^a. This power object satisfies properties similar to properties that the powerset satisfies in $\mathbb{S}\mathrm{et}$.
- **Epi-mono factorization.** Just as in $\mathbb{S}\mathrm{et}$, every map can be factored as an epi followed by a mono; that is, for every $f\colon a \longrightarrow b$, there is an object m_f, an epi $a \twoheadrightarrow m_f$, and a mono $m_f \rightarrowtail b$ such that

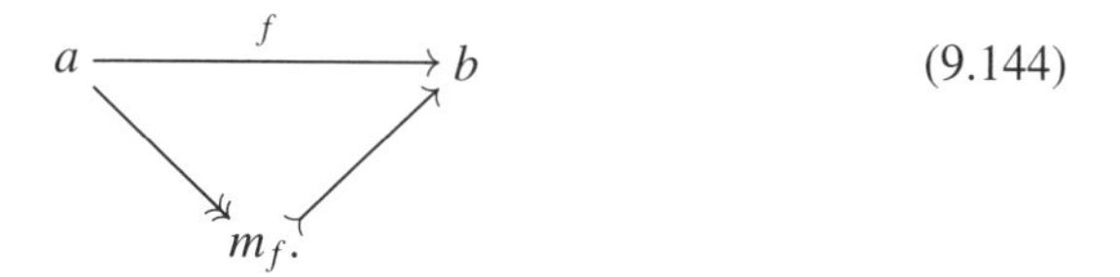

(9.144)

- **Logic**. In $\mathbb{S}\mathrm{et}$, the set $\{0,1\}$ is a Boolean algebra with maps $\vee\colon \{0,1\}\times\{0,1\} \longrightarrow \{0,1\}$, $\wedge\colon \{0,1\}\times\{0,1\} \longrightarrow \{0,1\}$, and $\neg\colon \{0,1\} \longrightarrow \{0,1\}$. We use this structure on $\{0,1\}$ to form the union, intersection, complement, etc. of subsets of any set. That is, the logic of $\{0,1\}$ is used to describe the logic of every set. In detail, for every set S, one looks at the subsets of S that correspond to maps $S \longrightarrow \{0,1\}$. Given two maps $f\colon S \longrightarrow \{0,1\}$ and $g\colon S \longrightarrow \{0,1\}$ we form their union by looking at the map $\vee \circ (f\times g)\colon S\times S \longrightarrow \{0,1\}\times\{0,1\} \longrightarrow \{0,1\}$. The intersection and the compliment are done similarly. In every elementary topos, it can be shown that the subobject classifier has the structure of a Heyting algebra. That is, there are the maps $\wedge\colon \Omega\times\Omega \longrightarrow \Omega$, $\vee\colon \Omega\times\Omega \longrightarrow \Omega$ and $\Rightarrow\colon \Omega\times\Omega \longrightarrow \Omega$, which satisfy the properties of being a Heyting algebra. These maps are then used to talk about the Heyting algebra of subobjects. In summary, Important Categorical Idea 2.4.4 in Chapter 2 taught us that for all $a \in \mathbb{E}$, the collection $Hom_{\mathbb{E}}(a,\Omega)$ inherits the structure of Ω. Thus, the Heyting structure of Ω provides a logic associated to the elementary topos $\mathbb{E}$.

There are many properties that we can request from an elementary topos. The following are some examples:

- **Well-pointed.** In sets, if there are two functions $f, g\colon S \longrightarrow R$ and f is not equal to g, then there must be an element $s \in S$ such that $f(s) \neq g(s)$. We can generalize this to any category with a terminal object 1. Let $f, g\colon a \longrightarrow b$. The category is well-pointed if $f \neq g$; and then there is a "point" $p\colon 1 \longrightarrow a$ such that $fp \neq gp$.
- **Locally small.** In general, for topos $\mathbb{E}$ and for element a, the collection $Sub_{\mathbb{E}}(a)$ might be a class, not a set. If we insist that $\mathbb{E}$ is locally small, then $Sub_{\mathbb{A}}(a)$ is a set with an order relation.
- **Boolean.** This means that Ω is a Boolean algebra. The logical system associated with such a topos is not just a Heyting algebra; it is also a Boolean algebra and satisfies the law of excluded middle.

Example 9.5.12. Choice. Another potential axiom that one can add to the an elementary topos is the axiom of choice. Consider an infinite collection of sets. A set theory "has choice" or "has the axiom of choice" if there is a way to pick one element of each of the sets. This is nicely explained with a cute example. Consider an infinite collection of pairs of shoes. It is easy to choose one element of each of the shoes: simply either choose all the left ones (or all the right ones). In contrast, consider an infinite collection of pairs of tube socks. How should you choose one of them? You cannot describe how to do it. If you nevertheless have the ability to do it, then your system has choice.

We can define choice easily in terms of arrows in a category. We consider an infinite collection of sets by looking at an onto set function $f\colon S \longrightarrow T$, with T being infinite. Then the collection of $f^{-1}(t)$ for all t is an infinite collection of sets. A way of choosing one of those elements for each of the sets is a function $c\colon T \longrightarrow S$ such that $f \circ c = id_T$. For each $t \in T$, the element is $c(t)$ is the choice from $f^{-1}(t)$.

Let us generalize to any category. A category $\mathbb{E}$ has choice— or the axiom of choice—if for any epi morphism $f\colon a \longrightarrow b$, there is a $c\colon b \longrightarrow a$ such that $f \circ c = id_b$.

Radu Diaconescu proved that any elementary topos that satisfies the axiom of choice is a Boolean topos. □

Example 9.5.13. Natural number object. Yet another property that a topos can have is a natural number object. In a topos with a natural number object, one has the notion of recursion. This is related to the category being able to deal with the notion of infinity. Let us start at the beginning, with a simple notion of recursion. Given a number f in $\mathbf{N}$ and a function $g\colon \mathbf{N} \longrightarrow \mathbf{N}$, one can describe a function $h\colon \mathbf{N} \longrightarrow \mathbf{N}$ such that

$$h(0) = f, \tag{9.145}$$

$$h(s(n)) = g(h(n)). \tag{9.146}$$

Category theorists have distilled this idea and written it in arrow notation. While reading this definition, think of the category $\mathbb{S}\mathrm{et}$, with $\mathbf{N}$ as the set of natural numbers. A natural number object in a category is a object $\mathbf{N}$ and two maps:

$$* \xrightarrow{\quad 0 \quad} \mathbf{N} \xrightarrow{\quad s \quad} \mathbf{N}, \tag{9.147}$$

where $*$ is the terminal object in the category. (Think of the 0 map as picking out the zero of the natural numbers and the s function as taking a natural number to its successor.) These two maps must satisfy the following universal property: for any $f\colon * \longrightarrow \mathbf{N}$ and $g\colon \mathbf{N} \longrightarrow \mathbf{N}$, there is a unique $h\colon \mathbf{N} \longrightarrow \mathbf{N}$ such that the following diagram commutes:

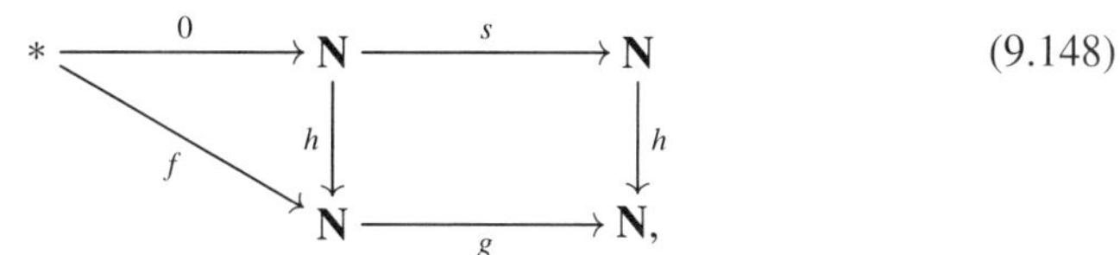

(9.148)

see, [32, 33, 180]. Saying that Diagram (9.148) commutes is the same as saying that h is defined by the simple recursion scheme.

We can use the same scheme to generalize to any object in the category. For any object A, and any maps $f\colon * \longrightarrow A$, $g\colon A \longrightarrow A$, there is an $h\colon \mathbf{N} \longrightarrow A$ that makes the following commute:

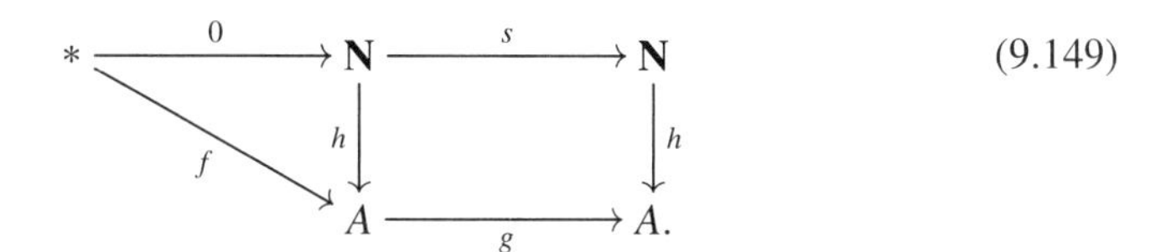

(9.149)

While this definition definitely works in the category of $\mathbb{S}\mathrm{et}$, there are many other categories that have natural number objects. For each category, we have to see what it means.

It pays to meditate on what this definition does. It takes the infinite set of natural numbers and asks the question: What can one do with natural numbers? The answer is that we can always find a successor. This is the essence of infinite. With this successor, we can discuss recursion. The notion of infinity is linked to the notion of a natural number object. From this, we can understand why the topos $\mathbb{S}\mathrm{et}$ has a natural number object while the topos $\mathbb{F}\mathrm{in}\mathbb{S}\mathrm{et}$ does not. All topoi of the form $\mathbb{S}\mathrm{et}^{\mathbb{A}^{op}}$ have a natural number object.

There is a need to discuss functions that have more than one input and hence need to consider a souped-up natural number object called a **parameterized natural number object**. The souped-up version of recursion is

$$h(\bar{x}, 0) = f(\bar{x}), \tag{9.150}$$

$$h(\bar{x}), s(n)) = g(\bar{x}), h(\bar{x}), n)). \tag{9.151}$$

In terms of categories, we need two objects in the category A and B, and for every $f: A \longrightarrow B$ and $g: A \times B \longrightarrow B$, there is an $h: A \times \mathbf{N} \longrightarrow B$ such that the following two squares commute:

$$\begin{array}{ccccccc} A \times * & \xrightarrow{\pi \times 0} & A \times \mathbf{N} & \qquad & A \times \mathbf{N} & \xrightarrow{\pi \times s} & A \times \mathbf{N} \\ \wr \downarrow & & \downarrow h & & \langle \pi_A^{A \times \mathbf{N}}, h \rangle \downarrow & & \downarrow h \\ A & \xrightarrow[f]{} & B & & A \times B & \xrightarrow[g]{} & B. \end{array} \tag{9.152}$$

□

Set theorists use the **Zermelo–Fraenkel (ZF)** axioms to describe sets. Sometimes they insist that their sets also have choice. The axioms of ZF with choice usually go by the name **ZFC**. Notice that a topos has morphisms. In contrast, a model of set theory does not have morphisms. Topos theorists have worked out exactly what properties an elementary topos needs so that the objects in that topos will satisfy the axioms of ZFC. The properties include Boolean, with a natural number object, locally small, and (of course) choice. Amazingly, there are many such toposes.

There is a notion of a morphism between elementary toposes, explored in Definition 9.5.14.

Definition 9.5.14. Given two elementary topoi $\mathbb{E}$ and $\mathbb{E}'$, a functor $F\colon \mathbb{E} \longrightarrow \mathbb{E}'$ is a **logical functor** if it preserves (i) the Cartesian closed structure, (ii) finite limits, and (iii) the subobject classifier. The collection of all elementary toposes and logical functors form a category called $\mathbb{ETopoi}$.

Geometry: Grothendieck Topoi

A Grothendieck topos is a vast generalization of a topological space. Before we define it, let us recall some history. Topologists realized that many properties of a topological space are characterized by the category of sheaves on the topological space. What are sheaves on a topological space? A sheaf on a topological space is a special type of **presheaf** on a topological space. What is a presheaf on a topological space? It is a way of assigning a set to each open set of a topological space. One can think of this as assigning local information to every part of the topological space. For a topological space T, we consider the preorder $O(T)$ of open sets on T, and a presheaf is a functor $P\colon O(T)^{op} \longrightarrow \mathbb{Set}$. For example, a presheaf might assign to every open set U the set of all real functions $U \longrightarrow \mathbf{R}$. The collection of all presheaves on T forms a category $\mathbb{Set}^{O(T)^{op}}$.

While presheaves are interesting in themselves, we are more interested in special types of presheaves where the various pairs of local information cohere with each other. This means that if there are two open sets U_1 and U_2 and they overlap as $U_3 = U_1 \cup U_2$, then the local information of U_3 should match up well with the local information of U_1 and U_2. A presheaf on a topological space T where such good matching occurs is called a **sheaf**. The subcategory of all sheaves on T is denoted as $Sh(T) \hookrightarrow \mathbb{Set}^{O(T)^{op}}$. It turns out that this inclusion functor has a left adjoint $S : \mathbb{Set}^{O(T)^{op}} \longrightarrow Sh(T)$, which takes every presheaf to a related sheaf. This functor is called **sheafification**. The main idea is that one can look at the category of ways that local information matches up and determine various characteristics of the original topological space.

The history continues with Grothendieck, who in the 1960s started applying these ideas to a branch of mathematics called "algebraic geometry." He wanted to characterize certain geometric structures using sheaves. Within algebraic geometry, the local parts of the studied shape does not form a preorder. Rather, they form a general category, which we write as $\mathbb{A}$. A presheaf for the shape is a functor $P\colon \mathbb{A}^{op} \longrightarrow \mathbb{Set}$. The collection of all presheaves forms the category $\mathbb{Set}^{\mathbb{A}^{op}}$. There are special types of presheaves called sheaves where the outputs cohere with the categorical structure of $\mathbb{A}$ (we are simplifiing here). The collection of all sheaves forms a category $Sh(\mathbb{A})$, and there is an inclusion functor $Sh(\mathbb{A}) \hookrightarrow \mathbb{S}\mathbb{e}\mathbb{t}^{\mathbb{A}^{op}}$. This inclusion also has a left adjoint, which is called sheafification.

Grothendick and his school were able to use these fancy category-theoretic tools to solve many problems in algebraic geometry. One should see [194] for the true historical motivation of these concepts.

Definition 9.5.15. A **Grothendieck topos** is a category $\mathbb{G}$ that includes into a category of presheaves $\mathbb{S}\mathbb{e}\mathbb{t}^{\mathbb{A}^{op}}$ for some category $\mathbb{A}$, and where the inclusion has a left adjoint (i.e., $\mathbb{G}$ is a full reflective subcategory of a category of presheaves). Notice that any category of presheaves is also a Grothendieck topos.

It can be shown that a Grothendiesk topos has finite limits, is Cartesian closed, and has a subobject classifier. This means that every Grothendieck topos is an elementary topos.

It can be shown that a Grothendieck topos is Cartesian closed and also has a classifying object. This means that every Grothendieck topos is an elementary topos. The reverse is not always true.

There is a story that brings all these parts together. We saw that Georg Cantor showed there are fewer natural numbers than there are subsets of natural numbers. This means that $|\mathbf{N}| < |\mathcal{P}(\mathbf{N})|$. Cantor then posed the following simple question: Is there a set X whose size is somewhere between $\mathbf{N}$ and $|\mathcal{P}(n)|$? Cantor conjectured that the answer was no. This statement that no such intermediate set exist, called the **continuum hypothesis**, takes the following form:

$$\text{If there is a set } X \text{ such that} |\mathbf{N}| \leq |X| \leq |\mathcal{P}(\mathbf{N})|, \text{ then either } X \cong \mathbf{N} \text{ or } X \cong \mathcal{P}(\mathbf{N}). \tag{9.153}$$

Kurt Gödel and Paul Cohen showed that the axioms of ZFC do not determine if this statement is true or false. That is, they showed that there are models of ZFC set theory where the continuium hypothesis is true and models where it is not true. In topos theory, this corresponds to the fact that one can construct Grothendiesk toposes where the continuum hypothesis is true and one can construct Grothendiesk toposes where the continuum hypothesis is false. This leads to the counterintuitive idea that there are many different collections of sets. Notice what was done here: ideas generated from topology and differential geometry were used to describe strange models of set theory. Only category theory can unite these different worlds.

There is a notion of a morphism between Grothendieck toposes.

Definition 9.5.16. Given two Grothendieck topoi $\mathbb{G}$ and $\mathbb{G}'$, a functor $F\colon \mathbb{G} \longrightarrow \mathbb{G}'$ is a **geometric functor** if it has a left adjoint $F^*\colon \mathbb{G}' \longrightarrow \mathbb{G}$ such that F^* preserves limits. The collection of all Grothendieck toposes and geometric functors forms a category called $\mathbb{GTopoi}$.

Other Toposes

We have only scratched the surface of topos theory. Different fields (including music theory discussed in [193]) have used topos theory in fundamental ways. Here are a few such fields:

- **The effective topos.** A computer can be programmed to enumerate an infinite list of numbers. However, there are some infinite lists where those elements not in the list cannot be enumerated. This has a feel of Heyting algebra, as opposed to a Boolean algebra. Researchers have formulated the effective topos as a way of discussing these issues. It is the beginning of computability theory, which is part of theoretical computer science. See [117, 118, 255] for an introduction.
- **The quantum topos.** This is a new branch of topos theory dealing with quantum physics. There is an important idea called the "Kochen-Specker theorem," which gives a mathematical proof showing that superposition is a very real phenomenon. This also has to do with the concept of contextuality, which means that a result of a measurement depends on how it is measured. See Section 7.3 of [272] for a very readable introduction to these ideas. All these ideas are shown to be describable by nontrivial sheaves, which are Grothendieck toposes. See [76] for a nice introduction. For a modern introduction to the algebra and math of quantum theory, see [152]. See also the citations at the end of Sections 10.2 and 10.3.
- **Higher toposes.** In recent years, Jacob Lurie, Charles Rezk, and others have investigated the notion of an ∞-topos. These are special types of ∞-categories that have higher-order properties of toposes. See [176] and a four-part YouTube lecture by Charles Rezk [218] for more.

Suggestions for Further Study

- There are many nice introductions to topos theory, including [183], Chapter 15 of [33], [170], and [119].
- Textbooks for topos theory are (in order of easiness to read): [91, 181, 42, 32], and [123].
- For papers on this topic, see part II of [151], much of [125], Section 8.8 of [21], [78], and [50].
- The major resources for the topic are [126], and [127].
- There are some nice articles on the history of topos theory, including [194], [35], and [179].
- This presentation gained much from [181].

9.6 Mini-Course: Homotopy Type Theory

Homotopy type theory (HoTT) uses tools and ideas from homotopy theory, logic, and computer science to classify and describe the structures and proofs of mathematics. HoTT is touted as a new and relevant foundation of mathematics. This theory has connections to programs called **proof assistants**, which might revolutionize the very practice of mathematics. Like topos theory, HoTT can be seen from various perspectives, and many a blind wise man might declare his unique view of HoTT. Once again, only the language of category theory can bring all these different ideas under one umbrella.

The first part of this mini-course will explain what type theory is. The second part will describe the innovations provided by homotopy theory and how parts of mathematics can be formulated with HoTT. The final part will be a discussion of how HoTT is used as a foundation of mathematics and its connection with proof assistants.

Type Theory

We begin by introducing type theory. In computer programs, data comes in many forms. There are integers, strings, reals, and others. All these are "types" of data. Functions or methods in programs take input of certain types and return output of certain types. This is helpful for computers (and programmers) to know what information they are dealing with. We will be discussing a vast generalization of these types, called **Martin-Löf dependent type theory** or **constructive type theory**.

We start with **type** A and **term** a. These will have different meanings as we progress. At some point, a **judgment** is made that says that term a is of type A and is written as $a:A$. In the beginning, we think of types as in a computer type and terms as variables, constants, or expressions of that type. For the first part of this presentation, there will not be a problem if you think of a type A as a set and term a as an element of the set. In this sense, $a:A$ in type theory is similar to $a \in A$ in set theory. Later, we also think of type A as a proposition and term a as a proof or witness of A being true. We will also think of type A as a topological space and term a as a point in the topological space.

Definition 9.6.1. Besides types, there are functions between types. For types A and B, we write the function $f\colon A \longrightarrow B$. Notice that if $a:A$ and $f\colon A \longrightarrow B$, then $f(a):B$. The collection of types and functions form a category denoted as $\mathbb{Type}$, and we write $A:\mathbb{Type}$ to describe type A.

We can ask what type is $\mathbb{Type}$? In other words, is $\mathbb{Type}:\mathbb{Type}$? To avoid problems like Russell's paradox (see page 115), we must say that this judgment is invalid. In fact, $\mathbb{Type}$ should be denoted as $\mathbb{Type}_1$ and there is another larger collection called $\mathbb{Type}_2$ with a judgment $\mathbb{Type}_1:\mathbb{Type}_2$. Every type in $\mathbb{Type}_1$ will be a type in $\mathbb{Type}_2$. We can go on to a whole sequence of increasing "universes" $\mathbb{Type}_1, \mathbb{Type}_2, \mathbb{Type}_3, \ldots$. However, for our current needs, we restrict our attention to $\mathbb{Type} = \mathbb{Type}_1$.

The category $\mathbb{Type}$ has much structure. It is Cartesian, co-Cartesian, and closed. This means that there are different ways of constructing types from previously defined types. These methods of constructing types are called **constructors**. In particular, given types A and B, we can construct $A \times B$, $A+B$ and $A \longrightarrow B$ called the **product type**, **coproduct type**, and **function type**, respectively. A typical judgment for a product type is $\langle a,b\rangle : A \times B$, where $a:A$ and $b:B$. A typical judgment for a coproduct type is $inLeft(x):A+B$ if either $x:A$ and inRight(x) for $x:B$. Given types A and B, the type $A \longrightarrow B$ corresponds to functions from A to B. We make a judgment that f is such an element by writing $f:A \longrightarrow B$. (This can be seen as a mathematical pun, as we are using the symbol : in two different ways. On the one hand, it is used in a definition of a function, and on the other hand, it is used as a judgment.)

Since $\mathbb{T}\mathrm{ype}$ has all finite products and coproducts, it has an empty product and an empty coproduct. The empty coproduct is the empty type 0. This type does not have any valid judgment $a:0$. The empty product is type 1. This type has exactly one valid judgment (i.e., $*:1$).

Other commonly used constructors are **dependent types**. These are types that depend on another type. Let A be a type and consider a family of types indexed by A. In other words, for every $x:A$, there is a type $B(x)$. Another way of thinking of this is that there is a function $f\colon A \longrightarrow \mathbb{T}\mathrm{ype}$ (which is a function in $\mathbb{T}\mathrm{ype}_2$) that takes terms of A and outputs a type. Given such a family of types, we can take their coproduct $\sum_{x:A} B(x)$. A typical term of this type is a pair $\langle x, b\rangle$ where $x:A$ and $b:B(x)$. Another constructor that one can use on such a dependent family is $\prod_{x:A} B(x)$. A typical term here is a collection $(b_{x_1}, b_{x_2}, b_{x_3}, \ldots)$, where $b_{x_i} : B(x_i)$. One can think of an element of $\prod_{x:A} B(x)$ as a function $f\colon A \longrightarrow \mathbb{T}\mathrm{ype}$ such that $f(x):B(x)$.

The austerity of type theory must be noted and appreciated. Type theory is about terms and types. That's it! Even maps between types are discussed as function types. And yet, despite this austerity, a large part of mathematics can be described in type theory.

Homotopy Type Theory

The main innovation of homotopy type theory is to add to type theory some topological and homotopical ideas. The first step is to think of a type as a topological space rather than as a set.[6] (Don't worry about the open sets of the topological spaces.) The terms of the types are points or elements of the topological spaces. For type A and term a, we say $a:A$ if a is a point in A.

For terms $a:A$ and $b:A$, a path f from a to b is a proof that a is the same as b. Let us look at some of the properties of these paths:

- Notice that for a given term $a:A$, there is a constant path from a to a. This corresponds to the fact that a is the same as a. (reflexive.)
- Given a path from $a:A$ to $b:A$, there is an inverse path from $b:A$ to $a:A$. This corresponds to the fact that if a is the same as b, then b is the same as a. (symmetric.)
- Given a path from $a:A$ to $b:A$ and a path from $b:A$ to $c:A$, there is a composition path from $a:A$ to $c:A$. This corresponds to the fact that if there is a proof that a is the same as b and a proof that b is the same as c, then there is a proof that a is the same as c. (transitive.)

The collection of all types $\mathbb{T}\mathrm{ype}$ also forms a topological space where each point is a topological space of a type. A function from one type to another $f\colon A \longrightarrow B$ is a continuous map of topological spaces.

Martin-Löf dependent type theory has a way of making types that becomes very interesting when you look at types as topological spaces. For type A and terms $a:A$ and $b:A$, there is an **identity type**, which we denote as $Id_A(a,b)$ or $a =_A b$. This is the type of proof that a is the same as b. If a is not the same as b, this type is empty. One should think of the terms of $Id_A(a,b)$ as paths from a to b. A term of $Id_A(a,b)$ can

[6] In actuality, each type is a homotopy equivalence class of a space.

be drawn as $a \longleftrightarrow b$. The notation $Id_A(a,b)$ should remind one of a Hom set in a category.

This is just the beginning of the story. Let $x : Id_A(a,b)$ and $y : Id_A(a,b)$ be proofs that a is the same as b. We ask if these proofs are the same. If there is a proof that x and y are the same, then there is a homotopy, or a two-dimensional path, from x to y. This is a path in the identity type

$$Id_{Id_A(a,b)}(x,y). \tag{9.154}$$

We write a term of this type as $x \Longleftrightarrow y$ This type can also be empty if x is not the same as y.

This can be continued. Let $p : Id_{Id_A(a,b)}(x,y)$ and $q : Id_{Id_A(a,b)}(x,y)$. We can ask if p and q are the same. A proof that they are the same is a term of the type

$$Id_{Id_{Id_A(a,b)}(x,y)}(r,s). \tag{9.155}$$

This is a three-dimensional path $p \Longleftrightarrow q$. Such a term is a proof of the sameness of two proofs (p and q) of the fact that two proofs (x and y) are the same proofs that a is the same as b. This is a homotopy between homotopies. As always, this process can go on forever as we can climb the ladder of dimensionality.

This process is very similar to what we saw in Remark 9.4.8, where we take an (∞,∞)-category and look at the Hom set of two elements. This Hom set is itself an (∞,∞)-category. In HoTT, think of a type as an infinite-dimensional topological space (or an ∞-groupoid). Then an identity type is like a Hom set. It too can be an infinite-dimensional topological space.

Notice how the language of category theory is uniting the geometric intuitions of homotopy theory and the logical structures of proofs.

As HoTT is a foundation of mathematics, there must be a way for HoTT to deal with such diverse fields as logic, sets, and algebra. In the next few paragraphs, we discuss these topics from an HoTT prospective.

We begin with logic. This is done by considering every proposition as a type. This doctrine, called **propositions-as-types** or **Curry-Howard correspondence**, is central to HoTT's view of logic. Essentially, there is an adjunction between types and propositions. The map $\mathbb{Type} \longrightarrow \mathbb{Prop}$ takes a type A to the proposition "A is inhabited." Type A is deemed true if it is inhabited, which means that there is some term a such that there is a valid judgment $a : A$. So long as there is at least one such judgment, it is true. In contrast, if there is not term where such a judgment can be made, then we say that type A is false. The map $\mathbb{Prop} \longrightarrow \mathbb{Type}$ takes a proposition P to the type of all proofs of P. We need at least one proof for it to be considered true.

Definition 9.6.2. A type is **contractible** if there is some term such that there is a path from every term to that original term. In type theory, we write A is contractible as follows:

$$isCont(A) :\equiv \sum_{a:A} \prod_{b:A} a =_A b. \tag{9.156}$$

Notice that $isCont(0)$ is not inhabited.

Definition 9.6.3. A **proposition** (also called a **mere proposition**) is a type where any two terms are isomorphic. In symbols, A is a proposition as follows:

$$isProp(A) :\equiv \prod_{a:A} \prod_{b:A} a =_A b. \tag{9.157}$$

If $isProp(A)$ is inhabited, then A is either uninhabited or contactable.

Using this, we can talk about the logical constants as follows:

- Truth. 1 corresponds to true because it is always inhabited by one term: $* : 1$.
- False. 0 corresponds to false because it is not inhabited by any term.

The logical operations in HoTT follow—as in all of category theory—from the morphisms of $\mathbb{Type}$. Let us look at the operations:

- Conjunction. $A \wedge B$ corresponds to the product type $A \times B$. This follows from the existence of the projection functions $A \times B \longrightarrow A$ and $A \times B \longrightarrow B$. The type $A \times B$ will be inhabited only when both A <u>and</u> B are inhabited.
- Implication. $A \Longrightarrow B$ corresponds to a morphism $f\colon A \longrightarrow B$. If x inhabits A (i.e., $x : A$), then $f(x)$ inhabits B (i.e., $f(x) : B$).
- Negation. $\neg A$ corresponds to the function type $A \longrightarrow 0$. Notice that if A is inhabited, then it is impossible for $A \longrightarrow 0$ to be inhabited because 0 is not inhabited.
- Universal quantifiers. $\forall_a P(a)$ corresponds to the dependant type $\prod_{a:A} P(a)$. Consider a function of types $P\colon A \longrightarrow \mathbb{Type}$. For all $a : A$, there is a $P(a) : \mathbb{Type}$. If every type $P(a)$ is inhabited, then so is $\prod_{a:A} P(a)$.

All these work regularly. The following two operations do not work well because propositions are not closed under them:

- Disjunction. $A \vee B$ corresponds to the coproduct type $A + B$. This follows from the injection functions $A \longrightarrow A + B$ and $B \longrightarrow A + B$. The type $A + B$ will be inhabited only when either A <u>or</u> B are inhabited.
- Existential quantifiers. $\exists_a P(a)$ corresponds to the dependent type $\sum_{a:A} P(a)$. Consider a function of types $P\colon A \longrightarrow \mathbb{Type}$. For all $a : A$, there is a $P(a) : \mathbb{Type}$. If there is a type $P(a)$ that is inhabited, then so is $\sum_{a:A} P(a)$.

The problem with disjunction is that *true* $\vee$ *true* is not a proposition because it will have two different valid judgments. There is a similar problem with existential quantifiers. HoTT offers various solutions to this problem, which we omit in this discussion.

Let us use some of this new language to discuss some common mathematical notions. Since the symbol $X = Y$ has a special meaning in HoTT, we shall use $X :\equiv Y$ to mean that X is defined as Y.

Recall that we met the **Axiom of Choice** in our discussion of topos theory earlier in this chapter. We can state the axiom in the language of HoTT. The axiom says that if there is a relation $R \subseteq A \times B$ such that for every a in A, there is at least one b in B (but usually more than one) that is related to a, then there is a single-valued function f that picks out one such b that is related to it:

$$AC :\equiv \prod_{A,B:\mathbb{T}\mathrm{ype}} \prod_{R:A\times B\longrightarrow \mathbb{T}\mathrm{ype}} \left[\left(\prod_{a:A}\sum_{b:B} R(a,b)\right) \longrightarrow \left(\sum_{f:A\longrightarrow B}\prod_{a:A} R(a,f(a))\right)\right]. \tag{9.158}$$

In the literature, a modification of this axiom is used.

The **Law of Excluded Middle** is an axiom that says that the logical system that we are working with is Boolean. This means that every proposition is either true or false and there is nothing in the middle. We state this axiom as a type as follows:

$$LEM :\equiv \prod_{A:\mathbb{T}\mathrm{ype}} [isProp(A) \longrightarrow (A + \neg A)]. \tag{9.159}$$

In constructive mathematics, it is always true that for any proposition A, $A \longrightarrow \neg\neg A$. In classical mathematics, it also goes the other way. That is, for any proposition A, $\neg\neg A \longrightarrow A$. It turns out that we can equivalently define this notion as

$$LEM :\equiv \prod_{A:\mathbb{T}\mathrm{ype}} [isProp(A) \longrightarrow (\neg\neg A \longrightarrow A)]. \tag{9.160}$$

In a system where this type is inhabited, the system is Boolean.[7]

For HoTT, a set is a collection of "blobs," where each blob is a contractible space (no holes in it). For a given type, A, the type $isSet(A)$ is defined as

$$isSet(A) :\equiv \prod_{x,y:A} \prod_{p,q:x=_A y} p = q. \tag{9.161}$$

There is a function $\pi_0 : \mathbb{T}\mathrm{ype} \longrightarrow \mathbb{S}\mathrm{et}$ that takes any type A, to the set of isomorphism classes $\pi_0(A)$ of A. In other words, π_0 takes a collection of all the points in the blobs to the set of blobs.

We can go further with this. A groupoid is a type such that between any two objects, there is a set of morphisms:

$$Groupoid :\equiv \prod_{G:\mathbb{T}\mathrm{ype}} \left[\prod_{x,y:G} isSet(x =_G y)\right]. \tag{9.162}$$

This is similar to the how the Hom sets of a category are a set. However, here the Hom sets are invertible. We can move on now to talk about 2-groupoids.

Before we move on to algebra, it pays to review and summarize what we have seen so far:

- A proposition: $isProp(A) :\equiv \prod_{a,b:A} isCont(a =_A b)$
- A set: $isSet(A) :\equiv \prod_{a,b:A} isProp(a =_A b)$
- A groupoid: $isGroupid(A) :\equiv \prod_{a,b:A} isSet(a =_A b)$
- A 2-groupoid: $is2\text{-}Groupid(A) :\equiv \prod_{a,b:A} isGroupoid(a =_A b)$

[7]There is a HoTT form of **Diaconescu's theorem** that says that any system that has the axiom of choice also has the law of excluded middle. To get this, we would have to modify AC to AC' and LEM to LEM'. This modification comes from a process called **propositional truncation**. The theorem then means that there is a morphism $AC' \longrightarrow LEM'$, and if AC' is inhabited, then so is LEM'. See Section 10.1 of [254] for details.

This goes on. We can define a hierarchy or stratification of levels of all types recursively as follows:

- Type A is of 0-level if it is contractible.
- Type A is of n-level if for all terms $a, b : A$, the type $Id_A(a, b)$ is of $n-1$-level.

This hierarchy is cumulative. In other words, if A is a type of n-level, then A is of $n+1$-level.

This hierarchy is related to the hierarchy of categories discussed in Section 9.4:

- -2-level are contractible spaces.[8]
- -1-level are empty or contractible spaces—truth values or propositions.
- 0-level are sets (i.e., $(0, 0)$-categories).
- 1-level are groupoids (i.e., $(1, 0)$-categories).
- 2-level are a 2-groupoid (i.e., $(2, 0)$-categories).

  ⋮
- n-level are n-groupoids (i.e., $(n, 0)$-categories).

  ⋮
- ∞-groupoids (i.e., $(\infty, 0)$-categories).

The proper way to think of type A is as an ∞-groupoid. This is in line with our thinking from Section 9.3 that every topological space is the same as an ∞-groupoid.

The list of (n, r)-categories that we saw in Section 9.4 can be extended as follows:

- $(-2, 0)$-categories are contractible categories; that is, categories $\mathbb{A}$ such that the unique functor $\mathbb{A} \longrightarrow \mathbf{1}$ is an equivalence of categories.
- $(-1, 0)$-categories are truth values or propositions.

How does HoTT deal with algebra? Let us consider a standard mathematical structure such as a group, and show how it is described by HoTT. As we saw in Definition 1.4.56, a group is a 4-tuple containing a set G, a unit element $e \in G$, an inverse $(\ )^{-1} : G \longrightarrow G$, and a multiplication $m : G \times G \longrightarrow G$. In HoTT, we need to describe each of the parts of a group as a type as follows:

- We need a type G that is a set. This means that $isSet(G)$ is inhabited.
- There is a unit element $e : G$.
- There is an inverse operation $(\ )^{-1} : G \longrightarrow G$.
- There is a multiplication $m : G \times G \longrightarrow G$. All this sums up to the following type:

$$\sum_{G:\mathbb{Type}_n} \sum_{e:G} \sum_{(\ )^{-1} : G \longrightarrow G} (G \times G \longrightarrow G). \tag{9.163}$$

  However, this is not good enough for HoTT. We need to add in proofs that these axioms are satisfied.

[8]The -2-level types do not correspond to *all* contractible spaces. Rather, they correspond only to constructable ones that can be described with type theory. It is similar for the rest of this list.

- The multiplication is associative means that the type

$$\prod_{a:G}\prod_{b:G}\prod_{c:G} m(a,m(b,c)) = m(m(a,b),c) \tag{9.164}$$

is inhabited, say by α. In other words, for each a, b, and c, α will provide a proof of $a(bc) = (ab)c$.
- The left identity rule can be written as

$$\prod_{a:G} m(a,e) = a, \tag{9.165}$$

and this type can be inhabited by λ_i. A similar proof is needed for the right identity proof (say ρ_i).
- The right unit rule is stated as

$$\prod_{a:G} m(a,(a)^{-1}) = e, \tag{9.166}$$

which can be inhabited by ρ_u. A similar proof is needed for the left unit rule (say λ_u).

In conclusion, a group is a 9-tuple: $(G, e, (\ )^{-1}, m, \alpha, \lambda_i, \rho_i, \lambda_u, \rho_u)$. The type that describes groups[9] is defined as

$$Group :\equiv \sum_{G:\mathbb{Type}}\ \sum_{e:G}\ \sum_{(\ )^{-1}:\ G\longrightarrow G} (isSet(G)) \times (G\times G \longrightarrow G)\times \tag{9.167}$$

$$\prod_{a:G}\prod_{b:G}\prod_{c:G} m(a,m(b,c)) = m(m(a,b),c) \times \text{other identities}). \tag{9.168}$$

Univalent Foundations

The centerpiece of HoTT is an idea called the "univalence axiom," proposed by Vladimir Voevodsky. We now slowly build up to this idea.

The axiom concerns when two types are equivalent. Remember that the set $\{0,1\}$, $\{False, True\}$, $\{x,y\}$, etc., are all different sets. But they are all isomorphic. What about two types? When are they equivalent? In other words, when do they describe the same structures?

[9]Some authors do not include $isSet(G)$ as part of the definition of a group. In that case, there is a subtle complexity that is worthy of more thought. Let us say that there are two different proofs that the multiplication m is associative. This will entail two different groups: $(G, e, (\ )^{-1}, m, \alpha, \lambda_i, \rho_i, \lambda_u, \rho_u)$ and $(G, e, (\ )^{-1}, m, \alpha', \lambda_i, \rho_i, \lambda_u, \rho_u)$. The problem is that most mathematicians would consider these two groups to be the same. Because of this, the type *Group* would describe a larger collection than the usual collection of groups that a mathematician deals with. One solution to this is to include in the definition of a group the fact that there is exactly one proof that the multiplication m is associative. In other words, $\prod_{a:G}\prod_{b:G}\prod_{c:G} m(a,m(b,c)) = m(m(a,b),c)$ is a true proposition. There are similar deep issues with the other identities.

There is a definition of equivalence of types that is similar to the definition of equivalence of functors as essentially surjective and fully faithful. For every type function $f\colon A \longrightarrow B$ and for every $b\colon B$, there is a type called the fiber of f at b, defined as follows:

$$f^{-1}(b) :\equiv \prod_{a:A} f(a) =_B b. \tag{9.169}$$

Notice that there might be many such a's, but they are all isomorphic to each other. Let us use these fibers to define what equivalence of types is.

Definition 9.6.4. Given two types A and B, they are equivalent if the following type is inhabited:

$$Equiv_{\mathbb{Type}}(A, B) :\equiv \sum_{f\colon A \longrightarrow B} \prod_{b:B} isCont(f^{-1}(b)). \tag{9.170}$$

There is an inclusion:

$$inc\colon Id_{\mathbb{Type}}(A, B) \hookrightarrow Equiv_{\mathbb{Type}}(A, B). \tag{9.171}$$

This means that if two types are the same, then they are definitely equivalent. This is very similar to the fact that the collection of isomorphism functors in $\mathbb{Cat}$ are included into the collection of equivalence functors in $\mathbb{Cat}$.

Now for the main point: the **univalence axiom** says that not only is $Id_{\mathbb{Type}}(A, B)$ included into $Equiv_{\mathbb{Type}}(A, B)$, but the inclusion is an equivalence of types. This is such an important idea that we will describe it and its consequences in several different ways. The univalence axiom means the following:

- If A and B are equivalent, then we can think of them as equal.
- Every equivalence of types is "up to homotopy" really an identity.
- Equality is equivalent to equivalence.
- There is a quasi-inverse to the *inc* map

$$(A \simeq B) \longrightarrow (A = B), \tag{9.172}$$

  which takes equivalent notions and considers them identical.
- Stated in symbols, this means

$$Id_{\mathbb{Type}}(A, B) \simeq Equiv_{\mathbb{Type}}(A, B), \tag{9.173}$$

  or, more succinctly,

$$(A = B) \simeq (A \simeq B). \tag{9.174}$$

- $Equiv_{\mathbb{Type}}(A =_{\mathbb{Type}} B, Equiv_{\mathbb{Type}}(A, B))$ is inhabited.

Keep in mind that the univalence axiom is about a type, and the main question is whether this type is inhabited or not. We do not want to get into the details of this, but there are certain models of HoTT where the univalence axiom is not true. When we think of types as general topological spaces, then the axiom need not be true. When we think of the types as Kan complexes, then the univalence axiom is true. Momentarily, we will see why this axiom is very important for the use of HoTT as a practical foundation of mathematics.

One can only be struck by the similarity between the univalence axiom and the core ideas in coherence theory. They are both centered about telling when two structures are essentially the same. Imagine a structure where the multiplication of three elements is defined as $m(x, m(y, z))$, and another structure where the multiplication of three elements is defined as $m(m(x, y), z)$. The terms for the descriptions of these two structures are equivalent. The univalence axiom tells us that you might think of these two structures as the same and the multiplication of three elements might be written as $m(x, y, z)$. These are exactly the ideas that motivated our discussion of coherence theory. Both the univalence axiom and coherence theory are about change and when changed structures are the same. Much work remains to understand the connection.

A word must be said here about the potential influence that HoTT might have on the future development of mathematics. Traditionally, when a mathematician proves a theorem, the community inspects the proof and either finds errors in the proof or accepts it. As of late, some proofs are too large and complex for reviewers to determine their validity. Errors are made and proofs go unverified. Part of the appeal of HoTT is to use its formalization of mathematics to ensure that mathematical proofs are correct.

The long-term goal is to set up large computer systems with every true statement of mathematics. These systems will be running programs that go by the name **proof assistant** and **proof verifier**. These programs help mathematicians get the kinks out of their proofs and make sure that they are correct. While these programs have been around for a while, they have received a new impetus from HoTT. A program currently in use that performs such tasks is **Coq**, developed by Thierry Coquand and his collaborators. When a new theorem and proof are stated, they will have to be stated in a formal language and entered into the computer. The computer will then go through the proof and determine if there are flaws in the proof or if the proof is acceptable.

Univalence will play a special role here. Many times, mathematicians do not necessarily present a structure in the same way. ("Mathematics is the art of giving the same name to different objects."—Henri Poincaré.) What is needed is a computer program with the ability to utilize the univalence axiom, which considers equivalent structures to be the same. While it will be painful to formalize the statement and the proof in the correct language, the benefit of knowing that there are no errors will be immense. Once a result is accepted, it will be added to the computer so it can be part of accepted mathematics. See [20, 204] for more about the role of proof assistants.

HoTT is lauded as a foundation of modern mathematics. By a foundation of mathematics, we mean an ontology (an entity that we can say all mathematical structures really are) and a set of tools to deal with those structures. Traditionally, mathematicians have used logic and set theory as a foundation.

The problem with logic as a foundation is that it is "clunky" and most mathematicians do not think of the structures that they are dealing with as statements in mathematical logic.

Set theory is a more popular foundation of mathematics. This school of thought believes that all mathematical structures can be stated in the langauge of set theory. There is a criticism of these ideas from Saunders Mac Lane. He supposedly stated that saying all structures are made out of sets is like saying all skyscrapers are made out of sand. It might be true, but it is not very helpful.

HoTT as a foundation brings many new ideas to the table. For one, it is a foundation that is not some abstract metaphysics. As we just outlined, it has a very practical application that will make mathematics less prone to error. Another very positive aspect of HoTT as a foundation is that it is stresses proofs, not just structures. Mathematics is not about true mathematical statements. It is about proven mathematical statements. HoTT formalizes such proofs and looks at how different proofs relate. Yet another positive aspect of HoTT is that it makes informal mathematics done by everyday mathematicians closer to formal mathematics. These are some of the reasons why many believe that HoTT is a superior foundation for mathematics and it will have an important role in the coming years.

Suggestions for Further Study

Here are some helpful introductory articles: [22, 94], and [204]. Be aware that HoTT is a fast-developing field and any exposition is quickly outdated.

For a more categorical view of dependent type theory, see [232].

There are many wonderful lectures on YouTube that explain various parts of HoTT. Gershom Bazerman has a nice lecture titled "Homotopy Type Theory: What's the Big Idea" [34]. André Joyal gave a beautiful talk at the New York City Category Theory Seminar titled "Homotopy Type Theory: A New Bridge Between Logic, Category Theory and Topology" [129]. There are also several wonderful and understandable lectures by Vladimir Voevodsky on YouTube that are worth watching.

The bible of HoTT is [254], a product of a special year at the Institute of Advanced Study in Princeton, New Jersey, devoted to the topic "Univalent Foundations of Mathematics." The book is very clear and does not have the feel of being written by a committee.

10

More Mini-Courses

Knowledge comes, but wisdom lingers.
Alfred Lord Tennyson
Locksley Hall, 1835

Now that our toolbox is chock-full of categorical ideas and structures, let us delve into several more fields of math and science. Realize how easy it is to learn various topics once we know basic category theory.

This chapter is organized as follows:

- Section 10.1: Knot Theory
- Section 10.2: Basic Quantum Theory
- Section 10.3: Quantum Computing

These topics were chosen for a few reasons. They are easy to understand with the category theory that we learned, they are well known, and we felt comfortable presenting them. While they are mostly disconnected and unrelated, all three topics have something in common. They all are about methods of assigning morphisms in various monoidal categories. The mini-courses can be described as follows:

- Knot theory concerns assigning to every link a special endomorphism $I \longrightarrow I$, where I is the unit element in a monoidal category.
- Basic quantum theory deals with the assignment of morphisms of the monoidal category of Hilbert spaces to the concepts of quantum theory.
- Quantum computing concerns algorithms that are simply the composition of morphisms in the monoidal category of Hilbert spaces.

Obviously, the devil is in the details, but it is nice to see this common theme.

The mini-courses are independent of each other and can be read in any order, except that it is advisable to read basic quantum theory before quantum computing.

There are a myriad of other mini-courses that we could have included. This is in line with the central theme of this text: category theory is a unifying language that makes many different topics easy to learn and understand. However, due to time and space considerations, we must limit ourselves to these three. There is an urge to write an entire book with only mini-courses of various topics. That book might cover the standard model in particle physics, linear logic, quantum gravity, linguistics, topological quantum field theory (QFT), parts of chemistry, anyon computation, machine learning, loop quantum gravity, and many other subjects. There are research papers that connect these areas to monoidal categories. Perhaps this is a project for the future.

10.1 Mini-Course: Knot Theory

In the nineteenth century, knot theory was part of recreational mathematics. Later, physicists thought that different atoms were really knotted vortex tubes of the ether. In the early part of the twentieth century, knot theory was taken up more seriously by topologists and combinitorists who classified knots with different properties. In 1984, the New Zealand mathematician Vaughan Jones introduced a new result that revolutionized knot theory. His ideas were related to some concepts in physics. This revolution is intimately related to monoidal categories and is the focus of this mini-course. Category theory is perfectly suited to unite topological, combinitorial, and physical concepts.

Basic Definitions and Examples

One makes a knot by taking a string and wrapping it around itself in different ways. The ends of the string are then fused together. Here, we are going to deal with mathematical strings, which have no thickness and can be as long as needed. One can think of a knot as a smooth injective map from the circle to three-dimensional real space (i.e., a smooth inclusion $S^1 \longrightarrow \mathbf{R}^3$). Two such maps can represent the same knot if you can deform one into the other without passing the string through itself. We say two such maps are **ambient isotopic**. Technically, ambient isotopy is an equivalence relation, and a knot is an equivalence class of this collection of maps.

Let us spread our wings and define a more general concept. A **link** consists of a finite number of closed strings that are wrapped around each other. We already met links in Section 7.2 where we saw that links are the tangles that start and end from the empty space ($\emptyset$). In other words, links are the elements of the hom set $Hom_{\mathbb{Tang}}(\emptyset, \emptyset)$. One can think of a link as a smooth injective map from n disjoint circles to three-dimensional space; that is,

$$S^1 \amalg S^1 \amalg S^1 \amalg \cdots \amalg S^1 \qquad \longrightarrow \qquad \mathbf{R}^3. \tag{10.1}$$

Two links are considered the same if there is a way to deform the strings of one to the strings of the other with no strings passing through each other. Technically, a link is an equivalence class of this collection of maps.

Since this book is written on a two-dimensional page, we will follow what we did with braids and talk about **link projections**. As with braids, there are many link projections for every link. Figure 10.1 contains some famous knot and link projections.

Just as we asked when two braid projections describe the same braid, so too we ask when two link projections describe the same link. Consider the following **Reidemeister moves**:

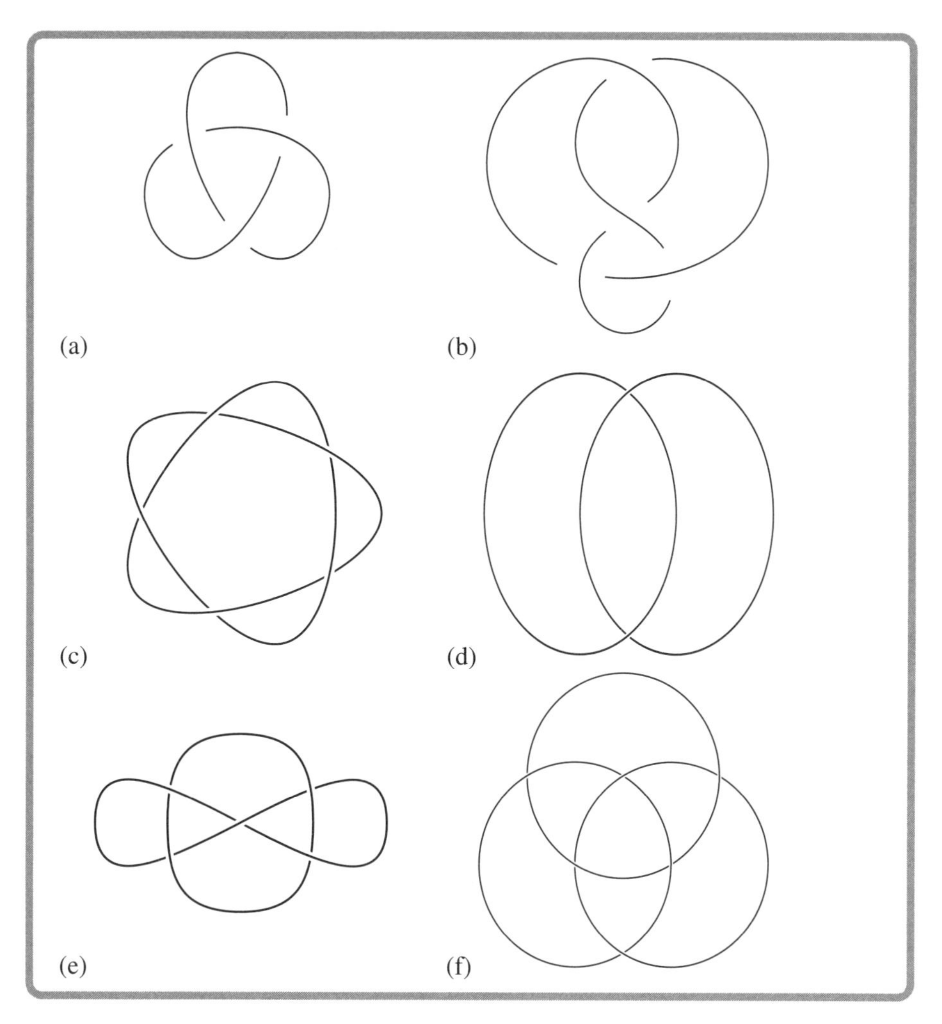

Figure 10.1. Several famous knots and link projections: (a) Trefoil knot; (b) figure-eight knot; (c) Solomon's seal knot; (d) Hopf link; (e) Whitehead link; (f) Borromean rings.

- Reidemeister I: $\Longleftrightarrow$ $\Longleftrightarrow$
- Reidemeister II: $\Longleftrightarrow$ $\Longleftrightarrow$
- Reidemeister III: $\Longleftrightarrow$ $\Longleftrightarrow$

These are relationships between small parts of link projections. When these small changes are made to a link projection, it is obvious that the entire link has not changed. Jones was dealing with some algebraic structure related to statistical mechanics when he noticed that this algebraic structure satisfied Reidermeister III. He concluded that the field that he was working in was related to knot theory.

Remark 10.1.1. Notice that these three types of moves are already familiar to those of us who have studied braided and rigid categories. Reidemeister I is exactly the middle line of Figure 7.4. Reidemeister II is the right side of Inequality 7.6. Reidemeister III is exactly the same as Equation 7.19. ♠

Just as we saw that there is a small group of relations that determine when two descriptions of permutations describe the same permutation, and there is a small group of relations that determine when two braid projection are the same braid, the Reidemeister moves are a small group of relations that determine when two link projections describe the same link. Two link projections represent the same link if they differ by a sequence of Reidemeister moves. This was proved in 1927 by Kurt Reidemeister, James Alexander, and Garland Briggs.

An example of asking when two link projections are the same is asking when a link projection represents the unlink. That is, when the link described by the link projection is the same as separated unknots. There exists an algorithm that can answer this question; however, it might take a very long time to finish. The reason for the complexity of the algorithm is that sometimes one must make the link have more crossings before it can have fewer crossings. In other words, sometimes the only way to reduce the complexity of a knot is to first increase its complexity.

Researchers love to classify knots and links. They do this with an **invariant**. This is a function *Inv* from the collection of links (or just knots) to a set of mathematical objects such as numbers, polynomials, or groups. Many times, an invariant is described by looking at properties of the link projection. This means that there is a function Inv^* from the collection of link projections to a set of mathematical objects, and this function will respect the Reidemeister rules. Another way to say this is that the following diagram commutes:

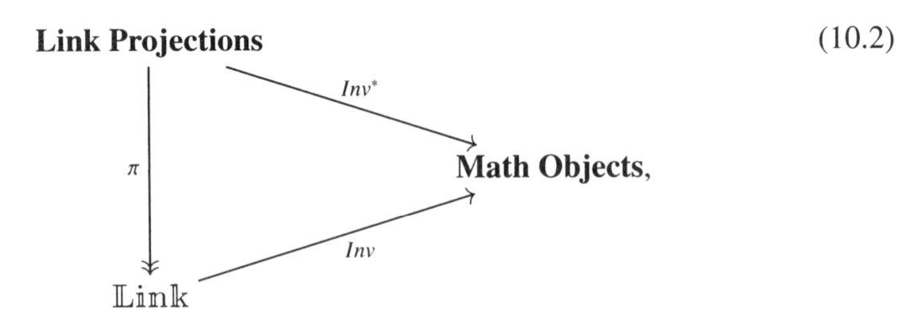

(10.2)

where π takes every link projection to the link that it describes. This means that the assignment of a mathematical object to the link projection is invariant (does not change) with respect to the various ways that the link is presented.

An invariant has the following property: given two links L_1 and L_2,

$$Inv(L_1) \neq Inv(L_2) \Longrightarrow L_1 \neq L_2. \tag{10.3}$$

There are rarer types of invariants called **complete invariants** which take different links to different mathematical objects; that is,

$$Inv(L_1) \neq Inv(L_2) \Longleftrightarrow L_1 \neq L_2. \tag{10.4}$$

Another way to say this is that *Inv* is a complete invariant if it is injective. Such invariants are the ultimate goal.

Here are some important invariants:

- The number of **components** of the link is a simple invariant.
- A knot lives in $\mathbf{R}^3$. Take the space $\mathbf{R}^3$ and cut out the knot. This is the space with empty tubes where the knot was. Look at the fundamental group of this space. This a powerful invariant of the knot and is called the **knot group**
- Gauss's **linking number** of a link tells how many times two links wrap around each other.
- The **crossing number** of a knot is the minimum number of crossings of any knot projection of the knot.
- The **unknotting number** of a knot is the minimum number of times that a strand must be passed through another strand to make it into the unknot.

The New Invariants

The revolution of Vaughan Jones introduced a whole slew of new, powerful invariants. These invariants are usually calculated using a so-called **skein relation**. (The definition of a skein is a loosely coiled string of yarn wound on a reel.) It is a geometric version of calculating something by recursion. Let us take a minute to remind ourselves of how recursion works. Remember that the Fibonacci numbers are given by the following recursion relation:

$$Fib(1) = 1, \quad Fib(2) = 1, \quad Fib(n+1) = Fib(n) + Fib(n-1). \tag{10.5}$$

To use this in calculating, say, $Fib(7)$, one first has to calculate $Fib(6)$ and $Fib(5)$. But for $Fib(6)$, one needs to calculate $Fib(5)$ and $Fib(4)$. And this goes on:

$$Fib(7) = Fib(6) + Fib(5) = (Fib(5) + Fib(4)) + (Fib(4) + Fib(3)) = \cdots \tag{10.6}$$

Notice that each term reduces to smaller terms. This goes on until all the terms are the boundary conditions $Fib(2) = 1$ and $Fib(1) = 1$.

Now let us examine a skein relation for one of the the new invariants, called the **Kauffman bracket**. Given a link L, one calculates the the Kauffman bracket of L. This is a polynomial that can have negative powers. Such polynomials are called **Laurent polynomials**. The Kauffman bracket is written as $\langle L \rangle$ and is calculated using one of the following schemes:

- The unknot rule. This is a boundary condition that says that the Kauffman bracket of the unknot is 1:

$$\langle \bigcirc \rangle = 1 \tag{10.7}$$

- The disjoint union rule. This is like a boundary condition. It tells what to do with one extra unknot:

$$\left\langle L \cup \bigcirc \right\rangle = (-A^2 - A^{-2})\langle L \rangle \tag{10.8}$$

Notice the A^{-2} is a negative power.

- The skein relation. This is the main rule, and it tells how to calculate the bracket by decomposing a crossing:

$$\left\langle \times \right\rangle = A \left\langle \asymp \right\rangle + A^{-1} \left\langle)(\right\rangle. \tag{10.9}$$

When one has a crossing in a small part of the link, it is decomposed in this way. The decomposed links will have fewer crossings. This goes on until one reaches a boundary condition.

The obvious question to ask is: In what order should the crossings be decomposed? The answer is that it does not make a difference. The result will be the same regardless of what order the crossings are decomposed. All roads lead to the same value of the invariant of the link. This is essentially a coherence theorem that says that it does not matter which operation is done in which order.

Here are several other invariants each given by a different skein relation. In all of them, the boundary case is that the unknot is 1:

- The first invariant goes back to James Alexander in the 1920s and is called the **Alexander polynomial**:

$$\Delta\left[\times\right] - \Delta\left[\times\right] = \frac{1}{\sqrt{1-x^2}}\Delta\left[\,)\;(\,\right]. \tag{10.10}$$

- In the 1960s, John Conway found another invariant, which is based on the Alexander polynomial and which is called the **Conway polynomial**:

$$\nabla\left[\times\right] - \nabla\left[\times\right] = x\nabla\left[\,)\;(\,\right]. \tag{10.11}$$

- The **Jones polynomial** is a Laurent polynomial, and it is given as follows:

$$xV\left[\times\right] - x^{-1}V\left[\times\right] = \left(\frac{1}{\sqrt{x}} - \sqrt{x}\right)V\left[\,)\;(\,\right]. \tag{10.12}$$

- The **HOMFLYPT polynomial** P got its name as an acronym of the last names of the eight people who introduced it: Jim Hoste, Adrian Ocneanu, Kenneth Millett, Peter J. Freyd, W. B. R. Lickorish, David N. Yetter, Józef H. Przytycki, and Paweł Traczyk. The polynomial has two variables, x and t:

$$xP\left[\times\right] - tP\left[\times\right] = P\left[\,)\;(\,\right]. \tag{10.13}$$

- There is also a **Jones polynomial in two variables**:

$$\frac{1}{\sqrt{\lambda}\sqrt{q}}\left[\times\right] - \sqrt{\lambda}\sqrt{q}X\left[\times\right] = \left(\frac{\sqrt{q}-1}{\sqrt{q}}\right)X\left[\,)\;(\,\right]. \tag{10.14}$$

All these polynomials are related to each other via various changes of variables and normalizations.

As an example of values for the various invariants, we give some of the polynomials for the Solomon's seal knot, as shown in Figure 10.1(c):

- The Alexander polynomial: $\Delta(x) = x^2 - x + 1 - x^{-1} + x^{-2}$
- The Conway polynomial: $\nabla(x) = x^4 + 3x^2 + 1$
- The Jones polynomial: $V(x) = x^2 + x^4 - x^5 + x^6 - x^7$
- The HOMFLYPT polynomial: $P(x,t) = t^4x^4 + t^2(-x^6 - 4x^4) + (3x^4 + 2x^6)$

(These calculations were taken from [260].)

There is another way to look at these new invariants that goes to the heart of monoidal category theory. One can perform an operation on a braid that turns it into a link. Take the braid and add in a strand that connects the top of the first strand with the bottom of the first strand. Think of the new strand going around the braid clockwise. Then take a second strand and connect the the top of the second strand of the first braid with the bottom strand of the braid. Continue along this way with all n strands of the braid. This forms the **closure of the braid** or the **Markov closure of the braid**. Some of those strands will be connected, but there are at most n components. The closure of a pure braid with n strands will be a link with n components.

We can think of this operation categorically by looking at $\mathbb{T}\mathrm{ang}$, the category of tangles. The following diagram will help:

(10.15)

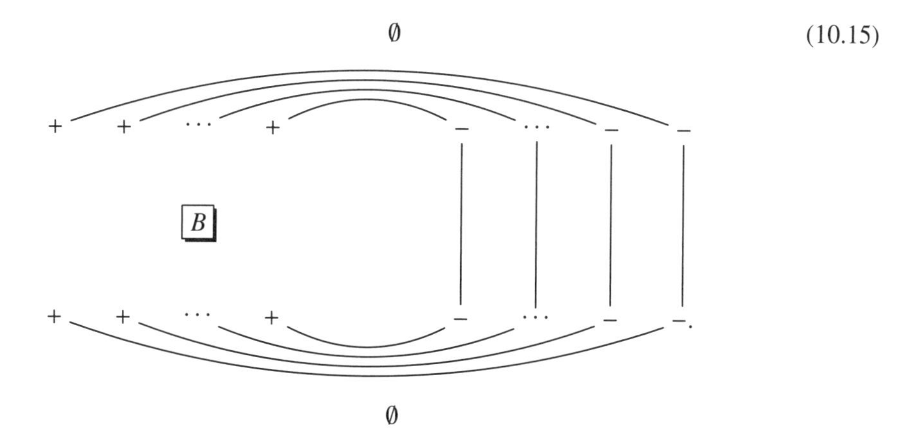

Consider an n strand braid B going from $++\cdots+$ to $++\cdots+$. The coevaluation η goes from the empty string of $+$ and $-$ to $++\cdots+\quad --\cdots-$. On the left side is the braid, and on the right side is the identity. We then use the evaluation ϵ to go back to the empty string. A simple version of this is

(10.16)

This gives us a function that takes a braid and outputs a link. In other words, it is a function from the category of braids to the category of links. Notice that this is not a functor because given two braids, B and B' of n strands, there is no reason why $trace(B' \circ B) = trace(B') \circ trace(B)$. We call this function **trace**. It is a theorem of Alexander that every link can be deformed so that all the activity happens in a braid and the right side of the closure of the braid has no activity. This means that the function $trace\colon \mathbb{Braid} \longrightarrow \mathbb{Link}$ is onto.

We can summarize what we have stated here by adding the trace function to Diagram 7.68 and getting the following result:

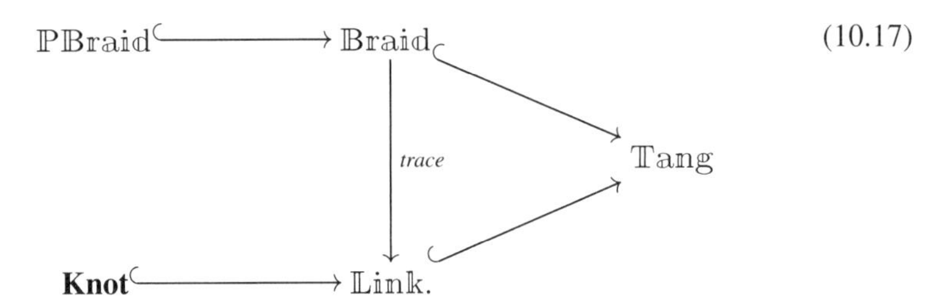

(10.17)

Since *trace* is surjective, we may ask when two braids represent the same link. The following two rules are called **Markov moves** (see Figure 10.2):

- For any n strand braids B and B', we have that $trace(B) = trace(B' \circ B \circ B'^{-1})$. The reason for this is that B' and B'^{-1} can travel around the closure and cancel each other out. This rule is called **conjugation**. Notice that since every B' is built of small σ_i that are the generators of braids, this rule is usually written as $trace(B) = trace(\sigma_i \circ B \circ \sigma^{-1})$.
- For any n strand braid, B, there is a $n+1$ strand braid $B \otimes 1$. If one twists the last two strands of this braid with a σ_n and then closes this bigger braid, the last two strands become one. Thus, we have shown that $trace(B) = trace(B\sigma_n)$. This rule is called **stabilization**.

In 1936, Andrei Markov proved that these two moves describe all such changes. In other words, two links are ambient isotopic if and only if any two corresponding braids differ by a finite sequence of Markov moves (conjugation and stabilization).

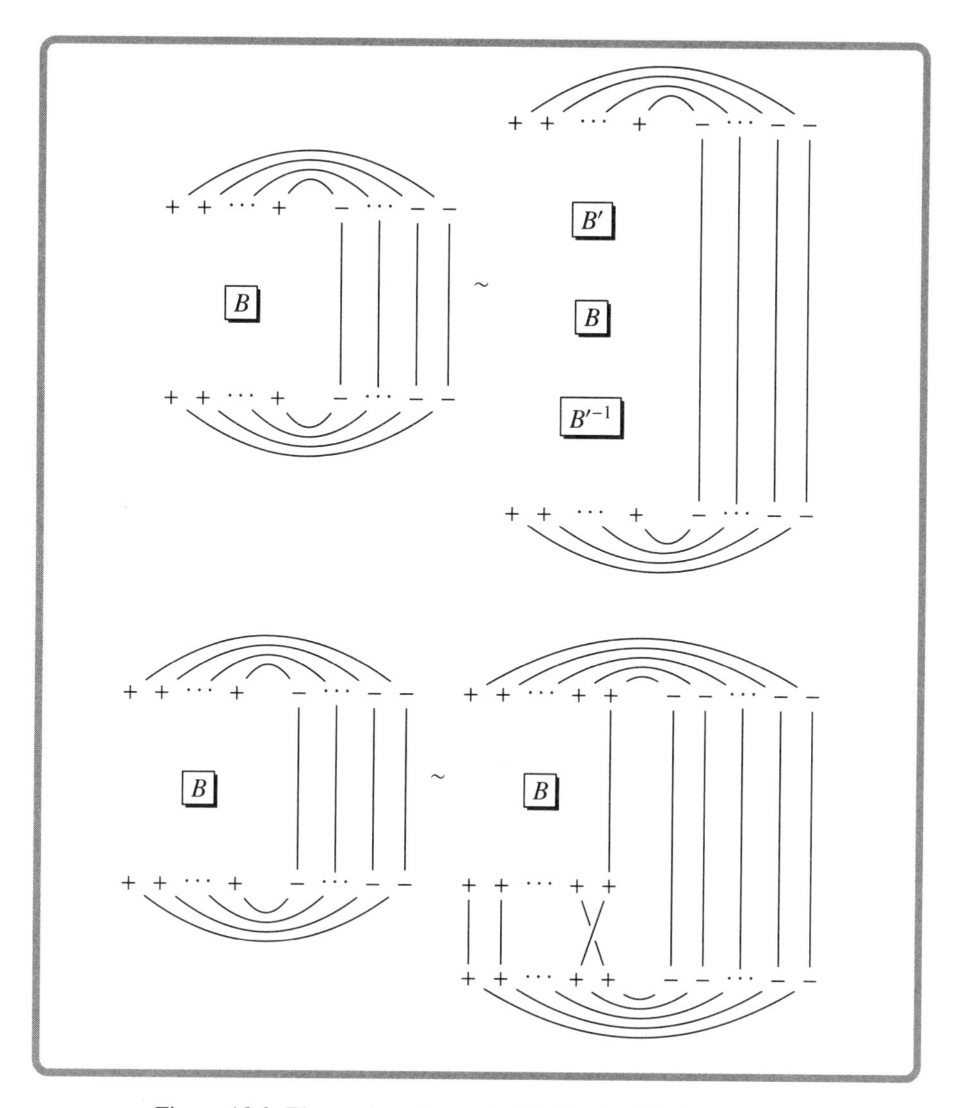

Figure 10.2. The conjugation and stabilization Markov moves.

The trace function satisfies another rule that is easy to see. Consider two braids, B and B', one after the other. The B' in Figure 10.3 can be swung around the closure of the braids to show that $trace(BB') = trace(B'B)$. This will correspond to the fact that the endomorphisms of the unit of a monoidal category is commutative.

The main point is that any link L is a closed braid that lives in the category $\mathbb{T}\mathrm{ang}$ as a morphism $L\colon \emptyset \longrightarrow \emptyset$. We saw in Theorem 7.3.9 that the category of ribbons is initial. Related to this is the fact that the category $\mathbb{T}\mathrm{ang}$ is initial in a certain sense. That means, for every rigid monoidal category $(\mathbf{C}\mathbb{V}\mathrm{ect}, \otimes, \mathbf{C}, \gamma)$, there is a a unique strong monoidal functor $F\colon (\mathbb{T}\mathrm{ang}, \amalg, \emptyset) \longrightarrow (\mathbf{C}\mathbb{V}\mathrm{ect}, \otimes, \mathbf{C}, \gamma)$ and we have $F(L)\colon F(\emptyset) \longrightarrow F(\emptyset)$. Since F takes unit objects to unit objects, we have $F(L)\colon \mathbf{C} \longrightarrow$

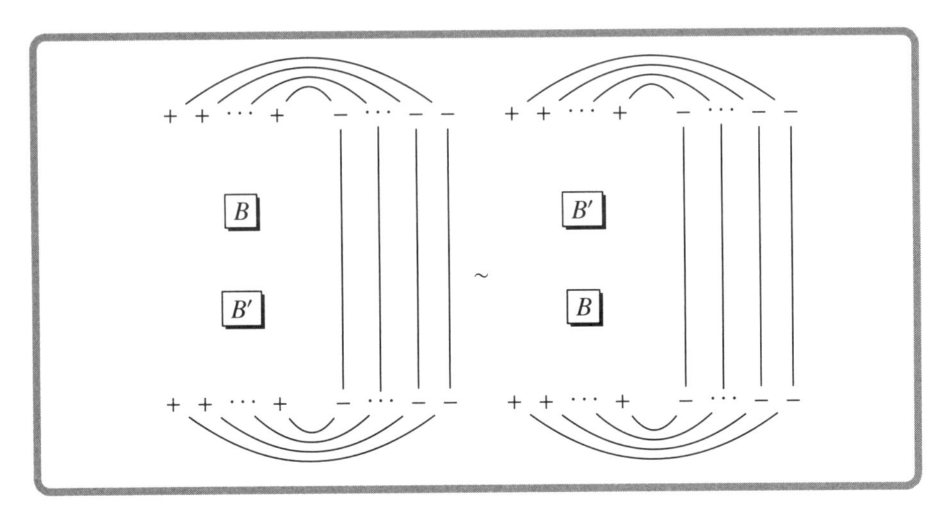

Figure 10.3. The trace of two braids is the same regardless of order.

C. This means that L is going to correspond to the complex number that corresponds to this function.

If γ in $\mathbf{C}\mathbb{V}\mathbb{ect}$ is the usual one (i.e., $a \otimes b \mapsto b \otimes a$), then this corresponds to all the strands just passing through each other, and F gives the number of components in the link. However different γ's in $\mathbf{C}\mathbb{V}\mathbb{ect}$ give different invariants. Researchers have worked out which categories of modules on quantum groups give rise to which knot invariant. In other words, for every new type of knot invariant, there is a quantum group and a rigid monoidal category of modules of that quantum group. The value of the invariant for a link is the same as the value $F(L) \colon \mathbf{C} \longrightarrow \mathbf{C}$.

Further Directions in Knot Theory

Knot theory is a very active area of research, with lot of new results continuously arising. Before we close, we would like to mention three other parts of knot theory that are related to this text.

What exactly does the Jones polynomial measure? Edward Witten showed that the Jones polynomial has a physical meaning. As we will see in Section 10.2, quantum theory is centered around probabilities. Many events in the quantum world can happen, but some events are more probable than others. The probability that some event will happen is given by a number called the "amplitude." Imagine a particle traveling between two points. With high probability, the particle will follow a straight line. With a lesser probability, it will go through space in a loopy way and travel like a knot around and around. If one thinks of a knot K as the orbit in spacetime of a charged particle, then the average amplitude of the particle following that path is V_K, the Jones polynomial of K. The version of quantum theory that deals with this is called "Chern-Simons theory for gauge fields," which was developed by Shiing-Shen Chern and James Simons in the 1970s. See [264] for an exposition of these ideas.

We saw in Remark 9.4.13 the importance of going up a dimension by categorification. Mikhail Khovanov took the Jones polynomial invariant and categorified it to form what is now called the **Khovanov homology** invariant. This is a very powerful invariant that can distinguish the unknot from all other knots. The invariant also has a deep relevant meaning in quantum theory. In particular, string theorists are always looking at higher-dimensional structures and how to incorporate higher dimensions. Read more about this in [264].

In 1989, Victor Vassiliev formulated the notion of a **Vassiliev invariant** or **finite-type invariant** of a knot. While it is beyond us to get into all the details of this powerful invariant, there are some beautiful homotopy theory intuitions that are worth meditating on (discussed in Section 9.3). Consider the infinite dimensional space of all smooth inclusion maps $S^1 \longrightarrow \mathbf{R}^3$. Think of this space as a giant box where every point in the box represents such a mapping. There is an equivalence relation on this space where two mappings are deemed equivalent if they can be deformed into each other. Now generalize a little. Consider all smooth maps $S^1 \longrightarrow \mathbf{R}^3$, which might have a finite number of double points. In other words, two different points in S^1 can go to the same point of $\mathbf{R}^3$. Think of such a double point as sitting between two opposite crossings as follows:

$$[\times] \quad [\times] \quad [\times]. \tag{10.18}$$

In other words, to go from one inclusion $S^1 \longrightarrow \mathbf{R}^3$ to another inclusion that might represent a different knot, one has to pass through a map $S^1 \longrightarrow \mathbf{R}^3$ with this double point. Now one should envision all smooth maps with at most a finite number of double points as a giant box that is partitioned into a grid of smaller boxes. Each smaller box has the infinite number of maps that correspond to a single knot. The walls of the partitions are the double-pointed maps that separate the knots from each other. Passing through a wall represents the passage from one crossing through a double point to the opposite point. Another way to say this is that one can convert one knot to another knot via a finite number of such crossing changes and the associated passages through the walls of the partition. A Vassiliev invariant takes into account this larger space of maps and is a better invariant because of that. This topic is definitely worthy of study if you are interested in homotopy theory.

Suggestions for Further Study

The most accessible entry point for every aspect of knots is Colin C. Adams's *The Knot Book* [13]. See also the books [136, 135, 137] and papers of Louis Kauffman. This presentation gained much from Chapter X of [134], V. G. Turaev's [253], and David Yetter's book [278].

10.2 Mini-Course: Basic Quantum Theory

The most fascinating field in all of modern science is quantum theory. It is the branch of physics that describes the vast majority of phenomena in the universe. The core ideas are counterintuitive and mysterious. In this mini-course, we outline some of the main ideas of quantum theory with the category theory that we have learned.

Unfortunately, quantum theory has a bad reputation as a topic that is extremely difficult to learn. This is patently false. If you got to this point of the book, you already have the mathematics necessary to understand and work with quantum theory. The mathematics and the manipulations are not hard. The difficulty is accepting that the universe works in such a counterintuitive way. We will not be offering any explanations here; we will just learn how quantum theory works, not why.

Contrary to popular opinion to learn quantum theory, one does not need to have a background in classical physics. Nor does one need much advanced mathematics. Rather, this mini-course only assumes that the reader knows basic category theory and the material in the mini-courses on linear algebra in Chapters 2 and 5. To a large extent, basic quantum theory can be understood as applied complex vector spaces. Even better: the quantum theory that we will employ only demands finite-dimensional complex vector spaces.

Our presentation is centered around several basic postulates of quantum theory. The postulates will describe physical phenomena with parts of $\mathbb{Hilb}$, the symmetric monoidal category of Hilbert spaces. Along the way, we describe some of the main concepts, theorems, and experiments of quantum theory.

A disclaimer is in order at this point. Quantum theory has been in existence for more than a century and is a huge field. We can only give a small taste of the beauty and profundity of this theory. There are literally thousands of textbooks on this topic. Also, we are not presenting all the wonderful connections between category theory and quantum theory that researchers have described. Nor are we employing all the structure that $\mathbb{Hilb}$ and $\mathbb{FDHilb}$ has. Rather, the goal here is to describe—in a few pithy pages—the central notions of quantum theory based on the language of monoidal category theory. It is my sincere hope that the reader goes on from here and learns some more about this amazing field. See the texts listed at the end of this section to go further with your studies.

Classical Systems

We start with simple systems that describe parts of the classical world. Such systems are needed as a stepping stone to get to quantum theory. We will see that the quantum world is analogous (but very different) to aspects of the classical world. This part might feel overly pedantic and somewhat obvious. We are presenting classical systems in this manner to make the comparison with quantum systems.

A classical system is described by a set S of states of the system. For example, the system can be a bunch of juggling balls, and then we can look at the set of states that the describe the balls. Each state has within it various properties of the system, such as the position, speed, and color of each of the balls. The system can be a group of people, and then the states will be the properties of those people. If the system is a machine, then the states will be the properties of the components of that machine.

Each state of the system will be described by an element s of set S, and it will contain all the relevant information of the system in that state. We category theorists would say that the state is described by a set map $\{*\} \longrightarrow S$.

There are several operations that one can do with a system: (i) one can measure properties of the system, (ii) one can change the system, and (iii) one can combine systems to form larger systems. Let us look at these three operations carefully.

When we measure some aspect of the system, we learn about some properties of that system. For example, in the case of the juggling balls, we could be interested in the color of the sixteenth ball. When we learn that the color of this ball is red, the system will be described by a smaller set $S' \subseteq S$ of states:

$$S' = \{s \in S : s \text{ is a state with the 16th ball being red}\}. \tag{10.19}$$

All the other properties of the system are still unknown. We are stating that every measurement of a system described by S corresponds to an inclusion function $S' \hookrightarrow S$.

Changes to the system can also be described by a function from the set to itself. In other words, one state can go to another state via a function $S \longrightarrow S$. If the changes to the system are reversible, then such a function will be an isomorphism.

Another process that one does is combine two systems. This corresponds to taking the Cartesian product of the two sets of states in $\mathbb{S}\mathrm{et}$. So if we have two systems described by S_1 and S_2, then we can form the set of states of the total system as $S_1 \times S_2$. The states of the system will consist of a state of S_1 and a state of S_2. In other words, every map $\{*\} \longrightarrow S_1 \times S_2$ determines—and is determined by—maps $\{*\} \longrightarrow S_1$ and $\{*\} \longrightarrow S_2$.

In our discussion of classical systems, we only mentioned sets and functions between sets. In other words, classical systems are modeled in the symmetric Cartesian category $(\mathbb{S}\mathrm{et}, \times, \{*\})$. Quantum systems are modeled in the symmetric monoidal category $(\mathbb{H}\mathrm{ilb}, \otimes, \mathbf{C})$.

Quantum Systems

Quantum systems are systems that deal with objects that follow the laws of quantum mechanics. Usually, these are subatomic particles or objects that are determined by subatomic particles (i.e., everything).

We saw that a classical system is described by an object in $\mathbb{S}\mathrm{et}$. For quantum systems, we have Postulate 1.

Postulate 1. A quantum system is modeled by a Hilbert space $\mathcal{H}$. In other words, for each quantum system, there is an assignment of an object of the category $\mathbb{H}\mathrm{ilb}$. ⨀

The Hilbert space will have all the possible states of the quantum system. We saw that a state of a classical system is described by a map $\{*\} \longrightarrow S$. Remember that the terminal object in $\mathbb{H}\mathrm{ilb}$ is $\mathbf{C}$ and a map $f\colon \mathbf{C} \longrightarrow \mathcal{H}$ can be described by its value on $1 \in \mathbf{C}$ because $f(c) = c \cdot f(1)$ (Remark 2.4.15). So a map $\mathbf{C} \longrightarrow \mathcal{H}$ corresponds to the ray of vectors in $\mathcal{H}$ that are nonzero multiples of $f(1)$.

Postulate 2. A state of a quantum system is modeled by a ray of vectors in the Hilbert space (i.e., a linear map $\mathbf{C} \longrightarrow \mathcal{H}$). ⨀

Physicists have a different nomenclature, which we will follow. Instead of using the word "vector," they use the phrase **wave function** or the word **ket**, for reasons that will be explained soon. Rather than writing a vector as v, they use Greek letters and enclose those letters with $|\ \ \rangle$. Thus, kets are written as $|\phi\rangle$ or $|\psi\rangle$, and a state is described by a ray of kets.

For the rest of this text, we focus our attention and consider only finite-dimensional Hilbert spaces. This means that we only talk about finite vectors, finite bases, and finite linear combinations. In other words, we can think of a ket as a finite column vector of complex numbers. We might write a ket as

$$|\phi\rangle = [3 + 2i,\ 6i,\ 12.3 - 7.1i,\ 0,\ 14]^T. \tag{10.20}$$

The next concept is the most important in all of quantum mechanics. It is also the most mysterious. In stark contrast to a classical system, where two states $f\colon \{*\} \longrightarrow S$ and $g\colon \{*\} \longrightarrow S$ stand alone, in a quantum system, we can use the vector space structure of the Hilbert space and actually add two states. Given $f\colon \mathbf{C} \longrightarrow \mathcal{H}$ and $g\colon \mathbf{C} \longrightarrow \mathcal{H}$, we can get another state, $f + g\colon \mathbf{C} \longrightarrow \mathcal{H}$. In other words, given kets $|\phi\rangle$ and $|\psi\rangle$, we can form ket $|\phi\rangle + |\psi\rangle$. In a sense, this new state corresponds to the system being in a state that is represented by $|\phi\rangle$ *and* $|\psi\rangle$. We will elaborate more on what this means in the coming pages. This phenomenon is called **superposition**. The word comes from the following idea: If the state of a system tells you the position of a particle, then the sum of a few states is saying that the system is in a few positions or a superposition. However, a superposition is not only about position. Any property that a quantum system deals with can be in a superposition. For example, a quantum object can be in a superposition of energies, spins, velocities, and other elements. This means that an arbitrary ket describes a state where the components of the system have many values of properties at the same time. This is mind blowing.

When two kets are finite-dimensional, one can add the components of the vectors. So if

$$|\phi\rangle = [3 + 2i,\ 6i,\ 12.3 - 7.1i,\ 0,\ 14]^T \qquad \text{and} \qquad |\psi\rangle = [17.2,\ 6 - 4i,\ 12.2,\ 16i,\ 0]^T, \tag{10.21}$$

then

$$|\phi\rangle + |\psi\rangle = [20.2 + 2i,\ 6 + 2i,\ 24.5 - 7.1i,\ 16i,\ 14]^T. \tag{10.22}$$

One can also use the scalar multiplication to form kets $c|\phi\rangle$ for any $c \in \mathbf{C}$. Notice that the state that corresponds to $|\phi\rangle$ is considered to be the same state as $c|\phi\rangle$ for any nonzero complex number c. They are in the same ray. Ket $|\phi\rangle$ and $c|\phi\rangle$ represent the same physical state but are said to have a different **phase**.[1] Because different phases are represented by the same physical state, we can conclude that the phase of a ket does not correspond to anything physical in the system.

The most famous way of physically demonstrating the notion of superposition is the well-known **double split experiment**. Imagine a light shining on a wall that has two thin slits that are near each other. When light passes through each slit separately, light shines on the other side of the wall. When light goes through both slits at the

[1] Physicists generally reserve the word "phase" for c, such that $|c| = 1$.

same time, there will be an interference pattern. This is explained by saying that light is a wave, and when the light goes through both slits, the waves cancel each other out in some locations and reinforce each other in others. This causes an interference pattern. There is nothing strange about this. But physicists take this further. They are able to send photons one at a time toward the slits. When they do this with many, many photons, they again see an interference pattern. How can that be? If the photon is going through the slit one at a time, who does it cancel out with? Who does it reinforce with? The answer is that the single photon goes through *both slits at the same time*. In other words, rather than the single photon going through one slit or the other, the photon is in a superposition of going through both slits. There are many other experiments that show the physical reality of superposition.

Every ket has a dual version, which is called a **bra**. The bra that corresponds to $|\phi\rangle \in \mathcal{H}$ is written as $\langle\phi| \in \mathcal{H}^*$. For a finite-dimensional ket, its bra is simply the adjoint row vector. In other words, $\langle\phi| = |\phi\rangle^\dagger$. For ket $|\phi\rangle = [3+2i,\ 6i,\ 12.3-7.1i,\ 0,\ 14]^T$, the bra would be $\langle\phi| = [3-2i,\ -6i,\ 12.3+7.1i,\ 0,\ 14]$. We use the inner product structure of a Hilbert space to compare two kets. For example, given two kets, $|\phi\rangle$ and $|\psi\rangle$, we compare them by taking the bra of $|\phi\rangle$ and performing the inner product $\langle\phi||\psi\rangle = \langle\phi|\psi\rangle$ to get a complex number. The complex number will describe how the two states are related. Together, they form a "bra-ket" or a "bracket" (hence the strange nomenclature). Notice that in the finite-dimensional case, the inner product is just the matrix multiplication of a column vector and a row vector, which is a single complex number. Categorically, a ket corresponds to a map $\mathbf{C} \longrightarrow \mathcal{H}$ and a bra corresponds to a map $\mathcal{H} \longrightarrow \mathbf{C}$. A bracket corresponds to a the composition $\mathbf{C} \longrightarrow \mathcal{H} \longrightarrow \mathbf{C}$. The value of the bracket is the image of $1 \in \mathbf{C}$.

One can also look at a ket-bra. This is a ket followed by a bra. For example, $|\psi\rangle\langle\phi|$. In the finite-dimensional case, this is a column multiplied by a row, which gives a matrix. Categorically, one can see this as the composition of $\mathcal{H} \longrightarrow \mathbf{C} \longrightarrow \mathcal{H}$ and hence gives an operator $\mathcal{H} \longrightarrow \mathcal{H}$. What does this operator do? If one applies this operator to the ket $|\phi\rangle$, one gets

$$|\psi\rangle\langle\phi|\ \ |\phi\rangle = |\psi\rangle\ \ \langle\phi|\phi\rangle = \langle\phi|\phi\rangle\ \ |\psi\rangle. \tag{10.23}$$

In other words, the operator $|\psi\rangle\langle\phi|$ takes $|\phi\rangle$ to a scalar multiple of $|\psi\rangle$.

While any ket of a ray describes the state, there are special kets that are helpful and are called **normalized** kets. Such a ket has norm 1. Any ket can be normalized as follows: For a given ket $|\phi\rangle$, find the length of $|\phi\rangle$. If the length of the ket is L, then the ket $|\phi'\rangle = \frac{|\phi\rangle}{L}$ is normalized. For example, the ket $|\phi\rangle = [2-3i, 1+2i]^T$ has length $L = \sqrt{|2-3i|^2 + |1+2i|^2} = 4.2426$. And the normalized ket is

$$\frac{[2-3i, 1+2i]^T}{4.2426} = [0.41714 - 0.70711i, 0.23570 + 0.47140i]^T. \tag{10.24}$$

The normalization of $1|\phi\rangle + 1|\psi\rangle$ is

$$\frac{1}{\sqrt{2}}|\phi\rangle + \frac{1}{\sqrt{2}}|\psi\rangle. \tag{10.25}$$

Normalized kets will be important in the coming discussion.

Observable and Measurements

In our mini-course on advanced linear operators, we highlighted two types of morphisms from $\mathcal{H}$ to itself: Hermitian operators and unitary operators The first type of operator will correspond to measuring a property of a system, while the second type of operator is about making changes to the system. Let us first consider making observations.

At some point, we would like to know a property of a system. Any physical property that we can learn about a system is called an **observable**.

Postulate 3. For a system that is represented by $\mathcal{H}$, each observable of the system corresponds to the Hermitian operator $A: \mathcal{H} \longrightarrow \mathcal{H}$. ⊙

Remember that for a Hermitian operator A, the eigenvectors for distinct eigenvalues are orthogonal. In particular, the eigenvectors for every Hermitian operator forms a basis that we call its **eigenbasis**. In quantum theory, the eigenvectors are called **eigenkets**. The eigenbasis will consist of the different possible outputs of the observable. The main point is that the state of the system will be expressed in different ways depending on what measurement is being made. In other words, the states of the system are expressed by different eigenbases. There is no essentail state of the system.

Example 10.2.1. Let us look at some examples of observables and their associated eigenkets:

- A particle can be on a line, and we can ask what its position is on that line.
- We can ask what the momentum of a particle is.
- Some quantum particles have a property called **spin**. We can measure the spin in different directions. With respect to a specific direction, certain particles can spin in one of two ways: up or down. We denote these two ways as $|\uparrow\rangle$ and $|\downarrow\rangle$. Each direction corresponds to a different observable. For example, the eigenbases for another observable is $|\nearrow, \swarrow\rangle$. Yet another eigenbasis is $|\leftarrow\rangle$ and $|\rightarrow\rangle$.
- Some particles have electric charges, and we can measure them with a basis that looks like this:
$$\ldots, |-1\rangle, |-0.5\rangle, |0\rangle, |0.5\rangle, |+1\rangle, \ldots. \tag{10.26}$$
- We can measure how many particles are in the system. The basis will be the particle numbers:
$$|0\rangle, |1\rangle, |2\rangle, |3\rangle, \ldots, |n\rangle, \ldots. \tag{10.27}$$

(This can be infinite-dimensional.) □

This notion of spin is shown with the **Stern-Gerlach experiment**. In this experiment, a stream of particles, usually silver, passes between two magnets. After it passes, the stream usually splits into different parts. The particles are being measured, and some have spin in this direction while others have spin in that direction. The discreteness of the outgoing streams show that quantum spin is different from classical spin. The rate of quantum spin cannot be any possible value. Rather, particles spin only in a certain discrete set of values.

For any observable, the Hermitian operator determines an orthonormal eigenbasis, such as

$$|\phi_1\rangle, |\phi_2\rangle, |\phi_3\rangle, \ldots, |\phi_n\rangle. \tag{10.28}$$

Since these are basis vectors, any ket $|\psi\rangle$ is a linear combination of the eigenkets. This means there is a unique sequence of complex numbers $c_1, c_2, \ldots, c_n$ such that

$$|\psi\rangle = c_1|\phi_1\rangle + c_2|\phi_2\rangle + c_3|\phi_3\rangle + \cdots + c_n|\phi_n\rangle = \sum c_i|\phi_i\rangle. \tag{10.29}$$

The complex number c_i is called the **amplitude** of that eigenvector. Let us compare $|\psi\rangle$ to one of the eigenkets, such as $|\phi_i\rangle$, as follows:

$$\langle\phi_i|\psi\rangle = \langle\phi_i| \ \ (c_1|\phi_1\rangle + c_2|\phi_2\rangle + c_3|\phi_3\rangle + \cdots + c_n|\phi_n\rangle) \ \ \rangle. \tag{10.30}$$

This amounts to

$$\langle\phi_i|\psi\rangle = c_1\langle\phi_i|\phi_1\rangle + c_2\langle\phi_i|\phi_2\rangle + c_3\langle\phi_i|\phi_3\rangle + \cdots + c_n\langle\phi_i|\phi_n\rangle. \tag{10.31}$$

Since the eigenbases is orthonormal, the only part that remains is

$$\langle\phi_i|\psi\rangle = c_i\langle\phi_i|\phi_i\rangle = c_i. \tag{10.32}$$

That is, when comparing any ket to a particular eigenket, we get the coefficient of that eigenket. This gives us a nice way of writing every ket:

$$|\psi\rangle = \langle\phi_1|\psi\rangle|\phi_1\rangle + \langle\phi_2|\psi\rangle|\phi_2\rangle + \langle\phi_3|\psi\rangle|\phi_3\rangle + \cdots + \langle\phi_n|\psi\rangle|\phi_n\rangle = \sum \langle\phi_i|\psi\rangle|\phi_i\rangle. \tag{10.33}$$

The central idea is that each ket can be expressed in different ways depending on how it is observed.

This idea, that different observables have different results, relates to one of the most famous concepts about quantum theory, called **wave-particle duality**. The notion comes from the oldest quantum observed phenomenon, namely, light. When we measure light going through the double slit experiment, we think of light as a wave. In contrast when light hits a certain type of metal, electrons are emitted from the metal. This happens only because light is acting like a particle and is absorbed by the electron. The higher energy allows the electron to escape the metal. So which is it? Is light a wave or a particle? The answer is that it depends how the light is being observed. The different ways of observing the light give us different phenomena.

What happens to a system after we observe some aspect of it?

Postulate 4. When an observation is made with a Hermitian operator $A\colon \mathcal{H} \longrightarrow \mathcal{H}$, the state of the system changes from $|\psi\rangle$ to $|\psi'\rangle$, which is a normalized projection onto the eigenspace of the observed eigenvector. ⊙

The system was in an superposition for some property and is now a position with respect to that property. This is called a **collapse of the wave function**. The new state will be like the old state, except that what is observed will change. If one does an observation with a Hermitian operator and then immediately observes it again withe

the same operator, the state will not change another time. This is because the second observation is looking at a state in an eigenstate already.

This brings to the forefront one of the deepest philosophical questions in quantum theory. Why should a ket that represents a superposition of states collapse into a nonsuperposition state when it is measured? What is it about measuring that changes it? This is called the **measurement problem**. In other words, why does the process of measuring change the physical state from a superposition to a position? Why don't we see objects in a superposition? There are several schools of thought with regard to the measurement problem. The current popular response to the measurement problem is the notion of **decoherence** (which is not easily related to the coherence theory of category theory). This says that a superposition collapses when the system interacts with another large measuring system that is not in a superposition. When a physical system that is in a superposition meets a measuring device, such as a human eye, then it joins that system and becomes a position. (There is some controversy as to whether decoherence actually solves the measurement problem.)

Two observables correspond to two Hermitian operators, which might have different eigenbases. That means when we use one observable on a state, we might get a different output than if we measure it with a different observable.

What happens if there are two or more observables? In detail, let us say that there are two observables that correspond to the Hermitian operators $A\colon \mathcal{H} \longrightarrow \mathcal{H}$ and $B\colon \mathcal{H} \longrightarrow \mathcal{H}$. We can choose to first measure the property that corresponds to A and then measure the property that corresponds to B. On the other hand, we can do it in the reverse order. Categorically, this gives us the square

$$\begin{array}{ccc} \mathcal{H} & \xrightarrow{A} & \mathcal{H} \\ {\scriptstyle B}\downarrow & & \downarrow{\scriptstyle B} \\ \mathcal{H} & \xrightarrow[A]{} & \mathcal{H}. \end{array} \tag{10.34}$$

This is written algebraically as

$$[A, B] = AB - BA = 0, \tag{10.35}$$

where [,] is called the **commutator**.

When this square commutes, we say that the observables **commute** or the observables are **compatible**. This corresponds to them having a **common eigenbasis**.

In contrast, if the diagram does not commute,[2] then we say that the observables are **incompatible**. This happens when the eigenbases are different. In that case, when you first measure with A and then measure with B, you will get the eigenbasis of B. In contrast, when you first measure with B and then with A, you will get a result in the eigenbasis of A. Niels Bohr called this phenomenon of having two incompatable observables **complementarity**. He pushed the notion that there is no objective value before a property is measured, and the value that it will eventually have depends on

[2]My thesis adviser, Alex Heller, used to say that the diagrams that do *not* commute are the interesting ones.

how it is measured. This dependency is called **contextuality**. The outcome depends on the context of the observable.

Certain incompatible observables—such as the position observable and the momentum observable—are called **complementary observables**. If their Hermitian operators are A and B, then

$$[A, B] = AB - BA = i\hbar. \tag{10.36}$$

This is a form of the **Heisenberg uncertainty principle**.

We saw how a measurement is made and what happens to the system after a measurement is made. Now we ask which of the possible outcomes will actually happen when a measurement is done. The answer is that the amplitudes of the state determine the probabilities of which state a ket will collapse into.

Postulate 5. Given an Hermitian operator $A: \mathcal{H} \longrightarrow \mathcal{H}$ whose eigenbases is

$$|\phi_1\rangle, |\phi_2\rangle, |\phi_3\rangle, \ldots, |\phi_n\rangle, \tag{10.37}$$

and a ket with the given amplitudes

$$|\psi\rangle = c_1|\phi_1\rangle + c_2|\phi_2\rangle + c_3|\phi_3\rangle + \cdots + c_n|\phi_n\rangle = \sum c_i|\phi_i\rangle, \tag{10.38}$$

the probability that it will collapse to the $|\phi_i\rangle$ state is given by the **Born rule**:

$$P(|\phi_i\rangle) = \frac{|c_i|^2}{|c_1|^2 + |c_2|^2 + |c_3|^2 + \cdots + |c_n|^2}. \tag{10.39}$$

If $|\psi\rangle$ is normalized, then $|c_1|^2 + |c_2|^2 + |c_3|^2 + \cdots + |c_n|^2 = 1$ and the Born rule shortens to

$$P(|\phi_i\rangle) = |c_i|^2. \tag{10.40}$$

⊙

Any two kets in in a ray describe the same state and hence give the same probabilities. Let us work this out. Consider two kets $|\psi\rangle$ and $c|\psi\rangle$. The later ket is

$$c|\psi\rangle \quad = \quad cc_1|\phi_1\rangle + cc_2|\phi_2\rangle + cc_3|\phi_3\rangle + \cdots + cc_n|\phi_n\rangle. \tag{10.41}$$

The probability that $c|\psi\rangle$ will collapse into the $|\phi_i\rangle$ state is

$$P(c|\phi\rangle) = \frac{|cc_i|^2}{|cc_1|^2 + |cc_2|^2 + |cc_3|^2 + \cdots + |cc_n|^2}, \tag{10.42}$$

which is

$$\frac{c|c_i|^2}{c|c_1|^2 + c|c_2|^2 + c|c_3|^2 + \cdots + c|c_n|^2} \tag{10.43}$$

or

$$\frac{c|c_i|^2}{c(|c_1|^2 + |c_2|^2 + |c_3|^2 + \cdots + |c_n|^2)}. \tag{10.44}$$

This is exactly Equation (10.39) once the c is canceled out.

There are those who are unhappy with the inherent probabilistic nature of quantum mechanics. They feel that the laws of physics should be deterministic and not probabilistic. Einstein famously disparaged the randomness of qunatum theory by saying that "God does not play dice." Such people believe that there are **hidden variables** that we do not know about. To them, if we took into account the hidden variables, then quantum theory would also be deterministic.

Dynamics

Quantum systems change. To deal with this, we have to add in the element of time. Rather than discussing the state of the system as $|\phi\rangle$, we have to talk about the state of the system at time t, which we denote as $|\phi(t)\rangle$. Now we can ask: How does the system change from $|\phi(t)\rangle$ to $|\phi(t')\rangle$, or how does it change from $|\phi(0)\rangle$ to $|\phi(t)\rangle$? We saw that in classical systems, reversible changes to a system are described by invertable maps $S \longrightarrow S$. For a quantum system, we have Postulate 6.

Postulate 6. A change of a quantum system (when it is not being measured or interacting with another system) corresponds to the unitary operator $U\colon \mathcal{H} \longrightarrow \mathcal{H}$. For every time t, there is a unitary operator $U_t\colon \mathcal{H} \longrightarrow \mathcal{H}$ such that

$$U_t|\phi(0)\rangle = |\phi(t)\rangle. \tag{10.45}$$

⊙

Remember that a unitary operator is a map U such that $UU^\dagger = U^\dagger U = id_{\mathcal{H}}$. Not only is a unitary map invertable, but it is easy to invert because the inverse is given by the dagger operation. This means that all changes described by a unitary operator can be easily undone, and the laws of quantum mechanics for such changes are reversible.

Some examples of unitary operators are rotation operators, which rotate parts of the system, and transition operators, which shift parts the system. We will see many more examples of unitary operators in the mini-course on quantum computing.

The fact that unitary operators preserve norms means that $|U|\phi\rangle| = ||\phi\rangle|$. This implies that the probabilities of the system do not change after applying the unitary operator.

There are rules as to how quantum systems change. There is a Hermitian operator called the **Hamiltonian**, denoted as $\hat{H}$. This is the observable of the total energy of the system. All changes to a quantum system have to satisfy the time-dependent **Schrödinger equation**:

$$i\hbar\frac{d}{dt}|\psi(t)\rangle = \hat{H}|\psi(t)\rangle. \tag{10.46}$$

Combined Systems

The universe consists of more than one quantum system. We must be able to construct larger systems from smaller systems.

Postulate 7. If there are two systems described by the Hilbert spaces $\mathcal{H}_1$ and $\mathcal{H}_2$, then the combined system is described by the Hilbert space $\mathcal{H}_1 \otimes \mathcal{H}_2$, where $\otimes$ is from the symmetric monoidal category structure of $\mathbb{H}\mathbb{i}\mathbb{l}\mathbb{b}$. ⊙

If $|\phi\rangle$ is a ket in $\mathcal{H}_1$ and $|\psi\rangle$ is a ket in $\mathcal{H}_2$, then there is a corresponding ket, written as

$$|\phi\rangle \otimes |\psi\rangle \qquad \text{or} \qquad |\phi \otimes \psi\rangle \qquad \text{or} \qquad |\phi\psi\rangle, \tag{10.47}$$

in $\mathcal{H}_1 \otimes \mathcal{H}_2$ that describes both their states.

Any two kets can be combined. Even superpositions can be combined. So if $|\phi\rangle$ is a superposition of $|\phi_0\rangle$ and $|\phi_1\rangle$ and $|\psi\rangle$ is a superposition of $|\psi_0\rangle$ and $|\psi_1\rangle$, then the tensor product is the superposition

$$|\phi, \psi\rangle = (|\phi_0\rangle + |\phi_1\rangle) \otimes (|\psi_0\rangle + |\psi_1\rangle) = \tag{10.48}$$

$$1|\phi_0\psi_0\rangle \quad + \quad 1|\phi_0\psi_1\rangle \quad + \quad 1|\phi_1\psi_0\rangle \quad + \quad 1|\phi_1\psi_1\rangle. \tag{10.49}$$

When this system is measured, any of the four possibilities are equally possible.

As we have seen, the category of Hilbert spaces is a symmetric monoidal category. This means that there is a twist map $\gamma\colon \mathcal{H} \otimes \mathcal{H}' \longrightarrow \mathcal{H}' \otimes \mathcal{H}$, which is defined on kets as $tw(|\phi\rangle \otimes |\psi\rangle) = |\psi\rangle \otimes |\phi\rangle$. This map is actually very important for particle physics. Suffice it to say that this map is at the center of the difference between matter particles called **fermions** and force-carrying particles called **bosons**.[3]

There is a need to generalize by combining more than two Hilbert spaces. There might be n systems described by n Hilbert spaces, $\mathcal{H}_1, \mathcal{H}_2, \ldots, \mathcal{H}_n$. The combined system is the Hilbert space $\mathcal{H}_1 \otimes \mathcal{H}_2 \otimes \cdots \otimes \mathcal{H}_n$. An example of this will occur when we discuss quantum computers later in this chapter. There, we will be interested in looking at n of the same systems that represent qubits (the quantum analog of bits), so if $\mathcal{H}$ is the Hilbert space that corresponds to one qubit, we construct the Hilbert space $\mathcal{H} \otimes \mathcal{H} \otimes \cdots \otimes \mathcal{H}$ (n times), which is also denoted as $\mathcal{H}^{\otimes n}$. This will correspond to n qubits.

There is a phenomenon called **entanglement**, which is related to combining systems and is shocking and counterintuitive. The easiest example of entanglement can be shown via the concept of spin for subatomic particles. Consider measuring particles in the $|\uparrow\rangle, |\downarrow\rangle$ basis. A single particle can be in a superposition $|\uparrow\rangle \ + \ |\downarrow\rangle$. Given two particles, where each is in a superposition, we have

$$(|\uparrow\rangle \ + \ |\downarrow\rangle) \otimes (|\uparrow\rangle \ + \ |\downarrow\rangle). \tag{10.50}$$

This describes four possible states of the two particles:

$$1|\uparrow\downarrow\rangle \quad + \quad 1|\uparrow\uparrow\rangle \quad + \quad 1|\downarrow\downarrow\rangle \quad + \quad 1|\downarrow\uparrow\rangle. \tag{10.51}$$

In the usual case, when we measure one particle, there are still two possibilities for the other particle.

---

[3]The Higgs boson is technically not a force-carrying boson.

Now let us look at a more interesting case. There are certain particles that do not have any spin at all, but these particles can decay into two particles that do have spin. The universe is governed by conservation rules. Just as there is a conservation of matter rule and a conservation of energy rule, there is also a conservation of spin rule. This means that the amount of spin in a system cannot change. So if a system starts off with no spin, then the amount of up spin must be the same as for the down spin. When a no-spin particle splits up, each of the daughter particles is in a superposition of spinning up and down at the same time. When one particle is measured, that particle collapses to either $|\uparrow\rangle$ or $|\downarrow\rangle$. The other particle must also automatically collapse the other way to preserve conservation of spin. This means that after measuring the spin on one of the particles, the state of the system can either be $|\uparrow\downarrow\rangle$ or $|\downarrow\uparrow\rangle$. After measurement, the system cannot be in the state $|\uparrow\uparrow\rangle$, nor can it be in the state $|\downarrow\downarrow\rangle$. Another way to say this is that before measurement, the state of the two particles is

$$1|\uparrow\downarrow\rangle \quad + \quad 0|\uparrow\uparrow\rangle \quad + \quad 0|\downarrow\downarrow\rangle \quad + \quad 1|\downarrow\uparrow\rangle \qquad = \qquad 1|\uparrow\downarrow\rangle \quad + \quad 1|\downarrow\uparrow\rangle. \tag{10.52}$$

These two particles are said to be entangled. Measuring one affects not only that particle, but also the other particle. The most amazing part of entanglement is that the two particles can be separated across the universe. Even though they are far away from each other, they are still considered a single system, and when one particle is measured, the universe ensures that the other particle will collapse to a spin in the opposite direction. This feature of one measurement instantly affecting a particle far away is called **nonlocality**, which Albert Einstein disparagingly called "spooky action at a distance."

Another way to describe entangled states is to compare them to unentangled states. Line (10.51) is unentangled and is the tensor of the states given in Line (10.50). In stark contrast, the entangled state described by Line (10.52) is not a tensor of two states. That is, Line (10.52) is a description of a two-particle system that cannot be separated into two systems. In terms of categories, entanglement means that there are more maps $\mathbf{C} \longrightarrow \mathcal{H}_1 \otimes \mathcal{H}_2$ than pairs of maps $\mathbf{C} \longrightarrow \mathcal{H}_1$ and $\mathbf{C} \longrightarrow \mathcal{H}_2$. Yet another way to say it is $\mathcal{H}_1 \otimes \mathcal{H}_2$, which contains entangled and unentangled states, is much larger than $\mathcal{H}_1 \times \mathcal{H}_2$ which only has unentangled states.

EPR Paradox. Entanglement plays a major role in the Einstein–Podolsky–Rosen (EPR)) paradox. In 1935 Albert Einstein, Boris Podolsky, and Nathan Rosen came up with a thought experiment that shows the consequence of entanglement. Imagine the two entangled particles going toward Ann and Bob, who are far apart. The particles are entangled as the superposition

$$|\uparrow\downarrow\rangle \; + \; |\downarrow\uparrow\rangle \qquad \text{and} \qquad |\nearrow\swarrow\rangle \; - \; |\swarrow\nearrow\rangle.^4 \tag{10.53}$$

Ann and Bob can each measure the spins of the particles in these two incompatible observables:

$$|\uparrow\rangle|\downarrow\rangle \qquad \text{and} \qquad |\nearrow\rangle|\swarrow\rangle. \tag{10.54}$$

[4] The minus is for technical reasons. The superposition is essentially the same as if there were a plus.

EPR argues that if one can determine the value of of an observable for a particle without disturbing the particle, then that value must be what it calls an "element of reality." Now Ann can measure in the up-down basis, and whichever direction she gets, she knows that the spin of Bob's particle will be in the opposite direction. So the up-down direction of Bob's particle must be an element of reality because Ann can determine it, and she does not touch Bob's particle. On the other hand, Ann can measure her spin in the diagonal direction, and, again, whichever direction she finds, Bob will find the opposite. She also will not have to touch Bob's particle to do this. That means that the diagonal direction of Bob's particle must be an element of reality. However, according to quantum mechanics, these observables cannot have well-defined values because the operators corresponding to them do not commute. Therefore, according to EPR, there are elements of reality that cannot be described by quantum mechanics, and, hence, quantum mechanics does not present a complete picture of the world.

Bell's Theorem. In the 1960s, John Bell took the EPR paradox one step further. He looked at two entangled particles. However, rather than measuring the two particles in two directions, he added the third direction:

$$|\leftarrow\rangle, |\rightarrow\rangle. \tag{10.55}$$

(See pages 196–201 of [272] for a readable account of all the details.) He was able to definitively show that if there are hidden variables in quantum mechanics, then the hidden variables are nonlocal. This means that regardless of the question of hidden variables, quantum mechanical phenomena are nonlocal. When a particle is measured here, the other particle far away will collapse from a superposition to a position. Although Bell proved his theorem using quantum theory and never mentioned experiments to show that it is true, in the 1980s, several researchers started working on experiments to prove Bell's theorem. For this work, some of the researchers were awarded the 2022 Nobel Prize in Physics.

Schrödinger's Cat. Another important consequence of entanglement is the Schrödinger's cat thought experiment. It is so famous that it does not need to be repeated here. Suffice it to say that the system of a subatomic particle is combined and entangled with the seemingly classical system of a cat in the box. The main point of the experiment is that the strangeness of subatomic particle superposition is intimately related to the vitality of the cat. The consequence of this is to firmly demonstrate that there really is no distinction between quantum systems and classical systems. All systems are quantum.

Further Directions in Quantum Theory

How is one to make sense of quantum theory? While no one argues with the math and the results of the experiments in quantum theory, what they mean is subject to much debate. There are many schools of thought as to how to interpret quantum theory, including the following:

- **Copenhagen interpretation**: This was a school started by Niels Bohr and was dominant for a long time. Our presentation mostly followed this school of thought.
- **Bohmian mechanics**: This school employs hidden variables and hence assumes quantum mechanics is deterministic,[5] not probabilistic. A criticism that was leveled against this theory is that it is extremely nonlocal. Now that Bell showed us that quantum mechanics is not local, this criticism is without merit. Bohmian mechanics was founded by David Bohm.
- **Multiverse interpretation**: This is a radical way to solve the measurement problem. It says that every time there is a measurement, the entire universe splits into many copies. Each copy has a different outcome. This school was founded by Hugh Everett III. A more modern interpretation in line with this school is that there are many universes, and there is exactly one gigantic wave function that describes the entire multiverse. When a measurement is taken, different parts of the single wave function collapse. See [54] for a clear presentation. A popular science book on this is [53].
- **Relational quantum theory**. This is a recent school promoted by Carlo Rovelli. The central idea is there are many different quantum systems, and they are all interacting with each other. One system has the observer, and another system has the observed.
- **Quantum baysianism**. This is a way of interpreting quantum theory in a more subjective sense. The wave function represents what an observer believes. This changes with more information. These ideas were recently developed, and its main proponents are Christopher Fuchs and Rüdiger Schack.
- **Ghirardi–Rimini–Weber (GRW) theory**. This interpretation, named after Giancarlo Ghirardi, Alberto Rimini, and Tullio Weber, replaces the collapse of the wave function upon measurement with a principle of spontaneous collapse as a fundamental property of the wave function. This additional postulate solves the measurement problem and explains the lack of macroscopic quantum phenomena, but it has been criticized for being ad hoc.

It is important to note that none of these schools of thought is scientifically right or wrong. This means that there are no scientific experiments that show that one school has the correct view and the others do not. Rather, all these interpretations have different views that agree and disagree with certain philosophical viewpoints. See [107] for a nice introduction and in-depth study of the various ways of interpreting quantum theory.

Quantum theory is more than a century old, and there are many branches of it. Here we just list some of the branches. **Quantum field theory (QFT)** describes ways that quantum mechanics extends from dealing with a finite number of particles to dealing with infinite fields of objects. There are many different types of fields, and hence many different types of quantum field theories:

[5] There are different schools of Bohmian mechanics with different beliefs. For more, see [153].

- **Quantum electrodynamics (QED)** is about the interactions between electrically charged particles and the electromagnetic field.
- **Quantum chromodynamics (QCD)** is concerned with the interactions between quarks and gluons, which are the building blocks of many subatomic particles.
- **Quantum gravity** attempts to unify quantum theory with gravity. There are many theories about such a unification.
- **Topological quantum field theory (TQFT)** is a way of dealing with the relationship of topological structures with quantum theory. This is usually discussed categorically as functors from categories of topological structures called **cobordisms** to categories of Hilbert space structures. See [25] for a beautiful categorical introduction.
- **Conformal field theory (CFT)** is a type of TQFT that studies systems that are invariant under morphisms called conformal transformations. String theory is part of CFT.

Two other areas of quantum theory that are very current are **quantum computing** and **quantum information theory**, which are discussed in the next mini-course.

Suggestions for Further Study

For an easy popular science introduction to quantum theory, may I humbly suggest Section 7.2 of my [272]. Two other popular science books for quantum theory are [199] and [107]. For a short history of the early development of quantum theory, see [84].

There are many introductory textbooks. From easier to harder, consider the following [60, 88, 262, 191, 213]. See [250] for a vector space approach. For a more algebraic approach, see [101] and [152]. Modern textbooks include [230], [72] or [152]. Special mention must be made of the classic introduction by Paul A. M. Dirac [64]. Even though it was first published in the 1930s, this book is still well worth reading.

There are many texts dedicated to understanding quantum theory from the perspective of category theory, including [110, 61].

10.3 Mini-Course: Quantum Computing

Quantum computing is a field that employs the features of quantum mechanics to make computers and information systems more efficient. This new field shows how one can use phenomena like superposition and entanglement to create computers that go faster and to improve communication.

Here is the intuition of how superposition can help. Consider the task of looking for an entry in an array that is not ordered in any way. The only way to do this with a classical computer is to go through every entry in the array one at a time. Now imagine a quantum computer that permits one to look at *all* the entries in the array at the same time. With such a superposition, the task will be completed much more quickly. The real story of quantum algorithms is not as simple as this, but it is a nice intuition.

Entanglement is used to connect distant objects and permit better communication and a type of teleportation.

There are many parts of quantum computing such as cryptography, teleportation, quantum theoretical computer science, and quantum information theory. In this mini-course, we will focus on quantum algorithms. These are algorithms that work on quantum computers, and they demand fewer operations than classical algorithms working on regular computers. This speed-up will have tremendous impact on the world we live in.

Quantum algorithms are described by morphisms in the category of complex vector space, $\mathbb{Hilb}$. For our purposes, we only need to know that $\mathbb{Hilb}$ is a monoidal category. This category has much more structure, which arises in other parts of quantum computing. At the end of the mini-course, there is a short discussion about an important result in quantum information theory called the "no-cloning theorem." We show the categorical reason for this result.

Parts of this exposition were shamelessly stolen and modified from Chapters 5 and 6 of [276].

Bits and Qubits

Quantum computers can be seen as a vast generalization of classical computers. Let us back up for a moment and look at how those familiar machines work. Classical computers are based on the notion of a **bit**, which is a system with two clearly defined states. For example, a switch can be up or down; a light can be on or off; we can write "true" or "false"; and a wire can have electricity go through it or not. We describe a bit as a two-dimensional Boolean matrix and write the two states as the following kets:

$$|0\rangle = \begin{matrix} 0 \\ 1 \end{matrix} \begin{bmatrix} 1 \\ 0 \end{bmatrix} \quad \text{and} \quad |1\rangle = \begin{matrix} 0 \\ 1 \end{matrix} \begin{bmatrix} 0 \\ 1 \end{bmatrix}. \tag{10.56}$$

To make things easier, sometimes we put the numbers of the rows outside the matrices.

More than one bit is usually needed to describe the state of a computer. Eight bits together are called a **byte**. A typical byte would look like this: 01101011. In terms of kets, we can write this as a tensor product of 8 bits:

$$|0\rangle \otimes |1\rangle \otimes |1\rangle \otimes |0\rangle \otimes |1\rangle \otimes |0\rangle \otimes |1\rangle \otimes |1\rangle. \tag{10.57}$$

In terms of matrices, this looks like

$$\begin{bmatrix} 1 \\ 0 \end{bmatrix}, \begin{bmatrix} 0 \\ 1 \end{bmatrix}, \begin{bmatrix} 0 \\ 1 \end{bmatrix}, \begin{bmatrix} 1 \\ 0 \end{bmatrix}, \begin{bmatrix} 0 \\ 1 \end{bmatrix}, \begin{bmatrix} 1 \\ 0 \end{bmatrix}, \begin{bmatrix} 0 \\ 1 \end{bmatrix}, \begin{bmatrix} 0 \\ 1 \end{bmatrix}. \tag{10.58}$$

We can also represent this byte as a $2^8 = 256$–row vector with a single 1 in the correct row:

$$\begin{array}{c} 00000000 \\ 00000001 \\ \vdots \\ 01101010 \\ 01101011 \\ 01101100 \\ \vdots \\ 11111110 \\ 11111111 \end{array} \begin{bmatrix} 0 \\ 0 \\ \vdots \\ 0 \\ 1 \\ 0 \\ \vdots \\ 0 \\ 0 \end{bmatrix} \tag{10.59}$$

Operations on the states of computers will be represented by matrices. That is, a matrix with Boolean entries is multiplied with representations of bits to change the bits. This will be the dynamics of the computers.

Qubits

Bits are great for classical computers. Now we have to generalize this idea for quantum computers. Within a quantum computer, a state is represented by a system that can be in a superposition of $|0\rangle$ and $|1\rangle$. Such a generalization is a **quantum bit** or **qubit**.

Qubits are implemented in several ways. Here are some examples:

- An electron might be in one of two orbits around the nucleus of an atom (ground state and excited state).
- A photon might be in one of two polarized states.
- A subatomic particle might have one of two spin directions.

In each of these cases, there is enough quantum indeterminacy and quantum superposition effects to represent a qubit. To date, the engineers have not come to any firm conclusions as to what works for large-scale quantum computers.

The qubits are in a state that is s superposition. When we measure the qubit, it will become either bit $|0\rangle$ or bit $|1\rangle$. We will use two complex numbers to describe how the qubit will act when measured. Formally, we represent a qubit as a 2-by-1 matrix with complex numbers:

$$\begin{array}{c} 0 \\ 1 \end{array} \begin{bmatrix} c_0 \\ c_1 \end{bmatrix}. \tag{10.60}$$

In other words, a qubit is a element of the complex vector space $\mathbf{C} \times \mathbf{C} = \mathbf{C}^2$. When we measure such a qubit, there is a $\frac{|c_0|^2}{|c_0|^2 + |c_1|^2}$ chance of it becoming $|0\rangle$ and a $\frac{|c_1|^2}{|c_0|^2 + |c_1|^2}$ chance of it becoming $|1\rangle$.[6]

[6] A categorical way of understanding qubits is as a Yoneda type generalization of a bit. In detail, the set of bits is $\{0, 1\}$. There is an embedding of $\{0, 1\}$ into $\mathbf{C}^{\{0,1\}}$. That is, we look at all functions from $\{0, 1\}$ into the complex numbers. The bit 0 is represented by the function that takes 0 to 1 and 1 to 0. The bit 1! is represented by the function that takes 0 to 0 and 1 to 1. The main idea is that the set $\mathbf{C}^{\{0,1\}}$ has a lot more structure than $\{0, 1\}$. It is a finite-dimensional Hilbert space.

Bits $|0\rangle$ and $|1\rangle$ should be seen as the canonical basis of $\mathbf{C}^2$. Thus, any qubit can be written as

$$\begin{bmatrix} c_0 \\ c_1 \end{bmatrix} = c_0 \cdot \begin{bmatrix} 1 \\ 0 \end{bmatrix} + c_1 \cdot \begin{bmatrix} 0 \\ 1 \end{bmatrix} = c_0|0\rangle + c_1|1\rangle. \tag{10.61}$$

Let us look at several ways of denoting different qubits.

Example 10.3.1.

- $\frac{1}{\sqrt{2}}\begin{bmatrix} 1 \\ 1 \end{bmatrix}$ can be written as $\begin{bmatrix} \frac{1}{\sqrt{2}} \\ \frac{1}{\sqrt{2}} \end{bmatrix} = \frac{1}{\sqrt{2}}|0\rangle + \frac{1}{\sqrt{2}}|1\rangle = \frac{|0\rangle + |1\rangle}{\sqrt{2}}$.

- $\frac{1}{\sqrt{2}}\begin{bmatrix} 1 \\ -1 \end{bmatrix}$ can be written as $\begin{bmatrix} \frac{1}{\sqrt{2}} \\ \frac{-1}{\sqrt{2}} \end{bmatrix} = \frac{1}{\sqrt{2}}|0\rangle - \frac{1}{\sqrt{2}}|1\rangle = \frac{|0\rangle - |1\rangle}{\sqrt{2}}$.

Both of these qubits have an equal chance of becoming $|0\rangle$ and $|1\rangle$

It is important to realize that

$$\frac{|0\rangle + |1\rangle}{\sqrt{2}} = \frac{|1\rangle + |0\rangle}{\sqrt{2}}. \tag{10.62}$$

These are both ways of denoting $\begin{bmatrix} \frac{1}{\sqrt{2}} \\ \frac{1}{\sqrt{2}} \end{bmatrix}$. In contrast,

$$\frac{|0\rangle - |1\rangle}{\sqrt{2}} \neq \frac{|1\rangle - |0\rangle}{\sqrt{2}}. \tag{10.63}$$

The left state is the vector $\begin{bmatrix} \frac{1}{\sqrt{2}} \\ \frac{-1}{\sqrt{2}} \end{bmatrix}$ and the right state is the vector $\begin{bmatrix} \frac{-1}{\sqrt{2}} \\ \frac{1}{\sqrt{2}} \end{bmatrix}$. However, the two kets are related as follows:

$$\frac{|0\rangle - |1\rangle}{\sqrt{2}} = (-1)\frac{|1\rangle - |0\rangle}{\sqrt{2}} \tag{10.64}$$

and represent the same state. This will be important in the coming algorithms. □

More than one qubit is needed. The complex vector space that deals with 2 qubits is $\mathbf{C}^2 \otimes \mathbf{C}^2 = \mathbf{C}^4$. A typical element might look like this:

$$\begin{array}{c} 0 \\ 1 \end{array}\begin{bmatrix} c_0 \\ c_1 \end{bmatrix} \otimes \begin{array}{c} 0 \\ 1 \end{array}\begin{bmatrix} c'_0 \\ c'_1 \end{bmatrix} \quad \text{or} \quad \begin{array}{c} 00 \\ 01 \\ 10 \\ 11 \end{array}\begin{bmatrix} c_{00} \\ c_{01} \\ c_{10} \\ c_{11} \end{bmatrix}. \tag{10.65}$$

Qubits can be written in many ways. When the first qubit is in state $|0\rangle$ and the second qubit is in state $|1\rangle$, we write it as

$$|0\rangle \otimes |1\rangle = |0 \otimes 1\rangle = |0, 1\rangle = \begin{array}{c} 00 \\ 01 \\ 10 \\ 11 \end{array}\begin{bmatrix} 0 \\ 1 \\ 0 \\ 0 \end{bmatrix}. \tag{10.66}$$

In stark contrast to Equation (10.62), we have the following inequality:

$$|0\rangle \otimes |1\rangle \neq |1\rangle \otimes |0\rangle. \tag{10.67}$$

The kets on the left have the first qubit in the zero state and the second qubit in the one state. The kets on the right have the first qubit in the one state and the second qubit in the zero state.

We can see entanglement from the qubit point of view. Consider a system where the states are described by these kets:

$$\frac{|11\rangle + |00\rangle}{\sqrt{2}} = \frac{1}{\sqrt{2}}|11\rangle + \frac{1}{\sqrt{2}}|00\rangle. \tag{10.68}$$

This means that the two qubits are entangled. That is, if we measure the first qubit and it is found in state $|1\rangle$, then we automatically know that the state of the second qubit is $|1\rangle$. Similarly, if we measure the first qubit and find it in state $|0\rangle$, then we know that the second qubit is also in state $|0\rangle$.[7]

When there are eight qubits, we have a **qubyte**. Such a state can be written as

$$\begin{array}{c} 00000000 \\ 00000001 \\ \vdots \\ 01101010 \\ 01101011 \\ 01101100 \\ \vdots \\ 11111110 \\ 11111111 \end{array}\begin{bmatrix} c_0 \\ c_1 \\ \vdots \\ c_{106} \\ c_{107} \\ c_{108} \\ \vdots \\ c_{254} \\ c_{255} \end{bmatrix} \tag{10.69}$$

[7]One can continue the past footnote and consider many qubits as the Hilbert space from many bits. Consider the set of n bits as $\{0, 1\}^{\times n}$. With this in mind, we can consider the Hilbert space of n qubits as

$$(\mathbf{C}^{\{0,1\}})^{\otimes n} \cong \mathbf{C}^{\{0,1\}^{\times n}}.$$

The Born rule tells us that when we measure such a qubyte, the chances of the system ending up in the state where there is a 1 in the ith row and 0's everywhere else is

$$\frac{|c_i|^2}{\sum_{j=0}^{255} |c_j|^2}.$$

From here, we can see why quantum computers are so special and cannot be mimicked by classical computers. To describe the state of 8 bits, we need eight Boolean digits. In contrast, to describe the state of 8 qubits, we need $2^8 = 256$ complex numbers. For very added qubit, we need to double the number of complex numbers. If we wanted to mimic a quantum computer with 64 qubits, we would need 18,446,744,073,709,551,616 complex numbers. This is beyond our resources. It was this exponential growth that was the impetus of Richard Feynman to start thinking of quantum computers. He wanted to model some quantum systems and realized that the exponential amount of resources needed is too much for a classical computer. Only a quantum computer has the ability to model quantum systems.

Classical Gates and Quantum Gates

We already encountered the usual logical gates (AND, OR, NOT, NAND, and NOR) in Example 2.1.57. We saw that logical gates combine to form circuits. The collection of circuits form the category $\mathbb{Circuit}$. In Example 4.1.58, we described a functor $MatrixDesc\colon \mathbb{Circuit} \longrightarrow Bool\mathbb{Mat}^{op}$, which describes every circuit as a matrix. In Example 6.1.17, we saw that the functor *MatrixDesc* respects parallel processing.

We will travel the same way for quantum gates. We will show that for every quantum gate, there is a corresponding unitary matrix. These quantum gates combine to form quantum circuits. The circuits and matrices cohere with each other.

Reversible Gates

Before we move from classical gates to quantum gates, there is an intermediate level called **reversible gates**. These are gates that are like classical gates because they do not use quantum effects, but they share a property of quantum gates in that they are reversible (i.e., their operation can be undone).

To understand reversible gates, we have to realize that most logical gates are not reversible. For example, if you look at the AND gate and we see that the output is 0, we do not know if the input was $(0,0)$, or $(0,1)$, or $(1,0)$. Similarly with the OR gate, if we get an output of 1, we do not know if the input was $(0,1)$, $(1,0)$, or $(1,1)$. In both cases, information was lost, and we cannot use the output to figure out what the input was.

In contrast, there is a classical gate that is reversible. If the **NOT gate** outputs 1, we know the input is 0. On the other hand, if the output of the NOT gate is 0, then we know the input is 1. We can use the output to get back the original input. The NOT gate is its own inverse. That is, if you use the NOT gate on the output of the NOT gate, you get back to the input. In terms of morphisms, this is

$$Bool \xrightarrow{NOT} Bool \xrightarrow{NOT} Bool. \quad (\text{composite} = id) \tag{10.70}$$

In terms of matrices, we can write this as

$$NOT \cdot NOT = \begin{bmatrix} 0 & 1 \\ 1 & 0 \end{bmatrix} \cdot \begin{bmatrix} 0 & 1 \\ 1 & 0 \end{bmatrix} = \begin{bmatrix} 1 & 0 \\ 0 & 1 \end{bmatrix} = id. \tag{10.71}$$

Let us introduce some more reversible gates. The **controlled-NOT gate** or the **CNOT** gate takes two Boolean variables and outputs two Boolean variable (i.e., $CNOT \colon Bool^2 \longrightarrow Bool^2$). It can be drawn as follows:

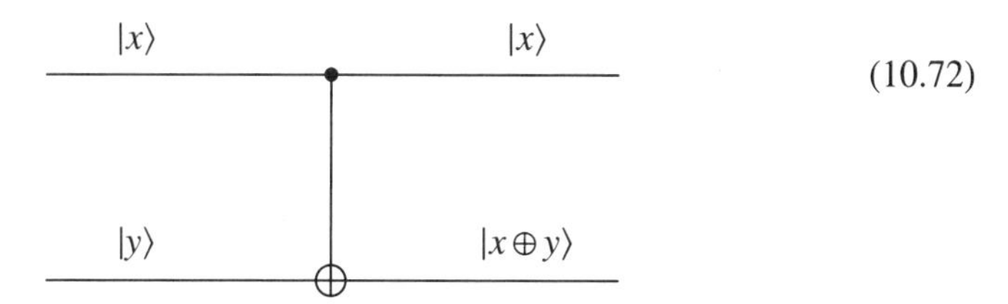

(10.72)

The top input is the control bit. It controls what the output will be. If $|x\rangle = |0\rangle$, then the bottom output of $|y\rangle$ is the same as the input. If $|x\rangle = |1\rangle$, then the bottom output will be the opposite. If we write the top qubit first and then the bottom qubit, then the controlled-NOT gate takes $|x, y\rangle$ to $|x, x \oplus y\rangle$ where $\oplus$ is the binary exclusive or (XOR) operation.

The matrix that corresponds to this reversible gate is

$$\begin{array}{c} \\ 00 \\ 01 \\ 10 \\ 11 \end{array} \begin{array}{c} \begin{array}{cccc} 00 & 01 & 10 & 11 \end{array} \\ \begin{bmatrix} 1 & 0 & 0 & 0 \\ 0 & 1 & 0 & 0 \\ 0 & 0 & 0 & 1 \\ 0 & 0 & 1 & 0 \end{bmatrix} \end{array}. \tag{10.73}$$

The controlled-NOT gate can be reversed by itself; that is,

$$Bool^2 \xrightarrow{CNOT} Bool^2 \xrightarrow{CNOT} Bool^2. \quad (\text{composite} = id) \tag{10.74}$$

We can see this as the circuit:

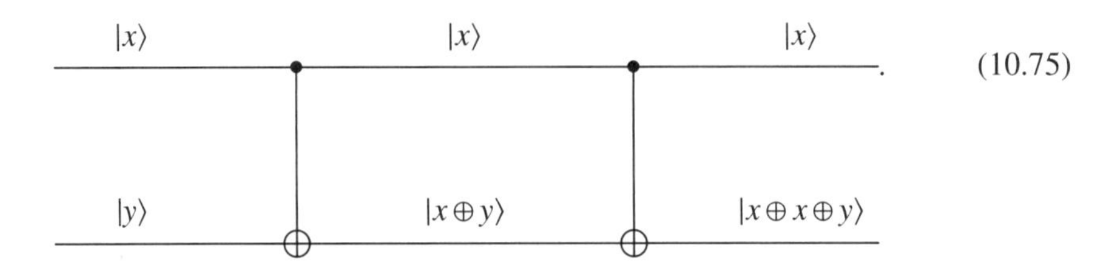

(10.75)

State $|x, y\rangle$ goes to $|x, x \oplus y\rangle$, which further goes to $|x, x \oplus (x \oplus y)\rangle$. This last state is equal to $|x, (x \oplus x) \oplus y\rangle$ because $\oplus$ is associative. Since $x \oplus x$ is always equal to 0, this state reduces to the original $|x, y\rangle$.

Another interesting reversible gate is the **Toffoli gate**, which operates on three Boolean variables:

$$TOFF: Bool^3 \longrightarrow Bool^3. \tag{10.76}$$

The circuit is depicted as

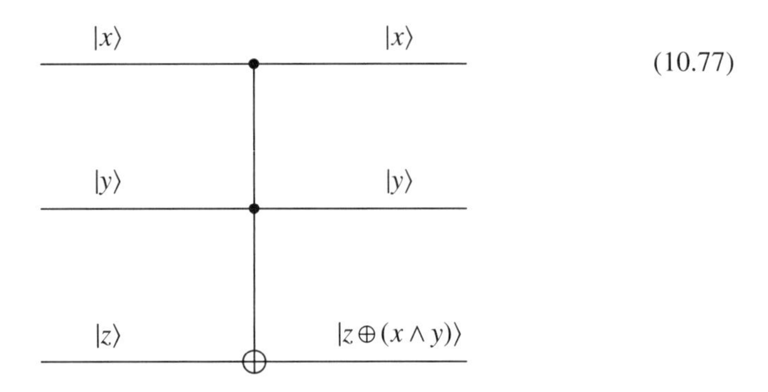

(10.77)

This is similar to the controlled-NOT gate, but with two controlling bits. The bottom bit flips only when *both* of the top two bits are in state $|1\rangle$. We can write this operation as taking state $|x, y, z\rangle$ to $|x, y, (z \oplus (x \wedge y))\rangle$.

The matrix that corresponds to the Toffoli gate is

$$\begin{array}{c} \\ 000 \\ 001 \\ 010 \\ 011 \\ 100 \\ 101 \\ 110 \\ 111 \end{array} \begin{array}{c} \begin{array}{cccccccc} 000 & 001 & 010 & 011 & 100 & 101 & 110 & 111 \end{array} \\ \left[\begin{array}{cccccccc} 1 & 0 & 0 & 0 & 0 & 0 & 0 & 0 \\ 0 & 1 & 0 & 0 & 0 & 0 & 0 & 0 \\ 0 & 0 & 1 & 0 & 0 & 0 & 0 & 0 \\ 0 & 0 & 0 & 1 & 0 & 0 & 0 & 0 \\ 0 & 0 & 0 & 0 & 1 & 0 & 0 & 0 \\ 0 & 0 & 0 & 0 & 0 & 1 & 0 & 0 \\ 0 & 0 & 0 & 0 & 0 & 0 & 0 & 1 \\ 0 & 0 & 0 & 0 & 0 & 0 & 1 & 0 \end{array} \right] \end{array}. \tag{10.78}$$

One reason why the Toffoli gate is interesting is that it is universal. In other words, one can form any logical gate with copies of the Toffoli gate. (We already saw that any logical circuit can be formed with NAND gates.) This means that one can make a reversible computer using only Toffoli gates. Such a computer, in theory, would neither use any energy nor give off any heat.

Exercise 10.3.2. Show that the Toffoli gate is its own inverse. ■

Another interesting reversible gate is the **Fredkin gate**. This gate also operates on three Boolean variables; that is:

$$FRED\colon Bool^3 \longrightarrow Bool^3. \tag{10.79}$$

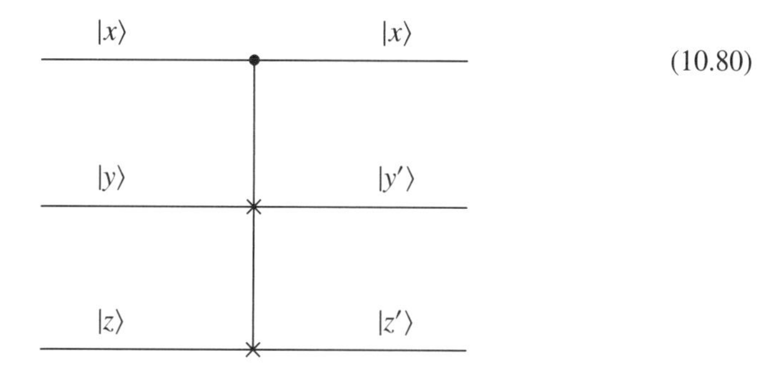

(10.80)

The top $|x\rangle$ input is the control input. The top output is always the same $|x\rangle$. If $|x\rangle$ is set to $|0\rangle$, then $|y'\rangle = |y\rangle$ and $|z'\rangle = |z\rangle$ (i.e., the values stay the same). If, on the other hand, the control $|x\rangle$ is set to $|1\rangle$, then the outputs are reversed: $|y'\rangle = |z\rangle$ and $|z'\rangle = |y\rangle$. In short, $|0, y, z\rangle \longmapsto |0, y, z\rangle$ and $|1, y, z\rangle \longmapsto |1, z, y\rangle$.

Exercise 10.3.3. Show that the Fredkin gate is its own inverse. ■

The matrix that corresponds to the Fredkin gate is

$$\begin{array}{c} \\ 000 \\ 001 \\ 010 \\ 011 \\ 100 \\ 101 \\ 110 \\ 111 \end{array}
\begin{array}{c}
\begin{array}{cccccccc} 000 & 001 & 010 & 011 & 100 & 101 & 110 & 111 \end{array} \\
\left[\begin{array}{cccccccc}
1 & 0 & 0 & 0 & 0 & 0 & 0 & 0 \\
0 & 1 & 0 & 0 & 0 & 0 & 0 & 0 \\
0 & 0 & 1 & 0 & 0 & 0 & 0 & 0 \\
0 & 0 & 0 & 1 & 0 & 0 & 0 & 0 \\
0 & 0 & 0 & 0 & 1 & 0 & 0 & 0 \\
0 & 0 & 0 & 0 & 0 & 0 & 1 & 0 \\
0 & 0 & 0 & 0 & 0 & 1 & 0 & 0 \\
0 & 0 & 0 & 0 & 0 & 0 & 0 & 1
\end{array}\right]
\end{array}. \tag{10.81}$$

The Fredkin gate is also universal and can be used to make reversible computers that do not use energy.

Quantum Gates

Quantum gates are unitary operators that act on qubits. They correspond to unitary matrices.

First, here are some examples. The following three matrices are called **Pauli matrices** and are very important because they arise everywhere in quantum mechanics and quantum computing:

$$X = \begin{bmatrix} 0 & 1 \\ 1 & 0 \end{bmatrix}; \quad Y = \begin{bmatrix} 0 & -i \\ i & 0 \end{bmatrix}; \quad Z = \begin{bmatrix} 1 & 0 \\ 0 & -1 \end{bmatrix}. \tag{10.82}$$

Note that the X matrix is our NOT matrix.

Another quantum gate is the **Hadamard gate**. The matrix associated to this operator is

$$H = \frac{1}{\sqrt{2}} \begin{array}{c} \\ 0 \\ 1 \end{array} \begin{array}{c} \begin{array}{cc} 0 & 1 \end{array} \\ \begin{bmatrix} 1 & 1 \\ 1 & -1 \end{bmatrix} \end{array}. \tag{10.83}$$

This matrix is fundamental for quantum algorithms because it places qubits into a superposition. In detail,

$$H|0\rangle = \begin{bmatrix} \frac{1}{\sqrt{2}} & \frac{1}{\sqrt{2}} \\ \frac{1}{\sqrt{2}} & -\frac{1}{\sqrt{2}} \end{bmatrix} \begin{bmatrix} 1 \\ 0 \end{bmatrix} = \begin{bmatrix} \frac{1}{\sqrt{2}} \\ \frac{1}{\sqrt{2}} \end{bmatrix} \tag{10.84}$$

and

$$H|1\rangle = \begin{bmatrix} \frac{1}{\sqrt{2}} & \frac{1}{\sqrt{2}} \\ \frac{1}{\sqrt{2}} & -\frac{1}{\sqrt{2}} \end{bmatrix} \begin{bmatrix} 0 \\ 1 \end{bmatrix} = \begin{bmatrix} \frac{1}{\sqrt{2}} \\ -\frac{1}{\sqrt{2}} \end{bmatrix}. \tag{10.85}$$

When you measure either of these states, the system will be found to be $|0\rangle$ 50 percent of the time and $|1\rangle$ 50 percent of the time.

It is helpful to see the Hadamard matrix as being defined as follows:

$$H = \frac{1}{\sqrt{2}} \begin{array}{c} \\ 0 \\ 1 \end{array} \begin{array}{c} \begin{array}{cc} 0 & \quad 1 \end{array} \\ \begin{bmatrix} (-1)^{0\wedge 0} & (-1)^{0\wedge 1} \\ (-1)^{1\wedge 0} & (-1)^{1\wedge 1} \end{bmatrix} \end{array}. \tag{10.86}$$

Notice that we are thinking of 0 and 1 as both Boolean values and numbers that are exponents. (Remember that $(-1)^0 = 1$ and $(-1)^1 = -1$.)

We do not only need to place 1 qubit in a superposition. We need to put many qubits into a superposition. Hence, we need to calculate $H \otimes H$. We can calculate it as follows:

$$H^{\otimes 2} = H \otimes H = \frac{1}{\sqrt{2}} \begin{array}{c} \\ 0 \\ 1 \end{array} \begin{array}{c} \begin{array}{cc} 0 & \quad 1 \end{array} \\ \begin{bmatrix} (-1)^{0\wedge 0} & (-1)^{0\wedge 1} \\ (-1)^{1\wedge 0} & (-1)^{1\wedge 1} \end{bmatrix} \end{array} \otimes \frac{1}{\sqrt{2}} \begin{array}{c} \\ 0 \\ 1 \end{array} \begin{array}{c} \begin{array}{cc} 0 & \quad 1 \end{array} \\ \begin{bmatrix} (-1)^{0\wedge 0} & (-1)^{0\wedge 1} \\ (-1)^{1\wedge 0} & (-1)^{1\wedge 1} \end{bmatrix} \end{array} \tag{10.87}$$

$$
=
$$

$$
\frac{1}{\sqrt{2}} * \frac{1}{\sqrt{2}}
\begin{array}{c}
\\ 00 \\ 01 \\ 10 \\ 11
\end{array}
\begin{array}{c}
\begin{array}{cccc} \quad 00 \quad & \quad 01 \quad & \quad 10 \quad & \quad 11 \quad \end{array} \\
\begin{bmatrix}
(-1)^{0\wedge 0} * (-1)^{0\wedge 0} & (-1)^{0\wedge 0} * (-1)^{0\wedge 1} & (-1)^{0\wedge 1} * (-1)^{0\wedge 0} & (-1)^{0\wedge 1} * (-1)^{0\wedge 1} \\
(-1)^{0\wedge 0} * (-1)^{1\wedge 0} & (-1)^{0\wedge 0} * (-1)^{1\wedge 1} & (-1)^{0\wedge 1} * (-1)^{1\wedge 0} & (-1)^{0\wedge 1} * (-1)^{1\wedge 1} \\
(-1)^{1\wedge 0} * (-1)^{0\wedge 0} & (-1)^{1\wedge 0} * (-1)^{0\wedge 1} & (-1)^{1\wedge 1} * (-1)^{0\wedge 0} & (-1)^{1\wedge 1} * (-1)^{0\wedge 1} \\
(-1)^{1\wedge 0} * (-1)^{1\wedge 0} & (-1)^{1\wedge 0} * (-1)^{1\wedge 1} & (-1)^{1\wedge 1} * (-1)^{1\wedge 0} & (-1)^{1\wedge 1} * (-1)^{1\wedge 1}
\end{bmatrix}
\end{array}.
\tag{10.88}
$$

When we multiply $(-1)^x * (-1)^y$, we are not interested in $(-1)^{x+y}$. Rather, we are interested in the parity of x and y. So we do not add x and y, but rather take their exclusive or ($\oplus$). This gives us

$$
H^{\otimes 2} = \frac{1}{2}
\begin{array}{c}
\\ 00 \\ 01 \\ 10 \\ 11
\end{array}
\begin{array}{c}
\begin{array}{cccc} \quad 00 \quad & \quad 01 \quad & \quad 10 \quad & \quad 11 \quad \end{array} \\
\begin{bmatrix}
(-1)^{0\wedge 0\oplus 0\wedge 0} & (-1)^{0\wedge 0\oplus 0\wedge 1} & (-1)^{0\wedge 1\oplus 0\wedge 0} & (-1)^{0\wedge 1\oplus 0\wedge 1} \\
(-1)^{0\wedge 0\oplus 1\wedge 0} & (-1)^{0\wedge 0\oplus 1\wedge 1} & (-1)^{0\wedge 1\oplus 1\wedge 0} & (-1)^{0\wedge 1\oplus 1\wedge 1} \\
(-1)^{1\wedge 0\oplus 0\wedge 0} & (-1)^{1\wedge 0\oplus 0\wedge 1} & (-1)^{1\wedge 1\oplus 0\wedge 0} & (-1)^{1\wedge 1\oplus 0\wedge 1} \\
(-1)^{1\wedge 0\oplus 1\wedge 0} & (-1)^{1\wedge 0\oplus 1\wedge 1} & (-1)^{1\wedge 1\oplus 1\wedge 0} & (-1)^{1\wedge 1\oplus 1\wedge 1}
\end{bmatrix}
\end{array}
\tag{10.89}
$$

$$
=
$$

$$
\frac{1}{2}
\begin{array}{c}
\\ 00 \\ 01 \\ 10 \\ 11
\end{array}
\begin{array}{c}
\begin{array}{cccc} 00 & 01 & 10 & 11 \end{array} \\
\begin{bmatrix}
1 & 1 & 1 & 1 \\
1 & -1 & 1 & -1 \\
1 & 1 & -1 & -1 \\
1 & -1 & -1 & 1
\end{bmatrix}
\end{array}.
\tag{10.90}
$$

Continuing along the same path, we have

$$
H^{\otimes 3} = \frac{1}{2\sqrt{2}}
\begin{array}{c}
\\ 000 \\ 001 \\ 010 \\ 011 \\ 100 \\ 101 \\ 110 \\ 111
\end{array}
\begin{array}{c}
\begin{array}{cccccccc} 000 & 001 & 010 & 011 & 100 & 101 & 110 & 111 \end{array} \\
\begin{bmatrix}
1 & 1 & 1 & 1 & 1 & 1 & 1 & 1 \\
1 & -1 & 1 & -1 & 1 & -1 & 1 & -1 \\
1 & 1 & -1 & -1 & 1 & 1 & -1 & -1 \\
1 & -1 & -1 & 1 & 1 & -1 & -1 & 1 \\
1 & 1 & 1 & 1 & -1 & -1 & -1 & -1 \\
1 & -1 & 1 & -1 & -1 & 1 & -1 & 1 \\
1 & 1 & -1 & -1 & -1 & -1 & 1 & 1 \\
1 & -1 & -1 & 1 & -1 & 1 & 1 & -1
\end{bmatrix}
\end{array}.
\tag{10.91}
$$

From this, we can write a general formula for $H^{\otimes n}$ as

$$
H^{\otimes n}[i, j] = \frac{1}{\sqrt{2^n}}(-1)^{\langle i,j \rangle},
\tag{10.92}
$$

where i and j are the row and column numbers in binary and $\langle i, j \rangle$ is taking the inner product in binary.

Now we need to evaluate functions. However, not every function is unitary. We remedy this with the following: For every function $f: V \longrightarrow V$, we can form a quantum gate

$$V \otimes V \xrightarrow{U_f} V \otimes V, \tag{10.93}$$

which is unitary and mimics what function f does. The circuit corresponding to this morphism is as follows:

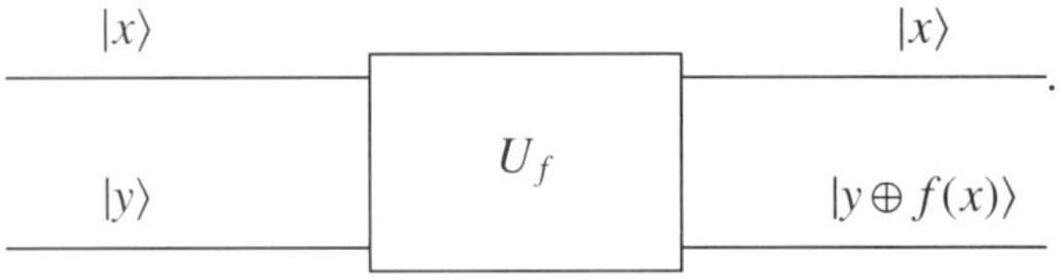

$|x\rangle$ $|x\rangle$

U_f

$|y\rangle$ $|y \oplus f(x)\rangle$ (10.94)

The top input, $|x\rangle$, will be the qubit value that one wishes to evaluate and the bottom input, $|y\rangle$, controls the output. The top output is the same as the input qubit $|x\rangle$, and the bottom output will be $|y \oplus f(x)\rangle$. This gate is its own inverse:

$$V \otimes V \xrightarrow{U_f} V \otimes V \xrightarrow{U_f} V \otimes V. \quad (\text{composite} = id) \tag{10.95}$$

We can see this with the following circuit:

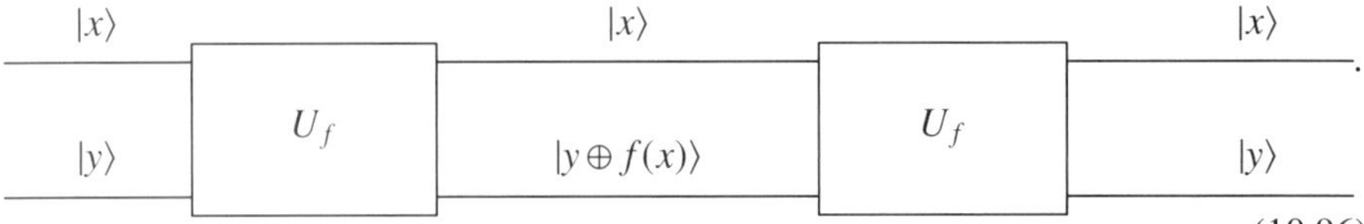

$|x\rangle$ $|x\rangle$ $|x\rangle$

U_f U_f

$|y\rangle$ $|y \oplus f(x)\rangle$ $|y\rangle$

(10.96)

State $|x, y\rangle$ goes to $|x, y \oplus f(x)\rangle$, which further goes to

$$|x, (y \oplus f(x)) \oplus f(x)\rangle = |x, y \oplus (f(x) \oplus f(x))\rangle = |x, y \oplus 0\rangle = |x, y\rangle. \tag{10.97}$$

We need to generalize this for functions with many inputs (i.e., $f: V^{\otimes n} \longrightarrow V$). We can form a quantum gate:

$$V^{\otimes n} \otimes V \xrightarrow{U_f} V^{\otimes n} \otimes V. \tag{10.98}$$

The circuit corresponding to this morphism is as follows:

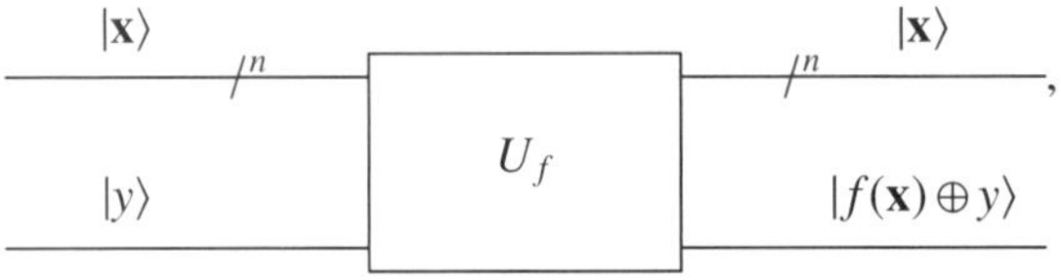

$|\mathbf{x}\rangle$ $/^n$ $/^n$ $|\mathbf{x}\rangle$

U_f

$|y\rangle$ $|f(\mathbf{x}) \oplus y\rangle$ (10.99)

with n qubits (denoted as $/^n$) as the top input and output. For the rest of this mini-course, a binary string is denoted by a boldface letter.

This needs to be further generalized for a function with many inputs and many outputs (i.e., $f\colon V^{\otimes n} \longrightarrow V^{\otimes n}$). For such a function, we can form a quantum gate:

$$V^{\otimes n} \otimes V^{\otimes n} \xrightarrow{U_f} V^{\otimes n} \otimes V^{\otimes n}. \tag{10.100}$$

The quantum gate U_f can be visualized as the following circuit:

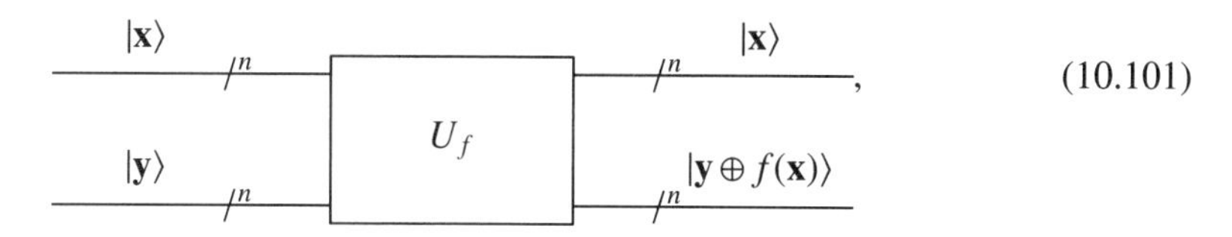

(10.101)

where $|\mathbf{x},\mathbf{y}\rangle$ goes to $|\mathbf{x},\mathbf{y}\oplus f(\mathbf{x})\rangle$. U_f is again its own inverse. Setting $\mathbf{y}=0^n$ would give us an easy way to evaluate $f(\mathbf{x})$.

There are still other quantum gates. One of the central features of computer science is an operation that is done only under certain conditions and not under others. This is equivalent to (an IF-THEN statement). If a certain (qu)bit is true, then a particular operation should be performed; otherwise, the operation is not performed. For every n-qubit unitary operation U, we can create a unitary $n+1$–qubit operation **controlled-U** or ${}^{C}U$. We depict it as a circuit as follows:

$|x\rangle$ $\quad$ $|x\rangle$ $\quad$ U $\quad$ (10.102)

This operation will perform the U operation if the top $|x\rangle$ input is a $|1\rangle$ and simply perform the identity operation if $|x\rangle$ is $|0\rangle$.

For the simple case of

$$U = \begin{bmatrix} a & b \\ c & d \end{bmatrix}, \tag{10.103}$$

the controlled-U operation can be seen to be

$${}^{C}U = \begin{bmatrix} 1 & 0 & 0 & 0 \\ 0 & 1 & 0 & 0 \\ 0 & 0 & a & b \\ 0 & 0 & c & d \end{bmatrix}. \tag{10.104}$$

This same construction works for large matrices.

Exercise 10.3.4. Show that the constructed ${}^{C}U$ works as it should when the top qubit is set to $|0\rangle$ or $|1\rangle$. ■

Exercise 10.3.5. Show that if U is unitary, then so is ${}^{C}U$. ■

Exercise 10.3.6. Show that the Toffoli gate is nothing more than $^C(^C\text{NOT})$. ■

There are many more quantum gates, but we will not list them here. There is one more gate that is decidedly not unitary but is still needed: the **measurement operation**. This is not a quantum gate and, in general, is not even reversible. This operation is usually performed at the end of a computation when we want to measure qubits (and find bits). We denote it as

$$\text{———}\boxed{\nearrow}. \tag{10.105}$$

Quantum Algorithms

With these qubits and quantum gates in our toolbox, we can move on to the core of this mini-course, which consists of five of the most famous quantum algorithms. We proceed from the simplest algorithm to the two most important: Grover's algorithm, which helps find a needle in a haystack more quickly than usual; and Shor's algorithm, which factors large numbers in polynomial time.

Just as an algorithm in a classical computer can be stated as a sequence of operations on bits, a quantum algorithm will be stated as a sequence of operators on complex vector spaces that represent qubits. These operators will be stated as a sequence of morphisms in $\mathbb{Hilb}$. We will then draw them as a circuit diagram. For each algorithm, we will show the state of the qubits as the algorithm progresses.

The algorithms have a certain common form. They start in a particular state, which is described as a linear map from $\mathbf{C}$ to a vector space. We then use Hadamard matrices to put the system in a superposition. A function or two is then applied to the system, and finally a measurement is made. Sometimes these operations have to be performed more than once.

The problems that these algorithms solve also have a common form. In all five cases, we are given a function as a black box. This means we do not know how the function is defined, and the only way we can determine information about the function is to evaluate it. Furthermore, we are assured or promised that this function has a special property. We are asked to determine some aspect of that special property.

Deutsch's Algorithm

We start with the world's simplest quantum algorithm and work our way toward more complicated ones. We go through the first three in agonizing detail and leave some of the details out of the last two algorithms.

Deutsch's algorithm solves a slightly contrived problem. This algorithm is concerned with functions from the set $\{0, 1\}$ to the set $\{0, 1\}$. There are four such functions, which might be visualized as

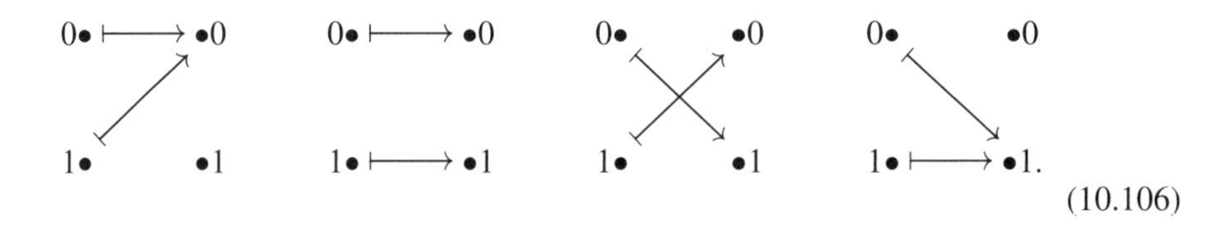

(10.106)

Call a function $f\colon \{0,1\} \longrightarrow \{0,1\}$, **balanced** if $f(0) \neq f(1)$, that is, it is one to one. In contrast, call a function **constant** if $f(0) = f(1)$. Of the four functions, two are balanced and two are constant.

Deutsch's algorithm solves the following problem: given a function $f\colon \{0,1\} \longrightarrow \{0,1\}$ as a black box, determine if the function is balanced or constant.

With a classical computer, one would have to evaluate f on one input, evaluate f on the second input, and finally compare the outputs. The following decision tree shows what a classical computer must do:

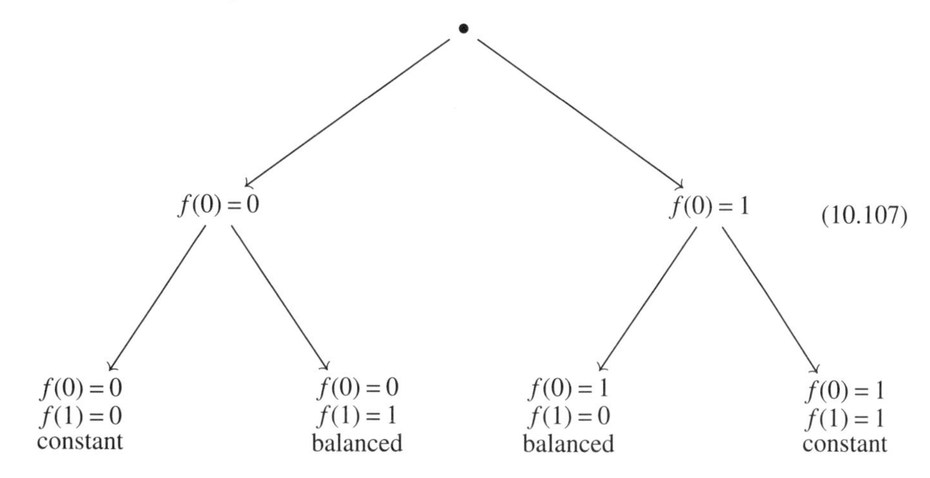

(10.107)

The point is that with a classical computer, f must be evaluated twice. With a quantum computer, this can be done with one function evaluation. The single function evaluation is done on a superposition of inputs.

Rather than just give the algorithm, we build toward it to see the intuition behind the algorithm.

First Try at Deutsch's Algorithm. The obvious way of solving this problem is putting the top input qubit into a superposition. In terms of morphisms in $\mathbb{Hilb}$, the algorithm is as follows:

$$\mathbf{C} \otimes \mathbf{C} \xrightarrow{|0\rangle \otimes |0\rangle} V \otimes V \xrightarrow{H \otimes id} V \otimes V \xrightarrow{U_f} V \otimes V \xrightarrow{id \otimes Ms} V \otimes V. \tag{10.108}$$

This forms the following circuit:

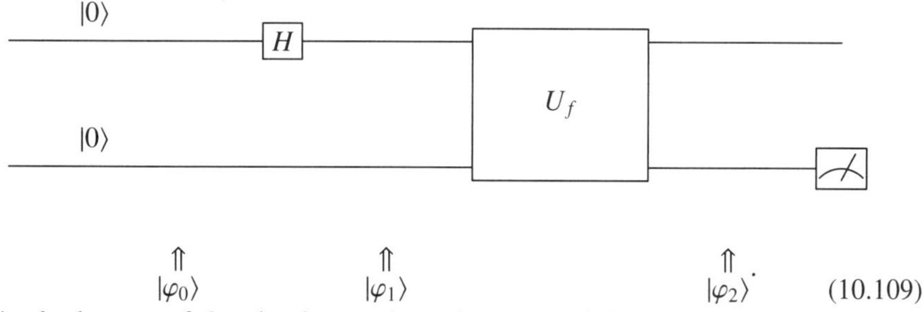

(10.109)

At the bottom of the circuit, we show the states of the system. Let us carefully examine the states of the system as they change through the algorithm:

- The system starts in

$$|\varphi_0\rangle = |0\rangle \otimes |0\rangle = |0, 0\rangle. \tag{10.110}$$

- We then apply the Hadamard matrix only to the top input—leaving the bottom input alone—to get

$$|\varphi_1\rangle = \left[\frac{|0\rangle + |1\rangle}{\sqrt{2}}\right]|0\rangle = \frac{|0, 0\rangle + |1, 0\rangle}{\sqrt{2}}. \tag{10.111}$$

- After applying U_f, we have

$$|\varphi_2\rangle = \frac{|0, f(0)\rangle + |1, f(1)\rangle}{\sqrt{2}}. \tag{10.112}$$

  If we measure the top qubit, there will be a 50 percent chance of finding it in state $|0\rangle$ and a 50 percent chance of finding it in state $|1\rangle$. Similarly, if we measure the bottom qubit, we do not gain anything. The obvious algorithm does not work. We need a better trick.

Second Try at Deutsch's Algorithm. Rather than putting the top input into a superposition, let us see what would happen if we put the bottom input into a superposition. In terms of morphisms in $\mathbb{Hilb}$, the algorithm is as follows:

$$V \otimes \mathbf{C} \xrightarrow{id \otimes |1\rangle} V \otimes V \xrightarrow{id \otimes H} V \otimes V \xrightarrow{U_f} V \otimes V \xrightarrow{id \otimes Ms} V \otimes V. \tag{10.113}$$

This forms the following circuit:

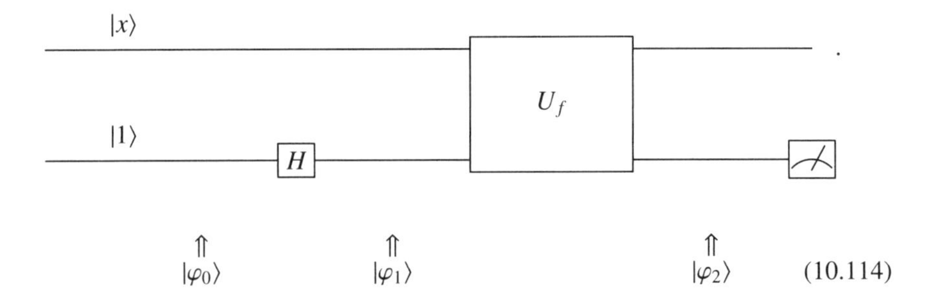

(10.114)

Let us look carefully at how the states of the qubits change:

- The system starts in

$$|\varphi_0\rangle = |x, 1\rangle. \tag{10.115}$$

- After the Hadamard operator is applied to the bottom qubit, we have

$$|\varphi_1\rangle = |x\rangle \left[\frac{|0\rangle - |1\rangle}{\sqrt{2}} \right] = \frac{|x,0\rangle - |x,1\rangle}{\sqrt{2}}. \tag{10.116}$$

- Applying U_f, we get

$$|\varphi_2\rangle = |x\rangle \left[\frac{|0 \oplus f(x)\rangle - |1 \oplus f(x)\rangle}{\sqrt{2}} \right] = |x\rangle \left[\frac{|f(x)\rangle - |\overline{f(x)}\rangle}{\sqrt{2}} \right], \tag{10.117}$$

where $\overline{f(x)}$ means the opposite of $f(x)$. Therefore, we have

$$|\varphi_2\rangle = \begin{cases} |x\rangle \left[\frac{|0\rangle - |1\rangle}{\sqrt{2}} \right] & \text{if } f(x) = 0 \\ \\ |x\rangle \left[\frac{|1\rangle - |0\rangle}{\sqrt{2}} \right] & \text{if } f(x) = 1 \end{cases}. \tag{10.118}$$

Remembering that $a - b = (-1)(b - a)$, we might write this as

$$|\varphi_2\rangle = (-1)^{f(x)}|x\rangle \left[\frac{|0\rangle - |1\rangle}{\sqrt{2}} \right]. \tag{10.119}$$

What would happen if we evaluate either the top or the bottom state? We do not gain any information if we measure the top qubit because it will be in state $|x\rangle$. Nor will measuring the bottom qubit be a help because after measuring, the bottom qubit will be either in state $|0\rangle$ or in state $|1\rangle$. We need something more.

Deutsch's Algorithm. The final algorithm is a combination of the first two tries and puts *both* the top and the bottom qubits into a superposition. We will also put the results of the top qubit through a Hadamard matrix. In terms of morphisms in $\mathbb{Hilb}$, the algorithm is as follows:

$$\mathbf{C} \otimes \mathbf{C} \xrightarrow{|0\rangle \otimes |1\rangle} V \otimes V \xrightarrow{H \otimes H} V \otimes V \xrightarrow{U_f} V \otimes V \xrightarrow{H \otimes id} V \otimes V \xrightarrow{Ms \otimes id} V \otimes V. \tag{10.120}$$

This forms the following circuit:

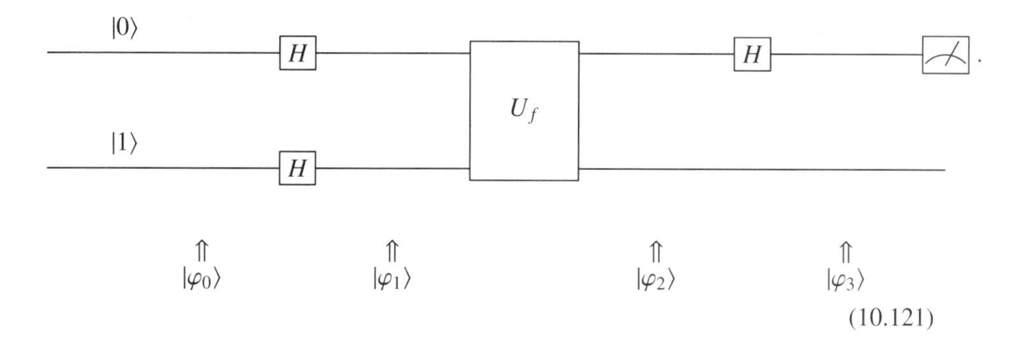

(10.121)

- We start with

$$|\varphi_0\rangle = |0, 1\rangle. \tag{10.122}$$

- After applying Hadamard operators to both qubits, we get

$$|\varphi_1\rangle = \left[\frac{|0\rangle + |1\rangle}{\sqrt{2}}\right]\left[\frac{|0\rangle - |1\rangle}{\sqrt{2}}\right] = \frac{+|0,0\rangle - |0,1\rangle + |1,0\rangle - |1,1\rangle}{2} = \begin{matrix} 00 \\ 01 \\ 10 \\ 11 \end{matrix}\begin{bmatrix} +\frac{1}{2} \\ -\frac{1}{2} \\ +\frac{1}{2} \\ -\frac{1}{2} \end{bmatrix}. \tag{10.123}$$

- We saw from our last attempt at solving this problem that when we put the bottom qubit into a superposition and apply U_f, we are in the superposition

$$(-1)^{f(x)}|x\rangle\left[\frac{|0\rangle - |1\rangle}{\sqrt{2}}\right]. \tag{10.124}$$

Now, with $|x\rangle$ in a superposition, we have

$$|\varphi_2\rangle = \left[\frac{(-1)^{f(0)}|0\rangle + (-1)^{f(1)}|1\rangle}{\sqrt{2}}\right]\left[\frac{|0\rangle - |1\rangle}{\sqrt{2}}\right]. \tag{10.125}$$

For example, if $f(0) = 1$ and $f(1) = 0$, the top qubit becomes

$$\frac{(-1)|0\rangle + (+1)|1\rangle}{\sqrt{2}} = (-1)\left[\frac{|0\rangle - |1\rangle}{\sqrt{2}}\right]. \tag{10.126}$$

For a general function f, let us look carefully at

$$(-1)^{f(0)}|0\rangle + (-1)^{f(1)}|1\rangle. \tag{10.127}$$

If f is constant, this becomes either

$$+1(|0\rangle + |1\rangle) \qquad \text{or} \qquad -1(|0\rangle + |1\rangle) \tag{10.128}$$

(depending on whether it is constantly 0 or constantly 1).

If f is balanced, it becomes either

$$+1(|0\rangle - |1\rangle) \qquad \text{or} \qquad -1(|0\rangle - |1\rangle) \tag{10.129}$$

(depending on which way it is balanced).

Summing up, we have that

$$|\varphi_2\rangle = \begin{cases} (\pm 1)\left[\frac{|0\rangle + |1\rangle}{\sqrt{2}}\right]\left[\frac{|0\rangle - |1\rangle}{\sqrt{2}}\right] & \text{if } f \text{ is constant.} \\ \\ (\pm 1)\left[\frac{|0\rangle - |1\rangle}{\sqrt{2}}\right]\left[\frac{|0\rangle - |1\rangle}{\sqrt{2}}\right] & \text{if } f \text{ is balanced.} \end{cases} \tag{10.130}$$

- Remembering that the Hadamard matrix is its own inverse, which takes $\frac{|0\rangle + |1\rangle}{\sqrt{2}}$ to $|0\rangle$ and takes $\frac{|0\rangle - |1\rangle}{\sqrt{2}}$ to $|1\rangle$, we apply the Hadamard matrix to the top qubit to get

$$|\varphi_3\rangle = \begin{cases} (\pm 1)|0\rangle\left[\frac{|0\rangle - |1\rangle}{\sqrt{2}}\right] & \text{if } f \text{ is constant.} \\ \\ (\pm 1)|1\rangle\left[\frac{|0\rangle - |1\rangle}{\sqrt{2}}\right] & \text{if } f \text{ is balanced.} \end{cases} \tag{10.131}$$

For example, if $f(0) = 1$ and $f(1) = 0$, then we get

$$|\varphi_3\rangle = -1|0\rangle\left[\frac{|0\rangle - |1\rangle}{\sqrt{2}}\right]. \tag{10.132}$$

Now, we simply measure the top qubit. If it is in state $|0\rangle$, then we know that f is a constant function; otherwise, it is a balanced function. This was all accomplished with only one function evaluation as opposed to the two evaluations that the classical algorithm demands. This is a speed-up that is not really significant. However the ideas learned in this algorithm are used for the other problems.

Deutsch-Jozsa Algorithm

Let us generalize the Deutsch algorithm to other functions. Rather than talking about functions $f\colon \{0,1\} \longrightarrow \{0,1\}$, let us talk about functions with a larger domain. Consider functions $f\colon \{0,1\}^n \longrightarrow \{0,1\}$, which accept a string of n zeros and 1s and outputs a zero or 1. The domain might be thought of as any natural number from 0 to $2^n - 1$.

We call a function $f\colon \{0,1\}^n \longrightarrow \{0,1\}$ **balanced** if exactly half of the inputs go to 0 (and the other half go to 1). Call a function **constant** if *all* the inputs go to 0 or *all* the inputs go to 1.

Exercise 10.3.7. How many functions are there from $\{0,1\}^n$ to $\{0,1\}$? How many of them are balanced? How many of them are constant? ■

The Deutsch-Jozsa algorithm solves the following problem: Suppose that you are given a function from $\{0,1\}^n$ to $\{0,1\}$ as a black box. Suppose further that you are assured that the function is either balanced or constant, and not anything else. Determine if the function is balanced or constant. Notice that when $n = 1$, this is exactly the problem that the Deutsch algorithm solved.

Classically, this algorithm can be solved by evaluating the function on different inputs. The best-case scenario is when the first two different inputs have different outputs, which assures us that the the function is balanced. In contrast, to be sure that the function is constant, one must evaluate the function on more than half the possible inputs. So the worst-case scenario requires $\frac{2^n}{2} + 1 = 2^{n-1} + 1$ function evaluations. Can we do better?

With the last algorithm, we solved the problem by entering into a superposition of two possible input states. In this algorithm, we solve the problem by entering a superposition of all 2^n possible input states. We saw that we can get such a superposition by multiplying $H^{\otimes n}$ by $|\mathbf{0}\rangle = |000\cdots 0\rangle$. In terms of morphisms in $\mathbb{Hilb}$, the algorithm is as follows:

$$\mathbf{C}^{\otimes n} \otimes \mathbf{C} \xrightarrow{|\mathbf{0}\rangle^{\otimes n} \otimes |1\rangle} V^{\otimes n} \otimes V \xrightarrow{H^{\otimes n} \otimes H} V^{\otimes n} \otimes V \xrightarrow{U_f} V^{\otimes n} \otimes V \xrightarrow{H^{\otimes n} \otimes id} V^{\otimes n} \otimes V \xrightarrow{Ms \otimes id} V^{\otimes n} \otimes V. \tag{10.133}$$

This forms the following circuit:

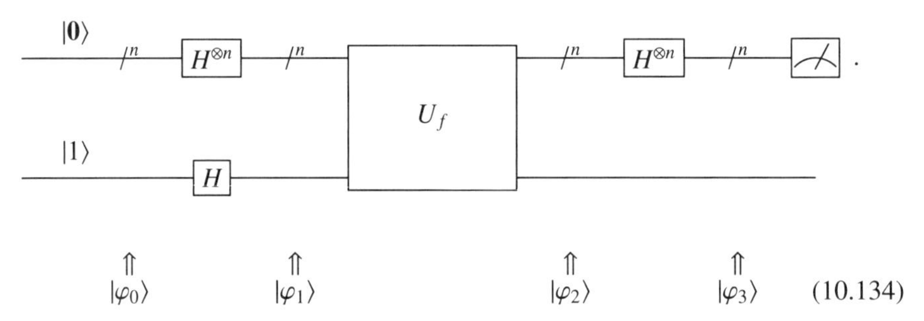

(10.134)

Let us go through all the states:

- We begin with

$$|\varphi_0\rangle = |\mathbf{0}, 1\rangle. \tag{10.135}$$

- After applying the Hadamard operators, we get

$$|\varphi_1\rangle = \left[\frac{\sum_{\mathbf{x} \in \{0,1\}^n} |\mathbf{x}\rangle}{\sqrt{2^n}}\right] \left[\frac{|0\rangle - |1\rangle}{\sqrt{2}}\right]. \tag{10.136}$$

- After applying the U_f operator, we have

$$|\varphi_2\rangle = \left[\frac{\sum_{\mathbf{x} \in \{0,1\}^n} (-1)^{f(\mathbf{x})} |\mathbf{x}\rangle}{\sqrt{2^n}}\right] \left[\frac{|0\rangle - |1\rangle}{\sqrt{2}}\right]. \tag{10.137}$$

- Finally, we apply $H^{\otimes n}$ to the top qubits, which are already in a superposition of different **x** states, to get a superposition of a superposition:

$$|\varphi_3\rangle = \left[\frac{\sum_{\mathbf{x}\in\{0,1\}^n} (-1)^{f(\mathbf{x})} \sum_{\mathbf{z}\in\{0,1\}^n} (-1)^{\langle \mathbf{z},\mathbf{x}\rangle} |\mathbf{z}\rangle}{2^n} \right] \left[\frac{|0\rangle - |1\rangle}{\sqrt{2}} \right]. \tag{10.138}$$

We can combine parts and add exponents to get

$$|\varphi_3\rangle = \left[\frac{\sum_{\mathbf{x}\in\{0,1\}^n} \sum_{\mathbf{z}\in\{0,1\}^n} (-1)^{f(\mathbf{x})} (-1)^{\langle \mathbf{z},\mathbf{x}\rangle} |\mathbf{z}\rangle}{2^n} \right] \left[\frac{|0\rangle - |1\rangle}{\sqrt{2}} \right] \tag{10.139}$$

$$=$$

$$\left[\frac{\sum_{\mathbf{x}\in\{0,1\}^n} \sum_{\mathbf{z}\in\{0,1\}^n} (-1)^{f(\mathbf{x}) \oplus \langle \mathbf{z},\mathbf{x}\rangle} |\mathbf{z}\rangle}{2^n} \right] \left[\frac{|0\rangle - |1\rangle}{\sqrt{2}} \right]. \tag{10.140}$$

Now the top qubits of state $|\varphi_3\rangle$ are measured. Rather than figuring out what we will get after measuring the top qubit, let us ask the following question: What is the probability that the top qubits of $|\varphi_3\rangle$ will collapse to state $|\mathbf{0}\rangle$? We can answer this by setting $\mathbf{z} = \mathbf{0}$ and realizing that $\langle \mathbf{z}, \mathbf{x}\rangle = \langle \mathbf{0}, \mathbf{x}\rangle = 0$ for all $\mathbf{x}$. In this case, we have reduced $|\varphi_3\rangle$ to

$$\left[\frac{\sum_{\mathbf{x}\in\{0,1\}^n} (-1)^{f(\mathbf{x})} |\mathbf{0}\rangle}{2^n} \right] \left[\frac{|0\rangle - |1\rangle}{\sqrt{2}} \right]. \tag{10.141}$$

So the probability of collapsing to $|\mathbf{0}\rangle$ is totally dependent on $f(\mathbf{x})$. If $f(\mathbf{x})$ is constant at 1, the top qubits become

$$\frac{\sum_{\mathbf{x}\in\{0,1\}^n} (-1)|\mathbf{0}\rangle}{2^n} = \frac{-(2^n)|\mathbf{0}\rangle}{2^n} = -1|\mathbf{0}\rangle. \tag{10.142}$$

If $f(\mathbf{x})$ is constant at 0, the top qubits become

$$\frac{\sum_{\mathbf{x}\in\{0,1\}^n} 1|\mathbf{0}\rangle}{2^n} = \frac{2^n|\mathbf{0}\rangle}{2^n} = +1|\mathbf{0}\rangle. \tag{10.143}$$

And finally, if f is balanced, then half of the **x**'s will cancel the other half and the top qubits will become

$$\frac{\sum_{\mathbf{x}\in\{0,1\}^n} (-1)^{f(\mathbf{x})} |\mathbf{0}\rangle}{2^n} = \frac{0|\mathbf{0}\rangle}{2^n} = 0|\mathbf{0}\rangle. \tag{10.144}$$

This means that when measuring the top qubits of $|\varphi_3\rangle$, we will get $|\mathbf{0}\rangle$ only if the function is constant. Because the coefficient is 0, we will never get $|\mathbf{0}\rangle$ if it is balanced. We will get $|\mathbf{1}\rangle$.

In conclusion, in the worst-case scenario, the problem is that a classical algorithm demands exponential function evaluations. In contrast, this quantum algorithm demands only one function evaluation. This is what is called an "exponential speed-up." However, the problem is a bit contrived.

Simons's Periodicity Algorithm

Simons's periodicity algorithm is about finding patterns in functions. We will use methods that we already learned with previous algorithms, but we will also employ other ideas. This algorithm is a combination of quantum procedures and classical procedures.

Suppose that we are given a function $f\colon \{0,1\}^n \longrightarrow \{0,1\}^n$ as a black box. We are further assured that there is a secret, hidden binary string $\mathbf{c} = c_0 c_1 c_2 \cdots c_{n-1}$, such that for all strings $\mathbf{x}, \mathbf{y} \in \{0,1\}^n$, we have

$$f(\mathbf{x}) = f(\mathbf{y}) \quad \text{if and only if} \quad \mathbf{x} = \mathbf{y} \oplus \mathbf{c}, \tag{10.145}$$

where $\oplus$ is the bitwise exclusive or operation. In other words, the values of f repeat themselves in some pattern, which is determined by $\mathbf{c}$. We call $\mathbf{c}$ the **period** of f. The goal of Simons's algorithm is to determine $\mathbf{c}$.

Example 10.3.8. Let us work out an example. Let $n = 3$. Consider $\mathbf{c} = 101$. Then we are going to have the following requirements on f:

- $000 \oplus 101 = 101$, hence $f(000) = f(101)$.
- $001 \oplus 101 = 100$, hence $f(001) = f(100)$.
- $010 \oplus 101 = 111$, hence $f(010) = f(111)$.
- $011 \oplus 101 = 110$, hence $f(011) = f(110)$.
- $100 \oplus 101 = 001$, hence $f(100) = f(001)$.
- $101 \oplus 101 = 000$, hence $f(101) = f(000)$.
- $110 \oplus 101 = 011$, hence $f(110) = f(011)$.
- $111 \oplus 101 = 010$, hence $f(111) = f(010)$. □

How would one solve this problem classically? We would have to evaluate f on different binary strings. After each evaluation, check to see if that output has already been found. If one finds two inputs $\mathbf{x}_1$ and $\mathbf{x}_2$ such that $f(\mathbf{x}_1) = f(\mathbf{x}_2)$, then we are assured that

$$\mathbf{x}_1 = \mathbf{x}_2 \oplus \mathbf{c} \tag{10.146}$$

and can obtain $\mathbf{c}$ by $\oplus$-ing both sides with $\mathbf{x}_2$:

$$\mathbf{x}_1 \oplus \mathbf{x}_2 = \mathbf{x}_2 \oplus \mathbf{c} \oplus \mathbf{x}_2 = \mathbf{c}. \tag{10.147}$$

If the function is a two-to-one function, then we will not have to evaluate more than half the inputs before we get a repeat. If we evaluate more than half the strings and still cannot find a match, then we know that f is one-to-one and $\mathbf{c} = 0^n$. So, in the worst case, $\frac{2^n}{2} + 1 = 2^{n-1} + 1$ function evaluations are needed. Can we do better?

The quantum part of Simons's algorithm basically consists of performing the following operations several times:

$$\mathbf{C}^{\otimes n} \otimes \mathbf{C}^{\otimes n} \xrightarrow{|0\rangle^{\otimes n} \otimes |0\rangle^{\otimes n}} V^{\otimes n} \otimes V^{\otimes n} \xrightarrow{H^{\otimes n} \otimes id^{\otimes n}} V^{\otimes n} \otimes V^{\otimes n} \xrightarrow{U_f} V^{\otimes n} \otimes V^{\otimes n}$$

$$\xrightarrow{H^{\otimes n} \otimes id^{\otimes n}} V^{\otimes n} \otimes V^{\otimes n} \xrightarrow{Ms \otimes id^{\otimes n}} V^{\otimes n} \otimes V^{\otimes n} \tag{10.148}$$

The circuit corresponding to these morphisms is as follows:

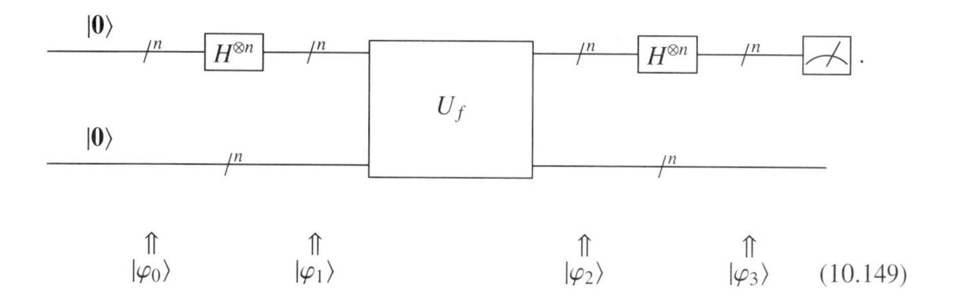

(10.149)

- We start at

$$|\varphi_0\rangle = |\mathbf{0}, \mathbf{0}\rangle. \tag{10.150}$$

- We then place the input in a superposition of all possible inputs We know that looks as follows:

$$|\varphi_1\rangle = \frac{\sum_{\mathbf{x} \in \{0,1\}^n} |\mathbf{x}, \mathbf{0}\rangle}{\sqrt{2^n}}. \tag{10.151}$$

- Evaluation of f on all these possibilities gives us

$$|\varphi_2\rangle = \frac{\sum_{\mathbf{x} \in \{0,1\}^n} |\mathbf{x}, f(\mathbf{x})\rangle}{\sqrt{2^n}}. \tag{10.152}$$

And finally, let us apply $H^{\otimes n}$ to the top output. We get

$$|\varphi_3\rangle = \frac{\sum_{\mathbf{x} \in \{0,1\}^n} \sum_{\mathbf{z} \in \{0,1\}^n} (-1)^{\langle \mathbf{z}, \mathbf{x} \rangle} |\mathbf{z}, f(\mathbf{x})\rangle}{2^n}. \tag{10.153}$$

From here on, we will not give much detail. When you measure the top qubits, you get a binary string $\mathbf{x}$, which has the following property:

$$\langle \mathbf{x}, \mathbf{c} \rangle = \mathbf{0}. \tag{10.154}$$

This procedure is run n times, each time getting another possible output. At the end, there will be n binary strings $\mathbf{x}_i$ such that $\langle \mathbf{x}_i, \mathbf{c} \rangle = \mathbf{0}$. What remains to be done is to solve a classical linear equations problem to find $\mathbf{c}$.

In conclusion, for a given periodic f, we can find the period $\mathbf{c}$ with n function evaluations. This is in contrast to the $2^{n-1}+1$ needed with the classical algorithm.

Grover's Algorithm

How do you find a needle in a haystack? You look at each piece of hay separately and check each one to see if it is the desired needle. That is not very efficient, however. The computer science version of this problem is about unordered arrays instead of haystacks. Given an unordered array of m elements, find a particular element. Classically, in the worst case, this takes m queries. On average, we will find the desired element in about $m/2$ queries. Can we do better?

Lov Grover's search algorithm does the job in $\sqrt{m}$ queries. Although this is not the exponential speedup of the Deutsch-Jozsa algorithm and Simons's algorithm, it is still very good. Grover's algorithm has many applications to database theory and other areas.

Since we have become quite adept at binary functions, let us look at the search problem from the point of view of binary functions. Imagine that we are given a function $f\colon \{0,1\}^n \longrightarrow \{0,1\}$ and we are assured that there is exactly one binary string $\mathbf{x_0}$ such that

$$f(\mathbf{x}) = \begin{cases} 1 & \text{if } \mathbf{x} = \mathbf{x_0}. \\ 0 & \text{if } \mathbf{x} \neq \mathbf{x_0}. \end{cases} \tag{10.155}$$

We are asked to find $\mathbf{x_0}$. Classically, in the worst case, we would have to evaluate all 2^n binary strings to find the desired $\mathbf{x_0}$. Grover's algorithm will demand only $\sqrt{2^n} = 2^{\frac{n}{2}}$ evaluations.

The naive way to solve this problem is to put $\mathbf{x}$ into a superposition of all 2^n possible values and evaluate that superposition with the U_f operator. Alas, this will not work. The output of U_f will be a superposition. When we measure this superposition, we will probably get 0. Very rarely will we get 1.

To solve this problem, Grover's algorithm uses a two-step process. One step is called **phase inversion**, which is described by the operator $U_f(id \otimes H)$. Without getting into the details, this changes the coefficient of $\mathbf{x_0}$. The next step highlights this changed coefficient with an **inversion about the mean**. Again, we omit the details, but this ensures that when the superposition is measured, it will collapse to the the highlighted position. This operator is described as the matrix $-id + 2A$ for some matrix A. These two steps have to be preformed $\sqrt{2^n}$ "times to" make sure that $\mathbf{x_0}$ is properly highlighted.

This is Grover's algorithm in pseudocode:

Input: A function $f\colon \{0,1\}^n \longrightarrow \{0,1\}$, where exactly one input has output 1.

Output: The single input that gives that output.

Step 1. Start with a state $|\mathbf{0}\rangle$.

Step 2. Apply $H^{\otimes n}$.

Step 3. Repeat $\sqrt{2^n}$ times:

Step 3a. Apply the phase inversion operator: $U_f(id \otimes H)$.

Step 3b. Apply the inversion about the mean operator$-id + 2A$.

Step 4. Measure the qubits.

The quantum part of the algorithm can be represented as the following sequence of morphisms in $\mathbb{Hilb}$:

$$V^{\otimes n} \otimes V \xrightarrow{id^{\otimes n} \otimes H} V^{\otimes n} \otimes V \xrightarrow{U_f} V^{\otimes n} \otimes V \xrightarrow{(-I+2A)\otimes id} V^{\otimes n} \otimes V \xrightarrow{M s \otimes id} V^{\otimes n} \otimes V. \tag{10.156}$$

In terms of a circuit, Grover's algorithm looks like this:

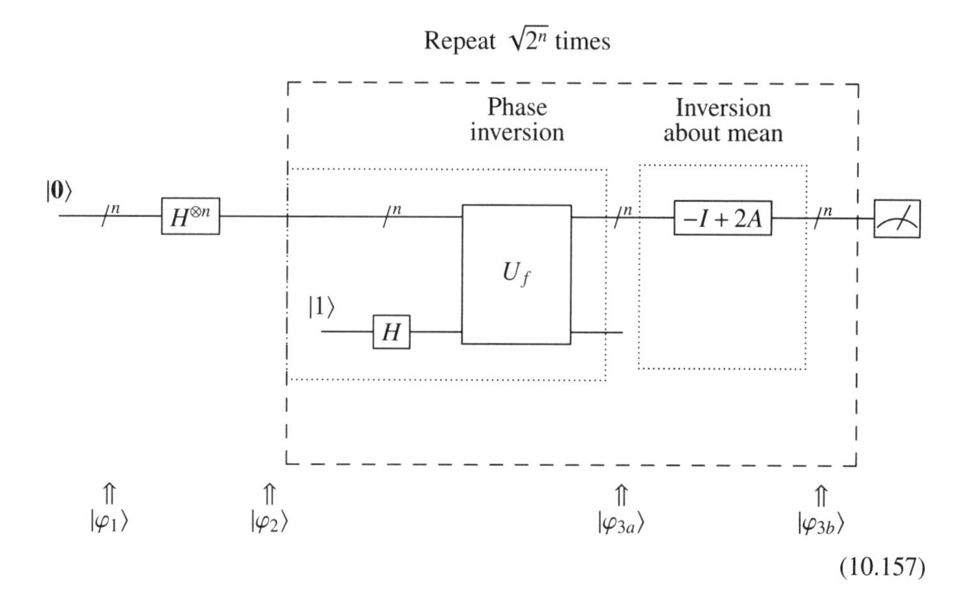

(10.157)

It would take us too far afield to show all the details of how this algorithm works. Suffice it to say that the $\sqrt{2^n}$ function evaluations are much more efficient than the 2^n function evaluations.

Shor's Algorithm

The problem of factoring integers is very significant. Much of internet security is based on the fact that it is hard to factor integers on classical computers. Peter Shor's amazing algorithm factors integers in polynomial time and really brings quantum computing into the limelight.

Shor's algorithm is based on the following idea: the factoring problem can be reduced to finding the period of a certain function. With Simons's periodicity algorithm, we learned how to find the period of a function. In this section, we employ some of those periodicity techniques to factor integers.

The number that we wish to factor is N. In practice, N is a large number. We assume that the given N is not a prime number but is a composite number. We can easily check to see if N is prime before we try to factor it because there is a deterministic, polynomial algorithm that determines if N is prime [14].

Before we go on to Shor's algorithm, we have to remind ourselves of basic number theory. We begin by looking at **modular arithmetic.** For a positive integer N and any integer a, we write $a \ Mod \ N$ for the remainder (or residue) of the quotient a/N. (For C/C++, and for Java programmers, *Mod* is recognizable as the percent operation.)

Example 10.3.9. Here are some examples:

- $7 \ Mod \ 15 = 7$ because $7/15 = 0$ remainder 7.
- $99 \ Mod \ 15 = 9$ because $99/15 = 6$ remainder 9.
- $199 \ Mod \ 15 = 4$ because $199/15 = 13$ remainder 4.
- $5317 \ Mod \ 371 = 123$ because $5{,}317/371 = 14$ remainder 123.
- $23374 \ Mod \ 371 = 1$ because $23{,}374/371 = 63$ remainder 1.
- $1446 \ Mod \ 371 = 333$ because $1{,}446/371 = 3$ remainder 333. □

We write

$$a \equiv a' \ Mod \ N \text{ if and only if } (a \ Mod \ N) = (a' \ Mod \ N), \tag{10.158}$$

or equivalently, if N is a divisor of $a - a'$, that is, $N|(a - a')$.

With *Mod* understood, we can start discussing Shor's algorithm. Let us randomly choose an integer a that is less than N but does not have a nontrivial factor in common with N. One can test for such a factor by performing Euclid's algorithm to calculate GCD(a, N). If the GCD is not 1, then we have found a factor of N and we are done. If the GCD is 1, then a is called **co-prime** to N and we can use it. We need to find the powers of a modulo N; that is,

$$a^0 \ Mod \ N, \ a^1 \ Mod \ N, \ a^2 \ Mod \ N, \ a^3 \ Mod \ N, \ \ldots \tag{10.159}$$

In other words, we need to find the values of the function

$$f_{a,N}(x) = a^x \ Mod \ N. \tag{10.160}$$

Some examples are in order.

Example 10.3.10. Let $N = 15$ and $a = 2$. A few simple calculations show that we get the following:

x	0	1	2	3	4	5	6	7	8	9	10	11	12	$\cdots$
$f_{2,15}(x)$	1	2	4	8	1	2	4	8	1	2	4	8	1	$\cdots$

(10.161)

For $a = 4$, we have

x	0	1	2	3	4	5	6	7	8	9	10	11	12	$\cdots$
$f_{4,15}(x)$	1	4	1	4	1	4	1	4	1	4	1	4	1	$\cdots$

(10.162)

For $a = 13$, we have

x	0	1	2	3	4	5	6	7	8	9	10	11	12	$\cdots$
$f_{13,15}(x)$	1	13	4	7	1	13	4	7	1	13	4	7	1	$\cdots$

(10.163)

□

Notice that the outputs of these functions are repetitive, that is, they have a period. Formally, the period is the smallest r such that

$$f_{a,N}(r) = a^r \ \ Mod \ \ N = 1. \tag{10.164}$$

It is a theorem of number theory that for any co-prime $a \leq N$, the function $f_{a,N}$ will output 1 for some $r < N$. After it hits 1, the sequence of numbers will simply repeat. If $f_{a,N}(r) = 1$, then

$$f_{a,N}(r+1) = f_{a,N}(1) \tag{10.165}$$

and, in general,

$$f_{a,N}(r+s) = f_{a,N}(s). \tag{10.166}$$

We use the period of $f_{a,N}$ to calculate the factors of N.

But how do we find the period? We employ similar techniques as Simons's periodicity algorithm. First we form the quantum gate that computes $U_{f_{a,N}}$. This gate corresponds to a morphism in $\mathbb{Hilb}$:

$$V^{\otimes n} \otimes V^{\otimes n} \xrightarrow{U_f} V^{\otimes n} \otimes V^{\otimes n}. \tag{10.167}$$

The inputs and outputs of this gate look like this:

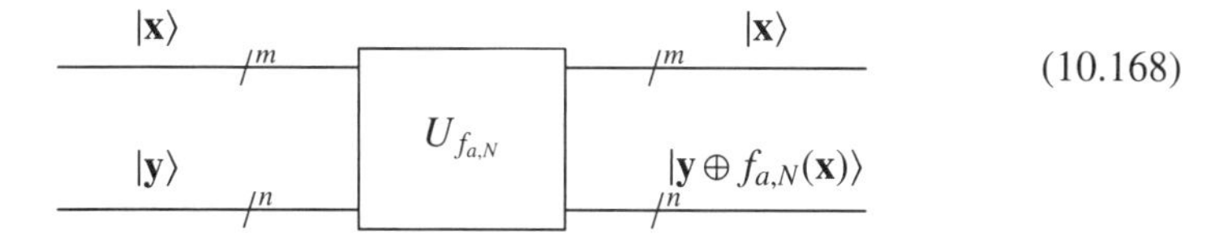

(10.168)

where $|\mathbf{x}, \mathbf{y}\rangle$ goes to $|\mathbf{x}, \mathbf{y} \oplus f_{a,N}(\mathbf{x})\rangle = |\mathbf{x}, \mathbf{y} \oplus (a^{\mathbf{x}} \ \ Mod \ \ N)\rangle$.

The $U_{f_{a,N}}$ operator is used in the quantum part of Shor's algorithm, which corresponds to the composition of the morphism in $\mathbb{Hilb}$:

$$\mathbf{C}^{\otimes m} \otimes \mathbf{C}^{\otimes n} \xrightarrow{|0\rangle^{\otimes n} \otimes |0\rangle^{\otimes m}} V^{\otimes n} \otimes V^{\otimes m} \xrightarrow{H^{\otimes m} \otimes id^{\otimes n}} V^{\otimes n} \otimes V^{\otimes m} \xrightarrow{U_f} V^{\otimes n} \otimes V^{\otimes n}.$$

$$\xrightarrow{id^{\otimes m} \otimes Ms} V^{\otimes n} \otimes V^{\otimes n} \xrightarrow{QFT^{\dagger} \otimes id^{\otimes n}} V^{\otimes n} \otimes V^{\otimes n} \tag{10.169}$$

In terms of a quantum circuit, this looks very similar to the circuit of Simons's periodicity algorithm:

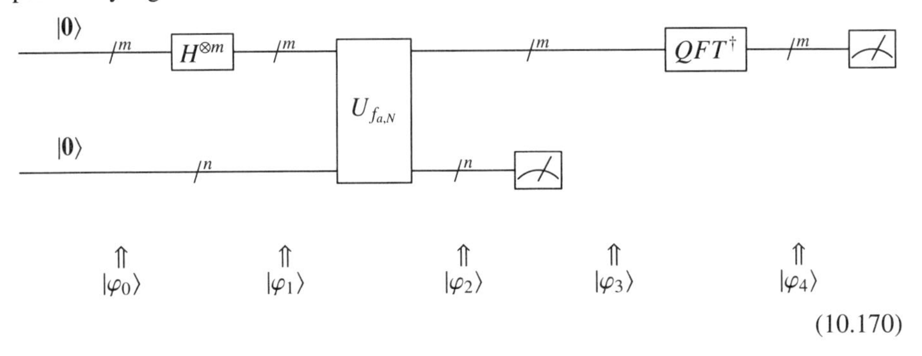

$$\Uparrow \quad |\varphi_0\rangle \qquad \Uparrow \quad |\varphi_1\rangle \qquad \Uparrow \quad |\varphi_2\rangle \qquad \Uparrow \quad |\varphi_3\rangle \qquad \Uparrow \quad |\varphi_4\rangle \tag{10.170}$$

The first few states are already familiar:

$$|\varphi_0\rangle = |\mathbf{0}_m, \mathbf{0}_n\rangle, \quad |\varphi_1\rangle = \frac{\sum_{\mathbf{x} \in \{0,1\}^m} |\mathbf{x}, \mathbf{0}_n\rangle}{\sqrt{2^m}},$$
$$|\varphi_2\rangle = \frac{\sum_{\mathbf{x} \in \{0,1\}^m} |\mathbf{x}, f_{a,N}(\mathbf{x})\rangle}{\sqrt{2^m}} = \frac{\sum_{\mathbf{x} \in \{0,1\}^m} |\mathbf{x}, a^{\mathbf{x}} \ Mod \ N\rangle}{\sqrt{2^m}}. \tag{10.171}$$

The rest of the states are a little bit more complicated then we need. The $QFT^{\dagger}$ is a **quantum Fourier transform**. This is like a Hadamard operator that gets the superposition results into the right shape. After applying this operator, we measure the period r of the $f_{a,N}$ function.

Knowledge of period r helps us find a factor of N. We need a period that is an even number. There is a theorem of number theory that tells us that for the majority of a, the period of $f_{a,N}$ is an even number. If, however, we do choose a such that the period is an odd number, simply throw that a away and choose another one. Once an even r is found, so that

$$a^r \equiv 1 \ Mod \ N, \tag{10.172}$$

we may subtract 1 from both sides of the equivalence to get

$$a^r - 1 \equiv 0 \ Mod \ N, \tag{10.173}$$

or equivalently,

$$N|(a^r - 1). \tag{10.174}$$

Remembering that $1 = 1^2$ and $x^2 - y^2 = (x+y)(x-y)$, we get that

$$N|(\sqrt{a^r}+1)(\sqrt{a^r}-1) \qquad \text{or} \qquad N|(a^{\frac{r}{2}}+1)(a^{\frac{r}{2}}-1). \tag{10.175}$$

(If r were odd, we would not be able to evenly divide by 2.) We must make sure that

$$a^{\frac{r}{2}} - 1 \neq 0 \;\; Mod \;\; N \tag{10.176}$$

because if they were equal, then $a^{\frac{r}{2}} \equiv 1 \;\; Mod \;\; N$, which would mean that $\frac{r}{2}$ was the period of a. But we already know that r is the period of a. If we find such an r, we must throw away that particular a and start over. If, however, a passes all these tests, then a factor of N is also a factor of either $(a^{\frac{r}{2}}+1)$ or $(a^{\frac{r}{2}}-1)$ or both. Either way, a factor for N can be found by looking at

$$\mathrm{GCD}((a^{\frac{r}{2}}+1), N) \qquad \text{and} \qquad \mathrm{GCD}((a^{\frac{r}{2}}-1), N). \tag{10.177}$$

Finding the GCD can be done with the classical Euclidean algorithm.

Let us put all the pieces of Shor's algorithm together and state it in pseudocode:

Input: A positive integer N.

Output: A factor p of N if it exists.

Step 1. Use a polynomial algorithm to determine if N is a prime number or a power of a prime number. If it is a prime number, declare that it is and exit. If it is a power of a prime number, declare that it is and exit.

Step 2. Randomly choose an integer a such that $1 < a < N$. Perform Euclid's algorithm to determine $\mathrm{GCD}(a, N)$. If the GCD is not 1, then we found a factor, and we return it and exit.

Step 3. Use the quantum part of the algorithm to find the period r of $f_{a,N}$.

Step 4. If r is odd or if $a^{\frac{r}{2}} \equiv -1 \;\; Mod \;\; N$, then return to Step 2 and choose another a.

Step 5. Use Euclid's algorithm to calculate $\mathrm{GCD}((a^{\frac{r}{2}}+1), N)$ and $\mathrm{GCD}((a^{\frac{r}{2}}-1), N)$. Return at least one of the nontrivial answers.

What is the complexity of Shor's algorithm? To determine this, one needs to have an in-depth analysis of the details of how $U_{f_{a,N}}$ and $QFT^{\dagger}$ are implemented. One would also need to know what percentage of times things can go wrong. For example, what percentage of a would $f_{a,N}$ have an odd period? Rather than going into the gory details, let us just state that Shor's algorithm works in

$$O((n^2 \log n)\,(\log \log n)) \tag{10.178}$$

number of steps, where n is the number of bits needed to represent the number N. This is polynomial in terms of n. This is in contrast to the best-known classical algorithms, which demand

$$O(e^{cn^{1/3}\,\log^{2/3} n}) \tag{10.179}$$

steps, where c is some constant. This is exponential in terms of n. Shor's quantum algorithm demands fewer operations.

Hidden Subgroup Problem. As is typical of category theory, there is a marvelous way of looking at most of the algorithms as instances of a single-arrow theoretic problem.

Definition 10.3.11. Consider group G and set S, which one should think of as a set of colors. Imagine that there is a set function $f\colon G \longrightarrow S$, which assigns to every $g \in G$ a color $f(g)$. We are "promised" or "assured" that this f has the following property: there is a subgroup H of G such that f gives the same color to all the members of each quotient class. This means that function f factors as

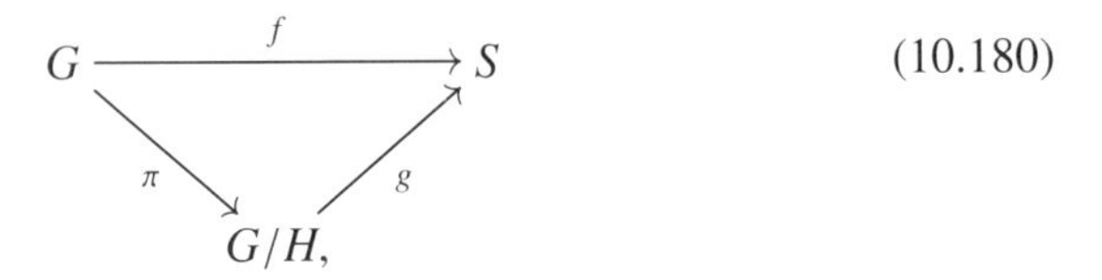

(10.180)

where π is the group homomorphism that takes every element of G to its quotient class and g assigns each quotient class a single color. The **hidden subgroup problem** asks one to determine H.

Remark 10.3.12. It is important to notice that the hidden subgroup problem is a simple problem in mathematics. Just set $H = f^{-1}(f(e))$, where e is the identity of the group. In other words, the solution is the collection of elements that go to the color of the identity. However, this is not a mathematics problem. It is a computer science problem. In general, we do not know—or have access to—the inverse of f. We are asking to solve the problem by only evaluating f. ♠

Let us look at the algorithms that we worked with and see them as instances of the hidden subgroup problem.

Example 10.3.13.

- **Deutsch's algorithm.** $G = \mathbf{Z}_2 = \{0, 1\}, S = \{0, 1\}$ and f is the given function. Group G only has two possible subgroups $H = \{e\}$ and $H = \mathbf{Z}_2$:

$$\begin{array}{ccc} \mathbf{Z}_2 & \xrightarrow{\quad f \quad} & \{0,1\} \\ & {\scriptstyle\pi}\searrow \quad \nearrow{\scriptstyle g} & \\ & \mathbf{Z}_2/H. & \end{array} \tag{10.181}$$

If $H = \{0\}$, then $f = g$ is balanced. In contrast, if $H = \mathbf{Z}_2$, then f is constant.

- **Deutsch-Jozsa's algorithm.** $G = \mathbf{Z}_2^n$ and $S = \{0, 1\}$, and we are assured that f is either balanced or constant:

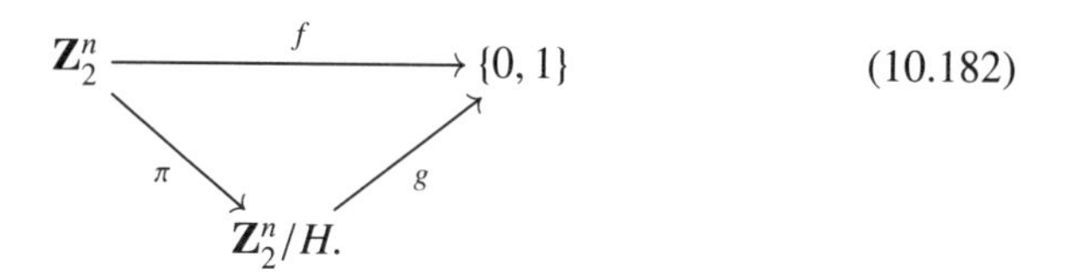

(10.182)

  If $H = \mathbf{Z}_2^n$, then f is constant. Otherwise, it is balanced.
- **Simons's periodicity algorithm.** $G = \mathbf{Z}_2^n$ and $S = \mathbf{Z}_2^n = \{0, 1\}^n$. We are given a map $f \colon \mathbf{Z}_2^n \longrightarrow \mathbf{Z}_2^n$ and assured that there is a $c \in \mathbf{Z}_2^n$ such that for all $x \in \mathbf{Z}_2^n$, we have $f(x) = f(x \oplus c)$. We are asked to find c:

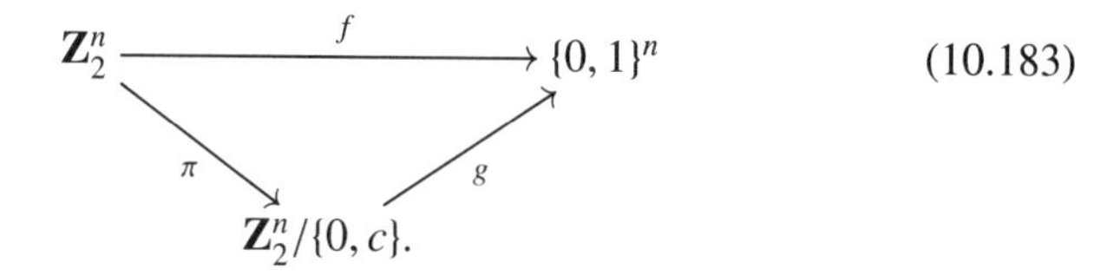

(10.183)

  We are searching for c so we can find $H = \{0, c\}$.
- **Shor's order-finding algorithm.** $G = \mathbf{Z}$ and $S = \{0, 1, 2, 3, \ldots N-1\}$, where N is a large number that is not prime. Let a be some number less than N such that $GCD(a, N) = 1$. Consider the function $f_{a,N}(x) = a^x MOD\ N$. The possible outputs of this set is $S = \{0, 1, 2, 3, \ldots N-1\}$. This function has a certain period r, which means that $f_{a,N}(r) = a^r MOD \quad N = 1$ and $f_{a,N}(x) = f_{a,N}(x + r)$. We are asked to find r:

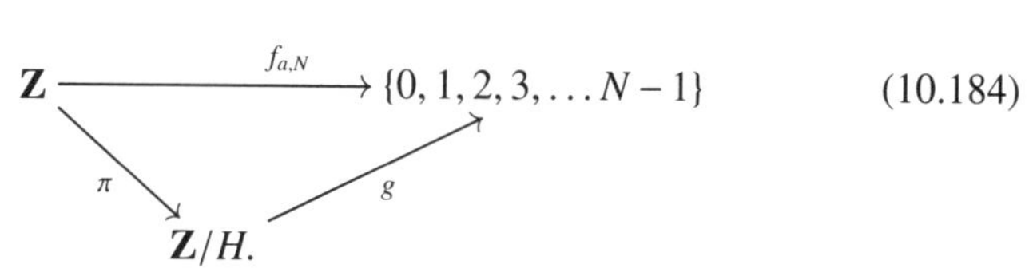

(10.184)

  The needed subgroup is $H = \{0, r, 2r, 3r, \ldots\}$; that is, the subgroup generated by r. Once we have this r, we can do a little classical computing to calculate the factors of N.

Notice that Grover's algorithm is missing from this list. This is because it's a different type of algorithm. □

We close with two fundamental limitations of quantum computing and quantum information theory. These ideas are called the **no-cloning theorem** and the **no-deleting theorem** and are related to the notion of conservation of information that is intertwined with quantum theory. We show that this theorem highlights the difference between a Cartesian category and a monoidal category.

The no-cloning theorem (see Section 3.5.4 of [263]) states that it is impossible to clone an exact quantum state. In other words, it is impossible to make a copy of an arbitrary quantum state without first destroying the original. In "computerese," this says that we can cut and paste a quantum state but we cannot *copy* and paste it. "Move is possible. Copy is impossible." This shows that quantum theory is resource sensitive.

What is the difficulty? Let us look what such a cloning operation would look like. Let $\mathcal{H}$ be a Hilbert space that represents a quantum system. As we intend to clone states in this system, we "double" this Hilbert space and work with $\mathcal{H}\otimes\mathcal{H}$. A potential cloning operation would be a linear map:

$$C\colon \mathcal{H}\otimes\mathcal{H} \longrightarrow \mathcal{H}\otimes\mathcal{H}, \tag{10.185}$$

which should take an arbitrary state $|x\rangle$ in the first system, and perhaps nothing in the second system, and clone $|x\rangle$; that is,

$$C(|x\rangle\otimes|0\rangle) = (|x\rangle\otimes|x\rangle). \tag{10.186}$$

Since C is a map in $\mathbb{Hilb}$, it must be linear. That is, it is linear for any quantum ket to include an arbitrary *superposition* of states. Suppose that we start with $|x\rangle + |y\rangle$. Cloning such a state would mean that

$$C\left((|x\rangle + |y\rangle)\otimes 0\right) = \left((|x\rangle + |y\rangle)\ \otimes\ (|x\rangle + |y\rangle)\right). \tag{10.187}$$

However, C must be linear, and hence, it must respect the addition in $\mathcal{H}\otimes\mathcal{H}$. So we have

$$C\left((|x\rangle + |y\rangle)\otimes 0\right) = C((|x\rangle\otimes 0) + (|y\rangle\otimes 0)) = C(|x\rangle\otimes 0) + C(|y\rangle\otimes 0) \tag{10.188}$$

$$= ((|x\rangle\otimes|x\rangle) + (|y\rangle\otimes|y\rangle)). \tag{10.189}$$

However,

$$((|x\rangle + |y\rangle)\ \otimes\ (|x\rangle + |y\rangle))\ \neq\ (|x\rangle\otimes|x\rangle) + (|y\rangle\otimes|y\rangle). \tag{10.190}$$

So C is not a linear map and hence is not permitted.

The no-cloning theorem is about a diagonal map. In Cartesian categories, there is a well-defined map $\Delta\colon a \longrightarrow a\times a$. In stark contrast, an arbitrary monoidal category like Hilbert space need not have a map $\Delta\colon \mathcal{H} \longrightarrow \mathcal{H}\otimes\mathcal{H}$. Another way to say this is that while the Δ morphism exists in the category of $\mathbb{Set}$, it does not exist in $\mathbb{Hilb}$ and therefore is not possible in quantum theory.

In contrast to cloning, there is no problem with **transporting** arbitrary quantum states from one system to another. Such a transporting operation would be a linear map:

$$T: \mathcal{H}\otimes\mathcal{H} \longrightarrow \mathcal{H}\otimes\mathcal{H}, \tag{10.191}$$

which should take an arbitrary state $|x\rangle$ in the first system and, say, nothing in the second system, and transport $|x\rangle$ to the second system, leaving nothing in the first system. That is,

$$T(|x\rangle\otimes|0\rangle) = (|0\rangle\otimes|x\rangle). \tag{10.192}$$

We do not run into the same problem as before if we transport a superposition of states. In detail,

$$T((|x\rangle + |y\rangle) \otimes |0\rangle) = T((|x\rangle \otimes 0) + (|y\rangle \otimes 0)) \tag{10.193}$$

$$= (T(|x\rangle \otimes |0\rangle) + T(|y\rangle \otimes |0\rangle)) = ((|0\rangle \otimes |x\rangle) + (|0\rangle \otimes |y\rangle)) = (|0 \otimes (|x\rangle + |y\rangle). \tag{10.194}$$

This is exactly what we would expect from a transporting operation.

In terms of categories, transporting says that in $\mathbb{Hilb}$, there is a well-defined braiding morphism $br\colon \mathcal{H} \otimes \mathcal{H} \longrightarrow \mathcal{H} \otimes \mathcal{H}$.

The **no-deleting theorem** (see, e.g., Exercise 3.5.8 of [263]) states that there is no way to delete a copy of a quantum state. In detail, a deleting morphism

$$D\colon \mathcal{H} \otimes \mathcal{H} \otimes \mathcal{H} \longrightarrow \mathcal{H} \otimes \mathcal{H} \otimes \mathcal{H} \tag{10.195}$$

would be defined as follows: For some quantum state $|\psi\rangle$ that we want to delete, we have

$$D(|\psi\rangle \otimes |\psi\rangle \otimes |\phi\rangle) = |\psi\rangle \otimes |0\rangle \otimes |\phi'\rangle, \tag{10.196}$$

where $|\phi\rangle$ is some start quantum state and $|\phi'\rangle$ is an ending quantum state. The problem again is that such a D morphism is not linear, and hence is not in $\mathbb{Hilb}$.

In terms of categories, this means that although in a Cartesian category, we have the following commutative diagram:

$$S \xrightarrow{\Delta} S \times S \xrightarrow{id \times !} S \times 1 \xrightarrow{\cong} S, \quad \text{(with an arc } S \xrightarrow{id} S \text{)} \tag{10.197}$$

where the morphism ! destroys the information in S, an analogous diagram with $\otimes$ instead of $\times$ does not necessarily exist.

In summation, the monoidal category $\mathbb{Hilb}$ is most definitely not a Cartesian category. For more information, see Important Categorical Idea 5.3.7. For more information, see also [25] for more along these lines.

Suggestions for Further Study

For more on quantum computing, may I humbly suggest [276], which is a clear introduction to the entire field. The details of all the algorithms are worked out with many examples and exercises. Although category theory is not mentioned in that text, the ideas of category theory are infused throughout it. For an interesting popular science book on quantum computing, and much much more, see [1].

For more on quantum theory and its connection to category theory, see [110]. Although [61] claims not to be a category theory book, it is full of interesting category theory ideas. There are also several very interesting papers by Abramsky and others that are important, such as [8, 7, 9, 6, 5]. These texts and papers use the full categorical structure of the category $\mathbb{Hilb}$.

Appendix A

Venn Diagrams

Legend

B's / A's	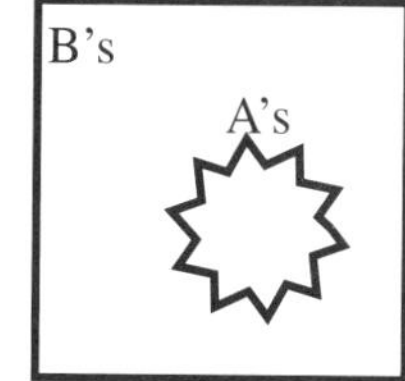 B's / A's
A's satisfy more axioms or requirements than B's. There is an inclusion functor from A's to B's.	A's have more operations or structure than B's. There is a forgetful functor from A's to B's.

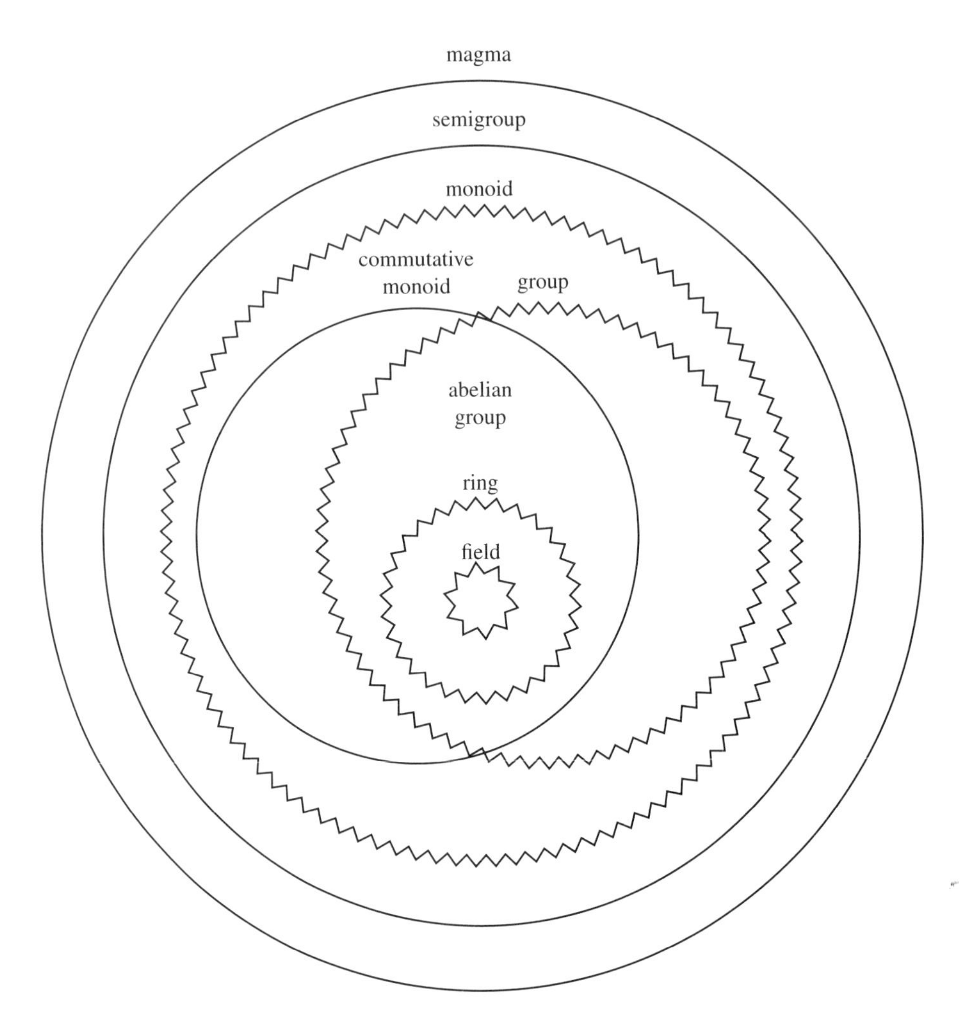

Figure A.1. Algebraic structures

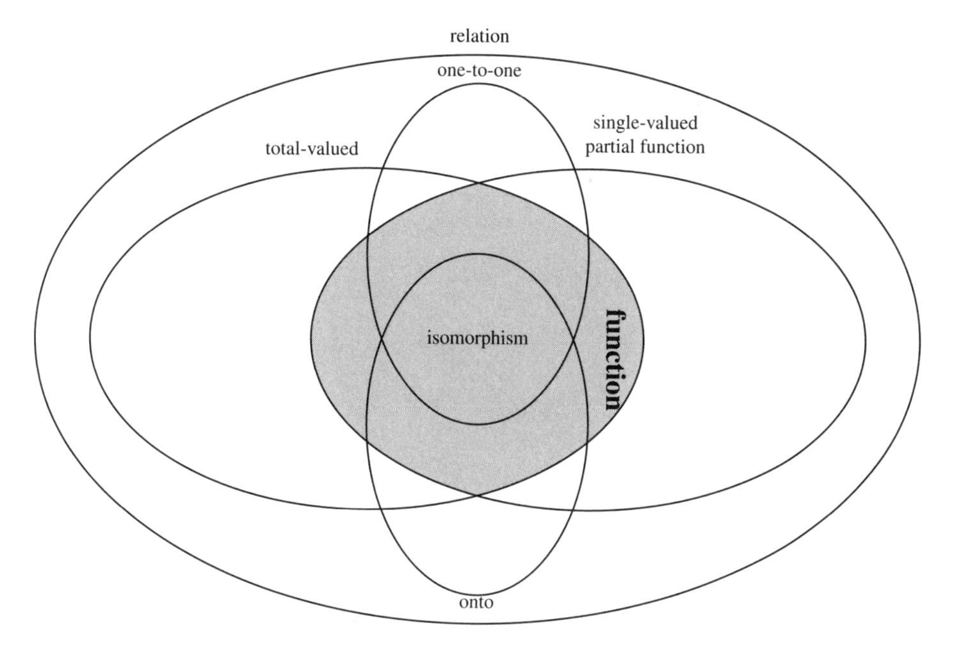

Figure A.2. Types of relations

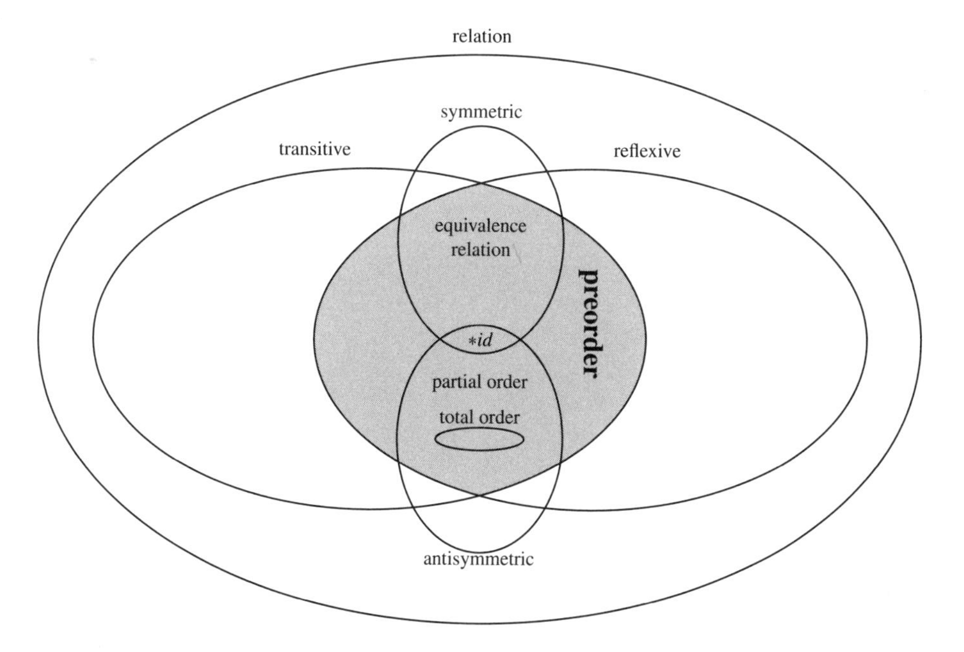

Figure A.3. Types of relations between a set and itself

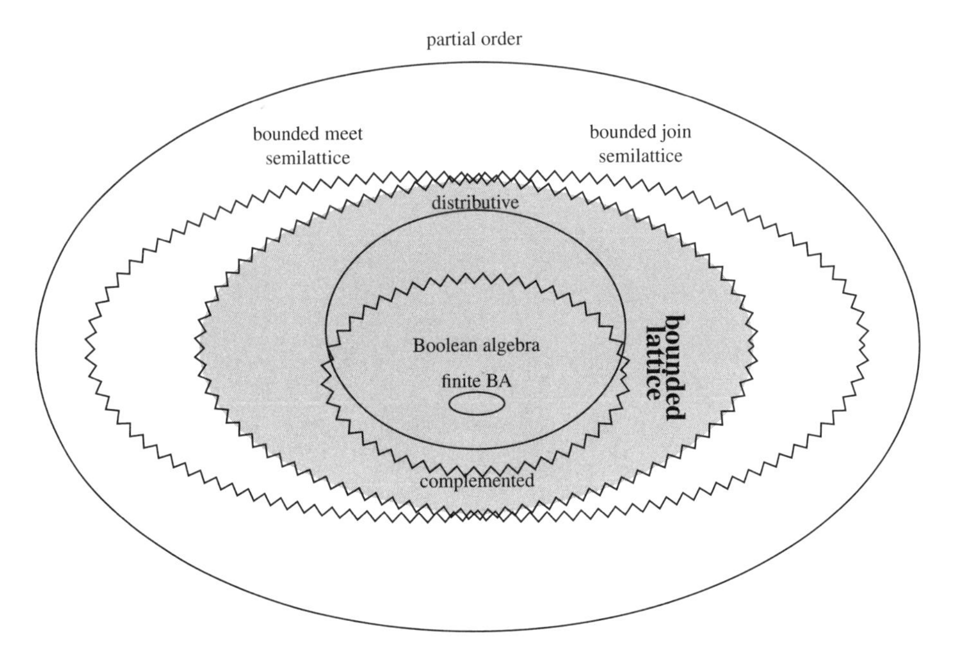

Figure A.4. Types of partial orders

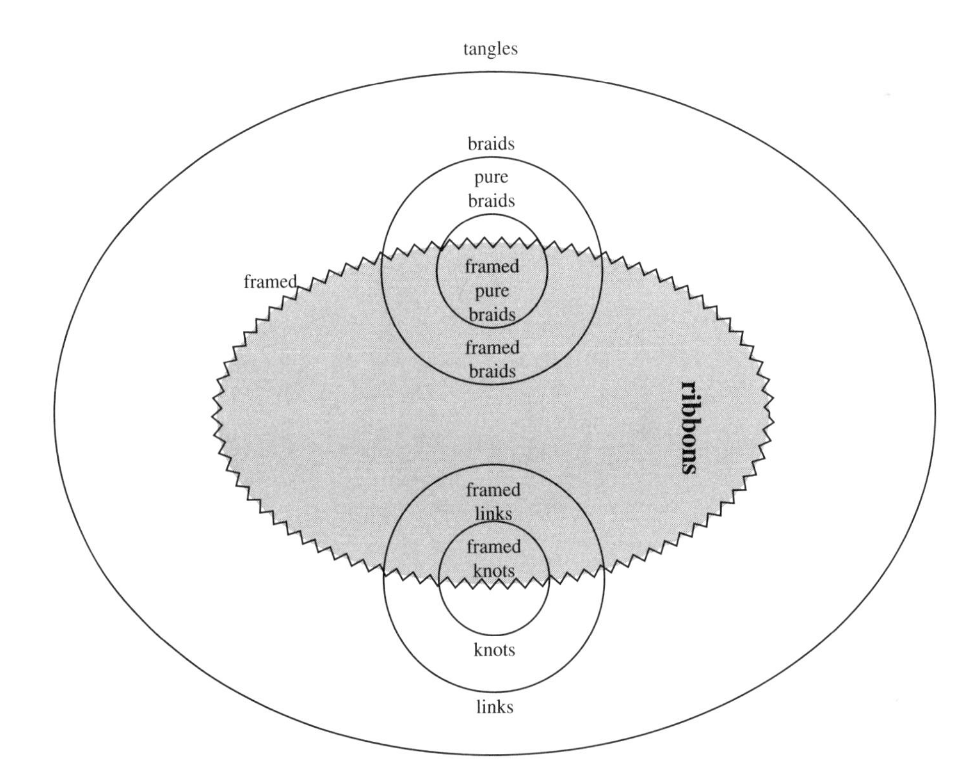

Figure A.5. Low-dimensional topological (oriented) structures

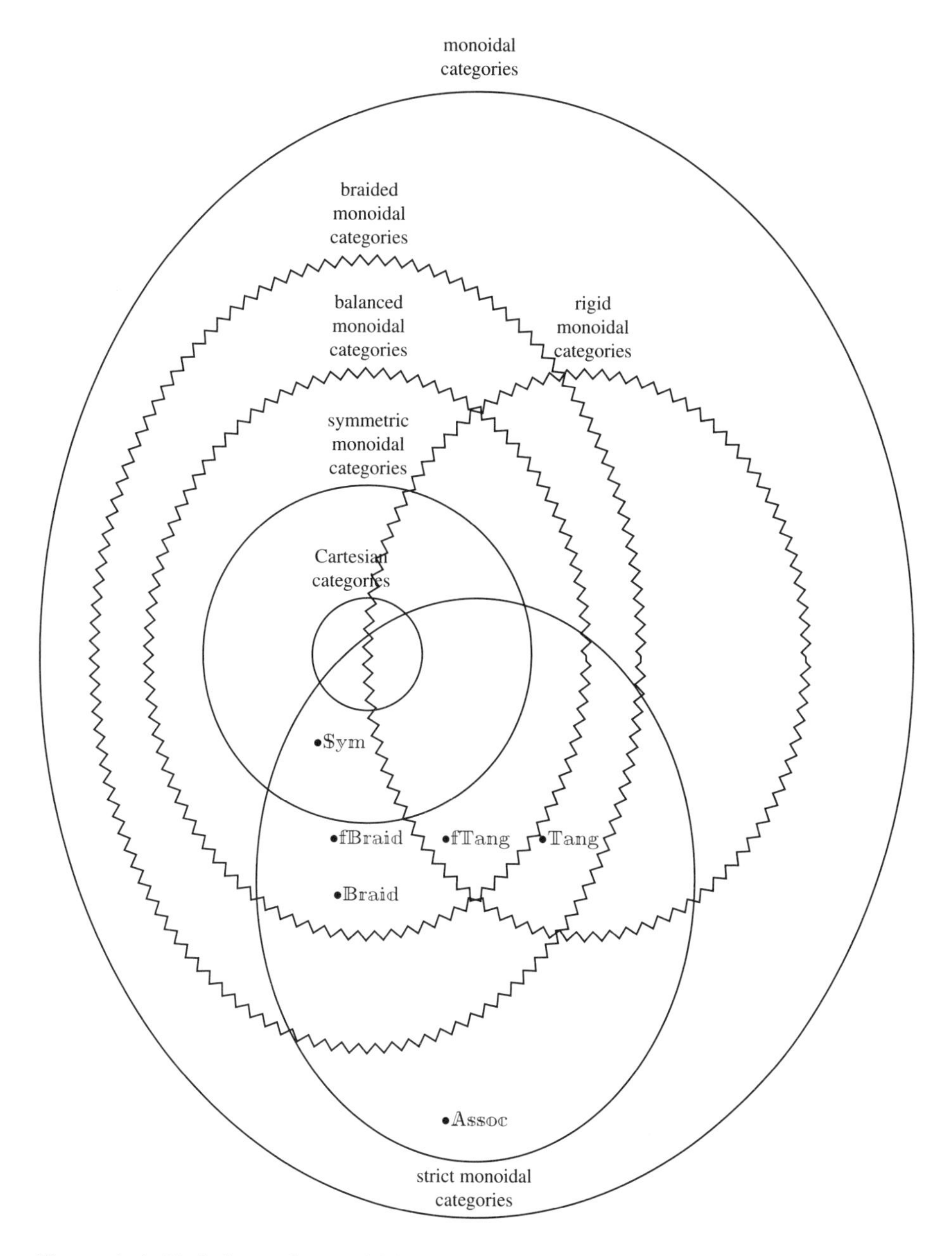

Figure A.6. Variations of monoidal categories and their paradigms. The left part has many notions of commutativity. The right part is about duality. The bottom is about associativity.

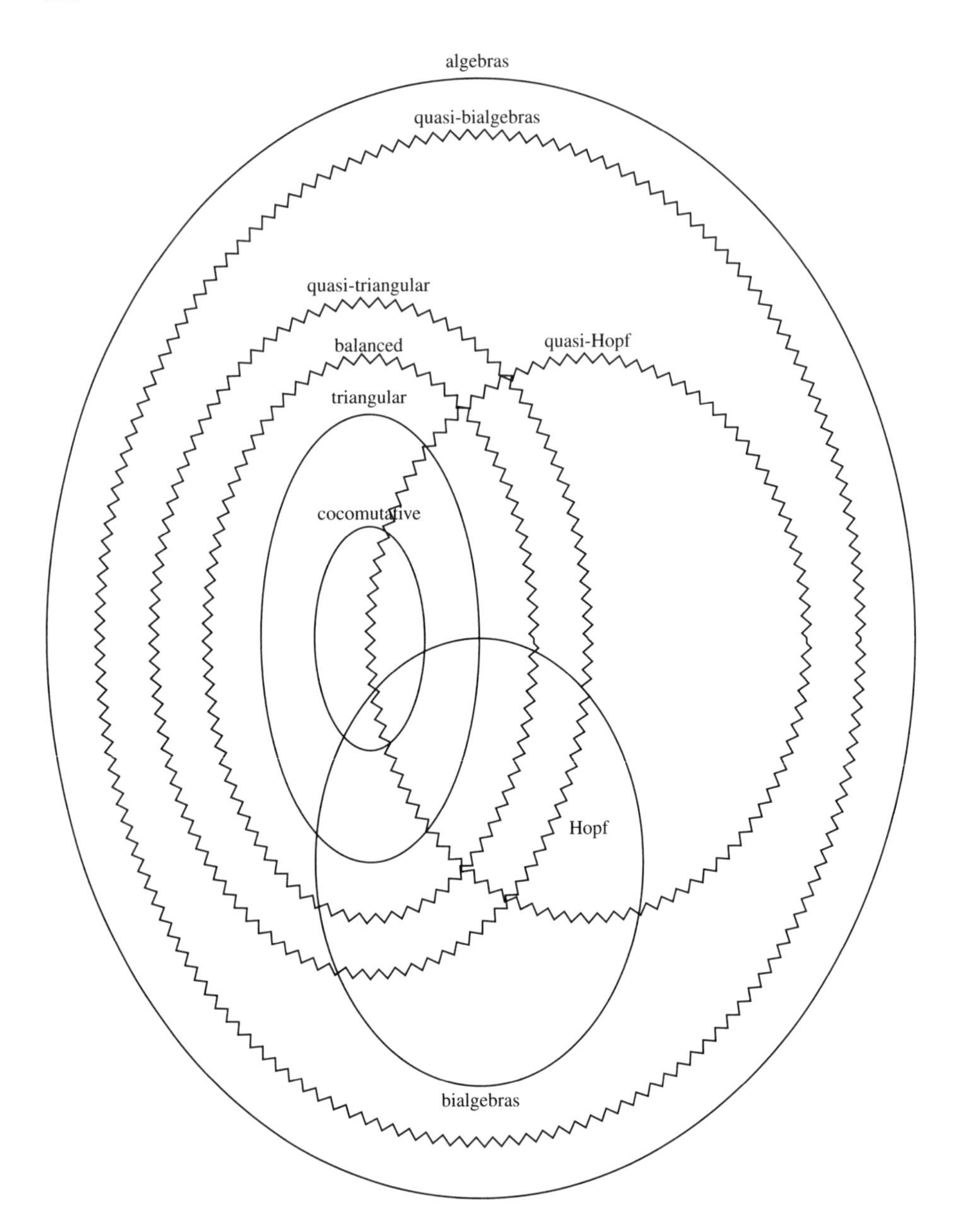

Figure A.7. Variations of algebras. The left part has many notions of cocommutativity. The right part is about duality. The bottom is about coassociativity.

Appendix B

Index of Categories

Category	Objects and Morphisms	Defined
Δ	The objects are totally ordered sets of numbers, and the morphisms are order-preserving functions.	9.3.5
$2\mathbb{C}\mathbb{a}\mathbb{t}$	2-categories and 2-functors.	9.4.4
$Bool\mathbb{M}\mathbb{a}\mathbb{t}$	Objects are natural numbers, and the morphisms are matrices with Boolean entries.	2.1.30
n-$\mathbb{M}\mathbb{a}\mathbb{n}\mathbb{i}\mathbb{f}$	n-dimensional manifolds and smooth maps.	2.1.29
$Nat\mathbb{I}\mathbb{s}\mathbb{o}$	Objects are sets $\{1, 2, 3, \ldots, n\}$ for all n, and morphisms are isomorphisms.	4.1.54
$Nat\mathbb{P}\mathbb{a}\mathbb{r}$	Objects are sets $\{1, 2, 3, \ldots, n\}$ for all n, and morphisms are partial functions.	4.1.53
$Nat\mathbb{R}\mathbb{e}\mathbb{l}$	Objects are sets $\{1, 2, 3, \ldots, n\}$ for all n, and morphisms are relations.	4.1.50
$Nat\mathbb{S}\mathbb{e}\mathbb{t}$	Objects are sets $\{1, 2, 3, \ldots, n\}$ for all n, and morphisms are functions.	4.1.53
$\mathbf{C}\mathbb{M}\mathbb{a}\mathbb{t}$	Objects are natural numbers, and the morphisms are matrices with complex number entries.	2.1.30
$\mathbf{C}\mathbb{V}\mathbb{e}\mathbb{c}\mathbb{t}$	Complex vector spaces and linear maps.	2.4.14
$\mathbf{K}\mathbb{F}\mathbb{D}\mathbb{V}\mathbb{e}\mathbb{c}\mathbb{t}$	Finite-dimensional $\mathbf{K}$ vector spaces and linear maps.	4.1.55
$\mathbf{K}\mathbb{M}\mathbb{a}\mathbb{t}$	Objects are natural numbers, and the morphisms are matrices with entries in the field $\mathbf{K}$.	2.1.30
$\mathbf{K}\mathbb{V}\mathbb{e}\mathbb{c}\mathbb{t}$	$\mathbf{K}$ vector spaces and linear maps.	2.4.14
$\mathbf{Q}\mathbb{M}\mathbb{a}\mathbb{t}$	Objects are natural numbers, and the morphisms are matrices with rational entries.	2.1.30
$\mathbf{R}\mathbb{M}\mathbb{a}\mathbb{t}$	Objects are natural numbers, and the morphisms are matrices with real number entries.	2.1.30

Category	Objects and Morphisms	Defined
ZMat	Objects are natural numbers, and the morphisms are matrices with integer entries.	2.1.30
AbGp	Abelian groups and group homomorphisms.	2.1.16
Assoc	Objects are associations, and there is a unique morphism between any two associations.	5.3.9
BClattice	Bounded complimented lattices and functions that preserve $\wedge$, $\vee$, 0, and 1.	6.4.1
BDlattice	Bounded distributive lattices and functions that preserve $\wedge$, $\vee$, 0, and 1.	6.4.1
BJslattice	Bounded join semilattices and functions that preserve $\vee$ and 0.	6.4.1
Blattice	Bounded lattices and functions that preserve $\wedge$, $\vee$, 0, and 1.	6.4.1
BMslattice	Bounded meet semilattices and functions that preserve $\wedge$ and 1.	6.4.1
BoolAlg	Boolean algebras and functions that preserve $\wedge$, $\vee$, 0, and 1.	6.4.1
Braid	Objects are the natural numbers, and the morphisms from n to n are the elements of the nth braid group.	7.1.6
CABoolAlg	Complete, atomic Boolean algebras and Boolean algebra homomorphisms.	6.4.7
CAT	Categories and functors.	4.1.22
Cat	Small categories and functors.	4.1.22
CGHS	Compactly generated Hausdorff spaces and continuous maps.	7.2.4
Circuit	Objects are natural numbers, and a morphism from m to n is a logical circuit with m inputs and n outputs.	2.1.57
Cocone(F)	For a functor $F\colon \mathbb{D} \longrightarrow \mathbb{B}$, the objects are natural transformations $F \Longrightarrow \Delta(b)$ for some $b \in B$. The morphisms are morphisms in $\mathbb{B}$ such that the appropriate diagrams commute.	4.6.1
ComMonoid	Commutative monoids and monoid homomorphisms.	2.1.16
CompFunc	Types and computable functions.	2.1.6
Cone(F)	For a functor $F\colon \mathbb{D} \longrightarrow \mathbb{B}$, the objects are natural transformations $\Delta(b) \Longrightarrow F$ for some $b \in B$. The morphisms are morphisms in $\mathbb{B}$ such that the appropriate diagrams commute.	4.6.1
Equiv	Equivalence relations and morphisms that respect the equivalence relations.	8.3.7
Esakia	Esakia spaces and their morphisms.	6.4.14
ETopoi	Elementary toposes and logical functors.	9.5.14
fBraid	Objects are natural numbers and the morphisms are framed braids.	7.3.2

Category	Objects and Morphisms	Defined
$\mathbb{FDHilb}$	Finite-dimensional Hilbert spaces and linear maps.	5.6.15
$\mathbb{Field}$	Fields and field homomorphisms.	2.1.16
$\mathbb{FinBoolAlg}$	Finite Boolean algebras and functions that preserve $\wedge$, $\vee$, 0, and 1.	6.4.1
$\mathbb{FinSet}$	Finite sets and functions between them.	2.3.2
$\mathbb{fPBraid}$	Objects are natural numbers and the morphisms are framed pure braids.	7.3.2
$\mathbb{Frame}$	Frame and frame maps.	6.4.18
$\mathbb{fTang}$	Objects are sequences of +'s and −'s, and and morphisms are framed tangles between them.	7.3.7
$\mathbb{Graph}$	Directed graphs and graph homomorphisms.	2.1.3
$\mathbb{Group}$	Groups and group homomorphisms.	2.1.3, 2.1.16
$\mathbb{GTopoi}$	Grothendieck toposes and geometrical functors.	9.5.16
$\mathbb{Hilb}$	Hilbert spaces and bounded linear maps.	5.6.15
$\mathbb{HeytAlg}$	Heyting algebras and functions that preserve $\Rightarrow$, $\wedge$, $\vee$, 0, and 1.	6.4.1
$\mathbb{hoTop}$	Objects are topological spaces, and morphisms are equivalence classes of continuous maps, where two maps are equivalent if there is a homotopy between them.	9.3.3
$\mathbb{hoTop}_*$	Objects are pointed topological spaces, and morphisms are equivalence classes of continuous maps that preserve the point, where two maps are equivalent if there is a homotopy between them.	9.3.3
$\mathbb{Lind}$	Objects are equivalence classes of propositions, and morphisms are entailments.	4.8.5
$\mathbb{Link}$	A single object $\emptyset$ and links between from $\emptyset$ to $\emptyset$.	7.2.16
$\mathbb{Locale}$	Locales and locale maps.	6.4.18
$\mathbb{Logic}$	The objects are natural numbers, and a morphism from m to n are equivalence classes of n-tuples of logical formulas where each formula uses at most m variables.	6.1.19
$\mathbb{Magma}$	Magmas and magma homomorphisms.	2.1.16
$\mathbb{Manif}$	Manifolds and smooth maps.	2.1.29
$\mathbb{Monad}(\mathbb{A})$	Monads on $\mathbb{A}$ and monad homomorphisms.	8.3.14
$\mathbb{Monoid}$	Monoids and monoid homomorphisms.	2.1.16
$\mathbb{MonCat}$	Monoidal categories and monoidal functors.	6.1.10
$\mathbb{NANDCircuit}$	Objects are natural numbers, and a morphism from m to n is a circuit that is only made of NAND gates with m inputs and n outputs.	2.3.4

Category	Objects and Morphisms	Defined
$\mathbb{OMat}$	Objects are natural numbers, and morphisms are orthogonal matrices.	5.6.19
$\mathbb{Operad}$	Operads and operad homomorphisms.	8.2.6
$\mathbb{Par}$	Sets and partial set functions.	2.1.47
$\mathbb{PBraid}$	Objects are the natural numbers, and the morphisms from n to n are the elements of the nth pure braid group.	7.1.7
$\mathbb{PO}$	Partial orders and order-preserving maps.	2.1.52
$\mathbb{Pred}(\bar{x})$	Let $\bar{x}$ be a set of variables. The objects are predicates that only use variables from $\bar{x}$, and morphisms are entailment.	4.8.9
$\mathbb{PreO}$	Preorder and order-preserving maps.	2.1.54
$\mathbb{Priestley}$	Priestley spaces and continuous maps that are order preserving.	6.4.16
$\mathbb{Proof}$	Propositions and proofs.	2.1.59
$\mathbb{Prop}$	Propositions and entailment.	2.1.7
$\mathbb{Rel}$	Sets and relations.	2.1.40
$\mathbb{Ring}$	Rings and ring homomorphisms.	2.1.16
$\mathbb{SemiGp}$	Semigroups and semigroup homomorphisms.	2.1.16
$\mathbb{Set}$	Sets and set functions.	2.1.1
$\mathbb{Set}^{\Delta^{op}}$	Simplicial sets and simplicial maps.	9.3.9
$\mathbb{SFrame}$	Spatial frames and frame maps.	6.4.20
$\mathbb{Sober}$	Sober spaces and continuous maps.	6.4.20
$\mathbb{sSet}$	Simplicial sets and simplicial maps.	9.3.9
$\mathbb{Stone}$	Stone spaces and continuous maps.	6.4.20
$\mathbb{StrMonCat}$	Strict monoidal categories and monoidal functors.	6.1.10
$\mathbb{SymMonCat}$	Symmetric monoidal categories and symmetric monoidal functors.	6.1.10
$\mathbb{StrSymMonCat}$	Strictly associative symmetric monoidal categories and strict symmetric monoidal functors.	6.1.10
$\mathbb{Sym}$	Objects are the natural numbers, and the morphisms from n to n are the elements of the nth symmetry group.	5.3.10
$\mathbb{Tang}$	Objects are sequences of +'s and −'s, and morphisms are tangles between them.	7.2.15
$\mathbb{Theory}$	Algebraic theories and theory morphisms.	8.1.13
$\mathbb{Top}$	Topological spaces and continuous maps.	2.1.25

Category	Objects and Morphisms	Defined
Type	Types and functions.	9.6.1
UHilb	Objects are Hilbert spaces, and morphisms are unitary, bounded, linear maps.	5.6.28
UMat	Objects are natural numbers, and morphisms are unitary matrices.	5.6.19
UFDHilb	Objects are finite-dimensional Hilbert spaces, and morphisms are unitary, bounded, linear maps.	5.6.28
Theory	Algebraic theories and theory morphisms.	8.1.13
Top	Topological spaces and continuous maps.	2.1.25
Type	Types and functions.	9.6.1
UHilb	Objects are Hilbert spaces, and morphisms are unitary, bounded, linear maps.	5.6.28
UMat	Objects are natural numbers, and morphisms are unitary matrices.	5.6.19
UFDHilb	Objects are finite-dimensional Hilbert spaces, and morphisms are unitary, bounded, linear maps.	5.6.28

2-Category	Objects, Morphisms, and 2-cells	Defined
$\overline{\mathbb{2Theory}}$	Algebraic 2-theories, 2-theory morphisms, and 2-theory natural transformations.	8.4.4
$\overline{\mathbb{Algebra}}$	Algebras, algebra homomorphisms, and algebra conjugates.	7.4.1
$\overline{\mathbb{BalMonCat}}$	Balanced monoidal categories, balanced monoidal functors, and their monoidal natural transformations.	7.3.3
$\overline{\mathbb{BraMonCat}}$	Braided monoidal categories, braided monoidal functors, and braided monoidal natural transformations.	7.1.8
$\overline{\mathbb{CAT}}$	Categories, functors, and natural transformations.	4.2.11
$\overline{\mathbb{Cat}}$	Small categories, functors, and natural transformations.	4.2.11
$\overline{\mathbb{Cat}}/\mathbf{C}\mathbb{Vect}$	Functors from small categories to $\mathbf{C}\mathbb{Vect}$, triangles of functors ending in $\mathbf{C}\mathbb{Vect}$, and natural transformations between such triangles.	7.4.1
$\overline{\mathbb{MonCat}}$	Monoidal categories, monoidal functors, and monoidal natural transformations.	6.1.12
$\overline{\mathbb{RibMonCat}}$.	Ribbon categories, their functors, and their natural transformations.	7.3.8
$\overline{\mathbb{RigidMonCat}}$	Rigid monoidal categories, rigid monoidal functors, and their monoidal natural transformations.	7.2.14
$\overline{\mathbb{StrBalMonCat}}$	Strictly associative, balanced monoidal categories, balanced monoidal functors, and their monoidal natural transformations.	7.3.3
$\overline{\mathbb{StrBraMonCat}}$	Strictly associative, braided monoidal categories, braided monoidal functors, and braided monoidal natural transformations.	7.1.9
$\overline{\mathbb{StrMonCat}}$	Strictly associative, monoidal categories, monoidal functors, and monoidal natural transformations.	6.1.12
$\overline{\mathbb{StrSymMonCat}}$	Strictly associative, symmetric monoidal categories, strict symmetric monoidal functors, and symmetric monoidal natural transformations.	6.1.12
$\overline{\mathbb{SymMonCat}}$	Symmetric monoidal categories, symmetric monoidal functors, and symmetric monoidal natural transformations.	6.1.12
$\overline{\mathbb{Top}}$	The objects are topological spaces, the morphisms are continuous maps, and the 2-cells are equivalence classes of homotopies between continuous maps.	9.3.3
$\overline{\mathbb{VCat}}$	For some monoidal category $\mathbb{V}$, $\mathbb{V}$-categories, $\mathbb{V}$-functors, and $\mathbb{V}$-natural transformations.	9.1.9

Appendix C

Suggestions for Further Study

This book is only the first step. The whole world of category theory is out there waiting for you. Here are several sources that can help you continue your explorations. Choose the topic that interests you the most.

- **Mathematics** Since category theory started as a branch of mathematics, most of the classical introductions to category theory are geared for people who know mathematics.
  - *Categories for the Working Mathematician* by Saunders Mac Lane [180]
    - This book was the "bible" of the subject for a long time. It is extremely clear and has a wealth of information.
  - *Category Theory: An Introduction* by Horst Herrlich and George E. Strecker [108]
  - *Handbook of Categorical Algebra* by Francis Borceux [40, 41, 42]
    - This three-volume work has much in it and is a valuable resource.
  - *Toposes, Triples, and Theories* by Michael Barr and Charles Wells [32]

  Here are some recent mathematics-based introductions to category theory:
  - *Category Theory* by Steve Awody [21]
  - *Category Theory in Context* by Emily Riehl [222]
  - *Basic Category Theory* by Tom Leinster [171]
- **Physics**
  - "This Week's Finds in Mathematical Physics," by John Baez
    - Probably the best way to learn about the relationship of physics and category theory is to leap into these blog posts. From 1993 to 2010, Baez described many wonderful ideas and connections between category theory and physics. In almost eighteen years, he produced about 300 posts. Each one is a delightful gem that is easy to read. It is well worth your time to study these. The website is at http://math.ucr.edu/home/baez/twf.html.

- "Physics, Topology, Logic and Computation: A Rosetta Stone," by John Baez and Mike Stay [24].
- *Mathematical Physics* by Robert Geroch [86]

- **Computing**
  - *Categories and Computer Science* by R. F. C. Walters [258]
  - *Basic Category Theory for Computer Scientists* by Benjamin C. Pierce [208]
  - *Category Theory for Computing Science* by Michael Barr and Charles Wells [33]
    - This great book has many topics and is available for free online.
  - *Category Theory for the Sciences* by David Spivak [243]
    - This is a nice introduction with connections to databases.
  - *Categories, Types, and Structures: An Introduction to Category Theory for the Working Computer Scientist* by Andrea Asperti and Giuseppe Longo [18]
  - *Arrows, Structures, and Functors: The Categorical Imperative* by Michael A. Arbib and Ernest G. Manes [17]

  The following three are geared more for logicians:

  - *Topoi: The Categorial Analysis of Logic* by Robert Goldblatt [91]
  - *Stone Spaces* by Peter T. Johnstone [125]
  - *Introduction to Higher-Order Categorical Logic* by Joachim Lambek and Philip Scott [151]

- **History and Philosophy**
  - "Category Theory," in the *Stanford Online Encyclopedia of Philosophy* by Jean-Pierre Marquis [188].
  - *Tool and Object: A History and Philosophy of Category Theory* by Ralf Krömer [150]
  - *From a Geometrical Point of View: A Study of the History and Philosophy of Category Theory* by Jean-Pierre Marquis [187]
  - *Axiomatic Method and Category Theory* by Andrei Rodin [224]

  Eugenia Cheng has two delightful popular science books on category theory:

  - *How to Bake Pi: An Edible Exploration of the Mathematics of Mathematics* [58]
  - *The Joy of Abstraction—An Exploration of Math, Category Theory, and Life* [59]. This is more than just a popular science book. It goes into a lot of detail and explains the information very well.

- **Journals** Here are a few journals that specialize in category theory:
  - *Theory and Applications of Categories* - TAC - http://www.tac.mta.ca/tac/
    - This is a free online journal that has also republished classic category theory papers and books that are out of print. These are called "TAC Reprints" and can be found at http://www.tac.mta.ca/tac/reprints/index.html.

- *Applied Categorical Structures*
  * http://www.springer.com/mathematics/journal/10485.
- *Mathematical Structures in Computer Science*
- *Categories and General Algebraic Structures with Applications*
  * http://www.cgasa.ir/.
- *Cahiers de Topologie et Géométrie Différentielle Catégoriques*
  * http://ehres.pagesperso-orange.fr/Cahiers/Ctgdc.htm.
- *Journal of Pure and Applied Algebra*
  * https://www.journals.elsevier.com/journal-of-pure-and-applied-algebra

- **Online Resources** Books and papers are not the only way to learn information. There are now many different resources online, including the following:
  - Wikipedia
  - nLab https://ncatlab.org/nlab/show/HomePage
    * This has many different entries on every concept related to category theory and higher category theory. (A warning is in order. nLab is like Wikipedia in that everyone can put in whatever data they want. Researchers have added many results. At the present time, there is not much deleting going on there. Many pages are made for experts, and the novice might get lost.)
  - The n-Category Cafe: A group blog on math, physics and philosophy at https://golem.ph.utexas.edu/category/.
    * This is a forum with many different announcements, discussions, and notes on various current topics.
  - Catsters, at https://www.youtube.com/user/TheCatsters/featured
    * This is a series of YouTube videos by Eugenia Cheng and Simon Willerton that teach various topics in category theory.

The world of category theory is now open to you. Go and explore!

Appendix D

Answers to Selected Problems

1.4.6. 4, 3, $12 = 4 \cdot 3$.

1.4.7. $m \cdot n \cdot p \cdot q \cdot r$.

1.4.8. When the two sets have nothing in common (i.e., when the intersection of the two sets is empty).

1.4.18. Let us just focus on the subset $\mathbf{Q} \subsetneq \mathbf{R}$. The characteristic function $\chi_{\mathbf{Q}} : \mathbf{R} \longrightarrow \{0, 1\}$ is defined as follows:

$$\chi_{\mathbf{Q}}(r) = \begin{cases} 1 & : r \text{ is rational.} \\ 0 & : r \text{ is not rational.} \end{cases}$$

The others are done similarly.

1.4.20. $(\sqrt{5}, -5)$.

1.4.23. The isomorphism will take a pair of real numbers $(r_1, r_2) \in \mathbf{R} \times \mathbf{R}$ to the complex number $r_1 + ir_2 \in \mathbf{C}$, where $i = \sqrt{-1}$.

1.4.24. Each of the following lines is a function:

$f(a) = 0$	$f(b) = 0$	$f(c) = 0$
$f(a) = 0$	$f(b) = 0$	$f(c) = 1$
$f(a) = 0$	$f(b) = 1$	$f(c) = 0$
$f(a) = 0$	$f(b) = 1$	$f(c) = 1$
$f(a) = 1$	$f(b) = 0$	$f(c) = 0$
$f(a) = 1$	$f(b) = 0$	$f(c) = 1$
$f(a) = 1$	$f(b) = 1$	$f(c) = 0$
$f(a) = 1$	$f(b) = 1$	$f(c) = 1$

1.4.29. There are two ways of associating the functions: $h \circ (g \circ f)$ and $(h \circ g) \circ f$. These two functions are the same. On input s of S, both functions have the value $h(g(f(s)))$.

1.4.37. This is essentially the content of Diagram (1.30).

1.4.52. We prove the contrapositive. If G is not weakly connected, then G will be able to split the graph into two parts, with no arrow from one part to the other part. There is, then, a function from the graph to this two-vertex graph, where the nodes of one part of the graph go to a and all the arrows of that part go to the single arrow f_a. The nodes of the other part go to b, and all the arrows of that part go to f_b. If the graph is weakly connected, then this partitioning would not be possible.

1.4.53.

- One-to-one graph homomorphisms from the snake graph, Diagram (1.52), to any graph will correspond to simple paths.
- A cycle can be described as a graph homomorphisms from ring graphs. These are graphs of the form

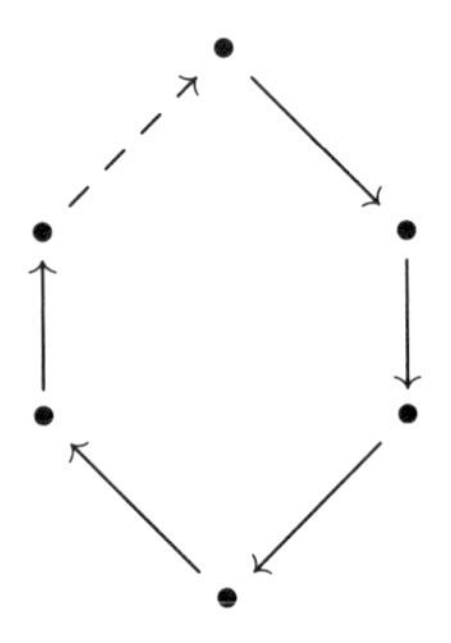

  where the dashed arrow is an ellipsis that possibly denotes many more arrows.
- One-to-one graph homomorphisms from the ring graphs will correspond to simple cycles.

1.4.54. Let $H\colon G \longrightarrow G'$ and $H'\colon G' \longrightarrow G''$ be graph homomorphisms. Then $H' \circ H : G \longrightarrow G''$ is a graph homomorphism. The fact that $H' \circ H$ preserves the sources of arrows amounts to the commutativity of the following diagram:

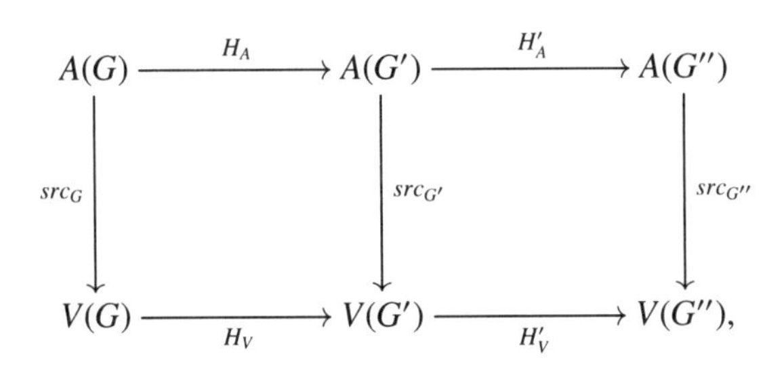

which is assured because each square commutes. A similar argument must be made to show that $H' \circ H$ preserves targets. The proof of the associativity of the composition of graph homomorphisms comes from the fact that graph homomorphisms are functions and is similar to the solution to Exercise 1.4.29.

1.4.55. The identity graph homomorphism is defined as $I_{G,A}(x) = x$ and $I_{G,V}(f) = f$. This graph homomorphism preserves the sources and targets of the arrows for trivial reasons. The fact that it acts like a unit to composition is because it is essentially two identity functions.

1.4.58. It is essentially the same diagram but change the $G \times \{*\}$ to $\{*\} \times G$.

1.4.59. It is essentially the same diagram with the $id \times (\ \ )^{-1}$ map switched to $(\ \ )^{-1} \times id$.

1.4.64. Let $f\colon G \longrightarrow G'$ and $f'\colon G' \longrightarrow G''$ be group homomorphisms. Then $(f' \circ f)(x) = f'(f(x))$. To show that composition preserves the group operations, do the following: for $x, y \in G$,

$$\begin{aligned}(f' \circ f)(x \star y) &= f'(f(x \star y)) = f'(f(x) \star' f(y)) = f'(f(x)) \star'' \\ f'(f(y)) &= (f' \circ f)(x) \star'' (f' \circ f)(y)\end{aligned}$$

and

$$(f' \circ f)(e) = f'(f(e)) = f'(e') = e''.$$

1.4.65. The identity trivially respects the group operations. The last part follows from the fact that group homomorphisms are simply functions that satisfy certain properties.

2.1.15. The fact that the identity function respects all the operations is trivially true. The fact that the identity function acts like a unit to composition follows from the fact that homomorphisms are simply special types of set functions and such functions satisfy this property.

2.1.22. Let $f\colon (X, \tau) \longrightarrow (Y, \sigma)$ and $g\colon (Y, \sigma) \longrightarrow (Z, \zeta)$ be two continuous maps. Now consider the function $g \circ f$ and $V \in \zeta$. Then $g^{-1}(V)$ is open in σ by the continuity of g, and $f^{-1}(g^{-1}(V))$ is open in τ by the continuity of f. Hence, $(g \circ f)^{-1} = f^{-1} \circ g^{-1}$ takes open sets to open sets.

2.1.23. This follows from the fact that continuous maps are special types of set functions, and we showed in Exercise 1.4.29 that the composition operation of set functions is associative.

2.1.24. The preimage of the identity function is the identity function, and it takes open sets to open sets. The fact that the identity acts like a unit to the composition follows from the fact that the identity function is a special type of set functions.

2.1.36. Given the relations $R\colon S \not\longrightarrow T$, $Q\colon T \not\longrightarrow U$ and $P\colon U \not\longrightarrow V$, the composition of all three is $P \circ Q \circ R$, which is defined as

$$\{(s, v) : \text{there exist } t \in T \text{ and } u \in U \text{ such that } (s, t) \in R,\ (t, u) \in Q, \text{ and } (u, v) \in P\}.$$

2.1.37. Let $f\colon S \longrightarrow T$ and $g\colon T \longrightarrow U$ be functions that are associated with relations $\hat{f}\colon S \not\longrightarrow T$ and $\hat{g}\colon T \not\longrightarrow U$, respectively. If $f(s) = t$ and $g(t) = u$, then $(s, t) \in \hat{f}$ and $(t, u) \in \hat{g}$. Hence, $(s, u) \in \widehat{g \circ f}$.

2.1.39. Let $(s,t) \in R$. For every $s \in S$, $(s,s) \in id_S$. The fact that (s,s) and (s,t) are in the relations ensures that (s,t) is in the composition. A similar argument works for the second identity.

2.1.45. We prove that single-valued is equivalent to $R \circ R^{-1} \subseteq id_T$ and leave the rest for the reader:

R is single-valued $\iff$ every $s \in S$ and $t, t' \in T$, $(s,t) \in R$ and $(s,t') \in R$ implies $t = t'$
$\iff$ every $s \in S$ and $t, t' \in T$, $(s,t) \in R$ and $(t', s) \in R^{-1}$ implies $t = t'$
$\iff$ every $t, t' \in T$, $(t', t) \in R \circ R^{-1}$ implies $t = t'$
$\iff$ every $t, t' \in T$, $(t', t) \in R \circ R^{-1}$ implies $(t', t) \subseteq Id_T$
$\iff$ $R \circ R^{-1} \subseteq Id_T$

2.1.50. Consider the order-preserving $f\colon (P, \leq) \longrightarrow (Q, \leq')$ and $f'\colon (Q, \leq') \longrightarrow (R, \leq'')$. The inequality $p \leq p'$ implies $f(p) \leq' f(p')$, which in turn implies $f'(f(p)) \leq'' f'(f(p'))$. The associativity follows because an order-preserving function is a special type of function.

2.1.51. If $p \leq p'$, then $id_P(p) = p \leq p' = id_P(p')$. The unit part follows because these are all simply special types of set functions.

2.1.62.

- $Mor(\mathbf{2}) = \{id_a, f, id_b\}$. $dom(id_a) = id_a, cod(id_a) = id_a; dom(f) = id_a, cod(f) = id_b; dom(id_b) = id_b, cod(id_b) = id_b$
- $Mor(\mathbf{2}_I) = \{id_a, f, g, id_b\}$. $dom(id_a) = id_a, cod(id_a) = id_a; dom(f) = id_a, cod(f) = id_b; dom(g) = id_b, cod(g) = id_a; dom(id_b) = id_b, cod(id_b) = id_b$
- $Mor(\mathbf{3}) = \{f, g, h, id_a id_b, id_c\}$. $dom(f) = id_a, cod(f) = id_b; dom(g) = id_b, cod(g) = id_c; dom(h) = id_a, cod(h) = id_c; dom(id_a) = id_a, cod(id_a) = id_a; dom(id_b) = id_b, cod(id_b) = id_b; dom(id_c) = id_c, cod(id_c) = id_c$

We leave the composition operation to the reader.

2.1.63. Just use the set of morphisms. Let the domain of morphisms be the identities of the sources of the arrows. Let the codomain of morphisms be the identities of the targets of the arrows.

2.1.64. Make the objects correspond to the identity morphisms.

2.2.4. Let $f\colon a \longrightarrow b$ and $f'\colon b \longrightarrow b'$ be monics. Since f' is monic, then

$$f' \circ (f \circ g) = f' \circ (f \circ h) \text{ implies } (f \circ g) = (f \circ h).$$

Since f is monic, then

$$f \circ g = f \circ h \text{ implies } g = h.$$

Combining the implications and using the associativity of maps, we have that

$$(f' \circ f) \circ g = (f' \circ f) \circ h \text{ implies } g = h.$$

2.2.5. Assume that f is a split monomorphism. If $f \circ g = f \circ h$, then by composition with f', we have $f' \circ f \circ g = f' \circ f \circ h$. This implies that $id \circ g = id \circ h$ and $g = h$.

2.2.7. If $f\colon S \longrightarrow T$ is onto, then by Theorem 1.4.32 we know that f has a right inverse. If we further assume that $g \circ f = h \circ f$, then precomposing with the right inverse gives us that $g = h$. If, on the other hand, f is epic,

$$S \xrightarrow{\ f\ } T \overset{g}{\underset{h}{\rightrightarrows}} \{0,1\} \tag{1}$$

then set $g\colon T \longrightarrow \{0,1\}$ as the characteristic function of the image of f (i.e., $g(t) = 1$ iff t is in the image of f) and h as the function that is constant at 1. Saying that $g \circ f = h \circ f$ implies $g = h$, which implies that every t is in the image of f.

2.2.8. This is similar to the solution of Exercise 2.2.4.

2.2.9. Assume that f is a split epimorphism. If $g \circ f = h \circ f$, then by composition with f', we have $g \circ f \circ f' = h \circ f \circ f'$. This implies that $g \circ id = h \circ id$ and $g = h$.

2.2.11. Assume that $f\colon a \longrightarrow b$ has two inverses, $g\colon b \longrightarrow a$ and $g'\colon b \longrightarrow a$. Then

$$g = g \circ id_b = g \circ (f \circ g') = (g \circ f) \circ g' = id_a \circ g' = g'.$$

2.2.12. The inverse of the identity is itself.

2.2.13. The composition of the inverses form the inverse to the composition.

2.2.14.

- Reflexive comes from Exercise 2.2.12, where we see that every object is isomorphic to itself (reflective).
- Symmetry comes from the fact that the inverse of an isomorphism is also an isomorphism.
- Transitivity comes from Exercise 2.2.13.

2.2.25. The proof is very similar to the proof of Theorem 2.2.23.

2.3.19. If x is an initial object in $\mathbb{A}$, then for every element $a \in \mathbb{A}$, there is a unique morphism $x \longrightarrow a$. This means that in $\mathbb{A}^{op}$, there is a unique morphism $a \longrightarrow x$ for every $a \in \mathbb{A}^{op}$.

2.3.22. The category **3** has three objects, while the category $\mathbf{2}_I$ has two objects. The Cartesian product has $3 \cdot 2 = 6$ objects. We can think of these six objects as split into two sides, and each side contains **3**. These two sides are connected by isomorphisms, as in the following:

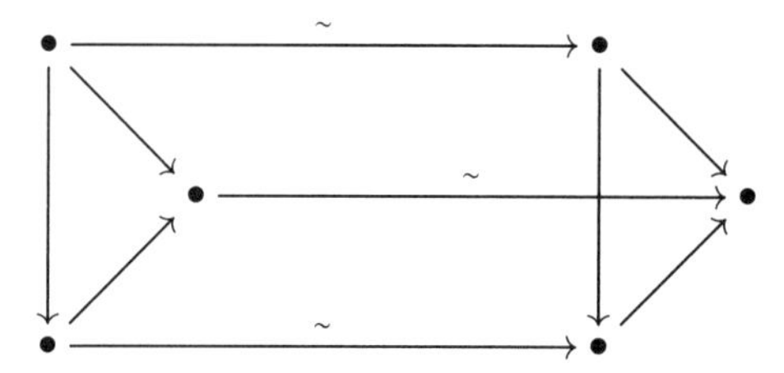

2.4.3. This is very similar to showing that *Func*(**N**, **C**) is a vector space.

2.4.8. Setting c and c' to be 1 in Equation (2.44) gives Equation (2.42). Setting $c'=0$ in Equation (2.44) gives Equation (2.43). The other direction is trivial.

2.4.10.

- $\pi_{1,3}((c_1,c_2,c_3)+(c_1',c_2',c_3'))=\pi_{1,3}(c_1+c_1',c_2+c_2',c_3+c_3')$
  $=(c_1+c_1',c_3+c_3')=\pi_{1,3}(c_1,c_2,c_3)+\pi_{1,3}(c_1',c_2',c_3')$
- $\pi_{1,3}(c\cdot(c_1,c_2,c_3))=\pi_{1,3}(cc_1,cc_2,cc_3)=(cc_1,cc_3)=c\cdot\pi_{1,3}(c_1,c_2,c_3)$

2.4.12. Let $T\colon V\longrightarrow V'$ and $T'\colon V'\longrightarrow V''$ be linear maps. Then their composition is linear because the following are both true:

- $T'(T(v+v'))=T'(T(v)+T(v'))=T'(T(v))+T'(T(v'))$
- $T'(T(c\cdot v))=T'(c\cdot(T(v)))=c\cdot(T'(T(v)))$

The composition operation is associative because linear maps are nothing more than special types of set functions.

2.4.13. The identity map does not change anything, so it clearly respects the operations. The fact that the identity linear maps are a unit to the composition follows from the fact that linear maps are simply special types of set functions.

2.4.22.

$$[c_0,c_1,c_2,\ldots,c_m]^T\longmapsto c_0+c_1x+c_2x^2+\cdots+c_mx^m.$$

2.4.23. Since S is finite, one can order the elements. You might as well think of S as $\{s_1,s_2,\ldots,s_m\}$. Then a function $f\colon S\longrightarrow\mathbf{C}$ can be uniquely associated with $[f(s_1),f(s_2),\ldots,f(s_m)]^T$.

2.4.26. Only the 0 vector space can have an injection into the 0 vector space.

2.4.29. This is not a simple proof. It can be found in any book on linear algebra. (Hint: show that you can replace the elements of one basis by the elements of the other basis.)

2.4.32. The canonical basis consists of special functions f_s for every $s\in S$ defined as

$$f_s(s')=\begin{cases}1 & : s=s'\\ 0 & : s\neq s'.\end{cases}$$

The dimension is the same as the size (i.e., cardinality) of S.

2.4.36. We have that there is some w and w' such that $v-v'=w$ and $u-u'=w'$. From this, we have $v=w+v'$ and $u=w'+u'$. Therefore,

$$[v]+[u]=[v+u]=[w+v'+w'+u']=[v'+u']=[v']+[u'].$$

The equation $c\cdot[v]=c\cdot[v']$ is done similarly.

3.1.7. By the universal property, there are morphisms $c\longrightarrow c'$ and $c'\longrightarrow c$, where each one commutes with the projection maps. When you compose these maps, you get maps $c\longrightarrow c$ and $c'\longrightarrow c'$, which also commute with the projection maps. Such maps are unique by the universal properties. We already know the identity maps have this property. This ensures that the two maps are the inverse of each other. One must still show that there is only one such isomorphism. This follows along the same line as the proof of Theorem 2.2.23.

3.1.21.

$$\begin{aligned}
\alpha' \circ \alpha &= \langle \pi_a^{(ab)c}, \langle \pi_b^{(ab)c}, \pi_c^{(ab)c} \rangle \rangle \circ \alpha && \text{by definition of } \alpha' \\
&= \langle \pi_a^{(ab)c}\alpha, \langle \pi_b^{(ab)c}, \pi_c^{(ab)c} \rangle \alpha \rangle && \text{by Equation (3.22)} \\
&= \langle \pi_a^{(ab)c}\alpha, \langle \pi_b^{(ab)c}\alpha, \pi_c^{(ab)c}\alpha \rangle \rangle && \text{by Equation (3.22)} \\
&= \\
&\langle \pi_a^{(ab)c} \langle \langle \pi_a^{a(bc)}, \pi_b^{a(bc)} \rangle, \pi_c^{a(bc)} \rangle, \langle \\
&\pi_b^{(ab)c} \langle \langle \pi_a^{a(bc)}, \pi_b^{a(bc)} \rangle, \pi_c^{a(bc)} \rangle, \\
&\pi_c^{(ab)c} \langle \langle \pi_a^{a(bc)}, \pi_b^{a(bc)} \rangle, \pi_c^{a(bc)} \rangle \rangle \rangle && \text{by definition of } \alpha \\
&= \langle \pi_a^{a(bc)}, \langle \pi_b^{a(bc)}, \pi_c^{a(bc)} \rangle \rangle && \text{by Equation (3.13)} \\
&= id_{a \times (b \times c)} && \text{by definition of } id.
\end{aligned}$$

We leave the $\alpha \circ \alpha'$ calculation for the reader to do.

3.1.24. Here, $br^{-1} \colon b \times a \longrightarrow a \times b$ is defined as $br^{-1} = \langle \pi_b^{ba}, \pi_a^{ba} \rangle$. Showing that they are inverse of each other means the following:

$$br^{-1} \circ br = \langle \pi_b^{ba}, \pi_a^{ba} \rangle \circ \langle \pi_b^{ab}, \pi_a^{ab} \rangle$$

$$= \langle \pi_b^{ba} \circ \langle \pi_b^{ab}, \pi_a^{ab} \rangle, \pi_a^{ba} \circ \langle \pi_b^{ab}, \pi_a^{ab} \rangle \rangle = \langle \pi_a^{ab}, \pi_b^{ab} \rangle = id_{a \times b}.$$

The other direction is similar.

3.1.26. Consider the following commutative diagram:

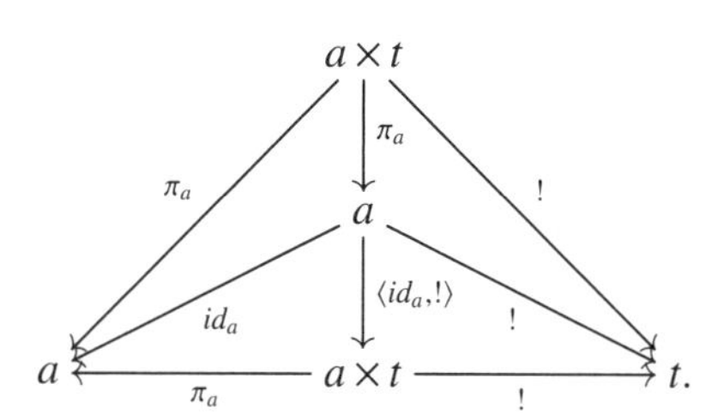

The isomorphism is $\langle id_a, ! \rangle$ with inverse π_a. By the universal property, $\pi_a \circ \langle id_a, ! \rangle = id_a$. Consider the map $\langle id_a, ! \rangle \circ \pi_a \colon a \times t \longrightarrow a \times t$. Notice that this map and $id_{a \times t}$ satisfy the universal property. By uniqueness, these maps are equal and we have shown $\langle id_a, ! \rangle$ and π_a are the inverse of each other.

3.1.31. Invert the arrows of the proof given in the solution to Exercise 3.1.7.

3.1.37.

$$\alpha = [inc_{a+b}inc_a, [inc_{a+b}inc_b, inc_c]].$$

3.2.1. A limit to diagram a is an object c with a map $\pi\colon c \longrightarrow a$, such that for any other object d and map $k\colon d \longrightarrow a$, there is a unique $f\colon d \longrightarrow c$. Set $d = a$ and $k = id_a$. This shows that $\pi \circ f = id_a$. The equality $f \circ \pi = id_c$ follows from the fact that there is only one morphism $c \longrightarrow c$, making the obvious diagram commute.

3.2.2. Consider the empty diagram that does not have any objects or morphisms. A limit of this diagram will be an object (call it i). There is no requirement for a map to the diagram because there are no objects in the diagram. The universal property of being a limit says that for any object a in the category (again, there is no requirement for a morphism from a), there is a unique morphism to i. The colimit is similar.

3.2.13. This is very similar to the proof of Theorems 3.2.10 and 3.2.11.

3.3.3. For any object $f\colon b \longrightarrow a$ of $\mathbb{A}/a$, f is the unique morphism that makes the triangle commute.

3.3.6. This is similar to the solution to Exercise 3.3.3.

4.1.13. In *Bool*, 1+1=1, while in **N**, 1+1=2.

4.1.16. All the rest have proper inclusions of Hom sets.

4.1.19. Given $F\colon \mathbb{A} \longrightarrow \mathbb{B}$ and $G\colon \mathbb{B} \longrightarrow \mathbb{C}$, we have to show that $G \circ F$ is a functor as follows:

- On object a in $\mathbb{A}$, $(G \circ F)(a) = G(F(a))$.
- For $f\colon a \longrightarrow a'$ and $f'\colon a' \longrightarrow a''$, we have

$$(G \circ F)(f' \circ f) = G(F(f' \circ f)) = G(F(f')) \circ G(F(f)) = (G \circ F)(f') \circ (G \circ F)(f).$$

- For a in $\mathbb{A}$,

$$(G \circ F)(id_a) = G(F(id_a)) = G(id_{F(a)}) = id_{G(F(id_a))} = id_{(G \circ F)(id_a)}.$$

4.1.20. This is like the composition of functions.

4.1.21. This is like the composition of the identity functions with other functions.

4.1.25. The reverse of the reverse of an arrow is the original arrow.

4.1.35. On objects of $\mathbb{A}$, functor F' is exactly the same as F. Consider $f\colon a \longrightarrow a'$ in $\mathbb{A}^{op}$. There is a corresponding $f'\colon a' \longrightarrow a$ in $\mathbb{A}$. The morphism $F(f)\colon F(a) \longrightarrow F(a')$ in $\mathbb{B}$ will correspond to $F'(f')\colon F(a') \longrightarrow F(a)$ in $\mathbb{B}^{op}$.

4.1.36. Assume that f is an epimorphism (i.e., for g_1 and g_2, $g_1 f = g_2 f$ implies that $g_1 = g_2$). The morphism $Hom(f, a)$ takes map g_1 to $g_1 f$. Let $g_1, g_2\colon S \longrightarrow Hom(b', a)$ be two set functions such that $Hom(f, a)g_1 = Hom(f, a)g_2$. This means that $g_1 f = g_2 f$, which implies that $g_1 = g_2$. The inverse implication is very similar.

4.1.37. Assume that f is a split monomorphism (i.e., there is a $f'\colon b' \longrightarrow b$ such that $f' f = id_b$). The morphism $Hom(f, a)$ takes map g_1 to $g_1 f$. The morphism $Hom(f, a)$ is surjective if, for all $g\colon b \longrightarrow a$, there is a $\widehat{g}\colon b' \longrightarrow a$ such that $Hom(f, a)\widehat{g} = \widehat{g} f = g$. Consider the following:

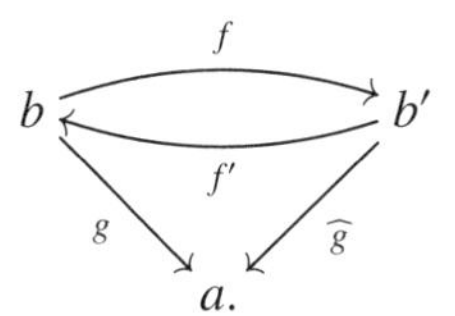

For any g, let $\widehat{g} = gf'$. Then $Hom(f,a)\widehat{g} = Hom(f,a)gf' = gf'f = g$. So $Hom(f,a)$ is an epimorphism.

For the other direction, set $a = b$, and then consider $Hom(f,b)\colon Hom(b',b) \longrightarrow Hom(b,b)$. Let this map be surjective/epimorphic, and then consider id_b in $Hom(b,b)$. The map f' such that $Hom(f,b)(f') = id_b$ will split f.

4.1.45. The inverse of $br_{\mathbb{A},\mathbb{B}}$ is $br_{\mathbb{B},\mathbb{A}}$.

4.1.48. A functor F picks out one object in $\mathbb{Set}$ (say, $F(*) = S$). For every $m\colon * \longrightarrow *$, we have $F(m\colon * \longrightarrow *)\colon F(*) \longrightarrow F(*)$. In other words, $F(m)\colon S \longrightarrow S$. The action that m does on s is the value of $F(m)$ at input s (i.e., $m \cdot s = F(m)(s)$). This functor respects composition in M; that is,

$$F(m \star m') = F(m) \circ F(m').$$

This translates into the first diagram. The functor respects the identity of M (i.e., $F(e) = id_{F(*)}$, translates into the second diagram).

4.1.51. We have to make sure that $F(R' \circ R) = F(R') \circ F(R)$. For this, we recall how relations compose in Equation (2.17). Basically, it comes down to there being a 1 in a row of one matrix and a 1 in the same column of another matrix. For all n, the value $F(id_{\bar{n}})$ is the n by n identity matrix.

4.1.54. A matrix will be in this image if there is exactly one 1 on each row and exactly one 1 on each column. Such matrices are called **permutation matrices**.

4.2.5. Draw three boxes and continue as expected.

4.2.9. Start by examining three composable natural transformations:

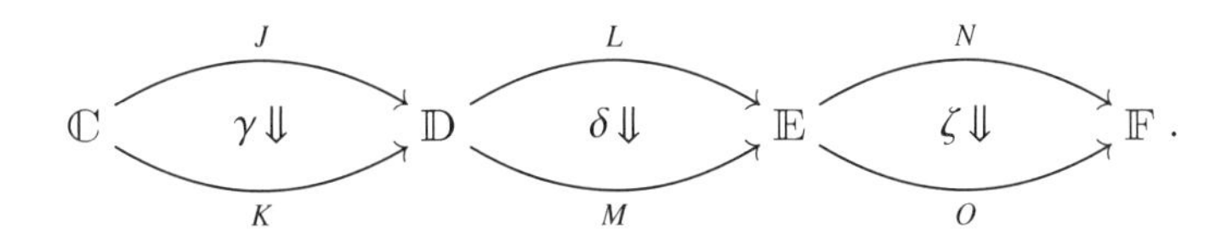

Simply write the definitions of the composition.

4.4.6. All you have to show is that Definition (IV) is equivalent to Definition (II), which is very similar to the fact that Definition (III) is equivalent to Definition (II).

4.4.26.

$$Hom_{\mathbb{B}^{op}}(b, L^{op}(a)) = Hom_{\mathbb{B}}(L^{op}(a), b) = Hom_{\mathbb{B}}(L(a), b) \cong$$

$$Hom_{\mathbb{A}}(a, R(b)) = Hom_{\mathbb{A}}(a, R^{op}(b)) = Hom_{\mathbb{A}^{op}}(R^{op}(b), a).$$

4.8.13. Consider the following adjunctions:

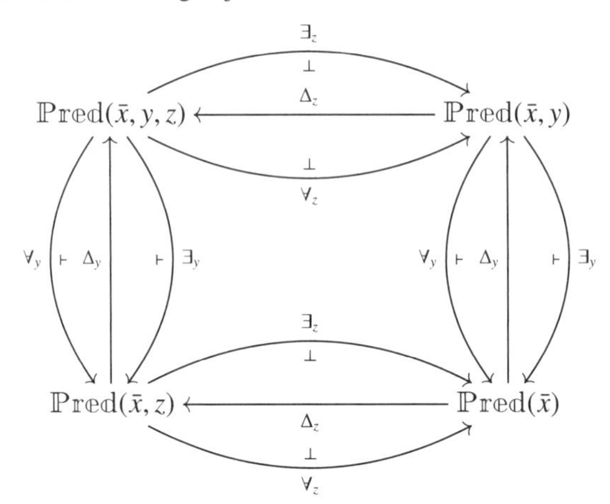

The two maps from the lower right to the upper left amount to $Q(\bar{x}) \mapsto \Delta_y\Delta_z Q(\bar{x}) = \Delta_z\Delta_y Q(\bar{x})$). This means that the two compositions of two right adjoints are isomorphic. Similarly, the composition of two left adjoints are isomorphic.

5.1.4. We will see that these two monoidal categories are isomorphic as monoidal categories. The string $1111\cdots 1$, where there are n copies of 1, goes to the number n.

5.2.6. All the operations are pointwise, and the structure is inherited from V and W.

5.2.8. The isomorphism is given by $(v,(w,x)) \longmapsto ((v,w),x)$. The existence of the isomorphism also follows from the fact that the projection maps satisfy the universal properties.

5.2.9. The isomorphism is given by $(v,0) \longmapsto v$. The existence of the isomorphism also follows from the fact that the projection maps satisfy the universal properties.

5.2.10. The isomorphism is given by $(v,w) \longmapsto (w,v)$. The existence of the isomorphism also follows from the fact that the projection maps satisfy the universal properties.

5.2.21. Use the fact that $\mathbf{K}\mathbb{M}\mathrm{at}^{op} \cong \mathbf{K}\mathbb{M}\mathrm{at}$, which we saw in Example 4.1.28 on page 149. To see that the universal property of being a coproduct is satisfied, consider the following:

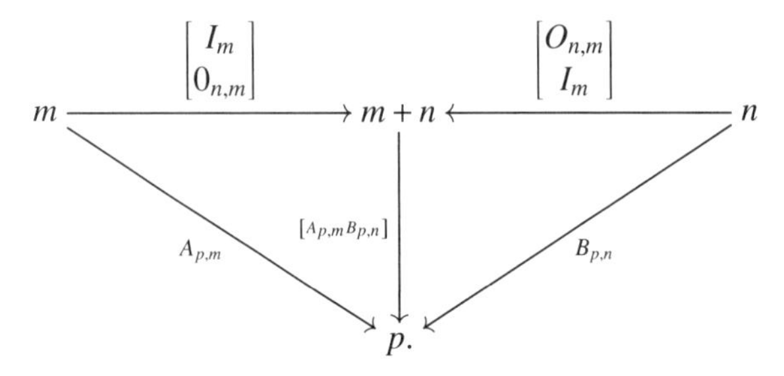

5.3.8. The proof is dual to the proof that every Cartesian category is a symmetric monoidal category.

5.6.3. We prove some of these properties as follows:

- $(c \cdot A)^{\dagger}[i, j] = \overline{(c \cdot A)[j, i]} = \overline{c} \cdot \overline{A[j, i]} = \overline{c} \cdot (A^{\dagger})[i, j]$.
- $(A \otimes B)^{\dagger}[i, j] = \overline{(A \otimes B)}[j, i] = (\overline{A} \otimes \overline{B})[j, i] = \overline{A}[j/m', i/n'] \cdot \overline{B}[j \; Mod \; m', i \; Mod \; n']$ $= A^{\dagger}[i/n', j/m'] \cdot B^{\dagger}[i \; Mod \; n', j \; Mod \; m'] = (A^{\dagger} \otimes B^{\dagger})[i, j]$.
- $Tr(A + B) = \sum_i (A + B)[i, i] = \sum_i (A[i, i] + B[i, i]) = \sum_i (A[i, i]) + \sum_i (B[i, i])$ $= Tr(A) + Tr(B)$.
- $Tr(c \cdot A) = \sum_i (c \cdot A)[i, i]) = c \cdot \sum_i A[i, i] = c \cdot Tr(A)$.
- $Tr(A \cdot B) = \sum_i AB[i, i] = \sum_i (\sum_k (A[i, k] B[k, i])) = \sum_k (\sum_i (B[i, k] A[k, i]))$ $= Tr(B \cdot A)$.
- $Tr(A \otimes B)$ will be the sum of the diagonal elements of $A \otimes B$. A little thought about the Kronecker product explains that the diagonal elements of $A \otimes B$ are obtained from the diagonal elements of A and B. So $Tr(A \otimes B) = (\sum_i (A[i, i]$ $(\sum_j B[i, i])) = (\sum_i (A[i, i])(\sum_j B[i, i]) = Tr(A) \cdot Tr(B)$.

5.6.22.

- The adjoint preserves vector addition:

$$\langle v, T^{\dagger}(w + w') \rangle = \langle T(v), w + w' \rangle = \langle T(v), w \rangle + \langle T(v), w' \rangle$$

$$= \langle v, T^{\dagger}(w) \rangle + \langle v, T^{\dagger}(w') \rangle = \langle v, T^{\dagger}(w) + T^{\dagger}(w') \rangle.$$

  Since there is only one adjoint map $T^{\dagger}$ that satisfies Equation (5.159) with $\langle T(v), w + w' \rangle$, we have that $T^{\dagger}(w + w') = T^{\dagger}(w) + T^{\dagger}(w')$.
- The adjoint preserves scalar multiplication:

$$\langle v, T^{\dagger}(c \cdot w) \rangle = \langle T(v), c \cdot w \rangle = \overline{c} \langle T(v), w \rangle = \overline{c} \langle v, T^{\dagger}(w) \rangle = \langle v, c \cdot T^{\dagger}(w) \rangle.$$

  Since there is only one adjoint map $T^{\dagger}$ that satisfies Equation (5.159) with $\langle T(v), c \cdot w \rangle$, we have that $T^{\dagger}(c \cdot w) = c \cdot T^{\dagger}(w)$.

5.6.23. We prove all the properties by using the uniqueness of the adjoint:

- $\langle v, (T + T')^{\dagger}(w) \rangle = \langle (T + T')(v), w \rangle = \langle T(v), w \rangle + \langle T'(v), w \rangle = \langle v, T^{\dagger}(w) \rangle$ $+ \langle v, T'^{\dagger}(w) \rangle = \langle v, T^{\dagger}(w) + T'^{\dagger}(w) \rangle$
- $\langle v, c \cdot T^{\dagger}(w) \rangle = \overline{c} \langle v, T^{\dagger}(w) \rangle = \overline{c} \langle T(v), w \rangle = \langle (\overline{c} \cdot T(v)), w \rangle = \langle v, (\overline{c} \cdot T^{\dagger})(w) \rangle$
- $\langle v, (T'^{\dagger} \circ T^{\dagger})(w) \rangle = \langle v, T'^{\dagger}(T^{\dagger}(w)) \rangle = \langle T'(v), T^{\dagger}(w) \rangle = \langle T(T'(v)), w \rangle =$ $\langle (T \circ T')(v), w \rangle = \langle v, (T \circ T')^{\dagger}(w) \rangle$
- $\langle v, (T \oplus T')^{\dagger}(w) \rangle = \langle (T \oplus T')(v), w \rangle = \langle T(v), w \rangle + \langle T'(v), w \rangle = \langle v, T^{\dagger}(w) \rangle$ $+ \langle v, T'^{\dagger}(w) \rangle = \langle v, T^{\dagger}(w) \oplus T'^{\dagger}(w) \rangle$
- $\langle v, (T \otimes T')^{\dagger}(w) \rangle = \langle (T \otimes T')(v), w \rangle = \langle T(v), w \rangle \cdot \langle T'(v), w \rangle = \langle v, T^{\dagger}(w) \rangle$ $\cdot \langle v, T'^{\dagger}(w) \rangle = \langle v, T^{\dagger}(w) \cdot T'^{\dagger}(w) \rangle$

5.6.27. Assume that T and T' are unitary. Then we have the following:

- $(T \circ T') \circ (T \circ T')^{\dagger} = (T \circ T') \circ (T'^{\dagger} \circ T^{\dagger}) = T \circ Id_V \circ T^{\dagger} = Id$
- $(T \oplus T') \circ (T \oplus T')^{\dagger} = (T \oplus T') \circ (T^{\dagger} \oplus T'^{\dagger}) = (T \circ T^{\dagger}) \oplus (T' \circ T'^{\dagger}) = Id \oplus Id = Id$
- $(T \otimes T') \circ (T \otimes T')^{\dagger} = (T \otimes T') \circ (T^{\dagger} \otimes T'^{\dagger}) = (T \circ T^{\dagger}) \otimes (T' \circ T'^{\dagger}) = Id \otimes Id = Id$

Bibliography

[1] Scott Aaronson. *Quantum computing since Democritus*. Cambridge University Press, Cambridge, 2013. 589

[2] S. Abramsky and N. Tzevelekos. Introduction to categories and categorical logic. In *New structures for physics*, vol. 813 of *Lecture Notes in Phys.*, pp. 3–94. Springer, Heidelberg, 2011. 224

[3] Samson Abramsky. A domain equation for bisimulation. *Inform. and Comput.*, 92(2): 161–218, 1991. 330

[4] Samson Abramsky. Domain theory in logical form. *Ann. Pure Appl. Logic*, 51 (1–2): 1–77, 1991. 330

[5] Samson Abramsky. Temperley-Lieb algebra: From knot theory to logic and computation via quantum mechanics. In *Mathematics of quantum computation and quantum technology*, Appl. Math. Nonlinear Sci. Ser., pp. 515–558. Chapman & Hall/CRC, Boca Raton, FL, 2008. 589

[6] Samson Abramsky. No-cloning in categorical quantum mechanics. In *Semantic techniques in quantum computation*, pp. 1–28. Cambridge University Press, Cambridge, 2010. 589

[7] Samson Abramsky and Bob Coecke. Abstract physical traces. *Theory Appl. Categ.*, 14(6): 111–124, 2005. 589

[8] Samson Abramsky and Bob Coecke. Categorical quantum mechanics. In *Handbook of quantum logic and quantum structures—quantum logic*, pp. 261–323. Elsevier/North-Holland, Amsterdam, 2009. 589

[9] Samson Abramsky and Chris Heunen. Operational theories and categorical quantum mechanics. In *Logic and algebraic structures in quantum computing*, vol. 45 of *Lect. Notes Log.*, pp. 88–122. Assoc. Symbol. Logic, La Jolla, CA, 2016. 589

[10] Samson Abramsky and Achim Jung. Domain theory. In *Handbook of logic in computer science*, vol. 3, pp. 1–168. Oxford University Press, New York, 1994. 330

[11] J. Adámek, J. Rosický, and E. M. Vitale. *Algebraic theories: A categorical introduction to general algebra*, vol. 184 of *Cambridge Tracts in Mathematics*. Cambridge University Press, Cambridge, 2011. 395

[12] Jiri Adamek, Stefan Milius, and Lawrence S. Moss. *Initial algebras, terminal coalgebras, and the theory of fixed points of functors*. Cambridge University Press, in preparation. 427

[13] Colin C. Adams. *The knot book: An elementary introduction to the mathematical theory of knots. Revised reprint of the 1994 original*. American Mathematical Society, Providence, RI, 2004. 380, 543

[14] M. Agrawal, N. Kayal, and N. Saxena. PRIMES in P. *Ann. of Math. 2*, 160(2): 781–793, 2004. 582

[15] Paolo Aluffi. *Algebra: Chapter 0*. In vol. 104 of *Graduate studies in mathematics*. American Mathematical Society, Providence, RI, 2009. 32, 42

[16] D. W. Anderson. Axiomatic homotopy theory. In *Algebraic topology, Waterloo, 1978 (Proc. Conf., Univ. Waterloo, Waterloo, Ontario, 1978)*, vol. 741 of *Lecture Notes in Math.*, pp. 520–547. Springer, Berlin, 1979. 500

[17] Michael A. Arbib and Ernest G. Manes. *Arrows, structures, and functors: The categorical imperative*. Academic Press, Inc. New York and London, 1975. 604

[18] Andrea Asperti and Giuseppe Longo. *Categories, types, and structures: An introduction to category theory for the working computer scientist*. Foundations of Computing Series. MIT Press, Cambridge, MA, 1991. 604

[19] Michael F. Atiyah. Duality in mathematics and physics. https://fme.upc.edu/ca/arxius/butlleti-digital/riemann/071218_conferencia_atiyah-d_article.pdf. Accessed March 15, 2022. 330

[20] Jeremy Avigad. The mechanization of mathematics. *Notices Amer. Math. Soc.*, 65(6): 681–690, 2018. 531

[21] Steve Awodey. *Category theory*, vol. 52 of *Oxford logic guides*, 2nd ed. Oxford University Press, Oxford, 2010. 67, 111, 224, 360, 427, 466, 522, 603

[22] Steve Awodey and Michael A. Warren. Homotopy theoretic models of identity types. *Math. Proc. Cambridge Philos. Soc.*, 146(1): 45–55, 2009. 532

[23] Sheldon Axler. *Linear algebra done right*, 2nd ed. Undergraduate Texts in Mathematics. Springer-Verlag, New York, 1997. 82, 279

[24] J. Baez and M. Stay. Physics, topology, logic and computation: A Rosetta stone. In *New structures for physics*, vol. 813 of *Lecture notes in physics*, pp. 95–172. Springer, Heidelberg, 2011. 6, 604

[25] John Baez. Quantum quandaries: A category-theoretic perspective. In *The structural foundations of quantum gravity*, pp. 240–265. Oxford University Press, Oxford, 2006. 557, 589

[26] John C. Baez and James Dolan. Higher-dimensional algebra III: n-categories and the algebra of opetopes, https://arxiv.org/abs/q-alg/9702014, 1997. Accessed January 5, 2024. 507

[27] John C. Baez and James Dolan. Categorification. In *Higher category theory*, vol. 230 of *Contemp. Math.*, pp. 1–36. American Mathematical Society, Providence, RI, 1998. 509

[28] John C. Baez and Aaron D. Lauda. A prehistory of n-categorical physics. In *Deep beauty*, pp. 13–128. Cambridge University Press, Cambridge, 2011. 266

[29] Dror Bar-Natan. On associators and the Grothendieck-Teichmuller group. I. *Selecta Math. (N.S.)*, 4(2): 183–212, 1998. 380

[30] Michael Barr, John F. Kennison, and R. Raphael. Isbell duality. *Theory Appl. Categ.*, 20(15): 504–542, 2008. 330

[31] Michael Barr, John F. Kennison, and R. Raphael. Isbell duality for modules. *Theory Appl. Categ.*, 22(17): 401–419, 2009. 330

[32] Michael Barr and Charles Wells. *Toposes, triples and theories*, vol. 278 of *Grundlehren der Mathematischen Wissenschaften [Fundamental principles of mathematical sciences]*. Springer-Verlag, New York, 1985. 67, 111, 393, 519, 522, 603

[33] Michael Barr and Charles Wells. *Category theory for computing science*, 3rd ed. Number 22. Les Publications CRM, Montreal, 2012. 7, 67, 111, 393, 427, 519, 522, 604

[34] Gershom Bazerman. Homotopy type theory: What's the big idea? 2014. https://www.youtube.com/watch?v=OupcXmLER7I. Accessed August 8, 2022. 532

[35] John L. Bell. The development of categorical logic. Available at https://publish.uwo.ca/~jbell/catlogprime.pdf. Accessed August 10, 2022. 522

[36] Jean Bénabou. Introduction to bicategories. In *Reports of the Midwest Category Seminar*, pp. 1–77. Springer, Berlin, 1967. 266

[37] R. Blackwell, G. M. Kelly, and A. J. Power. Two-dimensional monad theory. *J. Pure Appl. Algebra*, 59(1): 1–41, 1989. 427, 433

[38] J. M. Boardman and R. M. Vogt. *Homotopy invariant algebraic structures on topological spaces, Lecture Notes in Mathematics*. Springer Berlin Heidelberg, 2006. 511

[39] G. S. Boolos, J. P. Burgess, and R. C. Jeffrey. *Computability and logic*. Cambridge University Press, New York, 2007. 134, 140

[40] Francis Borceux. *Handbook of categorical algebra. Vol. 1: Basic category theory*, Vol. 50 of *Encyclopedia of mathematics and its applications*. Cambridge University Press, Cambridge, 1994. 67, 111, 176, 461, 603

[41] Francis Borceux. *Handbook of categorical algebra. Vol. 2: Categories and structures*, Vol. 51 of *Encyclopedia of mathematics and its applications*. Cambridge University Press, Cambridge, 1994. 395, 453, 454, 603

[42] Francis Borceux. *Handbook of categorical algebra. Vol. 3: Categories of sheaves*, Vol. 52 of *Encyclopedia of mathematics and its applications*. Cambridge University Press, Cambridge, 1994. 522, 603

[43] Tai-Danae Bradley. Warming up to enriched category theory, part 1. https://www.math3ma.com/blog/warming-up-to-enriched-category-theory-1. Accessed April 24, 2022. 454

[44] Tai-Danae Bradley. Warming up to enriched category theory, part 2. https://www.math3ma.com/blog/warming-up-to-enriched-category-theory-part-2. Accessed April 24, 2022. 454

[45] Tai-Danae Bradley. What is an operad?, 2017. https://www.math3ma.com/blog/what-is-an-operad-part-1. Accessed April 30, 2022. 409

[46] Tai-Danae Bradley. The yoneda perspective, 2017. https://www.math3ma.com/blog/the-yoneda-perspective. Accessed March 7, 2022. 208

[47] Tai-Danae Bradley, Tyler Bryson, and John Terilla. *Topology—a categorical approach*. MIT Press, Cambridge, MA, 2020. 500

[48] Spencer Breiner, Peter Denno, and Eswaran Subrahmanian. Categories for planning and scheduling. *Notices Amer. Math. Soc.*, 67(11): 1666–1677, 2020. 443, 445

[49] Cristian Calude. To halt or not to halt? That is the question. World Scientific, 2024. 126

[50] Olivia Caramello. *Theories, sites, toposes: Relating and studying mathematical theories through topos-theoretic "bridges."* Oxford University Press, Oxford, 2018. 512, 522

[51] Aurelio Carboni, Stephen Lack, and R. F. C. Walters. Introduction to extensive and distributive categories. *J. Pure and Appl. Algebra*, 84(2): 145–158, 1993. 101

[52] Gunnar Carlsson and Facundo Mémoli. Classifying clustering schemes. *Found. Comput. Math.*, 13(2): 221–252, 2013. 463

[53] Sean M. Carroll. *Something deeply hidden: Quantum worlds and the emergence of spacetime*. Penguin Random House, New York, 2020. 556

[54] Sean M. Carroll. Reality as a vector in Hilbert space. In *Quantum mechanics and fundamentality—naturalizing quantum theory between scientific realism and ontological indeterminacy*, vol. 460 of *Synth. Libr.*, pp. 211–224. Springer, Cham, Switzerland, 2022. 556

[55] Pierre Cartier. What is an operad? In *Surveys in modern mathematics*, vol. 321 of *London Math. Soc. Lecture Note Ser.*, pp. 283–291. Cambridge University Press, Cambridge, 2005. 409

[56] Sergio A. Celani and Ramon Jansana. Easkia duality and its extensions. In *Leo Esakia on duality in modal and intuitionistic logics*, vol. 4 of *Outst. Contrib. Log.*, pp. 63–98. Springer, Dordrecht, Netherlands, 2014. 330

[57] Vyjayanthi Chari and Andrew Pressley. *A guide to quantum groups*. Cambridge University Press, Cambridge, 1995. 380

[58] Eugenia Cheng. *How to bake π: An edible exploration of the mathematics of mathematics*. Basic Books, New York, 2015. 604

[59] Eugenia Cheng. *The joy of abstraction—an exploration of math, category theory, and life*. Cambridge University Press, Cambridge, 2022. 604

[60] Marvin Chester. *Primer of quantum mechanics*. Dover Publications, New York, 2003. 557

[61] Bob Coecke and Aleks Kissinger. *Picturing quantum processes: A first course in quantum theory and diagrammatic reasoning*. Cambridge University Press, 2017. 557, 589

[62] Edward B. Curtis. Simplicial homotopy theory. *Adv. in Math.*, 6: 107–209 (1971), 1971. 500

[63] Georgi Dimov, Elza Ivanova-Dimova, and Walter Tholen. Categorical extension of dualities: From Stone to de Vries and beyond, I. *Appl. Categ. Struct.*, 30(2): 287–329, 2022. 330

[64] P. A. M. Dirac. *The principles of quantum mechanics*. The International Series of Monographs on Physics. Oxford University Press, 1982. 557

[65] V. G. Drinfeld. On quasitriangular quasi-Hopf algebras and on a group that is closely connected with Gal($\overline{\mathbf{Q}}/\mathbf{Q}$). *Algebra i Analiz*, 2(4): 149–181, 1990. 380

[66] William G. Dwyer, Philip S. Hirschhorn, Daniel M. Kan, and Jeffrey H. Smith. *Homotopy limit functors on model categories and homotopical categories*, vol. 113 of *Mathematical Surveys and Monographs*. American Mathematical Society, Providence, RI, 2004. 500

[67] Sammy Eilenberg and Saunders Mac Lane. *Eilenberg–Mac Lane, Collected Works*. Academic Press, Orlando, FL, 1986. 1

[68] Samuel Eilenberg and Saunders Mac Lane. General theory of natural equivalences. *Trans. Amer. Math. Soc.*, 58(2): 231–294, 1945. 165, 207, 331

[69] Herbert B. Enderton. *A mathematical introduction to logic*. Academic Press, New York and London, 1972. 224

[70] D. B. A. Epstein. Functors between tensored categories. *Invent. Math.*, 1: 221–228, 1966. 312

[71] Pavel Etingof, Shlomo Gelaki, Dmitri Nikshych, and Victor Ostrik. *Tensor categories*, vol. 205 of *Mathematical Surveys and Monographs*. American Mathematical Society, Providence, RI, 2015. 251, 266, 312, 360, 452, 454

[72] Richard P. Feynman. *Feynman Lectures on Physics (3-Volume Set)*. Addison Wesley, Boston, 1963. 557

[73] Richard P. Feynman. *The Pleasure of Finding Things Out: The Best Short Works of Richard P. Feynman*. Basic Books, Helix Books ed., New York, 2005. ix

[74] Thomas M. Fiore. Pseudo limits, biadjoints, and pseudo algebras: Categorical foundations of conformal field theory. *Mem. Amer. Math. Soc.*, 182(860): x+171, 2006. 433

[75] Thomas M. Fiore, Po Hu, and Igor Kriz. Laplaza sets, or how to select coherence diagrams for pseudo algebras. *Adv. Math.*, 218(6): 1705–1722, 2008. 433

[76] Cecilia Flori. *A first course in topos quantum theory*, vol. 868 of *Lecture Notes in Physics*. Springer, Heidelberg, 2013. 522

[77] Brendan Fong and David I. Spivak. *An invitation to applied category theory: Seven sketches in compositionality*. Cambridge University Press, Cambridge, 2019. 224, 445, 451, 454

[78] Michael P. Fourman. The logic of topoi. In *Handbook of mathematical logic*, vol. 90 of *Stud. Logic Found. Math.*, pp. 1053–1090. North-Holland, Amsterdam, 1977. 224, 522

[79] J. B. Fraleigh. *A first course in abstract algebra*. Pearson Education, Reading, London, 2003. 32, 42

[80] Peter Freyd. *Abelian categories: An introduction to the theory of functors*. Harper's Series in Modern Mathematics. Harper & Row, New York, 1964. 453

[81] Peter Freyd. Properties invariant within equivalence types of categories. In *Algebra, topology, and category theory: A collection of papers in honor of Samuel Eilenberg*, pp. 55–61. 1976. 305

[82] Peter J. Freyd and Andre Scedrov. *Categories, allegories*, vol. 39 of *North-Holland Mathematical Library*. North-Holland, Amsterdam, 1990. 305

[83] Greg Friedman. Survey article: An elementary illustrated introduction to simplicial sets. *Rocky Mountain J. Math.*, 42(2): 353–423, 2012. 500

[84] George Gamow. *Thirty years that shook physics: The story of quantum theory*. Dover Publications, New York, 1985. 557

[85] Mai Gehrke. Duality in computer science. In *Proceedings of the 31st Annual ACM-IEEE Symposium on Logic in Computer Science (LICS 2016)*, p. 15. Association for Computing Machinery (ACM), New York, 2016. 330

[86] Robert Geroch. *Mathematical physics*. Chicago Lectures in Physics. University of Chicago Press, Chicago, 1985. 604

[87] Jimmie Gilbert and Linda Gilbert. *Linear algebra and matrix theory, second edition*. Thomson, Brooks/Cole, Belmont, CA, 2004. 82, 279

[88] Daniel T. Gillespie. *A quantum mechanics primer: An introduction to the formal theory of non-relativistic quantum mechanics*. John Wiley & Sons, Hoboken, NJ, 1974. 557

[89] Steven Givant and Paul Halmos. *Introduction to Boolean algebras*. Undergraduate Texts in Mathematics. Springer, New York, 2009. 330

[90] Paul G. Goerss and John F. Jardine. *Simplicial homotopy theory*. Modern Birkhäuser Classics. Birkhäuser Verlag, Basel, Switzerland, 2009. 500

[91] Robert Goldblatt. *Topoi: The categorial analysis of logic*. Number 98 in Studies in Logic and the Foundations of Mathematics. North-Holland, third edition, 1983. 224, 522, 604

[92] John W. Gray. 2-algebraic theories and triples. *Cahiers de topologie et géométrie différentielle. In Colloque sur l'algèbre des catégories. Amiens-1973. Résumés des conférences*, 14(2): 178–180, 1973. 431

[93] John W. Gray. *Formal category theory: Adjointness for 2-categories. Lecture Notes in Mathematics*, vol. 391. Springer-Verlag, Berlin and New York, 1974. 432

[94] Daniel R. Grayson. An introduction to univalent foundations for mathematicians. *Bull. Amer. Math. Soc. (N.S.)*, 55(4): 427–450, 2018. 532

[95] Alexander Grothendieck. Sur quelques points d'algèbre homologique, I. *Tohoku Math. J.*, 9(2): 119–221, 1957. 207

[96] Alexandre Grothendieck. Pursuing stacks (À la poursuite des champs), 1983. Available at https://thescrivener.github.io/PursuingStacks/ps-online.pdf. Accessed April 24, 2022. 499

[97] Alexandre Grothendieck. Les dérivateurs, 1991. Available at https://web-users.imj-prg.fr/~georges.maltsiniotis/groth/Derivateurs.html. Accessed April 24, 2022. 499

[98] Alexandre Grothendieck. Some aspects of homological algebra, 2011. Translated by Marcia Barr with with the help of Michael Barr as the mathematical reviser and TeX editor. https://www.math.mcgill.ca/barr/papers/gk.pdf. Accessed March 7, 2022. 207

[99] Paul R. Halmos. *Lectures on Boolean algebras*. Van Nostrand Mathematical Studies, No. 1. D. Van Nostrand Co., Princeton, NJ, 1963. 330

[100] Paul R. Halmos. *I want to be a mathematician: An automathography*. Mathematical Association of America,, Washington, DC, 1985. 33

[101] Keith Hannabuss. *An introduction to quantum theory*. Oxford, NY, 1997. 557

[102] Allen Hatcher. *Algebraic topology*. Cambridge University Press, Cambridge, 2002. 500

[103] Alex Heller. Homotopy theories. *Mem. Amer. Math. Soc.*, 71(383): vi+78, 1988. 499

[104] Alex Heller. Homological algebra and (semi)stable homotopy. *J. Pure Appl. Algebra*, 115(2): 131–139, 1997. 499

[105] Alex Heller. Stable homotopy theories and stabilization. *J. Pure Appl. Algebra*, 115(2): 113–130, 1997. 499

[106] Alex Heller. Semistability and infinite loop spaces. *J. Pure Appl. Algebra*, 154(1–3): pp. 213–220, 2000. 499

[107] Nick Herbert. *Quantum reality: Beyond the new physics*. Anchor Press, Garden City, NY, Anchor, 1987. 556, 557

[108] Horst Herrlich and George E. Strecker. *Category theory: An introduction*, 2nd ed., vol. 1 of *Sigma Series in Pure Mathematics*. Heldermann Verlag, Berlin, 1979. 67, 111, 603

[109] Kathryn Hess. Model categories in algebraic topology. Vol. 10, Applied Categorical Structures, pp. 195–220, 2002. 500

[110] Chris Heunen and Jamie Vicary. *Categories for quantum theory*, vol. 28 of *Oxford Graduate Texts in Mathematics*. Oxford University Press, Oxford, 2019. 312, 360, 557, 589

[111] John G. Hocking and Gail S. Young. *Topology, second edition*. Dover Publications, New York, 1988. 500

[112] Mark Hovey. *Model categories*, vol. 63 of *Mathematical surveys and monographs*. American Mathematical Society, Providence, RI, 1999. 498, 500

[113] Mark Hovey. Quillen model categories. *J. K-Theory*, 11(3): 469–478, 2013. 500

[114] P. Hu and I. Kriz. Conformal field theory and elliptic cohomology. *Adv. Math.*, 189(2): 325–412, 2004. 433

[115] Po Hu and Igor Kriz. Closed and open conformal field theories and their anomalies. *Comm. Math. Phys.*, 254(1): 221–253, 2005. 433

[116] Po Hu and Igor Kriz. On modular functors and the ideal Teichmüller tower. *Pure Appl. Math. Q.*, 1(3, Special Issue: In Memory of Armand Borel. Part 2): 665–682, 2005. 433

[117] J. M. E. Hyland. The effective topos. In *The L.E.J. Brouwer Centenary Symposium (Noordwijkerhout, 1981)*, vol. 110 of *Studies in Logic and the Foundations of Mathematics*, pp. 165–216. North-Holland, Amsterdam and New York, 1982. 522

[118] J. M. E. Hyland, P. T. Johnstone, and A. M. Pitts. Tripos theory. *Math. Proc. Cambridge Philos. Soc.*, 88(2): 205–231, 1980. 522

[119] Luc Illusie. What is … a topos? *Notices Amer. Math. Soc.*, 51(9): 1060–1061, 2004. 522

[120] P. S. Isaac, W. P. Joyce, and J. Links. An algebraic approach to symmetric premonoidal statistics. *J. Algebra Appl.*, 6(1): 49–69, 2007. 310

[121] Rick Jardine. Lectures on homotopy theory. Available at https://math.sci.uwo.ca/~jardine/HomTh/. Accessed September 1, 2022. 500

[122] Rick Jardine. Tōhoku, 2015. https://inference-review.com/article/tohoku. Accessed March 7, 2022. 207

[123] P. T. Johnstone. *Topos theory. London Mathematical Society Monographs*, vol. 10. Academic Press [Harcourt Brace Jovanovich, Publishers], London-New York, 1977. 522

[124] Peter T. Johnstone. The point of pointless topology. *Bull. Amer. Math. Soc. (N.S.)*, 8(1): 41–53, 1983. 155, 330

[125] Peter T. Johnstone. *Stone spaces*, vol. 3 of *Cambridge Studies in Advanced Mathematics*. Cambridge University Press, Cambridge, 1986. Reprint of the 1982 edition. 330, 522, 604

[126] Peter T. Johnstone. *Sketches of an elephant: A topos theory compendium. Vol. 1*, vol. 43 of *Oxford Logic Guides*. Clarendon Press and Oxford University Press, New York, 2002. 511, 522

[127] Peter T. Johnstone. *Sketches of an elephant: A topos theory compendium. Vol. 2*, vol. 44 of *Oxford Logic Guides*. Clarendon Press and Oxford University Press, Oxford, 2002. 511, 522

[128] André Joyal. Quasi-categories and kan complexes. *J. Pure Appl. Algebra*, 175(1): 207–222, 2002. Special volume celebrating the 70th birthday of Professor Max Kelly. 511

[129] André Joyal. Homotopy type theory: A new bridge between logic, category theory and topology, 2018. https://www.youtube.com/watch?v=MxClaWFiGKw. Accessed August 8, 2022. 532

[130] André Joyal and Ross Street. An introduction to Tannaka duality and quantum groups. In *Category theory (Como, 1990)*, vol. 1488 of *Lecture Notes in Mathematics*, pp. 413–492. Springer, Berlin, 1991. 330

[131] André Joyal and Ross Street. Braided tensor categories. *Adv. Math.*, 102(1): 20–78, 1993. 251, 266, 312, 356, 360

[132] K. H. Kamps and T. Porter. *Abstract homotopy and simple homotopy theory.* World Scientific Publishing, River Edge, NJ, 1997. 500

[133] Daniel M. Kan. Adjoint functors. *Trans. Amer. Math. Soc.*, 87: 294–329, 1958. 207

[134] Christian Kassel. *Quantum groups*, vol. 155 of *Graduate Texts in Mathematics.* Springer-Verlag, New York, 1995. 7, 251, 266, 312, 349, 355, 360, 380, 543

[135] Louis H. Kauffman. *On knots*, vol. 115 of *Annals of Mathematics Studies.* Princeton University Press, Princeton, NJ, 1987. 543

[136] Louis H. Kauffman. *Knots and physics*, vol. 53 of *Series on Knots and Everything*, 4th ed. World Scientific Publishing, Hackensack, NJ, 2013. 543

[137] Louis H. Kauffman and Sóstenes L. Lins. *Temperley-Lieb recoupling theory and invariants of* 3*-manifolds*, vol. 134 of *Annals of Mathematics Studies.* Princeton University Press, Princeton, NJ, 1994. 543

[138] G. M. Kelly. On MacLane's conditions for coherence of natural associativities, commutativities, etc. *J. Algebra*, 1: 397–402, 1964. 266

[139] G. M. Kelly. An abstract approach to coherence. In *Coherence in categories*, pp. 106–147. *Lecture Notes in Mathematics*, vol. 281. 1972. 266, 433

[140] G. M. Kelly. Coherence theorems for lax algebras and for distributive laws. In *Category Seminar (Proc. Sem., Sydney, 1972/1973)*, pp. 281–375. *Lecture Notes in Mathematics*, vol. 420, 1974. 266

[141] G. M. Kelly. On clubs and doctrines. In *Category Seminar (Proc. Sem., Sydney, 1972/1973)*, pp. 181–256. *Lecture Notes in Mathematics*, vol. 420, 1974. 433

[142] G. M. Kelly. On clubs and data-type constructors. In *Applications of categories in computer science (Durham, 1991)*, vol. 177 of *London Mathematics Society Lecture Note Series*, pp. 163–190. Cambridge University Press, Cambridge, 1992. 433

[143] G. M. Kelly. Basic concepts of enriched category theory. *Repr. Theory Appl. Categ.*, (10): vi+137, 2005. 454

[144] G. M. Kelly and M. L. Laplaza. Coherence for compact closed categories. *J. Pure Appl. Algebra*, 19: 193–213, 1980. 266

[145] G. M. Kelly and S. Mac Lane. Coherence in closed categories. *J. Pure Appl. Algebra*, 1(1): 97–140, 1971. 266

[146] G. M. Kelly and Saunders Mac Lane. Closed coherence for a natural transformation. In *Coherence in categories*, pp. 1–28, *Lecture Notes in Mathematics*, vol. 281. Springer, Berlin, 1972. 266

[147] G. M. Kelly. *Basic concepts of enriched category theory*, vol. 64 of *London Mathematical Society Lecture Note Series*. Cambridge University Press, Cambridge and New York, 1982. 454

[148] A. Kock and G. E. Reyes. Doctrines in categorical logic. In *Handbook of mathematical logic*, vol. 90 of *Stud. Logic Found. Math.*, pp. 283–313. North-Holland, Amsterdam, 1977. 224, 393

[149] Joachim Kock. Elementary remarks on units in monoidal categories. *Math. Proc. Cambridge Philos. Soc.*, 144(1): 53–76, 2008. 266

[150] Ralf Kromer. *Tool and object: A history and philosophy of category theory*, vol. 32 of *Science Networks. Historical Studies*. Birkhäuser Verlag, Basel, Switzerland, 2007. 207, 604

[151] Joachim Lambek and Philip Scott. *Introduction to higher order categorical logic*. Number 7 in *Cambridge Studies in Advanced Mathematics*. Cambridge University Press, Cambridge and New York, 1986. 224, 522, 604

[152] Klaas Landsman. *Foundations of quantum theory: From classical concepts to operator algebras*, vol. 188 of *Fundamental Theories of Physics*. Springer, Cham, Switzerland, 2017. 330, 522, 557

[153] Klaas Landsman. Bohmian mechanics is not deterministic. *Found. Phys.*, 52(4), July 2022. 556

[154] Serge Lang. *Introduction to linear algebra, second edition*. Springer, Berlin, 1986. 82, 279

[155] Finnur Larusson. The homotopy theory of equivalence relations, 2006. Available at https://arxiv.org/abs/math/0611344. Accessed July 7, 2022. 493

[156] Aaron D. Lauda and Joshua Sussan. An invitation to categorification. *Notices Amer. Math. Soc.*, 69(1): 11–21, 2022. 509

[157] Mark V. Lawson. Classical stone duality. https://www.macs.hw.ac.uk/~markl/1-stone-duality.pdf. Accessed June 7, 2022. 330

[158] F. William Lawvere. *Functional semantics of algebraic theories—thesis (Ph.D.)—Columbia University*. ProQuest LLC, Ann Arbor, MI, 1963. 381

[159] F. William Lawvere. Functorial semantics of algebraic theories. *Proc. Nat. Acad. Sci. U.S.A.*, 50: 869–872, 1963. 395

[160] F. William Lawvere. An elementary theory of the category of sets. *Proc. Nat. Acad. Sci. U.S.A.*, 52: 1506–1511, 1964. 83

[161] F. William Lawvere. Some algebraic problems in the context of functorial semantics of algebraic structures. In Saunders Mac Lane, ed., *Reports of the Midwest Category Seminar II*, number 61 in *Lecture Notes in Mathematics*, pp. 41–61. Springer-Verlag, Berlin, 1968. 395

[162] F. William Lawvere. Diagonal arguments and Cartesian closed categories. In *Category theory, homology theory and their applications, II (Battelle Institute Conference, Seattle, Wash., 1968, vol. two)*, pp. 134–145. Springer, Berlin, 1969. 125, 139

[163] F. William Lawvere. Functorial semantics of algebraic theories and some algebraic problems in the context of functorial semantics of algebraic theories. *Repr. Theory Appl. Categ.*, (5): 1–121, 2004. Reprinted from *Proc. Nat. Acad. Sci. U.S.A.* 50 (1963), 869–872 [MR0158921] and *Reports of the Midwest Category Seminar*. II, 41–61, Springer, Berlin, 1968 [MR0231882]. 381

[164] F. William Lawvere. Adjointness in foundations. *Repr. Theory Appl. Categ.*, (16): 1–16, 2006. Reprinted from *Dialectica* 23 (1969). 224

[165] F. William Lawvere and Robert Rosebrugh. *Sets for mathematics*. Cambridge University Press, Cambridge, 2003. 32, 139

[166] F. William Lawvere and Stephen H. Schanuel. *Conceptual mathematics: A first introduction to categories*, 2nd ed. Cambridge University Press, Cambridge, 2009. 32, 139

[167] Tom Leinster. The yoneda lemma: What's it all about? 2000. https://www.maths.ed.ac.uk/~tl/categories/yoneda.ps. Accessed March 7, 2022. 208

[168] Tom Leinster. A survey of definitions of n-category, 2001. https://arxiv.org/abs/math/0107188. Accessed January 7, 2024. 507

[169] Tom Leinster. *Higher operads, higher categories*, vol. 298 of *London Mathematical Society Lecture Note Series*. Cambridge University Press, Cambridge, 2004. 409

[170] Tom Leinster. An informal introduction to topos theory, 2010. https://arxiv.org/abs/1012.5647. Accessed January 7, 2024. 522

[171] Tom Leinster. *Basic category theory*, vol. 143 of *Cambridge Studies in Advanced Mathematics*. Cambridge University Press, Cambridge, 2014. 67, 111, 603

[172] F. E. J. Linton. Some aspects of equational categories. In *Proceedings of the Conference on Categorical Algebra (La Jolla, Calif., 1965)*, pp. 84–94. Springer, New York, 1966. 395

[173] Seymour Lipschutz. *Schaum's outline of theory and problems of set theory and related topics*. 2nd ed. McGraw-Hill, NY, 1998. 330

[174] Jean-Louis Loday, James D. Stasheff, and Alexander A. Voronov, eds. *Operads: Proceedings of Renaissance Conferences*, vol. 202 of *Contemporary Mathematics*. American Mathematical Society, Providence, RI, 1997. 409

[175] Jacob Lurie. What is an infinity-category? *Notices AMS*, 55(8): 949–950, 2008. 511

[176] Jacob Lurie. *Higher topos theory*, vol. 170 of *Annals of Mathematics Studies*. Princeton University Press, Princeton, NJ, 2009. 499, 510, 522

[177] Saunders Mac Lane. Natural associativity and commutativity. *Rice Univ. Stud.*, 49(4): 28–46, 1963. 266

[178] Saunders Mac Lane. Categorical algebra. *Bull. Amer. Math. Soc.*, 71: 40–106, 1965. 266

[179] Saunders Mac Lane. The development of mathematical ideas by collision: The case of categories and topos theory. In *Categorical topology and its relation to analysis, algebra and combinatorics (Prague, 1988)*, pp. 1–9. World Scientific Publishing, Teaneck, NJ, 1989. 522

[180] Saunders Mac Lane. *Categories for the working mathematician*, 2nd ed., vol. 5 of *Graduate Texts in Mathematics*. Springer-Verlag, New York, 1998. xi, 6, 67, 111, 224, 312, 381, 427, 431, 454, 460, 461, 466, 500, 519, 603

[181] Saunders Mac Lane and Ieke Moerdijk. *Sheaves in geometry and logic: A first introduction to topos theory*. Universitext. Springer-Verlag, New York, 1994. 329, 330, 360, 500, 522

[182] Saunders Mac Lane, ed. *Coherence in categories*. Lecture Notes in Mathematics, vol. 281. Springer-Verlag, Berlin and New York, 1972. 266

[183] Saunders Mac Lane. Sets, topoi, and internal logic in categories. In *Logic Colloquium '73 (Bristol, 1973)*, pp. 119–134. *Studies in Logic and the Foundations of Mathematics*, vol. 80. 1975. 522

[184] Shahn Majid. *Foundations of quantum group theory*. Cambridge University Press, Cambridge, 1995. 380

[185] Yu. I. Manin. *A course in mathematical logic for mathematicians*, 2nd ed., vol. 53 of *Graduate Texts in Mathematics*. Springer, New York, 2010. 140, 224, 408

[186] Martin Markl, Steve Shnider, and Jim Stasheff. *Operads in algebra, topology and physics*, vol. 96 of *Mathematical Surveys and Monographs*. American Mathematical Society, Providence, RI, 2002. 409

[187] Jean-Pierre Marquis. *From a geometrical point of view, a study of the history and philosophy of category theory*, vol. 14 of *Logic, epistemology, and the unity of science*. Springer, Dordrecht, 2009. 207, 604

[188] Jean-Pierre Marquis. Category theory. In Edward N. Zalta, ed., the *Stanford encyclopedia of philosophy*. Metaphysics Research Lab, Stanford University, Stanford, CA, winter 2015 edition, 2015. 207, 604

[189] Jean-Pierre Marquis. An historical perspective on duality and category theory: Hom is where the heart is. In *Duality in 19th- and 20th-century mathematical thinking*. Birkhauser, Basel, 2022. 207, 330

[190] Jean-Pierre Marquis and Gonzalo E. Reyes. The history of categorical logic: 1963–1977. In *Sets and extensions in the twentieth century*, vol. 6 of *Handb. Hist. Log.*, pp. 689–800. Elsevier/North-Holland, Amsterdam, 2012. 224

[191] J. L. Martin. *Basic quantum mechanics*. Oxford Physics Series. Oxford University Press, Oxford, 1982. 557

[192] J. Peter May. *Simplicial objects in algebraic topology*. Chicago Lectures in Mathematics. University of Chicago Press, Chicago, 1992. Reprint of the 1967 original. 500

[193] Guerino Mazzola. *The topos of music. I. Theory*. Computational Music Science. Springer, Cham, Switzerland, 2017. 521

[194] Colin McLarty. The uses and abuses of the history of topos theory. *British J. Philos. Sci.*, 41(3): 351–375, 1990. 512, 521, 522

[195] Elliott Mendelson. *Introduction to mathematical logic*. 4th ed. Chapman & Hall, London, 1997. 134, 140, 224

[196] Micah Miller. A primer on homotopy limits., 2018. https://www.youtube.com/watch?v=hF_T1JHaVpU. Accessed August 13, 2023. 500

[197] W. Keith Nicholson. *Linear algebra with applications*. 3rd ed. PWS Publishing, Boston, 1994. 82, 279

[198] Michael O'Nan. *Linear algebra*. 2nd ed. Harcourt Brace Jovanovich, New York, 1976. 82, 279

[199] Heinz R. Pagels. *The cosmic code: Quantum physics as the language of nature*. Simon and Schuster, New York, 1982. 557

[200] Prakash Panangaden. Dualities in mathematics: Analysis dressed up as algebra is dual to topology. part II: Gelfand duality. https://www.cs.mcgill.ca/~prakash/Talks/ssqs14-2.pdf. Accessed June 7, 2022. 330

[201] Prakash Panangaden. Dualities in mathematics: Locally compact abelian groups. part III: Pontryagin duality. https://www.cs.mcgill.ca/~prakash/Talks/ssqs14-3.pdf Accessed June 7, 2022. 330

[202] Prakash Panangaden. The mirror of mathematics, part I: Classical stone duality. https://www.cs.mcgill.ca/~prakash/Talks/ssqs14-1.pdf. Accessed June 7, 2022. 330

[203] Rohit Parikh. Existence and feasibility in arithmetic. *J. Sym. Logic*, 36: 494–508, 1971. 140

[204] Álvaro Pelayo and Michael A. Warren. Homotopy type theory and Voevodsky's univalent foundations. *Bull. Amer. Math. Soc. (N.S.)*, 51(4): 597–648, 2014. 531, 532

[205] Richard C. Penney. *Linear algebra, ideas and applications*. John Wiley & Sons, New York, 1998. 82, 279

[206] Paolo Perrone. Categorical probability, Markov categories, and the de Finetti Theorem, 2021. https://www.youtube.com/watch?v=qH9gvVVYVUM. Accessed August 13, 2023. 427

[207] Paolo Perrone. *Starting category theory*. World Scientific Publishing, Singapore, 2024. 427

[208] Benjamin C. Pierce. *Basic category theory for computer scientists*. Foundations of Computing Series. MIT Press, Cambridge, MA, 1991. 111, 604

[209] A. M. Pitts. *Categorical logic*. Technical Report 367, 94 pp, University of Cambridge Computer Laboratory, 1995. 224

[210] Andrew M. Pitts. Categorical logic. In *Handbook of logic in computer science, vol. 5*, pp. 39–128. Oxford University Press, New York, 2000. 224

[211] Henri Poincaré. *Science et Méthode*. 1908. https://archive.org/details/science etmthod00poin/mode/2up. Accessed September 18, 2022. vi

[212] Henri Poincaré. *Science and Method*. 1914. https://archive.org/details/science method00poinuoft. Accessed September 18, 2022. vi

[213] John Polkinghorne. *Quantum theory: A very short introduction*. Oxford University Press, Oxford, 2002. 557

[214] H.-E. Porst and W. Tholen. Concrete dualities. In *Category theory at work (Bremen, 1990)*, vol. 18 of *Res. Exp. Math.*, pp. 111–136. Heldermann, Berlin, 1991. 330

[215] Daniel Quillen. Rational homotopy theory. *Ann. Math. (2)*, 90: 205–295, 1969. 491

[216] Daniel G. Quillen. *Homotopical algebra*. Lecture Notes in Mathematics, No. 43. Springer-Verlag, Berlin and New York, 1967. 489, 491, 500

[217] Charles Rezk. A model category for categories, 2000. Available at https://faculty .math.illinois.edu/~rezk/cat-ho.dvi. Accessed April 26, 2022. 493

[218] Charles Rezk. Higher topos theory, 2021. https://www.youtube.com/watch? v=f3bdYrxTa1E. Accessed July 18, 2022. 522

[219] Birgit Richter. *From categories to homotopy theory*. Cambridge Studies in Advanced Mathematics. Cambridge University Press, Cambridge, 2020. 409, 454

[220] Emily Riehl. A leisurely introduction to simplicial sets. Available at https://math .jhu.edu/~eriehl/ssets.pdf. Accessed April 26, 2022. 500

[221] Emily Riehl. *Categorical homotopy theory*, vol. 24 of *New Mathematical Monographs*. Cambridge University Press, Cambridge, 2014. 500

[222] Emily Riehl. *Category Theory in Context*. Dover Publications, NY, 2016. 67, 111, 427, 461, 466, 603

[223] Emily Riehl and Dominic Verity. *Elements of ∞-category theory*. Cambridge Studies in Advanced Mathematics. Cambridge University Press, Cambridge, 2022. 510

[224] Andrei Rodin. *Axiomatic method and category theory*, vol. 364 of *Synthese Library*. Springer, Cham, Switzerland, 2014. 604

[225] Kenneth H. Rosen. *Discrete mathematics and its applications*. 5th ed. McGraw Hill, New York, 2003. 32

[226] Kenneth A. Ross and Charles R. B. Wright. *Discrete mathematics*. 5th ed. Prentice Hall, Upper Saddle River, NJ, 2003. 32, 224

[227] Joseph J. Rotman. *An introduction to algebraic topology*, vol. 119 of *Graduate Texts in Mathematics*. Springer-Verlag, New York, 1988. 478, 500

[228] Rudy Rucker. *Infinity and the mind: The science and philosophy of the infinite*. Birkhäuser, Boston, 1982. 139

[229] Neantro Saavedra Rivano. *Catégories Tannakiennes. Lecture Notes in Mathematics*, vol. 265. Springer-Verlag, Berlin and New York, 1972. 266

[230] Jun John Sakurai. *Modern quantum mechanics*. Rev. ed. Addison-Wesley Publishing, Reading, MA, 1994. 557

[231] Patrick Schultz, David I. Spivak, Christina Vasilakopoulou, and Ryan Wisnesky. Algebraic databases. *Theory Appl. Categ.*, 32(16): 547–619, 2017. 445

[232] Robert Seely. Locally Cartesian closed categories and type theory. *Math. Proc. Cambridge Phil. Soc.*, 95: 33–48, 1984. 532

[233] P. Selinger. A survey of graphical languages for monoidal categories. In *New structures for physics*, vol. 813 of *Lecture Notes in Physics*, pp. 289–355. Springer, Heidelberg, 2011. 266, 331, 360

[234] Jean-Pierre Serre. Motifs. No. 198–200, pp. 11, 333–349 (1992). 1991. Journées Arithmétiques, 1989 (Luminy, 1989). 141

[235] Dan Shiebler. Kan extensions in data science and machine learning. Available at https://arxiv.org/abs/2203.09018. Accessed July 4, 2022. 462

[236] Dan Shiebler, Bruno Gavranović, and Paul Wilson. Category theory in machine learning, 2021. https://arxiv.org/abs/2106.07032. Accessed January 7, 2024. 463

[237] Steven Shnider and Shlomo Sternberg. *Quantum groups: From coalgebras to Drinfeld algebras, A guided tour*. Graduate Texts in Mathematical Physics, II. International Press, Cambridge, MA, 1993. 380

[238] Mike Shulman. Derivators: Doing homotopy theory with 2-category theory, 2013. Available at http://web.science.mq.edu.au/groups/coact/seminar/ct2013/slides/introder.pdf. Accessed April 24, 2022. 499

[239] Mei Chee Shum. Tortile tensor categories. *J. Pure Appl. Algebra*, 93(1): 57–110, 1994. 360

[240] I. M. Singer and John A. Thorpe. *Lecture notes on elementary topology and geometry*. Scott, Foresman, Glenview, IL, 1967. 500

[241] Michael Sipser. *Introduction to the theory of computation*, 2nd ed. Thomson Course Technology, Boston, 2006. 140

[242] David I. Spivak. Functorial data migration. *Inform. Comput.*, 217: 31–51, 2012. 442, 445

[243] David I. Spivak. *Category theory for the sciences*. MIT Press, Cambridge, MA, 2014. 445, 604

[244] Jim Stasheff. What is ... an operad? *Notices Amer. Math. Soc.*, 51(6): 630–631, 2004. 409

[245] Ross Street. The formal theory of monads. *J. Pure Appl. Algebra*, 2(2): 149–168, 1972. 466

[246] Ross Street. The algebra of oriented simplexes. *J. Pure and Appl. Algebra*, (49): 283–335, 1987. 506

[247] Ross Street. *Quantum groups: A path to current algebra*, vol. 19 of *Australian Mathematical Society Lecture Series*. Cambridge University Press, Cambridge, 2007. 360, 380

[248] Ross Street. Monoidal categories in, and linking, geometry and algebra. *Bull. Belg. Math. Soc. Simon Stevin*, 19(5): 769–821, 2012. 360

[249] Ross Street and Robert Walters. Yoneda structures on 2-categories. *J. Algebra*, 50(2): 350–379, 1978. 466

[250] Anthony Sudbery. *Quantum mechanics and the particles of nature: An outline for mathematicians*. Cambridge University Press, Cambridge, 1986. 557

[251] Robert M. Switzer. *Algebraic topology—homotopy and homology*. Die Grundlehren der mathematischen Wissenschaften, Band 212. Springer-Verlag, New York and Heidelberg, 1975. 500

[252] Andy Tonks. Relating the associahedron and the permutohedron. In *Operads: Proceedings of Renaissance Conferences (Hartford, CT/Luminy, 1995)*, vol. 202 of *Contemp. Math.*, pp. 33–36. American Mathematical Society, Providence, RI, 1997. 404

[253] Vladimir G. Turaev. *Quantum invariants of knots and 3-manifolds*, 3rd ed., vol. 18 of *De Gruyter Studies in Mathematics*. De Gruyter, Berlin, 2016. 360, 543

[254] Univalent Foundations Program. *Homotopy type theory—univalent foundations of mathematics*. Univalent Foundations Program, Institute for Advanced Study (IAS), Princeton, NJ, 2013. 527, 532

[255] Jaap van Oosten. *Realizability: An introduction to its categorical side*, vol. 152 of *Studies in Logic and the Foundations of Mathematics*. Elsevier B. V., Amsterdam, 2008. 522

[256] Steven Vickers. *Topology via logic*, vol. 5 of *Cambridge Tracts in Theoretical Computer Science*. Cambridge University Press, Cambridge, 1989. 330

[257] Eric G. Wagner. Algebraic semantics. In *Handbook of Logic in Computer Science*, vol. 3, pp. 323–393. Oxford University Press, New York, 1994. 395

[258] R. F. C. Walters. *Categories and computer science*, vol. 28 of *Cambridge Computer Science Texts*. Cambridge University Press, Cambridge, 1991. 101, 604

[259] Charles A. Weibel. *An introduction to homological algebra*, vol. 38 of *Cambridge Studies in Advanced Mathematics*. Cambridge University Press, Cambridge, 1994. 453, 454

[260] Eric W. Weisstein. Solomon's seal knot. From MathWorld–A Wolfram Web Resource. https://mathworld.wolfram.com/SolomonsSealKnot.html. Accessed July 2, 2023. 539

[261] Charles Wells. Sketches: Outline with references, 1993. Available at https://ncatlab.org/nlab/files/Wells_Sketches.pdf. Accessed: July 16, 2023. 393

[262] Robert L White. *Basic quantum mechanics*. McGraw-Hill, New York, 1966. 557

[263] Mark M. Wilde. *Quantum information theory*, 2nd ed. Cambridge University Press, Cambridge, 2017. 587, 589

[264] Edward Witten. Knots and quantum theory, 2011. https://www.ias.edu/ideas/2011/witten-knots-quantum-theory. Accessed July 2, 2023. 542, 543

[265] Noson S. Yanofsky. Computability and complexity of categorical structures. Available at http://arxiv.org/pdf/1507.05305v1.pdf. Preprint. Accessed July 6, 2022. 32

[266] Noson S. Yanofsky. *Obstructions to coherence: Natural noncoherent associativity and tensor functors—thesis (Ph.D.)—City University of New York*. ProQuest LLC, Ann Arbor, MI, 1996. 309, 310

[267] Noson S. Yanofsky. Obstructions to coherence: Natural noncoherent associativity. *J. Pure Appl. Algebra*, 147(2): 175–213, 2000. 309

[268] Noson S. Yanofsky. The syntax of coherence. *Cahiers Topologie Géom. Différentielle Catég.*, 41(4): 255–304, 2000. 431, 432

[269] Noson S. Yanofsky. Coherence, homotopy and 2-theories. *K-Theory*, 23(3): 203–235, 2001. 431, 432, 500

[270] Noson S. Yanofsky. A universal approach to self-referential paradoxes, incompleteness and fixed points. *Bull. Symbolic Logic*, 9(3): 362–386, 2003. 139

[271] Noson S. Yanofsky. Towards a definition of an algorithm. *J. Logic Comput.*, 21(2): 253–286, 2011. 408

[272] Noson S. Yanofsky. *The outer limits of reason: What science, mathematics, and logic cannot tell us*. MIT Press, Cambridge, MA, 2013. 139, 140, 522, 555, 557

[273] Noson S. Yanofsky. Resolving paradoxes. *Philosophy Now*, 016: 10–12, 2015. 139

[274] Noson S. Yanofsky. *Theoretical computer science for the working category theorist*. Cambridge University Press, Cambridge, 2022. 6, 38, 507

[275] Noson S. Yanofsky. Paradoxes, contradictions, and the limits of science. *American Scientist*, pp. 166–173, May-June 2016. 139

[276] Noson S. Yanofsky and Mirco A. Mannucci. *Quantum computing for computer scientists*. Cambridge University Press, Cambridge, 2008. 558, 589

[277] Donald Yau. *Operads of wiring diagrams*, vol. 2192 of *Lecture Notes in Mathematics*. Springer, Cham, Switzerland, 2018. 409

[278] David N. Yetter. *Functorial knot theory*, vol. 26 of *Series on Knots and Everything*. World Scientific Publishing Co., Inc., River Edge, NJ, 2001. Categories of tangles, coherence, categorical deformations, and topological invariants. 360, 543

Index